Introduction to Coalgebra 59

Towards Mathematics of States and Observation

The area of coalgebra has emerged within theoretical computer science with a unifying claim: to be the mathematics of computational dynamics. It combines ideas from the theory of dynamical systems and from the theory of state-based computation. Although still in its infancy, it is an active area of research that generates wide interest.

Written by one of the founders of the field, this book acts as the first mature and accessible introduction to coalgebra. It provides clear mathematical explanations with many examples and exercises involving deterministic and non-deterministic automata, transition systems, streams, Markov chains and weighted automata. The theory is expressed in the language of category theory, which provides the right abstraction to make the similarity and duality between algebra and coalgebra explicit and which the reader is introduced to in a hands-on manner. The book will be useful to mathematicians and (theoretical) computer scientists and of interest to mathematical physicists, biologists and economists.

BART JACOBS is Professor of Software and Security Correctness at Radboud University Nijmegen in the Netherlands. His work has been supported by prestigious grants such as the Pionier grant from the Netherlands Organisation for Scientific Research and the Advanced Investigator Grant from the European Research Council. Jacobs is a member of Academia Europaea.

CAMBRIDGE TRACTS IN THEORETICAL COMPUTER SCIENCE 59

Titles in the series
A complete list of books in the series can be found at
www.cambridge.org/computerscience.
Recent titles include the following:

30. M. Anthony & N. Biggs *Computational Learning Theory*
31. T. F. Melham *Higher Order Logic and Hardware Verification*
32. R. Carpenter *The Logic of Typed Feature Structures*
33. E. G. Manes *Predicate Transformer Semantics*
34. F. Nielson & H. R. Nielson *Two-Level Functional Languages*
35. L. M. G. Feijs & H. B. M. Jonkers *Formal Specification and Design*
36. S. Mauw & G. J. Veltink (eds) *Algebraic Specification of Communication Protocols*
37. V. Stavridou *Formal Methods in Circuit Design*
38. N. Shankar *Metamathematics, Machines and Gödel's Proof*
39. J. B. Paris *The Uncertain Reasoner's Companion*
40. J. Desel & J. Esparza *Free Choice Petri Nets*
41. J.-J. Ch. Meyer & W. van der Hoek *Epistemic Logic for AI and Computer Science*
42. J. R. Hindley *Basic Simple Type Theory*
43. A. S. Troelstra & H. Schwichtenberg *Basic Proof Theory*
44. J. Barwise & J. Seligman *Information Flow*
45. A. Asperti & S. Guerrini *The Optimal Implementation of Functional Programming Languages*
46. R. M. Amadio & P.-L. Curien *Domains and Lambda-Calculi*
47. W.-P. de Roever & K. Engelhardt *Data Refinement*
48. H. Kleine Büning & T. Lettmann *Propositional Logic*
49. L. Novak & A. Gibbons *Hybrid Graph Theory and Network Analysis*
50. J. C. M. Baeten, T. Basten & M. A. Reniers *Process Algebra: Equational Theories of Communicating Processes*
51. H. Simmons *Derivation and Computation*
52. D. Sangiorgi & J. Rutten (eds) *Advanced Topics in Bisimulation and Coinduction*
53. P. Blackburn, M. de Rijke & Y. Venema *Modal Logic*
54. W.-P. de Roever et al. *Concurrency Verification*
55. Terese *Term Rewriting Systems*
56. A. Bundy et al. *Rippling: Meta-Level Guidance for Mathematical Reasoning*
57. A. M. Pitts *Nominal Sets*
58. S. Demri, V. Goranko & M. Lange *Temporal Logics in Computer Science*
59. B. Jacobs *Introduction to Coalgebra*

Introduction to Coalgebra

Towards Mathematics of States and Observation

BART JACOBS
Radboud University Nijmegen

CAMBRIDGE
UNIVERSITY PRESS

CAMBRIDGE
UNIVERSITY PRESS

University Printing House, Cambridge CB2 8BS, United Kingdom

One Liberty Plaza, 20th Floor, New York, NY 10006, USA

477 Williamstown Road, Port Melbourne, VIC 3207, Australia

4843/24, 2nd Floor, Ansari Road, Daryaganj, Delhi - 110002, India

79 Anson Road, #06-04/06, Singapore 079906

Cambridge University Press is part of the University of Cambridge.

It furthers the University's mission by disseminating knowledge in the pursuit of education, learning and research at the highest international levels of excellence.

www.cambridge.org
Information on this title: www.cambridge.org/9781107177895
10.1017/9781316823187

First published 2017

A catalogue record for this publication is available from the British Library

Library of Congress Cataloging in Publication data
Names: Jacobs, Bart.
Title: Introduction to coalgebra : towards mathematics of states and observation / Bart Jacobs.
Description: Cambridge : Cambridge University Press, [2017] |
Series: Cambridge tracts in theoretical computer science ; 59 |
Includes bibliographical references and indexes.
Identifiers: LCCN 2016023664 | ISBN 9781107177895 (Hardback)
Subjects: LCSH: Associative algebras. | Universal enveloping algebras. |
Algebra, Universal.
Classification: LCC QA251.5 .B36 2017 | DDC 512/.46–dc23 LC record available at
https://lccn.loc.gov/2016023664

ISBN 978-1-107-17789-5 Hardback

Contents

Preface

Mathematics is about the formal structures underlying counting, measuring, transforming etc. It has developed fundamental notions like number systems, groups, vector spaces (see e.g. [345]) and has studied their properties. In more recent decades also 'dynamical' features have become a subject of research. The emergence of computers has contributed to this development. Typically, dynamics involves a 'state of affairs', which can possibly be observed and modified. For instance, the contents of a tape of a Turing machine contribute to its state. Such a machine may thus have many possible states and can move from one state to another. Also, the combined contents of all memory cells of a computer can be understood as the computer's state. A user can observe part of this state via the screen (or via the printer) and modify this state by typing commands. In reaction, the computer can display certain behaviour. Describing the behaviour of such a computer system is a non-trivial matter. However, formal descriptions of such complicated systems are needed if we wish to reason formally about their behaviour. Such reasoning is required for the correctness or security of these systems. It involves a specification describing the required behaviour, together with a correctness proof demonstrating that a given implementation satisfies the specification.

Both mathematicians and computer scientists have introduced various formal structures to capture the essence of state-based dynamics, such as automata (in various forms), transition systems, Petri nets and event systems. The area of coalgebra has emerged within theoretical computer science with a unifying claim. It aims to be the mathematics of computational dynamics. It combines notions and ideas from the mathematical theory of dynamical systems and from the theory of state-based computation. The area of coalgebra is still in its infancy but promises a perspective on uniting, say, the theory of differential equations with automata and process theory and with biological and quantum computing, by providing an appropriate semantical basis with associated logic. The theory of coalgebras may be seen as one of the original contributions stemming from the area of theoretical computer science. The span of applications of coalgebras is still fairly limited but may in the future be extended to include dynamical phenomena in areas like physics, biology or economics – based for instance on the claim of Adleman (the father of DNA computing) that biological life can be equated with computation [32]; on [362], which gives a coalgebraic description of type spaces used in economics [211]; on [52], describing network dynamics that is common to all these areas; on [447], using coalgebras in biological modelling or on [7, 249], where coalgebras are introduced in quantum computing.

Coalgebras are of surprising simplicity. They consist of a state space, or set of states, say X, together with a structure map of the form $X \to F(X)$. The symbol F describes some expression involving X (a functor), capturing the possible outcomes of the structure map applied to a state. The map $X \to F(X)$ captures the dynamics in the form of a function acting on states. For instance, one can have F as powerset in $F(X) = \mathcal{P}(X)$ for non-deterministic computation $X \to \mathcal{P}(X)$, or $F(X) = \{\bot\} \cup X$ for possibly non-terminating computations $X \to \{\bot\} \cup X$. At this level of generality, algebras are described as the duals of coalgebras (or the other way round), namely as maps of the form $F(X) \to X$. One of the appealing aspects of this abstract view is the duality between structure (algebras) and behaviour (coalgebras).

Computer Science Is about Generated Behaviour

What is the essence of computing? What is the topic of the discipline of computer science? Answers that are often heard are 'data processing' or 'symbol manipulation'. Here we follow a more behaviouristic approach and describe the subject of computer science as *generated behaviour*. This is the behaviour that can be observed on the outside of a computer, for instance via a screen or printer. It arises in interaction with the environment, as a result of the computer executing instructions, laid down in a computer program. The aim of computer programming is to make a computer do certain things, i.e. to generate behaviour. By executing a program a computer displays behaviour that is ultimately produced by humans, as programmers.

This behaviouristic view allows us to understand the relation between computer science and the natural sciences: biology is about 'spontaneous' behaviour, and physics concentrates on lifeless natural phenomena, without autonomous behaviour. Behaviour of a system in biology or physics is often described as evolution, where evolutions in physics are transformational changes according to the laws of physics. Evolutions in biology seem to lack inherent directionality and predictability [174]. Does this mean that behaviour is deterministic in (classical) physics, and non-deterministic in biology? And that coalgebras of corresponding kinds capture the situation? At this stage the coalgebraic theory of modelling has not yet demonstrated its usefulness in those areas. Therefore this text concentrates on coalgebras in mathematics and computer science.

The behaviouristic view does help in answering questions like: can a computer think? Or: does a computer feel pain? All a computer can do is display thinking behaviour, or pain behaviour, and that is it. But it is good enough in interactions – think of the famous Turing test – because in the end we never know for sure if other people actually feel pain. We see only pain behaviour and are conditioned to associate such behaviour with certain internal states. But this association may not always work, for instance not in a different culture: in Japan it is common to touch one's ear after burning a finger; for Europeans this is non-standard pain behaviour. This issue of external behaviour versus internal states is nicely demonstrated in [350] where it turns out to be surprisingly difficult for a human to kill a 'Mark III Beast' robot once it starts displaying desperate survival behaviour with corresponding sounds, so that people easily attribute feelings to the machine and start to feel pity.

These wide-ranging considerations form the background for a theory about computational behaviour in which the relation between observables and internal states is of central importance.

The generated behaviour that we claim to be the subject of computer science arises by a computer executing a program according to strict operational rules. The behaviour is typically observed via the computer's input and output (I/O). More technically, the program can be understood as an element in an inductively defined set P of terms. This set forms a suitable (initial) algebra $F(P) \to P$, where the expression (or functor) F captures the signature of the operations for forming programs. The operational rules for the behaviour of programs are described by a coalgebra $P \to G(P)$, where the functor G captures the kind of behaviour that can be displayed – such as deterministic or probabilistic. In abstract form, generated computer behaviour amounts to the repeated evaluation of an (inductively defined) coalgebra structure on an algebra of terms. Hence the algebras (structure) and coalgebras (behaviour) that are studied systematically in this text form the basic matter at the heart of computer science.

One of the big challenges of computer science is to develop techniques for effectively establishing properties of generated behaviour. Often such properties are formulated positively as wanted, functional behaviour. But these properties may also be negative, such as in computer security, where unwanted behaviour must be excluded. However, an elaborate logical view about actual program properties within the combined algebraic/coalgebraic setting has not been fully elaborated yet.

Algebras and Coalgebras

The duality with algebras forms a source of inspiration and of opposition: there is a 'hate-love' relationship between algebra and coalgebra. First, there is a fundamental divide. Think of the difference between an inductively defined data type in a functional programming language (an algebra) and a class in an object-oriented programming language (a coalgebra). The data type is completely determined by its 'constructors': algebraic operations of the form $F(X) \to X$ going *into* the data type. The class however involves an internal state, given by the values of all the public and private fields of the class. This state can be observed (via the public fields) and can be modified (via the public methods). These operations of a class act on a state (or object) and are naturally described as 'destructors' pointing *out of* the class: they are of the coalgebraic form $X \to F(X)$.

Next, besides these differences between algebras and coalgebras there are also many correspondences, analogies and dualities, for instance between bisimulations and congruences or between initiality and finality. Whenever possible, these connections will be made explicit and will be exploited in the course of this work.

As already mentioned, ultimately, stripped to its bare minimum, a programming language involves both a coalgebra and an algebra. A program is a structured element of the algebra that arises (as so-called initial algebra) from the programming language that is being used. Each construct of the language corresponds to certain dynamics (behaviour), captured via a coalgebra. The program's behaviour is thus described by a coalgebra acting on the state space of the computer. This is the view underlying the so-called structural operational semantics. Coalgebraic behaviour is generated by an algebraically structured program. This is a simple, clear and appealing view. It turns out that this approach requires a certain level of compatibility between the algebras and coalgebras involved. It is expressed in terms of so-called distributive laws connecting algebra–coalgebra pairs. These laws appear in Chapter 5.

Coalgebras Have a Black Box State Space

Coalgebra is thus the study of states and their operations and properties. The set of states is best seen as a black box, to which one has limited access –

as with the states of a computer mentioned above. As already mentioned, the tension between what is actually inside and what can be observed externally is at the heart of the theory of coalgebras. Such tension also arises for instance in quantum mechanics where the relation between observables and states is a crucial issue [365]. Similarly, it is an essential element of cryptography that parts of data are not observable – via encryption or hashing. In a coalgebra it may very well be the case that two states are internally different but are indistinguishable as far as one can see with the available operations. In that case one calls the two states *bisimilar* or *observationally equivalent*. Bisimilarity is indeed one of the fundamental notions of the theory of coalgebras; see Chapter 3. Also important are *invariant* properties of states: once such a property holds, it continues to hold no matter which of the available operations is applied; see Chapter 6. Safety properties of systems are typically expressed as invariants. Finally, specifications of the behaviour of systems are conveniently expressed using assertions and modal operators such as: for all direct successor states (nexttime), for all future states (henceforth), for some future state (eventually); see also Chapter 6. This text describes these basic elements of the theory of coalgebras – bisimilarity, invariants and assertions. It is meant as an introduction to this new and fascinating field within theoretical computer science. The text is too limited in both size and aims to justify the grand unifying claims mentioned above. But it is hoped it will inspire and generate much further research in the area.

Brief Historical Perspective

Coalgebra does not come out of the blue. Below we shall sketch several, relatively independent, developments during the last few decades that appeared to have a common coalgebraic basis and that have contributed to the area of coalgebra as it stands today. This short sketch is of course far from complete.

1. **The categorical approach to mathematical system theory.** During the 1970s several people, notably Arbib, Manes and Goguen, and also Adámek, have analysed Kalman's [287] work on linear dynamical systems, in relation to automata theory. They realised that linearity does not really play a role in Kalman's famous results about minimal realisation and duality and that these results could be reformulated and proved more abstractly using elementary categorical constructions. Their aim was 'to place sequential machines and control systems in a unified framework' (abstract of [40]), by developing a notion of 'machine in a category' (see also [13, 14]). This led to general notions of state, behaviour, reachability, observability and

realisation of behaviour. However, the notion of coalgebra did not emerge explicitly in this approach, probably because the setting of modules and vector spaces from which this work arose provided too little categorical infrastructure (especially, no cartesian closure) to express these results purely coalgebraically.

2. **Non-well-founded sets.** Aczel [9] formed a next crucial step with his special set theory that allows infinitely descending $\in$-chains, because it used coalgebraic terminology right from the beginning. The development of this theory was motivated by the desire to provide meaning to concurrent processes with potentially infinite behaviour in Milner's calculus of communicating systems (CCS). Therefore, the notion of bisimulation from process theory played a crucial role. An important contribution of Aczel is that he showed how to treat bisimulation in a coalgebraic setting, especially by establishing the first link between proofs by bisimulations and finality of coalgebras; see also [12, 10].
3. **Data types of infinite objects.** The first systematic approach to data types in computing [165] relied on initiality of algebras. The elements of such algebraic structures are finitely generated objects. However, many data types of interest in computer science (and mathematics) consist of infinite objects, such as infinite lists or trees (or even real numbers). The use of (final) coalgebras in [462, 41, 192, 378] to capture such structures provided a next important step. Such infinite structures can be represented in functional programming languages (typically with lazy evaluation) or in logical programming languages [431, 189, 190].
4. **Initial and final semantics.** In the semantics of program and process languages it appeared that the relevant semantical domains carry the structure of a final coalgebra (sometimes in combination with an initial algebra structure [144, 132]). Especially in the metric space–based tradition (see e.g. [50]) this insight was combined with Aczel's techniques by Rutten and Turi. It culminated in the recognition that 'compatible' algebra–coalgebra pairs (called bialgebras) are highly relevant structures, described via distributive laws. The basic observation of [452, 451], further elaborated in [58], is that such laws correspond to specification formats for operational rules on (inductively defined) programs (see also [298]). These bialgebras satisfy elementary properties such as: observational equivalence (i.e. bisimulation with respect to the coalgebra) is a congruence (with respect to the algebra).
5. **Behavioural approaches in specification.** Reichel [397] was the first to use so-called behavioural validity of equations in the specification of algebraic structures that are computationally relevant. The basic idea is to

divide types (also called sorts) into 'visible' and 'hidden' ones. The latter are supposed to capture states and are not directly accessible. Equality is used only for the 'observable' elements of visible types. For elements of hidden types (or states) one uses behavioural equality instead: two elements x_1 and x_2 of hidden type are behaviourally equivalent if $t(x_1) = t(x_2)$ for each term t of visible type. This means that they are equal as far as can be observed. The idea is further elaborated in what has become known as hidden algebra [164] (see for instance also [152, 418, 68]) and has been applied to describe classes in object-oriented programming languages, which have an encapsulated state space. But it was later realised that behavioural equality is essentially bisimilarity in a coalgebraic context (see e.g. [339]), and it was again Reichel [399] who first used coalgebras for the semantics of object-oriented languages. Later on they have been applied also to actual programming languages such as Java [264, 245].

6. **Modal logic.** A more recent development is the connection between coalgebras and modal logics. In general, such logics qualify the truth conditions of statements concerning knowledge, belief and time. In computer science such logics are used to reason about the way programs behave and to express dynamical properties of transitions between states. Temporal logic is a part of modal logic which is particularly suitable for reasoning about (reactive) state-based systems, as argued for example in [389, 390], via its nexttime and lasttime operators. Since coalgebras give abstract formalisations of such state-based systems one expects a connection. It was Moss [359] who first associated a suitable modal logic to coalgebras – which inspired much subsequent work [403, 404, 322, 230, 243, 373, 317]; see [318] for an overview. The idea is that the role of equational formulas in algebra is played by modal formulas in coalgebra.

Position of This Text

There are several recent texts presenting a synthesis of several of the developments in the area of coalgebra [266, 453, 176, 411, 320, 374, 178, 16, 269]. This text is a first systematic presentation of the subject in the form of a book. Key points are: coalgebras are general dynamical systems; final coalgebras describe behaviour of such systems (often as infinite objects) in which states and observations coincide; bisimilarity expresses observational indistinguishability; the natural logic of coalgebras is modal logic; etc.

During the last decade a 'coalgebraic community' has emerged, centered around the workshops Coalgebraic Methods in Computer Science, see the proceedings [260, 267, 400, 109, 361, 180, 22, 151, 19, 262, 420, 82];

the conferences Coalgebra and Algebra in Computer Science (CALCO), see [130, 363, 326, 108, 209]; and the associated special journal issues [261, 268, 110, 181, 23, 20, 263, 375, 83]. Publications on coalgebras have also appeared outside these dedicated fora, in essentially all conference proceedings and journals in theoretical computer science – and also as a cover story in the *Communications of the ACM* [79]. This book is specifically not solely focused on the coalgebra community but tries to reach a wider audience. This means that the emphasis lies – certainly in the beginning – on explaining the theory via concrete examples and on motivation rather than on generality and (categorical) abstraction.

Coalgebra and Category Theory

Category theory is a modern, abstract mathematical formalism that emerged in the 1940s and 1950s in algebraic topology. It has become the preferred formalism in the area of semantics of datatypes and programming languages: it adequately captures the relevant phenomena and makes it possible to express similarities between different structures (such as sets, domains and metric spaces). The field of coalgebra requires the theory of categories already in the definition of the notion of coalgebra itself – since it uses the concept of a functor. However, the reader is not assumed to know category theory: in this text the intention is not to describe the theory of coalgebras in its highest form of generality, making systematic use of category theory right from the beginning. After all, this is only an introduction. Rather, the text starts from concrete examples and introduces the basics of category theory as it proceeds. Categories will thus be introduced gradually, without making it a proper subject matter. Hopefully, readers unfamiliar with category theory can thus pick up the basics along the way, seeing directly how it is used. Anyway, most of the examples that are discussed live in the familiar standard setting of sets and functions, so that it should be relatively easy to see the underlying categorical structures in a concrete setting. Thus, more or less familiar set-theoretic language is used in the first half of the book, but with a perspective on the greater generality offered by the theory of categories that is more prominent in the second half. In this way we hope to serve the readers without background in category theory, and at the same time offer the more experienced cognoscenti an idea of what is going on at a more abstract level – which they can find to a limited extent in the Exercises but to a greater extent in the literature. Clearly, this is a compromise which runs the risk of satisfying no one: the description may be too abstract for some, and too concrete for others. The hope is that it does have something to offer for everyone.

In the first half of the book (Chapters 1–3) the formalism of categories will not be very prominent, for instance, in the restriction to so-called polynomial functors which can be handled rather concretely. This is motivated by our wish to produce an introduction that is accessible to non-specialists. Certainly, the general perspective is always right around the corner and we hope it will be appreciated once this more introductory material has been digested. Certainly in the second half of the book, starting from Chapter 4, the language of category theory will be inescapable.

Often the theory of categories is seen as a very abstract part of mathematics that is not very accessible. However, this theory is essential in this text, for several good reasons.

1. It greatly helps to properly organise the relevant material on coalgebras.
2. Only by using categorical language can the duality between coalgebra and algebra be fully seen – and exploited.
3. Almost all of the literature on coalgebra uses category theory in one way or another. Therefore, an introductory text that wishes to properly prepare the reader for further study cannot avoid the language of categories.
4. Category helps you to structure your thinking and to ask relevant questions: ah, this mapping is a functor, so what structure does it preserve? Does it have an adjoint?

In the end, we think that coalgebras form a very basic and natural mathematical concept and that their identification is real step forward. Many people seem to be using coalgebras in various situations, without being aware of it. It is hoped that this text will make them aware and will contribute to a better understanding and exploitation of these situations. And it is hoped that many more such application areas will be identified, further enriching the theory of coalgebras.

Intended Audience

This text is written for everyone with an interest in the mathematical aspects of computational behaviour. This probably includes primarily mathematicians, logicians and (theoretical) computer scientists, but hopefully also an audience with a different background such as mathematical physics, biology or even economics. A basic level of mathematical maturity is assumed, for instance familiarity with elementary set theory and logic (and its notation). The examples in the text are taken from various areas. Each section is accompanied by a series of exercises, to facilitate teaching – typically at a late Bachelor's or early Master's level – and for testing one's own understanding in self-study.

Acknowledgements

An earlier version of this book has been on the web for quite some time. This generated useful feedback from many people. In fact, there are too many of them to mention individually here. Therefore I would like to thank everyone in the coalgebra community (and beyond) for their cooperation, feedback, help, advice, wisdom, insight, support and encouragement.

1
Motivation

This chapter tries to explain why coalgebras are interesting structures in mathematics and computer science. It does so via several examples. The notation used for these examples will be explained informally, as we proceed. The emphasis at this stage is not so much on precision in explanation but on transfer of ideas and intuitions. Therefore, for the time being we define a coalgebra – very informally – to be a function of the form

$$S \xrightarrow{\quad c \quad} \boxed{\;\cdots \quad S \quad \cdots\;} \,. \tag{1.1}$$

What we mean is: a coalgebra is given by a set S and a function c with S as domain and with a 'structured' codomain (result, output, the box $\boxed{\cdots}$), in which the domain S may occur again. The precise general form of these codomain boxes is not of immediate concern.

Some terminology: We often call S the *state space* or *set of states* and say that the coalgebra *acts on* S. The function c is sometimes called the *transition function* or *transition structure*. The idea that will be developed is that coalgebras describe general 'state-based systems' provided with 'dynamics' given by the function c. For a state $x \in S$, the result $c(x)$ tells us what the successor states of x are, if any. The codomain $\boxed{\cdots}$ is often called the *type* or *interface* of the coalgebra. Later we shall see that it is a *functor*.

A simple example of a coalgebra is the function

$$\mathbb{Z} \xrightarrow{\quad n \,\mapsto\, (n-1,\, n+1) \quad} \mathbb{Z} \times \mathbb{Z}$$

with state space $\mathbb{Z}$ occurring twice on the right-hand side. Thus the box or type of this coalgebra is: $\boxed{(-) \times (-)}$. The transition function $n \mapsto (n-1, n+1)$

may also be written using λ-notation as $\lambda n.\,(n-1, n+1)$ or as $\lambda n \in \mathbb{Z}.\,(n-1, n+1)$.

Another example of a coalgebra, this time with state space the set $A^{\mathbb{N}}$ of functions from $\mathbb{N}$ to some given set A, is

$$A^{\mathbb{N}} \xrightarrow{\quad \sigma \mapsto (\sigma(0),\, \lambda n.\, \sigma(n+1)) \quad} A \times A^{\mathbb{N}}.$$

In this case the box is $\boxed{A \times (-)}$. If we write σ as an infinite sequence $(\sigma_n)_{n \in \mathbb{N}}$ we may write this coalgebra as a pair of functions $\langle \mathsf{head}, \mathsf{tail} \rangle$ where

$$\mathsf{head}((\sigma_n)_{n\in\mathbb{N}}) = \sigma_0 \qquad \text{and} \qquad \mathsf{tail}((\sigma_n)_{n\in\mathbb{N}}) = (\sigma_{n+1})_{n\in\mathbb{N}}.$$

Many more examples of coalgebras will occur throughout this text.

This chapter is devoted to 'selling' and 'promoting' coalgebras. It does so by focusing on the following topics.

1. A representation as a coalgebra (1.1) is often very natural, from the perspective of state-based computation.
2. There are powerful 'coinductive' definition and proof principles for coalgebras.
3. There is a very natural (and general) temporal logic associated with coalgebras.
4. The coalgebraic notions are on a suitable level of abstraction, so that they can be recognised and used in various settings.

Full appreciation of this last point requires some familiarity with basic category theory. It will be provided in Section 1.4.

Remark 1.0.1 Readers with a mathematical background may be familiar with the notion of coalgebra as comonoid in vector spaces, dual to an algebra as a monoid. In that case one has a 'counit' map $V \to K$, from the carrier space V to the underlying field K, together with a 'comultiplication' $V \to V \otimes V$. These two maps can be combined into a single map $V \to K \times (V \otimes V)$ of the form (1.1), forming a coalgebra in the present sense. The notion of coalgebra used here is thus much more general than the purely mathematical one.

1.1 Naturalness of Coalgebraic Representations

We turn first to an area where coalgebraic representations as in (1.1) occur naturally and may be useful, namely programming languages – used for writing computer programs. What are programs, and what do they do? Well, programs are lists of instructions telling a computer what to do. Fair enough. But what are programs from a mathematical point of view? Put differently, what do programs mean?[1] One view is that programs are certain functions that take an input and use it to compute a certain result. This view does not cover all programs: certain programs, often called processes, are meant to be running forever, like operating systems, without really producing a result. But we shall follow the view of programs as functions for now. The programs we have in mind work not only on input, but also on what is usually called a state, for example for storing intermediate results. The effect of a program on a state is not immediately visible and is therefore often called the *side effect* of the program. One may think of the state as given by the contents of the memory in the computer that is executing the program. This is not directly observable.

Our programs should thus be able to modify a state, typically via an assignment like `i = 5` in a so-called imperative programming language. Such an assignment statement is interpreted as a function that turns a state x into a new, successor state x' in which the value of the identifier `i` is equal to `5`. Statements in such languages are thus described via suitable 'state transformer' functions. In simplest form, ignoring input and output, they map a state to a successor state, as in

$$S \xrightarrow{\quad \mathsf{stat} \quad} S, \tag{1.2}$$

where we have written S for the set of states. Its precise structure is not relevant. Often the set S of states is considered to be a 'black box' to which we do not have direct access, so that we can observe only certain aspects. For instance, the function $i \colon S \to \mathbb{Z}$ representing the above integer `i` allows us to observe the value of i. The value $i(x')$ should be 5 in the result state x' after evaluating the assignment `i = 5`, considered as a function $S \to S$, as in (1.2).

This description of statements as functions $S \to S$ is fine as first approximation, but one quickly realises that statements do not always terminate normally and produce a successor state. Sometimes they can 'hang' and continue to compute without ever producing a successor state. This typically happens because of an infinite loop, for example in a `while` statement or because of a recursive call without exit.

There are two obvious ways to incorporate such non-termination.

1. **Adjust the state space**. In this case one extends the state space S to a space $S_\bot \stackrel{\text{def}}{=} \{\bot\} \cup S$, where $\bot$ is a new 'bottom' element not occurring in

[1] This question comes up frequently when confronted with two programs – one possibly as a transformation from the other – which perform the same task in a different manner and which could thus be seen as the same program. But how can one make precise that they are the same?

S that is especially used to signal non-termination. Statements then become functions:

$$S_\perp \xrightarrow{\;\mathsf{stat}\;} S_\perp \quad \text{with the requirement} \quad \mathsf{stat}(\perp) = \perp.$$

The side-condition expresses the idea that once a statement hangs it will continue to hang.

The disadvantage of this approach is that the state space becomes more complicated and that we have to make sure that all statements satisfy the side-condition, namely that they preserve the bottom element $\perp$. But the advantage is that composition of statements is just function composition.

2. **Adjust the codomain**. The second approach keeps the state space S as it is but adapts the codomain of statements, as in

$$S \xrightarrow{\;\mathsf{stat}\;} S_\perp \quad \text{where, recall,} \quad S_\perp = \{\perp\} \cup S.$$

In this representation we easily see that in each state $x \in S$ the statement can either hang, when $\mathsf{stat}(x) = \perp$, or terminate normally, namely when $\mathsf{stat}(x) = x'$ for some successor state $x' \in S$. What is good is that there are no side-conditions anymore. But composition of statements cannot be defined via function composition, because the types do not match. Thus the types force us to deal explicitly with the propagation of non-termination: for these kind of statements $s_1, s_2 \colon S \to S_\perp$ the composition $s_1 \, ; s_2$, as a function $S \to S_\perp$, is defined via a case distinction (or pattern match) as

$$s_1 \, ; s_2 \;=\; \lambda x \in S. \begin{cases} \perp & \text{if } s_1(x) = \perp \\ s_2(x') & \text{if } s_1(x) = x'. \end{cases}$$

This definition is more difficult than function composition (as used in point 1 above), but it explicitly deals with the case distinction that is of interest, namely between non-termination and normal termination. Hence being forced to make these distinctions explicitly is maybe not so bad at all.

We push these same ideas a bit further. In many programming languages (such as Java [43]) programs may not only hang, but also terminate ‘abruptly’ because of an exception. An exception arises when some constraint is violated, such as a division by zero or an access `a[i]` in an array `a` which is a null-reference. Abrupt termination is fundamentally different from non-termination: non-termination is definitive and irrevocable, whereas a program can recover from abrupt termination via a suitable exception handler that restores normal termination. In Java this is done via a `try-catch` statement; see for instance [43, 173, 240].

Let us write E for the set of exceptions that can be thrown. Then there are again two obvious representations of statements that can terminate normally or abruptly or can hang.

1. **Adjust the state space**. Statements then remain endofunctions[2] on an extended state space:

$$\big(\{\bot\} \cup S \cup (S \times E)\big) \xrightarrow{\ \mathsf{stat}\ } \big(\{\bot\} \cup S \cup (S \times E)\big).$$

   The entire state space clearly becomes complicated now. But also the side-conditions are becoming non-trivial: we still want $\mathsf{stat}(\bot) = \bot$, and also $\mathsf{stat}(x, e) = (x, e)$, for $x \in S$ and $e \in E$, but the latter only for non-catch statements. Keeping track of such side-conditions may easily lead to mistakes. But on the positive side, composition of statements is still function composition in this representation.
2. **Adjust the codomain**. The alternative approach is again to keep the state space S as it is, but to adapt the codomain type of statements, namely as

$$S \xrightarrow{\ \mathsf{stat}\ } \big(\{\bot\} \cup S \cup (S \times E)\big). \tag{1.3}$$

   Now we do not have side-conditions and we can clearly distinguish the three possible termination modes of statements. This structured output type in fact forces us to make these distinctions in the definition of the composition $s_1 \,;\, s_2$ of two such statements $s_1, s_2 \colon S \to \{\bot\} \cup S \cup (S \times E)$, as in

$$s_1 \,;\, s_2 \;=\; \lambda x \in S.\begin{cases} \bot & \text{if } s_1(x) = \bot \\ s_2(x') & \text{if } s_1(x) = x' \\ (x', e) & \text{if } s_1(x) = (x', e). \end{cases}$$

   Thus, if s_1 hangs or terminates abruptly, then the subsequent statement s_2 is not executed. This is very clear in this second *coalgebraic* representation.

   When such a coalgebraic representation is formalised within the typed language of a theorem prover (as in [265]), the type checker of the theorem prover will make sure that appropriate case distinctions are made, according to the output type as in (1.3). See also [240] where Java's exception mechanism is described via such case distinctions, closely following the official language definition [173].

These examples illustrate that coalgebras as functions with structured codomains $\boxed{\cdots}$, as in (1.1), arise naturally and that the structure of the

[2] An endofunction is a function $A \to A$ from a set A to itself.

codomain indicates the kind of computations that can be performed. This idea will be developed further and applied to various forms of computation. For instance, non-deterministic statements may be represented via the powerset $\mathcal{P}$ as coalgebraic state transformers $S \rightarrow \mathcal{P}(S)$ with multiple result states. But there are many more such examples, involving for instance probability distributions on states.

(Readers familiar with computational monads [357] may recognise similarities. Indeed, in a computational setting there is a close connection between coalgebraic and monadic representations. Briefly, the monad introduces the computational structure, like composition and extension, whereas the coalgebraic view leads to an appropriate program logic. This is elaborated for Java in [264].)

Exercises

1.1.1 1. Prove that the composition operation ; as defined for coalgebras $S \rightarrow \{\perp\} \cup S$ is associative, i.e. satisfies $s_1 ;(s_2 ; s_3) = (s_1 ; s_2) ; s_3$, for all statements $s_1, s_2, s_3 \colon S \rightarrow \{\perp\} \cup S$.

Define a statement $\mathsf{skip} \colon S \rightarrow \{\perp\} \cup S$ which is a unit for composition ; i.e. which satisfies $(\mathsf{skip} ; s) = s = (s ; \mathsf{skip})$, for all $s \colon S \rightarrow \{\perp\} \cup S$.

2. Do the same for ; defined on coalgebras $S \rightarrow \{\perp\} \cup S \cup (S \times E)$. (In both cases, statements with an associative composition operation and a unit element form a monoid.)

1.1.2 Define also a composition monoid $(\mathsf{skip}, ;)$ for coalgebras $S \rightarrow \mathcal{P}(S)$.

1.2 The Power of the Coinduction

In this section we shall look at sequences – or lists or words, as they are also called. Sequences are basic data structures, both in mathematics and in computer science. One can distinguish finite sequences $\langle a_1, \ldots, a_n \rangle$ and infinite $\langle a_1, a_2, \ldots \rangle$ ones. The mathematical theory of finite sequences is well understood and a fundamental part of computer science, used in many programs (notably in the language Lisp). Definition and reasoning with finite lists is commonly done with induction. As we shall see, infinite lists require *coinduction*. Infinite sequences can arise in computing as the observable outcomes of a program that runs forever. Also, in functional programming, they can occur as so-called lazy lists, as in the languages Haskell [71] or Clean [381]. Modern extensions of logical programming languages have support for infinite sequences [431, 189]. (Logic programming languages themselves can also be described coalgebraically; see [78, 107, 305, 306, 80, 81].)

In the remainder of this section we shall use an arbitrary but fixed set A and wish to look at both finite $\langle a_1, \ldots, a_n \rangle$ and infinite $\langle a_1, a_2, \ldots \rangle$ sequences of elements a_i of A. The set A may be understood as a parameter, and our sequences are thus parametrised by A, or, put differently, are polymorphic in A.

We shall develop a slightly unusual and abstract perspective on sequences. It treats sequences not as completely given at once but as arising in a local, step-by-step manner. This coalgebraic approach relies on the following basic fact. It turns out that the set of both finite and infinite sequences enjoys a certain 'universal' property, namely that it is a *final* coalgebra (of suitable type). We shall explain what this means and how this special property can be exploited to define various operations on sequences and to prove properties about them. A special feature of this universality of the final coalgebra of sequences is that it avoids making the (global) distinction between finiteness and infiniteness for sequences.

First some notation. We write $A^\star$ for the set of *finite* sequences $\langle a_1, \ldots, a_n \rangle$ (or lists or words) of elements $a_i \in A$, and $A^{\mathbb{N}}$ for the set of infinite ones: $\langle a_1, a_2, \ldots \rangle$. The latter may also be described as functions $a_{(-)} \colon \mathbb{N} \to A$, which explains the exponent notation in $A^{\mathbb{N}}$. Sometimes, the infinite sequences in $A^{\mathbb{N}}$ are called *streams*. Finally, the set of both finite and infinite sequences A^∞ is then the (disjoint) union $A^\star \cup A^{\mathbb{N}}$.

The set of sequences A^∞ carries a coalgebra or transition structure, which we simply call next. It tries to decompose a sequence into its head and tail, if any. Hence one may understand next as a partial function. But we describe it as a total function which possibly outputs a special element $\perp$ for undefined:

$$
\begin{array}{ccc}
A^\infty & \xrightarrow{\qquad\qquad \mathsf{next} \qquad\qquad} & \{\perp\} \cup (A \times A^\infty) \\
\sigma & \longmapsto & \begin{cases} \perp & \text{if } \sigma \text{ is the empty sequence } \langle\rangle \\ (a, \sigma') & \text{if } \sigma = a \cdot \sigma' \text{ with head } a \in A \text{ and tail } \sigma' \in A^\infty. \end{cases}
\end{array}
\tag{1.4}
$$

The type of the coalgebra is thus $\boxed{\{\perp\} \cup (A \times (-))}$, as in (1.1), with A^∞ as state space that is plugged in the hole $(-)$ in the box. The successor of a state $\sigma \in A^\infty$, if any, is its tail sequence, obtained by removing the head.

The function next captures the external view on sequences: it tells what can be *observed* about a sequence σ, namely whether or not it is empty, and if not, what its head is. By repeated application of the function next all observable elements of the sequence appear. This 'observational' approach is fundamental in coalgebra.

A first point to note is that this function next is an isomorphism: its inverse next^{-1} sends $\perp$ to the empty sequence $\langle\rangle$ and a pair $(a, \tau) \in A \times A^\infty$ to the sequence $a \cdot \tau$ obtained by prefixing a to τ.

The following result describes a crucial 'finality' property of sequences that can be used to characterise the set A^∞. Indeed, as we shall see later in Lemma 2.3.3, final coalgebras are unique, up-to-isomorphism.

Proposition 1.2.1 (Finality of sequences) *The coalgebra* $\mathsf{next}\colon A^\infty \to \{\perp\} \cup A \times A^\infty$ *from (1.4) is final among coalgebras of this type: for an arbitrary coalgebra* $c\colon S \to \{\perp\} \cup (A \times S)$ *on a set* S *there is a unique 'behaviour' function* $\mathrm{beh}_c\colon S \to A^\infty$ *which is a homomorphism of coalgebras. That is, for each* $x \in S$, *both*

- *if* $c(x) = \perp$, *then* $\mathsf{next}(\mathrm{beh}_c(x)) = \perp$.
- *if* $c(x) = (a, x')$, *then* $\mathsf{next}(\mathrm{beh}_c(x)) = (a, \mathrm{beh}_c(x'))$.

Both these two points can be combined in a commuting diagram, namely as

$$\begin{array}{ccc} \{\perp\} \cup (A \times S) & \xrightarrow{\ \mathrm{id} \cup (\mathrm{id} \times \mathrm{beh}_c)\ } & \{\perp\} \cup (A \times A^\infty) \\ {\scriptstyle c}\uparrow & & \uparrow{\scriptstyle \cong\ \mathsf{next}} \\ S & \xrightarrow[\ \mathrm{beh}_c\ ]{} & A^\infty \end{array}$$

where the function $\mathrm{id} \cup (\mathrm{id} \times \mathrm{beh}_c)$ *on top maps* $\perp$ *to* $\perp$ *and* (a, x) *to* $(a, \mathrm{beh}_c(x))$.

In general, we shall write dashed arrows, like above, for maps that are uniquely determined. In the course of this chapter we shall see that a general notion of homomorphism between coalgebras (of the same type) can be defined by such commuting diagrams.

Proof The idea is to obtain the required behaviour function $\mathrm{beh}_c\colon S \to A^\infty$ via repeated application of the given coalgebra c as follows.

$$\mathrm{beh}_c(x) = \begin{cases} \langle\rangle & \text{if } c(x) = \perp \\ \langle a\rangle & \text{if } c(x) = (a, x') \wedge c(x') = \perp \\ \langle a, a'\rangle & \text{if } c(x) = (a, x') \wedge c(x') = (a', x'') \wedge c(x'') = \perp \\ \vdots & \end{cases}$$

Doing this formally requires some care. We define for $n \in \mathbb{N}$ an iterated version $c^n\colon S \to \{\perp\} \cup A \times S$ of c as

$$\begin{aligned} c^0(x) &= c(x) \\ c^{n+1}(x) &= \begin{cases} \perp & \text{if } c^n(x) = \perp \\ c(y) & \text{if } c^n(x) = (a, y). \end{cases} \end{aligned}$$

Obviously, $c^n(x) \neq \bot$ implies $c^m(x) \neq \bot$, for $m < n$. Thus we can define

$$\mathrm{beh}_c(x) = \begin{cases} \langle a_0, a_1, a_2, \ldots \rangle & \text{if } \forall n \in \mathbb{N}.\, c^n(x) \neq \bot, \text{ and } c^i(x) = (a_i, x_i) \\ \langle a_0, \ldots, a_{m-1} \rangle & \text{if } m \in \mathbb{N} \text{ is the least number with } c^m(x) = \bot, \\ & \text{and } c^i(x) = (a_i, x_i), \text{ for } i < m. \end{cases}$$

We check the two conditions for homomorphism from the proposition above.

- If $c(x) = \bot$, then the least m with $c^m(x) = \bot$ is 0, so that $\mathrm{beh}_c(x) = \langle\rangle$, and thus also $\mathsf{next}(\mathrm{beh}_c(x)) = \bot$.
- If $c(x) = (a, x')$, then we distinguish two cases:
  - If $\forall n \in \mathbb{N}.\, c^n(x) \neq \bot$, then $\forall n \in \mathbb{N}.\, c^n(x') \neq \bot$, and $c^{i+1}(x) = c^i(x')$. Let $c^i(x') = (a_i, x_i)$, then

$$\begin{aligned} \mathsf{next}(\mathrm{beh}_c(x)) &= \mathsf{next}(\langle a, a_0, a_1, \ldots \rangle) \\ &= (a, \langle a_0, a_1, \ldots \rangle) \\ &= (a, \mathrm{beh}_c(x')). \end{aligned}$$

  - If m is least with $c^m(x) = \bot$, then $m > 0$ and $m - 1$ is the least k with $c^k(x') = \bot$. For $i < m - 1$ we have $c^{i+1}(x) = c^i(x')$, and thus by writing $c^i(x') = (a_i, x_i)$, we get as before:

$$\begin{aligned} \mathsf{next}(\mathrm{beh}_c(x)) &= \mathsf{next}(\langle a, a_0, a_1, \ldots, a_{m-2} \rangle) \\ &= (a, \langle a_0, a_1, \ldots, a_{m-2} \rangle) \\ &= (a, \mathrm{beh}_c(x')). \end{aligned}$$

Finally, we still need to prove that this behaviour function beh_c is the unique homomorphism from c to next. Thus, assume also $g\colon S \to A^\infty$ is such that $c(x) = \bot \Rightarrow \mathsf{next}(g(x)) = \bot$ and $c(x) = (a, x') \Rightarrow \mathsf{next}(g(x)) = (a, g(x'))$. We then distinguish:

- $g(x)$ is infinite, say $\langle a_0, a_1, \ldots \rangle$. Then one shows by induction that for all $n \in \mathbb{N}$, $c^n(x) = (a_n, x_n)$, for some x_n. This yields $\mathrm{beh}_c(x) = \langle a_0, a_1, \ldots \rangle = g(x)$.
- $g(x)$ is finite, say $\langle a_0, \ldots, a_{m-1} \rangle$. Then one proves that for all $n < m$, $c^n(x) = (a_n, x_n)$, for some x_n, and $c^m(x) = \bot$. So also now, $\mathrm{beh}_c(x) = \langle a_0, \ldots, a_{m-1} \rangle = g(x)$. ❑

Before exploiting this finality result we illustrate the behaviour function.

Example 1.2.2 (Decimal representations as behaviour) So far we have considered sequence coalgebras parametrised by an arbitrary set A. In this example we take a special choice, namely $A = \{0, 1, 2, 3, 4, 5, 6, 7, 8, 9\}$, the set of decimal digits. We wish to define a coalgebra (or machine) which generates decimal representations of real numbers in the unit interval $[0, 1) \subseteq \mathbb{R}$. Notice that this may give rise to both finite sequences ($\frac{1}{8}$ should yield the sequence $\langle 1, 2, 5\rangle$, for 0.125) and infinite ones ($\frac{1}{3}$ should give $\langle 3, 3, 3, \ldots\rangle$ for $0.333\ldots$).

The coalgebra we are looking for computes the first decimal of a real number $r \in [0, 1)$. Hence it should be of the form

$$[0, 1) \xrightarrow{\;\mathsf{nextdec}\;} \{\bot\} \cup \Big(\{0, 1, 2, 3, 4, 5, 6, 7, 8, 9\} \times [0, 1)\Big)$$

with state space $[0, 1)$. How to define nextdec? Especially, when does it stop (i.e. return $\bot$), so that a finite sequence is generated? Well, the decimal representation 0.125 may be identified with 0.12500000... with a tail of infinitely many zeros. Clearly, we wish to map such infinitely many zeros to $\bot$. Fair enough, but it does have as consequence that the real number $0 \in [0, 1)$ gets represented as the empty sequence.

A little thought brings us to the following:

$$\mathsf{nextdec}(r) = \begin{cases} \bot & \text{if } r = 0 \\ (d, 10r - d) & \text{otherwise, with } d \leq 10r < d + 1 \text{ for } d \in A. \end{cases}$$

Notice that this function is well defined, because in the second case the successor state $10r - d$ is within the interval $[0, 1)$.

According to the previous proposition, this nextdec coalgebra gives rise to a behaviour function:

$$[0, 1) \xrightarrow{\;\mathsf{beh}_{\mathsf{nextdec}}\;} \Big(\{0, 1, 2, 3, 4, 5, 6, 7, 8, 9\}\Big)^{\infty}.$$

In order to understand what it does, i.e. which sequences are generated by nextdec, we consider two examples.

Starting from $\frac{1}{8} \in [0, 1)$ we get

$$\begin{aligned} \mathsf{nextdec}(\tfrac{1}{8}) &= (1, \tfrac{1}{4}) \text{ because } 1 \leq \tfrac{10}{8} < 2 \text{ and } \tfrac{10}{8} - 1 = \tfrac{1}{4} \\ \mathsf{nextdec}(\tfrac{1}{4}) &= (2, \tfrac{1}{2}) \text{ because } 2 \leq \tfrac{10}{4} < 3 \text{ and } \tfrac{10}{4} - 2 = \tfrac{1}{2} \\ \mathsf{nextdec}(\tfrac{1}{2}) &= (5, 0) \text{ because } 5 \leq \tfrac{10}{2} < 6 \text{ and } \tfrac{10}{2} - 5 = 0 \\ \mathsf{nextdec}(0) &= \bot. \end{aligned}$$

Thus the resulting nextdec-behaviour on $\frac{1}{8}$ is $\langle 1, 2, 5\rangle$, i.e. $\mathsf{beh}_{\mathsf{nextdec}}(\frac{1}{8}) = \langle 1, 2, 5\rangle$. Indeed, in decimal notation we write $\frac{1}{8} = 0.125$.

Next, when we run nextdec on $\frac{1}{9} \in [0, 1)$ we see that

$$\mathsf{nextdec}(\tfrac{1}{9}) = (1, \tfrac{1}{9}) \text{ because } 1 \leq \tfrac{10}{9} < 2 \text{ and } \tfrac{10}{9} - 1 = \tfrac{1}{9}.$$

The function nextdec thus immediately loops on $\frac{1}{9}$, giving an infinite sequence $\langle 1, 1, 1, \ldots\rangle$ as behaviour. This corresponds to the fact that we can identify $\frac{1}{9}$ with the infinite decimal representation $0.11111\ldots$.

One sees in the proof of Proposition 1.2.1 that manipulating sequences via their elements is cumbersome and requires us to distinguish between finite and infinite sequences. However, the nice thing about the finality property of A^∞ is that we do not have to work this way anymore. This property states two important aspects, namely *existence* and *uniqueness* of a homomorphism $S \to A^\infty$ into the set of sequences, provided we have a coalgebra structure on S. These two aspects give us two principles:

- **A coinductive definition principle.** The existence aspect tells us how to obtain functions $S \to A^\infty$ into A^∞. If 'recursion' is the appropriate term for definition by induction, then the existence property at hand may be called 'corecursion'.
- **A coinductive proof principle.** The uniqueness aspect tells us how to prove that two functions $f, g \colon S \to A^\infty$ are equal, namely by showing that they are both homomorphisms from a single coalgebra $c \colon S \to \{\bot\} \cup (A \times S)$ to the final coalgebra $\mathsf{next} \colon A^\infty \to \{\bot\} \cup (A \times A^\infty)$.

Coinduction is thus the use of finality – just as induction is the use of initiality, as will be illustrated in Section 2.4. We shall see several examples of the use of these definition and proof principles for sequences in the remainder of this section.

Notation. The previous proposition shows us that coalgebras $c \colon S \to \{\bot\} \cup (A \times S)$ can be understood as generators of sequences, namely via the resulting behaviour function $\mathrm{beh}_c \colon S \to A^\infty$. Alternatively, these coalgebras can be understood as certain automata. The behaviour of a state $x \in S$ of this automaton is then the resulting sequence $\mathrm{beh}_c(x) \in A^\infty$. These sequences $\mathrm{beh}_c(x)$ show only the external behaviour and need not tell everything about states.

Given this behaviour-generating perspective on coalgebras, it will be convenient to use a transition-style notation. For a state $x \in S$ of an arbitrary coalgebra $c \colon S \to \{\bot\} \cup (A \times S)$ we shall often write

$$x \nrightarrow \quad \text{if } c(x) = \bot \qquad \text{and} \qquad x \xrightarrow{a} x' \quad \text{if } c(x) = (a, x'). \tag{1.5}$$

In the first case there is no transition starting from the state x: the automaton c halts immediately at x. In the second case one can do a c-computation starting with x; it produces an observable element $a \in A$ and results in a successor state x'.

This transition notation can be used in particular for the final coalgebra $\mathsf{next} \colon A^\infty \to \{\bot\} \cup (A \times A^\infty)$. In that case, for $\sigma \in A^\infty$, $\sigma \nrightarrow$ means that the sequence σ is empty. In the second case $\sigma \xrightarrow{a} \sigma'$ expresses that the sequence σ can do an a-step to σ', and hence that $\sigma = a \cdot \sigma'$.

Given this new notation we can reformulate the two homomorphism requirements from Proposition 1.2.1 as two implications:

- $x \nrightarrow \implies \mathrm{beh}_c(x) \nrightarrow$
- $x \xrightarrow{a} x' \implies \mathrm{beh}_c(x) \xrightarrow{a} \mathrm{beh}_c(x')$.

In the tradition of operational semantics, such implications can also be formulated as rules:

$$\frac{x \nrightarrow}{\mathrm{beh}_c(x) \nrightarrow} \qquad\qquad \frac{x \xrightarrow{a} x'}{\mathrm{beh}_c(x) \xrightarrow{a} \mathrm{beh}_c(x')}. \tag{1.6}$$

Such rules thus describe implications: (the conjunction of) what is above the line implies what is below.

In the remainder of this section we consider examples of the use of coinductive definition and proof principles for sequences.

Evenly Listed Elements from a Sequence

Our first aim is to take a sequence $\sigma \in A^\infty$ and turn it into a new sequence $\mathsf{evens}(\sigma) \in A^\infty$ consisting only of the elements of σ at even positions. Step by step we will show how such a function $\mathsf{evens}\colon A^\infty \to A^\infty$ can be defined within a coalgebraic framework, using finality.

Our informal description of $\mathsf{evens}(\sigma)$ can be turned into three requirements:

- If $\sigma \nrightarrow$, then $\mathsf{evens}(\sigma) \nrightarrow$, i.e. if σ is empty, then $\mathsf{evens}(\sigma)$ should also be empty.
- If $\sigma \xrightarrow{a} \sigma'$ and $\sigma' \nrightarrow$, then $\mathsf{evens}(\sigma) \xrightarrow{a} \sigma'$. Thus if σ is the singleton sequence $\langle a \rangle$, then also $\mathsf{evens}(\sigma) = \langle a \rangle$. Notice that by the previous point we could equivalently require $\mathsf{evens}(\sigma) \xrightarrow{a} \mathsf{evens}(\sigma')$ in this case.
- If $\sigma \xrightarrow{a} \sigma'$ and $\sigma' \xrightarrow{a'} \sigma''$, then $\mathsf{evens}(\sigma) \xrightarrow{a} \mathsf{evens}(\sigma'')$. This means that if σ has head a and tail σ', which in its turn has head a' and tail σ'', i.e. if $\sigma = a \cdot a' \cdot \sigma''$, then $\mathsf{evens}(\sigma)$ should have head a and tail $\mathsf{evens}(\sigma'')$, i.e. then $\mathsf{evens}(\sigma) = a \cdot \mathsf{evens}(\sigma'')$. Thus, the intermediate head at odd position is skipped. This is repeated 'coinductively': as long as needed.

As in (1.6) above we can write these three requirements as rules:

$$\frac{\sigma \nrightarrow}{\mathsf{evens}(\sigma) \nrightarrow} \qquad \frac{\sigma \xrightarrow{a} \sigma' \qquad \sigma' \nrightarrow}{\mathsf{evens}(\sigma) \xrightarrow{a} \mathsf{evens}(\sigma')} \qquad \frac{\sigma \xrightarrow{a} \sigma' \qquad \sigma' \xrightarrow{a'} \sigma''}{\mathsf{evens}(\sigma) \xrightarrow{a} \mathsf{evens}(\sigma'')}. \tag{1.7}$$

One could say that these rules give an 'observational description' of the sequence $\mathsf{evens}(\sigma)$: they describe what we can observe about $\mathsf{evens}(\sigma)$ in terms of what we can observe about σ. For example, if $\sigma = \langle a_0, a_1, a_2, a_3, a_4\rangle$ we can compute

$$\begin{aligned} \mathsf{evens}(\sigma) &= a_0 \cdot \mathsf{evens}(\langle a_2, a_3, a_4\rangle) \\ &= a_0 \cdot a_2 \cdot \mathsf{evens}(\langle a_4\rangle) \\ &= a_0 \cdot a_2 \cdot a_4 \cdot \langle\rangle \\ &= \langle a_0, a_2, a_4\rangle. \end{aligned}$$

Now that we have a reasonable understanding of the function $\mathsf{evens}\colon A^\infty \to A^\infty$ we will see how it arises within a coalgebraic setting. In order to define it coinductively, following the finality mechanism of Proposition 1.2.1, we need to have a suitable coalgebra structure e on the domain A^∞ of the function evens, as in a diagram:

$$\begin{array}{ccc} \{\bot\} \cup (A \times A^\infty) & \xrightarrow{\ \mathrm{id} \cup (\mathrm{id} \times \mathrm{beh}_e)\ } & \{\bot\} \cup (A \times A^\infty) \\ {\scriptstyle e}\uparrow & & \uparrow{\scriptstyle \cong\ \mathsf{next}} \\ A^\infty & \xrightarrow[\ \mathsf{evens} = \mathrm{beh}_e\ ]{} & A^\infty \end{array}$$

That is, for $\sigma \in A^\infty$,

- if $e(\sigma) = \bot$, then $\mathsf{evens}(\sigma) \not\to$
- if $e(\sigma) = (a, \sigma')$, then $\mathsf{evens}(\sigma) \xrightarrow{a} \mathsf{evens}(\sigma')$.

Combining these two points with the above three rules (1.7) we see that the coalgebra e must be

$$e(\sigma) = \begin{cases} \bot & \text{if } \sigma \not\to \\ (a, \sigma') & \text{if } \sigma \xrightarrow{a} \sigma' \text{ with } \sigma' \not\to \\ (a, \sigma'') & \text{if } \sigma \xrightarrow{a} \sigma' \text{ and } \sigma' \xrightarrow{a'} \sigma''. \end{cases}$$

This function e thus tells what can be observed immediately, if anything, and what will be used in the recursion (or corecursion, if you like). It contains the same information as the above three rules. In the terminology used earlier: the coalgebra or automaton e generates the behaviour of evens.

Remark 1.2.3 The coalgebra $e\colon A^\infty \to \{\bot\} \cup (A \times A^\infty)$ illustrates the difference between states and observables. Consider an arbitrary sequence $\sigma \in A^\infty$ and write $\sigma_1 = a \cdot a_1 \cdot \sigma$ and $\sigma_2 = a \cdot a_2 \cdot \sigma$, where $a, a_1, a_2 \in A$ with $a_1 \neq a_2$. These $\sigma_1, \sigma_2 \in A^\infty$ are clearly different states of the coalgebra $e\colon A^\infty \to \{\bot\} \cup (A \times A^\infty)$, but they have the same behaviour: $\mathsf{evens}(\sigma_1) = a \cdot \mathsf{evens}(\sigma) =$

$\mathsf{evens}(\sigma_2)$, where $\mathsf{evens} = \mathrm{beh}_e$. Such observational indistinguishability of the states σ_1, σ_2 is called bisimilarity, written as $\sigma_1 \,\underline{\leftrightarrow}\, \sigma_2$, and will be studied systematically in Chapter 3. Bisimilarity depends on the operations that are available. In this illustration we assume that we have only the coalgebra e.

Oddly Listed Elements from a Sequence

Next we would like to have a similar function $\mathsf{odds}\colon A^\infty \to A^\infty$ which extracts the elements at odd positions. We leave formulation of the appropriate rules to the reader, and claim this function odds can be defined coinductively via the behaviour-generating coalgebra $o\colon A^\infty \to \{\bot\} \cup (A \times A^\infty)$ given by:

$$o(\sigma) = \begin{cases} \bot & \text{if } \sigma \not\to \text{ or } \sigma \xrightarrow{a} \sigma' \text{ with } \sigma' \not\to \\ (a', \sigma'') & \text{if } \sigma \xrightarrow{a} \sigma' \text{ and } \sigma' \xrightarrow{a'} \sigma''. \end{cases} \tag{1.8}$$

Thus, we take $\mathsf{odds} = \mathrm{beh}_o$ to be the behaviour function resulting from o, following the finality principle of Proposition 1.2.1. Hence $o(\sigma) = \bot \Rightarrow \mathsf{odds}(\sigma) \not\to$ and $o(\sigma) = (a, \sigma') \Rightarrow \mathsf{odds}(\sigma) \xrightarrow{a} \mathsf{odds}(\sigma')$. This allows us to compute

$$\begin{aligned} \mathsf{odds}(\langle a_0, a_1, a_2, a_3, a_4\rangle) &= a_1 \cdot \mathsf{odds}(\langle a_2, a_3, a_4\rangle) \\ &\qquad \text{since } o(\langle a_0, a_1, a_2, a_3, a_4\rangle) = (a_1, \langle a_2, a_3, a_4\rangle) \\ &= a_1 \cdot a_3 \cdot \mathsf{odds}(\langle a_4\rangle) \\ &\qquad \text{since } o(\langle a_2, a_3, a_4\rangle) = (a_3, \langle a_4\rangle) \\ &= a_1 \cdot a_3 \cdot \langle\rangle \\ &\qquad \text{since } o(\langle a_4\rangle) = \langle\rangle \\ &= \langle a_1, a_3\rangle. \end{aligned}$$

At this point the reader may wonder: why not define odds via evens, using an appropriate tail function? We shall prove that this gives the same outcome, using coinduction.

Lemma 1.2.4 *One has*

$$\mathsf{odds} = \mathsf{evens} \circ \mathsf{tail},$$

where the function $\mathsf{tail}\colon A^\infty \to A^\infty$ *is given by:*

$$\mathsf{tail}(\sigma) = \begin{cases} \sigma & \text{if } \sigma \not\to \\ \sigma' & \text{if } \sigma \xrightarrow{a} \sigma'. \end{cases}$$

Proof In order to prove that the two functions $\mathsf{odds}, \mathsf{evens} \circ \mathsf{tail}\colon A^\infty \to A^\infty$ are equal one needs to show by Proposition 1.2.1 that they are both homomorphisms for the same coalgebra structure on A^∞. Since odds arises by definition from the function o in (1.8), it suffices to show that $\mathsf{evens} \circ \mathsf{tail}$ is also a homomorphism from o to next. This involves two points:

- If $o(\sigma) = \bot$, there are two subcases, both yielding the same result:
  - If $\sigma \not\rightarrow$, then $\mathsf{evens}(\mathsf{tail}(\sigma)) = \mathsf{evens}(\sigma) \not\rightarrow$.
  - If $\sigma \xrightarrow{a} \sigma'$ and $\sigma' \not\rightarrow$, then $\mathsf{evens}(\mathsf{tail}(\sigma)) = \mathsf{evens}(\sigma') \not\rightarrow$.
- Otherwise, if $o(\sigma) = (a', \sigma'')$, because $\sigma \xrightarrow{a} \sigma'$ and $\sigma' \xrightarrow{a'} \sigma''$, then we have $\mathsf{evens}(\mathsf{tail}(\sigma)) = \mathsf{evens}(\sigma') \xrightarrow{a'} \mathsf{evens}(\mathsf{tail}(\sigma''))$ since:
  - If $\sigma'' \not\rightarrow$, then $\mathsf{evens}(\sigma') \xrightarrow{a'} \mathsf{evens}(\sigma'') = \mathsf{evens}(\mathsf{tail}(\sigma''))$.
  - If $\sigma'' \xrightarrow{a''} \sigma'''$, then $\mathsf{evens}(\sigma') \xrightarrow{a'} \mathsf{evens}(\sigma''') = \mathsf{evens}(\mathsf{tail}(\sigma''))$. □

Such equality proofs using uniqueness may be a bit puzzling at first. But they are very common in category theory and in many other areas of mathematics dealing with universal properties. Later, in Section 3.4 we shall see that such proofs can also be done via bisimulations. This is a common proof technique in process theory – and in coalgebra, of course.

Merging Sequences

In order to further familiarise the reader with the way in which the 'coinductive game' is played, we consider merging two sequences, via a binary operation of the form $\mathsf{merge}\colon A^\infty \times A^\infty \to A^\infty$. We want $\mathsf{merge}(\sigma, \tau)$ to alternatingly take one element from σ and from τ, starting with σ. In terms of rules:

$$\frac{\sigma \not\rightarrow \qquad \tau \not\rightarrow}{\mathsf{merge}(\sigma,\tau) \not\rightarrow} \qquad\qquad \frac{\sigma \not\rightarrow \qquad \tau \xrightarrow{a} \tau'}{\mathsf{merge}(\sigma,\tau) \xrightarrow{a} \mathsf{merge}(\sigma,\tau')}$$

$$\frac{\sigma \xrightarrow{a} \sigma'}{\mathsf{merge}(\sigma,\tau) \xrightarrow{a} \mathsf{merge}(\tau,\sigma')}.$$

Notice the crucial reversal of arguments in the last rule.

Thus, the function $\mathsf{merge}\colon A^\infty \times A^\infty \to A^\infty$ is defined coinductively as the behaviour beh_m of the coalgebra

$$(A^\infty \times A^\infty) \xrightarrow{\quad m \quad} \{\bot\} \cup \big(A \times (A^\infty \times A^\infty)\big)$$

given by

$$m(\sigma,\tau) = \begin{cases} \bot & \text{if } \sigma \not\rightarrow \text{ and } \tau \not\rightarrow \\ (a, (\sigma, \tau')) & \text{if } \sigma \not\rightarrow \text{ and } \tau \xrightarrow{a} \tau' \\ (a, (\tau, \sigma')) & \text{if } \sigma \xrightarrow{a} \sigma'. \end{cases}$$

At this stage we can combine all of the coinductively defined functions so far in the following result. It says that the merge of the evenly listed and oddly listed elements in a sequence is equal to the original sequence. At first, this may seem obvious, but recall that our sequences may be finite or infinite, so there is some work to do. The proof is again an exercise in coinductive reasoning using uniqueness. It does not involve a global distinction between finite and infinite but proceeds by local, single step reasoning.

Lemma 1.2.5 *For each sequence $\sigma \in A^\infty$,*

$$\mathsf{merge}(\mathsf{evens}(\sigma), \mathsf{odds}(\sigma)) = \sigma.$$

Proof Let us write $f \colon A^\infty \to A^\infty$ as shorthand for the left-hand side $f(\sigma) = \mathsf{merge}(\mathsf{evens}(\sigma), \mathsf{odds}(\sigma))$. We need to show that f is the identity function. Since the identity function $\mathrm{id}_{A^\infty} \colon A^\infty \to A^\infty$ is a homomorphism from next to next – i.e. $\mathrm{id}_{A^\infty} = \mathrm{beh}_{\mathsf{next}}$ – it suffices to show that also f is such a homomorphism $\mathsf{next} \to \mathsf{next}$. This involves two points:

- If $\sigma \nrightarrow$, then $\mathsf{evens}(\sigma) \nrightarrow$ and $\mathsf{odds}(\sigma) \nrightarrow$; this means that we also have $\mathsf{merge}(\mathsf{evens}(\sigma), \mathsf{odds}(\sigma)) \nrightarrow$ and thus $f(\sigma) \nrightarrow$.
- If $\sigma \xrightarrow{a} \sigma'$, then we distinguish two cases and prove $f(\sigma) \xrightarrow{a} f(\sigma')$ in both, using Lemma 1.2.4.
  - If $\sigma' \nrightarrow$, then $\mathsf{evens}(\sigma) \xrightarrow{a} \mathsf{evens}(\sigma')$ and thus

$$\begin{aligned} f(\sigma) &= \mathsf{merge}(\mathsf{evens}(\sigma), \mathsf{odds}(\sigma)) \\ &\xrightarrow{a} \mathsf{merge}(\mathsf{odds}(\sigma), \mathsf{evens}(\sigma')) \\ &\qquad = \mathsf{merge}(\mathsf{evens}(\mathsf{tail}(\sigma)), \mathsf{evens}(\mathsf{tail}(\sigma'))) \\ &\qquad = \mathsf{merge}(\mathsf{evens}(\sigma'), \mathsf{odds}(\sigma')) \\ &\qquad = f(\sigma'). \end{aligned}$$

  - If $\sigma' \xrightarrow{a'} \sigma''$, then $\mathsf{evens}(\sigma) \xrightarrow{a} \mathsf{evens}(\sigma'')$, and one can derive $f(\sigma) \xrightarrow{a} f(\sigma')$ as before. □

We have seen sequences of elements of an arbitrary set A. Things become more interesting when the set A has some algebraic structure, for instance addition or (also) multiplication. Such structure can then be transferred via coinductive definitions to final coalgebras of sequences, leading to what may be called *stream calculus*; see [412].

This completes our introduction to coinduction for sequences. What we have emphasised is that the coalgebraic approach using finality does not consider sequences as a whole via their elements but concentrates on the local, one-step behaviour via head and tail (if any). This makes definitions and reasoning easier – even though the reader may need to see more examples and get more experience to fully appreciate this point. But there is already a clear analogy with induction, which also uses single steps instead of global ones. The formal analogy between induction and coinduction will appear in Section 2.4.

More coinductively defined functions for sequences can be found in [215]. A broader palette of examples is given in [312, 278].

Exercises

1.2.1 Compute the nextdec-behaviour of $\frac{1}{7} \in [0, 1)$ as in Example 1.2.2.

1.2.2 Formulate appropriate rules for the function $\mathsf{odds}: A^\infty \to A^\infty$ in analogy with the rules (1.7) for evens.

1.2.3 Use coinduction to define the empty sequence $\langle\rangle \in A^\infty$ as a map $\langle\rangle: \{\bot\} \to A^\infty$.

Fix an element $a \in A$, and similarly define the infinite sequence $\vec{a}: \{\bot\} \to A^\infty$ consisting only of as.

1.2.4 Compute the outcome of $\mathsf{merge}(\langle a_0, a_1, a_2\rangle, \langle b_0, b_1, b_2, b_3\rangle)$.

1.2.5 Is the merge operation associative, i.e. is $\mathsf{merge}(\sigma, \mathsf{merge}(\tau, \rho))$ the same as $\mathsf{merge}(\mathsf{merge}(\sigma, \tau), \rho)$? Give a proof or a counterexample. Is there is neutral element for merge?

1.2.6 Show how to define an alternative merge function which alternatingly takes *two* elements from its argument sequences.

1.2.7
1. Define three functions $\mathsf{ex}_i: A^\infty \to A^\infty$, for $i = 0, 1, 2$, which extract the elements at positions $3n + i$.
2. Define $\mathsf{merge3}: A^\infty \times A^\infty \times A^\infty \to A^\infty$ satisfying the equation $\mathsf{merge3}(\mathsf{ex}_0(\sigma), \mathsf{ex}_1(\sigma), \mathsf{ex}_2(\sigma)) = \sigma$, for all $\sigma \in A^\infty$.

1.2.8 Consider the sequential composition function $\mathsf{comp}: A^\infty \times A^\infty \to A^\infty$ for sequences, described by the three rules:

$$\frac{\sigma \not\to \qquad \tau \not\to}{\mathsf{comp}(\sigma, \tau) \not\to} \qquad \frac{\sigma \not\to \qquad \tau \xrightarrow{a} \tau'}{\mathsf{comp}(\sigma, \tau) \xrightarrow{a} \mathsf{comp}(\sigma, \tau')}$$

$$\frac{\sigma \xrightarrow{a} \sigma'}{\mathsf{comp}(\sigma, \tau) \xrightarrow{a} \mathsf{comp}(\sigma', \tau)}.$$

1. Show by coinduction that the empty sequence $\langle\rangle = \mathsf{next}^{-1}(\bot) \in A^\infty$ is a unit element for comp, i.e. that $\mathsf{comp}(\langle\rangle, \sigma) = \sigma = \mathsf{comp}(\sigma, \langle\rangle)$.
2. Prove also by coinduction that comp is associative, and thus that sequences carry a monoid structure.

1.2.9 Consider two sets A, B with a function $f: A \to B$ between them. Use finality to define a function $f^\infty: A^\infty \to B^\infty$ that applies f element-wise. Use uniqueness to show that this mapping $f \mapsto f^\infty$ is 'functorial' in the sense that $(\mathrm{id}_A)^\infty = \mathrm{id}_{A^\infty}$ and $(g \circ f)^\infty = g^\infty \circ f^\infty$.

1.2.10 Use finality to define a map $\mathrm{st}: A^\infty \times B \to (A \times B)^\infty$ that maps a sequence $\sigma \in A^\infty$ and an element $b \in B$ to a new sequence in $(A \times B)^\infty$ by adding this b at every position in σ. (This is an example of a 'strength' map; see Exercise 2.5.4.)

1.3 Generality of Temporal Logic of Coalgebras

This section will illustrate the important coalgebraic notion of invariant, and use it to introduce temporal operators like $\Box$ for henceforth and $\Diamond$ for eventually. These operators are useful for expressing various interesting properties about states of a coalgebra. As we shall see later in Section 6.4, they can be defined for general coalgebras. But here we shall introduce them in more concrete situations – although we try to suggest the more general perspective. First, the sequences from the previous Section 1.2 will be reconsidered, and next, the statements from Section 1.1 will be used to form a rudimentary notion of class, with associated temporal operators $\Box$ and $\Diamond$ for expressing safety and liveness properties.

1.3.1 Temporal Operators for Sequences

Consider a fixed set A and an arbitrary 'A-sequence' coalgebra $c: S \to \{\bot\} \cup (A \times S)$ with state space S. We will be interested in properties of states, expressed via predicates/subsets $P \subseteq S$. For a state $x \in S$ we shall often write $P(x)$ for $x \in P$, and then say that the predicate P holds for x. Such a property $P(x)$ may for instance be: 'the behaviour of x is an infinite sequence'.

For an arbitrary predicate $P \subseteq S$ we shall define several new predicates, namely $\bigcirc P \subseteq S$ for 'nexttime' P, $\Box P \subseteq S$ for 'henceforth' P and $\Diamond P \subseteq S$ for 'eventually' P. These temporal operators $\bigcirc$, $\Box$, $\Diamond$ are all defined with respect to an arbitrary coalgebra $c: S \to \{\bot\} \cup (A \times S)$ as above. In order to make this dependence on the coalgebra c explicit we could write $\bigcirc_c P$, $\Box_c P$ and $\Diamond_c P$. But usually it is clear from the context which coalgebra is meant.

All these temporal operators $\bigcirc$, $\Box$, $\Diamond$ talk about future states obtained via transitions to successor states, i.e. via successive applications of the coalgebra. The nexttime operator $\bigcirc$ is most fundamental because it talks about single transitions. The other two, $\Box$ and $\Diamond$, involve multiple steps (zero or more) and are defined in terms of $\bigcirc$. For a sequence coalgebra $c\colon S \to \{\bot\} \cup (A \times S)$ with a predicate $P \subseteq S$ on its state space we define a new predicate $\bigcirc P \subseteq S$, for 'nexttime P', as

$$\begin{aligned}(\bigcirc P)(x) &\iff \forall a \in A.\, \forall x' \in S.\, c(x) = (a, x') \Rightarrow P(x') \\ &\iff \forall a \in A.\, \forall x' \in S.\, x \xrightarrow{a} x' \Rightarrow P(x').\end{aligned} \tag{1.9}$$

In words:

> The predicate $\bigcirc P$ holds for those states x, all of whose successor states x', if any, satisfy P. Thus, $(\bigcirc P)(x)$ indeed means that nexttime after x, P holds.

This simple operator $\bigcirc$ turns out to be fundamental, for example in defining the following notion.

Definition 1.3.1 A predicate P is a (sequence) **invariant** if $P \subseteq \bigcirc P$.

An invariant P is thus a predicate such that if P holds for a state x, then also $\bigcirc P$ holds of x. The latter means that P holds in successor states of x. Hence, if P holds for x, it holds for successors of x. This means that once P holds, P will continue to hold, no matter which transitions are taken. Or, once inside P, one cannot get out.

In general, invariants are important predicates in the study of state-based systems. They often express certain safety or data integrity properties which are implicit in the design of a system; for example, the pressure in a tank will not rise above a certain safety level. An important aspect of formally establishing the safety of systems is to prove that certain crucial predicates are actually invariants.

A concrete example of an invariant on the state space A^∞ of the final sequence coalgebra $\mathsf{next}\colon A^\infty \xrightarrow{\cong} \{\bot\} \cup (A \times A^\infty)$ is the property 'σ is a finite sequence'. Indeed, if σ is finite, and $\sigma \xrightarrow{a} \sigma'$, then also σ' is finite.

Certain predicates $Q \subseteq S$ on the state space of a coalgebra are thus invariants. Given an arbitrary predicate $P \subseteq S$, we can consider those subsets $Q \subseteq P$ which are invariants. The greatest among these subsets plays a special role.

Definition 1.3.2 Let $P \subseteq S$ be an arbitrary predicate on the state space S of a sequence coalgebra.

1. We define a new predicate $\square P \subseteq S$, for **henceforth** P, to be the greatest invariant contained in P. That is,

$$(\square P)(x) \iff \exists Q \subseteq S.\, Q \text{ is an invariant} \wedge Q \subseteq P \wedge Q(x).$$

   More concretely, $(\square P)(x)$ means that all successor states of x satisfy P.

2. And $\Diamond P \subseteq S$, for **eventually** P, is defined as

$$\Diamond P = \neg \Box \neg P,$$

where, for an arbitrary predicate $U \subseteq S$, the negation $\neg U \subseteq S$ is $\{x \in S \mid x \notin U\}$. Hence

$$(\Diamond P)(x) \iff \forall Q \subseteq S. (Q \text{ is an invariant } \wedge Q \subseteq \neg P) \Rightarrow \neg Q(x).$$

Thus, $(\Diamond P)(x)$ says that some successor state of x satisfies P.

The way these temporal operators $\Box$ and $\Diamond$ are defined may seem somewhat complicated at first, but will turn out to be at the right level of abstraction. As we shall see later in Section 6.4, the same formulation in terms of invariants works much more generally for coalgebras of different types (and not just for sequence coalgebras): the definition is 'generic' or 'polytypic'.

In order to show that the abstract formulations in the definition indeed capture the intended meaning of $\Box$ and $\Diamond$ as 'for all future states' and 'for some future state', we prove the following result.

Lemma 1.3.3 *For an arbitrary sequence coalgebra $c: S \to \{\bot\} \cup (A \times S)$, consider its iterations $c^n: S \to \{\bot\} \cup (A \times S)$, for $n \in \mathbb{N}$, as defined in the proof of Proposition 1.2.1. Then, for $P \subseteq S$ and $x \in S$,*

$$\begin{aligned}
(\Box P)(x) &\iff P(x) \wedge (\forall n \in \mathbb{N}. \forall a \in A. \forall y \in S. c^n(x) = (a, y) \Rightarrow P(y)) \\
(\Diamond P)(x) &\iff P(x) \vee (\exists n \in \mathbb{N}. \exists a \in A. \exists y \in S. c^n(x) = (a, y) \wedge P(y)).
\end{aligned}$$

Proof Since the second equivalence follows by purely logical manipulations from the first one, we shall prove only the first.

($\Rightarrow$) Assume $(\Box P)(x)$, i.e. $Q(x)$ for some invariant $Q \subseteq P$. By induction on $n \in \mathbb{N}$ one gets $c^n(x) = (a, y) \Rightarrow Q(y)$. But then also $P(y)$, for all such y in $c^n(x) = (a, y)$.

($\Leftarrow$) The predicate $\{x \in S \mid P(x) \wedge \forall n \in \mathbb{N}. \forall a \in A. \forall y \in S. c^n(x) = (a, y) \Rightarrow P(y)\}$ is an invariant contained in P. Hence it is contained in $\Box P$. ❑

Example 1.3.4 Consider an arbitrary sequence coalgebra $c: S \to \{\bot\} \cup (A \times S)$. We give three illustrations of the use of temporal operators $\Box$ and $\Diamond$ to express certain properties about states $x \in S$ of this coalgebra c.

1. Recall the termination predicate $(-) \not\rightarrow$ introduced in (1.5): $x \not\rightarrow$ means $c(x) = \bot$. Now consider the predicate $\Diamond((-) \not\rightarrow) \subseteq S$. It holds for those states which are eventually mapped to $\bot$, i.e. for those states whose behaviour is a finite sequence in $A^\star \subseteq A^\infty$.

2. In a similar way we can express that an element $a \in A$ occurs in the behaviour of a state $x \in S$. This is done as

$$\begin{aligned}\mathsf{Occ}(a) &= \Diamond(\{y \in S \mid \exists y' \in S.\, c(y) = (a, y')\}) \\ &= \Diamond(\{y \in S \mid \exists y' \in S.\, y \xrightarrow{a} y'\}).\end{aligned}$$

   One may wish to write $a \in x$ as a more intuitive notation for $x \in \mathsf{Occ}(a)$. It means that there is a future state of x which can do an a-step, i.e. that a occurs somewhere in the behaviour sequence of the state x.
3. Now assume our set A carries an order $\leq$. Consider the predicate:

$$\begin{aligned}&\mathsf{LocOrd}(x) \\ &\iff \forall a, a' \in A.\, \forall x', x'' \in S.\, c(x) = (a, x') \wedge c(x') = (a', x'') \Rightarrow a \leq a' \\ &\iff \forall a, a' \in A.\, \forall x', x'' \in S.\, x \xrightarrow{a} x' \wedge x' \xrightarrow{a'} x'' \Rightarrow a \leq a'.\end{aligned}$$

   Thus, LocOrd holds for x if the first two elements of the behaviour of x, if any, are related by $\leq$. Then,

$$\mathsf{GlobOrd} = \Box\, \mathsf{LocOrd}$$

   holds for those states whose behaviour is an ordered sequence: the elements appear in increasing order.

Next we wish to illustrate how to reason with these temporal operators. We show that an element occurs in the merge of two sequences if and only if it occurs in at least one of the two sequences. Intuitively this is clear, but technically it is not entirely trivial. The proof makes essential use of invariants.

Lemma 1.3.5 *Consider for an element $a \in A$ the occurrence predicate $a \in (-) = \mathsf{Occ}(a) \subseteq A^\infty$ from the previous example, for the final coalgebra $\mathsf{next}\colon A^\infty \xrightarrow{\cong} \{\bot\} \cup (A \times A^\infty)$ from Proposition 1.2.1. Then, for sequences $\sigma, \tau \in A^\infty$,*

$$a \in \mathsf{merge}(\sigma, \tau) \iff a \in \sigma \vee a \in \tau,$$

where $\mathsf{merge}\colon A^\infty \times A^\infty \to A^\infty$ is the merge operator introduced in the previous section.

Proof (⇒) Assume $a \in \mathsf{merge}(\sigma, \tau)$ but neither $a \in \sigma$ nor $a \in \tau$. The latter yields two invariants $P, Q \subseteq A^\infty$ with $P(\sigma)$, $Q(\tau)$ and $P, Q \subseteq \neg\{\rho \mid \exists \rho'.\, \rho \xrightarrow{a} \rho'\}$. These inclusions mean that sequences in P or Q cannot do an a-step.

In order to derive a contradiction we form a new predicate:

$$R = \{\mathsf{merge}(\alpha, \beta) \mid \alpha, \beta \in P \cup Q\}.$$

Clearly, $R(\mathsf{merge}(\sigma, \tau))$. The only transitions that a sequence $\mathsf{merge}(\alpha, \beta) \in R$ can do are

1. $\mathsf{merge}(\alpha,\beta) \xrightarrow{b} \mathsf{merge}(\alpha,\beta')$ because $\alpha \not\rightarrow$ and $\beta \xrightarrow{b} \beta'$.
2. $\mathsf{merge}(\alpha,\beta) \xrightarrow{b} \mathsf{merge}(\beta,\alpha')$ because $\alpha \xrightarrow{b} \alpha'$.

In both cases the successor state is again in R, so that R is an invariant. Also, sequences in R cannot do an a-step. The predicate R thus disproves the assumption $a \in \mathsf{merge}(\sigma,\tau)$.

($\Leftarrow$) Assume, without loss of generality, $a \in \sigma$ but not $a \in \mathsf{merge}(\sigma,\tau)$. Thus there is an invariant $P \subseteq \neg\{\rho \mid \exists \rho'.\rho \xrightarrow{a} \rho'\}$ with $P(\mathsf{merge}(\sigma,\tau))$. We now take

$$Q = \{\alpha \mid \exists \beta.\, P(\mathsf{merge}(\alpha,\beta)) \vee P(\mathsf{merge}(\beta,\alpha))\}.$$

Clearly $Q(\sigma)$. In order to show that Q is an invariant, assume an element $\alpha \in Q$ with a transition $\alpha \xrightarrow{b} \alpha'$. There are then several cases.

1. If $P(\mathsf{merge}(\alpha,\beta))$ for some β, then $\mathsf{merge}(\alpha,\beta) \xrightarrow{b} \mathsf{merge}(\beta,\alpha')$, so that $\alpha' \in Q$, because $P(\mathsf{merge}(\beta,\alpha'))$, and also $b \neq a$.
2. If $P(\mathsf{merge}(\beta,\alpha))$ for some β, then there are two further cases:
   a. If $\beta \not\rightarrow$, then $\mathsf{merge}(\beta,\alpha) \xrightarrow{b} \mathsf{merge}(\beta,\alpha')$, so that $\alpha' \in Q$, and $b \neq a$.
   b. If $\beta \xrightarrow{c} \beta'$, then $\mathsf{merge}(\beta,\alpha) \xrightarrow{c} \mathsf{merge}(\alpha,\beta') \xrightarrow{b} \mathsf{merge}(\alpha',\beta')$. Thus $P(\mathsf{merge}(\alpha',\beta'))$, so that $\alpha' \in Q$, and also $b \neq a$.

These cases also show that Q is contained in $\neg\{\rho \mid \exists \rho'.\rho \xrightarrow{a} \rho'\}$. This contradicts the assumption that $a \in \sigma$. ❑

This concludes our first look at temporal operators for sequences, from a coalgebraic perspective.

1.3.2 Temporal Operators for Classes

A class in an object-oriented programming language encapsulates data with associated operations, called ‘methods’ in this setting. They can be used to access and manipulate the data. These data values are contained in so-called fields or attributes. Using the representation of methods as statements with exceptions E as in Section 1.1 we can describe the operations of a class as a collection of attributes and methods, acting on a state space S:

$$\begin{array}{rcl} \mathsf{at}_1 : S & \longrightarrow & D_1 \\ & \vdots & \\ \mathsf{at}_n : S & \longrightarrow & D_n \\ \mathsf{meth}_1 : S & \longrightarrow & \{\bot\} \cup S \cup (S \times E) \\ & \vdots & \\ \mathsf{meth}_m : S & \longrightarrow & \{\bot\} \cup S \cup (S \times E). \end{array} \tag{1.10}$$

These attributes at_i give the data value $\mathsf{at}_i(x) \in D_i$ in each state $x \in S$. Similarly, each method meth_j can produce a successor state, either normally or exceptionally, in which the attributes have possibly different values. Objects, in the sense of object-oriented programming (not of category theory), are thus identified with states.

For such classes, as for sequences, coalgebraic temporal logic provides a tailor-made nexttime operator $\bigcirc$. For a predicate $P \subseteq S$, we have $\bigcirc P \subseteq S$, defined on $x \in S$ as

$$\begin{aligned}(\bigcirc P)(x) \iff \forall j \leq m.\ &(\forall y \in S.\, \mathsf{meth}_j(x) = y \Rightarrow P(y)) \wedge \\ &(\forall y \in S.\, \forall e \in E.\, \mathsf{meth}_j(x) = (y, e) \Rightarrow P(y)).\end{aligned}$$

Thus, $(\bigcirc P)(x)$ means that P holds in each possible successor state of x, resulting from normal or abrupt termination.

From this point on we can follow the pattern used above for sequences. A predicate $P \subseteq S$ is a **class invariant** if $P \subseteq \bigcirc P$. Also: $\Box P$ is the greatest invariant contained in P, and $\Diamond P = \neg\Box\neg P$. Predicates of the form $\Box P$ are so-called **safety properties** expressing that 'nothing bad will happen': P holds in all future states. And predicates $\Diamond P$ are **liveness properties** saying that 'something good will happen': P holds in some future state.

A typical example of a safety property is: this integer field `i` will always be non-zero (so that it is safe to divide by `i`), or: this array `a` will always be a non-null reference and have length greater than 1 (so that we can safely access `a[0]` and `a[1]`).

Such temporal properties are extremely useful for reasoning about classes. As we have tried to indicate, they arise quite naturally and uniformly in a coalgebraic setting.

Exercises

1.3.1 The nexttime operator $\bigcirc$ introduced in (1.9) is the so-called **weak** nexttime. There is an associated **strong** nexttime, given by $\neg\bigcirc\neg$. Note the difference between weak and strong nexttime for sequences.

1.3.2 Prove that the 'truth' predicate that always holds is a (sequence) invariant. And if P_1 and P_2 are invariants, then so is the intersection $P_1 \cap P_2$. Finally, if P is an invariant, then so is $\bigcirc P$.

1.3.3
1. Show that $\Box$ is an interior operator, i.e. satisfies: $\Box P \subseteq P$, $\Box P \subseteq \Box\Box P$, and $P \subseteq Q \Rightarrow \Box P \subseteq \Box Q$.
2. Prove that a predicate P is an invariant if and only if $P = \Box P$.

1.3.4 Recall the finite behaviour predicate $\Diamond(- \not\rightarrow)$ from Example 1.3.4.1 and show that it is an invariant: $\Diamond(- \not\rightarrow) \subseteq \bigcirc\Diamond(- \not\rightarrow)$. *Hint*: For an invariant Q, consider the predicate $Q' = (\neg(-) \not\rightarrow) \cap (\bigcirc Q)$.

1.3.5 Let $(A, \leq)$ be a complete lattice, i.e. a poset in which each subset $U \subseteq A$ has a join $\bigvee U \in A$. It is well known that each subset $U \subseteq A$ then also has a meet $\bigwedge U \in A$, given by $\bigwedge U = \bigvee\{a \in A \mid \forall b \in U. a \leq b\}$.

Let $f : A \to A$ be a monotone function: $a \leq b$ implies $f(a) \leq f(b)$. Recall, e.g. from [119, chapter 4] that such a monotone f has both a least fixed point $\mu f \in A$ and a greatest fixed point $\nu f \in A$ given by the formulas:

$$\mu f = \bigwedge\{a \in A \mid f(a) \leq a\}, \qquad \nu f = \bigvee\{a \in A \mid a \leq f(a)\}.$$

Now let $c : S \to \{\perp\} \cup (A \times S)$ be an arbitrary sequence coalgebra, with associated nexttime operator $\bigcirc$.

1. Prove that $\bigcirc$ is a monotone function $\mathcal{P}(S) \to \mathcal{P}(S)$, i.e. that $P \subseteq Q$ implies $\bigcirc P \subseteq \bigcirc Q$, for all $P, Q \subseteq S$.
2. Check that $\Box P \in \mathcal{P}(S)$ is the greatest fixed point of the function $\mathcal{P}(S) \to \mathcal{P}(S)$ given by $U \mapsto P \cap \bigcirc U$.
3. Define for $P, Q \subseteq S$ a new predicate $P \,\mathcal{U}\, Q \subseteq S$, for '$P$ until Q' as the least fixed point of $U \mapsto Q \cup (P \cap \neg\bigcirc\neg U)$. Check that 'until' is indeed a good name for $P \,\mathcal{U}\, Q$, since it can be described explicitly as

$$\begin{aligned} P \,\mathcal{U}\, Q = \{x \in S \mid \exists n \in \mathbb{N}. \exists x_0, x_1, \ldots, x_n \in S.\\ x_0 = x \wedge (\forall i < n. \exists a. x_i \xrightarrow{a} x_{i+1}) \wedge Q(x_n)\\ \wedge \forall i < n. P(x_i)\}. \end{aligned}$$

Hint: Don't use the fixed point definition μ, but first show that this subset is a fixed point, and then that it is contained in an arbitrary fixed point.

(The fixed point definitions that we described above are standard in temporal logic; see e.g. [128, 3.24–3.25]. The above operation $\mathcal{U}$ is what is called the 'strong' until. The 'weak' one does not have the negations $\neg$ in its fixed-point description in point 3.)

1.4 Abstractness of the Coalgebraic Notions

In this final section of the chapter we wish to consider the different settings in which coalgebras can be studied. Proper appreciation of the level of generality of coalgebras requires a certain familiarity with the theory of categories. Category theory is a special area that studies the fundamental structures used within mathematics. It is based on the very simple notion of an arrow between objects. Category theory is sometimes described as abstract nonsense, but it is often useful because it provides an abstract framework in which similarities between seemingly different notions become apparent. It has become a standard tool in theoretical computer science, especially in the semantics of programming languages. In particular, the categorical description of fixed points, both of recursive functions and of recursive types, captures the relevant 'universal' properties that are used in programming and reasoning with these constructs. This categorical approach to fixed points forms one of the starting points for the use of category theory in the study of algebras and coalgebras.

For this reason we need to introduce the fundamental notions of category and functor, simply because a bit of category theory helps enormously in presenting the theory of coalgebras and in recognising the common structure underlying many examples. Readers who wish to learn more about categories may consider introductory texts such as [46, 37, 113, 461, 379, 57, 330, 332], or more advanced ones such as [344, 85, 346, 239, 443].

In the beginning of this chapter we have described a coalgebra in (1.1) as a function of the form $\alpha\colon S \to \boxed{\cdots\ S\ \cdots}$ with a structured output type in which the state space S may occur. Here we shall describe such a result type as an expression $F(S) = \boxed{\cdots\ S\ \cdots}$ involving S. Shortly we shall see that F is a functor. A coalgebra is then a map of the form $\alpha\colon S \to F(S)$. It can thus be described in an arrow-theoretic setting, as given by a category.

Definition 1.4.1 A **category** is a mathematical structure consisting of objects with arrows between them, that can be composed.

More formally, a category $\mathbb{C}$ consists of a collection $\mathsf{Obj}(\mathbb{C})$ of objects and a collection $\mathsf{Arr}(\mathbb{C})$ of arrows (also called maps or morphisms). Usually we write $X \in \mathbb{C}$ for $X \in \mathsf{Obj}(\mathbb{C})$. Each arrow in $\mathbb{C}$, written as $X \xrightarrow{f} Y$ or as $f\colon X \to Y$, has a domain object $X \in \mathbb{C}$ and a codomain object $Y \in \mathbb{C}$. These objects and arrows carry a composition structure.

1. For each pair of maps $f\colon X \to Y$ and $g\colon Y \to Z$ there is a composition map $g \circ f\colon X \to Z$. This composition operation $\circ$ is associative: if $h\colon Z \to W$, then $h \circ (g \circ f) = (h \circ g) \circ f$.
2. For each object $X \in \mathbb{C}$ there is an identity map $\mathrm{id}_X\colon X \to X$, such that id is neutral element for composition $\circ$: for $f\colon X \to Y$ one has $f \circ \mathrm{id}_X = f = \mathrm{id}_Y \circ f$. Often, the subscript X in id_X is omitted when it is clear from the context.

Ordinary sets with functions between them form an obvious example of a category, for which we shall write **Sets**. Although **Sets** is a standard example, it is important to realise that a category may be a very different structure. In particular, an arrow in a category need not be a function.

We give several standard examples, and leave it to the reader to check that the requirements of a category hold for all of them.

Example 1.4.2 1. Consider a monoid M with composition operation $+$ and unit element $0 \in M$. This M can also be described as a category with one object, say $\star$, and with arrows $\star \to \star$ given by elements $m \in M$. The identity arrow is then $0 \in M$, and composition of arrows $m_1 : \star \to \star$ and $m_2 : \star \to \star$ is $m_1 + m_2 : \star \to \star$. The associativity and identity requirements required for a category are precisely the associativity and identity laws of the monoid.

2. Here is another degenerate example: a preorder consists of a set D with a reflexive and transitive order relation $\leq$. It corresponds to a category in which there is at most one arrow between each pair of objects. Indeed, the preorder $(D, \leq)$ can be seen as a category with elements $d \in D$ as objects and with an arrow $d_1 \to d_2$ if and only if $d_1 \leq d_2$.
3. Many examples of categories have certain mathematical structures as objects, and structure-preserving functions between them as morphisms. Examples are:

   a. **Mon**, the category of monoids with monoid homomorphisms (preserving composition and unit).
   b. **Grp**, the category of groups with group homomorphisms (preserving composition and unit, and thereby also inverses).
   c. **PreOrd**, the category of preorders with monotone functions (preserving the order). Similarly, there is a category **PoSets** with posets as objects, and also with monotone functions as morphisms.
   d. **Dcpo**, the category of directed complete partial orders (dcpos) with continuous functions between them (preserving the order and directed joins $\bigvee$).
   e. **Sp**, the category of topological spaces with continuous functions (whose inverse image preserves open subsets).
   f. **Met**, the category of metric spaces with non-expansive functions between them. Consider two objects (M_1, d_1) and (M_2, d_2) in **Met**, where $d_i : M_i \times M_i \to [0, \infty)$ is a distance function on the set M_i. A morphism $(M_1, d_1) \to (M_2, d_2)$ in **Met** is defined as a function $f : M_1 \to M_2$ between the underlying sets satisfying $d_2(f(x), f(y)) \leq d_1(x, y)$, for all $x, y \in M_1$.

4. An example that we shall frequently use is the category **SetsRel** of sets and relations. Its objects are ordinary sets, and its morphisms $X \to Y$ are relations $R \subseteq X \times Y$. Composition of $R \colon X \to Y$ and $S \colon Y \to Z$ in **SetsRel** is given by relational composition:

$$S \circ R = \{(x, z) \in X \times Z \mid \exists y \in Y.\, R(x, y) \wedge S(y, z)\}. \tag{1.11}$$

The identity morphism $X \to X$ in **SetsRel** is the equality relation (also called diagonal or identity relation) $\mathrm{Eq}(X) \subseteq X \times X$ given by $\mathrm{Eq}(X) = \{(x, x) \mid x \in X\}$.

A category is thus a very general mathematical structure, with many possible instances. In the language of categories one can discuss standard mathematical notions, such as mono-/epi-/iso-morphism, product, limit, etc. For example, an isomorphism in a category $\mathbb{C}$ is a morphism $f \colon X \to Y$ for which there is a (necessarily unique) morphism $g \colon Y \to X$ in the opposite direction with $f \circ g = \mathrm{id}_Y$ and $g \circ f = \mathrm{id}_X$. If there is such an isomorphism, one often writes $X \cong Y$. Such general categorical definitions then have meaning in every example of a category. For instance, it yields a notion of isomorphism for groups, posets, topological spaces, etc.

In a category a morphism from an object to itself, that is, a morphism of the form $f \colon X \to X$, will be called an **endomorphism** or simply an **endomap**. For instance, a relation $R \subseteq X \times X$ on a single set X – instead of a relation $S \subseteq X \times Y$ on two sets X, Y – will be called an endorelation, since it forms an endomap $X \to X$ in the above category **SetsRel**.

Categorical properties are expressed in terms of morphisms, often drawn as diagrams. Two fundamental aspects are commutation and uniqueness.

- **Commutation**: An equation in category theory usually has the form $f_1 \circ \cdots \circ f_n = g_1 \circ \cdots \circ g_m$, for certain morphisms f_i, g_j. Such an equation can be expressed in a commuting diagram as

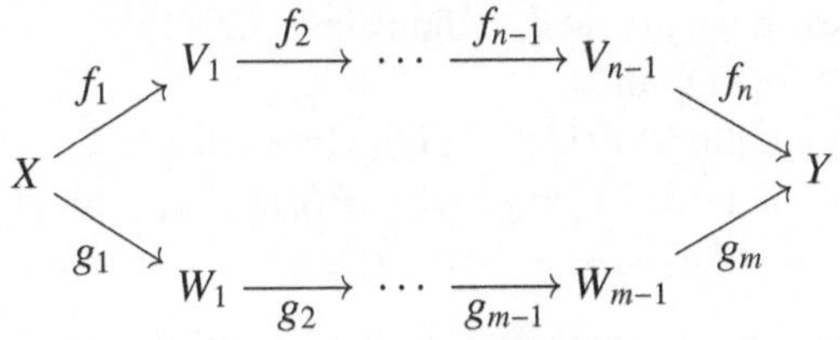

  Extracting such an equation from a commuting diagram by following two paths is an example of what is called *diagram chasing*.
- **Uniqueness**: A frequently occurring formulation is: for every +++ there is a unique morphism $f \colon X \to Y$ satisfying ***. Such uniqueness is

often expressed by writing a dashed arrow, $f: X \dashrightarrow Y$, especially in a diagram.

As we have already seen in Section 1.2, uniqueness is a powerful reasoning principle: one can derive an equality $f_1 = f_2$ for two morphisms $f_1, f_2 : X \to Y$ by showing that they both satisfy ***. Often, this property *** can be established via diagram chasing, i.e. by following paths in a diagram (both for f_1 and for f_2).

Both commutation and uniqueness will be frequently used in the course of this book.

For future use we mention two ways to construct new categories from old.

- Given a category $\mathbb{C}$, one can form what is called the **opposite** category $\mathbb{C}^{op}$ which has the same objects as $\mathbb{C}$, but the arrows reversed. Thus $f: X \to Y$ in $\mathbb{C}^{op}$ if and only if $f: Y \to X$ in $\mathbb{C}$. Composition $g \circ f$ in $\mathbb{C}^{op}$ is then $f \circ g$ in $\mathbb{C}$.
- Given two categories $\mathbb{C}$ and $\mathbb{D}$, we can form the **product** category $\mathbb{C} \times \mathbb{D}$. Its objects are pairs of objects (X, Y) with $X \in \mathbb{C}$ and $Y \in \mathbb{D}$. A morphism $(X, Y) \to (X', Y')$ in $\mathbb{C} \times \mathbb{D}$ consists of a pair of morphisms $X \to X'$ in $\mathbb{C}$ and $Y \to Y'$ in $\mathbb{D}$. Identities and compositions are obtained componentwise.

The above example categories of monoids, groups, etc. indicate that structure-preserving mappings are important in category theory. There is also a notion of such a mapping between categories, called functor. It preserves the relevant structure.

Definition 1.4.3 Consider two categories $\mathbb{C}$ and $\mathbb{D}$. A **functor** $F: \mathbb{C} \to \mathbb{D}$ consists of two mappings $\mathsf{Obj}(\mathbb{C}) \to \mathsf{Obj}(\mathbb{D})$ and $\mathsf{Arr}(\mathbb{C}) \to \mathsf{Arr}(\mathbb{D})$, both written as F, such that:

1. F preserves domains and codomains: if $f: X \to Y$ in $\mathbb{C}$, then $F(f): F(X) \to F(Y)$ in $\mathbb{D}$.
2. F preserves identities: $F(\mathrm{id}_X) = \mathrm{id}_{F(X)}$ for each $X \in \mathbb{C}$.
3. F preserves composition: $F(g \circ f) = F(g) \circ F(f)$, for all maps $f: X \to Y$ and $g: Y \to Z$ in $\mathbb{C}$.

For each category $\mathbb{C}$ there is a trivial 'identity' functor $\mathrm{id}_{\mathbb{C}} : \mathbb{C} \to \mathbb{C}$, mapping $X \mapsto X$ and $f \mapsto f$. Also, for each object $A \in \mathbb{C}$ there are functors which map everything to A. They can be defined as functors $A: \mathbb{D} \to \mathbb{C}$, for an arbitrary category $\mathbb{D}$. This constant functor maps any object $X \in \mathbb{D}$ to A, and any morphism f in $\mathbb{D}$ to the identity map $\mathrm{id}_A : A \to A$.

Further, given two functors $F\colon \mathbb{C} \to \mathbb{D}$ and $G\colon \mathbb{D} \to \mathbb{E}$, there is a composite functor $G \circ F\colon \mathbb{C} \to \mathbb{E}$. It is given by $X \mapsto G(F(X))$ and $f \mapsto G(F(f))$. Often we simply write $GF = G \circ F$, and similarly $GF(X) = G(F(X))$.

Example 1.4.4 1. Consider two monoids $(M, +, 0)$ and $(N, \cdot, 1)$ as categories, as in Example 1.4.2.1. A functor $f\colon M \to N$ is then the same as a monoid homomorphism: it preserves the composition operation and unit element.

2. Similarly, consider two preorders $(D, \leq)$ and $(E, \sqsubseteq)$ as categories, as in Example 1.4.2.3. A functor $f\colon D \to E$ is then nothing but a monotone function: $x \leq y$ implies $f(x) \sqsubseteq f(y)$.
3. Frequently occurring examples of functors are so-called **forgetful** functors. They forget part of the structure of their domain. For instance, there is a forgetful functor **Mon** $\to$ **Sets** mapping a monoid $(M, +, 0)$ to its underlying set M, and mapping a monoid homomorphism f to f, considered as a function between sets. Similarly, there is a forgetful functor **Grp** $\to$ **Mon** mapping groups to monoids by forgetting their inverse operation.
4. There is a 'graph' functor **Sets** $\to$ **SetsRel**. It maps a set X to X itself, and a function $f\colon X \to Y$ to the corresponding graph relation $\mathrm{Graph}(f) \subseteq X \times Y$ given by $\mathrm{Graph}(f) = \{(x, y) \mid f(x) = y\}$.
5. Recall from Section 1.2 that sequence coalgebras were described as functions of the form $c\colon S \to \{\bot\} \cup (A \times S)$. Their codomain can be described via a functor $\mathsf{Seq}\colon$ **Sets** $\to$ **Sets**. It maps a set X to the set $\{\bot\} \cup (A \times X)$. And it sends a function $f\colon X \to Y$ to a function $\{\bot\} \cup (A \times X) \to \{\bot\} \cup (A \times Y)$ given by

$$\bot \longmapsto \bot \qquad \text{and} \qquad (a, x) \longmapsto (a, f(x)).$$

We leave it to the reader to check that Seq preserves compositions and identities. We do note that the requirement that the behaviour function $\mathrm{beh}_c\colon S \to A^\infty$ from Proposition 1.2.1 is a homomorphism of coalgebras can now be described via commutation of the following diagram:

$$\begin{array}{ccc}
\mathsf{Seq}(S) & \xrightarrow{\mathsf{Seq}(\mathrm{beh}_c)} & \mathsf{Seq}(A^\infty) \\
{\scriptstyle c}\uparrow & & \uparrow{\scriptstyle \cong\ \mathsf{next}} \\
S & \xrightarrow[\mathrm{beh}_c]{} & A^\infty
\end{array}$$

In this book we shall be especially interested in **endofunctors**, i.e. in functors $\mathbb{C} \to \mathbb{C}$ from a category $\mathbb{C}$ to itself. In many cases this category $\mathbb{C}$ will simply be **Sets**, the category of sets and functions. Often we say that a mapping $A \mapsto G(A)$ of sets to sets is **functorial** if it can be extended in a more or less obvious way to a mapping $f \mapsto G(f)$ on functions such that G becomes a functor $G\colon \mathbf{Sets} \to \mathbf{Sets}$. We shall see many examples in the next chapter.

We can now introduce coalgebras in full generality.

Definition 1.4.5 Let $\mathbb{C}$ be an arbitrary category, with an endofunctor $F\colon \mathbb{C} \to \mathbb{C}$.

1. An F-**coalgebra**, or just a **coalgebra** when F is understood, consists of an object $X \in \mathbb{C}$ together with a morphism $c\colon X \to F(X)$. As before, we often call X the state space, or the carrier of the coalgebra, and c the transition or coalgebra structure.
2. A **homomorphism of coalgebras**, or a **map of coalgebras** or a **coalgebra map**, from one coalgebra $c\colon X \to F(X)$ to another coalgebra $d\colon Y \to F(Y)$ consists of a morphism $f\colon X \to Y$ in $\mathbb{C}$ which commutes with the structures, in the sense that the following diagram commutes:

$$\begin{array}{ccc} F(X) & \xrightarrow{\;F(f)\;} & F(Y) \\ {\scriptstyle c}\uparrow & & \uparrow{\scriptstyle d} \\ X & \xrightarrow[\;f\;]{} & Y \end{array}$$

3. F-coalgebras with homomorphisms between them form a category, which we shall write as $\mathbf{CoAlg}(F)$. It comes with a forgetful functor $\mathbf{CoAlg}(F) \to \mathbb{C}$, mapping a coalgebra $X \to F(X)$ to its state space X, and a coalgebra homomorphism f to f.

The abstractness of the notion of coalgebra lies in the fact that it can be expressed in any category. So we not only need to talk about coalgebras in **Sets**, as we have done so far, but can also consider coalgebras in other categories. For instance, one can have coalgebras in **PreOrd**, the category of preorders. In that case, the state space is a preorder, and the coalgebra structure is a monotone function. Similarly, a coalgebra in the category **Mon** of monoids has a monoid as state space, and a structure which preserves this monoid structure. We can even have a coalgebra in a category $\mathbf{CoAlg}(F)$ of coalgebras. We briefly mention some examples, without going into details.

- Real numbers (and also Baire and Cantor space) are described in [378, theorem 5.1] as final coalgebras (via continued fractions; see also [366]) of an endofunctor on the category **PoSets**.
- So-called descriptive general frames (special models of modal logic) appear in [316] as coalgebras of the Vietoris functor on the category of Stone spaces.

- At several places in this book we shall see coalgebra of endofunctors other than sets. For instance, Exercise 1.4.6 mentions invariants as coalgebras of endofunctors on poset categories, and Example 2.3.10 and Exercise 2.3.7 describe streams with their topology as final coalgebra in the category of topological spaces. Section 5.3 introduces traces of suitable coalgebras via coalgebra homomorphism to a final coalgebra in the category **SetsRel** of sets with relations as morphisms.

In the next few chapters we shall concentrate on coalgebras in **Sets**, but occasionally this more abstract perspective will be useful.

Exercises

1.4.1 Let $(M, +, 0)$ be a monoid, considered as a category. Check that a functor $F \colon M \to \mathbf{Sets}$ can be identified with a **monoid action**: a set X together with a function $\mu \colon X \times M \to X$ with $\mu(x, 0) = x$ and $\mu(x, m_1 + m_2) = \mu(\mu(x, m_2), m_1)$.

1.4.2 Check in detail that the opposite $\mathbb{C}^{\mathrm{op}}$ and product $\mathbb{C} \times \mathbb{D}$ are indeed categories.

1.4.3 Assume an arbitrary category $\mathbb{C}$ with an object $I \in \mathbb{C}$. We form a new category $\mathbb{C}/I$, the so-called **slice category** over I, with

objects maps $f \colon X \to I$ with codomain I in $\mathbb{C}$

morphisms from $X \xrightarrow{f} I$ to $Y \xrightarrow{g} I$ are morphisms $h \colon X \to Y$ in $\mathbb{C}$ for which the following diagram commutes:

$$\begin{array}{ccc} X & \xrightarrow{\;h\;} & Y \\ & {\scriptstyle f}\searrow \quad \swarrow{\scriptstyle g} & \\ & I & \end{array}$$

1. Describe identities and composition in $\mathbb{C}/I$, and verify that $\mathbb{C}/I$ is a category.
2. Check that taking domains yields a functor $\mathrm{dom} \colon \mathbb{C}/I \to \mathbb{C}$.
3. Verify that for $\mathbb{C} = \mathbf{Sets}$, a map $f \colon X \to I$ may be identified with an I-indexed family of sets $(X_i)_{i \in I}$, namely where $X_i = f^{-1}(i)$. What do morphisms in $\mathbb{C}/I$ correspond to, in terms of such indexed families?

1.4.4 Recall that for an arbitrary set A we write $A^\star$ for the set of finite sequences $\langle a_0, \ldots, a_n \rangle$ of elements $a_i \in A$.

1. Check that $A^\star$ carries a monoid structure given by concatenation of sequences, with the empty sequence $\langle\rangle$ as neutral element.
2. Check that the assignment $A \mapsto A^\star$ yields a functor **Sets** $\to$ **Mon** by mapping a function $f\colon A \to B$ between sets to the function $f^\star\colon A^\star \to B^\star$ given by $\langle a_0, \ldots, a_n\rangle \mapsto \langle f(a_0), \ldots, f(a_n)\rangle$. (Be aware of what needs to be checked: $f^\star$ must be a monoid homomorphism, and $(-)^\star$ must preserve composition of functions and identity functions.)
3. Prove that $A^\star$ is the **free monoid on** A: there is the singleton-sequence insertion map $\eta\colon A \to A^\star$ which is universal among all mappings of A into a monoid. The latter means that for each monoid $(M, 0, +)$ and function $f\colon A \to M$ there is a unique monoid homomorphism $g\colon A^\star \to M$ with $g \circ \eta = f$.

1.4.5 Recall from (1.3) the statements with exceptions of the form $S \to \{\bot\} \cup S \cup (S \times E)$.

1. Prove that the assignment $X \mapsto \{\bot\} \cup X \cup (X \times E)$ is functorial, so that statements are coalgebras for this functor.
2. Show that all the operations $\mathsf{at}_1, \ldots, \mathsf{at}_n, \mathsf{meth}_1, \ldots, \mathsf{meth}_m$ of a class as in (1.10) can also be described as a single coalgebra, namely of the functor:

$$X \mapsto D_1 \times \cdots \times D_n \times \underbrace{(\{\bot\} \cup X \cup (X \times E)) \times \cdots \times (\{\bot\} \cup X \cup (X \times E))}_{m \text{ times}}\ .$$

1.4.6 Recall the nexttime operator $\bigcirc$ for a sequence coalgebra $c\colon S \to \mathsf{Seq}(S) = \{\bot\} \cup (A \times S)$ from the previous section. Exercise 1.3.5.1 says that it forms a monotone function $\mathcal{P}(S) \to \mathcal{P}(S)$ – with respect to the inclusion order – and thus a functor. Check that invariants are precisely the $\bigcirc$-coalgebras!

2
Coalgebras of Polynomial Functors

The previous chapter has introduced several examples of coalgebras and has illustrated basic coalgebraic notions such as behaviour and invariance (for those examples). This chapter will go deeper into the study of the area of coalgebra, introducing some basic notions, definitions and terminology. It will first discuss several fundamental set theoretic constructions, such as products, coproducts, exponents and powersets in a suitably abstract (categorical) language. These constructs are used to define a collection of elementary functors, the so-called polynomial functors. As will be shown in Section 2.2, this class of functors is rich enough to capture many examples of interesting coalgebras, including deterministic and non-deterministic automata. One of the attractive features of polynomial functors is that almost all of them have a final coalgebra – except when the (non-finite) powerset occurs. The unique map into a final coalgebra will appear as behaviour morphism, mapping a state to its behaviour. The two last sections of this chapter, Sections 2.4 and 2.5, provide additional background information, namely on algebras (as duals of coalgebras) and on adjunctions. The latter form a fundamental categorical notion describing back-and-forth translations that occur throughout mathematics.

2.1 Constructions on Sets

This section describes familiar constructions on sets, such as products, coproducts (disjoint unions), exponents and powersets. It does so in order to fix notation and to show that these operations are functorial, i.e. give rise to functors. This latter aspect is maybe not so familiar. Functoriality is essential for properly developing the theory of coalgebras; see Definition 1.4.5.

These basic constructions on sets are instances of more general constructions in categories. We shall give a perspective on these categorical formulations, but we do not overemphasise this point. Readers without much familiarity with the theory of categories may then still follow the development, and readers who are quite comfortable with categories will recognise this wider perspective anyway.

Products

We recall that for two arbitrary sets X, Y the product $X \times Y$ is the set of pairs

$$X \times Y = \{(x, y) \mid x \in X \wedge y \in Y\}.$$

There are then obvious projection functions $\pi_1 : X \times Y \to X$ and $\pi_2 : X \times Y \to Y$ by $\pi_1(x, y) = x$ and $\pi_2(x, y) = y$. Also, for functions $f : Z \to X$ and $g : Z \to Y$ there is a tuple (or pairing) function $\langle f, g\rangle : Z \to X \times Y$ given by $\langle f, g\rangle(z) = (f(z), g(z)) \in X \times Y$ for $z \in Z$. Here are some basic equations which are useful in computations:

$$\begin{aligned} \pi_1 \circ \langle f, g\rangle &= f & \qquad \langle \pi_1, \pi_2\rangle &= \mathrm{id}_{X\times Y} \\ \pi_2 \circ \langle f, g\rangle &= g & \qquad \langle f, g\rangle \circ h &= \langle f \circ h, g \circ h\rangle. \end{aligned} \tag{2.1}$$

The latter equation holds for functions $h : W \to Z$.

Given these equations it is not hard to see that the product operation gives rise to a bijective correspondence between pairs of functions $Z \to X$, $Z \to Y$ on the one hand, and functions $Z \to X \times Y$ into the product on the other. Indeed, given two functions $Z \to X$, $Z \to Y$, one can form their pair $Z \to X \times Y$. And in the reverse direction, given a function $Z \to X \times Y$, one can post-compose with the two projections π_1 and π_2 to get two functions $Z \to X$, $Z \to Y$. The above equations help show that these operations are each other's inverses. Such a bijective correspondence is conveniently expressed by a 'double rule', working in two directions:

$$\frac{Z \longrightarrow X \qquad Z \longrightarrow Y}{\overline{\overline{\quad Z \longrightarrow X \times Y \quad}}}\,. \tag{2.2}$$

Interestingly, the product operation $(X, Y) \mapsto X \times Y$ applies not only to sets, but also to functions: for functions $f : X \to X'$ and $g : Y \to Y'$ we can define a function $f \times g$, namely:

$$X \times Y \xrightarrow{\;f \times g\;} X' \times Y' \quad \text{given by} \quad (x, y) \longmapsto (f(x), g(y)). \tag{2.3}$$

Notice that the symbol $\times$ is overloaded: it is used both on sets and on functions. This product function $f \times g$ can also be described in terms of projections and pairing as $f \times g = \langle f \circ \pi_1, g \circ \pi_2 \rangle$. It is easily verified that the operation $\times$ on functions satisfies

$$\mathrm{id}_X \times \mathrm{id}_Y = \mathrm{id}_{X \times Y} \quad \text{and} \quad (f \circ h) \times (g \circ k) = (f \times g) \circ (h \times k).$$

These equations express that the product $\times$ is *functorial*: it applies not only to sets, but also to functions, and it does so in such a way that identity maps and compositions are preserved (see Definition 1.4.3). The product operation $\times$ is a functor $\mathbf{Sets} \times \mathbf{Sets} \to \mathbf{Sets}$, from the product category $\mathbf{Sets} \times \mathbf{Sets}$ of $\mathbf{Sets}$ with itself, to $\mathbf{Sets}$.

Products of sets form an instance of the following general notion of product in a category.

Definition 2.1.1 Let $\mathbb{C}$ be a category. The **product** of two objects $X, Y \in \mathbb{C}$ is a new object $X \times Y \in \mathbb{C}$ with two projection morphisms

$$X \xleftarrow{\ \pi_1\ } X \times Y \xrightarrow{\ \pi_2\ } Y,$$

which are universal: for each pair of maps $f\colon Z \to X$ and $g\colon Z \to Y$ in $\mathbb{C}$ there is a *unique* tuple morphism $\langle f, g \rangle\colon Z \to X \times Y$ in $\mathbb{C}$, making the following diagram commute.

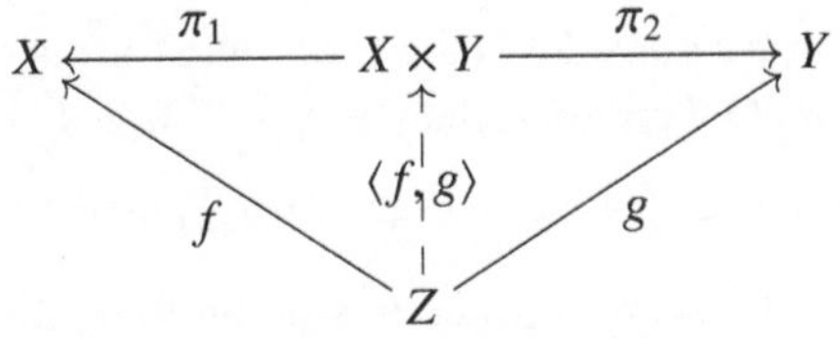

The first two equations from (2.1) clearly hold for this abstract definition of product. The other two equations in (2.1) follow by using the uniqueness property of the tuple.

Products need not exist in a category, but if they exist they are determined up-to-isomorphism: if there is another object with projections $X \xleftarrow{p_1} X \otimes Y \xrightarrow{p_2} Y$ satisfying the above universal property, then there is a unique isomorphism $X \times Y \xrightarrow{\cong} X \otimes Y$ commuting with the projections. Similar results can be proven for the other constructs in this section.

What we have described is the product $X \times Y$ of *two* sets/objects X, Y. For a given X, we shall write $X^n = X \times \cdots \times X$ for the n-fold product (also known as **power**). The special case where $n = 0$ involves the empty product X^0, which is called a final or terminal object.

Definition 2.1.2 A **final object** in a category $\mathbb{C}$ is an object, usually written as $1 \in \mathbb{C}$, such that for each object $X \in \mathbb{C}$ there is a unique morphism $!_X\colon X \to 1$ in $\mathbb{C}$.

Not every category has a final object, but **Sets** does. Any singleton set is final. We choose one, and write it as $1 = \{*\}$. Notice then that elements of a set X can be identified with functions $1 \to X$. Hence we could forget about membership $\in$ and talk only about arrows.

When a category has binary products $\times$ and a final object 1, one says that the category has **finite products**: for each finite list $X_1, \ldots, X_n$ of objects one can form the product $X_1 \times \cdots \times X_n$. The precise bracketing in this expression is not relevant, because products are associative (up-to-isomorphism); see Exercise 2.1.8 below.

One can generalise these finite products to arbitrary, **set-indexed products**. For an index set I, and a collection $(X_i)_{i \in I}$ of I-indexed objects there is a notion of I-indexed product. It is an object $X = \prod_{i \in I} X_i$ with projections $\pi_i : X \to X_i$, for $i \in I$, which are universal as in Definition 2.1.1: for an arbitrary object Y and an I-indexed collection $f_i : Y \to X_i$ of morphisms there is a unique map $f = \langle f_i \rangle_{i \in I} : Y \to X$ with $\pi_i \circ f = f_i$, for each $i \in I$. In the category **Sets** such products exist and may be described as

$$\prod_{i \in I} X_i = \{t : I \to \bigcup_{i \in I} X_i \mid \forall i \in I.\, t(i) \in X_i\}. \tag{2.4}$$

Coproducts

The next construction we consider is the coproduct (or disjoint union, or sum) $+$. For sets X, Y we write their coproduct as $X + Y$. It is defined as

$$X + Y = \{(x, 1) \mid x \in X\} \cup \{(y, 2) \mid y \in Y\}.$$

The components 1 and 2 serve to force this union to be disjoint. These 'tags' enable us to recognise the elements of X and of Y inside $X + Y$. Instead of projections as above we now have 'coprojections' $\kappa_1 : X \to X + Y$ and $\kappa_2 : Y \to X + Y$ going in the other direction. One puts $\kappa_1(x) = (x, 1)$ and $\kappa_2(y) = (y, 2)$. And instead of tupling we now have 'cotupling' (sometimes called 'source tupling' or 'pattern matching'): for functions $f : X \to Z$ and $g : Y \to Z$ there is a cotuple function $[f, g] : X + Y \to Z$ going out of the coproduct, defined by case distinction:

$$[f, g](w) = \begin{cases} f(x) & \text{if } w = (x, 1) \\ g(y) & \text{if } w = (y, 2). \end{cases}$$

There are standard equations for coproducts, similar to those for products (2.1):

$$\begin{array}{rclcrcl} [f, g] \circ \kappa_1 & = & f & \qquad & [\kappa_1, \kappa_2] & = & \mathrm{id}_{X+Y} \\ [f, g] \circ \kappa_2 & = & g & & h \circ [f, g] & = & [h \circ f, h \circ g]. \end{array} \tag{2.5}$$

Earlier we described the essence of products in a bijective correspondence; see (2.2). There is a similar correspondence for coproducts, but with all arrows reversed:

$$\frac{X \longrightarrow Z \qquad Y \longrightarrow Z}{X + Y \longrightarrow Z} \; . \tag{2.6}$$

(The horizontal bar in (2.6) is printed as a double line.)

This duality between products and coproducts can be made precise in categorical language; see Exercise 2.1.3 below.

So far we have described the coproduct $X + Y$ on sets. We can extend it to functions in the following way. For $f: X \to X'$ and $g: Y \to Y'$ there is a function $f + g: X + Y \to X' + Y'$ by

$$(f + g)(w) = \begin{cases} (f(x), 1) & \text{if } w = (x, 1) \\ (g(y), 2) & \text{if } w = (y, 2). \end{cases} \tag{2.7}$$

Equivalently, we could have defined: $f + g = [\kappa_1 \circ f, \kappa_2 \circ g]$. This operation $+$ on functions preserves identities and composition:

$$\mathrm{id}_X + \mathrm{id}_Y = \mathrm{id}_{X+Y} \qquad \text{and} \qquad (f \circ h) + (g \circ k) = (f + g) \circ (h + k).$$

Thus, coproducts yield a functor $+$: **Sets** $\times$ **Sets** $\to$ **Sets**, like products.

Coproducts in **Sets** satisfy some additional properties. For example, the coproduct is disjoint, in the sense that $\kappa_1(x) \neq \kappa_2(y)$, for all x, y. Also, the coprojections cover the coproduct: every element of a coproduct is of the form either $\kappa_1(x)$ or $\kappa_1(y)$. Further, products distribute over coproducts; see Exercise 2.1.7 below.

We should emphasise that a coproduct $+$ is very different from ordinary union $\cup$. For example, $\cup$ is idempotent: $X \cup X = X$, but there is not even an isomorphism between $X+X$ and X (if $X \neq \emptyset$). Union is an operation on subsets, whereas coproduct is an operation on sets.

Also the coproduct $+$ in **Sets** is an instance of a more general categorical notion of coproduct.

Definition 2.1.3 The **coproduct** of two objects X, Y in a category $\mathbb{C}$ is a new object $X + Y \in \mathbb{C}$ with two coprojection morphisms

$$X \xrightarrow{\quad \kappa_1 \quad} X + Y \xleftarrow{\quad \kappa_2 \quad} Y$$

satisfying a universal property: for each pair of maps $f: X \to Z$ and $g: Y \to Z$ in $\mathbb{C}$ there is a *unique* cotuple morphism $[f, g]: X + Y \to Z$ in $\mathbb{C}$, making the following diagram commute:

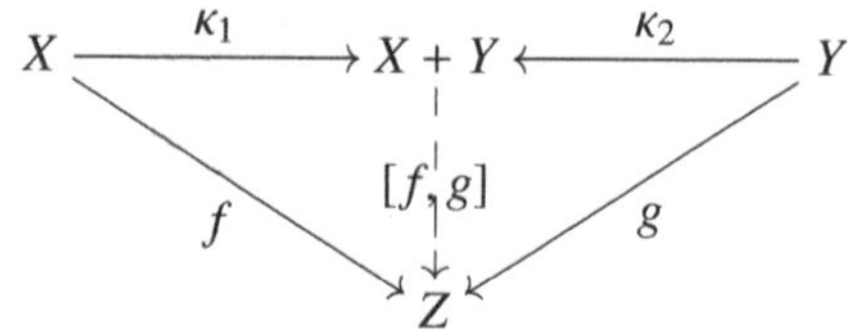

As for products, Equation (2.5) can be derived. Also, there is a notion of empty coproduct.

Definition 2.1.4 An **initial** object 0 in a category $\mathbb{C}$ has the property that for each object $X \in \mathbb{C}$ there a unique morphism $!_X : 0 \to X$ in $\mathbb{C}$.

The exclamation mark ! is often used to describe uniqueness, as in unique existence $\exists!$. Hence we use it to describe both maps $0 \to X$ out of initial objects and maps $X \to 1$ into final objects (see Definition 2.1.2). Usually this does not lead to confusion.

In **Sets** the empty set 0 is initial: for each set X there is precisely one function $0 \to X$, namely the empty function (the function with the empty graph). In **Sets** one has the additional property that each function $X \to 0$ is an isomorphism. This makes $0 \in$ **Sets** a so-called strict initial object.

As for products, one says that a category has finite coproducts when it has binary coproducts + together with an initial object 0. In that case one can form coproducts $X_1 + \cdots + X_n$ for any finite list of objects X_i. Taking the n-fold coproduct of the same object X yields what is called the **copower**, written as $n \cdot X = X + \cdots + X$. Also, a **set-indexed coproduct**, for a set I and a collection $(X_i)_{i \in I}$ of I-indexed objects, may exist. If so, it is an object $X = \coprod_{i \in I} X_i$ with coprojections $\kappa_i : X_i \to X$, for $i \in I$, which are universal: for an arbitrary object Y and a collection of maps $f_i : X_i \to Y$ there is a unique morphism $f = [f_i]_{i \in I} : \coprod_{i \in I} X_i \to Y$ with $f \circ \kappa_i = f_i$, for each $i \in I$. In the category **Sets** such coproducts are disjoint unions, like finite coproducts, but with tags from the index set I, as in

$$\coprod_{i \in I} X_i = \{(i, x) \mid i \in I \wedge x \in X_i\}. \tag{2.8}$$

Whereas products are well known, coproducts are relatively unfamiliar. But from a purely categorical perspective, they are essentially the same and not more difficult than products, because they are their duals (see Exercise 2.1.3 below). In a non-categorical setting however, the cotuple $[f, g]$ is a bit complicated, because it involves variable binding and pattern matching: in a term calculus one can write $[f, g](z)$ for instance as

$$\begin{array}{l} \textsf{CASES } z \textsf{ OF} \\ \quad \kappa_1(x) \longmapsto f(x) \\ \quad \kappa_2(y) \longmapsto g(y). \end{array}$$

Notice that the variables x and y are bound: they are mere place-holders, and their names are not relevant. Functional programmers are quite used to such cotuple definitions by pattern matching.

Another reason why coproducts are not so standard in mathematics is probably that in many algebraic structures coproducts coincide with products; in that case one speaks of biproducts. This is for instance the case for (commutative) monoids/groups and vector spaces and complete lattices; see Exercise 2.1.6. Additionally, in many continuous structures coproducts do not exist (e.g. in categories of domains).

However, within the theory of coalgebras coproducts play an important role. They occur in the construction of many functors F, used to describe coalgebras (namely as F-coalgebras; see Definition 1.4.5), in order to capture different output options, such as normal and abrupt termination in Section 1.1. But additionally, one can form new coalgebras from existing ones via coproducts. This will be illustrated next. It will be our first purely categorical construction. Therefore, it is elaborated in some detail.

Proposition 2.1.5 *Let $\mathbb{C}$ be a category with finite coproducts $(0, +)$, and let F be an arbitrary endofunctor $\mathbb{C} \to \mathbb{C}$. The category* $\mathbf{CoAlg}(F)$ *of F-coalgebras then also has finite coproducts, given by*

$$\textit{initial coalgebra:} \begin{pmatrix} F(0) \\ \uparrow ! \\ 0 \end{pmatrix} \qquad \textit{coproduct coalgebra:} \begin{pmatrix} F(X+Y) \\ \uparrow \\ X+Y \end{pmatrix} \tag{2.9}$$

where the map $X+Y \to F(X{+}Y)$ on the right is the cotuple $[F(\kappa_1)\circ c, F(\kappa_2)\circ d]$, assuming coalgebras $X \xrightarrow{c} F(X)$ and $Y \xrightarrow{d} F(Y)$.

This result generalises to arbitrary (set-indexed) coproducts $\coprod_{i\in I} X_i$ (see Exercise 2.1.13) and also to coequalisers (see Exercise 2.1.14) (and thus to all colimits).

Proof It is important to distinguish between reasoning in the two different categories at hand, namely $\mathbb{C}$ and $\mathbf{CoAlg}(F)$. For the above map $0 \to F(0)$ to be an initial object in $\mathbf{CoAlg}(F)$ we have to show that there is a unique map to any object in $\mathbf{CoAlg}(F)$. This means, for an arbitrary coalgebra $c\colon X \to F(X)$ there must be a unique map f in $\mathbf{CoAlg}(F)$ of the form:

$$\begin{array}{ccc} F(0) & \xrightarrow{F(f)} & F(X) \\ {\scriptstyle !}\uparrow & & \uparrow{\scriptstyle c} \\ 0 & \xrightarrow[f]{} & X \end{array}$$

Since this map f must also be a map $0 \to X$ in $\mathbb{C}$, by initiality of $0 \in \mathbb{C}$, it can be only the unique map $f = \,!\colon 0 \to X$. We still have to show that for $f = \,!$ the above diagram commutes. But this follows again by initiality of 0: there can be only a single map $0 \to F(X)$ in $\mathbb{C}$. Hence both composites $c \circ f$ and $F(f) \circ !$ must be the same.

Next, in order to see that the coalgebra on the right in (2.9) is a coproduct in **CoAlg**(F), we precisely follow Definition 2.1.3. We have to have two coprojections in **CoAlg**(F), for which we take

$$\begin{array}{ccccc} F(X) & \xrightarrow{F(\kappa_1)} & F(X+Y) & \xleftarrow{F(\kappa_2)} & F(Y) \\ {\scriptstyle c}\uparrow & & \uparrow{\scriptstyle [F(\kappa_1) \circ c,\, F(\kappa_2) \circ d]} & & \uparrow{\scriptstyle d} \\ X & \xrightarrow[\kappa_1]{} & X+Y & \xleftarrow[\kappa_2]{} & Y \end{array}$$

It is almost immediately clear that these κ_1, κ_2 are indeed homomorphisms of coalgebras. Next, according to Definition 2.1.3 we must show that one can do cotupling in **CoAlg**(F). So assume two homomorphisms f, g of coalgebras:

$$\begin{array}{ccccc} F(X) & \xrightarrow{F(f)} & F(W) & \xleftarrow{F(g)} & F(Y) \\ {\scriptstyle c}\uparrow & & {\scriptstyle e}\uparrow & & \uparrow{\scriptstyle d} \\ X & \xrightarrow[f]{} & W & \xleftarrow[g]{} & Y \end{array}$$

These f, g are by definition also morphisms $X \to W, Y \to W$ in $\mathbb{C}$. Hence we can take their cotuple $[f, g]\colon X + Y \to W$ in $\mathbb{C}$, since by assumption $\mathbb{C}$ has coproducts. What we need to show is that this cotuple $[f, g]$ is also a map in **CoAlg**(F), in

$$\begin{array}{ccc} F(X+Y) & \xrightarrow{F([f,g])} & F(W) \\ {\scriptstyle [F(\kappa_1) \circ c,\, F(\kappa_2) \circ d]}\uparrow & & \uparrow{\scriptstyle e} \\ X+Y & \xrightarrow[{[f,g]}]{} & W \end{array}$$

This follows by using the coproduct equations (2.5):

$$\begin{aligned} & F([f,g]) \circ [F(\kappa_1) \circ c,\ F(\kappa_2) \circ d] \\ & = [F([f,g]) \circ F(\kappa_1) \circ c,\ F([f,g]) \circ F(\kappa_2) \circ d] && \text{by (2.5)} \\ & = [F([f,g] \circ \kappa_1) \circ c,\ F([f,g] \circ \kappa_2) \circ d] && \text{since } F \text{ is a functor} \\ & = [F(f) \circ c,\ F(g) \circ d] && \text{by (2.5)} \\ & = [e \circ f,\ e \circ g] && f, g \text{ are coalgebra maps} \\ & = e \circ [f,g] && \text{see (2.5).} \end{aligned}$$

Now we know that $[f,g]$ is a map in **CoAlg**(F). Clearly it satisfies $[f,g] \circ \kappa_1 = f$ and $[f,g] \circ \kappa_2 = g$ in **CoAlg**(F) because composition in **CoAlg**(F) is the same as in $\mathbb{C}$. Finally, Definition 2.1.3 requires that this $[f,g]$ is the unique map in **CoAlg**(F) with this property. But this follows because $[f,g]$ is the unique such map in $\mathbb{C}$. □

Thus, coproducts form an important construct in the setting of coalgebras.

Exponents

Given two sets X and Y one can consider the set

$$Y^X = \{f \mid f \text{ is a total function } X \to Y\}.$$

This set Y^X is sometimes called the function space, or exponent of X and Y. As in products and coproducts, it comes equipped with some basic operations. There is an evaluation function $\text{ev}\colon Y^X \times X \to Y$, which sends the pair (f, x) to the function application $f(x)$. And for a function $f\colon Z \times X \to Y$ there is an abstraction function $\Lambda(f)\colon Z \to Y^X$, which maps $z \in Z$ to the function $x \mapsto f(z,x)$ that maps $x \in X$ to $f(z,x) \in Y$. Some basic equations are

$$\begin{aligned} \text{ev} \circ (\Lambda(f) \times \text{id}_X) &= f \\ \Lambda(\text{ev}) &= \text{id}_{Y^X} \\ \Lambda(f) \circ h &= \Lambda(f \circ (h \times \text{id}_X)). \end{aligned} \tag{2.10}$$

Again, the essence of this construction can be summarised concisely in the form of a bijective correspondence, sometimes called currying.

$$\frac{Z \times X \longrightarrow Y}{\overline{\quad Z \longrightarrow Y^X \quad}}. \tag{2.11}$$

Notice that if the set X is finite, say with n elements, then Y^X is isomorphic to the n-fold product $Y \times \cdots \times Y = Y^n$.

We have seen that both the product $\times$ and the coproduct $+$ give rise to functors **Sets** $\times$ **Sets** $\to$ **Sets**. The situation for exponents is more subtle, because of the so-called contravariance in the first argument. This leads to an exponent functor $\mathbf{Sets}^{\mathrm{op}} \times \mathbf{Sets} \to \mathbf{Sets}$, involving an opposite category for its first argument. We will show how this works.

For two maps $k \colon X \to U$ in $\mathbf{Sets}^{\mathrm{op}}$ and $h \colon Y \to V$ in **Sets** we need to define a function $h^k \colon Y^X \to V^U$ between exponents. The fact that $k \colon X \to U$ is a morphism in $\mathbf{Sets}^{\mathrm{op}}$ means that it really is a function $k \colon U \to X$. Therefore we can define h^k on a function $f \in Y^X$ as

$$h^k(f) = h \circ f \circ k. \tag{2.12}$$

This yields indeed a function in V^U. Functoriality also means that identities and compositions must be preserved. For identities this is easy:

$$(\mathrm{id}^{\mathrm{id}})(f) = \mathrm{id} \circ f \circ \mathrm{id} = f.$$

But for preservation of composition we have to remember that composition in an opposite category is reversed:

$$\begin{aligned}
(h_2^{k_2} \circ h_1^{k_1})(f) = h_2^{k_2}(h_1^{k_1}(f)) &= h_2^{k_2}(h_1 \circ f \circ k_1) \\
&= h_2 \circ h_1 \circ f \circ k_1 \circ k_2 \\
&= (h_2 \circ h_1)^{(k_1 \circ k_2)}(f) \\
&= (h_2 \circ_{\mathbb{C}} h_1)^{(k_2 \circ_{\mathbb{C}^{\mathrm{op}}} k_1)}(f).
\end{aligned}$$

We conclude this discussion of exponents with the categorical formulation.

Definition 2.1.6 Let $\mathbb{C}$ be a category with products $\times$. The **exponent** of two objects $X, Y \in \mathbb{C}$ is a new object $Y^X \in \mathbb{C}$ with an evaluation morphism

$$Y^X \times X \xrightarrow{\;\mathrm{ev}\;} Y$$

such that for each map $f \colon Z \times X \to Y$ in $\mathbb{C}$ there is a *unique* abstraction morphism $\Lambda(f) \colon Z \to Y^X$ in $\mathbb{C}$, making the following diagram commute:

$$\begin{array}{ccc}
Y^X \times X & \xrightarrow{\;\mathrm{ev}\;} & Y \\
{\scriptstyle \Lambda(f) \times \mathrm{id}_X} \uparrow & \nearrow {\scriptstyle f} & \\
Z \times X & &
\end{array}$$

The following notions are often useful. A **cartesian closed category**, or **CCC** for short, is a category with finite products and exponents. And a **bicartesian closed category**, or **BiCCC**, is a CCC with finite coproducts. As we have seen, **Sets** is a BiCCC.

Powersets

For a set X we write $\mathcal{P}(X) = \{U \mid U \subseteq X\}$ for the set of (all) subsets of X. In more categorical style we shall also write $U \hookrightarrow X$ or $U \rightarrowtail X$ for $U \subseteq X$. These subsets will also be called predicates. Therefore, we sometimes write $U(x)$ for $x \in U$, and say in that case that U holds for x. The powerset $\mathcal{P}(X)$ is naturally ordered by inclusion: $U \subseteq V$ iff $\forall x \in X. x \in U \Rightarrow x \in V$. This yields a poset $(\mathcal{P}(X), \subseteq)$, with (arbitrary) meets given by intersection $\bigcap_{i \in I} U_i = \{x \in X \mid \forall i \in I. x \in U_i\}$, (arbitrary) joins by unions $\bigcup_{i \in I} U_i = \{x \in X \mid \exists i \in I. x \in U_i\}$ and negation by complement $\neg U = \{x \in X \mid x \notin U\}$. In brief, $(\mathcal{P}(X), \subseteq)$ is a complete Boolean algebra. Of special interest is the truth predicate $\top_X = (X \subseteq X)$, which always holds, and the falsity predicate $\bot_X = (\emptyset \subseteq X)$, which never holds. The two-element set $\{\bot, \top\}$ of Booleans is thus the powerset $\mathcal{P}(1)$ of the final object 1.

Relations may be seen as special cases of predicates. For example, a (binary) relation R on sets X and Y is a subset $R \subseteq X \times Y$ of the product set, i.e. an element of the powerset $\mathcal{P}(X \times Y)$. We shall use the following notations interchangeably:

$$R(x, y), \qquad (x, y) \in R, \qquad xRy.$$

Relations, like predicates, can be ordered by inclusion. The resulting poset $(\mathcal{P}(X \times Y), \subseteq)$ is again a complete Boolean algebra. It also contains a truth relation $\top_{X \times Y} \subseteq X \times Y$, which always holds, and a falsity relation $\bot_{X \times Y} \subseteq X \times Y$, which never holds.

Reversal and composition are two basic constructions on relations. For a relation $R \subseteq X \times Y$ we shall write $R^\dagger \subseteq Y \times X$ for the reverse relation given by $yR^\dagger x$ iff xRy. If we have another relation $S \subseteq Y \times Z$ we can describe the composition of relations $S \circ R$ as a new relation $(S \circ R) \subseteq X \times Z$, via $x(S \circ R)z$ iff $\exists y \in Y. R(x, y) \wedge S(y, z)$, as already described in (1.11).

Often we are interested in 'endo' relations $R \subseteq X \times X$ on a single set X. Of special interest then is the equality relation $\mathrm{Eq}(X) \subseteq X \times X$ given by $\mathrm{Eq}(X) = \{(x, y) \in X \times X \mid x = y\} = \{(x, x) \mid x \in X\}$. As we saw in Example 1.4.2.4, sets and relations form a category **SetsRel**, with equality relations as identity maps. The reversal operation $(-)^\dagger$ yields a functor $\mathbf{SetsRel}^{\mathrm{op}} \to \mathbf{SetsRel}$ that is the identity on objects and satisfies $R^{\dagger\dagger} = R$. It makes **SetsRel** into what is called a dagger category. Such categories are used for reversible computations, like in quantum computing; see e.g. [8].

The powerset operation $X \mapsto \mathcal{P}(X)$ is also functorial. For a function $f\colon X \to Y$ there is a function $\mathcal{P}(f)\colon \mathcal{P}(X) \to \mathcal{P}(Y)$ given by so-called **direct image**: for $U \subseteq X$,

$$\begin{aligned} \mathcal{P}(f)(U) &= \{f(x) \mid x \in U\} \\ &= \{y \in Y \mid \exists x \in X.\, f(x) = y \wedge x \in U\}. \end{aligned} \tag{2.13}$$

Alternative notation for this direct image is $f[U]$ or $\coprod_f(U)$. In this way we may describe the powerset as a functor $\mathcal{P}$: **Sets** → **Sets**.

It turns out that one can also describe powerset as a functor $\mathcal{P}$: $\mathbf{Sets}^{\mathrm{op}} \to$ **Sets** with the opposite of the category of sets as domain. In that case a function $f: X \to Y$ yields a map $f^{-1}: \mathcal{P}(Y) \to \mathcal{P}(X)$, which is commonly called **inverse image**: for $U \subseteq Y$,

$$f^{-1}(U) = \{x \mid f(x) \in U\}. \tag{2.14}$$

The powerset operation with this inverse image action on morphisms is sometimes called the contravariant powerset. But standardly we shall consider the 'covariant' powersets with direct images as functor **Sets** → **Sets**. We shall frequently encounter these direct $\coprod_f$ and inverse f^{-1} images. They are related by a Galois connection:

$$\frac{\coprod_f(U) \subseteq V}{\overline{\overline{U \subseteq f^{-1}(V)}}}. \tag{2.15}$$

See also in Exercise 2.1.12 below.

We have seen bijective correspondences characterising products, coproducts, exponents and images. There is also such a correspondence for powersets:

$$\frac{X \longrightarrow \mathcal{P}(Y)}{\overline{\overline{\text{relations} \subseteq Y \times X}}}. \tag{2.16}$$

This leads to a more systematic description of a powerset as a so-called relation classifier. There is a special inhabitation $\in \subseteq Y \times \mathcal{P}(Y)$, given by $\in(y, U) \Leftrightarrow y \in U$. For any relation $R \subseteq Y \times X$ there is then a relation classifier, or characteristic function, $\mathsf{char}(R): X \to \mathcal{P}(Y)$ mapping $x \in X$ to $\{y \in Y \mid R(y, x)\}$. This map $\mathsf{char}(R)$ is the unique function $f: X \to \mathcal{P}(Y)$ with $\in(y, f(x)) \Leftrightarrow R(y, x)$, i.e. with $(\mathrm{id} \times f)^{-1}(\in) = R$.

This formalisation of this special property in categorical language yields so-called power objects. The presence of such objects is a key feature of 'toposes'. The latter are categorical set-like universes, with constructive logic. They form a topic that goes beyond the introductory material covered in this text. The interested reader is referred to the extensive literature on toposes [279, 281, 166, 56, 346, 85].

Finally, we often need the *finite* powerset $\mathcal{P}_{\mathrm{fin}}(X) = \{U \in \mathcal{P}(X) \mid U \text{ is finite}\}$. It also forms a functor **Sets** → **Sets**.

Injections and Surjections (in Sets)

A function $f: X \to Y$ is called injective (or an injection, a monomorphism or mono, for short, or a monic map) if $f(x) = f(x')$ implies $x = x'$. In that case we often write $f: X \rightarrowtail Y$ or $f: X \hookrightarrow Y$ in case X is a subset of Y. A surjection (or surjective function, epimorphism or epi or an epic map) is a map written as $f: X \twoheadrightarrow Y$ such that for each $y \in Y$ there is an $x \in X$ with $f(x) = y$. Injectivity and surjectivity can be formulated categorically (see Exercise 2.5.8 later on) and then appear as dual notions. In the category **Sets** these functions have some special 'splitting' properties that we shall describe explicitly because they are used from time to time.

The standard formulation of the axiom of choice (AC) says that for each collection $(X_i)_{i \in I}$ of non-empty sets there is a choice function $c: I \to \bigcup_{i \in I} X_i$ with $c(i) \in X_i$ for each $i \in I$. It is used for instance to see that the set-theoretic product $\prod_{i \in I} X_i$ from (2.4) is a non-empty set in case each X_i is non-empty.

An equivalent, more categorical, formulation of the axiom of choice is: every surjection $f: X \twoheadrightarrow Y$ has a section (also called splitting), a function $s: Y \to X$ in the reverse direction with $f \circ s = \mathrm{id}$. This s thus chooses an element $s(y)$ in the non-empty set $f^{-1}(y) = \{x \in X \mid f(x) = y\}$. Notice that such a section is an injection.

For injections there is a comparable splitting result. Assume $f: X \rightarrowtail Y$ in **Sets**, where $X \neq \emptyset$. Then there is a function $g: Y \to X$ with $g \circ f = \mathrm{id}$. This g is obtained as follows. Since $X \neq \emptyset$ we may assume an element $x_0 \in X$, and use it in

$$g(y) = \begin{cases} x & \text{if there is a (necessarily unique) element } x \text{ with } f(x) = y \\ x_0 & \text{otherwise.} \end{cases}$$

Notice that this g is a surjection $Y \twoheadrightarrow X$.

These observations will often be used in the following form.

Lemma 2.1.7 *Let F: **Sets** $\to$ **Sets** be an arbitrary functor.*

1. *If $f: X \to Y$ is surjective, then so is $F(f): F(X) \to F(Y)$.*
2. *If $f: X \to Y$ is injective and X is non-empty, then $F(f)$ is also injective.*

Proof 1. If $f: X \to Y$ is surjective, then, by the axiom of choice, there is a splitting $s: Y \to X$ with $f \circ s = \mathrm{id}_Y$. Hence $F(f) \circ F(s) = F(f \circ s) = F(\mathrm{id}_Y) = \mathrm{id}_{F(Y)}$. Thus $F(f)$ has a splitting (right inverse) and is thus surjective.

2. In the same way, as we have seen, for injective functions $f: X \to Y$ with $X \neq \emptyset$, there is a $g: Y \to X$ with $g \circ f = \mathrm{id}_X$. Thus $F(g) \circ F(f) = \mathrm{id}_{F(X)}$, so that $F(f)$ is injective. ❑

Exercises

2.1.1 Verify in detail the bijective correspondences (2.2), (2.6), (2.11) and (2.16).

2.1.2 Consider a poset $(D, \leq)$ as a category. Check that the product of two elements $d, e \in D$, if it exists, is the meet $d \wedge e$. And a coproduct of d, e, if it exists, is the join $d \vee e$.

Similarly, show that a final object is a top element $\top$ (with $d \leq \top$, for all $d \in D$) and that an initial object is a bottom element $\bot$ (with $\bot \leq d$, for all $d \in D$).

2.1.3 Check that a product in a category $\mathbb{C}$ is the same as a coproduct in $\mathbb{C}^{\text{op}}$.

2.1.4 Fix a set A and prove that assignments $X \mapsto A \times X$, $X \mapsto A + X$ and $X \mapsto X^A$ are functorial and give rise to functors **Sets** $\to$ **Sets**.

2.1.5 Prove that the category **PoSets** of partially ordered sets and monotone functions is a BiCCC. The definitions on the underlying sets X of a poset $(X, \leq)$ are like for ordinary sets but should be equipped with appropriate orders.

2.1.6 Consider the category **Mon** of monoids with monoid homomorphisms between them.

1. Check that the singleton monoid 1 is both an initial and a final object in **Mon**; this is called a zero object.
2. Given two monoids $(M_1, +_1, 0_1)$ and $(M_2, +_2, 0_2)$, one defines a product monoid $M_1 \times M_2$ with componentwise addition $(x, y) + (x', y') = (x +_1 x', y +_2 y')$ and unit $(0_1, 0_2)$. Prove that $M_1 \times M_2$ is again a monoid, which forms a product in the category **Mon** with the standard projection maps $M_1 \xleftarrow{\pi_1} M_1 \times M_2 \xrightarrow{\pi_2} M_2$.
3. Note that there are also coprojections $M_1 \xrightarrow{\kappa_1} M_1 \times M_2 \xleftarrow{\kappa_2} M_2$, given by $\kappa_1(x) = (x, 0_2)$ and $\kappa_2(y) = (0_1, y)$, which are monoid homomorphisms and which make $M_1 \times M_2$ at the same time the coproduct of M_1 and M_2 in **Mon** (and hence a biproduct). *Hint*: Define the cotuple $[f, g]$ as $x \mapsto f(x) + g(x)$.

2.1.7 Show that in **Sets** products distribute over coproducts, in the sense that the canonical maps

$$(X \times Y) + (X \times Z) \xrightarrow{[\mathrm{id}_X \times \kappa_1, \mathrm{id}_X \times \kappa_2]} X \times (Y + Z)$$

$$0 \xrightarrow{\quad !\quad} X \times 0$$

are isomorphisms. Categories in which this is the case are called **distributive**; see [101] for more information on distributive categories in general and see [182] for an investigation of such distributivities in categories of coalgebras.

2.1.8 1. Consider a category with finite products $(\times, 1)$. Prove that there are isomorphisms:

$$X \times Y \cong Y \times X, \quad (X \times Y) \times Z \cong X \times (Y \times Z), \quad 1 \times X \cong X.$$

2. Similarly, show that in a category with finite coproducts $(+, 0)$ one has

$$X + Y \cong Y + X, \quad (X + Y) + Z \cong X + (Y + Z), \quad 0 + X \cong X.$$

(This means that both the finite product and coproduct structure in a category yield so-called *symmetric monoidal* structure. See [344, 85] for more information.)

3. Next, assume that our category also has exponents. Prove that

$$X^0 \cong 1, \qquad X^1 \cong X, \qquad 1^X \cong 1.$$

And also that

$$Z^{X+Y} \cong Z^X \times Z^Y, \quad Z^{X \times Y} \cong (Z^Y)^X, \quad (X \times Y)^Z \cong X^Z \times Y^Z.$$

2.1.9 Show that the finite powerset also forms a functor $\mathcal{P}_{\text{fin}} : \mathbf{Sets} \to \mathbf{Sets}$.

2.1.10 Check that

$$\mathcal{P}(0) \cong 1, \qquad \mathcal{P}(X + Y) \cong \mathcal{P}(X) \times \mathcal{P}(Y).$$

And similarly for the finite powerset $\mathcal{P}_{\text{fin}}$ instead of $\mathcal{P}$. This property says that $\mathcal{P}$ and $\mathcal{P}_{\text{fin}}$ are 'additive'; see [111].

2.1.11 Notice that a powerset $\mathcal{P}(X)$ can also be understood as exponent 2^X, where $2 = \{0, 1\}$. Check that the exponent functoriality gives rise to the contravariant powerset $\mathbf{Sets}^{\text{op}} \to \mathbf{Sets}$.

2.1.12 Consider a function $f : X \to Y$. Prove that

1. The direct image $\mathcal{P}(f) = \coprod_f : \mathcal{P}(X) \to \mathcal{P}(Y)$ preserves all joins and that the inverse image $f^{-1}(-) : \mathcal{P}(Y) \to \mathcal{P}(X)$ preserves not only joins but also meets and negation (i.e. all the Boolean structure).
2. There is a Galois connection $\coprod_f(U) \subseteq V \Longleftrightarrow U \subseteq f^{-1}(V)$, as claimed in (2.15).
3. There is a product function $\prod_f : \mathcal{P}(X) \to \mathcal{P}(Y)$ given by $\prod_f(U) = \{y \in Y \mid \forall x \in X.\, f(x) = y \Rightarrow x \in U\}$, with a Galois connection $f^{-1}(V) \subseteq U \Longleftrightarrow V \subseteq \prod_f(U)$.

2.1.13 Assume a category $\mathbb{C}$ has arbitrary, set-indexed coproducts $\coprod_{i \in I} X_i$. Demonstrate, as in the proof of Proposition 2.1.5, that the category **CoAlg**(F) of coalgebras of a functor $F \colon \mathbb{C} \to \mathbb{C}$ then also has such coproducts.

2.1.14 For two parallel maps $f, g \colon X \to Y$ between objects X, Y in an arbitrary category $\mathbb{C}$ a **coequaliser** $q \colon Y \to Q$ is a map in a diagram

$$X \underset{g}{\overset{f}{\rightrightarrows}} Y \xrightarrow{\quad q \quad} Q$$

with $q \circ f = q \circ g$ in a 'universal way': for an arbitrary map $h \colon Y \to Z$ with $h \circ f = h \circ g$ there is a unique map $k \colon Q \to Z$ with $k \circ q = h$.

1. An **equaliser** in a category $\mathbb{C}$ is a coequaliser in $\mathbb{C}^{\mathrm{op}}$. Formulate explicitly what an equaliser of two parallel maps is.
2. Check that in the category **Sets** the set Q can be defined as the quotient Y/R, where $R \subseteq Y \times Y$ is the least equivalence relation containing all pairs $(f(x), g(x))$ for $x \in X$.
3. Returning to the general case, assume a category $\mathbb{C}$ has coequalisers. Prove that for an arbitrary functor $F \colon \mathbb{C} \to \mathbb{C}$ the associated category of coalgebras **CoAlg**(F) also has coequalisers, as in $\mathbb{C}$: for two parallel homomorphisms $f, g \colon X \to Y$ between coalgebras $c \colon X \to F(X)$ and $d \colon Y \to F(Y)$ there is by universality an induced coalgebra structure $Q \to F(Q)$ on the coequaliser Q of the underlying maps f, g, yielding a diagram of coalgebras

$$\begin{pmatrix} F(X) \\ \uparrow c \\ X \end{pmatrix} \underset{g}{\overset{f}{\rightrightarrows}} \begin{pmatrix} F(Y) \\ \uparrow d \\ Y \end{pmatrix} \xrightarrow{\quad q \quad} \begin{pmatrix} F(Q) \\ \uparrow \\ Q \end{pmatrix}$$

with the appropriate universal property in **CoAlg**(F): for each coalgebra $e \colon Z \to F(Z)$ with homomorphism $h \colon Y \to Z$ satisfying $h \circ f = h \circ g$ there is a unique homomorphism of coalgebras $k \colon Q \to Z$ with $k \circ q = h$.

2.2 Polynomial Functors and Their Coalgebras

Earlier in Definition 1.4.5 we have seen the general notion of a coalgebra as a map $X \to F(X)$ in a category $\mathbb{C}$, where F is a functor $\mathbb{C} \to \mathbb{C}$. Here, in this section and in much of the rest of this text, we shall concentrate on a more restricted situation: as category $\mathbb{C}$ we use the category **Sets** of ordinary sets and functions. And as functors $F \colon$ **Sets** $\to$ **Sets** we shall use so-called polynomial functors, such as $F(X) = A + (B \times X)^C$. These are functors built up inductively from certain simple basic functors, using products, coproducts, exponents and powersets for forming new functors. There are three reasons for this restriction to polynomial functors.

1. Polynomial functors are concrete and easy to grasp.
2. Coalgebras of polynomial functors include many of the basic examples; they suffice for the time being.
3. Polynomial functors allow definitions by induction, for many of the notions that we shall be interested in – notably relation lifting and predicate lifting in the next two chapters. These inductive definitions are easy to use and can be introduced without any categorical machinery.

This section contains the definition of polynomial functor and also many examples of such functors and of their coalgebras.

Definition 2.2.1 We define three collections of functors **Sets** → **Sets**, namely SPF, EPF, and KPF, for *simple*, *exponent* and *Kripke* polynomial functors, as in:

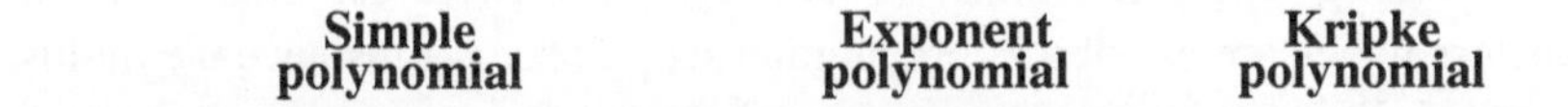

1. The collection SPF of **simple polynomial functors** is the least class of functors **Sets** → **Sets** satisfying the following four clauses.
   i The identity functor **Sets** → **Sets** is in SPF.
   ii For each set A, the constant functor A: **Sets** → **Sets** is in SPF. Recall that it maps every set X to A and every function f to the identity id_A on A.
   iii If both F and G are in SPF, then so is the product functor $F \times G$, defined as $X \mapsto F(X) \times G(X)$. On functions it is defined as $f \mapsto F(f) \times G(f)$; see (2.3).
   iv If we have a non-empty set I and an I-indexed collection of functors F_i in SPF, then the set-indexed coproduct $\coprod_{i \in I} F_i$ is also in SPF. This new functor maps a set X to the I-indexed coproduct $\coprod_{i \in I} F_i(X) = \{(i, u) \mid i \in I \wedge u \in F_i(X)\}$. It maps a function $f: X \to Y$ to the mapping $(i, u) \mapsto (i, F_i(f)(u))$.

2. The collection EPF of **exponent polynomial functors** has the above four clauses, and additionally:

   v. For each set A, if F in EPF, then so is the 'constant' exponent F^A defined as $X \mapsto F(X)^A$. It sends a function $f\colon X \to Y$ to the function $F(f)^A = F(f)^{\mathrm{id}_A}$, which maps $h\colon A \to F(X)$ to $F(f) \circ h\colon A \to F(Y)$; see (2.12).

3. The class KPF of **Kripke polynomial functors** is the superset of SPF defined by the above clauses (i)–(v), with 'SPF' replaced by 'KPF', plus one additional rule:

   vi. If F is in KPF, then so is the powerset $\mathcal{P}(F)$, defined as $X \mapsto \mathcal{P}(F(X))$ on sets and as $f \mapsto \mathcal{P}(F(f))$ on functions; see (2.13).

   Occasionally, we shall say that a functor F is a *finite* KPF. This means that all the powersets $\mathcal{P}$ occurring in F are actually finite powersets $\mathcal{P}_{\mathrm{fin}}$.

We notice that exponents $(-)^A$ for *finite* sets A are already included in simple polynomial functors via iterated products $F_1 \times \cdots \times F_n$. The collection EPF is typically used to capture coalgebras (or automata) with infinite sets of inputs, given as exponents; see Section 2.2.3 below. The collection KPF is used for non-deterministic computations via powersets; see Section 2.2.4.

The above clauses yield a reasonable collection of functors to start from, but we could of course have included some more constructions in our definition of polynomial functor – such as iterations via initial and final (co)algebras (see Exercise 2.3.8 and e.g. [214, 402, 275]), as used in the experimental programming language Charity [105, 103, 102]. There are thus interesting functors which are out of the 'polynomial scope'; see for instance the multiset or probability distribution functors from Section 4.1 or 'dependent' polynomial functors in Exercise 2.2.6. However, the above clauses suffice for many examples, for the time being.

The coproducts that are used to construct simple polynomial functors are arbitrary, set-indexed coproducts. Frequently we shall use binary versions $F_1 + F_2$, for an index set $I = \{1, 2\}$. But we like to go beyond such finite coproducts, for instance in defining the **list functor** $F^\star$, given as infinite coproduct of products:

$$F^\star = \textstyle\coprod_{n\in\mathbb{N}} F^n \quad \text{where} \quad F^n = \underbrace{F \times \cdots \times F}_{n \text{ times}}. \tag{2.17}$$

Thus, if F is the identity functor, then $F^\star$ maps a set X to the set of lists:

$$X^\star = 1 + X + (X \times X) + (X \times X \times X) + \cdots .$$

The collection SPF of simple polynomial functors is reasonably stable in the sense that it can be characterised in various ways. Below we give one such alternative characterisation; the other one is formulated later on, in Theorem 4.7.8, in terms of preservation properties. The characterisation below uses 'arities' as commonly used in universal algebra to capture the number of arguments in a primitive function symbol. For instance, addition + has arity 2, and minus − has arity 1. These arities will be used more systematically in Section 6.6 to associate a term calculus with a simple polynomial functor.

Definition 2.2.2 An **arity** is given by a set I and a function $\#\colon I \to \mathbb{N}$. It determines a simple polynomial functor $F_\#\colon \mathbf{Sets} \to \mathbf{Sets}$, namely

$$F_\#(X) \stackrel{\text{def}}{=} \coprod_{i \in I} X^{\# i} = \{(i, \vec{x}) \mid i \in I \text{ and } \vec{x} \in X^{\# i}\}. \tag{2.18}$$

We often call such an $F_\#$ an **arity functor**.

In the style of universal algebra one describes the operations of a group via an index set $I = \{\mathsf{s}, \mathsf{m}, \mathsf{z}\}$, with symbols for sum, minus and zero, with obvious arities $\#(\mathsf{s}) = 2$, $\#(\mathsf{m}) = 1$, $\#(\mathsf{z}) = 0$. The associated functor $F_\#$ sends X to $(X \times X) + X + 1$. In general, these arity functors have a form that clearly is 'polynomial'.

Proposition 2.2.3 *The collections of simple polynomial functors and arity functors are isomorphic.*

Proof By construction an arity functor $F_\# = \coprod_{i \in I}(-)^{\# i}$ is a simple polynomial functor, so we concentrate on the converse. We show that each simple polynomial functor F is an arity functor, with index set $F(1)$, by induction on the structure of F.

- If F is the identity functor $X \mapsto X$ we take $I = F(1) = 1$ and $\# = 1\colon 1 \to \mathbb{N}$. Then $F_\#(X) = \coprod_{i \in 1} X^{\# i} \cong \coprod_{i \in 1} X^1 \cong X = F(X)$.

  If F is a constant functor $X \mapsto A$, then we choose as arity the map $\#\colon A \to \mathbb{N}$, which is constantly 0. Then $F_\#(X) = \coprod_{a \in A} X^{\# a} = \coprod_{a \in A} X^0 \cong \coprod_{a \in A} 1 \cong A = F(X)$.
- If F is a product $X \mapsto F_1(X) \times F_2(X)$, we may assume arities $\#_j\colon I_j \to \mathbb{N}$ for $j \in \{1, 2\}$. We now define $\#\colon I_1 \times I_2 \to \mathbb{N}$ as $\#(i_1, i_2) = \#_1 i_1 + \#_2 i_2$. Then:

$$\begin{aligned} F_\#(X) &= \coprod_{(i_1,i_2) \in I_1 \times I_2} X^{(\#_1 i_1 + \#_2 i_2)} \\ &\cong \coprod_{i_1 \in I_1} \coprod_{i_2 \in I_2} X^{\#_1 i_1} \times X^{\#_2 i_2} \\ &\cong \coprod_{i_1 \in I_1} X^{\#_1 i_1} \times \coprod_{i_2 \in I_2} X^{\#_2 i_2} \\ &\qquad \text{since } Y \times (-) \text{ preserves coproducts (see also Exercise 2.1.7)} \\ &\stackrel{\text{(IH)}}{\cong} F_1(X) \times F_2(X) \\ &= F(X). \end{aligned}$$

- If F is a coproduct $X \mapsto \coprod_{j\in J} F_j(X)$, we may assume arities $\#_j \colon I_j \to \mathbb{N}$ by the induction hypothesis. The cotuple $\# = [\#_j]_{j\in J} \colon \coprod_{j\in j} I_j \to \mathbb{N}$ then does the job:

$$\begin{aligned} F_{\#}(X) &= \coprod_{(j,i)\in\coprod_{j\in J} I_j} X^{\#(j,i)} \\ &\cong \coprod_{j\in J}\coprod_{i\in I_j} X^{\#_j(i)} \\ &\overset{\text{(IH)}}{\cong} \coprod_{j\in J} F_j(X) \\ &= F(X). \end{aligned}$$ ❑

The arities $\# \colon I \to \mathbb{N}$ that we use here are *single-sorted* arities. They can be used to capture operations of the form $n \to 1$, with n inputs, all of the same sort, and a single output, of this same sort. But multi-sorted (or multi-typed) operations such as $\mathsf{even}\colon \mathbb{N} \to \mathsf{Bool} = \{\mathsf{true}, \mathsf{false}\}$ are out of scope. More generally, given a set of sorts/types S, one can also consider multi-sorted arities as functions $\# \colon I \to S^+ = S^* \times S$. A value $\#(i) = (\langle s_1, \ldots s_n\rangle, t)$ then captures a function symbol with type $s_1 \times \cdots \times s_n \to t$, taking n inputs of sort $s_1, \ldots, s_n$ to an output sort t. Notice that the (single-sorted) arities that we use here are a special case, when the set of sorts S is a singleton $1 = \{0\}$, since $1^\star \cong \mathbb{N}$.

In the remainder of this section we shall see several instances of (simple, exponent and Kripke) polynomial functors. This includes examples of fundamental mathematical structures that arise as coalgebras of such functors.

2.2.1 Statements and Sequences

In Chapter 1 we have used program statements (in Section 1.1) and sequences (in Section 1.2) as motivating examples for the study of coalgebras. We briefly review these examples using the latest terminology and notation.

Recall that statements were introduced as functions acting on a state space S, with different output types depending on whether these statements could hang or terminate abruptly because of an exception. These two representations were written as

$$S \longrightarrow \{\bot\} \cup S, \qquad\qquad S \longrightarrow \{\bot\} \cup S \cup (S \times E).$$

Using the notation from the previous section we now write these as:

$$S \longrightarrow 1 + S, \qquad\qquad S \longrightarrow 1 + S + (S \times E).$$

And so we recognise these statements as coalgebras

$$S \longrightarrow F(S), \qquad\qquad S \longrightarrow G(S)$$

of the simple polynomial functors:

$$F = 1 + \mathrm{id} = \big(X \longmapsto 1 + X\big) \quad \text{and} \quad G = 1 + \mathrm{id} + (\mathrm{id} \times E) = \big(X \longmapsto 1 + X + (X \times E)\big).$$

Thus, these functors determine the kind of computations.

Sequence coalgebras, for a fixed set A, were described in Section 1.2 as functions:

$$S \longrightarrow \{\bot\} \cup (A \times S)$$

i.e. as coalgebras:

$$S \longrightarrow 1 + (A \times S)$$

of the simple polynomial functor $1 + (A \times \mathrm{id})$. This functor was called Seq in Example 1.4.4.5. Again, the functor determines the kind of computations: either fail or produce an element in A together with a successor state. We could change this a bit and drop the fail option. In that case, each state yields an element in A with a successor state. This different kind of computation is captured by a different polynomial functor, namely by $A \times \mathrm{id}$. A coalgebra of this functor is a function

$$S \longrightarrow A \times S,$$

as briefly mentioned in the introduction to Chapter 1 (before Section 1.1). Its behaviour will be an infinite sequence of elements of A: since there is no fail option, these behaviour sequences do not terminate. In the next section we shall see how to formalise this as: infinite sequences $A^{\mathbb{N}}$ form the final coalgebra of this functor $A \times \mathrm{id}$.

2.2.2 Trees

We shall continue this game of capturing different kinds of computation via different polynomial functors. Trees form a good illustration because they occur in various forms. We recall that in computer science trees are usually written upside-down.

Let us start by fixing an arbitrary set A, elements of which will be used as labels in our trees. Binary trees are most common. They arise as outcomes of computations of coalgebras:

$$S \longrightarrow A \times S \times S$$

of the simple polynomial functor $A \times \mathrm{id} \times \mathrm{id}$. Indeed, given a state $x \in S$, a one-step computation yields a triple (a_0, x_1, x_2) of an element $a_0 \in A$ and two successor states $x_1, x_2 \in S$. Continuing the computation with both x_1 and x_2 yields two more elements in A, and four successor states, etc. This yields for each $x \in S$ an infinite binary tree with one label from A at each node:

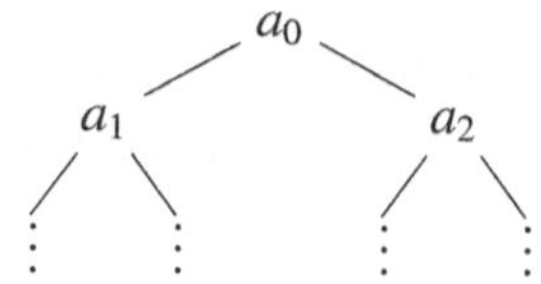

In a next step we could consider *ternary* trees as behaviours of coalgebras:

$$S \longrightarrow A \times S \times S \times S$$

of the simple polynomial functor $A \times \mathrm{id} \times \mathrm{id} \times \mathrm{id}$. By a similar extension one can get quaternary trees, etc. These are all instances of finitely branching trees, arising from coalgebras

$$S \longrightarrow A \times S^\star$$

of the Kripke polynomial functor $A \times \mathrm{id}^\star$. Each state $x \in S$ is now mapped to an element in A with a finite sequence $\langle x_1, \ldots x_n \rangle$ of successor states – with which one continues to observe the behaviour of x.

We can ask if the behaviour trees should always be infinitely deep. Finiteness must come from the possibility that a computation fails and yields no successor state. This can be incorporated by considering coalgebras

$$S \longrightarrow 1 + (A \times S \times S)$$

of the simple polynomial functor $1 + (A \times \mathrm{id} \times \mathrm{id})$. Notice that the resulting behaviour trees may be finite in one branch, but infinite in the other. There is nothing in the shape of the polynomial functor that will guarantee that the whole behaviour will actually be finite.

A minor variation is in replacing the set 1 for non-termination by another set B, in

$$S \longrightarrow B + (A \times S \times S).$$

The resulting behaviour trees will have elements from A at their (inner) nodes and from B at the leaves.

2.2.3 Deterministic Automata

Automata are very basic structures in computing, used in various areas: language theory, text processing, complexity theory, parallel and distributed computing, circuit theory, etc. Their state-based, dynamical nature makes them canonical examples of coalgebras. There are many versions of automata, but here, in this section, we shall concentrate on the two most basic ones: deterministic and non-deterministic. For more information on the extensive theory of automata, see e.g. [42, 127, 416, 30], and on coalgebras and automata see e.g. [412, 457, 319, 196, 429].

A deterministic automaton is usually defined as consisting of a set S of states, a set A of labels (or actions or letters of an alphabet), a transition function $\delta\colon S \times A \to S$ and a set $F \subseteq S$ of final states. Sometimes, also an initial state $x_0 \in S$ is considered part of the structure, but here it is not. Such an automaton is called deterministic because for each state $x \in S$, and input $a \in A$, there is precisely one successor state $x' = \delta(x, a) \in S$.

First, we shall massage these ingredients into coalgebraic shape. Via the bijective correspondence (2.11) for exponents, the transition function $S \times A \to S$ can also be understood as a map $S \to S^A$. And the subset $F \subseteq S$ of final states corresponds to a characteristic function $S \to \{0, 1\}$. These two functions $S \to S^A$ and $S \to \{0, 1\}$ can be combined into a single function,

$$S \xrightarrow{\langle \delta, \epsilon \rangle} S^A \times \{0, 1\}, \tag{2.19}$$

using the product correspondences (2.2). Thus, deterministic automata with A as a set of labels are coalgebras of the exponent polynomial functor $\mathrm{id}^A \times \{0, 1\}$.

An example of such an automaton with input set $A = \{0, 1\}$ is

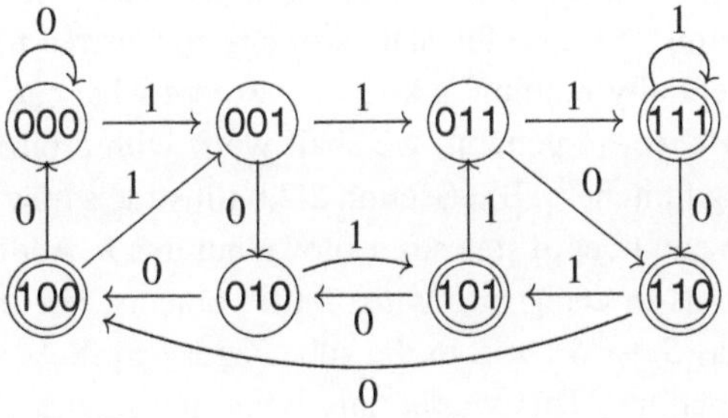

(2.20)

The states are the numbers $0, 1, \ldots, 7$ in binary notation, giving as state space:

$$S = \{000, 001, 010, 011, 100, 101, 110, 111\}.$$

Using the standard convention, final states are doubly encircled. In our coalgebraic representation (2.19) this is captured by the function $\epsilon\colon S \to \{0, 1\}$ defined by

$$\begin{aligned}\epsilon(111) = \epsilon(100) = \epsilon(101) = \epsilon(110) = 1\\ \epsilon(000) = \epsilon(001) = \epsilon(011) = \epsilon(010) = 0.\end{aligned}$$

The transition function $\delta\colon S \to S^{\{0,1\}}$ is

$$\begin{array}{llll}
\delta(000)(0) = 000 & \delta(001)(0) = 010 & \delta(011)(0) = 110 & \delta(111)(0) = 110 \\
\delta(000)(1) = 001 & \delta(001)(1) = 011 & \delta(011)(1) = 111 & \delta(111)(1) = 111 \\
\delta(100)(0) = 000 & \delta(010)(0) = 100 & \delta(101)(0) = 010 & \delta(110)(0) = 111 \\
\delta(100)(1) = 001 & \delta(010)(1) = 101 & \delta(101)(1) = 011 & \delta(110)(1) = 101.
\end{array}$$

It is clear that the graphical representation (2.20) is more pleasant to read than these listings. Shortly we discuss what this automaton does.

First we shall stretch the representation (2.19) a little bit and replace the set $\{0, 1\}$ of observable values by an arbitrary set B of outputs. Thus, what shall call a **deterministic automaton** is a coalgebra

$$S \xrightarrow{\langle \delta, \epsilon \rangle} S^A \times B \tag{2.21}$$

of the exponent polynomial functor $\mathrm{id}^A \times B$. We recall that if the set A of inputs is finite, we have a *simple* polynomial functor.

This kind of coalgebra, or deterministic automaton, $\langle \delta, \epsilon \rangle\colon S \to S^A \times B$, thus consists of a **transition function** $\delta\colon S \to S^A$ and an **observation function** $\epsilon\colon S \to B$. For such an automaton, we shall frequently use a transition notation $x \xrightarrow{a} x'$ for $x' = \delta(x)(a)$. Also, we introduce a notation for observation: $x \downarrow b$ stands for $\epsilon(x) = b$. Finally, a combined notation is sometimes useful: $(x \downarrow b) \xrightarrow{a} (x' \downarrow b')$ means three things at the same time: $x \downarrow b$, $x \xrightarrow{a} x'$ and $x' \downarrow b'$.

Often one considers automata with a *finite* state space. This is not natural in a coalgebraic setting, because the state space is considered to be a black box, about which essentially nothing is known – except what can be observed via the operations. Hence, in general, we shall work with arbitrary state spaces, without assuming finiteness. But Section 2.2.6 illustrates how to model n-state systems, when the number of states is a known number n. Additionally, one can consider coalgebras in categories with 'finite' objects. For instance, one can look at coalgebras $S \to S^A \times B$ in the subcategory **FinSets** $\hookrightarrow$ **Sets** of finite sets and (all) functions. This works only when the sets A, B are both finite. More generally, one can work in categories of finitely presented objects [351] or of finitely generated objects [352].

Let $x \in S$ be an arbitrary state of such a coalgebra/deterministic automaton $\langle \delta, \epsilon \rangle\colon S \to S^A \times B$. Applying the output function $\epsilon\colon S \to B$ to x yields an immediate observation $\epsilon(x) \in B$. For each element $a_1 \in A$ we can produce a successor state $\delta(x)(a_1) \in S$; it also gives rise to an immediate observation $\epsilon(\delta(x)(a_1)) \in B$, and for each $a_2 \in A$ a successor state $\delta(\delta(x)(a_1))(a_2) \in S$, etc. Thus, for each finite sequence $\langle a_1, \ldots, a_n \rangle \in A^\star$ we can observe an element $\epsilon(\delta(\cdots \delta(x)(a_1) \cdots)(a_n)) \in B$. Everything we can possibly observe about the state x is obtained in this way, namely as a function $A^\star \to B$. Such behaviours will form the states of the final coalgebra; see Proposition 2.3.5 in the next section.

For future reference we shall be a bit more precise about lists of inputs. Behaviours can best be described via an iterated transition function

$$S \times A^\star \xrightarrow{\delta^*} S \qquad \text{defined as} \qquad \begin{cases} \delta^*(x, \langle\rangle) = x \\ \delta^*(x, a \cdot \sigma) = \delta^*(\delta(x)(a), \sigma). \end{cases} \tag{2.22}$$

This iterated transition function δ^* gives rise to the multiple-step transition notation: $x \xrightarrow{\sigma}{}^* y$ stands for $y = \delta^*(x, \sigma)$ and means that y is the (non-immediate) successor state of x obtained by applying the inputs from the sequent $\sigma \in A^\star$, from left to right.

The behaviour $\mathrm{beh}(x)\colon A^\star \to B$ of a state $x \in S$ is then obtained as the function that maps a finite sequence $\sigma \in A^\star$ of inputs to the observable output

$$\mathrm{beh}(x)(\sigma) \stackrel{\text{def}}{=} \epsilon(\delta^*(x, \sigma)) \in B. \tag{2.23}$$

An element of the set $B^{A^\star}$ of all such behaviours can be depicted as a rooted tree with elements from the set A of inputs as labels, and elements from the set B of outputs as nodes. For example, a function $\varphi \in B^{A^\star}$ can be described as an infinite tree:

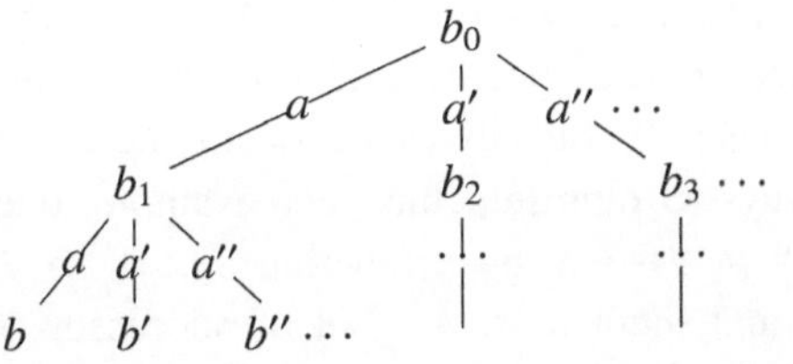

where

$$\begin{aligned} \varphi(\langle\rangle) &= b_0 \\ \varphi(\langle a\rangle) &= b_1 \\ \varphi(\langle a'\rangle) &= b_2 \\ \varphi(\langle a''\rangle) &= b_3 \\ &\text{etc.} \\ \varphi(\langle a, a\rangle) &= b \\ \varphi(\langle a, a'\rangle) &= b' \\ \varphi(\langle a, a''\rangle) &= b'' \\ &\text{etc.} \end{aligned}$$

If the behaviour $\mathrm{beh}(x)$ of a state x looks like this, then one can immediately observe $b_0 = \epsilon(x)$, observe $b_1 = \epsilon(\delta(x)(a))$ after input a, observe $b = \epsilon(\delta(\delta(x)(a))(a)) = \epsilon(\delta^*(x, \langle a, a\rangle))$ after inputting a twice, etc. Thus, there is an edge $b \xrightarrow{a} b'$ in the tree if and only if there are successor states y, y' of x with $(y \downarrow b) \xrightarrow{a} (y' \downarrow b')$. In the next section we shall see (in Proposition 2.3.5) that these behaviours in $B^{A^\star}$ themselves carry the structure of a deterministic automaton.

In the illustration (2.20) one has, for a list of inputs $\sigma \in \{0, 1\}^\star$,

$$\mathrm{beh}(000)(\sigma) = 1 \iff \sigma \text{ contains a 1 in the third position from its end.}$$

In order to see the truth of this statement, notice that the names of the states are chosen in such a way that their bits represent the last three bits that have been consumed so far from the input string σ. In Corollary 2.3.6.2 we shall see more generally that these behaviour maps beh capture the language accepted by the automaton.

Here is a very special way to obtain deterministic automata. A standard result (see e.g. [222, 8.7]) in the theory of differential equations says that unique solutions to such equations give rise to monoid actions; see Exercises 1.4.1 and 2.2.9. Such a monoid action may be of the form $X \times \mathbb{R}_{\geq 0} \to X$, where X is the set of states and $\mathbb{R}_{\geq 0}$ is the set of non-negative reals with monoid structure $(+, 0)$ describing the input (which may be understood as time). In this context monoid actions are sometimes called *flows*, motions, solutions or trajectories; see Exercise 2.2.11 for an example.

Exercise 2.2.12 below gives an account of linear dynamical systems which is very similar to the approach to deterministic automata that we described above. It is based on the categorical analysis by Arbib and Manes [35, 36, 39, 38, 40] of Kalman's [287] module-theoretic approach to linear systems.

Lemma 2.2.4 contains a description of what coalgebra homomorphisms are for automata as in (2.21).

2.2.4 Non-deterministic Automata and Transition Systems

Deterministic automata have a transition *function* $\delta\colon S \times A \to S$. For non-deterministic automata one replaces this function by a *relation*. A state can then have multiple successor states – which is the key aspect of non-determinism. There are several, equivalent ways to represent this. For example, one can replace the transition function $S \times A \to S$ by a function $S \times A \to \mathcal{P}(S)$, yielding for each state $x \in S$ and input $a \in A$ a set of successor states. Of course, this function can also be written as $S \to \mathcal{P}(S)^A$, using currying. This is the same as using a transition relation, commonly written as an arrow: $\to \subseteq S \times A \times S$. Or alternatively, one can use a function $S \to \mathcal{P}(A \times S)$. All this amounts to the same, because of the bijective correspondences from Section 2.1:

$$
\begin{array}{c}
S \longrightarrow \mathcal{P}(S)^A \\
\hline\hline
S \times A \longrightarrow \mathcal{P}(S) \\
\hline\hline
\text{relations} \subseteq (S \times A) \times S \\
\hline\hline
\text{relations} \subseteq S \times (A \times S) \\
\hline\hline
S \longrightarrow \mathcal{P}(A \times S)
\end{array}
\qquad
\begin{array}{l}
(2.11) \\
(2.16) \\
\\
(2.16)
\end{array}
$$

We have a preference for the first representation and will thus use functions $\delta\colon S \to \mathcal{P}(S)^A$ as non-deterministic transition structures.

(It should be noted that if we use the finite powerset $\mathcal{P}_{\text{fin}}$ instead of the ordinary one $\mathcal{P}$, then there are no such bijective correspondences, and there is then a real choice of representation.)

Standardly, non-deterministic automata are also considered with a subset $F \subseteq S$ of final states. As for deterministic automata, we like to generalise this subset to an observation function $S \to B$. This leads us to the following definition. A **non-deterministic automaton** is a coalgebra

$$S \xrightarrow{\langle \delta, \epsilon \rangle} \mathcal{P}(S)^A \times B \tag{2.24}$$

of a Kripke polynomial functor $\mathcal{P}(\text{id})^A \times B$, where A is the set of its inputs and B the set of its outputs or observations. As before, the coalgebra consists of a **transition function** $\delta\colon S \to \mathcal{P}(S)^A$ and an **observation function** $\epsilon\colon S \to B$.

As illustration we give a non-deterministic version of the (deterministic) automaton in (2.20) trying to find a 1 in the third position from the end of the input string. As is often the case, the non-deterministic version is much simpler. We use as inputs $A = \{0, 1\}$ and outputs $B = \{0, 1\}$, for observing final states, in

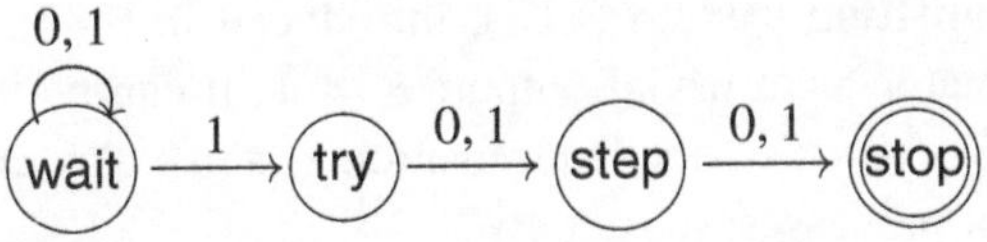

(2.25)

More formally, the state space is $S = \{\mathsf{wait}, \mathsf{try}, \mathsf{step}, \mathsf{stop}\}$ and $\epsilon\colon S \to \{0, 1\}$ outputs 1 only on the state stop. The transition function $\delta\colon S \to \mathcal{P}(S)^{\{0,1\}}$ is

$$\begin{aligned}
\delta(\mathsf{wait})(0) &= \{\mathsf{wait}\} & \qquad \delta(\mathsf{try})(0) &= \{\mathsf{step}\}\\
\delta(\mathsf{wait})(1) &= \{\mathsf{wait}, \mathsf{try}\} & \qquad \delta(\mathsf{try})(1) &= \{\mathsf{step}\}\\
\delta(\mathsf{step})(0) &= \{\mathsf{stop}\} & \qquad \delta(\mathsf{stop})(0) &= \emptyset\\
\delta(\mathsf{step})(1) &= \{\mathsf{stop}\} & \qquad \delta(\mathsf{stop})(1) &= \emptyset.
\end{aligned}$$

Clearly, we need sets of states to properly describe this automaton coalgebraically.

For non-deterministic automata, as in (2.24), we shall use the same notation as for deterministic ones: $x \xrightarrow{a} x'$ stands for $x' \in \delta(x)(a)$, and $x \downarrow b$ means $\epsilon(x) = b$. New notation is $x \stackrel{a}{\nrightarrow}$, which means $\delta(x)(a) = \emptyset$; i.e. there is no successor state x' such that x can do an a-step $x \xrightarrow{a} x'$ to x'. In general, for a given x and a there may be multiple (many or no) x' with $x \xrightarrow{a} x'$, but there is precisely one b with $x \downarrow b$. Also in the non-deterministic case we use the combined notation $(x \downarrow b) \xrightarrow{a} (x' \downarrow b')$ for $x \downarrow b$ and $x \xrightarrow{a} x'$ and $x' \downarrow b'$.

As for deterministic automata we wish to define the behaviour of a state from transitions $(x \downarrow b) \xrightarrow{a} (x' \downarrow b')$ by omitting the states and keeping the inputs and outputs. But in the non-deterministic case things are not so easy because there may be multiple successor states. More technically, there are no final coalgebras for functors $\mathcal{P}(\mathrm{id})^A \times B$ describing non-deterministic automata; see Proposition 2.3.8 and the discussion in the preceding paragraph. However, if we restrict ourselves to the *finite* powerset $\mathcal{P}_{\mathrm{fin}}$, final coalgebras do exist. For general non-deterministic automata (2.24), with the proper (non-finite) powerset, one can consider other forms of behaviour, such as traces; see Section 5.3.

Now that we have some idea of what a non-deterministic automaton is, namely a coalgebra $S \to \mathcal{P}(S)^A \times B$, we can introduce various transition systems as special cases.***

- An **unlabelled transition system** (UTS) is a non-deterministic automaton with trivial inputs and outputs: $A = 1$ and $B = 1$. It is thus nothing else but a relation $\to \subseteq S \times S$ on a set of states. UTSs are very basic dynamical systems, but they are important for instance as a basis for model checking: automatic state exploration for proving or disproving properties about systems; see for instance [347, 128].
- A **labelled transition system** (LTS), introduced in [383], is a non-deterministic automaton with trivial output: $B = 1$. It can be identified with a relation $\to \subseteq S \times A \times S$. Labelled transition systems play an important role in the theory of processes, see e.g. [67].
- A **Kripke structure** is a non-deterministic automaton of the form $S \longrightarrow \mathcal{P}(S) \times \mathcal{P}(\mathsf{AtProp})$ with trivial input ($A = 1$) and special output: $B = \mathcal{P}(\mathsf{AtProp})$, for a set AtProp of atomic propositions. The transition function $\delta\colon S \to \mathcal{P}(S)$ thus corresponds to an unlabelled transition system. And the observation function function $\epsilon\colon S \to \mathcal{P}(\mathsf{AtProp})$ tells for each state which of the atomic propositions are true (in that state). Such a function, when written as $\mathsf{AtProp} \to \mathcal{P}(S)$, is often called a valuation. Kripke structures are fundamental in the semantics of modal logic; see e.g. [73] or [129]. The latter reference describes Kripke structures for 'multiple agents', that is, as coalgebras $S \longrightarrow \mathcal{P}(S) \times \cdots \times \mathcal{P}(S) \times \mathcal{P}(\mathsf{AtProp})$ with multiple transition functions.

When we consider (non-)deterministic automata as coalgebras, we get an associated notion of (coalgebra) homomorphism for free. As we shall see next, such a homomorphism both preserves and reflects the transitions. The proofs are easy and are left to the reader.

Lemma 2.2.4 *1. Consider two deterministic automata $X \to X^A \times B$ and $Y \to Y^A \times B$ as coalgebras. A function $f\colon X \to Y$ is then a homomorphism of coalgebras if and only if*

i. $x \downarrow b \implies f(x) \downarrow b$;
ii. $x \xrightarrow{a} x' \implies f(x) \xrightarrow{a} f(x')$.

2. Similarly, for two non-deterministic automata $X \to \mathcal{P}(X)^A \times B$ and $Y \to \mathcal{P}(Y)^A \times B$ a function $f\colon X \to Y$ is a homomorphism of coalgebras if and only if

i. $x \downarrow b \implies f(x) \downarrow b$;
ii. $x \xrightarrow{a} x' \implies f(x) \xrightarrow{a} f(x')$;
iii. $f(x) \xrightarrow{a} y \implies \exists x' \in X.\ f(x') = y \wedge x \xrightarrow{a} x'$.

Such a function f is sometimes called a zig-zag morphism; see [63]. □

Note that the analogue of point (iii) in (2) trivially holds for deterministic automata. Later, in Chapter 4, we shall see that if we replace the powerset functor $\mathcal{P}$ in the coalgebraic description $X \to \mathcal{P}(X)^A \times B$ of non-deterministic automata by another functor, then we obtain a coalgebraic description of for instance probabilistic or weighted automata.

2.2.5 Context-Free Grammars

A **context-free grammar** (CFG) is a basic formalism in computer science to describe the syntax of programming languages via so-called production rules. These rules are of the form $v \longrightarrow \sigma$, where v is a non-terminal symbol and σ is a finite list of both terminal and non-terminal symbols. If we write V for the set of non-terminals, and A for the terminals, then a CFG is a coalgebra of the form

$$V \xrightarrow{\quad g \quad} \mathcal{P}((V + A)^\star).$$

A CFG is thus a coalgebra of the Kripke polynomial functor $X \longmapsto \mathcal{P}((X{+}A)^\star)$. It sends each non-terminal v to a set $g(v) \subseteq (V + A)^\star$ of right-hand sides $\sigma \in g(v)$ in productions $v \longrightarrow \sigma$.

A word $\tau \in A^\star$ – with terminals only – can be generated by a CFG g if there is a non-terminal $v \in V$ from which τ arises by applying rules repeatedly. The collection of all such strings is the language that is generated by the grammar.

A simple example is the grammar with $V = \{v\}$, $A = \{a, b\}$ and $g(v) = \{\langle\rangle, avb\}$. It thus involves two productions $v \longrightarrow \langle\rangle$ and $v \longrightarrow a \cdot v \cdot b$. This grammar generates the language of words $a^n b^n$ consisting of a number of as followed by an equal number of bs. Such a grammar is often written in Backus–Naur form (BNF) as $v ::= \langle\rangle \mid avb$.

A derivation of a word imposes a structure on the word that is generated. This structure may be recognised in an arbitrary word in a process called **parsing**. This is one of the very first things a compiler does (after lexical analysis); see for instance [33]. Example 5.3.2 describes the parsed language associated with a CFG via a trace semantics defined by coinduction.

2.2.6 Turing-Style Machines

Turing machines are paradigmatic models of computation that are of fundamental importance in the theory of computation (see e.g. [367]), in particular for the concept of computation itself. Turing machines are special automata involving a head moving over a tape that can read and write symbols, following certain rules determined by a finite state machine. Turing machines come in different flavours, with different kinds of tapes or different kinds of computation (e.g. (non-)deterministic or probabilistic, or even of quantum kind). Here we give a sketch of how to describe them coalgebraically, following [249].

The coalgebraic representation requires some care because a Turing machines involves two kinds of states, namely 'register' states given by the contents of the tape and 'steering' states in the finite state machine that controls the head. The register states may be represented as an infinite list $D^{\mathbb{N}}$ of data cells, on which certain operations (reading and writing) may be performed. We abstract from all this and represent it simply as a coalgebra $S \to F(S)$, where S is the set of register states. In case we know that there are n steering states, the resulting Turing-style machine can be represented as a coalgebra:

$$S \longrightarrow \big(F(n \cdot S)\big)^n. \tag{2.26}$$

Recall that $X^n = X \times \cdots \times X$ is the n-fold power and $n \cdot X = X + \cdots + X$ is the n-fold copower. This representation (2.26) gives for each steering state a separate coalgebra $S \to F(n \cdot S)$ telling how the machine moves to successor states (of both register and steering kind).

We consider a concrete example of a non-deterministic Turing machine that accepts all strings $\sigma \in \{a, b, c\}^\star$ where σ contains a c that is preceded or followed by ab. It can be described diagrammatically as

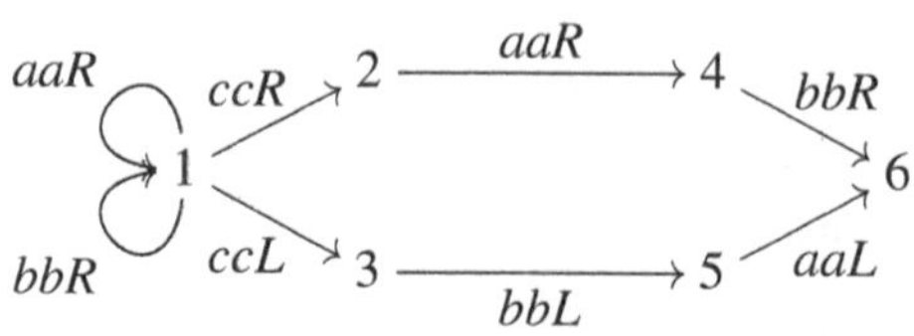

As usual, the label xyL means: if you read symbol x at the current location, then write y and move left. Similarly, xyR involves a move to the right.

We model such a machine with a tape with symbols from the alphabet $\Sigma = \{a, b, c\}$. The tape stretches in two dimensions, and so we use the integers $\mathbb{Z}$ as index. Thus the type $\mathbb{T}$ of tapes is given by $\mathbb{T} = \Sigma^{\mathbb{Z}} \times \mathbb{Z}$, consisting of pairs (t, p) where $t\colon \mathbb{Z} \to \Sigma = \{a, b, c\}$ is the tape itself and $p \in \mathbb{Z}$ the current position of the head.

Following the description (2.26) we represent the above non-deterministic Turing machine with six states as a coalgebra:

$$\mathbb{T} \longrightarrow \big(\mathcal{P}(6 \cdot \mathbb{T})\big)^6 . \tag{2.27}$$

Below we describe this coalgebra explicitly by enumerating the input cases, where the six coprojections $\kappa_i\colon \mathbb{T} \to 6 \cdot \mathbb{T} = \mathbb{T} + \mathbb{T} + \mathbb{T} + \mathbb{T} + \mathbb{T} + \mathbb{T}$ describe the successor states.

$$(t, p) \longmapsto \begin{cases} 1 \longmapsto \{\kappa_1(t, p+1) \mid t(p) = a \text{ or } t(p) = b\} \cup \\ \qquad\quad \{\kappa_2(t, p+1) \mid t(p) = c\} \cup \{\kappa_3(t, p-1) \mid t(p) = c\} \\ 2 \longmapsto \{\kappa_4(t, p+1) \mid t(p) = a\} \\ 3 \longmapsto \{\kappa_5(t, p-1) \mid t(p) = b\} \\ 4 \longmapsto \{\kappa_6(t, p+1) \mid t(p) = b\} \\ 5 \longmapsto \{\kappa_6(t, p-1) \mid t(p) = a\} \\ 6 \longmapsto \emptyset. \end{cases} \tag{2.28}$$

Notice that we need the powerset $\mathcal{P}$ to capture the non-determinism (especially at state 1).

Later on, in Exercise 5.1.6, we shall see that the functor $X \mapsto (\mathcal{P}(6 \cdot X))^6$ used in (2.27) carries a monad structure that makes it possible to compose this Turing coalgebra with itself, as in Exercise 1.1.2, so that transitions can be iterated.

2.2.7 Non-Well-Founded Sets

Non-well-founded sets form a source of examples of coalgebras which have been of historical importance in the development of the area. Recall that in ordinary set theory there is a foundation axiom (see e.g. [141, chapter II, §5]) stating that there are no infinite descending $\in$-chains $\cdots \in x_2 \in x_1 \in x_0$. This foundation axiom is replaced by an anti-foundation axiom in [140] and in [9], allowing for non-well-founded sets. The second reference [9] received much attention; it formulated an anti-foundation axiom as: every graph has a unique decoration. This can be reformulated easily in coalgebraic terms, stating that the universe of non-well-founded sets is a final coalgebra for the (special) powerset functor $\wp$ on the category of classes and functions:

$$\wp(X) = \{U \subseteq X \mid U \text{ is a small set}\}.$$

Incidentally, the 'initial algebra' (see Section 2.4) of this functor is the ordinary universe of well-founded sets; see [453, 451] for details.

Aczel developed his non-well-founded set theory in order to provide a semantics for Milner's theory CCS [354] of concurrent processes. An important contribution of this work is the link it established between the proof principle in process theory based on bisimulations (going back to [354, 371]) and coinduction as proof principle in the theory of coalgebras – which will be described in Section 3.4. This work formed a source of much inspiration in the semantics of programming languages [453] and also in logic [60].

Exercises

2.2.1 Check that a polynomial functor which does not contain the identity functor is constant.

2.2.2 Describe the kind of trees that can arise as behaviours of coalgebras:

1. $S \longrightarrow A + (A \times S)$.
2. $S \longrightarrow A + (A \times S) + (A \times S \times S)$.

2.2.3 Check, using Exercise 2.1.10, that non-deterministic automata $X \to \mathcal{P}(X)^A \times 2$ can equivalently be described as transition systems $X \to \mathcal{P}(1 + (A \times X))$. Work out the correspondence in detail.

2.2.4 Describe the arity # for the functors

1. $X \mapsto B + (X \times A \times X)$.
2. $X \mapsto A_0 \times X^{A_1} \times (X \times X)^{A_2}$, for finite sets A_1, A_2.

2.2.5 Check that *finite* arity functors correspond to simple polynomial functors in the construction of which all constant functors $X \mapsto A$ and exponents X^A have finite sets A.

2.2.6 Consider an indexed collection of sets $(A_i)_{i \in I}$ and define the associated 'dependent' polynomial functor **Sets** $\to$ **Sets** by

$$X \longmapsto \coprod_{i \in I} X^{A_i} = \{(i, f) \mid i \in I \wedge f \colon A_i \to X\}.$$

1. Prove that we get a functor in this way; obviously, by Proposition 2.2.3, each polynomial functor is of this form, for a finite set A_i.
2. Check that all simple polynomial functors are dependent – by finding suitable collections $(A_i)_{i \in I}$ for each of them.

(These functors are studied as 'containers' in the context of so-called W-types in dependent type theory for well-founded trees; see for instance [3, 2, 368].)

2.2.7 Recall from (2.13) and (2.14) that the powerset functor $\mathcal{P}$ can be described both as a covariant functor **Sets** $\to$ **Sets** and as a contravariant one $2^{(-)}$: $\mathbf{Sets}^{\mathrm{op}} \to$ **Sets**. In the definition of Kripke polynomial functors we use the powerset $\mathcal{P}$ covariantly. The functor $\mathcal{N} = \mathcal{P}\mathcal{P}$: **Sets** $\to$ **Sets** is obtained by using the contravariant powerset functor twice – yielding a covariant, but not Kripke polynomial functor. Coalgebras of this so-called neighbourhood functor are used in [423, 358] as models of a special modal logic (see also [194, 193] for the explicitly coalgebraic view).

1. Describe the action $\mathcal{N}(f): \mathcal{N}(X) \to \mathcal{N}(Y)$ of a function $f: X \to Y$.
2. Try to see a coalgebra $c: X \to \mathcal{N}(X)$ as the setting of a two–player game, with the first player's move in state $x \in X$ given by a choice of a subset $U \in c(x)$ and the second player's reply by a choice of successor state $x' \in U$.

2.2.8
1. Notice that the behaviour function $\mathrm{beh}: S \to B^{A^\star}$ from (2.23) for a deterministic automaton satisfies:

$$\begin{aligned} \mathrm{beh}(x)(\langle\rangle) &= \epsilon(x) \\ &= b && \text{where } x \downarrow b \\ \mathrm{beh}(x)(a \cdot \sigma) &= \mathrm{beh}(\delta(x)(a))(\sigma) \\ &= \mathrm{beh}(x')(\sigma) && \text{where } x \xrightarrow{a} x'. \end{aligned}$$

2. Consider a homomorphism $f: X \to Y$ of coalgebras/deterministic automata from $X \to X^A \times B$ and $Y \to Y^A \times B$ and prove that for all $x \in X$,

$$\mathrm{beh}_2(f(x)) = \mathrm{beh}_1(x).$$

2.2.9 Check that the iterated transition function $\delta^*: S \times A^\star \to S$ of a deterministic automaton is a monoid action – see Exercise 1.4.1 – for the free monoid structure on $A^\star$ from Exercise 1.4.4.

2.2.10 Note that a function space S^S carries a monoid structure given by composition. Show that the iterated transition function δ^* for a deterministic automaton, considered as a monoid homomorphism $A^\star \to S^S$, is actually obtained from δ by freeness of $A^\star$ – as described in Exercise 1.4.4.

2.2.11 Consider a very simple differential equation of the form $df/dy = -Cf$, where $C \in \mathbb{R}$ is a fixed positive constant. The solution is usually described as $f(y) = f(0) \cdot e^{-Cy}$. Check that it can be described as a monoid action $\mathbb{R} \times \mathbb{R}_{\geq 0} \to \mathbb{R}$, namely $(x, y) \mapsto xe^{-Cy}$, where $\mathbb{R}_{\geq 0}$ is the monoid of non-negative real numbers with addition $+, 0$.

2.2.12 Let **Vect** be the category with finite-dimensional vector spaces over the real numbers $\mathbb{R}$ (or some other field) as objects, and with linear transformations between them as morphisms. This exercise describes the basics of linear dynamical systems, in analogy with deterministic automata. It does require some basic knowledge of vector spaces.

1. Prove that the product $V \times W$ of (the underlying sets of) two vector spaces V and W is at the same time a product and a coproduct in **Vect** – the same phenomenon as in the category of monoids; see Exercise 2.1.6. Show also that the singleton space 1 is both an initial and a final object. And notice that an element x in a vector space V may be identified with a linear map $\mathbb{R} \to V$.
2. A **linear dynamical system** [287] consists of three vector spaces: S for the state space, A for input and B for output, together with three linear transformations: an input map $G: A \to S$, a dynamics $F: S \to S$ and an output map $H: S \to B$. Note how the first two maps can be combined via cotupling into one transition function $S \times A \to S$, as used for deterministic automata. Because of the possibility of decomposing the transition function in this linear case into two maps $A \to S$ and $S \to S$, these systems are called *decomposable* by Arbib and Manes [35]. (But this transition function $S \times A \to S$ is not bilinear (i.e. linear in each argument separately), so it does not give rise to a map $S \to S^A$ to the vector space S^A of linear transformations from A to S. Hence we do not have a purely coalgebraic description $S \to S^A \times B$ in this linear setting.)
3. For a vector space A, consider, in the notation of [35], the subset of infinite sequences:

$$A^{\S} = \{\alpha \in A^{\mathbb{N}} \mid \text{only finitely many } \alpha(n) \text{ are non-zero}\}.$$

Equip the set $A^{\S}$ with a vector space structure, such that the insertion map $\mathsf{in}: A \to A^{\S}$, defined as $\mathsf{in}(a) = (a, 0, 0, 0, \ldots)$, and shift map $\mathsf{sh}: A^{\S} \to A^{\S}$, given as $\mathsf{sh}(\alpha) = (0, \alpha(0), \alpha(1), \ldots)$, are linear transformations. (This vector space $A^{\S}$ may be understood as the space of polynomials over A in one variable. It can be defined as the infinite coproduct $\coprod_{n \in \mathbb{N}} A$ of $\mathbb{N}$-copies of A – which is also called a *copower* and written as $\mathbb{N} \cdot A$;

see [344, III, 3]. It is the analogue in **Vect** of the set of finite sequences $B^\star$ for $B \in$ **Sets**. This will be made precise in Exercise 2.4.8.)

4. Consider a linear dynamical system $A \xrightarrow{G} S \xrightarrow{F} S \xrightarrow{H} B$ as above and show that the analogue of the behaviour $A^\star \to B$ for deterministic automata (see also [37, 6.3]) is the linear map $A^\S \to B$ defined as

$$(a_0, a_2, \ldots, a_n, 0, 0, \ldots) \longmapsto \sum_{i \leq n} HF^i Ga_i.$$

This is the standard behaviour formula for linear dynamical systems; see e.g. [287, 38]. (This behaviour map can be understood as starting from the 'default' initial state $0 \in S$. If one wishes to start from an arbitrary initial state $x \in S$, one gets the formula

$$(a_0, a_2, \ldots, a_n, 0, 0, \ldots) \longmapsto HF^{(n+1)}x + \sum_{i \leq n} HF^i Ga_i.$$

It is obtained by consecutively modifying the state x with inputs $a_n, a_{n-1}, \ldots, a_0$.)

2.3 Final Coalgebras

In the previous chapter we have seen the special role that is played by the final coalgebra $A^\infty \to 1 + (A \times A^\infty)$ of sequences, both for defining functions into sequences and for reasoning about them – with 'coinduction'. Here we shall define finality in general for coalgebras and investigate this notion more closely – which leads for example to language acceptance by automata; see Corollary 2.3.6.2. This general definition will allow us to use coinductive techniques for arbitrary final coalgebras. The underlying theme is that in final coalgebras there is no difference between states and their behaviours.

In system-theoretic terms, final coalgebras are of interest because they form so-called minimal representations: they are canonical realisations containing all the possible behaviours of a system. It is this idea that we have already tried to suggest in the previous section when discussing various examples of coalgebras of polynomial functors.

Once we have a final coalgebra (for a certain functor), we can map a state from an arbitrary coalgebra (of the same functor) to its behaviour. This induces a useful notion of equivalence between states, namely equality of the associated behaviours. As we shall see later (in Section 3.4), this is bisimilarity.

We shall start in full categorical generality – building on Definition 1.4.5 – but we quickly turn to concrete examples in the category of sets. Examples of final coalgebras in other categories – such as categories of domains or of metric spaces – may be found in many places, such as [143, 144, 145, 433, 5, 453, 133, 132, 218, 86], but also in Section 5.3.

Definition 2.3.1 Let $\mathbb{C}$ be an arbitrary category with an endofunctor $F: \mathbb{C} \to \mathbb{C}$. A **final F-coalgebra** is simply a final object in the associated category $\mathbf{CoAlg}(F)$ of F-coalgebras. Thus, it is a coalgebra $\zeta: Z \to F(Z)$ such that for any coalgebra $c: X \to F(X)$ there is a unique homomorphism $\mathrm{beh}_c: X \to Z$ of coalgebras, as in

$$\begin{pmatrix} F(X) \\ \uparrow c \\ X \end{pmatrix} \xrightarrow{\ \mathrm{beh}_c\ } \begin{pmatrix} F(Z) \\ \uparrow \zeta \\ Z \end{pmatrix} \qquad \text{i.e.} \qquad \begin{array}{ccc} F(X) & \xrightarrow{F(\mathrm{beh}_c)} & F(Z) \\ {\scriptstyle c}\uparrow & & \uparrow{\scriptstyle \zeta} \\ X & \xrightarrow[\mathrm{beh}_c]{} & Z \end{array}$$

As before, the dashed arrow notation is used for uniqueness. What we call a behaviour map beh_c is sometimes called an *unfold* or a *coreduce* of c.

A common mistake is to read in this definition 'for any *other* coalgebra $c: X \to F(X)$' instead of 'for any coalgebra $c: X \to F(X)$'. It is important that we can take ζ for this c; the resulting map beh_ζ is then the identity, by uniqueness. This will for instance be used in the proof of Lemma 2.3.3 below.

Recall from Section 1.2 the discussion (after Example 1.2.2) about the two aspects of unique existence of the homomorphism into the final coalgebra, namely (1) existence (used as coinductive/corecursive definition principle) and (2) uniqueness (used as coinductive proof principle). In the next section we shall see that ordinary induction – from a categorical perspective – also involves such a unique existence property. At this level of abstraction there is thus a perfect duality between induction and coinduction.

This unique existence is in fact all we need about final coalgebras. What these coalgebras precisely look like – what their elements are in **Sets** – is usually not relevant. Nevertheless, in order to become more familiar with this topic of final coalgebras, we shall describe several examples concretely. More general theory about the existence of final coalgebras is developed in Section 4.6.

But first we shall look at two general properties of final coalgebras.

Lemma 2.3.2 *A final coalgebra, if it exists, is determined up to isomorphism.*

Proof This is in fact a general property of final objects in a category: if 1 and 1′ are both a final object in the same category, then there are unique maps $f\colon 1 \to 1'$ and $g\colon 1' \to 1$ by finality. Thus we have two maps $1 \to 1$, namely the composition $g \circ f$ and of course the identity id_1. But then they must be equal by finality of 1. Similarly, by finality of 1′ we get $f \circ g = \mathrm{id}_{1'}$. Therefore $1 \cong 1'$. □

In view of this result we often talk about *the* final coalgebra of a functor, if it exists.

Earlier, in the beginning of Section 1.2 we have seen that the final coalgebra map $A^\infty \to 1 + (A \times A^\infty)$ for sequences is an isomorphism. This turns out to be a general phenomenon, as observed by Lambek [327]: a final F-coalgebra is a fixed point for the functor F. The proof of this result is a nice exercise in diagrammatic reasoning.

Lemma 2.3.3 *A final coalgebra $\zeta\colon Z \to F(Z)$ is necessarily an isomorphism $\zeta\colon Z \xrightarrow{\cong} F(Z)$.*

Proof The first step towards constructing an inverse of $\zeta\colon Z \to F(Z)$ is to apply the functor $F\colon \mathbb{C} \to \mathbb{C}$ to the final coalgebra $\zeta\colon Z \to F(Z)$, which yields again a coalgebra, namely $F(\zeta)\colon F(Z) \to F(F(Z))$. By finality we get a homomorphism $f\colon F(Z) \to Z$ as in

$$
\begin{array}{ccc}
F(F(Z)) & \xrightarrow{\;F(f)\;} & F(Z) \\
{\scriptstyle F(\zeta)}\uparrow & & \uparrow{\scriptstyle \zeta} \\
F(Z) & \xrightarrow[\;f\;]{} & Z
\end{array}
$$

The aim is to show that this f is the inverse of ζ. We first consider the composite $f \circ \zeta\colon Z \to Z$ and show that it is the identity. We do so by first observing that the identity $Z \to Z$ is the unique homomorphism $\zeta \to \zeta$. Therefore, it suffices to show that $f \circ \zeta$ is also a homomorphism $\zeta \to \zeta$. This follows from an easy diagram chase:

$$
\begin{array}{ccccc}
F(Z) & \xrightarrow{\;F(\zeta)\;} & F(F(Z)) & \xrightarrow{\;F(f)\;} & F(Z) \\
{\scriptstyle \zeta}\uparrow & & {\scriptstyle F(\zeta)}\uparrow & & \uparrow{\scriptstyle \zeta} \\
Z & \xrightarrow{\;\zeta\;} & F(Z) & \xrightarrow{\;f\;} & Z
\end{array}
\qquad
Z \xrightarrow{\;\mathrm{beh}_\zeta \,=\, \mathrm{id}_Z\;} Z
$$

The rectangle on the right is the one defining f, and thus commutes by definition. And the one on the left obviously commutes. Therefore the outer rectangle commutes. This says that $f \circ \zeta$ is a homomorphism $\zeta \to \zeta$, and thus allows us to conclude that $f \circ \zeta = \mathrm{id}_Z$.

But now we are done since the reverse equation $\zeta \circ f = \mathrm{id}_{F(Z)}$ follows from a simple computation:

$$\begin{aligned} \zeta \circ f &= F(f) \circ F(\zeta) && \text{by definition of } f \\ &= F(f \circ \zeta) && \text{by functoriality of } F \\ &= F(\mathrm{id}_Z) && \text{as we just proved} \\ &= \mathrm{id}_{F(Z)} && \text{because } F \text{ is a functor.} \end{aligned}$$ ❑

An immediate negative consequence of this fixed point property of final coalgebras is the following. It clearly shows that categories of coalgebras need not have a final object.

Corollary 2.3.4 *The powerset functor $\mathcal{P}$:* **Sets** $\to$ **Sets** *does not have a final coalgebra.*

Proof A standard result of Cantor (proved by so-called diagonalisation; see e.g. [116, theorem 15.10] or [92, theorem 1.10]) says that there cannot be an injection $\mathcal{P}(X) \rightarrowtail X$, for any set X. This excludes a final coalgebra $X \xrightarrow{\cong} \mathcal{P}(X)$. ❑

As we shall see later in this section, the *finite* powerset functor does have a final coalgebra. But first we shall look at some easier examples. The following result, occurring for example in [342, 399, 237], is simple but often useful. It will be used in Section 4.6 to prove more general existence results for final coalgebras.

Proposition 2.3.5 *Fix two sets A and B and consider the polynomial functor* $\mathrm{id}^A \times B$ *whose coalgebras are deterministic automata. The final coalgebra of this functor is given by the set of behaviour functions $B^{A^\star}$, with structure*

$$B^{A^\star} \xrightarrow{\quad \zeta = \langle \zeta_1, \zeta_2 \rangle \quad} (B^{A^\star})^A \times B$$

given by

$$\zeta_1(\varphi)(a) = \lambda\sigma \in A^\star.\, \varphi(a \cdot \sigma) \qquad \textit{and} \qquad \zeta_2(\varphi) = \varphi(\langle\rangle).$$

Proof We have to show that for an arbitrary coalgebra/deterministic automaton $\langle \delta, \epsilon \rangle \colon X \to X^A \times B$ there is a unique homomorphism of coalgebras $X \to B^{A^\star}$. For this we take of course the behaviour function $\mathrm{beh} \colon X \to B^{A^\star}$ from the previous section, defined in (2.23) by $\mathrm{beh}(x) = \lambda\sigma \in A^\star.\, \epsilon(\delta^*(x, \sigma))$.

We have to prove that it is the unique function making the following diagram commute.

$$\begin{array}{ccc}
X^A \times B & \xrightarrow{\mathrm{beh}^{\mathrm{id}_A} \times \mathrm{id}_B} & (B^{A^\star})^A \times B \\
\langle \delta, \epsilon \rangle \uparrow & & \uparrow \zeta = \langle \zeta_1, \zeta_2 \rangle \\
X & \xrightarrow[\mathrm{beh}]{} & B^{A^\star}
\end{array}$$

We prove commutation first. It amounts to two points; see Lemma 2.2.4.1.

$$\begin{aligned}
(\zeta_1 \circ \mathrm{beh})(x)(a) &= \zeta_1(\mathrm{beh}(x))(a) \\
&= \lambda\sigma.\, \mathrm{beh}(x)(a \cdot \sigma) \\
&= \lambda\sigma.\, \mathrm{beh}(\delta(x)(a))(\sigma) \qquad \text{see Exercise 2.2.8.1} \\
&= \mathrm{beh}(\delta(x)(a)) \\
&= \mathrm{beh}^{\mathrm{id}_A}(\delta(x))(a) \\
&= (\mathrm{beh}^{\mathrm{id}_A} \circ \delta)(x)(a) \\
(\zeta_2 \circ \mathrm{beh})(x) &= \mathrm{beh}(x)(\langle\rangle) \\
&= \epsilon(\delta^*(x, \langle\rangle)) \\
&= \epsilon(x).
\end{aligned}$$

Next we have to prove uniqueness. Assume that $f\colon X \to B^{A^\star}$ is also a homomorphism of coalgebras. Then one can show, by induction on $\sigma \in A^\star$, that for all $x \in X$ one has $f(x)(\sigma) = \mathrm{beh}(x)(\sigma)$:

$$\begin{aligned}
f(x)(\langle\rangle) &= \zeta_2(f(x)) \\
&= \epsilon(x) && \text{since } f \text{ is a homomorphism} \\
&= \mathrm{beh}(x)(\langle\rangle) \\
f(x)(a \cdot \sigma) &= \zeta_1(f(x))(a)(\sigma) \\
&= f(\delta(x)(a))(\sigma) && \text{since } f \text{ is a homomorphism} \\
&= \mathrm{beh}(\delta(x)(a))(\sigma) && \text{by induction hypothesis} \\
&= \mathrm{beh}(x)(a \cdot \sigma) && \text{see Exercise 2.2.8.1.}
\end{aligned}$$ ❑

There are two special cases of this general result that are worth mentioning explicitly.

Corollary 2.3.6 *Consider the above final coalgebra* $B^{A^\star} \xrightarrow{\cong} (B^{A^\star})^A \times B$ *of the deterministic automata functor* $\mathrm{id}^A \times B$.

1. *When* $A = 1$, *so that* $A^\star = \mathbb{N}$, *the resulting functor* $\mathrm{id} \times B$ *captures stream coalgebras* $X \to X \times B$. *Its final coalgebra is the set* $B^{\mathbb{N}}$ *of infinite sequences (streams) of elements of* B, *with* (tail, head) *structure,*

$$B^{\mathbb{N}} \xrightarrow{\cong} B^{\mathbb{N}} \times B \qquad \textit{given by} \qquad \varphi \longmapsto (\,\lambda n \in \mathbb{N}.\, \varphi(n+1),\ \varphi(0)\,),$$

as described briefly towards the end of the introduction to Chapter 1.

2. *When* $B = 2 = \{0,1\}$ *describes final states of the automaton, the final coalgebra* $B^{A^\star}$ *is the set* $\mathcal{P}(A^\star)$ *of languages over the alphabet A, with structure*

$$\mathcal{P}(A^\star) \xrightarrow{\;\cong\;} \mathcal{P}(A^\star)^A \times \{0,1\}$$

given by

$$L \longmapsto (\lambda a \in A.\, L_a,\ \textsf{if}\ \langle\rangle \in L\ \textsf{then}\ 1\ \textsf{else}\ 0),$$

where L_a *is the so-called a-derivative, introduced by Brzozowski [89], and defined as*

$$L_a = \{\sigma \in A^\star \mid a \cdot \sigma \in L\}.$$

Given an arbitrary automaton $\langle \delta, \epsilon\rangle \colon X \to X^A \times \{0,1\}$ *of this type, the resulting behaviour map* $\mathrm{beh}_{\langle\delta,\epsilon\rangle} \colon X \to \mathcal{P}(A^\star)$ *thus describes the language* $\mathrm{beh}_{\langle\delta,\epsilon\rangle}(x) \subseteq A^\star$ ***accepted*** *by this automaton with* $x \in X$ *considered as initial state.*

Both these final coalgebras $A^{\mathbb{N}}$ and $\mathcal{P}(A^\star)$ are studied extensively by Rutten; see [410, 412, 413] and Example 3.4.5 later on.

Example 2.3.7 The special case of (1) in the previous result is worth mentioning, where B is the set $\mathbb{R}$ of real numbers (and $A = 1$). We then get a final coalgebra $\mathbb{R}^{\mathbb{N}} \xrightarrow{\cong} (\mathbb{R}^{\mathbb{N}})^{\mathbb{N}} \times \mathbb{R}$ of streams of real numbers. Recall that a function $f \colon \mathbb{R} \to \mathbb{R}$ is called **analytic** if it possesses derivatives of all orders and agrees with its Taylor series in the neighbourhood of every point. Let us write $\mathcal{A}$ for the set of such analytic functions. It carries a coalgebra structure:

$$\begin{array}{ccc} \mathcal{A} & \xrightarrow{\quad d \quad} & \mathcal{A} \times \mathbb{R} \\ f & \longmapsto & (f', f(0)). \end{array}$$

Here we write f' for the derivative of f. The induced coalgebra homomorphism $\mathrm{beh}_d \colon \mathcal{A} \to \mathbb{R}^{\mathbb{N}}$ maps an analytic function f to the stream of derivatives at 0, i.e. to $\textsf{Taylor}(f) = \mathrm{beh}_d(f) = (f(0), f'(0), f''(0), \ldots, f^{(n)}(0), \ldots) \in \mathbb{R}^{\mathbb{N}}$. The output values (in $\mathbb{R}$) in this stream are of course the coefficients in the Taylor series expansion of f in

$$f(x) = \sum_{n=0}^{\infty} \frac{f^{(n)}(0)}{n!} x^n.$$

This shows that beh_d is an isomorphism – and thus that $\mathcal{A}$ can also be considered as the final coalgebra of the functor $(-) \times \mathbb{R}$.

The source of this 'coinductive view on calculus' is [376]. It contains a further elaboration of these ideas. Other coalgebraic 'next' or 'tail' operations are also studied as derivatives in [412, 413], with the familiar derivative notation $(-)'$ to describe for instance the tail of streams. We may then write $\mathsf{Taylor}(f)' = \mathsf{Taylor}(f')$, so that taking Taylor series commutes with derivatives.

Corollary 2.3.4 implies that there is no final coalgebra for the non-deterministic automata functor $\mathcal{P}(\mathrm{id})^A \times B$. However if we restrict ourselves to the finite powerset functor $\mathcal{P}_{\mathrm{fin}}$ there is a final coalgebra. At this stage we shall be very brief, basically limiting ourselves to the relevant statement. It is a special case of Theorem 2.3.9, the proof of which will be given later, in Section 4.6.

Proposition 2.3.8 (After [55]) *Let A and B be arbitrary sets. Consider the finite Kripke polynomial functor $\mathcal{P}_{\mathit{fin}}(\mathrm{id})^A \times B$ whose coalgebras are image finite non-deterministic automata. This functor has a final coalgebra.*

So far in this section we have seen several examples of final coalgebras. One might wonder which polynomial functors possess a final coalgebra. The following result gives an answer. Its proof will be postponed until later in Section 4.6, because it requires some notions that have not been introduced yet.

Theorem 2.3.9 *Each finite Kripke polynomial functor* **Sets** $\rightarrow$ **Sets** *has a final coalgebra.*

As argued before, one does not need to know what a final coalgebra looks like in order to work with it. Its states coincide with its behaviours, so a purely behaviouristic view is justified: unique existence properties are sufficiently strong to use it as a black box. See for instance Exercise 2.3.4 below.

A good question raised explicitly in [411] is: which kind of functions can be defined with coinduction? Put another way: is there, in analogy with the collection of recursively defined functions (on the natural numbers), a reasonable collection of *corecursively* defined functions? This question is still largely open.

2.3.1 Beyond Sets

So far we have concentrated on functors F: **Sets** $\rightarrow$ **Sets** on the category of sets and functions. Indeed, most of the examples in this book will arise from such functors. It is important however to keep the broader (categorical) perspective in mind and realise that coalgebras are also of interest in other universes. Good examples appear in Section 5.3 where traces of suitable coalgebras are described via coalgebras in the category **SetsRel** of sets with relations as morphisms. At this stage we shall consider a single example in the category **Sp** of topological spaces and continuous functions between them.

Example 2.3.10 The set $2^{\mathbb{N}} = \mathcal{P}(\mathbb{N})$, for $2 = \{0, 1\}$, of infinite sequences of bits (or subsets of $\mathbb{N}$) carries a topology yielding the well-known 'Cantor space'; see [432, Section 2.3]. Alternatively, this space can be represented as the intersection of a descending chain of intervals (2^{n+1} separate pieces at stage n) of the real interval $[0, 1]$; see below or see or any textbook on topology, for instance [87]:

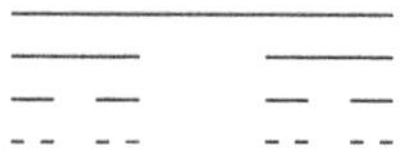

Starting from the unit interval $[0, 1]$ one keeps in step one the left and right thirds, given by the closed subintervals $[0, \frac{1}{3}]$ and $[\frac{2}{3}, 1]$. These can be described as 'left' and 'right', or as '0' and '1'. In a next step one again keeps the left and right thirds, giving us four closed intervals $[0, \frac{1}{9}]$, $[\frac{2}{9}, \frac{1}{3}]$, $[\frac{2}{3}, \frac{7}{9}]$ and $[\frac{8}{9}, 1]$; they can be referred to via the two-bit words 00, 01, 10, 11, respectively. And so on. The Cantor set is then defined as the (countable) intersection of all these intervals. It is a closed subspace of $[0, 1]$, obtained as an intersection of closed subspaces. It is also totally disconnected.

Elements of the Cantor set can be identified with infinite streams $2^{\mathbb{N}}$, seen as consecutive 'left' or 'right' choices. The basic opens of $2^{\mathbb{N}}$ are the subsets $\uparrow\sigma = \{\sigma \cdot \tau \mid \tau \in 2^{\mathbb{N}}\}$ of infinite sequences starting with σ, for $\sigma \in 2^{\star}$ a finite sequence.

Recall that **Sp** is the category of topological spaces with continuous maps between them. This category has coproducts, given as in **Sets**, with topology induced by the coprojections. In particular, for an arbitrary topological space X the coproduct $X + X$ carries a topology in which subsets $U \subseteq X + X$ are open if and only if both $\kappa_1^{-1}(U)$ and $\kappa_2^{-1}(U)$ are open in X. The Cantor space can then be characterised as the final coalgebra of the endofunctor $X \mapsto X + X$ on **Sp**.

A brief argument goes as follows. Corollary 2.3.6.1 tells that the set of streams $2^{\mathbb{N}}$ is the final coalgebra of the endofunctor $X \mapsto X \times 2$ on **Sets**. There is an obvious isomorphism $X \times 2 \cong X + X$ of sets (see Exercise 2.1.7), which is actually also an isomorphism of topological spaces if one considers the set 2 with the discrete topology (in which every subset is open); see Exercise 2.3.7 for more details.

But one can of course also check the finality property explicitly in the category **Sp** of topological spaces. The final coalgebra $\zeta\colon 2^{\mathbb{N}} \xrightarrow{\cong} 2^{\mathbb{N}} + 2^{\mathbb{N}}$ is given on $\sigma \in 2^{\mathbb{N}}$ by

$$\zeta(\sigma) = \begin{cases} \kappa_1 \mathsf{tail}(\sigma) & \text{if } \mathsf{head}(\sigma) = 0 \\ \kappa_2 \mathsf{tail}(\sigma) & \text{if } \mathsf{head}(\sigma) = 1. \end{cases}$$

It is not hard to see that both ζ and ζ^{-1} are continuous. For an arbitrary coalgebra $c\colon X \to X + X$ in **Sp** we can describe the unique homomorphism of coalgebra $\mathrm{beh}_c\colon X \to 2^{\mathbb{N}}$ on $x \in X$ and $n \in \mathbb{N}$ as

$$\mathrm{beh}_c(x)(n) = \begin{cases} 0 & \text{if } \exists y.\, c(([\mathrm{id}, \mathrm{id}] \circ c)^n(x)) = \kappa_1 y \\ 1 & \text{if } \exists y.\, c(([\mathrm{id}, \mathrm{id}] \circ c)^n(x)) = \kappa_2 y. \end{cases}$$

Again, this map is continuous. More examples and theory about coalgebras and such fractal-like structure may be found in [331] and [205].

Exercises

2.3.1 Check that a final coalgebra of a monotone endofunction $f\colon X \to X$ on a poset X, considered as a functor, is nothing but a greatest fixed point. (See also Exercise 1.3.5.)

2.3.2 For arbitrary sets A, B, consider the (simple polynomial) functor $X \mapsto (X \times B)^A$. Coalgebras $X \to (X \times B)^A$ of this functor are often called Mealy machines.

1. Check that Mealy machines can equivalently be described as deterministic automata $X \to X^A \times B^A$ and that the final Mealy machine is B^{A^+}, by Proposition 2.3.5, where $A^+ \hookrightarrow A^\star$ is the subset of non-empty finite sequences. Describe the final coalgebra structure $B^{A^+} \to (B^{A^+} \times B)^A$ explicitly.
2. Fix a set A and consider the final coalgebra $A^{\mathbb{N}}$ of the stream functor $A \times (-)$. Define by finality an 'alternation' function $\mathsf{alt}\colon A^{\mathbb{N}} \times A^{\mathbb{N}} \to A^{\mathbb{N}}$, such that

$$\mathsf{alt}(a_0 a_1 a_2 \cdots,\ b_0 b_1 b_2 \cdots) = a_0 b_1 a_2 b_3 \cdots$$

Prove by coinduction

$$\mathsf{alt}(\sigma, \mathsf{alt}(\tau_1, \rho)) = \mathsf{alt}(\sigma, \mathsf{alt}(\tau_2, \rho)).$$

Thus, in such a combination the middle argument is irrelevant.

3. Consider the set Z of so-called causal stream functions, given by

$$Z = \Big\{\psi\colon A^{\mathbb{N}} \to B^{\mathbb{N}} \;\Big|\; \forall n \in \mathbb{N}.\, \forall \alpha, \alpha' \in A^{\mathbb{N}}.\; (\forall i \le n.\, \alpha(i) = \alpha'(i)) \Rightarrow \psi(\alpha)(n) = \psi(\alpha')(n) \Big\}.$$

For such a causal stream function ψ, the output $\psi(\alpha)(n) \in B$ is thus determined by the first $n + 1$ elements $\alpha(0), \ldots, \alpha(n) \in A$.

Prove that Z yields an alternative description of the final Mealy automaton, via the structure map $\zeta\colon Z \to (Z \times B)^A$ given by

$$\zeta(\psi)(a) = (\,\lambda\alpha \in A^{\mathbb{N}}.\,\lambda n.\,\psi(a \cdot \alpha)(n+1),\ \psi(\lambda n \in \mathbb{N}.\,a)(0)\,)$$

where $a \cdot \alpha$ is prefixing to $\alpha \in A^{\mathbb{N}}$, considered as infinite sequence.

For more information on Mealy machines, see [84, 195].

2.3.3 Assume a category $\mathbb{C}$ with a final object $1 \in \mathbb{C}$. Call a functor $F\colon \mathbb{C} \to \mathbb{C}$ affine if it preserves the final object: the map $F(1) \to 1$ is an isomorphism. Prove that the inverse of this map is the final F-coalgebra, i.e. the final object in the category **CoAlg**(F). (Only a few of the functors F that we consider are affine; examples are the identity functor the non-empty powerset functor, or the distribution functor $\mathcal{D}$ from Section 4.1. Affine functions occur especially in probabilistic computation.)

2.3.4 Let Z be the (state space of the) final coalgebra of the binary tree functor $X \mapsto 1 + (A \times X \times X)$. Define by coinduction a mirror function $\mathsf{mir}\colon Z \to Z$ which (deeply) exchanges the subtrees. Prove, again by coinduction, that $\mathsf{mir} \circ \mathsf{mir} = \mathrm{id}_Z$. Can you tell what the elements of Z are?

2.3.5 Recall the decimal representation coalgebra $\mathsf{nextdec}\colon [0,1) \to 1 + (\{0, 1, \ldots, 9\} \times [0,1))$ from Example 1.2.2, with its behaviour map $\mathrm{beh}_{\mathsf{nextdec}}\colon [0,1) \to \{0, 1, \ldots, 9\}^{\infty}$. Prove that this behaviour map is a split mono: there is a map e in the reverse direction for which we have $e \circ \mathrm{beh}_{\mathsf{nextdec}} = \mathrm{id}_{[0,1)}$. (The behaviour map is not an isomorphism, because both numbers 5 and $49999\cdots$, considered as sequences in $\{0, 1, \ldots, 9\}^{\infty}$, represent the fraction $\frac{1}{2} \in [0,1)$. See other representations as continued fractions in [378] or [366] which do yield isomorphisms.)

2.3.6 This exercise is based on [237, lemma 5.4].

1. Fix three sets A, B, C, and consider the simple polynomial functor

$$X \longmapsto (C + (X \times B))^A.$$

Show that its final coalgebra can be described as the set of functions:

$$Z = \{\varphi \in (C + B)^{A^+} \mid \forall \sigma \in A^+.\, \forall c \in C.\, \varphi(\sigma) = \kappa_1(c) \Rightarrow \\ \forall \tau \in A^\star.\, \varphi(\sigma \cdot \tau) = \kappa_1(c)\}.$$

Once such functions $\varphi \in Z$ hit C, they keep this value in C. Here we write A^+ for the subset of $A^\star$ of non-empty finite sequences.

The associated coalgebra structure $\zeta\colon Z \xrightarrow{\cong} (C + (Z \times B))^A$ is given by

$$\zeta(\varphi)(a) = \begin{cases} \kappa_1(c) & \text{if } \varphi(\langle a \rangle) = \kappa_1(c) \\ \kappa_2(b, \varphi') & \text{if } \varphi(\langle a \rangle) = \kappa_2(b) \text{ where } \varphi'(\sigma) = \varphi(a \cdot \sigma). \end{cases}$$

2. Check that the fact that the set B^∞ of both finite and infinite sequences is the final coalgebra of the functor $X \mapsto 1 + (X \times B)$ is a special case of this.
3. Generalise the result in (1) to functors of the form

$$X \longmapsto (C_1 + (X \times B_1))^{A_1} \times \cdots \times (C_n + (X \times B_n))^{A_n}$$

using this time as state space of the final coalgebra the set

$$\begin{aligned} \Big\{ \varphi \in (C + B)^{A^+} \,\Big|\, & \forall \sigma \in A^\star.\, \forall i \leq n. \\ & \forall a \in A_i.\, \varphi(\sigma \cdot \kappa_i(a)) \in \kappa_1[\kappa_i[C_i]] \vee \varphi(\sigma \cdot \kappa_i(a)) \in \kappa_2[\kappa_i[B_i]] \\ & \wedge \forall c \in C_i.\, \sigma \neq \langle\rangle \wedge \varphi(\sigma) = \kappa_1(\kappa_i(c)) \\ & \qquad \Rightarrow \forall \tau \in A^\star.\, \varphi(\sigma \cdot \tau) = \kappa_1(\kappa_i(c)) \Big\} \end{aligned}$$

where $A = A_1 + \cdots + A_n$, $B = B_1 + \cdots + B_n$ and $C = C_1 + \cdots + C_n$.
4. Show how classes as in (1.10) fit into this last result. *Hint*: Use that $S + (S \times E) \cong (S \times 1) + (S \times E) \cong S \times (1 + E)$, using distributivity from Exercise 2.1.7.

2.3.7 For a topological space A consider the set $A^\mathbb{N}$ of streams with the product topology (the least topology that makes the projections $\pi_n\colon A^\mathbb{N} \to A$ continuous).

1. Check that the head $A^\mathbb{N} \to A$ and tail $A^\mathbb{N} \to A^\mathbb{N}$ operations are continuous.
2. Prove that the functor $A \times (-)\colon \mathbf{Sp} \to \mathbf{Sp}$ has $A^\mathbb{N}$ as final coalgebra.
3. Show that in the special case where A carries the discrete topology (in which every subset is open) the product topology on $A^\mathbb{N}$ is given by basic open sets $\uparrow\sigma = \{\sigma \cdot \tau \mid \tau \in A^\mathbb{N}\}$, for $\sigma \in A^\star$ as in Example 2.3.10.

2.3.8 1. Note that the assignment $A \mapsto A^\mathbb{N}$ yields a functor $\mathbf{Sets} \to \mathbf{Sets}$.
2. Prove the general result: consider a category $\mathbb{C}$ with a functor $F\colon \mathbb{C} \times \mathbb{C} \to \mathbb{C}$ in two variables. Assume that for each object $A \in \mathbb{C}$, the functor $F(A, -)\colon \mathbb{C} \to \mathbb{C}$ has a final coalgebra $Z_A \xrightarrow{\cong} F(A, Z_A)$. Prove that the mapping $A \mapsto Z_A$ extends to a functor $\mathbb{C} \to \mathbb{C}$.

2.4 Algebras

So far we have talked much about coalgebras. One way to introduce coalgebras is as duals of algebras. We shall do this the other way around and introduce algebras (in categorical formulation) as duals of coalgebras. This reflects of course the emphasis in this book.

There are many similarities (or dualities) between algebras and coalgebras which are often useful as guiding principles. But one should keep in mind that there are also significant differences between algebra and coalgebra. For example, in a computer science setting, algebra is mainly of interest for dealing with *finite* data elements – such as finite lists or trees – using induction as main definition and proof principle. A key feature of coalgebra is that it deals with potentially infinite data elements and with appropriate state-based notions and techniques for handling these objects. Thus, algebra is about construction, whereas coalgebra is about deconstruction – understood as observation and modification.

This section will introduce the categorical definition of algebras for a functor, in analogy with coalgebras of a functor in Definition 1.4.5. It will briefly discuss initiality for algebras, as dual of finality, and will illustrate that it amounts to ordinary induction. Also, it will mention several possible ways of combining algebras and coalgebras. A systematic approach to such combinations using distributive laws will appear in Chapter 5.

We start with the abstract categorical definition of algebras, which is completely dual (i.e. with reversed arrows) to what we have seen for coalgebras.

Definition 2.4.1 Let $\mathbb{C}$ be an arbitrary category, with an endofunctor $F\colon \mathbb{C} \to \mathbb{C}$.

1. An F**-algebra**, or just an **algebra** for F, consists of a 'carrier' object $X \in \mathbb{C}$ together with a morphism $a\colon F(X) \to X$, often called the constructor, or operation.
2. A **homomorphism of algebras**, or a **map of algebras** or **algebra map**, from one algebra $a\colon F(X) \to X$ to another coalgebra $b\colon F(Y) \to Y$ consists of a morphism $f\colon X \to Y$ in $\mathbb{C}$ which preserves the operations:

$$\begin{array}{ccc} F(X) & \xrightarrow{\;F(f)\;} & F(Y) \\ {\scriptstyle a}\big\downarrow & & \big\downarrow{\scriptstyle b} \\ X & \xrightarrow[\;f\;]{} & Y \end{array}$$

This yields a category, for which we shall write $\mathbf{Alg}(F)$.

3. An **initial** F-algebra is an initial object in the category $\mathbf{Alg}(F)$: it is an algebra $\alpha\colon F(A) \to A$ such that for any F-algebra $b\colon F(X) \to X$ there is a unique homomorphism of algebras int_b from α to b in

$$\begin{pmatrix} F(A) \\ \downarrow \alpha \\ A \end{pmatrix} \xrightarrow{\ \mathrm{int}_b\ } \begin{pmatrix} F(B) \\ \downarrow b \\ B \end{pmatrix} \quad \text{i.e.} \quad \begin{array}{ccc} F(A) & \xrightarrow{F(\mathrm{int}_b)} & F(X) \\ \alpha \downarrow & & \downarrow b \\ X & \xrightarrow[\mathrm{int}_b]{} & X \end{array}$$

We call this map int_b an interpretation map for the algebra b; sometimes it is also called a *fold* or a *reduce* of b.

Certain maps can be seen both as algebra and as coalgebra. For example, a map $S \times A \to S$ is an algebra – of the functor $X \mapsto X \times A$ – but can also be regarded as a coalgebra $S \to S^A$, after currying (2.11). But for other maps there is more clarity: maps $A \to S$ are algebras, which can be used to construct elements in S from elements in A. And maps $S \to A$ are coalgebras, which provide observations in A about elements in S.

The functor F in the above definition corresponds to what is traditionally called a signature. In simple form a (single-sorted) signature consists of a number of operations op_i, say for $1 \leq i \leq n$, each with their own arity $k_i \in \mathbb{N}$. An algebra for such a signature is then a set S with interpretations $[\![\mathsf{op}_i]\!]\colon S^{k_i} \to S$. Using the coproduct correspondence (2.6), they may be combined to a single cotuple operation,

$$S^{k_1} + \cdots + S^{k_n} \xrightarrow{\ [\,[\![\mathsf{op}_i]\!], \ldots, [\![\mathsf{op}_i]\!]\,]\ } S, \tag{2.29}$$

forming an algebra of the simple polynomial functor $X \mapsto X^{k_1} + \cdots + X^{k_n}$. This functor thus captures the number and types of the operations – that is, the signature. In fact, we have already seen the more general description of signatures via arities $\#\colon I \to \mathbb{N}$ and the associated simple polynomial functor $F_\# = \coprod_{i \in I} (-)^{\# i}$, namely in Definition 2.2.2.

A rule of thumb is: data types are algebras, and state-based systems are coalgebras. But this does not always give a clear-cut distinction. For instance, is a stack a data type or does it have a state? In many cases, however, this rule of thumb works: natural numbers are algebras (as we are about to see), and machines are coalgebras. Indeed, the latter have a state that can be observed and modified.

Initial algebras are special, just like final coalgebras. Initial algebras (in **Sets**) can be built as so-called term models: they contain everything that can be built from the operations themselves, and nothing more. Similarly, we saw that final coalgebras consist of observations only. The importance of initial algebras in computer science was first emphasised in [165]. For example, if F is the signature functor for some programming language, the initial algebra $F(P) \to P$ may be considered as the set of programs, and arbitrary algebras $F(D) \to D$ may be seen as denotational models. Indeed, the resulting unique homomorphism $\mathrm{int} = [\![-]\!] \colon P \to D$ is the semantical interpretation function. It is *compositional* by construction, because it commutes with the operations of the programming language.

Example 2.4.2 In mathematics many algebraic structures are single-sorted (or single-typed). Examples are monoids, groups, rings, etc. In these structures there is only one carrier set involved. Some structures, such as vector spaces, involve two sorts, namely scalars and vectors. In this case one speaks of a multisorted or typed algebra. Still, in mathematics vector spaces are often described as single-sorted, by keeping the field involved fixed.

In computer science most of the examples are multi-sorted. For instance, a specification of some data structure, say a queue, involves several sorts (or types): the type D of data on the queue, the type Q of the queue itself, and possibly some other types like $\mathbb{N}$ for the length of the queue. Type theory has developed into an area of its own, studying various calculi for types and terms.

Most programming languages are typed, involving various types like `nat`, `int`, `float`, `bool`, and possibly also additional user-definable types. We shall describe a toy typed programming language and show how its signature can be described via a functor. Interestingly, this will not be a functor on the standard category **Sets**, but a functor on the product category $\mathbf{Sets}^3 = \mathbf{Sets} \times \mathbf{Sets} \times \mathbf{Sets}$. The 3-fold product $\mathbf{Sets}^3$ is used because our toy language contains only three types, namely N for numbers, B for booleans (true and false) and S for program statements.

We assume that our programming language involves the following operations on numbers and booleans:

$\mathtt{n}\colon N$	for $0 \leq \mathtt{n} < 2^{16}$	
$\mathtt{v}\colon N$	for variables $\mathtt{v} \in V$	
$+\colon N \times N \to N$	for addition	
$*\colon N \times N \to N$	for multiplication	
$\mathtt{true}, \mathtt{false}\colon B$	truth values	
$\wedge\colon B \times B \to B$	for conjunction	
$\neg\colon B \to B$	for negation	
$=\colon N \times N \to B$	for equality	
$\leq\colon N \times N \to B$	for comparison.	

Here we use a (fixed) set V of variables. Note that the two operations $=$ and $\leq$ involve both types N and B. Also statements involve multiple types:

$$\begin{array}{rll} \texttt{skip}\colon & S & \text{the program that does nothing} \\ \texttt{;}\colon & S \times S \to S & \text{sequential program composition} \\ \texttt{-:=-}\colon & V \times N \to S & \text{for assignment} \\ \texttt{if-then-else-fi}\colon & B \times S \times S \to S & \text{for the conditional statement} \\ \texttt{while-do-od}\colon & B \times S \to S & \text{for repetition.} \end{array}$$

Following the (single-sorted) signature description as in (2.29) we could capture the constants n, sum + and product $*$ via a functor $X \mapsto 2^{16} + (X \times X) + (X \times X)$. But we need to capture all operations at the same time. Thus we use a functor $F\colon \mathbf{Sets}^3 \to \mathbf{Sets}^3$ in three variables, where the first one, written as X, will be used for the type of naturals, the second one Y for booleans and the third one Z for program statements. The functor F is then given by the following expression:

$$\begin{aligned} F\big(X, Y, Z\big) = \big(&2^{16} + V + (X \times X) + (X \times X), \\ &(X \times X) + (X \times X) + 2 + (Y \times Y) + Y, \\ &1 + (Z \times Z) + (V \times X) + (Y \times Z \times Z) + (Y \times Z)\big). \end{aligned}$$

An algebra $F(A, B, C) \to (A, B, C)$ in $\mathbf{Sets}^3$ for this functor is given by a triple of sets $(A, B, C) \in \mathbf{Sets}^3$ interpreting naturals, booleans and statements, and a triple of functions $(a, b, c)\colon F(A, B, C) \to (A, B, C)$ giving interpretations of the operations. For instance the function b in the middle is given by a 5-cotuple of the form

$$(A \times A) + (A \times A) + 2 + (B \times B) + B \xrightarrow{\; b = [b_1, b_2, b_3, b_4, b_5] \;} B$$

where $b_3\colon 2 \to B$ interprets the two booleans (since $2 = 1 + 1 = \{0, 1\}$) and $b_5\colon B \to B$ interprets negation; similarly for the maps a, c in this algebra. We see that signature functors for specific languages can become quite complicated. But at an abstract level we work with arbitrary (polynomial) functors, and such complexity remains hidden.

The initial F-algebra is given by three sets containing all syntactically constructed expressions for naturals, booleans and statements, using the above operations. This will not be proven, but the examples below illustrate how initial algebras consist of terms.

In this context of program language semantics, a final coalgebra is also understood as a canonical operational model of behaviours. The unique homomorphism to the final coalgebra may be seen as an operational interpretation function which is *fully abstract*, in the sense that objects have the same interpretation if and only if they are observationally indistinguishable. This relies on Theorem 3.4.1; see [451–453] for more information.

Compositionality for such models in final coalgebras is an important issue; see Section 5.5.

Because of the duality in the definitions of algebras and coalgebras, certain results can be dualised. For example, Lambek's fixed point result for final coalgebras (Lemma 2.3.3) also holds for initial algebras. The proof is the same, but with arrows reversed.

Lemma 2.4.3 *An initial algebra $F(A) \to A$ of a functor $F\colon \mathbb{C} \to \mathbb{C}$ is an isomorphism $F(A) \xrightarrow{\cong} A$ in $\mathbb{C}$.* ❑

The unique existence in the definition of initiality (again) has two aspects, namely existence corresponding to recursion, or definition by induction, and uniqueness, corresponding to proof by induction. This will be illustrated in two examples.

The natural numbers $\mathbb{N}$ form a trivial but important example of an initial algebra. We shall consider it in some detail, relating the familiar description of induction to the categorical one based on initiality.

Example 2.4.4 (Natural numbers) According to Peano, the most basic operations on the natural numbers $\mathbb{N}$ are zero $0 \in \mathbb{N}$ and successor $\mathsf{S}\colon \mathbb{N} \to \mathbb{N}$. Using that an element of $\mathbb{N}$ can be described as an arrow $1 \to \mathbb{N}$, together with the coproduct correspondence (2.6), these two maps can be combined into an algebra

$$1 + \mathbb{N} \xrightarrow{\quad [0, \mathsf{S}] \quad} \mathbb{N}$$

of the simple polynomial function $F(X) = 1 + X$. This functor adds a new point $* \in 1$ to an arbitrary set X. It is easy to see that the algebra $[0, \mathsf{S}]$ is an isomorphism, because each natural number is either zero or a successor. The isomorphism $[0, \mathsf{S}]\colon 1 + \mathbb{N} \xrightarrow{\cong} \mathbb{N}$ is in fact the initial algebra of the functor $F(X) = 1 + X$. Indeed, for an arbitrary F-algebra $[a, g]\colon 1 + X \to X$ with carrier X, consisting of functions $a\colon 1 \to X$ and $g\colon X \to X$, there is a unique homomorphism of algebras $f = \mathrm{int}_{[a,g]}\colon \mathbb{N} \to X$ with

$$\begin{array}{ccc} 1 + \mathbb{N} & \xrightarrow{\mathrm{id}_1 + f} & 1 + X \\ {\scriptstyle [0,\mathsf{S}]}\downarrow{\scriptstyle \cong} & & \downarrow{\scriptstyle [a,g]} \\ \mathbb{N} & \xrightarrow[f]{} & X \end{array} \qquad \text{i.e. with} \qquad \begin{cases} f \circ 0 = a \\ f \circ \mathsf{S} = g \circ f. \end{cases}$$

In ordinary mathematical language, this says that f is the unique function with $f(0) = a$, and $f(n + 1) = g(f(n))$. This indeed determines f completely, by induction. Thus, ordinary natural number induction implies initiality. We shall illustrate the reverse implication.

The initiality property of $[0, \mathsf{S}]\colon 1 + \mathbb{N} \xrightarrow{\cong} \mathbb{N}$ is strong enough to prove all of Peano's axioms. This was first shown in [142] (see also [345] or [166]), where the initiality is formulated as 'natural numbers object'. As an example, we shall consider the familiar induction rule: for a predicate $P \subseteq \mathbb{N}$,

$$\frac{P(0) \qquad \forall n \in \mathbb{N}.\, P(n) \implies P(n+1)}{\forall n \in \mathbb{N}.\, P(n)}.$$

In order to show how the validity of this induction rule follows from initiality, let us write i for the inclusion function $P \hookrightarrow \mathbb{N}$. We then note that the assumptions of the induction rule state that the zero and successor functions restrict to P, as in

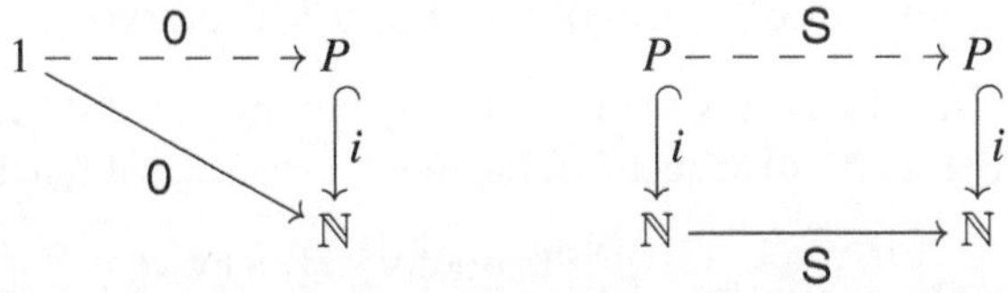

Thus, the subset P itself carries an algebra structure $[0, \mathsf{S}]\colon 1 + P \to P$. Therefore, by initiality of $\mathbb{N}$ we get a unique homomorphism $j\colon \mathbb{N} \to P$. Then we can show $i \circ j = \mathrm{id}_{\mathbb{N}}$, by uniqueness:

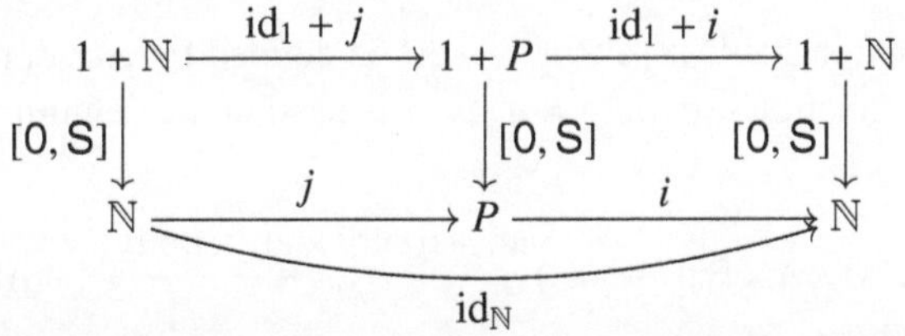

The rectangle on the left commutes by definition of j, and the one on the right by the previous two diagrams. The fact that $i \circ j = \mathrm{id}_{\mathbb{N}}$ now yields $P(n)$, for all $n \in \mathbb{N}$.

Example 2.4.5 (Binary trees) Fix a set A of labels. A signature for A-labelled binary trees may be given with two operations:

nil	for the empty tree,
$\mathsf{node}(b_1, a, b_2)$	for the tree constructed from subtrees b_1, b_2 and label $a \in A$.

Thus, the associated signature functor is $T(X) = 1 + (X \times A \times X)$. The initial algebra will be written as $\mathsf{BinTree}(A)$, with operation

$$1 + (\mathsf{BinTree}(A) \times A \times \mathsf{BinTree}(A)) \xrightarrow[\cong]{[\mathsf{nil}, \mathsf{node}]} \mathsf{BinTree}(A).$$

This carrier set $\mathsf{BinTree}(A)$ can be obtained by considering all terms that can be formed from the constructor nil via repeated application of the node operation $\mathsf{node}(\mathsf{nil}, a, \mathsf{nil})$, $\mathsf{node}(\mathsf{nil}, a', \mathsf{node}(\mathsf{nil}, a, \mathsf{nil}))$, etc. We do not describe it in too much detail because we wish to use it abstractly. Our aim is to traverse such binary trees and collect their labels in a list. We consider two obvious ways to do this – commonly called *inorder* traversal and *preorder* traversal – resulting in two functions $\mathsf{iotrv}, \mathsf{potrv} \colon \mathsf{BinTree}(A) \to A^\star$. These functions can be defined inductively as

$$\begin{aligned} \mathsf{iotrv}(\mathsf{nil}) &= \langle\rangle \\ \mathsf{iotrv}(\mathsf{node}(b_1, a, b_2)) &= \mathsf{iotrv}(b_1) \cdot a \cdot \mathsf{iotrv}(b_2) \\ \mathsf{potrv}(\mathsf{nil}) &= \langle\rangle \\ \mathsf{potrv}(\mathsf{node}(b_1, a, b_2)) &= a \cdot \mathsf{potrv}(b_1) \cdot \mathsf{potrv}(b_2). \end{aligned}$$

They can be defined formally via initiality, by putting two different T-algebra structures on the set $A^\star$ of sequences: the inorder transversal function arises in

$$\begin{array}{ccc} 1+(\mathsf{BinTree}(A) \times A \times \mathsf{BinTree}(A)) & \xrightarrow{\mathrm{id}_1 + (\mathsf{iotrv} \times \mathrm{id}_A \times \mathsf{iotrv})} & 1+(A^\star \times A \times A^\star) \\ \downarrow{\scriptstyle [\mathsf{nil}, \mathsf{node}]} & & \downarrow{\scriptstyle [\langle\rangle, g]} \\ \mathsf{BinTree}(A) & \xrightarrow[\mathsf{iotrv} = \mathrm{int}_{[\langle\rangle, g]}]{} & A^\star \end{array}$$

where the function $g \colon A^\star \times A \times A^\star \to A^\star$ is defined by $g(\sigma, a, \tau) = \sigma \cdot a \cdot \tau$. Similarly, the function $h(\sigma, a, \tau) = a \cdot \sigma \cdot \tau$ is used in the definition of preorder traversal:

$$\begin{array}{ccc} 1+(\mathsf{BinTree}(A) \times A \times \mathsf{BinTree}(A)) & \xrightarrow{\mathrm{id}_1 + (\mathsf{potrv} \times \mathrm{id}_A \times \mathsf{potrv})} & 1+(A^\star \times A \times A^\star) \\ \downarrow{\scriptstyle [\mathsf{nil}, \mathsf{node}]} & & \downarrow{\scriptstyle [\langle\rangle, h]} \\ \mathsf{BinTree}(A) & \xrightarrow[\mathsf{potrv} = \mathrm{int}_{[\langle\rangle, h]}]{} & A^\star \end{array}$$

What we see is that the familiar pattern matching in inductive definitions fits perfectly in the initiality scheme, where the way the various patterns are handled corresponds to the algebra structure on the codomain of the function that is being defined.

It turns out that many such functional programs can be defined elegantly via initiality (and also finality); this style that is called 'origami programming' in [153]. Moreover, via uniqueness one can establish various properties about these programs. This has developed into an area of its own, which is usually referred to as the Bird–Meertens formalism – with its own peculiar notation and terminology; see [72] for an overview.

In Example 2.4.2 we have seen how functors on categories other than **Sets** can be useful, namely for describing multi-sorted signatures. The next example is another illustration of the usefulness of descriptions in terms of functors.

Example 2.4.6 (Refinement types) For a fixed set A we can form the functor $F(X) = 1 + (A \times X)$. The initial algebra of this functor is the set $A^\star$ of finite lists of elements of A, with algebra structure given by the empty list nil and the prefix operation cons in

$$1 + (A \times A^\star) \xrightarrow[\cong]{[\mathsf{nil}, \mathsf{cons}]} A^\star.$$

Via this initiality one can define for instance the length function $\mathsf{len}\colon A^\star \to \mathbb{Z}$. For some special reasons we take the integers $\mathbb{Z}$, instead of the natural numbers $\mathbb{N}$, as codomain type (as is common in computer science); the reasons will become clear later on. A definition of length by initiality requires that the set $\mathbb{Z}$ be equipped with an appropriate F-coalgebra structure:

$$\begin{array}{ccc}
1 + A \times A^\star & \xrightarrow{\mathrm{id}_1 + \mathrm{id}_A \times \mathsf{len}} & 1 + A \times \mathbb{Z} \\
{\scriptstyle [\mathsf{nil}, \mathsf{cons}]}\big\downarrow{\scriptstyle \cong} & & \big\downarrow{\scriptstyle [0, \mathsf{S} \circ \pi_2]} \\
A^\star & \xrightarrow[\mathsf{len}]{} & \mathbb{Z}
\end{array}$$

Commutation of this diagram corresponds to the expected defining equations for list-length:

$$\mathsf{len}(\mathsf{nil}) = 0 \quad \text{and} \quad \mathsf{len}(\mathsf{cons}(a, \sigma)) = \mathsf{S}(\mathsf{len}(\sigma)) = 1 + \mathsf{len}(\sigma).$$

One can also define an append map $\mathsf{app}\colon A^\star \times A^\star \to A^\star$ that concatenates two lists producing a new one, but this requires some care: by initiality one can define only maps of the form $A^\star \to X$, when X carries an F-algebra structure. The easiest way out is to fix a list $\tau \in A^\star$ and to define $\mathsf{app}(-, \tau)\colon A^\star \to A^\star$ by initiality in

$$\begin{array}{ccc}
1 + A \times A^\star & \xrightarrow{\mathrm{id}_1 + \mathrm{id}_A \times \mathsf{app}(-, \tau)} & 1 + A \times A^\star \\
{\scriptstyle [\mathsf{nil}, \mathsf{cons}]}\big\downarrow{\scriptstyle \cong} & & \big\downarrow{\scriptstyle [\tau, \mathsf{cons}]} \\
A^\star & \xrightarrow[\mathsf{app}(-, \tau)]{} & A^\star
\end{array}$$

Alternatively, one can define the append map app via currying, namely as map $A^\star \to (A^\star)^{A^\star}$. This will be left to the interested reader. But one can also use the stronger principle of 'induction with parameters'; see Exercise 2.5.5.

Now assume the set A carries a binary operation $+\colon A \times A \to A$. We would like to use it to define an operation sum on lists, namely as

$$\mathsf{sum}(\langle a_1, \ldots, a_n\rangle, \langle b_1, \ldots, b_n\rangle) = \langle a_1 + b_1, \ldots, a_n + b_n\rangle.$$

But this works only for lists of the same length!

Hence what we would like to have is a special type of lists of a certain length n. It can be formed as an inverse image:

$$\mathsf{len}^{-1}(\{n\}) = \{\sigma \in A^\star \mid \mathsf{len}(\sigma) = n\} = \begin{cases} A^n & \text{if } n \geq 0 \\ \emptyset & \text{otherwise.} \end{cases}$$

The type of the function sum can now be described accurately as $A^n \times A^n \to A^n$. We now consider the map $\mathsf{len}\colon A^\star \to \mathbb{Z}$ as a map in a slice category $\mathbf{Sets}/\mathbb{Z}$. Recall from Exercise 1.4.3 that this map $\mathsf{len}\colon A^\star \to \mathbb{Z}$ can be identified with the indexed collection $(\mathsf{len}^{-1}(\{n\}))_{n\in\mathbb{Z}}$ given by inverse images. An interesting question is: can we also define such maps like sum on dependent types like $\mathsf{len}^{-1}(\{n\})$ by initiality? This is relevant for dependently typed programming languages like *Agda*.[1]

In [45] a solution is given. We sketch the essentials, in quite general terms. Let $F\colon \mathbb{C} \to \mathbb{C}$ be an arbitrary functor on a category $\mathbb{C}$. Assume an F-algebra $b\colon F(B) \to B$; it plays the role of $\mathbb{Z}$ above, and the elements $x \in B$ are used to form a refinement of F's initial algebra. But first we transform F from an endofunctor on $\mathbb{C}$ into an endofunctor on the slice category $\mathbb{C}/B$. Intuitively this slice category contains objects of $\mathbb{C}$ indexed by B; see Exercise 1.4.3. Recall that objects of $\mathbb{C}/B$ are arrows with codomain B, and morphisms are commuting triangles.

We now define a functor $F^b\colon \mathbb{C}/B \to \mathbb{C}/B$ by

$$(X \xrightarrow{f} B) \longmapsto (F(X) \xrightarrow{F(f)} F(B) \xrightarrow{b} B)$$

$$\left(\begin{array}{c} X \xrightarrow{h} Y \\ {\scriptstyle f}\searrow \;\; \swarrow{\scriptstyle g} \\ B \end{array}\right) \longmapsto \left(\begin{array}{c} F(X) \xrightarrow{\quad F(h) \quad} F(Y) \\ {\scriptstyle F(f)}\searrow \qquad\qquad \swarrow{\scriptstyle F(g)} \\ F(B) = F(B) \\ {\scriptstyle b}\searrow \;\; \swarrow{\scriptstyle b} \\ B \end{array}\right)$$

Next we assume an initial F-algebra $\alpha\colon F(A) \xrightarrow{\cong} A$. The algebra b gives rise to a homomorphism $\mathrm{int}_b\colon A \to B$ by initiality. This map is an object in the slice category $\mathbb{C}/B$. It can be understood as what is called a **refinement type** since it corresponds to an indexed family $(A_x)_{x\in B}$, where

[1] See http://wiki.portal.chalmers.se/agda.

$A_x = \text{int}_b^{-1}(\{x\}) = \{a \in A \mid \text{int}_b(a) = x\}$. In fact, this type is also an initial algebra, but in a slice category, as observed in [45].

Claim. The refinement type $\text{int}_b : A \to B$ in $\mathbb{C}/B$ carries the initial algebra of the functor $F^b : \mathbb{C}/B \to \mathbb{C}/B$ defined above.

The truth of this claim is not hard to see but requires that we carefully keep track of the category that we work in (namely $\mathbb{C}$ or $\mathbb{C}/B$). We first have to produce an algebra structure $F^b(\text{int}_b) \to \text{int}_b$ in the slice category $\mathbb{C}/B$. When we work out what it can be we see that the map $\alpha : F(A) \xrightarrow{\cong} A$ is the obvious candidate, on the left in

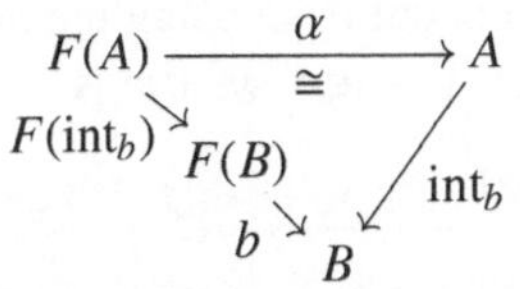

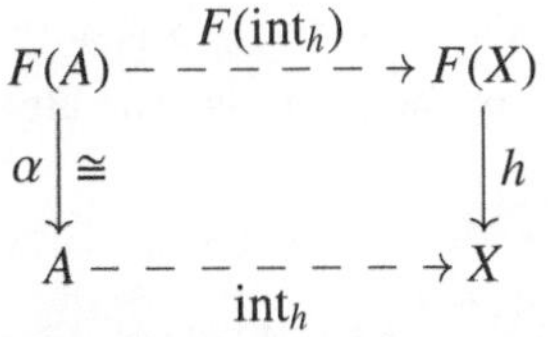

Next, if we have an arbitrary F^b-coalgebra, say given by a map $h: F^b(X \xrightarrow{f} B) \longrightarrow (X \xrightarrow{f} B)$ in the slice category $\mathbb{C}/B$, then this h is a map $h: F(X) \to X$ in $\mathbb{C}$ satisfying $f \circ h = b \circ F(h)$. By initiality we get an algebra homomorphism $\text{int}_h : A \to X$ as on the right above. We claim that the resulting interpretation map $\text{int}_h : A \to X$ is a morphism $\text{int}_b \to f$ in $\mathbb{C}/B$. This means that $f \circ \text{int}_h = \text{int}_b$, which follows by a uniqueness argument in

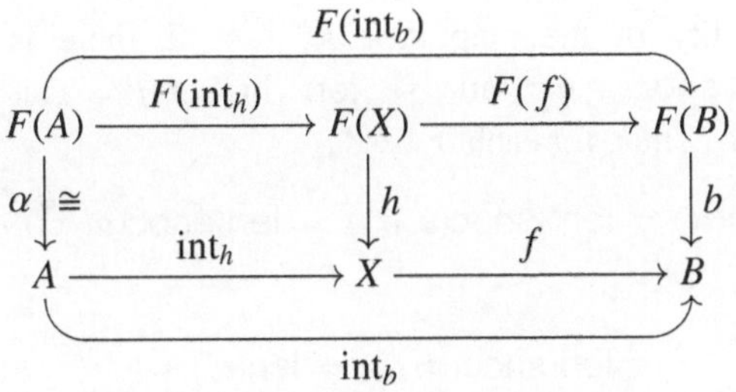

What we finally need to show is that we have a commuting diagram in $\mathbb{C}/B$ of the form

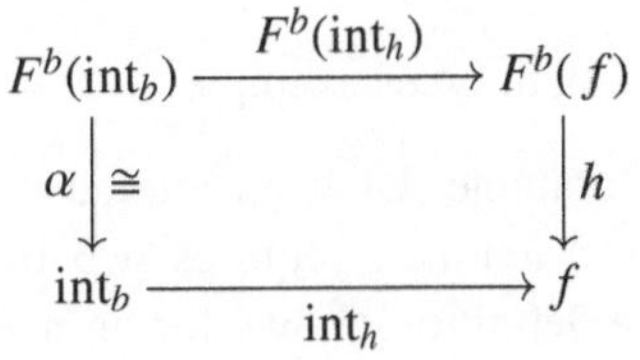

Commutation of this square in $\mathbb{C}/B$ amounts to commutation of the corresponding diagram in $\mathbb{C}$, which is the case by definition of int_h. The proof of uniqueness of this homomorphism of F^b-algebras $\mathrm{int}_b \to f$ is left to the reader.

At this stage we return to the length map $\mathsf{len}\colon A^\star \to \mathbb{Z}$ defined in the beginning of this example. We now use initiality in the slice category $\mathbf{Sets}/\mathbb{Z}$ to define a 'reverse cons' operation $\mathsf{snoc}\colon A^n \times A \to A^{n+1}$ that adds an element at the end of a list. We do so, as before, by fixing a parameter $a \in A$ and defining $\mathsf{snoc}(-, a)\colon A^n \to A^{n+1}$. We need to come up with an algebra of the endofunctor F^{len} on $\mathbf{Sets}/\mathbb{Z}$, where $F = 1 + (A \times -)$. The claim above says that $\mathsf{len}\colon A^\star \to \mathbb{Z}$ is the initial F^{len}-algebra.

As carrier of the algebra we take the object $\mathsf{len}'\colon A^\star \to \mathbb{Z}$ in $\mathbf{Sets}/\mathbb{Z}$ given by $\mathsf{len}'(\sigma) = \mathsf{len} - 1$. It is at this point that we benefit from using the integers instead of the naturals. The algebra map $F^{\mathsf{len}}(\mathsf{len}') \to \mathsf{len}'$ we use is

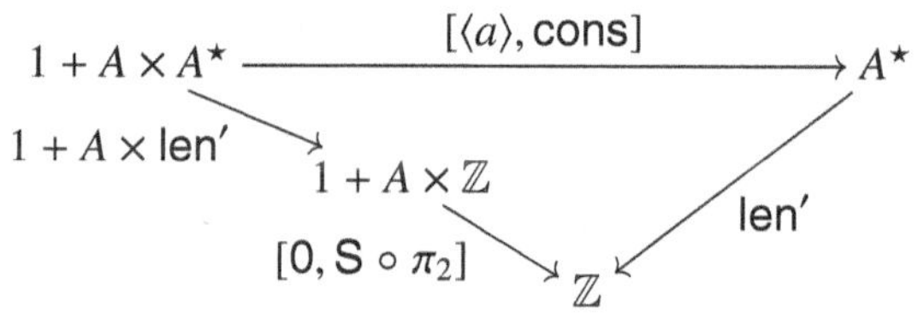

Here we use $\langle a \rangle = \mathsf{cons}(a, \mathsf{nil})$ for the singleton list. The triangle commutes because

$$\begin{aligned}
\mathsf{len}'(\langle a \rangle) &= \mathsf{len}(\langle a \rangle) - 1 \;=\; 1 - 1 \;=\; 0 \\
\mathsf{len}'(\mathsf{cons}(a, \sigma)) &= \mathsf{len}(\mathsf{cons}(a, \sigma)) - 1 \;=\; \mathsf{len}(\sigma) + 1 - 1 \;=\; \mathsf{len}(\sigma) - 1 + 1 \\
&= \mathsf{S}(\mathsf{len}'(\sigma)) \;=\; (\mathsf{S} \circ \pi_2 \circ (A \times \mathsf{len}'))(a, \sigma).
\end{aligned}$$

Hence by initiality of the map $\mathsf{len}\colon A^\star \to \mathbb{Z}$ there is unique algebra homomorphism $\mathsf{snoc}(-, a)\colon \mathsf{len} \to \mathsf{len}'$ in $\mathbf{Sets}/\mathbb{Z}$. This means: $\mathsf{len}' \circ \mathsf{snoc}(-, a) = \mathsf{len}$. Thus, for each $\sigma \in A^\star$,

$$\mathsf{len}(\sigma) \;=\; \mathsf{len}'(\mathsf{snoc}(a, \sigma)) \;=\; \mathsf{len}(\mathsf{snoc}(a, \sigma)) - 1,$$

i.e.

$$\mathsf{len}(\mathsf{snoc}(\sigma, a)) \;=\; \mathsf{len}(\sigma) + 1.$$

Alternatively, for each $n \in \mathbb{N}$ one has a typing:

$$A^n = \mathsf{len}^{-1}(\{n\}) \xrightarrow{\;\mathsf{snoc}\;} \mathsf{len}^{-1}(\{n + 1\}) = A^{n+1}\,.$$

This concludes our refinement types example.

As we have seen in Example 2.4.4, an inductive definition of a function $f\colon \mathbb{N} \to X$ on the natural numbers requires two functions, $a\colon 1 \to X$ and $g\colon X \to X$. A **recursive** definition allows for an additional parameter, via a function $h: X \times \mathbb{N} \to X$, determining $f(n+1) = h(f(n), n)$. The next result shows that recursion can be obtained via induction. A more general approach via comonads may be found in [455]. An alternative way to add parameters to induction occurs in Exercise 2.5.5.

Proposition 2.4.7 (Recursion) *An arbitrary initial algebra $\alpha: F(A) \xrightarrow{\cong} A$ satisfies the following recursion property: for each map $h: F(X \times A) \to X$ there is a unique map $f: A \to X$ making the following diagram commute:*

$$\begin{array}{ccc} F(A) & \xrightarrow{F\langle f, \mathrm{id}\rangle} & F(X \times A) \\ {\scriptstyle\alpha}\big\downarrow{\scriptstyle\cong} & & \big\downarrow{\scriptstyle h} \\ A & \xrightarrow[f]{} & X \end{array}$$

Proof We first turn $h: F(X \times A) \to X$ into an algebra on $X \times A$, namely by taking $h' = \langle h, \alpha \circ F(\pi_2)\rangle: F(X \times A) \to X \times A$. It gives by initiality rise to a unique map $k: A \to X \times A$ with $k \circ \alpha = h' \circ F(k)$. Then $\pi_2 \circ k = \mathrm{id}$ by uniqueness of algebra maps $\alpha \to \alpha$:

$$\pi_2 \circ k \circ \alpha = \pi_2 \circ h' \circ F(k) = \alpha \circ F(\pi_2) \circ F(k) = \alpha \circ F(\pi_2 \circ k).$$

Hence we take $f = \pi_1 \circ k$. □

Adding parameters to the formulation of induction gives another strengthening; see Exercise 2.5.5.

So far we have seen only algebras of functors describing fairly elementary datatypes. We briefly mention some other less trivial applications.

- Joyal and Moerdijk [286] introduce the notion of a *Zermelo–Fraenkel algebra*, or *ZF-algebra*, with two operations for union and singleton. The usual system of Zermelo–Fraenkel set theory can then be characterised as the free ZF-algebra. Such a characterisation can be extended to ordinals.
- Fiore, Plotkin and Turi [136] describe how variable binding – such as $\lambda x.\, M$ in the lambda calculus – can also be captured via initial algebras, namely of suitable functors on categories of presheaves. An alternative approach, also via initiality, is described by Gabbay and Pitts [148], using set theory with atoms (or *urelements*). See also [270].
- So-called name-passing systems, like the π-calculus, are also modelled as coalgebras, namely for endofunctors on categories of presheaves, see [137, 97, 349, 435].

In the remainder of this section we shall briefly survey several ways to combine algebras and coalgebras. Such combinations are needed in descriptions of more complicated systems involving data as well as behaviour. Later on, in Section 5.5, we shall study bialgebras more systematically, for operational semantics.

Additionally, so-called binary methods are problematic in a coalgebraic setting. These are (algebraic) operations like $X \times X \to X$ in which the state space X occurs multiple times (positively) in the domain. The name 'binary method' comes from object-oriented programming, where the status of such methods is controversial due to the typing problems they can cause; see [88]. They are also problematic in coalgebra, because they cannot be described as a coalgebra $X \to F(X)$. However, one may ask whether it makes sense in system theory to have an operation acting on two states, so that the problematic character of their representation need not worry us very much. But see Exercise 2.4.10 for a possible way to handle binary methods in a coalgebraic manner.

2.4.1 Bialgebras

A **bialgebra** consists of an algebra-coalgebra pair $F(X) \to X \to G(X)$ on a common state space X, where $F, G \colon \mathbb{C} \to \mathbb{C}$ are two endofunctors on the same category $\mathbb{C}$. We shall investigate them more closely in Section 5.5 using so-called distributive laws, following the approach of [452, 451, 58, 298].

Such structures are used in [69] to distinguish certain observer operations as coalgebraic operations within algebraic specification and to use these in a duality between reachability and observability (following the result of Kalman; see Exercise 2.5.15).

2.4.2 Dialgebras

A **dialgebra** consists of a mapping $F(X) \to G(X)$ where F and G are two functors $\mathbb{C} \to \mathbb{C}$. This notion generalises both algebras (for $G = \mathrm{id}$) and coalgebras (for $F = \mathrm{id}$). It was introduced in [191] and further studied for example in [138] in combination with 'laws' as a general concept of equation. In [392] a collection of such dialgebras $F_i(X) \to G_i(X)$ is studied, in order to investigate the commonalities between algebra and coalgebra, especially related to invariants and bisimulations.

2.4.3 Hidden Algebras

Hidden algebra is introduced in [164] as the 'theory of everything' in software engineering, combining the paradigms of object-oriented, logic, constraint and functional programming. A hidden algebra does not formally combine algebras and coalgebras, as in bi-/di-algebras, but it uses an algebraic syntax to handle essentially coalgebraic concepts, such as behaviour and observational equivalence. A key feature of the syntax is the partition of sorts (or types) into visible and hidden ones. Typically, data structures are visible, and states are hidden. Elements of the visible sorts are directly observable and comparable, but observations about elements of the hidden sorts can be made only via the visible ones. This yields a notion of behavioural equivalence – first introduced by Reichel – expressed in terms of equality of contexts of visible sort. This is in fact bisimilarity, from a coalgebraic perspective; see [339, 98].

Because of the algebraic syntax of hidden algebras one cannot represent typically coalgebraic operations. But one can mimic them via subsorts. For instance, a coalgebraic operation $X \longrightarrow 1 + X$ can be represented as a 'partial' function $S \longrightarrow X$ for a subsort $S \subseteq X$. Similarly, an operation $X \longrightarrow X + X$ can be represented via two operations, $S_1 \longrightarrow X$ and $S_2 \longrightarrow X$, for subsorts $S_1, S_2 \subseteq X$ with $S_1 \cup S_2 = X$ and $S_1 \cap S_2 = \emptyset$. This quickly gets out of hand for more complicated operations, like the methods meth_i from (1.10), so that the naturality of coalgebraic representations is entirely lost. On the positive side, hidden algebras can handle binary methods without problems – see also Exercise 2.4.10.

2.4.4 Coalgebras as Algebras

In general, there is reasonable familiarity with algebra, but not (yet) with coalgebra. Therefore, people like to understand coalgebras as if they were algebras. Indeed, there are many connections. For example, in [55] it is shown that quite often the final coalgebra $Z \xrightarrow{\cong} F(Z)$ of a functor F can be understood as a suitable form of completion of the initial algebra $F(A) \xrightarrow{\cong} A$ of the same functor. Examples are the extended natural numbers $\mathbb{N} \cup \{\infty\}$ from Exercise 2.4.1 below, as completion of the initial algebra $\mathbb{N}$ of the functor $X \mapsto 1+X$, and the set A^∞ of finite and infinite sequences as completion of the initial algebra $A^\star$ with finite sequences only, of the functor $X \mapsto 1 + (A \times X)$.

Also, since a final coalgebra $\zeta\colon Z \xrightarrow{\cong} F(Z)$ is an isomorphism, one can consider its inverse $\zeta^{-1}\colon F(Z) \xrightarrow{\cong} Z$ as an algebra. This happens for example in the formalisation of 'coinductive types' (final coalgebras) in the theorem prover Coq[2] (and used for instance in [112, 225]). However, this may lead to rather complicated formulations of coinduction, distracting from the coalgebraic essentials. Therefore we like to treat coalgebra as a field of its own, and not in an artificial way as part of algebra.

2 See https://coq.inria.fr/.

By now coalgebraic constructs are appearing in many languages for programming and proving; see e.g. [189, 415, 448, 1, 277, 278].

Exercises

2.4.1 Check that the set $\mathbb{N} \cup \{\infty\}$ of extended natural numbers is the final coalgebra of the functor $X \mapsto 1 + X$, as a special case of finality of A^∞ for $X \mapsto 1 + (A \times X)$. Use this fact to define appropriate addition and multiplication operations $(\mathbb{N} \cup \{\infty\}) \times (\mathbb{N} \cup \{\infty\}) \to \mathbb{N} \cup \{\infty\}$; see [411].

2.4.2 Define an appropriate size function $\mathsf{BinTree}(A) \to \mathbb{N}$ by initiality.

2.4.3 Consider the obvious functions $\mathsf{even}, \mathsf{odd} \colon \mathbb{N} \to 2 = \{0, 1\}$.

1. Describe the two algebra structures $1 + 2 \to 2$ for the functor $F(X) = 1 + X$ that define even and odd by initiality.
2. Define a single algebra $1 + (2 \times 2) \to 2 \times 2$ that defines the pair $\langle \mathsf{even}, \mathsf{odd} \rangle$ by mutual recursion, that is via $\mathsf{even}(n + 1) = \mathsf{odd}(n)$.

2.4.4 Show, dually to Proposition 2.1.5, that finite products $(1, \times)$ in a category $\mathbb{C}$ give rise to finite products in a category **Alg**(F) of algebras of a functor $F \colon \mathbb{C} \to \mathbb{C}$.

2.4.5 Define addition $+ \colon \mathbb{N} \times \mathbb{N} \to \mathbb{N}$, multiplication $\cdot \colon \mathbb{N} \times \mathbb{N} \to \mathbb{N}$ and exponentiation $(-)^{(-)} \colon \mathbb{N} \times \mathbb{N} \to \mathbb{N}$ by initiality. *Hint*: Use the correspondence (2.11) and define these operations as function $\mathbb{N} \to \mathbb{N}^{\mathbb{N}}$. Alternatively, one can use Exercise 2.5.5 from the next section.

2.4.6 Complete the proof of Proposition 2.4.7.

2.4.7 ('Rolling lemma') Assume two endofunctors $F, G \colon \mathbb{C} \to \mathbb{C}$ on the same category. Let the composite $FG \colon \mathbb{C} \to \mathbb{C}$ have an initial algebra $\alpha \colon FG(A) \xrightarrow{\cong} A$.

1. Prove that also the functor $GF \colon \mathbb{C} \to \mathbb{C}$ has an initial algebra, with $G(A)$ as carrier.
2. Formulate and prove a dual result, for final coalgebras.

2.4.8 This exercise illustrates the analogy between the set $A^\star$ of finite sequences in **Sets** and the vector space $A^\S$ in **Vect**, following Exercise 2.2.12. Recall from Example 2.4.6 that the set $A^\star$ of finite lists of elements of A is the initial algebra of the functor $X \mapsto 1 + (A \times X)$.

1. For an arbitrary vector space A, this same functor $1 + (A \times \mathrm{id})$, but considered as an endofunctor on **Vect**, can be rewritten as

$$
\begin{aligned}
1 + (A \times X) &\cong A \times X && \text{because } 1 \in \mathbf{Vect} \text{ is initial object}\\
&\cong A + X && \text{because } \times \text{ and } + \text{ are the same in } \mathbf{Vect}.
\end{aligned}
$$

Prove that the initial algebra of this functor $A + \mathrm{id} \colon \mathbf{Vect} \to \mathbf{Vect}$ is the vector space $A^{\S}$ with insertion and shift maps $\mathsf{in} \colon A \to A^{\S}$ and $\mathsf{sh} \colon A^{\S} \to A^{\S}$ defined in Exercise 2.2.12.3.

2. Check that the behaviour formula from Exercise 2.2.12.4 for a system $A \xrightarrow{F} X \xrightarrow{G} X \xrightarrow{H} B$ is obtained as $H \circ \mathrm{int}_{[F,G]} \colon A^{\S} \to B$ using initiality.
3. Show that the assignment $A \mapsto A^{\S}$ yields a functor $\mathbf{Vect} \to \mathbf{Vect}$. *Hint*: Actually, this is a special case of the dual of Exercise 2.3.8.

2.4.9 Use Exercise 2.1.10 to show that there is a commuting diagram of isomorphisms

$$
\begin{array}{ccc}
\mathcal{P}(A^\star)^A \times 2 & \xrightarrow{\quad\cong\quad} & \mathcal{P}(1 + (A \times A^\star)) \\
& {\scriptstyle \langle D,E\rangle}\nwarrow\cong \qquad \cong\swarrow{\scriptstyle \mathcal{P}([\mathsf{nil},\mathsf{cons}])} & \\
& \mathcal{P}(A^\star) &
\end{array}
$$

where the coalgebra $\langle D, E\rangle$ is the final Brzozowski deterministic automaton structure from Corollary 2.3.6.2, and $[\mathsf{nil}, \mathsf{cons}]$ is the initial list algebra (from Example 2.4.6).

2.4.10 Suppose we have a 'binary method', say of the form $m \colon X \times X \times A \longrightarrow 1 + (X \times B)$. There is a well-known trick from [143] to use a functorial description also in such cases, namely by separating positive and negative occurrences. This leads to a functor of the form $F \colon \mathbb{C}^{\mathrm{op}} \times \mathbb{C} \to \mathbb{C}$, which in this case would be $(Y, X) \mapsto (1 + (X \times B))^{Y \times A}$.

In general, for a functor $F \colon \mathbb{C}^{\mathrm{op}} \times \mathbb{C} \to \mathbb{C}$ define an F-coalgebra to be a map of the form $c \colon X \to F(X, X)$. A homomorphism from $c \colon X \to F(X, X)$ to $d \colon Y \to F(Y, Y)$ is then a map $f \colon X \to Y$ in $\mathbb{C}$ making the following pentagon commute:

$$
\begin{array}{ccccc}
& & F(X,Y) & & \\
& {\scriptstyle F(\mathrm{id}_X, f)}\nearrow & & \nwarrow{\scriptstyle F(f, \mathrm{id}_Y)} & \\
F(X,X) & & & & F(Y,Y) \\
{\scriptstyle c}\uparrow & & & & \uparrow{\scriptstyle d} \\
X & & \xrightarrow[\quad f \quad]{} & & Y
\end{array}
$$

1. Elaborate what this means for the above example $(Y, X) \mapsto (1 + (X \times B))^{Y \times A}$.

2. Prove that these coalgebras and their homomorphisms form a category.

(Such generalised coalgebras are studied systematically in [444].)

2.5 Adjunctions, Cofree Coalgebras, Behaviour-Realisation

The concept of an adjunction may be considered as one of the basic contributions of the theory of categories. It consists of a pair of functors going in opposite direction,

$$\mathbb{C} \underset{G}{\overset{F}{\rightleftarrows}} \mathbb{D}$$

satisfying certain properties. An adjunction is a fundamental notion, occurring in many, many situations. Identifying an adjunction is useful, because it captures much information and yields additional insights such as the 'left' adjoint functor preserves coproducts, and the 'right' adjoint preserves products. This is the good news. The bad news is that the notion of adjunction is considered to be a difficult one. It certainly requires some work and experience to fully appreciate its usefulness. Much of this text can be read without knowledge of adjunctions. However, there are certain results which can best be organised in terms of adjunctions. Therefore we include an introduction to adjunctions and apply it to three standard examples in the theory of coalgebras and algebras, namely behaviour-realisation, cofree coalgebras and liftings of adjunctions to categories of (co)algebras.

We start with 'baby adjunctions', namely Galois connections. Consider two posets C and D as categories, with monotone functions $f: C \to D$ and $g: D \to C$ between them, in opposite direction. They can be regarded as functors by Example 1.4.4.2. These functions form a **Galois connection** if for each $x \in C$ and $y \in D$ one has $f(x) \leq y$ in D if and only if $x \leq g(y)$ in C. We like to write this as a bijective correspondence:

$$\frac{f(x) \;\leq\; y}{\overline{\overline{\;x \;\leq\; g(y)\;}}}\,.$$

In this situation one calls f the left or lower adjoint, and g the right or upper adjoint.

With these correspondences it is easy to show that the left adjoint f preserves joins $\vee$ that exist in C, i.e. that $f(x_1 \vee x_2) = f(x_1) \vee f(x_2)$. This is done by showing that $f(x_1 \vee x_2)$ is the least upperbound in D of $f(x_1)$ and $f(x_2)$: for any $y \in D$,

$$\begin{array}{c} f(x_1 \vee x_2) \;\le\; y \\ \hline\hline x_1 \vee x_2 \;\le\; g(y) \\ \hline\hline \begin{array}{ccc} x_1 \;\le\; g(y) & \qquad & x_2 \;\le\; g(y) \\ \cline{1-1}\cline{3-3} \cline{1-1}\cline{3-3} f(x_1) \;\le\; y & & f(x_2) \;\le\; y \end{array} \end{array} .$$

This says that $f(x_1 \vee x_2) \le y$ if and only if both $f(x_1) \le y$ and $f(x_2) \le y$, and thus that $f(x_1 \vee x_2)$ is the join of $f(x_1)$ and $f(x_2)$.

The notion of adjunction is defined more generally for functors between categories. It involves a bijective correspondence like above, together with certain technical naturality conditions. These side conditions make the definition a bit complicated but are usually trivial in concrete examples. Therefore, the bijective correspondence is what should be kept in mind.

Definition 2.5.1 Consider two categories $\mathbb{C}$ and $\mathbb{D}$ with functors $F\colon \mathbb{C} \to \mathbb{D}$ and $G\colon \mathbb{D} \to \mathbb{C}$ between them. These functors form an **adjunction**, written as $F \dashv G$, if for all objects $X \in \mathbb{C}$ and $Y \in \mathbb{D}$ there are bijective correspondences between morphisms

$$\begin{array}{cc} F(X) \longrightarrow Y & \quad \text{in } \mathbb{D} \\ \hline\hline X \longrightarrow G(Y) & \quad \text{in } \mathbb{C} \end{array}$$

which are 'natural' in X and Y. In this case one says that F is the left adjoint, and G the right adjoint. The map $X \to G(Y)$ corresponding to $F(X) \to Y$ (and vice-versa) is sometimes called the transpose.

We write this correspondences as functions $\psi\colon \mathbb{D}(F(X), Y) \xrightarrow{\cong} \mathbb{C}(X, G(Y))$. The naturality requirement then says that for morphisms $f\colon X' \to X$ in $\mathbb{C}$ and $g\colon Y \to Y'$ in $\mathbb{D}$ one has, for $h\colon F(X) \to Y$,

$$G(g) \circ \psi(h) \circ f \;=\; \psi(g \circ h \circ F(f)).$$

Since morphisms in poset categories are so trivial, the naturality requirement disappears in the definition of a Galois connection.

There are several equivalent ways to formulate the concept of an adjunction, using for example unit and counit natural transformations, or freeness (for left adjoints) or cofreeness (for right adjoints). We shall describe one alternative, namely the unit-and-counit formulation, in Exercise 2.5.7, and refer the interested reader to [344, chapter IV] for more information. At this stage we are mainly interested in a workable formulation.

We continue with an elementary example. It illustrates that adjunctions involve canonical translations back and forth, which can be understood as minimal (or, more technically, free) constructions for left adjoints, and as maximal (or cofree) for right adjoints. Earlier, Exercise 1.4.4 described the set of finite sequences $A^\star$ as a *free* monoid on a set A. Below in Exercise 2.5.1 this will be rephrased in terms of an adjunction; it gives a similar minimality phenomenon.

Example 2.5.2 (From sets to preorders) Recall from Example 1.4.2.2 the definition of the category **PreOrd** with preorders $(X, \leq)$ as objects – where $\leq$ is a reflexive and transitive order on X – and monotone functions between them as morphisms. There is then an obvious forgetful functor $U\colon \mathbf{PreOrd} \to \mathbf{Sets}$, given by $(X, \leq) \mapsto X$. It maps a preorder to its underlying set and forgets about the order. This example shows that the forgetful functor U has both a left and a right adjoint.

The left adjoint $D\colon \mathbf{Sets} \to \mathbf{PreOrd}$ sends an arbitrary set A to the 'discrete' preorder $D(A) = (A, \mathrm{Eq}(A))$ obtained by equipping A with the equality relation $\mathrm{Eq}(A) = \{(a, a) \mid a \in A\}$. A key observation is that for a preorder $(X, \leq)$ any function $f\colon A \to X$ is automatically a monotone function $(A, \mathrm{Eq}(A)) \to (X, \leq)$. This means that there is a trivial (identity) bijective correspondence:

$$\frac{D(A) = (A, \mathrm{Eq}(A)) \xrightarrow{\;f\;} (X, \leq) \qquad \text{in } \mathbf{PreOrd}}{A \xrightarrow[\;f\;]{} X = U(X, \leq) \qquad \text{in } \mathbf{Sets}.}$$

This yields an adjunction $D \dashv U$.

The forgetful functor $U\colon \mathbf{PreOrd} \to \mathbf{Sets}$ has not only a left adjoint, via the discrete, minimal order $=$, but also a right adjoint, involving the indiscrete, maximal order: there is a functor $I\colon \mathbf{Sets} \to \mathbf{PreOrd}$ that equips a set A with the indiscrete preorder $I(A) = (A, \top_{A\times A})$ with 'truth' relation $\top_{A\times A} = \{(a, a') \mid a, a' \in A\} = (A \times A \subseteq A \times A)$, in which all elements of A are related. Then, for a preorder $(X, \leq)$, any function $X \to A$ is automatically a monotone function $(X, \leq) \to (A, \top_{A\times A})$. Hence we again have trivial correspondences:

$$\frac{(X, \leq) \xrightarrow{\;f\;} (A, \top_{A\times A}) = I(A) \qquad \text{in } \mathbf{Sets}}{U(X, \leq) = X \xrightarrow[\;f\;]{} A \qquad \text{in } \mathbf{PreOrd}}$$

yielding an adjunction $U \dashv I$.

This situation 'discrete $\dashv$ forgetful $\dashv$ indiscrete' is typical; see Exercise 2.5.6 below.

Next we shall consider examples of adjunctions in the context of coalgebras. The first result describes 'cofree' coalgebras: it gives a canonical construction of a coalgebra from an arbitrary set. Its proof relies on Theorem 2.3.9, which is as yet unproven. The proof shows that actually checking that one has an adjunction can be quite a bit of work. Indeed, the notion of adjunction combines much information.

Proposition 2.5.3 *For a finite Kripke polynomial functor $F\colon \mathbf{Sets} \to \mathbf{Sets}$, the forgetful functor $U\colon \mathbf{CoAlg}(F) \to \mathbf{Sets}$ has a right adjoint. It sends a set A to the carrier of the final coalgebra of the functor $A \times F(-)$.*

Proof Given F, consider the functor $F'\colon \mathbf{Sets} \to \mathbf{Sets}$ given by $A \mapsto A \times F(-)$. Then F' is also a finite Kripke polynomial functor. Hence, by Theorem 2.3.9 it has a final coalgebra $\zeta^A = \langle \zeta_1^A, \zeta_2^A \rangle\colon \widehat{A} \xrightarrow{\cong} A \times F(\widehat{A})$, which we use to define an F-coalgebra:

$$G(A) \stackrel{\text{def}}{=} \big(\widehat{A} \xrightarrow{\zeta_2^A} F(\widehat{A})\big).$$

We first show that G extends to a functor $\mathbf{Sets} \to \mathbf{CoAlg}(F)$. Let $f\colon A \to B$ therefore be an arbitrary function. We define G on f as a function $G(f)\colon \widehat{A} \to \widehat{B}$ by finality, in

$$\begin{array}{ccc}
B \times F(\widehat{A}) & \xrightarrow{\mathrm{id}_B \times F(G(f))} & B \times F(\widehat{B}) \\
{\scriptstyle f \times \mathrm{id}_{F(\widehat{A})}} \uparrow & & \\
A \times F(\widehat{A}) & & \uparrow {\scriptstyle \cong\ \zeta^B} \\
{\scriptstyle \zeta^A} \uparrow {\scriptstyle \cong} & & \\
\widehat{A} & \xrightarrow[G(f)]{} & \widehat{B}
\end{array}$$

Clearly,

$$\begin{aligned}
\zeta_2^B \circ G(f) &= \pi_2 \circ \zeta^B \circ G(f) \\
&= \pi_2 \circ (\mathrm{id}_B \times F(G(f))) \circ (f \times \mathrm{id}_{F(\widehat{A})}) \circ \zeta^A \\
&= F(G(f)) \circ \pi_2 \circ \zeta^A \\
&= F(G(f)) \circ \zeta_2^A.
\end{aligned}$$

Hence $G(f)$ is a homomorphism of coalgebras $G(A) \to G(B)$, as required. By uniqueness one shows that G preserves identities and compositions. This requires a bit of work but is not really difficult.

The adjunction $U \dashv G$ that we want requires a (natural) bijective correspondence:

$$U\begin{pmatrix} F(X) \\ c\uparrow \\ X \end{pmatrix} = X \xrightarrow{\quad g \quad} A \qquad \text{in } \textbf{Sets}$$
$$\overline{\overline{\qquad\qquad\qquad\qquad}}$$
$$\begin{pmatrix} F(X) \\ c\uparrow \\ X \end{pmatrix} \xrightarrow[h]{} \begin{pmatrix} F(\widehat{A}) \\ \uparrow\zeta_2^A \\ \widehat{A} \end{pmatrix} = G(A) \qquad \text{in } \textbf{CoAlg}(F).$$

That is,

$$\frac{X \xrightarrow{\quad g \quad} A}{\begin{array}{ccc} F(X) & \xrightarrow{\quad F(h) \quad} & F(\widehat{A}) \\ c\uparrow & & \uparrow\zeta_2^A \\ X & \xrightarrow[\quad h \quad]{} & \widehat{A} \end{array}}.$$

It is obtained as follows.

- Given a function $g\colon X \to A$, we can define $\overline{g}\colon X \to \widehat{A}$ by finality:

$$\begin{array}{ccc} A \times F(X) & \xrightarrow{\ \mathrm{id}_A \times F(\overline{g})\ } & A \times F(\widehat{A}) \\ \langle g, c\rangle \Big\uparrow & & \cong \Big\uparrow \zeta^A = \langle \zeta_1^A, \zeta_2^A \rangle \\ X & \xrightarrow[\quad \overline{g} \quad]{} & \widehat{A} \end{array}$$

  Then $\overline{g}$ is a homomorphism of coalgebras $c \to G(A)$ since

$$\zeta_2^A \circ \overline{g} = \pi_2 \circ (\mathrm{id}_A \times F(\overline{g})) \circ \langle g, c\rangle = F(\overline{g}) \circ \pi_2 \circ \langle g, c\rangle = F(\overline{g}) \circ c.$$

- Conversely, given a homomorphism $h\colon c \to G(A)$, take $\overline{h} = \zeta_1^A \circ h\colon X \to A$.

In order to complete the proof we still have to show bijectivity (i.e. $\overline{\overline{g}} = g$ and $\overline{\overline{h}} = h$) and naturality. The latter is left to the interested reader, but

$$\overline{\overline{g}} = \zeta_1^A \circ \overline{g} = \pi_1 \circ \zeta^A \circ \overline{g} = \pi_1 \circ (\mathrm{id}_A \times F(\overline{g})) \circ \langle g, c\rangle = \pi_1 \circ \langle g, c\rangle = g.$$

The second equation $\overline{\overline{h}} = h$ follows by uniqueness: $\overline{\overline{h}}$ is by construction the unique homomorphism k with $\zeta^A \circ k = (\mathrm{id}_A \times F(k)) \circ \langle \overline{h}, c\rangle$, i.e. with $\zeta^A \circ k = (\mathrm{id}_A \times F(k)) \circ \langle \zeta_1^A \circ h, c\rangle$. Since h is by assumption a homomorphism of coalgebras $c \to G(A)$, it also satisfies this condition. ❑

Dual to this result it makes sense to consider for algebras the *left* adjoint to a forgetful functor $\textbf{Alg}(F) \to \textbf{Sets}$. This gives so-called **free algebras**. They can be understood as term algebras built on top of a given collection of variables.

Since an adjunction $F \dashv G$ involves two translations $X \mapsto F(X)$ and $Y \mapsto G(Y)$ in opposite directions, one can translate objects 'back and forth', namely as $X \mapsto GF(X)$ and $Y \mapsto FG(Y)$. There are then canonical 'comparison' maps $\eta\colon X \to GF(X)$, called **unit**, and $\varepsilon\colon FG(Y) \to Y$, called **counit**.

In the context of the adjunction from the previous theorem, the unit η is a homomorphism from a coalgebra $X \to F(X)$ to the cofree coalgebra $\widehat{X} \to F(\widehat{X})$ on its carrier. It is obtained by finality. And the counit ε maps the carrier $\widehat{A}$ of a cofree coalgebra back to the original set A. It is ζ_1^A in the notation of the proof.

Using the notation $\psi\colon \mathbb{D}(F(X), Y) \xrightarrow{\cong} \mathbb{C}(X, G(Y))$ from Definition 2.5.1, the unit $\eta_X\colon X \to GF(X)$ and $\varepsilon_Y\colon FG(Y) \to Y$ maps can be defined as

$$\eta_X = \psi(\mathrm{id}_{F(X)}), \qquad \varepsilon_Y = \psi^{-1}(\mathrm{id}_{G(Y)}). \tag{2.30}$$

Using the naturality of ψ one easily checks that for arbitrary $f\colon X \to X'$ in $\mathbb{C}$ and $g\colon Y \to Y'$ in $\mathbb{D}$, the following diagrams commute:

$$\begin{array}{ccc} X & \xrightarrow{\;\eta_X\;} & GF(X) \\ {\scriptstyle f}\downarrow & & \downarrow{\scriptstyle GF(f)} \\ X' & \xrightarrow[\;\eta_{X'}\;]{} & GF(X') \end{array} \qquad\qquad \begin{array}{ccc} FG(Y) & \xrightarrow{\;\varepsilon_Y\;} & Y \\ {\scriptstyle FG(g)}\downarrow & & \downarrow{\scriptstyle g} \\ FG(Y') & \xrightarrow[\;\varepsilon_{Y'}\;]{} & Y' \end{array}$$

This leads to the following fundamental notion in category theory: a natural transformation. It is a mapping between functors.

Definition 2.5.4 Let $\mathbb{C}$, $\mathbb{D}$ be two categories, with two (parallel) functors $H, K\colon \mathbb{C} \to \mathbb{D}$ between them. A **natural transformation** α from H to K consists of a collection of morphisms $\alpha_X\colon H(X) \to K(X)$ in $\mathbb{D}$, indexed by objects $X \in \mathbb{C}$ satisfying the naturality requirement: for each morphism $f\colon X \to Y$ in $\mathbb{C}$, the following rectangle commutes:

$$\begin{array}{ccc} H(X) & \xrightarrow{\;\alpha_X\;} & K(X) \\ {\scriptstyle H(f)}\downarrow & & \downarrow{\scriptstyle K(f)} \\ H(Y) & \xrightarrow[\;\alpha_Y\;]{} & K(Y) \end{array}$$

In this case one often writes $\alpha\colon H \Rightarrow K$, with double arrow.

The next basic result illustrates the relevance of natural transformations in the theory of algebras and coalgebras.

Proposition 2.5.5 *Consider a natural transformation $\alpha\colon H \Rightarrow K$ as defined above, for two endofunctors $H, K\colon \mathbb{C} \to \mathbb{C}$. This α induces translation* *functors between the corresponding categories of algebras and coalgebras, in commuting triangles:*

$$\begin{array}{ccc} \mathbf{Alg}(K) & \longrightarrow & \mathbf{Alg}(H) \\ & \searrow \quad \swarrow & \\ & \mathbb{C} & \end{array} \qquad\qquad \begin{array}{ccc} \mathbf{CoAlg}(H) & \longrightarrow & \mathbf{CoAlg}(K) \\ & \searrow \quad \swarrow & \\ & \mathbb{C} & \end{array}$$

These (horizontal) functors are given on objects by pre- and post-composition:

$$\begin{pmatrix} K(X) \\ \downarrow b \\ X \end{pmatrix} \longmapsto \begin{pmatrix} H(X) \\ \downarrow b \circ \alpha_X \\ X \end{pmatrix} \qquad\qquad \begin{pmatrix} H(X) \\ \uparrow c \\ X \end{pmatrix} \longmapsto \begin{pmatrix} H(X) \\ \uparrow \alpha_X \circ c \\ X \end{pmatrix}$$

On morphisms these functors are simply $f \mapsto f$.

Proof We shall write $F\colon \mathbf{Alg}(K) \to \mathbf{Alg}(H)$ for the mapping defined above. It is a well-defined functor, since if $f\colon X \to X'$ is a map of K-algebras $b \to b'$, then f is also a map of H-algebras $F(b) \to F(b')$, since

$$\begin{aligned} F(b') \circ H(f) &= b' \circ \alpha_X \circ H(f) \\ &= b' \circ K(f) \circ \alpha_X && \text{by naturality} \\ &= f \circ b \circ \alpha_X && \text{since } f \text{ is a map of } K\text{-algebras} \\ &= f \circ F(b). \end{aligned}$$

□

For each set X there is an obvious map $\rho_X\colon X^\star \to \mathcal{P}_{\text{fin}}(X)$ mapping a sequence $\langle x_1, \ldots, x_n\rangle$ to the set $\{x_1, \ldots, x_n\}$ of elements involved – which may have size less than n in case duplicate elements occur. This operation is 'natural', also in a technical sense: for a function $f\colon X \to Y$ one has $\mathcal{P}_{\text{fin}}(f) \circ \rho_X = \rho_Y \circ f^\star$, since

$$\begin{aligned} (\rho_Y \circ f^\star)(\langle x_1, \ldots, x_n\rangle) &= \rho_Y(\langle f(x_1), \ldots, f(x_n)\rangle) \\ &= \{f(x_1), \ldots, f(x_n)\} \\ &= \mathcal{P}_{\text{fin}}(f)(\{x_1, \ldots, x_n\}) \\ &= (\mathcal{P}_{\text{fin}}(f) \circ \rho_X)(\langle x_1, \ldots, x_n\rangle). \end{aligned}$$

In the reverse direction $\mathcal{P}_{\text{fin}}(X) \to X^\star$ one can always choose a way to turn a finite set into a list, but there is no natural way to do this. More examples and non-examples are described at the end of Section 4.1. The idea is that natural transformations describe uniform mappings, such as given for instance by terms; see Exercise 2.5.17.

With this definition and notation we can write the unit and counit of an adjunction, from (2.30), as natural transformations $\eta\colon \mathrm{id}_{\mathbb{C}} \Rightarrow GF$ and $\varepsilon\colon FG \Rightarrow \mathrm{id}_{\mathbb{D}}$. The closely related notion of **equivalence** of categories is sometimes useful. It is an adjunction in which both the unit and counit are isomorphisms. This is a weaker notion than isomorphism of categories.

We shall consider another example of an adjunction which is typical in a coalgebraic setting, namely a so-called *behaviour-realisation* adjunction. Such adjunctions were first recognised in the categorical analysis of mathematical system theory; see [160–162]. Here we give a simple version using the deterministic automata introduced in Section 2.2. This requires two extensions of what we have already seen: (1) homomorphisms between these automata which allow variation in the input and output sets and (2) automata with initial states.

Definition 2.5.6 We write **DA** for a category of deterministic automata. It has the following:

objects deterministic automata $\langle \delta, \epsilon \rangle \colon X \to X^A \times B$ with an initial state $x_0 \in X$

morphisms from $\langle X \xrightarrow{\langle \delta, \epsilon \rangle} X^A \times B, x_0 \in X \rangle$ to $\langle Y \xrightarrow{\langle \delta', \epsilon' \rangle} Y^C \times D, y_0 \in Y \rangle$ consist of a triple of functions:

$$A \xleftarrow{\;f\;} C, \qquad B \xrightarrow{\;g\;} D, \qquad X \xrightarrow{\;h\;} Y,$$

with for all $x \in X$ and $c \in C$,

$$\begin{aligned} \delta'(h(x))(c) &= h(\delta(x)(f(c))) \\ \epsilon'(h(x)) &= g(\epsilon(x)) \\ h(x_0) &= y_0. \end{aligned}$$

The identity morphisms in **DA** are of the form (id, id, id). The composition of (f_1, g_1, h_1) followed by (f_2, g_2, h_2) is $(f_1 \circ f_2, g_2 \circ g_1, h_2 \circ h_1)$. Note the reversed order in the first component.

The first two equations express that h is a coalgebra homomorphism from the automaton $\langle \mathrm{id}_X^f \circ \delta, g \circ \epsilon \rangle \colon X \to X^C \times D$, translated via (f, g), to the automaton $\langle \delta', \epsilon' \rangle \colon X \to X^C \times D$. Here we use the exponent notation id_X^f for functions from (2.12). The third equation simply says that the initial state is preserved.

We also introduce a category of deterministic behaviours. Its objects are elements of the final coalgebras for deterministic automata; see Proposition 2.3.5.

Definition 2.5.7 Form the category **DB** with

objects functions of the form $\varphi \colon A^\star \to B$

morphisms from $A^\star \xrightarrow{\varphi} B$ to $C^\star \xrightarrow{\psi} D$ are pairs of functions

$$A \xleftarrow{\;f\;} C, \qquad B \xrightarrow{\;g\;} D$$

with

$$g \circ \varphi \circ f^\star = \psi.$$

That is, for all $\langle c_1, \ldots, c_n\rangle \in C^\star$,

$$g(\varphi(\langle f(c_1), \ldots, f(c_n)\rangle)) = \psi(\langle c_1, \ldots, c_n\rangle).$$

The maps $(\mathrm{id}, \mathrm{id})$ are identities in **DB**. And composition in **DB** is given by $(f_2, g_2) \circ (f_1, g_1) = (f_1 \circ f_2, g_2 \circ g_1)$.

The behaviour-realisation adjunction for deterministic automata yields a canonical way to translate back-and-forth between deterministic automata and behaviours. It looks as follows.

Proposition 2.5.8 *There are a behaviour functor $\mathcal{B}$ and a realisation functor $\mathcal{R}$ in an adjunction:*

$$\mathbf{DA} \underset{\mathcal{R}}{\overset{\mathcal{B}}{\rightleftarrows}} \mathbf{DB} \qquad \mathcal{B} \dashv \mathcal{R}$$

Proof We sketch the main lines and leave many details to the interested reader. The behaviour functor $\mathcal{B}\colon \mathbf{DA} \to \mathbf{DB}$ maps an automaton $\langle \delta, \epsilon\rangle\colon X \to X^A \times B$ with initial state $x_0 \in X$ to the behaviour $\mathrm{beh}_{\langle\delta,\epsilon\rangle}(x_0) \in B^{A^\star}$ of this initial state; see Proposition 2.3.5. On morphisms, it is $\mathcal{B}(f, g, h) = (f, g)$. This is well defined because $g \circ \mathrm{beh}(x) \circ f^\star = \mathrm{beh}'(h(x))$, as is checked easily by induction.

In the other direction, the realisation functor $\mathcal{R}\colon \mathbf{DB} \to \mathbf{DA}$ sends a behaviour $\psi\colon C^\star \to D$ to the final deterministic automaton $\zeta\colon D^{C^\star} \xrightarrow{\cong} (D^{C^\star})^C \times D$ described in Proposition 2.3.5, with ψ as initial state. The associated behaviour function beh_ζ is then given by $\mathrm{beh}_\zeta(\varphi) = \varphi$. On morphisms, $\mathcal{R}$ is defined as $\mathcal{R}(f, g) = (f, g, g^{f^\star})$.

The adjunction $\mathcal{B} \dashv \mathcal{R}$ is established by the bijective correspondence:

$$\frac{\mathcal{B}\Big(X \xrightarrow{\langle\delta,\epsilon\rangle} X^A \times B, x_0 \in X\Big) \xrightarrow{\quad (f,g) \quad} \Big(C^\star \xrightarrow{\;\psi\;} D\Big)}{\Big(X \xrightarrow{\langle\delta,\epsilon\rangle} X^A \times B, x_0 \in X\Big) \xrightarrow[\quad (f,g,h) \quad]{} \mathcal{R}\Big(C^\star \xrightarrow{\;\psi\;} D\Big)}.$$

This correspondence exists because the function h below the double line is uniquely determined by finality in the following diagram:

$$\begin{array}{ccc}
X^C \times D & \xrightarrow{\quad h^{\mathrm{id}_C} \times \mathrm{id}_D \quad} & (D^{C^\star})^C \times D \\
{\scriptstyle \mathrm{id}_X^f \times g}\uparrow & & \\
X^A \times B & & \uparrow{\scriptstyle \zeta}\ \cong \\
{\scriptstyle \langle \delta, \epsilon \rangle}\uparrow & & \\
X & \xrightarrow[\quad h \quad]{} & D^{C^\star}
\end{array}$$

It satisfies $h(x_0) = \psi$ if and only if $g \circ \mathrm{beh}_{\langle \delta,\epsilon \rangle}(x_0) \circ f^\star = \psi$. □

Several alternative versions of this behaviour-realisation adjunction for deterministic automata are described in the exercises below. Such adjunctions have also been studied in more abstract settings; see for example [48] and [289].

One interesting aspect of the behaviour-realisation adjunction is that it provides a setting in which to (categorically) study various process operators. For instance, one can consider several ways to put deterministic automata in parallel. For automata $M_i = \langle X_i \xrightarrow{\langle \delta_i, \epsilon_i \rangle} X_i^{A_i} \times B_i, x_i \in X \rangle$, one can define new automata:

$$\begin{aligned}
M_1 \mid M_2 &= \langle X_1 \times X_2 \xrightarrow{\langle \delta_|, \epsilon_| \rangle} (X_1 \times X_2)^{A_1 + A_2} \times (B_1 \times B_2), (x_0, x_1) \rangle \\
&\qquad \text{where } \delta_|(y_1, y_2)(\kappa_1(a)) = (\delta_1(y_1)(a), y_2) \\
&\qquad \phantom{\text{where }} \delta_|(y_1, y_2)(\kappa_2(a)) = (y_1, \delta_2(y_2)(a)) \\
&\qquad \phantom{\text{where }} \epsilon_|(y_1, y_2) = (\epsilon_1(y_1), \epsilon_2(y_2)). \\
M_1 \otimes M_2 &= \langle X_1 \times X_2 \xrightarrow{\langle \delta_\otimes, \epsilon_\otimes \rangle} (X_1 \times X_2)^{A_1 \times A_2} \times (B_1 \times B_2), (x_0, x_1) \rangle \\
&\qquad \text{where } \delta_\otimes(y_1, y_2)(a_1, a_2) = (\delta_1(y_1)(a), \delta_2(y_2)(a)) \\
&\qquad \phantom{\text{where }} \epsilon_\otimes(y_1, y_2) = (\epsilon_1(y_1), \epsilon_2(y_2)).
\end{aligned}$$

The first composition of automata involves transitions in each component automaton separately, whereas the second composition combines transitions.

Both these definitions yield a so-called symmetric monoidal structure on the category **DA** of deterministic automata. Similar structure can be defined on the associated category **DB** of behaviours, in such a way that the behaviour functor $\mathcal{B}\colon \mathbf{DA} \to \mathbf{DB}$ preserves this structure. Thus, complex behaviour can be obtained from more elementary building blocks.

There is a line of research studying such 'process categories' with operations that are commonly used in process calculi as appropriate categorical structure; see for instance [216, 290, 213, 417, 314, 467, 53, 207, 202, 204].

This section concludes with a result about lifting adjunctions to categories of algebras; see [218, theorem 2.14]. It is purely categorical abstract nonsense and does not talk about concrete categories or functors. In dual form it yields a lifting to categories of coalgebras.

Theorem 2.5.9 *Consider a natural transformation $\alpha\colon SF \Rightarrow FT$ in a situation:*

$$\begin{array}{ccc} \mathbb{A} & \xrightarrow{S} & \mathbb{A} \\ {\scriptstyle F}\uparrow & \overset{\alpha}{\Searrow} & \uparrow{\scriptstyle F} \\ \mathbb{B} & \xrightarrow[T]{} & \mathbb{B} \end{array} \tag{2.31}$$

It induces a lifting of F to a functor $\mathbf{Alg}(F)$ *in*

$$\begin{array}{ccc} \mathbf{Alg}(T) & \xrightarrow{\quad\mathbf{Alg}(F)\quad} & \mathbf{Alg}(S) \\ (T(X) \xrightarrow{a} X) & \longmapsto & (SF(X) \xrightarrow{\alpha_X} FT(X) \xrightarrow{F(a)} F(X)). \end{array}$$

If α is an isomorphism, then a right adjoint G to F induces a right adjoint $\mathbf{Alg}(G)$ *to* $\mathbf{Alg}(F)$ *in*

$$\begin{array}{ccc} \mathbb{A} & \xleftarrow{\quad U \quad} & \mathbf{Alg}(S) \\ F\big\uparrow \dashv \big\downarrow G & & \mathbf{Alg}(F)\big\uparrow \dashv \big\downarrow \mathbf{Alg}(G) \\ \mathbb{B} & \xleftarrow[\quad U \quad]{} & \mathbf{Alg}(T) \end{array}$$

where the Us are forgetful functors. The functor $\mathbf{Alg}(G)$ *arises from $\beta\colon TG \Rightarrow GS$, the adjoint transpose of $FTG \cong SFG \overset{S\varepsilon}{\Longrightarrow} S$.*

Proof Of course the functor $\mathbf{Alg}(F)$ is defined on morphisms as $f \mapsto F(f)$. We check that this is well defined: for a homomorphism of T-algebras $(T(X) \xrightarrow{a} X) \xrightarrow{f} (T(Y) \xrightarrow{b} Y)$ we obtain that $F(f)$ is a homomorphism between the corresponding S-algebras in

$$\begin{array}{ccc} SF(X) & \xrightarrow{SF(f)} & SF(Y) \\ {\scriptstyle \alpha_X}\big\downarrow & & \big\downarrow{\scriptstyle \alpha_Y} \\ FT(X) & \xrightarrow{FT(f)} & FT(Y) \\ {\scriptstyle F(a)}\big\downarrow & & \big\downarrow{\scriptstyle F(b)} \\ F(X) & \xrightarrow[F(f)]{} & F(Y) \end{array}$$

The upper square commutes by naturality of α, and the lower square because f is a homomorphism of T-algebras.

Next we assume that α is an isomorphism, and write β_Y, where $Y \in \mathbb{A}$, for the following map in $\mathbb{B}$:

$$\beta_Y = \Big(TG(Y) \xrightarrow{\eta_{TG(Y)}} GFTG(Y) \xrightarrow{G(\alpha^{-1}_{G(Y)})} GSFG(Y) \xrightarrow{GS(\varepsilon_Y)} GS(Y)\Big).$$

Following the construction of the functor $\mathbf{Alg}(F)$, this natural transformation β induces a functor $\mathbf{Alg}(G)\colon \mathbf{Alg}(S) \to \mathbf{Alg}(S)$ by

$$(S(Y) \xrightarrow{b} Y) \longmapsto (TG(Y) \xrightarrow{\beta_Y} GS(Y) \xrightarrow{G(b)} G(Y)).$$

Now we have to establish the bijective correspondence for $\mathbf{Alg}(F) \dashv \mathbf{Alg}(G)$. It takes the form

$$\frac{\mathbf{Alg}(F)\begin{pmatrix} T(X) \\ a\downarrow \\ X \end{pmatrix} = \begin{pmatrix} SF(X) \\ \downarrow F(a)\circ\alpha \\ F(X) \end{pmatrix} \xrightarrow{\quad f \quad} \begin{pmatrix} S(Y) \\ \downarrow b \\ Y \end{pmatrix}}{\begin{pmatrix} T(X) \\ a\downarrow \\ X \end{pmatrix} \xrightarrow[\quad g \quad]{} \begin{pmatrix} TGY \\ \downarrow G(b)\circ\beta \\ GY \end{pmatrix} = \mathbf{Alg}(G)\begin{pmatrix} S(Y) \\ \downarrow b \\ Y \end{pmatrix}}.$$

This correspondence is the one of the adjunction $F \dashv G$. We need only to check that the transposes are again homomorphisms of algebras. We shall do so for the map f above the lines, and leave the other case (for g) to the interested reader.

Assume therefore that f is a homomorphism of S-algebras as indicated above the lines. This means that $b \circ Sf = f \circ F(a) \circ \alpha_X$. The transpose $\overline{f} = G(f) \circ \eta_X\colon X \to G(Y)$ is then a homomorphism of T-algebras:

$$\begin{aligned}
&(G(b) \circ \beta_Y) \circ T(\overline{f}) \\
&= G(b) \circ GS(\varepsilon_Y) \circ G(\alpha^{-1}_{G(Y)}) \circ \eta_{TG(Y)} \circ TG(f) \circ T(\eta_X) \\
&= G(b) \circ GS(\varepsilon_Y) \circ G(\alpha^{-1}_{G(Y)}) \circ GFTG(f) \circ GFT(\eta_X) \circ \eta_{T(X)} \\
&\qquad \text{by naturality of } \eta \\
&= G(b) \circ GS(\varepsilon_Y) \circ GSFG(f) \circ GSF(\eta_X) \circ G(\alpha^{-1}_X) \circ \eta_{T(X)} \\
&\qquad \text{by naturality of } \alpha \text{ (and thus also of } \alpha^{-1}) \\
&= G(b) \circ GS(f) \circ GS(\varepsilon_{F(X)}) \circ GSF(\eta_X) \circ G(\alpha^{-1}_X) \circ \eta_{T(X)} \\
&\qquad \text{by naturality of } \varepsilon \\
&= G(b) \circ GS(f) \circ G(\alpha^{-1}_X) \circ \eta_{T(X)} \\
&\qquad \text{by one of the triangular identities; see Exercise 2.5.7} \\
&= G(f) \circ GF(a) \circ G(\alpha_X) \circ G(\alpha^{-1}_X) \circ \eta_{T(X)} \\
&\qquad \text{by assumption about } f \\
&= G(f) \circ \eta_X \circ a \\
&\qquad \text{by naturality of } \eta \\
&= \overline{f} \circ a.
\end{aligned}$$

□

Those readers in search of even more abstraction may want to check that the above pair of functors (S, T) with the natural transformations (α, β) forms a 'map of adjunctions' $(F \dashv G) \to (F \dashv G)$; see [344, IV, 7] for the definition of this notion. The construction $\mathbf{Alg}(F) \dashv \mathbf{Alg}(G)$ may then be understood as algebra of this endomorphism (of adjunctions), in a suitably abstract sense.

This section concludes with a relatively long series of exercises, mostly because adjunctions offer a perspective that leads to many more results. In a particular situation they can concisely capture the essentials of back-and-forth translations.

Exercises

2.5.1 Recall from Exercise 1.4.4 that the assignment $A \mapsto A^\star$ gives a functor $(-)^\star : \mathbf{Sets} \to \mathbf{Mon}$ from sets to monoids. Show that this functor is left adjoint to the forgetful functor $\mathbf{Mon} \to \mathbf{Sets}$ which maps a monoid $(M, +, 0)$ to its underlying set M and forgets the monoid structure.

2.5.2 Check that the bijective correspondences

$$\frac{\frac{\frac{X \longrightarrow \mathcal{P}(Y)}{\bullet \subseteq X \times Y}}{\bullet \subseteq Y \times X}}{Y \longrightarrow \mathcal{P}(X)}$$

induced by the correspondence (2.16) give rise to an adjunction of the form $\mathbf{Sets}^{\mathrm{op}} \leftrightarrows \mathbf{Sets}$.

2.5.3 The intention of this exercise is to show that the mapping $\# \mapsto F_\#$ from arities to functors is functorial. This requires some new notation and terminology. Let Endo(**Sets**) be the category with endofunctors $F : \mathbf{Sets} \to \mathbf{Sets}$ as objects and natural transformations between them as morphisms. One writes $\mathbf{Sets}/\mathbb{N}$ for the slice category over $\mathbb{N}$; see Exercise 1.4.3.

Prove that mapping $\# \mapsto F_\#$ from (2.18) yields a functor $\mathbf{Sets}/\mathbb{N} \to$ Endo(**Sets**). (This functor $\mathbf{Sets}/\mathbb{N} \to$ Endo(**Sets**) restricts to a 'full and faithful' functor via a suitable restriction of the category Endo(**Sets**); see [4, theorem 3.4]. This means that morphisms $F_{\#_1} \to F_{\#_2}$ in this restricted category are in one-to-one correspondence with morphisms of arities $\#_1 \to \#_2$.)

2.5.4 This exercise describes 'strength' for endofunctors on **Sets**. In general, this is a useful notion in the theory of datatypes [104, 105] and of computations [357]; see Section 5.2 for a systematic description.

Let $F\colon \mathbf{Sets} \to \mathbf{Sets}$ be an arbitrary functor. Consider for sets X, Y the strength map

$$\begin{array}{rcl} F(X) \times Y & \xrightarrow{\;\mathrm{st}_{X,Y}\;} & F(X \times Y) \\ (u, y) & \longmapsto & F(\lambda x \in X.\, (x, y))(u). \end{array} \tag{2.32}$$

1. Prove that this yields a natural transformation $F(-) \times (-) \stackrel{\mathrm{st}}{\Longrightarrow} F((-) \times (-))$, where both the domain and codomain are functors $\mathbf{Sets} \times \mathbf{Sets} \to \mathbf{Sets}$.
2. Describe this strength map for the list functor $(-)^\star$ and for the powerset functor $\mathcal{P}$.

2.5.5 The following strengthening of induction is sometimes called 'induction with parameters'. It is different from recursion, which also involves an additional parameter; see Proposition 2.4.7..

Assume a functor F with a strength natural transformation as in the previous exercise and with initial algebra $\alpha\colon F(A) \xrightarrow{\cong} A$. Let P be a set (or object) for parameters. Prove that for each map $h\colon F(X) \times P \to X$ there is a unique $f\colon A \times P \to X$ making the following diagram commute:

$$\begin{array}{ccc} F(A) \times P & \xrightarrow{\;\langle F(f) \circ \mathrm{st}, \pi_2 \rangle\;} & F(X) \times P \\ {\scriptstyle \alpha \times \mathrm{id}}\downarrow{\scriptstyle \cong} & & \downarrow{\scriptstyle h} \\ A \times P & \xrightarrow[\;f\;]{} & X \end{array}$$

Hint: First turn h into a suitable algebra $h'\colon F(X^P) \to X^P$ via strength.

Use this mechanism to define the append map $\mathsf{app}\colon A^\star \times A^\star \to A^\star$ from Example 2.4.6.

2.5.6 Show that the forgetful functor $U\colon \mathbf{Sp} \to \mathbf{Sets}$ from topological spaces to sets has both a left adjoint (via the discrete topology on a set, in which every subset is open) and a right adjoint (via the indiscrete topology, with only the empty set and the whole set itself as open sets).

2.5.7 Assume two functors $F\colon \mathbb{C} \to \mathbb{D}$ and $G\colon \mathbb{D} \to \mathbb{C}$ in opposite directions, with natural transformations $\eta\colon \mathrm{id}_{\mathbb{C}} \Rightarrow GF$ and $\varepsilon\colon FG \Rightarrow$

$\mathrm{id}_{\mathbb{D}}$. Define functions $\psi\colon \mathbb{D}(F(X), Y) \longrightarrow \mathbb{C}(X, G(Y)$ by $\psi(f) = G(f) \circ \eta_X$.

1. Check that such ψs are natural.
2. Prove that these ψs are isomorphisms if and only if the following **triangular identities** hold:

$$G(\varepsilon) \circ \eta G = \mathrm{id}, \qquad \varepsilon F \circ F(\eta) = \mathrm{id}.$$

2.5.8 A morphism $m\colon X' \to X$ in a category $\mathbb{D}$ is called a **monomorphism** (or **mono**, for short), written as $m\colon X' \rightarrowtail X$, if for each parallel pair of arrows $f, g\colon Y \to X'$, $m \circ f = m \circ g$ implies $f = g$.

1. Prove that the monomorphisms in **Sets** are precisely the injective functions.
2. Let $G\colon \mathbb{D} \to \mathbb{C}$ be a right adjoint. Show that if m is a monomorphism in $\mathbb{D}$, then so is $G(m)$ in $\mathbb{C}$.

Dually, an **epimorphism** (or **epi**, for short) in $\mathbb{C}$ is an arrow written as $e\colon X' \twoheadrightarrow X$ such that for all maps $f, g\colon X \to Y$, if $e \circ f = e \circ g$, then $f = g$.

3. Show that the epimorphism in **Sets** are the surjective functions. *Hint*: For an epi $X \twoheadrightarrow Y$, choose two appropriate maps $Y \to 1+Y$.
4. Prove that left adjoints preserve epimorphism.

2.5.9 Notice that the existence of final coalgebras for finite polynomial functors (Theorem 2.3.9) that is used in the proof of Proposition 2.5.3 is actually a special case of this proposition. *Hint*: Consider the right adjoint at the final set 1.

2.5.10 Assume an endofunctor $F\colon \mathbb{C} \to \mathbb{C}$. Prove that there are natural transformations

$$\mathbf{CoAlg}(F) \underset{FU}{\overset{U}{\rightrightarrows}} \mathbb{C} \;(\Downarrow) \qquad \text{and} \qquad \mathbf{Alg}(F) \underset{U}{\overset{FU}{\rightrightarrows}} \mathbb{C} \;(\Downarrow)$$

where U is the forgetful functor. (In 2-categorical terminology these maps form inserters; see e.g. [218, appendix].)

2.5.11 (Hughes) Let $\mathbb{C}$ be an arbitrary category with products $\times$ and let $F, H\colon \mathbb{C} \to \mathbb{C}$ be two endofunctors on $\mathbb{C}$. Assume that cofree F-coalgebras exist, i.e. that the forgetful functor $U\colon \mathbf{CoAlg}(F) \to \mathbb{C}$ has a right adjoint G – as in Proposition 2.5.3. Prove then that there is an isomorphism of categories of coalgebras,

$$\mathbf{CoAlg}(F \times H) \xrightarrow{\cong} \mathbf{CoAlg}(GHU),$$

where $\mathbf{CoAlg}(GHU)$ is a category of coalgebras on coalgebras, for the functor composition $GHU\colon \mathbf{CoAlg}(F) \to \mathbb{C} \to \mathbb{C} \to \mathbf{CoAlg}(F)$.

2.5.12 Consider two adjoint endofunctors as in

$$F \circlearrowright \mathbb{C} \circlearrowleft G \qquad \text{with} \qquad F \dashv G.$$

Prove that we then get an isomorphism of categories

$$\mathbf{Alg}(F) \xrightarrow{\cong} \mathbf{CoAlg}(G)$$

between the associated categories of algebras and coalgebras.
(Remark: As noted in [36, theorem 5.7], when $\mathbb{C} = \mathbf{Sets}$ the only such adjunctions $F \dashv G$ are product-exponent adjunctions $X \times (-) \dashv (-)^X$ as in (2.11). The argument goes as follows. In **Sets**, each object A can be written as coproduct $\coprod_{a\in A} 1$ of singletons. A left adjoint F must preserve such coproducts, so that $F(A) \cong \coprod_{a\in A} F(1) \cong F(1) \times A$. But then $G(-) \cong F(1) \Rightarrow (-)$, by uniqueness of adjoints.)

2.5.13 Theorem 2.5.9 deals with lifting adjunctions to categories of algebras. Check that its 'dual' version for coalgebras is (see [218]) the following:

For functors $T\colon \mathbb{B} \to \mathbb{B}$, $G\colon \mathbb{B} \to \mathbb{A}$, $S\colon \mathbb{A} \to \mathbb{A}$, a natural transformation $\alpha\colon GS \Rightarrow TG$ induces a functor $\mathbf{CoAlg}(G)\colon \mathbf{CoAlg}(S) \to \mathbf{CoAlg}(T)$. Furthermore, if α is an isomorphism, then a left adjoint $F \dashv G$ induces a left adjoint $\mathbf{CoAlg}(F) \dashv \mathbf{CoAlg}(G)$. Does it require a new proof?

2.5.14 A deterministic automaton $\langle\delta, \epsilon\rangle\colon X \to X^A \times B$ is called **observable** if its behaviour function $\mathrm{beh} = \lambda x.\,\lambda\sigma.\,\epsilon(\delta^*(x,\sigma))\colon X \to B^{A^\star}$ from Proposition 2.3.5 is injective. Later, in Corollary 3.4.3 we shall see that this means that bisimilar states are equal.

If this automaton comes equipped with an initial state $x_0 \in X$ one calls the automaton **reachable** if the function $\delta^*(x_0, -)\colon A^\star \to X$ from Section 2.2 is surjective. This means that every state can be reached from the initial state x_0 via a suitable sequence of inputs. The automaton is called **minimal** if it is both observable and reachable.

The realisation construction $\mathcal{R}\colon \mathbf{DB} \to \mathbf{DA}$ from Proposition 2.5.8 clearly yields an observable automaton since the resulting behaviour function is the identity. An alternative construction, the **Nerode realisation**, gives a minimal automaton. It is obtained from a behaviour $\psi\colon C^\star \to D$ as follows. Consider the equivalence relation $\equiv_\psi \subseteq C^\star \times C^\star$ defined by

$$\sigma \equiv_\psi \sigma' \iff \forall \tau \in C^\star.\, \psi(\sigma \cdot \tau) = \psi(\sigma' \cdot \tau).$$

We take the quotient $C^\star/\equiv_\psi$ as state space; it is defined as factorisation:

$$\begin{array}{ccc} C^\star & \xrightarrow{\quad \sigma \longmapsto \lambda\tau.\,\psi(\sigma\cdot\tau) \quad} & D^{C^\star} \\ & \searrow\!\!\!\!\searrow \qquad \nearrow\!\!\!\!\!\!\!\rightarrowtail & \\ & C^\star/\equiv_\psi & \end{array}$$

This quotient carries an automaton structure with transition function $\delta_\psi \colon (C^\star/\equiv_\psi) \to (C^\star/\equiv_\psi)^C$ given by $\delta_\psi([\sigma])(c) = [\sigma \cdot c]$, observation function $\epsilon_\psi \colon (C^\star/\equiv_\psi) \to D$ defined as $\epsilon_\psi([\sigma]) = \psi(\sigma)$, and initial state $[\langle\rangle] \in C^\star/\equiv_\psi$.

1. Check that this Nerode realisation $\mathcal{N}(C^\star \xrightarrow{\psi} D)$ is indeed a minimal automaton, forming a subautomaton/subcoalgebra $C^\star/\equiv_\psi \rightarrowtail D^{C^\star}$ of the final coalgebra.

Write **RDA** for the 'subcategory' of **DA** with reachable automata as objects, and morphisms (f, g, h) as in **DA** but with f a *surjective* function between the input sets. Similarly, let **RDB** be the subcategory of **DB** with the same objects but with morphisms (f, g) where f is surjective.

2. Check that the behaviour functor $\mathcal{B}\colon$ **DA** $\to$ **DB** from Proposition 2.5.8 restricts to a functor $\mathcal{B}\colon$ **RDA** $\to$ **RDB**, and show that it has Nerode realisation $\mathcal{N}$ yields a functor **RDA** $\to$ **RDB** in the opposite direction.
3. Prove that there is an adjunction $\mathcal{B} \dashv \mathcal{N}$.
4. Let **MDA** be the 'subcategory' of **RDA** with minimal automata as objects. Check that the adjunction in the previous point restricts to an equivalence of categories **MDA** $\simeq$ **RDB**. Thus, (states of) minimal automata are in fact (elements of) final coalgebras. (This result comes from [161, 162]; see also [14].)

2.5.15 This exercise and the next one continue the description of linear dynamical systems from Exercise 2.2.12. Here we look at the duality between reachability and observability. First a preliminary result.

1. Let B be an arbitrary vector space. Prove that the final coalgebra in **Vect** of the functor $X \mapsto B \times X$ is the set of infinite sequences $B^{\mathbb{N}}$, with obvious vector space structure, and with coalgebra structure $\langle \mathsf{hd}, \mathsf{tl} \rangle \colon B^{\mathbb{N}} \xrightarrow{\cong} B \times B^{\mathbb{N}}$ given by head and tail.

Call a linear dynamical system $A \xrightarrow{F} X \xrightarrow{G} X \xrightarrow{H} B$ **reachable** if the induced function $\mathrm{int}_{[F,G]} \colon A^{\S} \to X$ in the diagram on the left below is surjective (or equivalently, an epimorphism in **Vect**).

$$
\begin{array}{ccc}
A + A^{\S} & \xrightarrow{\mathrm{id}_A + \mathrm{int}_{[F,G]}} & A + X \\
{\scriptstyle [\mathsf{in},\mathsf{sh}]}\downarrow\cong & & \downarrow{\scriptstyle [F,G]} \\
A^{\S} & \xrightarrow[\mathrm{int}_{[F,G]}]{} & X
\end{array}
\qquad\qquad
\begin{array}{ccc}
B \times X & \xrightarrow{\mathrm{id}_B \times \mathrm{beh}_{\langle H,G\rangle}} & B \times B^{\mathbb{N}} \\
\uparrow{\scriptstyle \langle H,G\rangle} & & \cong\uparrow{\scriptstyle \langle \mathsf{hd},\mathsf{tl}\rangle} \\
X & \xrightarrow[\mathrm{beh}_{\langle H,G\rangle}]{} & B^{\mathbb{N}}
\end{array}
$$

Similarly, call this system **observable** if the map $\mathrm{beh}_{\langle H,G\rangle} \colon X \to B^{\mathbb{N}}$ on the right is injective (equivalently, a monomorphism in **Vect**). And call the system **minimal** if it is both reachable and observable.

2. Prove that $A \xrightarrow{F} X \xrightarrow{G} X \xrightarrow{H} B$ is reachable in **Vect** if and only if $B \xrightarrow{H} X \xrightarrow{G} X \xrightarrow{F} A$ is observable in $\mathbf{Vect}^{\mathrm{op}}$.

(Kalman's duality result [287, chapter 2] is now an easy consequence in *finite-dimensional* vector spaces, where the adjoint operator $(-)^*$ makes **Vect** isomorphic to $\mathbf{Vect}^{\mathrm{op}}$ – where V^* is of course the 'dual' vector space of linear maps to the underlying field. This result says that $A \xrightarrow{F} X \xrightarrow{G} X \xrightarrow{H} B$ is reachable if and only if $B^* \xrightarrow{H^*} X^* \xrightarrow{G^*} X^* \xrightarrow{F^*} A^*$ is observable. See also [36]. There is also a duality result for bialgebras in [69].)

2.5.16 This exercise sketches an adjunction capturing Kalman's minimal realisation [287, chapter 10] for linear dynamical systems, in analogy with the Nerode realisation, described in Exercise 2.5.14. This categorical version is based on [35, 36].

Form the category **RLDS** of reachable linear dynamical systems. Its objects are such reachable systems $A \xrightarrow{F} X \xrightarrow{G} X \xrightarrow{H} B$ in **Vect**. And its morphisms from $A \xrightarrow{F} X \xrightarrow{G} X \xrightarrow{H} B$ to $C \xrightarrow{F'} Y \xrightarrow{G'} Y \xrightarrow{H'} D$ are triples of functions $C \overset{f}{\twoheadrightarrow} A$, $B \xrightarrow{g} D$ and $X \xrightarrow{h} Y$ with

$$
\begin{array}{ccccccc}
A & \xrightarrow{F} & X & \xrightarrow{G} & X & \xrightarrow{H} & B \\
{\scriptstyle f}\Uparrow & & \downarrow{\scriptstyle h} & & \downarrow{\scriptstyle h} & & \downarrow{\scriptstyle g} \\
C & \xrightarrow[F']{} & Y & \xrightarrow[G']{} & Y & \xrightarrow[H']{} & D
\end{array}
$$

Note that f is required to be a surjection/epimorphism.

Also, there is a category **RLB** with linear maps $\varphi\colon A^{\S} \to B$ as objects. A morphism $(A^{\S} \overset{\varphi}{\to} B) \longrightarrow (C^{\S} \overset{\psi}{\to} D)$ is a pair of linear maps $f\colon C \twoheadrightarrow A$ and $g\colon B \to D$ with $g \circ \varphi \circ f^{\S} = \psi$ – where $f^{\S}$ results from the functoriality of $(-)^{\S}$; see Exercise 2.4.8.4.

1. Demonstrate that the behaviour formula from Exercises 2.2.12.4 and 2.4.8.3 yields a behaviour functor $\mathcal{B}\colon \mathbf{RLDS} \to \mathbf{RLB}$, given by $(F, G, H) \longmapsto H \circ \mathrm{int}_{[F,G]}$.
2. Construct a functor $\mathcal{K}\colon \mathbf{RLB} \to \mathbf{RLDS}$ in the reverse direction in the following way. Assume a behaviour $\psi\colon C^{\S} \to D$, and form the behaviour map $b = \mathrm{beh}_{\langle\psi,\mathsf{sh}\rangle}\colon C^{\S} \to D^{\mathbb{N}}$ below, using the finality from the previous exercise:

$$\begin{array}{ccc} D \times C^{\S} & \overset{\mathrm{id}_D \times b}{\dashrightarrow} & D \times D^{\mathbb{N}} \\ {\scriptstyle \langle\psi,\mathsf{sh}\rangle}\uparrow & & \cong\uparrow{\scriptstyle \langle\mathsf{hd},\mathsf{tl}\rangle} \\ C^{\S} & \underset{b}{\dashrightarrow} & D^{\mathbb{N}} \end{array}$$

The image $\mathrm{Im}(b)$ of this behaviour map can be written as

$$\Big(C^{\S} \xrightarrow{\mathrm{beh}_{\langle\psi,\mathsf{sh}\rangle}} D^{\mathbb{N}}\Big) = \Big(C^{\S} \overset{e}{\twoheadrightarrow} \mathrm{Im}(b) \overset{m}{\rightarrowtail} D^{\mathbb{N}}\Big).$$

It is not hard to see that the tail function $\mathsf{tl}\colon B^{\mathbb{N}} \to B^{\mathbb{N}}$ restricts to $\mathsf{tl}'\colon \mathrm{Im}(b) \to \mathrm{Im}(b)$ via diagonal fill-in:

$$\begin{array}{ccc} C^{\S} & \overset{e}{\twoheadrightarrow} & \mathrm{Im}(b) \\ {\scriptstyle e\,\circ\,\mathsf{sh}}\downarrow & \overset{\mathsf{tl}'}{\swarrow} & \downarrow{\scriptstyle \mathsf{tl}\,\circ\, m} \\ \mathrm{Im}(b) & \underset{m}{\rightarrowtail} & D^{\mathbb{N}} \end{array}$$

Hence one can define a linear dynamical system as

$$\mathcal{K}\Big(C^{\S} \xrightarrow{\ \psi\ } D\Big) \overset{\text{def}}{=} \Big(C \xrightarrow{e\,\circ\,\mathsf{in}} \mathrm{Im}(b) \xrightarrow{\mathsf{tl}'} \mathrm{Im}(b) \xrightarrow{\mathsf{hd}\,\circ\, m} D\Big).$$

Prove that this gives a minimal realisation in an adjunction $\mathcal{B} \dashv \mathcal{K}$.

2.5.17 This exercise describes the so-called terms-as-natural-transformations view which originally stems from [329]. It is elaborated in a coalgebraic context in [325].

Let $H\colon \mathbf{Sets} \to \mathbf{Sets}$ be a (not necessarily polynomial) functor, with free algebras, given by a left adjoint F to the forgetful functor

$U\colon \mathbf{Alg}(H) \to \mathbf{Sets}$. Let X be an arbitrary set whose elements are considered as variables. Elements of the carrier $UF(X)$ of the free algebra on X can then be seen as terms containing free variables from X. Show that there is a bijective correspondence:

$$\frac{\text{terms } t \in UF(X)}{\text{natural transformations } \tau\colon U^X \Longrightarrow U} \, .$$

Hint: The component of the natural transformation at a specific algebra $HA \to A$ is the mapping which takes a valuation $\rho\colon X \to A$ of the variables in A to an interpretation $[\![t]\!]^A_\rho$ of the term t in the algebra A. Naturality then says that for a homomorphism $f\colon A \to B$ of algebras, one has the familiar equation $f([\![t]\!]^A_\rho) = [\![t]\!]^B_{f \circ \rho}$.

3
Bisimulations

The operation c of a coalgebra $c\colon X \to F(X)$ gives us information about its states in X. It may allow us to observe certain things, and it may allow us to 'modify states' or to 'move to successor states'. Typically for coalgebras, we can observe and modify, but we have no means for constructing new states. The behaviour of a state x is all that we can observe about x, either directly or indirectly (via its successor states). This behaviour is written as $\mathrm{beh}(x)$, where beh is the unique map to the final coalgebra (if any), as introduced in Definition 2.3.1.

In this situation it may happen that two states have the same behaviour. In that case we cannot distinguish them with the operations (of the coalgebras) that we have at our disposal. The two states need not be equal then, since the operations may give only limited access to the state space, and certain aspects may not be observable. When two states x, y are observationally indistinguishable, they are called *bisimilar*. This is written as $x \mathrel{\underline{\leftrightarrow}} y$.

The bisimilarity relation, for a given coalgebra (or pair of coalgebras), is introduced as the union of all bisimulations. A bisimulation is a relation on state spaces, which is maintained by coalgebra transitions and leads to equal observations. Bisimulations were first introduced by Park [371] for automata, as mutual simulations – building on an earlier notion of simulation between programs [353]. Park proved that if the initial states of two deterministic automata are related by a bisimulation, then they accept the same sets of inputs (see also Corollary 3.4.4 below). Indeed, bisimulations form a crucial tool for stepwise reasoning – as in induction arguments.

Bisimilarity is the main topic of the present chapter, but only for coalgebras of Kripke polynomial functors. In the next chapter more general functors will be considered, but for reasons of simplicity we prefer to first study the concrete situation for polynomial functors. Bisimulation will be introduced – like congruence – via a technique called relation lifting. These liftings can be defined by induction on the structure of polynomial functors. In a related manner the notion of invariance will arise in the Chapter 6 via predicate lifting. The first part of this chapter concentrates on some basic properties of relation lifting and of bisimulation relations; these properties will be used frequently. The coinduction proof principle in Section 3.4 is a basic result in the theory of coalgebras. It says that two states have the same behaviour if and only if they are contained in a bisimulation relation. Coinduction via bisimulations is illustrated in a simple process calculus in Section 3.5.

3.1 Relation Lifting, Bisimulations and Congruences

This section will introduce the technique of relation lifting from [217, 218] and use it to define the notions of bisimulation for coalgebras, and congruence for algebras. Many elementary results about relation lifting are provided. Alternative ways for introducing bisimulations will be discussed later in Section 3.3.

We start by motivating the need for relation lifting. Consider a sequence coalgebra $c\colon X \to 1 + (A \times X)$, as in Section 1.2. A bisimulation for this coalgebra is a relation $R \subseteq X \times X$ on its state space which is 'closed under c'. What this means is that if R holds for states x, y, then either both x and y have no successor states (i.e. $c(x) = \kappa_1(*) = c(y)$), or they both have successor states which are again related by R and their observations are the same: $c(x) = \kappa_2(a, x')$, $c(y) = \kappa_2(b, y')$, with $a = b$ and $R(x', y')$.

One way to express such closure properties uniformly is via a 'lifting' of R from a relation on X to a relation R' on $1 + (A \times X)$ so that this closure can be expressed simply as

$$R(x, y) \Longrightarrow R'(c(x), c(y)).$$

This idea works if we take $R' \subseteq (1 + (A \times X)) \times (1 + (A \times X))$ to be

$$R' = \{\kappa_1(*), \kappa_1(*)\} \cup \{(\kappa_2(a, x), \kappa_2(b, y)) \mid a = b \wedge R(x, y)\}.$$

The general idea of relation lifting applies to a polynomial functor F. It is a transformation of a relation $R \subseteq X \times X$ to a relation $R' \subseteq F(X) \times F(X)$, which will be defined by induction on the structure of F. We shall use the notation $\mathrm{Rel}(F)(R)$ for R' above. Briefly, the lifted relation $\mathrm{Rel}(F)(R)$ uses equality on occurrences of constants A in F, and R on occurrences of the state space X, as suggested in

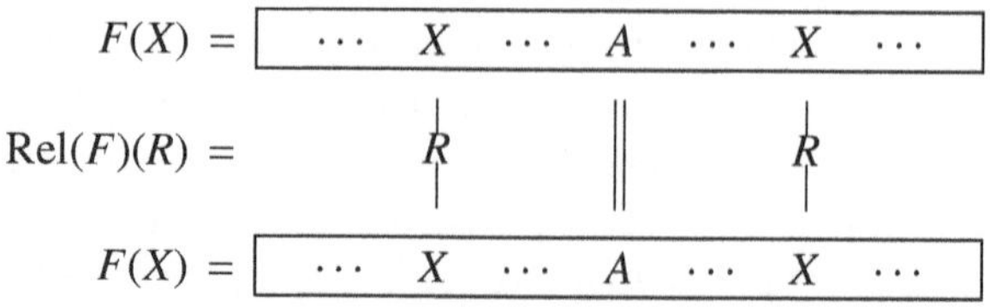

Actually, it will be convenient to define relation lifting slightly more generally and to allow different state spaces. Thus, it applies to relations $R \subseteq X \times Y$ and yields a relation $\mathrm{Rel}(F)(R) \subseteq F(X) \times F(Y)$.

Definition 3.1.1 (Relation lifting) Let $F\colon \mathbf{Sets} \to \mathbf{Sets}$ be a polynomial functor, and let X, Y be arbitrary sets. The mapping $\mathrm{Rel}(F)$ which sends a relation $R \subseteq X \times Y$ to a 'lifted' relation $\mathrm{Rel}(F)(R) \subseteq F(X) \times F(Y)$ is defined by induction on the structure of the functor F, following the points in Definition 2.2.1.

1. If F is the identity functor, then

$$\mathrm{Rel}(F)(R) = R.$$

2. If F is a constant functor $X \mapsto A$, then

$$\mathrm{Rel}(F)(R) = \mathrm{Eq}(A) = \{(a, a) \mid a \in A\}.$$

3. If F is a product $F_1 \times F_2$, then

$$\begin{aligned}\mathrm{Rel}(F)(R) = \{ ((u_1, u_2), (v_1, v_2)) \mid {} & \mathrm{Rel}(F_1)(R)(u_1, v_1) \wedge \\ & \mathrm{Rel}(F_2)(R)(u_2, v_2) \}.\end{aligned}$$

4. If F is a set-indexed coproduct $\coprod_{i \in I} F_i$, then:

$$\mathrm{Rel}(F)(R) = \bigcup_{i \in I} \{(\kappa_i(u), \kappa_i(v)) \mid \mathrm{Rel}(F_i)(R)(u, v)\}.$$

5. If F is an exponent G^A, then

$$\mathrm{Rel}(F)(R) = \{(f, g) \mid \forall a \in A.\, \mathrm{Rel}(G)(R)(f(a), g(a))\}.$$

6. If F is a powerset $\mathcal{P}(G)$, then

$$\begin{aligned}\mathrm{Rel}(F)(R) = \{ (U, V) \mid {} & \forall u \in U.\, \exists v \in V.\, \mathrm{Rel}(G)(R)(u, v) \wedge \\ & \forall v \in V.\, \exists u \in U.\, \mathrm{Rel}(G)(R)(u, v) \}.\end{aligned}$$

This same formula will be used in case F is a *finite* powerset $\mathcal{P}_{\mathrm{fin}}(G)$.

In the beginning of Section 3.3 we shall see that relation lifting can also be defined directly via images. The above inductive definition may seem more cumbersome but gives us a better handle on the different cases. Also, it better emphasises the relational aspects of lifting and the underlying logical infrastructure (such as finite conjunctions and disjunctions, and universal and existential quantification). This is especially relevant in more general settings, such as in [218, 45, 150].

Relation lifting with respect to a functor is closely related to so-called logical relations. These are collections of relations $(R_\sigma)_\sigma$ indexed by types σ, in such a way that $R_{\sigma\to\tau}$ and $R_{\sigma\times\tau}$ are determined by R_σ and R_τ. Similarly, we use collections of relations $(\mathrm{Rel}(F)(R))_F$ indexed by polynomial functors F, which are also structurally determined. Logical relations were originally introduced in the context of semantics of (simply) typed lambda calculus ([442, 147, 382]); see [356, chapter 8] for an overview. They are used for instance for definability, observational equivalence and data refinement.

In the next section we shall see various elementary properties of relation lifting. But first we show what it is used for: bisimulation for coalgebras and congruence for algebras. The definitions we use are *generic* or *polytypic*, in the sense that they apply uniformly to (co)algebras of an arbitrary polynomial functor.

Definition 3.1.2 Let $F\colon \mathbf{Sets} \to \mathbf{Sets}$ be a polynomial functor.

1. A **bisimulation** for coalgebras $c\colon X \to F(X)$ and $d\colon Y \to F(Y)$ is a relation $R \subseteq X \times Y$ which is 'closed under c and d':

$$(x, y) \in R \Longrightarrow (c(x), d(y)) \in \mathrm{Rel}(F)(R)$$

for all $x \in X$ and $y \in Y$. Equivalently:

$$R \subseteq (c \times d)^{-1}(\mathrm{Rel}(F)(R))$$

or, by (2.15),

$$\coprod_{c\times d}(R) \subseteq \mathrm{Rel}(F)(R).$$

2. A **congruence** for algebras $a\colon F(X) \to X$ and $b\colon F(Y) \to Y$ is a relation $R \subseteq X \times Y$ which is also 'closed under a and b':

$$(u, v) \in \mathrm{Rel}(F)(R) \Longrightarrow (a(u), b(v)) \in R.$$

That is,

$$\mathrm{Rel}(F)(R) \subseteq (a \times b)^{-1}(R) \qquad \text{or equivalently} \qquad \coprod_{a\times b}(\mathrm{Rel}(F)(R)) \subseteq R.$$

Often we are interested in bisimulations $R \subseteq X \times X$ on a *single* coalgebra $c\colon X \to F(X)$. We then use the definition with $d = c$. Similarly for congruences.

Notice that we require only that a congruence is closed under the (algebraic) operations and not that it is an equivalence relation. This minor deviation from standard terminology is justified by the duality we obtain between bisimulations and congruences. We shall use the following explicit terminology.

Definition 3.1.3 A **bisimulation equivalence** is a bisimulation on a single coalgebra which is an equivalence relation. Similarly, a **congruence equivalence** is a congruence on a single algebra which is an equivalence relation.

We continue with several examples of the notions of bisimulation and congruence for specific functors.

Bisimulations for Deterministic Automata

Consider a deterministic automaton $\langle \delta, \epsilon \rangle \colon X \to X^A \times B$. As we have seen in Section 2.2, it is a coalgebra for the functor $F = \mathrm{id}^A \times B$. Relation lifting for this functor yields for a relation $R \subseteq X \times X$ a new relation $\mathrm{Rel}(F)(R) \subseteq (X^A \times B) \times (X^A \times B)$, given by

$$\mathrm{Rel}(F)(R)\big((f_1, b_1), (f_2, b_2)\big) \iff \forall a \in A.\, R(f_1(a), f_2(a)) \wedge b_1 = b_2.$$

Thus, a relation $R \subseteq X \times X$ is a bisimulation with respect to the (single) coalgebra $\langle \delta, \epsilon \rangle \colon X \to X^A \times B$ if, for all $x, y \in X$,

$$R(x, y) \Longrightarrow \mathrm{Rel}(F)(R)\big((\delta(x), \epsilon(x)), (\delta(y), \epsilon(y))\big).$$

That is,

$$R(x, y) \Longrightarrow \forall a \in A.\, R(\delta(x)(a), \delta(y)(a)) \wedge \epsilon(x) = \epsilon(y).$$

That is, in transition notation:

$$R(x, y) \Longrightarrow \begin{cases} x \xrightarrow{a} x' \wedge y \xrightarrow{a} y' \text{ implies } R(x', y') \\ x \downarrow b \wedge y \downarrow c \text{ implies } b = c. \end{cases}$$

Thus, once two states are in a bisimulation R, they remain in R and give rise to the same direct observations. This makes them observationally indistinguishable.

Bisimulations for Non-deterministic Automata

Next, consider a non-deterministic automaton $\langle \delta, \epsilon \rangle \colon X \to \mathcal{P}(X)^A \times B$, as coalgebra for the functor $F = \mathcal{P}(\mathrm{id})^A \times B$. Relation lifting for this functor is slightly more complicated: it sends a relation $R \subseteq X \times X$ to the relation $\mathrm{Rel}(F)(R) \subseteq (\mathcal{P}(X)^A \times B) \times (\mathcal{P}(X)^A \times B)$ given by

$$\begin{aligned} \mathrm{Rel}(F)(R)((f_1, b_1), (f_2, b_2)) \iff & \forall a \in A.\ \forall x \in f_1(a).\, \exists y \in f_2(a).\, R(x, y) \\ & \wedge \forall y \in f_2(a).\, \exists x \in f_1(a).\, R(x, y) \\ & \wedge b_1 = b_2. \end{aligned}$$

Thus, $R \subseteq X \times X$ is a bisimulation if for all $x, y \in X$ with $R(x, y)$,

- $x \xrightarrow{a} x'$ implies there is a y' with $y \xrightarrow{a} y'$ and $R(x', y')$;
- $y \xrightarrow{a} y'$ implies there is an x' with $x \xrightarrow{a} x'$ and $R(x', y')$;
- $x \downarrow b$ and $y \downarrow c$ implies $b = c$.

This corresponds to the standard notion of bisimulation used in the theory of automata/transition systems.

Congruences for Monoids

Recall that a monoid is a set M carrying an associative operation $+\colon M \times M \to M$ with a unit element $0 \in M$. These two operations $+$ and 0, but not the relevant monoid equations, can be captured as an algebra:

$$1 + (M \times M) \xrightarrow{\quad [0,+] \quad} M$$

of the functor $F(X) = 1 + (X \times X)$. Relation lifting for F is described by

$$\mathrm{Rel}(F)(R) = \{(\kappa_1(*), \kappa_1(*))\} \cup \{(\kappa_2(x, x'), \kappa_2(y, y')) \mid R(x, y) \wedge R(x', y')\}.$$

Hence a relation $R \subseteq M \times M$ on the carrier of the monoid is a congruence if:

$$\mathrm{Rel}(F)(R)(u, v) \implies R([0, +](u), [0, +](v)).$$

This amounts to

$$R(0, 0) \qquad \text{and} \qquad R(x, y) \wedge R(x', y') \implies R(x + x', y + y').$$

Thus a congruence, like a bisimulation, is closed under the operations.

Congruences in a Binary Induction Proof Principle

We have already discussed the usual 'unary' induction proof principle for natural numbers in Example 2.4.4, expressed in terms of predicates, which are assumed to be closed under the operations. Later, in Section 6.1 we shall encounter it in full generality, stating that every invariant on an initial algebra is the truth predicate.

There is also a less well-known binary version of the induction proof principle, expressed in terms of congruences. It was first formulated as such for the natural numbers in [414] and further generalised in [218]. It also appeared in the derivations of induction and coinduction principles in [385] in the context of a formal logic for parametric polymorphism.

At this stage we only formulate this binary version, and postpone the proof. It can be given in various ways (see Exercises 3.3.2 and 6.2.2) but requires some properties of relation lifting which are still to come.

Theorem 3.1.4 (Binary induction proof principle) *Every congruence on an initial algebra contains the equality relation.*

This binary version of induction is the dual of a coinduction principle; see Corollary 3.4.2.

Bisimulations as Congruences

The so-called structural operational semantics (SOS) introduced by Plotkin is a standard technique in the semantics of programming languages to define the operational behaviour of programs. The latter are seen as elements of the initial algebra $F(\mathsf{Prog}) \xrightarrow{\cong} \mathsf{Prog}$ of a suitable functor F describing the signature of operations of the programming language, as in Example 2.4.2. A transition relation is then defined on top of the set of programs Prog, as the least relation closed under certain rules. This transition structure may be understood as coalgebra $\mathsf{Prog} \to G(\mathsf{Prog})$, for an appropriate functor G – which is often the functor $\mathcal{P}(\mathrm{id})^A$ for labelled transition systems (see [452, 451]); the transition structure is then given by transitions $p \xrightarrow{a} q$ describing an a-step between programs $p, q \in \mathsf{Prog}$.

The transition structure gives rise to certain equivalences for programs, such as bisimilarity (see below), trace equivalence or other equivalences; see [158]. These equivalences are typically bisimulation equivalences. An important issue in this setting is: are these bisimulation equivalences also *congruences* for the given algebra structure $F(\mathsf{Prog}) \xrightarrow{\cong} \mathsf{Prog}$? This is a basic requirement to make the equivalence a reasonable one for the kind of programs under consideration, because congruence properties are essential in reasoning with the equivalence. In this setting given by a bialgebra $F(\mathsf{Prog}) \to \mathsf{Prog} \to G(\mathsf{Prog})$, the two fundamental notions of this section (bisimulation and congruence) are thus intimately related. This situation will be investigated further in Section 5.5 in relation to distributive laws.

There is a whole line of research about suitable syntactic formats for SOS rules, ensuring that certain bisimulation equivalences are also congruences. See [175] for a basic reference. We shall use a more categorical perspective, first in Section 3.5, and later in Section 5.5, following [452, 451, 58]; see [298] for an overview of the coalgebraic approach.

Definition 3.1.5 Let $X \xrightarrow{c} F(X)$ and $Y \xrightarrow{d} F(Y)$ be two coalgebras of a polynomial functor F. The **bisimilarity** relation $\underline{\leftrightarrow}$ is the union of all bisimulations:

$$x \underline{\leftrightarrow} y \iff \exists R \subseteq X \times Y.\, R \text{ is a bisimulation for } c \text{ and } d, \text{ and } R(x, y).$$

As a result of Proposition 3.2.6.3 in the next section, this union is a bisimulation itself, so that $\underline{\leftrightarrow}$ can be characterised as the greatest bisimulation.

Sometimes we write ${}_c\underline{\leftrightarrow}_d$ for $\underline{\leftrightarrow}$ to make the dependence on the coalgebras c and d explicit.

Bisimilarity formalises the idea of observational indistinguishability. It will be an important topic in the remainder of this chapter.

Exercises

3.1.1 Use the description (2.17) of a list functor $F^\star$ to show that:

$$\mathrm{Rel}(F^\star)(R) = \{(\langle u_1, \ldots, u_n\rangle, \langle v_1, \ldots, v_n\rangle) \mid \forall i \leq n.\, \mathrm{Rel}(F)(R)(u_i, v_i)\}.$$

3.1.2 Unfold the definition of bisimulation for various kind of tree coalgebras, like $X \to 1 + (A \times X \times X)$ and $X \to (A \times X)^\star$.

3.1.3 Do the same for classes in object-oriented languages (see (1.10)), described as coalgebras of a functor in Exercise 2.3.6.3.

3.1.4 Note that the operations of a vector space V (over $\mathbb{R}$), namely zero, addition, inverse and scalar multiplication, can be captured as an algebra $1 + (V \times V) + V + (\mathbb{R} \times V) \longrightarrow V$. Investigate then what the associated notion of congruence is.

3.1.5 We have described relation lifting on a coproduct functor $F = F_1 + F_2$ in Definition 3.1.1 as

$$\mathrm{Rel}(F_1 + F_2)(R) = \coprod_{\kappa_1 \times \kappa_1}(\mathrm{Rel}(F_1)(R)) \cup \coprod_{\kappa_2 \times \kappa_2}(\mathrm{Rel}(F_2)(R)).$$

Prove that it can also be defined in terms of products $\prod$ and intersection $\cap$ as

$$\mathrm{Rel}(F_1 + F_2)(R) = \prod_{\kappa_1 \times \kappa_1}(\mathrm{Rel}(F_1)(R)) \cap \prod_{\kappa_2 \times \kappa_2}(\mathrm{Rel}(F_2)(R)),$$

where for a function $f \colon X \to Y$ the map $\prod_f \colon \mathcal{P}(X) \to \mathcal{P}(Y)$ is described in Exercise 2.1.12.

3.1.6 In this text we concentrate on *bi*simulations. There is also the notion of simulation, which can be defined via an order on a functor; see [446, 232]. For a functor $F \colon \mathbf{Sets} \to \mathbf{Sets}$ such an order consists of a functor $\sqsubseteq$ as in the diagram below.

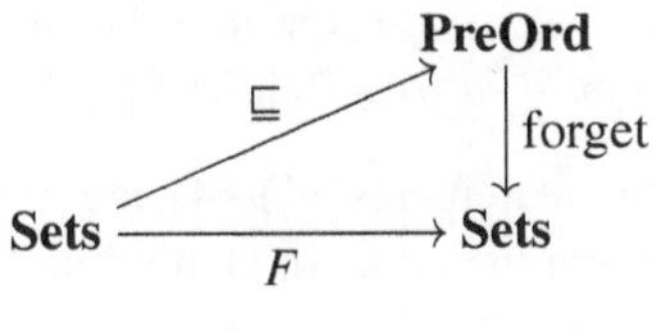

Given such an order we define 'lax' relation lifting $\mathrm{Rel}_{\sqsubseteq}(F)$ as $R \mapsto \sqsubseteq \circ \mathrm{Rel}(F)(R) \circ \sqsubseteq$. A relation $R \subseteq X \times Y$ is then a **simulation** on coalgebras $X \xrightarrow{c} F(X)$, $Y \xrightarrow{d} F(Y)$ if $R \subseteq (c \times d)^{-1}(\mathrm{Rel}_{\sqsubseteq}(F)(R))$. Similarity is then the union of all simulations.

1. Investigate what it means to have an order as described in the above diagram.
2. Describe on the functor $L = 1 + (A \times (-))$ a 'flat' order, and on the powerset functor $\mathcal{P}$ the inclusion order, as in the diagram. Check what the associated notions of simulation are.
3. Prove that similarity on the final coalgebra A^∞ of the functor L with order as in (2) is the prefix order given by $\sigma \leq \tau$ iff $\sigma \cdot \rho = \tau$ for some $\rho \in A^\infty$; see [232, example 5.7].

3.2 Properties of Bisimulations

This section is slightly technical, and possibly also slightly boring. It starts by listing various elementary properties of relation lifting, and subsequently uses these properties to prove standard results about bisimulations and bisimilarity.

First there are three lemmas about relation lifting.

Lemma 3.2.1 *Let F:* **Sets** $\rightarrow$ **Sets** *be a polynomial functor. Relation lifting* $\mathrm{Rel}(F)$ *with respect to F satisfies the following basic properties.*

1. *It preserves the equality relation:*

$$\mathrm{Rel}(F)(\mathrm{Eq}(X)) = \mathrm{Eq}(F(X)).$$

2. *It commutes with reversal of relations:*

$$\mathrm{Rel}(F)(R^\dagger) = \mathrm{Rel}(F)(R)^\dagger.$$

3. *It is monotone:*

$$R \subseteq S \implies \mathrm{Rel}(F)(R) \subseteq \mathrm{Rel}(F)(S).$$

4. *It preserves relation composition*

$$\mathrm{Rel}(F)(R \circ S) = \mathrm{Rel}(F)(R) \circ \mathrm{Rel}(F)(S).$$

5. *It preserves reflexivity, symmetry and transitivity, and thus, if R is an equivalence relation, then so is* $\mathrm{Rel}(F)(R)$.

Proof The first four statements (1)–(4) are proved by induction on the structure of F, following the cases in Definition 3.1.1. The case in (4) where

F is an exponent G^A requires the axiom of choice (AC), as will be illustrated: assume, as induction hypothesis (IH), that the functor G preserves composition of relations, then so does the exponent G^A, since:

$$\begin{aligned}
&\mathrm{Rel}(G^A)(R \circ S)(f, g) \\
&\iff \forall a \in A.\, \mathrm{Rel}(G)(R \circ S)(f(a), g(a)) \\
&\overset{\text{(IH)}}{\iff} \forall a \in A.\, (\mathrm{Rel}(G)(R) \circ \mathrm{Rel}(G)(S))(f(a), g(a)) \\
&\iff \forall a \in A.\, \exists z.\, \mathrm{Rel}(G)(R)(f(a), z) \wedge \mathrm{Rel}(G)(S)(z, g(a)) \\
&\overset{\text{(AC)}}{\iff} \exists h.\, \forall a \in \mathrm{Rel}(G)(R)(f(a), h(a)) \wedge \mathrm{Rel}(G)(S)(h(a), g(a)) \\
&\iff \exists h.\, \mathrm{Rel}(G^A)(R)(f, h) \wedge \mathrm{Rel}(G^A)(S)(h, g) \\
&\iff (\mathrm{Rel}(G^A)(R) \circ \mathrm{Rel}(G^A)(S))(f, g).
\end{aligned}$$

The last statement (5) follows from the previous ones:

- If the relation R is reflexive, i.e. satisfies $\mathrm{Eq}(X) \subseteq R$, then $\mathrm{Eq}(F(X)) = \mathrm{Rel}(F)(\mathrm{Eq}(X)) \subseteq \mathrm{Rel}(F)(R)$, so that $\mathrm{Rel}(F)(R)$ is also reflexive.
- If R is symmetric, that is, if $R \subseteq R^{-1}$, then $\mathrm{Rel}(F)(R) \subseteq \mathrm{Rel}(F)(R^{-1}) = \mathrm{Rel}(F)(R)^{-1}$, so that $\mathrm{Rel}(F)(R)$ is symmetric as well.
- If R is transitive, i.e. $R \circ R \subseteq R$, then $\mathrm{Rel}(F)(R) \circ \mathrm{Rel}(F)(R) = \mathrm{Rel}(F)(R \circ R) \subseteq \mathrm{Rel}(F)(R)$, so that $\mathrm{Rel}(F)(R)$ is also transitive. □

We proceed with a similar lemma, about relation lifting and (inverse and direct) images.

Lemma 3.2.2 *Let $F\colon$ **Sets** $\to$ **Sets** be a Kripke polynomial functor again, and let $f\colon X \to Z$ and $g\colon Y \to W$ be arbitrary functions.*

1. *Relation lifting commutes with inverse images: for $R \subseteq Z \times W$,*

$$\mathrm{Rel}(F)((f \times g)^{-1}(R)) = (F(f) \times F(g))^{-1}(\mathrm{Rel}(F)(R)).$$

2. *Relation lifting also commutes with direct images: for $R \subseteq X \times Y$,*

$$\mathrm{Rel}(F)(\textstyle\coprod_{f \times g}(R)) = \coprod_{F(f) \times F(g)}(\mathrm{Rel}(F)(R)).$$

Proof Both equations are proved by induction on the structure of F. We leave (1) to the reader. Once (1) is established, it can be used to prove the direction ($\supseteq$) of (2), using the Galois connection relating direct and inverse images in (2.15):

$$\begin{aligned}
&\textstyle\coprod_{F(f) \times F(g)}(\mathrm{Rel}(F)(R)) \subseteq \mathrm{Rel}(F)(\coprod_{f \times g}(R)) \\
&\iff \mathrm{Rel}(F)(R) \subseteq (F(f) \times F(g))^{-1}\mathrm{Rel}(F)(\textstyle\coprod_{f \times g}(R)) \\
&\qquad\qquad\qquad\quad = \mathrm{Rel}(F)((f \times g)^{-1} \textstyle\coprod_{f \times g}(R)).
\end{aligned}$$

This second inclusion is obtained from monotonicity of relation lifting (see Lemma 3.2.1(3)), using that $R \subseteq (f \times g)^{-1} \coprod_{f \times g}(R)$.

The inclusion ($\subseteq$) of (2) is proved by induction on the structure of the functor F. This requires the axiom of choice to handle the exponent functor case, as in the proof of point (4) in the previous lemma. The powerset case $F = \mathcal{P}G$ is most complicated, and will be described in detail.

$$
\begin{aligned}
&\mathrm{Rel}(\mathcal{P}G)(\textstyle\coprod_{f\times g}(R))(U, V) \\
&\iff \forall x \in U.\, \exists y \in V.\, \mathrm{Rel}(G)(\textstyle\coprod_{f\times g}(R))(x, y) \\
&\qquad\quad \wedge\ \forall y \in V.\, \exists x \in U.\, \mathrm{Rel}(G)(\textstyle\coprod_{f\times g}(R))(x, y) \\
&\overset{\text{(IH)}}{\iff} \forall x \in U.\, \exists y \in V.\ \textstyle\coprod_{G(f)\times G(g)} \mathrm{Rel}(G)(R)(x, y) \\
&\qquad\quad \wedge\ \forall y \in V.\, \exists x \in U.\ \textstyle\coprod_{G(f)\times G(g)} \mathrm{Rel}(G)(R)(x, y) \\
&\iff \forall x \in U.\, \exists y \in V.\, \exists u, v.\, G(f)(u) = x \wedge G(g)(v) = y \\
&\qquad\qquad \wedge\ \mathrm{Rel}(G)(R)(u, v) \\
&\qquad\quad \wedge\ \forall y \in V.\, \exists x \in U.\, \exists u, v.\, G(f)(u) = x \wedge G(g)(v) = y \\
&\qquad\qquad \wedge\ \mathrm{Rel}(G)(R)(u, v) \\
&\Longrightarrow \forall u \in U'.\, \exists v \in V'.\, \mathrm{Rel}(G)(R)(u, v) \\
&\qquad\quad \wedge\ \forall v \in V'.\, \exists u \in U'.\, \mathrm{Rel}(G)(R)(u, v), \quad \text{where} \\
&\qquad\qquad U' = \{u \mid G(f)(u) \in U \wedge \exists v.\, G(g)(v) \in V \wedge \mathrm{Rel}(G)(R)(u, v)\} \\
&\qquad\qquad V' = \{v \mid G(g)(v) \in V \wedge \exists u.\, G(f)(u) \in U \wedge \mathrm{Rel}(G)(R)(u, v)\} \\
&\iff \exists U', V'.\, \mathcal{P}(G(f))(U') = U \wedge \mathcal{P}(G(f))(V') \\
&\qquad\qquad\qquad\qquad\qquad\qquad = V \wedge \mathrm{Rel}(\mathcal{P}(G))(R)(U', V') \\
&\iff \textstyle\coprod_{\mathcal{P}(G(f))\times\mathcal{P}(G(g))}(\mathrm{Rel}(\mathcal{P}G)(R))(U, V).
\end{aligned}
$$

□

Below we show in a diagram why $\mathrm{Rel}(F)$ is called relation lifting. The term ‘relator’ is often used in this context for the lifting of F; see for instance [446] (and Definition 4.4.5, where a more general description of the situation is given). Additionally, the next result shows that bisimulations are coalgebras themselves. It involves a category **Rel** with relations *as objects*. This **Rel** should not be confused with the category **SetsRel**, from Example 1.4.2.4, which has relations *as morphisms*.

Definition 3.2.3 We write **Rel** for the category with binary relations $R \subseteq X \times Y$ as objects. A morphism $(R \subseteq X \times Y) \longrightarrow (S \subseteq U \times V)$ consists of two functions $f\colon X \to U$ and $g\colon Y \to V$ with $R(x, y) \implies S(f(x), g(y))$ for all $x \in X, y \in Y$. The latter amounts to the existence of the necessarily unique dashed map in:

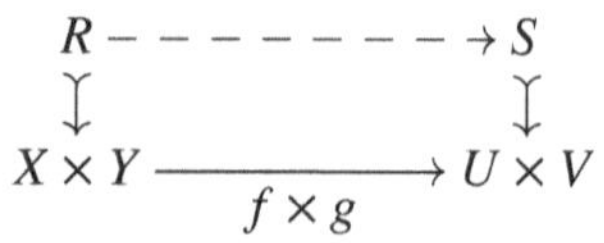

Equivalently, $R \subseteq (f\times g)^{-1}(S)$, or $\coprod_{f\times g}(R) \subseteq S$, by the correspondence (2.15).

Lemma 3.2.4 *Consider a Kripke polynomial functor F.*

1. *Relation lifting forms a functor:*

$$\begin{array}{ccc} \mathbf{Rel} & \xrightarrow{\mathrm{Rel}(F)} & \mathbf{Rel} \\ \downarrow & & \downarrow \\ \mathbf{Sets} \times \mathbf{Sets} & \xrightarrow[F \times F]{} & \mathbf{Sets} \times \mathbf{Sets} \end{array}$$

where the vertical arrows are the obvious forgetful functors.

2. *A bisimulation is a* Rel(F)*-coalgebra in this category* **Rel**. *Similarly, a congruence is a* Rel(F)*-algebra in* **Rel**.

Proof 1. We show that if a pair $(f, g)\colon R \to S$ is a morphism in **Rel** – where $f\colon X \to U$ and $g\colon Y \to V$ – then $(F(f), F(g))\colon \mathrm{Rel}(F)(R) \to \mathrm{Rel}(F)(S)$ is also a morphism in **Rel**. From the inclusion $R \subseteq (f \times g)^{-1}(S)$ we obtain by monotony and preservation of inverse images by relation lifting:

$$\mathrm{Rel}(F)(R) \subseteq \mathrm{Rel}(F)((f \times g)^{-1}(S)) = (F(f) \times F(g))^{-1}\mathrm{Rel}(F)(S).$$

This means that $(F(f), F(g))$ is a morphism $\mathrm{Rel}(F)(R) \to \mathrm{Rel}(F)(S)$ in **Rel**.

2. A Rel(F)-coalgebra $R \to \mathrm{Rel}(F)(R)$ in **Rel**, for $R \subseteq X \times Y$, consists of two underlying maps $c\colon X \to F(X)$ and $d\colon Y \to F(Y)$ with

$$\begin{array}{ccc} R & \dashrightarrow & \mathrm{Rel}(F)(R) \\ \downarrow & & \downarrow \\ X \times Y & \xrightarrow[c \times d]{} & F(X) \times F(Y) \end{array}$$

This says that R is a relation which is closed under the F-coalgebras c, d, i.e. that R is a bisimulation for c, d.

In the same way congruences are Rel(F)-algebras in the category **Rel**. ❑

The next result lists several useful preservation properties of relation lifting.

Lemma 3.2.5 *Relation lifting* Rel(F) *for a given Kripke polynomial functor* $F\colon$ **Sets** $\to$ **Sets** *preserves the following:*

1. ***Kernel*** *relations, given for an arbitrary function* $f\colon X \to Y$ *by:*

$$\begin{aligned} \mathrm{Ker}(f) &= \{(x_1, x_2) \in X \times X \mid f(x_1) = f(x_2)\} \\ &= (f \times f)^{-1}(\mathrm{Eq}(Y)) \end{aligned}$$

in

$$\mathrm{Rel}(F)(\mathrm{Ker}(f)) = \mathrm{Ker}(F(f)).$$

2. ***Graph*** *relations given for* $f\colon X \to Y$ *by:*

$$\begin{aligned}\mathrm{Graph}(f) &= \{(x,y) \in X \times Y \mid f(x) = y\}\\ &= (f \times \mathrm{id}_Y)^{-1}(\mathrm{Eq}(Y))\end{aligned}$$

in

$$\mathrm{Rel}(F)(\mathrm{Graph}(f)) = \mathrm{Graph}(F(f)).$$

3. ***Images*** *of tuples: for* $X \xleftarrow{f} Z \xrightarrow{g} Y$,

$$\begin{aligned}\mathrm{Im}(\langle f,g\rangle) &= \{(x,y) \in X \times Y \mid \exists z \in Z.\, f(z) = x \wedge g(z) = y\}\\ &= \textstyle\coprod_{f\times g}(\mathrm{Eq}(Z))\end{aligned}$$

in

$$\mathrm{Rel}(F)(\mathrm{Im}(\langle f,g\rangle)) = \mathrm{Im}(\langle F(f), F(g)\rangle).$$

4. ***Pullback*** *relations of spans: for* $X \xrightarrow{f} Z \xleftarrow{g} Y$,

$$\begin{aligned}\mathrm{Eq}(f,g) &= \{(x,y) \in X \times Y \mid f(x) = g(y)\}\\ &= (f \times g)^{-1}(\mathrm{Eq}(Z))\end{aligned}$$

in

$$\mathrm{Rel}(F)(\mathrm{Eq}(f,g)) = \mathrm{Eq}(F(f), F(g)).$$

Proof 1. By the results from the previous two lemmas:

$$\begin{aligned}\mathrm{Rel}(F)(\mathrm{Ker}(f)) &= \mathrm{Rel}(F)((f \times f)^{-1}(\mathrm{Eq}(Y)))\\ &= (F(f) \times F(f))^{-1}(\mathrm{Rel}(F)(\mathrm{Eq}(Y)))\\ &= (F(f) \times F(f))^{-1}(\mathrm{Eq}(F(Y)))\\ &= \mathrm{Ker}(F(f)).\end{aligned}$$

2. Similarly,

$$\begin{aligned}\mathrm{Rel}(F)(\mathrm{Graph}(f)) &= \mathrm{Rel}(F)((f \times \mathrm{id}_Y)^{-1}(\mathrm{Eq}(Y)))\\ &= (F(f) \times \mathrm{id}_{F(Y)})^{-1}(\mathrm{Rel}(F)(\mathrm{Eq}(Y)))\\ &= (F(f) \times \mathrm{id}_{F(Y)})^{-1}(\mathrm{Eq}(F(Y)))\\ &= \mathrm{Graph}(F(f)).\end{aligned}$$

3. And

$$\begin{aligned}\mathrm{Rel}(F)(\mathrm{Im}(\langle f,g\rangle)) &= \mathrm{Rel}(F)(\textstyle\coprod_{f\times g}(\mathrm{Eq}(Z)))\\ &= \textstyle\coprod_{F(f)\times F(g)}(\mathrm{Rel}(F)(\mathrm{Eq}(Z)))\\ &= \textstyle\coprod_{F(f)\times F(g)}(\mathrm{Eq}(F(Z)))\\ &= \mathrm{Im}(\langle F(f), F(g)\rangle)\end{aligned}$$

4. Finally,

$$\begin{aligned}
\mathrm{Rel}(F)(\mathrm{Eq}(f,g)) &= \mathrm{Rel}(F)((f\times g)^{-1}(\mathrm{Eq}(Z))) \\
&= (F(f)\times F(g))^{-1}(\mathrm{Rel}(F)(\mathrm{Eq}(Z))) \\
&= (F(f)\times F(g))^{-1}(\mathrm{Eq}(F(Z))) \\
&= \mathrm{Eq}(F(f),F(g)).
\end{aligned}$$

□

Once these auxiliary results are in place, the next two propositions establish a series of standard facts about bisimulations and bisimilarity; see [411, section 5]. We begin with closure properties.

Proposition 3.2.6 *Assume coalgebras* $X \xrightarrow{c} F(X)$, $X' \xrightarrow{c'} F(X')$, *and* $Y \xrightarrow{d} F(Y)$, $Y' \xrightarrow{d'} F(Y')$ *of a Kripke polynomial functor* F. *Bisimulations are closed under the following:*

1. *Reversal: if* $R \subseteq X \times Y$ *is a bisimulation, then so is* $R^{\dagger} \subseteq Y \times X$.
2. *Composition: if* $R \subseteq X \times X'$ *and* $S \subseteq X' \times Y$ *are bisimulations, then so is* $(S \circ R) \subseteq X \times Y$.
3. *Arbitrary unions: if* $R_i \subseteq X \times Y$ *is a bisimulation for each* $i \in I$, *then so is* $\bigcup_{i\in I} R_i \subseteq X \times Y$.
4. *Inverse images: for homomorphisms* $f\colon X \to Y$ *and* $f'\colon X' \to Y'$, *if* $R \subseteq Y \times Y'$ *is a bisimulation, then so is* $(f \times f')^{-1}(R) \subseteq X \times X'$.
5. *Direct images: for homomorphisms* $f\colon X \to Y$ *and* $f'\colon X' \to Y'$, *if* $R \subseteq X \times X'$ *is a bisimulation, then so is* $\coprod_{f\times f'}(R) \subseteq Y \times Y'$.

Proof 1. If the relation R is a bisimulation, i.e. if $R \subseteq (c\times d)^{-1}(\mathrm{Rel}(F)(R))$, then

$$\begin{aligned}
R^{\dagger} &\subseteq \left((c\times d)^{-1}(\mathrm{Rel}(F)(R))\right)^{\dagger} \\
&= (d\times c)^{-1}\left(\mathrm{Rel}(F)(R)^{\dagger}\right) \\
&= (d\times c)^{-1}(\mathrm{Rel}(F)(R^{\dagger})) \quad \text{by Lemma 3.2.1.2.}
\end{aligned}$$

2. Assume $R \subseteq (c\times e)^{-1}(\mathrm{Rel}(F)(R))$ and $S \subseteq (e\times d)^{-1}(\mathrm{Rel}(F)(S))$. If $(x,y) \in (S \circ R)$, say with $R(x,w)$ and $S(w,y)$, then by assumption $(c(x),e(w)) \in \mathrm{Rel}(F)(R)$ and $(e(w),d(y)) \in \mathrm{Rel}(F)(S)$. Hence $(c(x),d(y)) \in (\mathrm{Rel}(F)(S) \circ \mathrm{Rel}(F)(R)) = \mathrm{Rel}(F)(S \circ R)$, by Lemma 3.2.1.4. We have thus proved the inclusion $(S \circ R) \subseteq (c\times d)^{-1}$ $(\mathrm{Rel}(F)(S \circ R)$, and thus that $S \circ R$ is a bisimulation.
3. Assume $R_i \subseteq (c\times d)^{-1}(\mathrm{Rel}(F)(R_i))$, for each $i \in I$. Then

$$\begin{aligned}
\textstyle\bigcup_{i\in I} R_i &\subseteq \textstyle\bigcup_{i\in I}(c\times d)^{-1}(\mathrm{Rel}(F)(R_i)) \\
&= (c\times d)^{-1}\left(\textstyle\bigcup_{i\in I}\mathrm{Rel}(F)(R_i)\right) && \text{inverse image preserves unions} \\
&\subseteq (c\times d)^{-1}\left(\mathrm{Rel}(F)(\textstyle\bigcup_{i\in I} R_i)\right) && \text{by monotony of relation lifting (and of inverse images).}
\end{aligned}$$

4. If $R \subseteq (d \times d')^{-1}(\mathrm{Rel}(F)(R))$, then

$$\begin{aligned}
&(f \times f')^{-1}(R) \\
&\subseteq (f \times f')^{-1}(d \times d')^{-1}(\mathrm{Rel}(F)(R)) \\
&= (c \times c')^{-1}(F(f) \times F(f'))^{-1}(\mathrm{Rel}(F)(R)) && f, f' \text{ are homomorphisms} \\
&= (c \times c')^{-1}(\mathrm{Rel}(F)((f \times f')^{-1}(R))) && \text{by Lemma 3.2.2.1.}
\end{aligned}$$

5. If $\coprod_{c\times c'}(R) \subseteq \mathrm{Rel}(F)(R)$, then:

$$\begin{aligned}
&\textstyle\coprod_{d\times d'} \coprod_{f\times f'}(R) \\
&= \textstyle\coprod_{F(f)\times F(f')} \coprod_{c\times c'}(R) && \text{since } f, f' \text{ are homomorphisms} \\
&\subseteq \textstyle\coprod_{F(f)\times F(f')}(\mathrm{Rel}(F)(R)) && \text{by assumption} \\
&= \mathrm{Rel}(F)(\textstyle\coprod_{f\times f'}(R)) && \text{by Lemma 3.2.2.2.}
\end{aligned}$$

□

Proposition 3.2.7 *Let $X \xrightarrow{c} F(X)$, $Y \xrightarrow{d} F(Y)$ and $Z \xrightarrow{e} F(Z)$ be three coalgebras of a Kripke polynomial functor F.*

1. *An arbitrary function $f: X \to Y$ is a homomorphism of coalgebras if and only if its graph relation* Graph(f) *is a bisimulation.*
2. *The equality relation* Eq(X) *on X is a bisimulation equivalence. More generally, for a homomorphism $f: X \to Y$, the kernel relation* Ker(f) *is a bisimulation equivalence.*
3. *For two homomorphisms $X \xleftarrow{f} Z \xrightarrow{g} Y$ the image of the tuple* $\mathrm{Im}(\langle f, g\rangle) \subseteq X \times Y$ *is a bisimulation.*
4. *For two homomorphisms $X \xrightarrow{f} Z \xleftarrow{g} Y$ in the opposite direction, the pullback relation* $\mathrm{Eq}(f, g) \subseteq X \times Y$ *is a bisimulation.*

Proof By using Lemma 3.2.1.

1. Because

$$\begin{aligned}
&\mathrm{Graph}(f) \text{ is a bisimulation} \\
&\Longleftrightarrow \mathrm{Graph}(f) \subseteq (c \times d)^{-1}(\mathrm{Rel}(F)(\mathrm{Graph}(f))) \\
&\qquad\qquad = (c \times d)^{-1}(\mathrm{Graph}(F(f))) && \text{by Lemma 3.2.5.2} \\
&\Longleftrightarrow \forall x, y.\, f(x) = y \Rightarrow F(f)(c(x)) = d(y) \\
&\Longleftrightarrow \forall x.\, F(f)(c(x)) = d(f(x)) \\
&\Longleftrightarrow f \text{ is a homomorphism of coalgebras from } c \text{ to } d.
\end{aligned}$$

2. The fact that equality is a bisimulation follows from Lemma 3.2.1.1. Further, the kernel Ker(f) is a bisimulation because it can be written as Graph(f) $\circ$ Graph(f)$^{-1}$, which is a bisimulation by point (1) and by Proposition 3.2.6.1, 3.2.6.2.
3. We wish to prove an inclusion:

$$\mathrm{Im}(\langle f, g\rangle) \subseteq (c \times d)^{-1}(\mathrm{Rel}(F)(\mathrm{Im}(\langle f, g\rangle))) = (c \times d)^{-1}(\mathrm{Im}(\langle F(f), F(g)\rangle)),$$

where we used Lemma 3.2.5.3 for the last equation. This means that we have to prove: for each $z \in Z$ there is a $w \in F(Z)$ with $c(f(z)) = F(f)(w)$ and $d(f(z)) = F(g)(w)$. But clearly we can take $w = e(z)$.

4. We now seek an inclusion:

$$\mathrm{Eq}(f, g) \subseteq (c \times d)^{-1}(\mathrm{Rel}(F)(\mathrm{Eq}(f, g))) = (c \times d)^{-1}(\mathrm{Eq}(F(f), F(g))),$$

via Lemma 3.2.5.4. But this amounts to $f(x) = g(y) \Rightarrow F(f)(c(x)) = G(f)(d(y))$, for all $x \in X, y \in Y$. The implication holds because both f and g are homomorphisms. ❑

We can now also establish some elementary properties of bisimilarity.

Proposition 3.2.8 *Let* $F\colon \mathbf{Sets} \to \mathbf{Sets}$ *be a Kripke polynomial functor, with coalgebras* $X \stackrel{c}{\to} F(X)$, $X' \stackrel{c'}{\to} F(X')$, *and* $Y \stackrel{d}{\to} F(Y)$, $Y' \stackrel{d'}{\to} F(Y')$.

1. *The bisimilarity relation* ${}_c\underline{\leftrightarrow}_d \subseteq X \times Y$ *is a bisimulation; it is the greatest among all bisimulations between c and d.*
2. $({}_c\underline{\leftrightarrow}_d)^\dagger \subseteq {}_d\underline{\leftrightarrow}_c$ *and* ${}_c\underline{\leftrightarrow}_e \circ {}_e\underline{\leftrightarrow}_d \subseteq {}_c\underline{\leftrightarrow}_d$.
3. *The bisimilarity relation* ${}_c\underline{\leftrightarrow}_c \subseteq X \times X$ *for a single coalgebra is a bisimulation equivalence.*
4. *For homomorphisms of coalgebras* $f\colon X \to Y$ *and* $f'\colon X' \to Y'$ *one has, for* $x \in X, x' \in X'$,

$$f(x)\ {}_d\underline{\leftrightarrow}_{d'}\ f'(x') \Longleftrightarrow x\ {}_c\underline{\leftrightarrow}_{c'}\ x'.$$

Proof 1. By Proposition 3.2.6.2.

2. The first inclusion follows from Proposition 3.2.6.1 and the second one from Proposition 3.2.6.2.
3. The bisimilarity relation ${}_c\underline{\leftrightarrow}_c \subseteq X \times X$ is reflexive, because the equality relation $\mathrm{Eq}(X) \subseteq X \times X$ is a bisimulation; see Proposition 3.2.6.2. Symmetry and transitivity of ${}_c\underline{\leftrightarrow}_c$ follow from point (2).
4. Since ${}_d\underline{\leftrightarrow}_{d'} \subseteq Y \times Y'$ is a bisimulation, so is $(f \times f')^{-1}({}_d\underline{\leftrightarrow}_{d'}) \subseteq X \times X'$, by Proposition 3.2.6.4. Hence $(f \times f')^{-1}({}_d\underline{\leftrightarrow}_{d'}) \subseteq {}_c\underline{\leftrightarrow}_{c'}$, which corresponds to the implication $(\Longrightarrow)$.

   Similarly we obtain an inclusion $\coprod_{f \times f'}({}_c\underline{\leftrightarrow}_{c'}) \subseteq {}_d\underline{\leftrightarrow}_{d'}$ from Proposition 3.2.6.5, which yields $(\Longleftarrow)$. ❑

Exercises

3.2.1 1. Prove that relation lifting $\mathrm{Rel}(F)$ for an *exponent* polynomial functor F (hence without powersets $\mathcal{P}$) preserves non-empty intersections of relations: for $I \neq \emptyset$,

$$\mathrm{Rel}(F)(\textstyle\bigcap_{i\in I} R_i) = \bigcap_{i\in I} \mathrm{Rel}(F)(R_i).$$

2. Assume now that F is now a simple polynomial functor (also without exponents $(-)^A$, for A infinite). Prove that relation lifting $\mathrm{Rel}(F)$ preserves unions of ascending chains of relations: if $S_0 \subseteq S_1 \subseteq S_2 \subseteq \cdots$, then

$$\mathrm{Rel}(F)(\textstyle\bigcup_{n\in\mathbb{N}} S_n) = \bigcup_{n\in\mathbb{N}} \mathrm{Rel}(F)(S_n).$$

Which additional closure properties hold for bisimulations for coalgebras of such functors?

3.2.2 1. Check that if $\leq$ is a preorder on a set X, then $\mathrm{Rel}(F)(\leq)$ is also a preorder on $F(X)$.

2. Prove the same with ‘poset’ instead of ‘preorder’, for an exponent polynomial functor F (without powerset).

3. Prove that a Galois connection $f \dashv g$ in

$$(X, \leq_X) \underset{g}{\overset{f}{\rightleftarrows}} (Y, \leq_Y)$$

yields a ‘lifted’ Galois connection $F(f) \dashv F(g)$ in

$$(F(X), \mathrm{Rel}(F)(\leq_X)) \underset{F(g)}{\overset{F(f)}{\rightleftarrows}} (F(Y), \mathrm{Rel}(F)(\leq_Y)).$$

3.2.3 Check that, in analogy with Proposition 3.2.6, congruences are closed under inverses, composition, arbitrary intersections and inverse and direct images.

3.2.4 Prove the following analogue of Proposition 3.2.7. Let F be a polynomial functor, with algebras $F(X) \xrightarrow{a} X$, $F(Y) \xrightarrow{b} Y$ and $F(Z) \xrightarrow{c} Z$.

1. A function $f\colon X \to Y$ is a homomorphism of algebras if and only if its graph relation $\mathrm{Graph}(f) \subseteq X \times Y$ is a congruence.
2. The kernel relation $\mathrm{Ker}(f)$ of an algebra homomorphism $f\colon X \to Y$ is always a congruence equivalence.
3. The image $\mathrm{Im}(\langle f, g\rangle) \subseteq X \times Y$ of a pair of algebra homomorphisms $X \xleftarrow{f} Z \xrightarrow{g} Y$ is a congruence.
4. The pullback relation $\mathrm{Eq}(f, g) \subseteq X \times Y$ of a span of algebra homomorphisms $X \xrightarrow{f} Z \xleftarrow{g} Y$ is a congruence.

3.2.5 Let $R \subseteq X \times X$ be an arbitrary relation on the state space of a coalgebra, and let $\overline{R}$ be the least equivalence relation containing R. Prove that if R is a bisimulation, then $\overline{R}$ is a bisimulation equivalence. *Hint*: Write $\overline{R}$ as union of iterated compositions R^n for $n \in \mathbb{N}$.

3.2.6 Check that lax relation lifting as introduced in Exercise 3.1.6 forms a functor in

$$\begin{array}{ccc}
\mathbf{Rel} & \xrightarrow{\ \mathrm{Rel}_{\sqsubseteq}(F)\ } & \mathbf{Rel} \\
\downarrow & & \downarrow \\
\mathbf{Sets} \times \mathbf{Sets} & \xrightarrow[\ F \times F\ ]{} & \mathbf{Sets} \times \mathbf{Sets}
\end{array}$$

and that simulations are coalgebras of this functor $\mathrm{Rel}_{\sqsubseteq}(F)$ – as in Lemma 3.2.4.

3.2.7 This exercise describes a simple characterisation from [154] of when a function is definable by induction. The analogue for coinduction is in Exercise 6.2.4.

Consider an initial algebra $F(A) \xrightarrow{\cong} A$ of a polynomial functor F, where $A \neq \emptyset$. Prove that a function $f : A \to X$ is defined by initiality (i.e. $f = \mathrm{int}_a$ for some algebra $a : T(X) \to X$ on its codomain) if and only its kernel $\mathrm{Ker}(f)$ is a congruence.

Hint: Extend the induced map $F(A)/\mathrm{Ker}(F(f)) \to X$ along the inclusion $F(A)/\mathrm{Ker}(F(f)) \rightarrowtail F(X)$ by using an arbitrary element in $F(A)/\mathrm{Ker}(F(f))$ obtained from $A \neq \emptyset$.

3.3 Bisimulations as Spans and Cospans

This section continues the investigation of bisimulations and focusses specifically on the relation between the definition of bisimulation used here, given in terms of relation lifting, and an earlier definition given by Aczel and Mendler; see [9, 12]. One of the main results is that these definitions are equivalent, as will be shown below in Theorem 3.3.2.

The first lemma below establishes an important technical relationship which forms the basis for the subsequent theorem. The lemma uses that relations can be considered as sets themselves. From a logical perspective this involves a form of comprehension; see e.g. [239, chapter 4, section 6] or [218, 45, 150].

Lemma 3.3.1 *Let* $F : \mathbf{Sets} \to \mathbf{Sets}$ *be a Kripke polynomial functor and* $R \subseteq X \times Y$ *be an arbitrary relation, written via explicit functions* $\langle r_1, r_2 \rangle : R \rightarrowtail X \times Y$. *The lifted relation* $\mathrm{Rel}(F)(R)$, *considered as a set, is a* **retract** *of* $F(R)$*:*

there are functions $\alpha\colon \mathrm{Rel}(F)(R) \to F(R)$ *and* $\beta\colon F(R) \to \mathrm{Rel}(F)(R)$ *with* $\beta \circ \alpha = \mathrm{id}_{\mathrm{Rel}(F)(R)}$. *Moreover, these* α *and* β *make the following triangle commute:*

$$\begin{array}{ccc}
\mathrm{Rel}(F)(R) & \underset{\beta}{\overset{\alpha}{\rightleftarrows}} & F(R) \\
& \searrow \qquad \swarrow {\scriptstyle \langle F(r_1), F(r_2)\rangle} & \\
& F(X) \times F(Y) &
\end{array}$$

This means that $\mathrm{Rel}(F)(R) \rightarrowtail F(X) \times F(Y)$ *is the image of* $F(R) \to F(X) \times F(Y)$.

Proof The functions α and β are constructed by induction on the structure of F. In the two base cases where F is the identity functor or a constant functor, α and β are each other's inverses. We shall consider two induction steps, for product and powerset.

If F is a product $F_1 \times F_2$, we assume appropriate functions $\alpha_i\colon \mathrm{Rel}(F_i)(R) \to F_i(R)$ and $\beta_i\colon F_i(R) \to \mathrm{Rel}(F_i)(R)$, for $i = 1, 2$. The aim is to construct functions α, β in

$$\left(\begin{array}{r}\{((u_1,u_2),(v_1,v_2)) \mid \mathrm{Rel}(F_1)(R)(u_1,v_1) \\ \wedge\ \mathrm{Rel}(F_2)(R)(u_2,v_2)\}\end{array}\right) \underset{\beta}{\overset{\alpha}{\rightleftarrows}} F_1(R) \times F_2(R).$$

The definitions are obvious:

$$\begin{aligned}
\alpha((u_1,u_2),(v_1,v_2)) &= (\,\alpha_1(u_1,v_1),\ \alpha_2(u_2,v_2)\,) \\
\beta(w_1,w_2) &= (\,(\pi_1\beta_1(w_1), \pi_1\beta_2(w_2)),\ (\pi_2\beta_1(w_1), \pi_2\beta_2(w_2))\,).
\end{aligned}$$

If F is a powerset $\mathcal{P}(F_1)$, we may assume functions $\alpha_1 : \mathrm{Rel}(F_1)(R) \to F_1(R)$ and $\beta_1 : F_1(R) \to \mathrm{Rel}(F_1)(R)$ as in the lemma. We have to construct

$$\left(\begin{array}{r}\{(U,V) \mid \forall u \in U.\, \exists v \in V.\, \mathrm{Rel}(F_1)(R)(u,v) \\ \wedge\ \forall v \in V.\, \exists u \in U.\, \mathrm{Rel}(F_1)(R)(u,v)\}\end{array}\right) \underset{\beta}{\overset{\alpha}{\rightleftarrows}} \mathcal{P}(F_1(R)).$$

In this case we define:

$$\begin{aligned}
\alpha(U,V) &= \{\alpha_1(u,v) \mid u \in U \wedge v \in V \wedge \mathrm{Rel}(F_1)(R)(u,v)\}, \\
\beta(W) &= (\{\pi_1\beta_1(w) \mid w \in W\}, \{\pi_2\beta_1(w) \mid w \in W\}).
\end{aligned}$$

Then, using that $\beta_1 \circ \alpha_1 = \mathrm{id}$ holds by assumption, we compute:

$$\begin{aligned}
(\beta \circ \alpha)(U,V) &= (\{\pi_1\beta_1\alpha_1(u,v) \mid u \in U \wedge v \in V \wedge \mathrm{Rel}(F)(R)(u,v)\}, \\
&\qquad \{\pi_2\beta_1\alpha_1(u,v) \mid u \in U \wedge v \in V \wedge \mathrm{Rel}(F)(R)(u,v)\}) \\
&= (\{u \in U \mid \exists v \in V.\, \mathrm{Rel}(F_1)(R)(u,v)\}, \\
&\qquad \{v \in V \mid \exists u \in U.\, \mathrm{Rel}(F_1)(R)(u,v)\}) \\
&= (U,V).
\end{aligned}$$

❑

The formulation of bisimulation that we are using here relies on relation lifting; see Definition 3.1.2 and Lemma 3.2.4.2, where bisimulations for coalgebras of a functor F are described as $\mathrm{Rel}(F)$-coalgebras. This comes from [218]. An earlier definition was introduced by Aczel and Mendler; see [9, 12]. With the last lemma we can prove the equivalence of these definitions.

Theorem 3.3.2 *Let $X \xrightarrow{c} F(X)$ and $Y \xrightarrow{d} F(Y)$ be two coalgebras of a Kripke polynomial functor F. A relation $\langle r_1, r_2\rangle \colon R \rightarrowtail X \times Y$ is a bisimulation for c and d if and only if R is an* ***Aczel–Mendler bisimulation****: R itself is the carrier of some coalgebra $e \colon R \to F(R)$, making the legs r_i homomorphisms of coalgebras, as in*

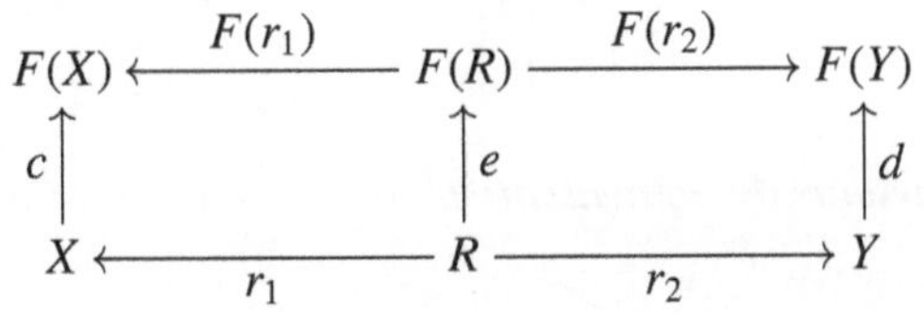

Thus, briefly: R carries a $\mathrm{Rel}(F)$*-coalgebra in the category* **Rel** *if and only if R carries an F-coalgebra in* **Sets** *making the diagram commute.*

Proof In Lemma 3.2.4.2 we already saw that $R \subseteq X \times Y$ is a bisimulation according to Definition 3.1.2 if and only if it carries a $\mathrm{Rel}(F)$-coalgebra; that is, if and only if the function $c \times d \colon X \times Y \to F(X) \times F(Y)$ restricts to a necessarily unique function $f \colon R \to \mathrm{Rel}(F)(R)$, making the square on the left below commute:

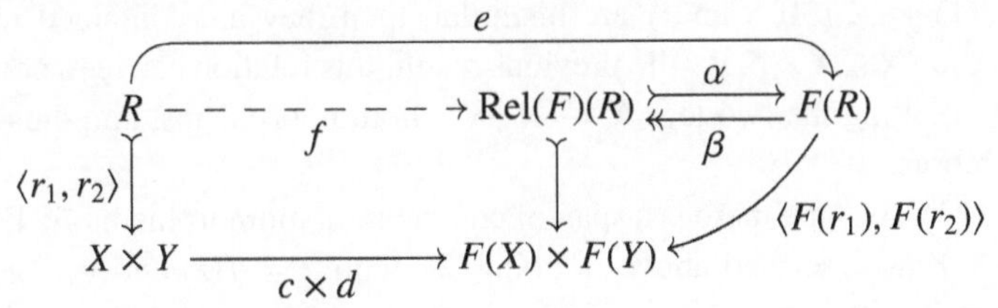

The functions α, β in the triangle on the right form the retract from the previous lemma. Then, if R is a bisimulation according to Definition 3.1.2, there is a function f as indicated, so that $\alpha \circ f \colon R \to F(R)$ yields an F-coalgebra on R making the legs r_i homomorphisms of coalgebras. Conversely, if there is a coalgebra $e \colon R \to F(R)$ making the r_i homomorphisms, then $\beta \circ e \colon R \to \mathrm{Rel}(F)(R)$ shows that $R \subseteq (c \times d)^{-1}(\mathrm{Rel}(F)(R))$. ❑

This result gives rise to another two alternative characterisations of bisimilarity. For non-polynomial functors on categories other than **Sets** these different descriptions may diverge; see [436] for more information.

Theorem 3.3.3 *Assume two elements $x \in X$ and $y \in Y$ of coalgebras $X \xrightarrow{c} F(X)$ and $Y \xrightarrow{d} F(Y)$ of a Kripke polynomial functor $F\colon \mathbf{Sets} \to \mathbf{Sets}$. The following statements are equivalent:*

1. *x, y are bisimilar: $x \;{}_{c}\!\underline{\leftrightarrow}_{d}\; y$.*
2. *There is a span of coalgebra homomorphisms:*

$$\begin{array}{ccc} & \bullet & \\ {}^{f}\swarrow & & \searrow^{g} \\ \begin{pmatrix} F(X) \\ \uparrow c \\ X \end{pmatrix} & & \begin{pmatrix} F(Y) \\ \uparrow d \\ Y \end{pmatrix} \end{array} \qquad \textit{with } x = f(z) \textit{ and } y = g(z), \textit{ for some element } z.$$

3. *x, y are **behaviourally equivalent**: there is a cospan of coalgebra homomorphisms:*

$$\begin{array}{ccc} \begin{pmatrix} F(X) \\ \uparrow c \\ X \end{pmatrix} & & \begin{pmatrix} F(Y) \\ \uparrow d \\ Y \end{pmatrix} \\ {}_{h}\searrow & & \swarrow_{k} \\ & \bullet & \end{array} \qquad \textit{with } h(x) = k(y).$$

For the proof we recall from Exercise 2.1.14 that the category **Sets** has coequalisers and that categories of coalgebras $\mathbf{CoAlg}(F)$ of endofunctors of **Sets** then also have such coequalisers.

Proof (1) ⇒ (2). If x and y are bisimilar, then they are contained in some bisimulation $R \subseteq X \times Y$. By the previous result, this relation carries a coalgebra structure making the two legs $X \leftarrow R \to Y$ homomorphisms, and thus a span of coalgebras.

(2) ⇒ (3). Assume there is a span of coalgebra homomorphisms $f\colon W \to X$, $g\colon W \to Y$ as described above in point (2), with $x = f(z)$ and $y = g(z)$, for some $z \in W$. One obtains a cospan as in (3) by taking the pushout of f, g in the category $\mathbf{CoAlg}(F)$. This notion of pushout has not been discussed yet, but it can be described in terms of notions that we have seen. We form the coequaliser in $\mathbf{CoAlg}(F)$ in

$$\begin{pmatrix} F(W) \\ \uparrow \\ W \end{pmatrix} \underset{\kappa_2 \circ g}{\overset{\kappa_1 \circ f}{\rightrightarrows}} \begin{pmatrix} F(X+Y) \\ \uparrow \\ X+Y \end{pmatrix} \overset{q}{\twoheadrightarrow} \begin{pmatrix} F(Q) \\ \uparrow \\ Q \end{pmatrix}$$

where the coalgebra in the middle on $X + Y$ is $[F(\kappa_1) \circ c, F(\kappa_2) \circ d]$, as in Proposition 2.1.5. Then we take $h = q \circ \kappa_1$ and $k = q \circ \kappa_2$, so that:

$$\begin{aligned} h(x) = q(\kappa_1 x) &= q(\kappa_1 f(z)) \\ &= q(\kappa_2 g(z)) \qquad \text{since } q \text{ is coequaliser} \\ &= q(\kappa_2 y) = k(y). \end{aligned}$$

(3) ⇒ (1). Assuming a cospan h, k of coalgebra homomorphisms with $h(x) = k(y)$ we take the pullback relation $\mathrm{Eq}(h,k) \subseteq X \times Y$. It is a bisimulation by Proposition 3.2.7 and it contains x, y by assumption. Hence $x \;{}_c\!\underline{\leftrightarrow}_d\; y$. □

We postpone a discussion of the different formulations of the notion of bisimulation to the end of this section. At this stage we shall use the new 'Aczel–Mendler' bisimulations in the following standard result – specifically in its point (2) – about monos and epis in categories of coalgebras.

Theorem 3.3.4 *Let $F\colon$ **Sets** → **Sets** be a Kripke polynomial functor.*

1. *For a coalgebra $c\colon X \to F(X)$ and a bisimulation equivalence $R \subseteq X \times X$, the quotient set X/R carries a unique quotient coalgebra structure, written as $c/R\colon X/R \to F(X/R)$, making the canonical quotient map $[-]_R\colon X \twoheadrightarrow X/R$ a homomorphism of coalgebras, as in*

$$\begin{array}{ccc} F(X) & \xrightarrow{F([-]_R)} & F(X/R) \\ {\scriptstyle c}\uparrow & & \uparrow{\scriptstyle c/R} \\ X & \underset{[-]_R}{\twoheadrightarrow} & X/R \end{array}$$

2. *A homomorphism of coalgebras $f\colon X \to Y$ from $X \overset{c}{\to} F(X)$ to $Y \overset{d}{\to} F(Y)$ is a monomorphism/epimorphism in the category $\mathbf{CoAlg}(F)$ if and only if f is an injective/surjective function between the underlying sets.*
3. *Every homomorphism of coalgebras factors as an epimorphism followed by a monomorphism in $\mathbf{CoAlg}(F)$. This factorisation is essentially unique because of the following 'diagonal-fill-in' property. For each commuting square of coalgebra homomorphisms as below, there is a unique diagonal homomorphism making both triangles commute:*

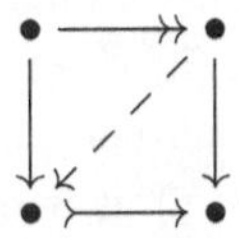

This means that monomorphisms and epimorphisms in a category of coalgebras **CoAlg**(F) *form a so-called factorisation system; see [56, 36, 146] and Section 4.3.*

As an aside, the result (2) that a monomorphism between coalgebras is an injective function holds for Kripke polynomial functors but not for arbitrary functors, as shown in [184].

Proof 1. It suffices to prove that $F([-]_R) \circ c$ is constant on R. But this is obvious:

$$\begin{aligned} R &\subseteq (c \times c)^{-1}(\mathrm{Rel}(F)(R)) && \text{since } R \text{ is a bisimulation} \\ &= (c \times c)^{-1}(\mathrm{Rel}(F)(\mathrm{Ker}([-]_R))) \\ &= (c \times c)^{-1}(\mathrm{Ker}(F([-]_R))) && \text{by Lemma 3.2.5.1} \\ &= \mathrm{Ker}(F([-]_R) \circ c). \end{aligned}$$

2. It is standard that if a coalgebra homomorphism f is injective/surjective, then it is a monomorphism/epimorphism in **Sets** – see also Exercise 2.5.8 – and hence also in **CoAlg**(F). Conversely, assume first that f is a monomorphism in **CoAlg**(F). The kernel $\langle r_1, r_2\rangle \colon \mathrm{Ker}(f) \rightarrowtail X \times X$ is a bisimulation by Proposition 3.2.7.2. Hence it carries an F-coalgebra structure by the previous theorem, making the r_i homomorphisms. From $f \circ r_1 = f \circ r_2$, we can conclude that $r_1 = r_2$, since f is a monomorphism in **CoAlg**(F). But $r_1 = r_2$ yields that f is injective:

$$\begin{aligned} f(x_1) = f(x_2) &\Longrightarrow (x_1, x_2) \in \mathrm{Ker}(f) \\ &\Longrightarrow x_1 = r_1(x_1, x_2) = r_2(x_1, x_2) = x_2. \end{aligned}$$

Next, assume that f is an epimorphism in **CoAlg**(F). There is a short categorical argument that tells that f is then an epi in **Sets**, and thus surjective: the forgetful functor **CoAlg**$(F) \to$ **Sets** creates colimits; see [411, proposition 4.7]. However, we spell out the argument. That f is epi in **CoAlg**(F) can be reformulated as: the following diagram is a coequaliser in **CoAlg**(F):

$$\begin{pmatrix} F(X) \\ \uparrow c \\ X \end{pmatrix} \underset{\kappa_2 \circ f}{\overset{\kappa_1 \circ f}{\rightrightarrows}} \begin{pmatrix} F(X+Y) \\ \uparrow \\ X+Y \end{pmatrix} \xrightarrow{[\mathrm{id}, \mathrm{id}]} \!\!\!\!\rightarrow \begin{pmatrix} F(Y) \\ \uparrow d \\ Y \end{pmatrix}$$

where the coalgebra in the middle is $[F(\kappa_1) \circ d, F(\kappa_2) \circ d]$, as in Proposition 2.1.5. Since in **Sets** a map is surjective if and only if it is an epimorphism – see Exercise 2.5.8 or any basic textbook in category theory – it suffices to show that f is an epimorphism in **Sets**. Thus we construct the coequaliser $q \colon Y + Y \twoheadrightarrow Q$ in **Sets** of $\kappa_1 \circ f, \kappa_2 \circ f \colon X \to Y + Y$.

By Exercise 2.1.14 this quotient set Q carries a unique coalgebra $Q \to F(Q)$ making q not only a homomorphism in **CoAlg**(F) but also a coequaliser of the homomorphisms $\kappa_1 \circ f, \kappa_2 \circ f$. Since coequalisers are determined up-to-isomorphism there is an isomorphism $\varphi\colon Y \xrightarrow{\cong} Q$ in **CoAlg**(F) with $\varphi \circ [\mathrm{id}, \mathrm{id}] = q$. This means that the codiagonal $[\mathrm{id}, \mathrm{id}]$ is also the coequaliser of $\kappa_1 \circ f, \kappa_2 \circ f$ in **Sets**, and thus that f is an epimorphism in **Sets** (and hence surjective).

3. Given a homomorphism of coalgebras $f\colon X \to Y$, we know by Proposition 3.2.7.2 that the kernel $\mathrm{Ker}(f)$ is a bisimulation equivalence. Hence by point (1) the quotient $X/\mathrm{Ker}(f)$ carries a coalgebra structure and yields a standard factorisation (in **Sets**):

$$\Big(X \xrightarrow{\;f\;} Y\Big) = \Big(X \overset{e}{\twoheadrightarrow} X/\mathrm{Ker}(f) \overset{m}{\rightarrowtail} Y\Big).$$

The map $e\colon X \twoheadrightarrow X/\mathrm{Ker}(f)$ is by definition of the unique coalgebra structure c' on $X/\mathrm{Ker}(f)$ – see point (1) – a homomorphism of coalgebras in

$$\begin{array}{ccccc}
F(X) & \xrightarrow{\;F(e)\;} & F(X/\mathrm{Ker}(f)) & \xrightarrow{\;F(m)\;} & F(Y) \\
{\scriptstyle c}\uparrow & & \uparrow{\scriptstyle c' = c/\mathrm{Ker}(f)} & & \uparrow{\scriptstyle d} \\
X & \underset{e}{\twoheadrightarrow} & X/\mathrm{Ker}(f) & \underset{m}{\rightarrowtail} & Y
\end{array}$$

Not only the square on the left, but also the one on the right commutes; using that e is an epimorphism: $d \circ m = F(m) \circ c'$ follows from

$$\begin{aligned}
d \circ m \circ e &= d \circ f \\
&= F(f) \circ c \\
&= F(m) \circ F(e) \circ c \\
&= F(m) \circ c' \circ e.
\end{aligned}$$

Thus, $m\colon X/\mathrm{Ker}(f) \rightarrowtail Y$ is also a homomorphism of coalgebras. Hence we have a factorisation $X \twoheadrightarrow X/\mathrm{Ker}(f) \rightarrowtail Y$ of f in the category **CoAlg**(F).

Regarding the diagonal-fill-in property, the diagonal is defined via the surjectivity of the top arrow. Hence this diagonal is a homomorphism of coalgebras. □

3.3.1 Comparing Definitions of Bisimulation

We started this chapter by introducing bisimulations via the logical technique of relation lifting, and later showed equivalence to quite different formulations in terms of (co)spans (Theorems 3.3.2 and 3.3.3). We shall discuss some differences between these 'logical' and 'span-based' approaches.

1. The logical approach describes bisimulations as relations with a special *property*, whereas the (co)span-based approaches rely on the existence of certain (coalgebra) *structure*. This aspect of the logical approach is more appropriate, because it is in line with the idea that bisimulations are special kinds of relations. If, in the Aczel–Mendler approach, the coalgebra structure is necessarily unique, existence of this structure also becomes a property. But uniqueness is neither required nor guaranteed. This is slightly unsatisfactory.
2. Relation lifting has been defined by induction on the structure of Kripke polynomial functors. Therefore, the logical approach (so far) applies only to such a limited collection of functors. Later, in Section 4.4 we will extend such lifting to functors on categories equipped with a factorisation system. But (co)span-based approaches apply more generally, without such factorisation structure. However, the dependence on such structure may also be seen as an advantage, as will be argued next.
3. Relation lifting is a logical technique which is not restricted to the standard classical logic of sets but may be defined for more general (categorical) logics, in terms of factorisation systems (see Section 4.4) or in terms of 'indexed' or 'fibred' preorders; see [218, 239]. For instance, one may wish to consider topological spaces with different logics, for instance with predicates given by the subsets which are closed, open, or both (clopen). Each of those preorders of predicates has different algebraic/logical properties. Thus, the logical approach is more general (or flexible) in another, logical dimension.
4. The cospan formulation (or behavioural equivalence) is essentially equivalent to equality on the final coalgebra. But the convenient aspect of behavioural equivalence is that it makes also makes sense in contexts where there is no final coalgebra.
5. Bisimulation relations can be constructed iteratively, by adding pairs until the relation is appropriately closed. This is done for instance in tools such as CIRC [338]. Such techniques are not available for behavioural equivalence.

There is no clear answer as to which approach is 'the best'. In more general situations – functors other than the polynomial ones, on categories other than **Sets** – the various notions may diverge (see [436]) and one may be better than the other. For instance, in coalgebraic modal logic the cospan-based approach, using behavioural equivalence (see Theorem 3.3.3.3), seems to work best. We have chosen to start from the logical approach because we consider it to be more intuitive and easier to use in concrete examples. However, from now on we shall freely switch between the different approaches and use whatever is most convenient in a particular situation.

In Chapter 6 on invariants we shall encounter the same situation. There is a logical definition based on 'predicate lifting'. It leads to a notion of invariant, which, in the classical logic of sets, is equivalent to the notion of subcoalgebra; see Theorem 6.2.5. The latter is again defined in terms of structure and applies to more general functors.

3.3.2 Congruences and Spans

We conclude this section with two basic results for algebra. The first one is an analogue of the Aczel–Mendler formulation of bisimulations (see Theorem 3.3.2) for congruences on algebras. The second one involves quotient algebras; it requires an auxiliary technical result about quotients in **Sets**.

Theorem 3.3.5 *Assume two algebras $a\colon F(X) \to X$ and $b\colon F(Y) \to Y$ of a Kripke polynomial functor F. A relation $R \subseteq X \times Y$ is a congruence relation if and only it carries a (necessarily unique) algebra structure $F(R) \to R$ itself, making the two legs $\langle r_1, r_2\rangle\colon R \rightarrowtail X \times Y$ of the inclusion homomorphisms of algebras, as in*

$$\begin{array}{ccccc} F(X) & \xleftarrow{\;F(r_1)\;} & F(R) & \xrightarrow{\;F(r_2)\;} & F(Y) \\ {\scriptstyle a}\downarrow & & \downarrow & & \downarrow{\scriptstyle b} \\ X & \xleftarrow[\;r_1\;]{} & R & \xrightarrow[\;r_2\;]{} & Y \end{array}$$

Proof Recall the retract (α, β), with $\beta \circ \alpha = \mathrm{id}$, from Lemma 3.3.1 in

$$\begin{array}{ccccc} F(R) & \underset{\beta}{\overset{\alpha}{\rightleftarrows}} & \mathrm{Rel}(F)(R) & \overset{f}{\dashrightarrow} & R \\ & {\scriptstyle \langle F(r_1), F(r_2)\rangle}\searrow & \downarrow & & \downarrow{\scriptstyle \langle r_1, r_2\rangle} \\ & & F(X) \times F(Y) & \xrightarrow[\;a \times b\;]{} & X \times Y \end{array}$$

Assume R is a congruence, say via $f\colon \mathrm{Rel}(F)(R) \to R$ as above. Then $f \circ \alpha\colon F(R) \to R$ is an algebra such that the r_i are homomorphisms. Conversely, if an algebra $c\colon F(R) \to R$ making the r_i morphisms of algebras exists, then $c \circ \beta$ ensures $\mathrm{Rel}(F)(R) \leq (a \times b)^{-1}(R)$. ❑

Theorem 3.3.4 describes coalgebras on quotients in the category **Sets**. Below we use some special properties of sets, like the axiom of choice and the fact that each equivalence relation is 'effective': it is equal to the kernel of its quotient map.

Lemma 3.3.6 *Let $F\colon \mathbf{Sets} \to \mathbf{Sets}$ be an arbitrary endofunctor. If $q\colon X \twoheadrightarrow X/R$ is the quotient map of an equivalence relation $\langle r_1, r_2\rangle\colon R \hookrightarrow X \times X$, then we have a coequaliser diagram:*

$$F(R) \underset{F(r_2)}{\overset{F(r_1)}{\rightrightarrows}} F(X) \xrightarrow{\quad F(q) \quad} F(X/R).$$

Proof The quotient map $q\colon X \twoheadrightarrow X/R$ sends an element $x \in X$ to its equivalence class $q(x) = [x]_R = \{x' \in X \mid R(x, x')\}$ with respect to the relation R. It is surjective, so by the axiom of choice we may assume a section $s\colon X/R \to X$ with $q \circ s = \mathrm{id}$ – as noted at the end of Section 2.1. This section s chooses a representative in each equivalence class. Since $q(x) = q(s(q(x)))$ we get $R(x, s(q(x)))$, for each $x \in X$. Thus we can define a function $d\colon X \to R$ by $d(x) = (x, s(q(x)))$; by construction it satisfies $r_1 \circ d = \mathrm{id}$ and $r_2 \circ d = s \circ q$.

We now turn to the diagram in the lemma. First, we know that the map $F(e)$ is surjective, because it has $F(s)$ as a section: $F(q) \circ F(s) = F(q \circ s) = \mathrm{id}$. Next, assume we have a map $f\colon F(X) \to Y$ with $f \circ F(r_1) = f \circ F(r_2)$. Then

$$\begin{aligned} f = f \circ \mathrm{id}_{F(X)} &= f \circ F(r_1) \circ F(d) \\ &= f \circ F(r_2) \circ F(d) && \text{by assumption about } f. \\ &= f \circ F(s) \circ F(q) \end{aligned}$$

This shows that $f' = f \circ F(s)\colon F(X/R) \to Y$ is the required mediating map: $f' \circ F(q) = f \circ F(s) \circ F(q) = f$. Clearly, f' is unique with this property, because $F(q)$ is an epimorphism (i.e. is surjective). ❑

Proposition 3.3.7 *Let $R \subseteq X \times X$ be a congruence equivalence relation* (i.e. *both a congruence and an equivalence relation) on an algebra $F(X) \to X$. The quotient X/R carries a unique algebra structure $a/R\colon F(X/R) \to X/R$ making the quotient map $q\colon X \to X/R$ a homomorphism of algebras.*

Proof Since R is a quotient we may assume by Theorem 3.3.5 an algebra $c\colon F(R) \to R$, as on the left in

$$\begin{array}{ccccc} F(R) & \underset{F(r_2)}{\overset{F(r_1)}{\rightrightarrows}} & F(X) & \xrightarrow{\quad F(q) \quad} & F(X/R) \\ {\scriptstyle c}\downarrow & & \downarrow{\scriptstyle a} & & \downarrow{\scriptstyle a/R} \\ R & \underset{r_2}{\overset{r_1}{\rightrightarrows}} & X & \xrightarrow[\quad q \quad]{} & X/R \end{array}$$

The algebra structure a/R on the right exists because the top row is a coequaliser by Lemma 3.3.6 and

$$\begin{aligned} q \circ a \circ F(r_1) &= q \circ r_1 \circ c \\ &= q \circ r_2 \circ c \qquad \text{since } q \text{ is coequaliser.} \\ &= q \circ a \circ F(r_2) \end{aligned}$$

❑

The last quotient result is of course well known in universal algebra. Here we obtain it in a uniform manner, for many functors (seen as signatures).

Exercises

3.3.1 Assume a homomorphism of coalgebras $f\colon X \to Y$ has two factorisations $X \twoheadrightarrow U \rightarrowtail Y$ and $X \twoheadrightarrow V \rightarrowtail Y$. Prove that the diagonal-fill-in property of Theorem 3.3.4.3 yields a unique isomorphism $U \xrightarrow{\cong} V$ commuting with the mono- and epimorphisms and with the coalgebra structures.

3.3.2 Use Theorem 3.3.5 to prove the binary induction proof principle in Theorem 3.1.4: every congruence on an initial algebra is reflexive.

3.3.3 Assume $f\colon X \to Y$ is an epimorphism in a category **CoAlg**(F), for a Kripke polynomial functor F on **Sets**. Prove that f is the coequaliser in **CoAlg**(F) of its own kernel pair $p_1, p_2\colon \mathrm{Ker}(f) \to X$. *Hint*: Use that this property holds in **Sets** and lift it to **CoAlg**(F).

3.3.4 Formally, a **quotient** of an object X in a category is an equivalence class of epimorphisms $X \twoheadrightarrow \bullet$. Two epis $e\colon X \twoheadrightarrow U$ and $d\colon X \twoheadrightarrow V$ are equivalent when there is a necessarily unique isomorphism $h\colon U \to V$ with $h \circ e = d$. As for subobjects – which are equivalence classes of monomorphisms – we often confuse a quotient with an epi that represents it.

We now concentrate on the category **Sets** and form a category **Quot** with

objects quotients $X \twoheadrightarrow U$

morphisms from $(X \overset{e}{\twoheadrightarrow} U)$ to $(Y \overset{d}{\twoheadrightarrow} V)$ that are maps $f\colon X \to Y$ for which there is a necessarily unique map:

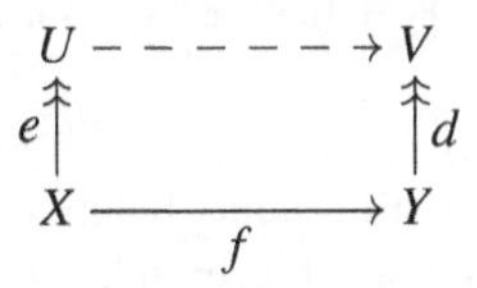

Use that quotients are effective in **Sets** to prove that there is an isomorphism of categories:

$$\begin{array}{ccc} \mathbf{Quot} & \xrightarrow{\;\cong\;} & \mathbf{EqRel} \\ & \searrow \quad \swarrow & \\ & \mathbf{Sets} & \end{array}$$

where **EqRel** $\hookrightarrow$ **Rel** is the subcategory of equivalence relations $R \rightarrowtail X \times X$.

3.4 Bisimulations and the Coinduction Proof Principle

We have already seen that states of final coalgebras coincide with behaviours and that bisimilarity captures observational indistinguishability. Hence the following fundamental result does not come as a surprise: states are bisimilar if and only if they have the same behaviour, i.e. become equal when mapped to the final coalgebra. This insight has been important in the development of the field of coalgebra.

Theorem 3.4.1 ([453, 411]) *Let* $F\colon \mathbf{Sets} \to \mathbf{Sets}$ *be a (finite) polynomial functor which has a final coalgebra* $\zeta\colon Z \xrightarrow{\cong} F(Z)$*. Let* $c\colon X \to F(X)$ *and* $d\colon Y \to F(Y)$ *be arbitrary coalgebras, with associated behaviour homomorphisms* $\mathrm{beh}_c\colon X \to Z$ *and* $\mathrm{beh}_d\colon Y \to Z$ *given by finality. Two states* $x \in X$ *and* $y \in Y$ *are then bisimilar if and only if they have the same behaviour:*

$$x \; {}_{c}{\underline{\leftrightarrow}}_{d} \; y \iff \mathrm{beh}_c(x) = \mathrm{beh}_d(y).$$

In particular, bisimilarity ${}_{\zeta}{\underline{\leftrightarrow}}_{\zeta} \subseteq Z \times Z$ *on the final coalgebra is equality.*

Proof ($\Leftarrow$) This is easy since we know by Proposition 3.2.7.4 that the pullback relation $\mathrm{Eq}(\mathrm{beh}_c, \mathrm{beh}_d) = \{(x, y) \mid \mathrm{beh}_c(x) = \mathrm{beh}_d(y)\} \subseteq X \times Y$ of two homomorphisms is a bisimulation. Hence it is included in the greatest bisimulation ${}_{c}{\underline{\leftrightarrow}}_{d}$.

($\Rightarrow$) The bisimilarity relation ${}_{c}{\underline{\leftrightarrow}}_{d}$ is itself a bisimulation, so it carries by Theorem 3.3.2 a coalgebra structure $e\colon ({}_{c}{\underline{\leftrightarrow}}_{d}) \to F({}_{c}{\underline{\leftrightarrow}}_{d})$ making the two legs r_i of the relation $\langle r_1, r_2\rangle : ({}_{c}{\underline{\leftrightarrow}}_{d}) \rightarrowtail X \times Y$ homomorphisms. By finality we then get $\mathrm{beh}_c \circ r_1 = \mathrm{beh}_d \circ r_2$, yielding the required result. ❑

This result gives rise to an important proof method for establishing that two states have the same behaviour. This method is often referred to as the coinduction proof principle and goes back to [354]. It corresponds to the uniqueness part of the unique existence property of behaviour maps in Definition 2.3.1.

Corollary 3.4.2 (Coinduction proof principle) *Two states have the same behaviour if and only if there is a bisimulation that contains them.*

Consequently, every bisimulation on a final coalgebra is contained in the equality relation. □

The second formulation in this corollary is sometimes called 'internal full abstractness'. It is the dual of the binary induction principle in Theorem 3.1.4, stating that on an initial algebra each congruence contains the equality relation (i.e. is reflexive).

As we shall see in Example 3.4.5 below, it may sometimes require a bit of ingenuity to produce an appropriate bisimulation. The standard way to find such a bisimulation is to start with the given equation as relation and close it off with successor states until no new elements appear. In that case one has only 'circularities'. Formalising this approach led to what is sometimes called circular rewriting; see e.g. [163], implemented in the tool CIRC [338].

Corollary 3.4.3 *Call a coalgebra* $c\colon X \to F(X)$ ***observable*** *if its bisimilarity relation* ${}_c\!\underline{\leftrightarrow}_c$ *is equality on X. This is equivalent to a generalisation of the formulation used in Exercise 2.5.14 for deterministic automata: the associated behaviour map* $\mathrm{beh}_c\colon X \to Z$ *to the final coalgebra* $Z \xrightarrow{\cong} F(Z)$*, if any, is injective.* □

Observable coalgebras are called *simple* in [411]. Coalgebras can always be forced to be observable via quotienting; see Exercise 3.4.1 below.

Bisimilarity is closely related to equivalence of automata, expressed in terms of equality of accepted languages (see Corollary 2.3.6). This result, going back to [371], will be illustrated next.

Corollary 3.4.4 *Consider two deterministic automata* $\langle \delta_i, \epsilon_i \rangle\colon S_i \to S_i^A \times \{0,1\}$ *with initial states* $s_i \in S_i$*, for* $i = 1, 2$*. These states* s_1, s_2 *are called* ***equivalent*** *if they accept the same language. The states* s_1 *and* s_2 *are then equivalent if and only if they are bisimilar.*

Proof The accepted languages are given by the behaviours $\mathrm{beh}_{\langle \delta_i, \epsilon_i \rangle}(s_i) \in \mathcal{P}(A^\star)$ of the initial states; see Corollary 2.3.6.2. □

Early on in Section 1.2 we already saw examples of coinductive reasoning for sequences. Here we continue to illustrate coinduction with (regular) languages.

Example 3.4.5 (Equality of regular languages [408]) In Corollary 2.3.6.2 we have seen that the set $\mathcal{L}(A) = \mathcal{P}(A^\star)$ of languages over an alphabet A forms a final coalgebra, namely for the deterministic automaton functor $X \mapsto X^A \times \{0, 1\}$. We recall that the relevant coalgebra structure on $\mathcal{L}(A)$ is given on a language $L \subseteq A^\star$ by

$$L \xrightarrow{a} L_a \qquad \text{where } L_a = \{\sigma \in A^\star \mid a \cdot \sigma \in L\} \text{ is the } a\text{-derivative of } L$$
$$L \downarrow 1 \iff \langle\rangle \in L \text{ which may simply be written as } L \downarrow.$$

The subset $\mathcal{R}(A) \subseteq \mathcal{L}(A)$ of so-called **regular languages** is built up inductively from constants

$$0 = \emptyset, \qquad 1 = \{\langle\rangle\}, \qquad \{a\}, \text{ for } a \in A, \text{ usually written simply as } a,$$

and the three operations of **union**, **concatenation** and **Kleene star**:

$$\begin{aligned} K + L &= K \cup L \\ KL &= \{\sigma \cdot \tau \mid \sigma \in K \wedge \tau \in L\} \\ K^* &= \textstyle\bigcup_{n \in \mathbb{N}} K^n, \quad \text{where } K^0 = 1 \text{ and } K^{n+1} = KK^n. \end{aligned}$$

See also Exercise 3.4.4 below. For example, the regular language $a(a + b)^*b$ consists of all (finite) words consisting of letters a, b only, that start with an a and end with a b. Regular languages can be introduced in various other ways, for example as the languages accepted by deterministic and non-deterministic automata with a finite state space (via what is called Kleene's theorem [294], or [430] for coalgebraic versions), or as the languages generated by regular grammars. Regular languages (or expressions) are used in many situations, such as lexical analysis (as patterns for tokens) or text editing and retrieval (for context searches). Regular expressions (see Exercise 3.4.4) are often used as search strings in a Unix/Linux environment; for example in the command `grep`, for 'general regular expression parser'.

An important topic is proving equality of regular languages. There are several approaches, namely via unpacking the definitions, via algebraic reasoning using a complete set of laws (see [309] and also [246]) or via minimalisation of associated automata. A fourth, coinductive approach is introduced in [408] using bisimulations. It is convenient and will be illustrated here.

Recall from Section 3.1 that a relation $R \subseteq \mathcal{L}(A) \times \mathcal{L}(A)$ is a bisimulation if for all languages L, K,

$$R(L, K) \Longrightarrow \begin{cases} R(L_a, K_a) & \text{for all } a \in A \\ L\downarrow \text{ iff } K\downarrow. \end{cases}$$

The coinduction proof principle then says, for $L, K \in \mathcal{L}(A)$,

$$L = K \iff \text{there is a bisimulation } R \subseteq \mathcal{L}(A) \times \mathcal{L}(A) \text{ with } R(L, K). \qquad (3.1)$$

We will use this same principle for the subset $\mathcal{R}(A) \subseteq \mathcal{L}(A)$ of regular languages. This is justified by the fact that the inclusion map $\mathcal{R}(A) \hookrightarrow \mathcal{L}(A)$ is the unique homomorphism of coalgebras obtained by finality in

$$\begin{array}{ccc} \mathcal{R}(A)^A \times 2 & \dashrightarrow & \mathcal{L}(A)^A \times 2 \\ \uparrow & & \uparrow \cong \\ \mathcal{R}(A) & \dashrightarrow & \mathcal{L}(A) = \mathcal{P}(A^\star) \end{array}$$

This works because the coalgebra structure on $\mathcal{L}(A)$ restricts to $\mathcal{R}(A)$. This coalgebra on $\mathcal{R}(A)$ is given by the following rules for (Brzozowski) derivatives L_a and termination $L\downarrow$:

$$\begin{array}{rcll} 0_a & = & 0 & \neg(0\downarrow) \\ 1_a & = & 0 & 1\downarrow \\ b_a & = & \begin{cases} 1 & \text{if } b = a \\ 0 & \text{otherwise} \end{cases} & \neg(b\downarrow) \\ (K+L)_a & = & K_a + L_a & K+L\downarrow \text{ iff } K\downarrow \text{ or } L\downarrow \\ (KL)_a & = & \begin{cases} K_aL + L_a & \text{if } K\downarrow \\ K_aL & \text{otherwise} \end{cases} & KL\downarrow \text{ iff } K\downarrow \wedge L\downarrow \\ (K^*)_a & = & K_aK^* & K^*\downarrow. \end{array}$$

We shall illustrate the use of the coinduction proof principle (3.1) for establishing equality of regular languages via two examples.

1. For an arbitrary element a in the alphabet A one has

$$(1+a)^* = a^*.$$

As candidate bisimulation R in (3.1) we take

$$R = \{((1+a)^*, a^*)\} \cup \{(0,0)\}.$$

The termination requirement obviously holds, so we concentrate on derivatives. First, for a itself:

$$((1+a)^*)_a = (1+a)_a\,(1+a)^* = (1_a + a_a)\,(1+a)^* = (0+1)\,(1+a)^* = (1+a)^*$$

and similarly

$$(a^*)_a = a_a\,a^* = 1\,a^* = a^*.$$

Hence the pair of a-derivatives $(((1+a)^*)_a, (a^*)_a) = ((1+a)^*, a^*)$ is again in the relation R. Likewise, $(0_a, 0_a) = (0,0) \in R$. And for an element $b \neq a$ we similarly have $(((1+a)^*)_b, (a^*)_b) = (0,0) \in R$. This shows that R is a bisimulation, and completes the proof. The reader may wish to compare it with an alternative proof of equality, using the definition of Kleene star $(-)^*$.

2. Next we restrict ourselves to an alphabet $A = \{a, b\}$ consisting of two (different) letters only. Consider the two languages

$$\begin{aligned} E(b) &= \{\sigma \in A^\star \mid \sigma \text{ contains an even number of } bs\} \\ O(b) &= \{\sigma \in A^\star \mid \sigma \text{ contains an odd number of } bs\}. \end{aligned}$$

(We consider $0 \in \mathbb{N}$ to be even.) Using the definitions of derivative and termination, it is not hard to see that

$$\begin{array}{lll} E(b)_a = E(b) & E(b)_b = O(b) & E(b)\downarrow \\ O(b)_a = O(b) & O(b)_b = E(b) & \neg\, O(b)\downarrow. \end{array}$$

Our aim is to prove the equality

$$E(b) = a^* + a^*b(a + ba^*b)^*ba^*$$

via coinduction. This requires by (3.1) a bisimulation R containing both sides of the equation. We take

$$\begin{aligned} R = \{(E(b), K)\} \cup \{(O(b), L)\} \text{ where} \\ K &= a^* + a^*b(a + ba^*b)^*ba^* \quad \text{(the right-hand side of the equation)} \\ L &= (a + ba^*b)^*ba^*. \end{aligned}$$

The computations that show that R is a bisimulation use the above computation rules for derivatives plus some obvious properties of the regular operations (such as associativity, commutativity, $X + 0 = X$ and $1X = X$):

$$\begin{aligned} K_a &= (a^*)_a + (a^*)_a b(a + ba^*b)^*ba^* + (b(a + ba^*b)^*ba^*)_a \\ &= a^* + a^*b(a + ba^*b)^*ba^* + b_a(a + ba^*b)^*ba^* \\ &= K + 0(a + ba^*b)^*ba^* \\ &= K \\ K_b &= (a^*)_b + (a^*)_b b(a + ba^*b)^*ba^* + (b(a + ba^*b)^*ba^*)_b \\ &= 0 + 0b(a + ba^*b)^*ba^* + b_b(a + ba^*b)^*ba^* \\ &= 0 + 0 + 1(a + ba^*b)^*ba^* \\ &= L \\ L_a &= ((a + ba^*b)^*)_a ba^* + (ba^*)_a \\ &= (a + ba^*b)_a(a + ba^*b)^*ba^* + b_a a^* \\ &= (a_a + b_a a^*b)(a + ba^*b)^*ba^* + 0a^* \\ &= (1 + 0a^*b)(a + ba^*b)^*ba^* + 0 \\ &= L \\ L_b &= ((a + ba^*b)^*)_b ba^* + (ba^*)_b \\ &= (a + ba^*b)_b(a + ba^*b)^*ba^* + b_b a^* \\ &= (a_b + b_b a^*b)(a + ba^*b)^*ba^* + 1a^* \\ &= a^*b(a + ba^*b)^*ba^* + a^* \\ &= K. \end{aligned}$$

This shows that $(U, V) \in R$ implies both $(U_a, V_a) \in R$ and $(U_b, V_b) \in R$. Further:

$$\begin{aligned} K\downarrow &\Longleftrightarrow a^* \downarrow \;\vee\; a^*b(a+ba^*b)^*ba^* \downarrow \\ &\Longleftrightarrow \text{true} \\ L\downarrow &\Longleftrightarrow (a+ba^*b)^* \downarrow \;\wedge\; ba^* \downarrow \\ &\Longleftrightarrow \text{true} \;\wedge\; b\downarrow \;\wedge\; a^* \downarrow \\ &\Longleftrightarrow \text{true} \;\wedge\; \text{false} \;\wedge\; \text{true} \\ &\Longleftrightarrow \text{false}. \end{aligned}$$

This shows that R is a bisimulation. As a result we obtain $E(b) = K$, as required, but also $O(b) = L$.

This concludes the example. For more information, see [408, 410, 412].

There are many more examples of coinductive reasoning in the literature, in various areas: non-well-founded sets [9, 60], processes [355], functional programs [172], streams [412, 215] (with analytic functions as special case [376]), datatypes [214], domains [133, 380] etc.

Exercises

3.4.1 Check that for an arbitrary coalgebra $c\colon X \to F(X)$ of a Kripke polynomial functor, the induced quotient coalgebra $c/\underline{\leftrightarrow}\colon X/\underline{\leftrightarrow} \to F(X/\underline{\leftrightarrow})$ is observable – using both Theorem 3.3.4.1 and Proposition 3.2.8.3. Is the mapping $c \mapsto c/\underline{\leftrightarrow}$ functorial?

Note that since the canonical map $[-]\colon X \twoheadrightarrow X/\underline{\leftrightarrow}$ is a homomorphism, its graph is a bisimulation. Hence a state $x \in X$ is bisimilar to its equivalence class $[x] \in X/\underline{\leftrightarrow}$. This means that making a coalgebra observable does not change the behaviour.

3.4.2
1. ([411, theorem 8.1]) Prove that a coalgebra c is observable (or simple) if and only if it has no proper quotients: every epimorphism $c \twoheadrightarrow d$ is an isomorphism. *Hint*: Consider the kernel of such a map.
2. Conclude that there is at most one homomorphism to an observable coalgebra.

3.4.3 Prove the following analogue of Theorem 3.4.1 for two arbitrary algebras $a\colon F(X) \to X$ and $b\colon F(Y) \to Y$ of a Kripke polynomial functor F, with an initial algebra $F(A) \xrightarrow{\cong} A$.

Two elements $x \in X$ and $y \in Y$ are interpretations $x = \text{int}_a(t)$ and $y = \text{int}_b(t)$ of the same element $t \in A$ if and only if the pair (x, y) is in each congruence relation $R \subseteq X \times Y$.

Conclude that for coalgebras c, d and algebras a, b:

$$\begin{aligned}
\mathrm{Eq}(\mathrm{beh}_c, \mathrm{beh}_d) &\stackrel{\text{def}}{=} (\mathrm{beh}_c \times \mathrm{beh}_d)^{-1}(\mathrm{Eq}) \\
&= \bigcup\{R \mid R \text{ is a bisimulation on the carriers of } c, d\} \\
\mathrm{Im}(\langle \mathrm{int}_a, \mathrm{int}_b\rangle) &\stackrel{\text{def}}{=} \coprod_{\mathrm{int}_a \times \mathrm{int}_b}(\mathrm{Eq}) \\
&= \bigcap\{R \mid R \text{ is a congruence on the carriers of } a, b\}.
\end{aligned}$$

3.4.4 Fix an alphabet A and consider the simple polynomial functor

$$\mathcal{R}(X) = 1 + 1 + A + (X \times X) + (X \times X) + X.$$

1. Show that the initial algebra RE of $\mathcal{R}$ is the set of **regular expressions**, given by the BNF syntax:

$$E := 0 \mid 1 \mid a \mid E + E \mid EE \mid E^*$$

where $a \in A$.

2. Define an interpretation map $\mathrm{int}\colon RE \to \mathcal{P}(A^\star) = \mathcal{L}(A)$ by initiality, whose image contains precisely the regular languages. In order to do so one needs an $\mathcal{R}$-algebra structure on the final coalgebra $\mathcal{L}(A)$; see also [246].

3.4.5 Use coinduction to prove the equation

$$(a + b)^* = (a^* b)^* a^* \qquad \text{for alphabet } A = \{a, b, c\}.$$

3.4.6 (From [408]) Prove the following equality of regular languages (over the alphabet $\{a, b\}$) by coinduction:

$$((b^* a)^* ab^*)^* = 1 + a(a + b)^* + (a + b)^* aa(a + b)^*.$$

3.4.7 Prove that the language $K = a^* + a^* b(a + ba^* b)^* ba^*$ of words with an even numbers of bs from Example 3.4.5 is the language that is accepted by the following finite deterministic automaton

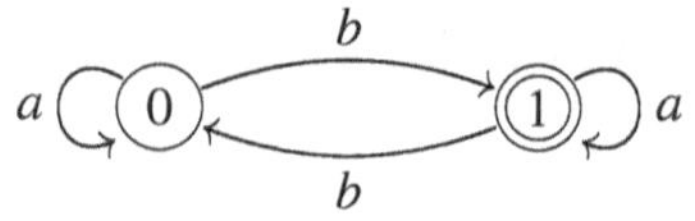

with 1 both as initial and as final state. More formally, this automaton is $\langle \delta, \epsilon \rangle\colon \{0, 1\} \to \{0, 1\}^{\{a,b\}} \times 2$ with

$$\begin{array}{lll}
\delta(0)(a) = 0 & \delta(0)(b) = 1 & \epsilon(0) = 0 \\
\delta(1)(a) = 1 & \delta(1)(b) = 0 & \epsilon(1) = 1.
\end{array}$$

3.5 Process Semantics

This section will introduce a semantics for processes using final coalgebras for the finite powerset functor $\mathcal{P}_{\text{fin}}$. They capture the behaviour of so-called finitely branching transition systems. This section forms an illustration of many of the ideas we have seen so far, such as behavioural interpretations via finality and compositional interpretations via initiality. Also, we shall see how the coalgebraic notion of bisimilarity forms a congruence – an algebraic notion. The material in this section builds on [453, 417], going back to [407]. It will be put in a broader context via distributive laws in Section 5.5.

A first, non-trivial question is: what is a process? Usually one understands it as a running program. Thus, a sequential program, transforming input to output, when in operation forms a process. But typical examples of processes are programs that are meant to be running 'forever', such as operating systems or controllers. Often they consist of several processes that run in parallel, with appropriate synchronisation between them. The proper way to describe such processes is not via input-output relations, as for sequential programs. Rather, one looks at their behaviour, represented as suitable (infinite) trees.

Let us start with the kind of example that is often used to introduce processes. Suppose we wish to describe a machine that can change 5- and 10-euro notes into 1- and 2-euro coins. We shall simply use '5' and '10' as input labels. And as output labels we use pairs (i, j) to describe the return of i 2-euro coins and j 1-euro coins. Also there is a special output action empty that indicates that the machine does not have enough coins left. Our abstract description will not determine which combination of coins is returned, but only gives the various options as a non-deterministic choice. Pictorially this yields a 'state-transition' diagram as in Figure 3.1. Notice that the machine can make a '5' or '10' transition only if it can return a corresponding change. Otherwise, it can do only an 'empty' step.

In this section we shall describe such transition systems as coalgebras, in particular as coalgebras of the functor $X \mapsto \mathcal{P}_{\text{fin}}(X)^A$, for various sets A of 'labels' or 'actions'. As usual, for states $s, s' \in S$ and actions $a \in A$ we write $s \xrightarrow{a} s'$ for $s' \in c(s)(a)$, where $c\colon S \to \mathcal{P}_{\text{fin}}(S)^A$ is our coalgebra. Note that for each state $s \in S$ and input $a \in A$ there are only finitely many successor states s' with $s \xrightarrow{a} s'$. Therefore, such transition systems are often called **finitely branching**.

In the example of Figure 3.1, the set of labels is

$$
\begin{aligned}
E = \{&5, 10, (2,1), (1,3), (0,5), (5,0), (4,2), (3,4), (2,6), \\
&(1,8), (0,10), \mathsf{empty}\}.
\end{aligned}
$$

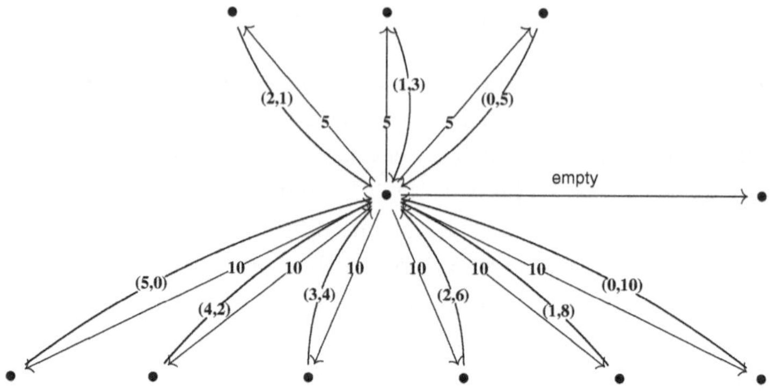

Figure 3.1 Transition diagram of a machine for changing 5- and 10-euro notes into coins.

And the set of states is

$$S = \{s_0, s_1, s_2, s_3, s_4, s_5, s_6, s_7, s_8, s_9, s_{10}\},$$

with 'change' coalgebra structure $\mathsf{ch}\colon S \to \mathcal{P}_{\text{fin}}(S)^E$ given by

$$\begin{array}{rcl} \mathsf{ch}(s_0) & = & \lambda a \in E. \begin{cases} \{s_1, s_2, s_3\} & \text{if } a = 5 \\ \{s_4, s_5, s_6, s_7, s_8, s_9\} & \text{if } a = 10 \\ \{s_{10}\} & \text{if } a = \mathsf{empty} \\ \emptyset & \text{otherwise} \end{cases} \\ \mathsf{ch}(s_1) & = & \lambda a \in E. \begin{cases} \{s_0\} & \text{if } a = (2,1) \\ \emptyset & \text{otherwise} \end{cases} \\ \mathsf{ch}(s_2) & = & \lambda a \in E. \begin{cases} \{s_0\} & \text{if } a = (1,3) \\ \emptyset & \text{otherwise} \end{cases} \\ & \text{etc.} & \end{array} \qquad (3.2)$$

Since the functor $X \mapsto \mathcal{P}_{\text{fin}}(X)^A$ is a finite Kripke polynomial functor, we know by Theorem 2.3.9 that it has a final coalgebra. In this section we shall write this final coalgebra as

$$Z_A \xrightarrow[\cong]{\zeta_A} \mathcal{P}_{\text{fin}}(Z_A)^A, \qquad (3.3)$$

where the set of inputs A is a parameter. We do not really care what these sets Z_A actually look like, because we shall use only their universal properties and do not wish to depend on a concrete representation. However, concrete descriptions in terms of certain finitely branching trees modulo bisimilarity may be given; see [453, section 4.3] (following [55]). In this context we shall call the elements of carrier Z_A of the final coalgebra **processes**.

Our change machine coalgebra ch from Figure 3.1 with its set of actions E thus gives rise to a behaviour map:

$$\begin{array}{ccc} \mathcal{P}_{\text{fin}}(S)^E & \xrightarrow{\mathcal{P}_{\text{fin}}(\mathsf{beh}_{\mathsf{ch}})^E} & \mathcal{P}_{\text{fin}}(Z_E)^E \\ \mathsf{ch}\uparrow & & \cong\uparrow\zeta_E \\ S & \xrightarrow[\mathsf{beh}_{\mathsf{ch}}]{} & Z_E \end{array} \tag{3.4}$$

The behaviour function $\mathsf{beh}_{\mathsf{ch}}$ turns the concrete states $s_0, \ldots, s_{10}$ of our change machine into abstract states of the final coalgebra Z_E – i.e. into processes – with the same behaviour.

3.5.1 Process Descriptions

Several languages have been proposed in the literature to capture processes, such as algebra of communicating processes (ACP) [67, 139], calculus of communicating systems (CCS) [354, 355] or communicating sequential processes (CSP) [223]. The goal of such languages is to study processes via axiom systems in which notions such as sequential and parallel execution, alternative choice and communication are formalised by means of algebraic operations and equations. It is not our aim to go into precise syntax and into the various differences between these formalisms. Instead we concentrate on the semantics and describe only a few basic operations on processes, showing how they can be interpreted in a final coalgebra of the form (3.3). As an example, the above change machine could be described via a (recursive) process equation:

$$\begin{aligned} \mathsf{CH} = \; & 5 \cdot (2,1) \cdot \mathsf{CH} + 5 \cdot (1,3) \cdot \mathsf{CH} + 5 \cdot (0,5) \cdot \mathsf{CH} + \\ & 10 \cdot (5,0) \cdot \mathsf{CH} + 10 \cdot (4,2) \cdot \mathsf{CH} + 10 \cdot (3,4) \cdot \mathsf{CH} + \\ & 10 \cdot (2,6) \cdot \mathsf{CH} + 10 \cdot (1,8) \cdot \mathsf{CH} + 10 \cdot (0,10) \cdot \mathsf{CH} + \\ & \mathsf{empty} \cdot \mathsf{0}. \end{aligned} \tag{3.5}$$

Process algebras can be understood more generally as providing a convenient syntax for describing various kinds of transition systems; see Example 3.5.1 below.

In the following we fix an arbitrary set A of actions, and we consider the associated final coalgebra $Z_A \xrightarrow{\cong} \mathcal{P}_{\text{fin}}(Z_A)^A$ as in (3.3). We shall describe a collection of operations (such as + and · above) on processes, understood as elements of Z_A.

The Null Process

One can define a trivial process which does nothing. It is commonly denoted as $\mathsf{0}$ and is defined as

$$\mathsf{0} \stackrel{\text{def}}{=} \zeta_A^{-1}(\lambda a \in A.\, \emptyset).$$

This means that there are no successor states of the process $\mathsf{0} \in Z_A$ – since by construction the set of successors $\zeta_A(\mathsf{0})(a)$ is empty, for each label $a \in A$.

Sum of Two Processes

Given two processes $z_1, z_2 \in Z_A$, one can define a sum process $z_1 + z_2 \in Z_A$ via the union operation $\cup$ on subsets:

$$z_1 + z_2 \stackrel{\text{def}}{=} \zeta_A^{-1}(\lambda a \in A.\, \zeta_A(z_1)(a) \cup \zeta_A(z_2)(a)).$$

Then:

$$\begin{aligned}(z_1 + z_2) \xrightarrow{a} w &\iff w \in \zeta_A(z_1 + z_2)(a) \\ &\iff w \in \zeta_A(z_1)(a) \cup \zeta_A(z_2)(a) \\ &\iff w \in \zeta_A(z_1)(a) \;\text{ or }\; w \in \zeta_A(z_2)(a) \\ &\iff z_1 \xrightarrow{a} w \;\text{ or }\; z_2 \xrightarrow{a} w.\end{aligned}$$

This means that there is an a-transition out of the sum process $z_1 + z_2$ if and only if there is an a-transition out of either z_1 or z_2. In the process literature one usually encounters the following two rules:

$$\frac{z_1 \xrightarrow{a} w}{z_1 + z_2 \xrightarrow{a} w} \qquad\qquad \frac{z_2 \xrightarrow{a} w}{z_1 + z_2 \xrightarrow{a} w}. \tag{3.6}$$

It is not hard to see that the structure $(Z_A, +, 0)$ is a commutative monoid with idempotent operation: $z + z = z$ for all $z \in Z_A$.

Prefixing of Actions

Given a process $z \in Z_A$ and an action $b \in A$ there is a process $b \cdot z$ which first performs b and then continues with z. It can be defined as

$$\begin{aligned}b \cdot z &\stackrel{\text{def}}{=} \zeta_A^{-1}(\lambda a \in A.\, \mathsf{if}\ a = b\ \mathsf{then}\ \{z\}\ \mathsf{else}\ \emptyset) \\ &= \zeta_A^{-1}(\lambda a \in A.\, \{z \mid a = b\}).\end{aligned}$$

We then have, for $w \in Z_A$ and $b \in A$,

$$\begin{aligned}(b \cdot z) \xrightarrow{a} w &\iff w \in \zeta_A(b \cdot z)(a) \\ &\iff w \in \{z \mid a = b\} \\ &\iff a = b \wedge w = z.\end{aligned}$$

This gives the standard rule:

$$\frac{}{b \cdot z \xrightarrow{b} z}. \tag{3.7}$$

Example 3.5.1 (Change machine, continued) Having defined prefixing and sum we can verify the validity of the change machine equation (3.5), for the interpretation $\mathsf{CH} = \mathrm{beh}_{\mathsf{ch}}(s_0) \in Z_E$ from (3.2). First we note that

$$\begin{aligned}
(2,1)\cdot\mathsf{CH} &= \zeta_E^{-1}\big(\lambda a \in E.\,(\text{if } a = (2,1) \text{ then } \{\mathrm{beh}_{\mathsf{ch}}(s_0)\} \text{ else } \emptyset)\big)\\
&= \zeta_E^{-1}\big(\lambda a \in E.\,\mathcal{P}_{\mathrm{fin}}(\mathrm{beh}_{\mathsf{ch}})(\text{if } a = (2,1) \text{ then } \{s_0\} \text{ else } \emptyset)\big)\\
&= \zeta_E^{-1}\big(\lambda a \in E.\,\mathcal{P}_{\mathrm{fin}}(\mathrm{beh}_{\mathsf{ch}})(\mathsf{ch}(s_1)(a))\big)\\
&= \zeta_E^{-1}\big(\lambda a \in E.\,\zeta_E(\mathrm{beh}_{\mathsf{ch}}(s_1))(a)\big)\\
&\qquad\qquad \text{since } \mathrm{beh}_{\mathsf{ch}} \text{ is a map of coalgebras in (3.4)}\\
&= \zeta_E^{-1}\big(\zeta_E(\mathrm{beh}_{\mathsf{ch}}(s_1))\big)\\
&= \mathrm{beh}_{\mathsf{ch}}(s_1).
\end{aligned}$$

Similar equations can be derived for the states $s_2, \ldots, s_9$. And for s_{10} we have

$$\begin{aligned}
\mathrm{beh}_{\mathsf{ch}}(s_{10}) &= \zeta_E^{-1}\big(\lambda a \in E.\,\zeta_E(\mathrm{beh}_{\mathsf{ch}}(s_{10}))(a)\big)\\
&= \zeta_E^{-1}\big(\lambda a \in E.\,\mathcal{P}_{\mathrm{fin}}(\mathrm{beh}_{\mathsf{ch}})(\mathsf{ch}(s_{10})(a))\big)\\
&= \zeta_E^{-1}\big(\lambda a \in E.\,\mathcal{P}_{\mathrm{fin}}(\mathrm{beh}_{\mathsf{ch}})(\emptyset)\big)\\
&= \zeta_E^{-1}\big(\lambda a \in E.\,\emptyset\big)\\
&= \mathsf{0}.
\end{aligned}$$

Finally we can check that Equation (3.5) holds for the behaviour $\mathsf{CH} = \mathrm{beh}_{\mathsf{ch}}(s_0)$. The calculation is described in Figure 3.2, for each action $a \in E$.

This example illustrates an approach for verifying that a transition system on an arbitrary state space satisfies a certain process equation:

1. Map the transition system to the final coalgebra via the behaviour map beh.
2. Check the equation for $\mathrm{beh}(s_0)$, where s_0 is a suitable (initial) state, using the above interpretations of the process combinators $+$, $\mathsf{0}$, $\cdot$ etc. on the final coalgebra.

3.5.2 A Simple Process Algebra

In the previous subsection we have seen several process combinators, described as functions on a terminal coalgebra $Z_A \xrightarrow{\cong} \mathcal{P}_{\mathrm{fin}}(Z_A)^A$. Next we shall consider these basic combinators as constructors of a very simple process language, often called basic process algebra (BPA); see [139, 451]. In the spirit of this text, the language of finite terms will be described as an initial algebra.

$$
\begin{aligned}
\zeta_E(\mathsf{CH})(a) &= \zeta_E(\mathrm{beh}_{\mathsf{ch}}(s_{10}))(a) \\
&= \mathcal{P}_{\mathrm{fin}}(\mathrm{beh}_{\mathsf{ch}})(\mathsf{ch}(s_0)(a)) \qquad \text{by (3.4)} \\
&= \begin{cases} \{\mathrm{beh}_{\mathsf{ch}}(s_1), \mathrm{beh}_{\mathsf{ch}}(s_2), \mathrm{beh}_{\mathsf{ch}}(s_3)\} & \text{if } a = 5 \\ \{\mathrm{beh}_{\mathsf{ch}}(s_4), \mathrm{beh}_{\mathsf{ch}}(s_5), \mathrm{beh}_{\mathsf{ch}}(s_6), \mathrm{beh}_{\mathsf{ch}}(s_7), \mathrm{beh}_{\mathsf{ch}}(s_8), \mathrm{beh}_{\mathsf{ch}}(s_9)\} & \text{if } a = 10 \\ \{\mathrm{beh}_{\mathsf{ch}}(s_{10})\} & \text{if } a = \mathsf{empty} \end{cases} \\
&= \{\, \mathrm{beh}_{\mathsf{ch}}(s_1), \mathrm{beh}_{\mathsf{ch}}(s_2), \mathrm{beh}_{\mathsf{ch}}(s_3) \mid a = 5\} \\
&\qquad \cup \{\, \mathrm{beh}_{\mathsf{ch}}(s_4), \mathrm{beh}_{\mathsf{ch}}(s_5), \mathrm{beh}_{\mathsf{ch}}(s_6), \mathrm{beh}_{\mathsf{ch}}(s_7), \mathrm{beh}_{\mathsf{ch}}(s_8), \mathrm{beh}_{\mathsf{ch}}(s_9) \mid a = 10\} \\
&\qquad \cup \{\, \mathsf{ch}(s_{10}) \mid a = \mathsf{empty}\} \\
&= \{\, \mathrm{beh}_{\mathsf{ch}}(s_1) \mid a = 5\} \cup \{\, \mathrm{beh}_{\mathsf{ch}}(s_2) \mid a = 5\} \cup \{\, \mathrm{beh}_{\mathsf{ch}}(s_3) \mid a = 5\} \\
&\qquad \cup \{\, \mathrm{beh}_{\mathsf{ch}}(s_4) \mid a = 10\} \cup \{\, \mathrm{beh}_{\mathsf{ch}}(s_5) \mid a = 10\} \cup \{\, \mathrm{beh}_{\mathsf{ch}}(s_6) \mid a = 10\} \\
&\qquad \cup \{\, \mathrm{beh}_{\mathsf{ch}}(s_7) \mid a = 10\} \cup \{\, \mathrm{beh}_{\mathsf{ch}}(s_8) \mid a = 10\} \cup \{\, \mathrm{beh}_{\mathsf{ch}}(s_9) \mid a = 10\} \cup \{\, \mathsf{ch}(s_{10}) \mid a = \mathsf{empty}\} \\
&= \zeta_E(5 \cdot (2,1) \cdot \mathsf{CH})(a) \cup \zeta_E(5 \cdot (1,3) \cdot \mathsf{CH})(a) \cup \zeta_E(5 \cdot (0,5) \cdot \mathsf{CH})(a) \\
&\qquad \cup \zeta_E(10 \cdot (5,0) \cdot \mathsf{CH})(a) \cup \zeta_E(10 \cdot (4,2) \cdot \mathsf{CH})(a) \cup \zeta_E(10 \cdot (3,4) \cdot \mathsf{CH})(a) \\
&\qquad \cup \zeta_E(10 \cdot (2,6) \cdot \mathsf{CH})(a) \cup \zeta_E(10 \cdot (1,8) \cdot \mathsf{CH})(a) \cup \zeta_E(10 \cdot (0,10) \cdot \mathsf{CH})(a) \cup \zeta_E(\mathsf{empty} \cdot 0)(a) \\
&= \zeta_E\Big(5 \cdot (2,1) \cdot \mathsf{CH} + 5 \cdot (1,3) \cdot \mathsf{CH} + 5 \cdot (0,5) \cdot \mathsf{CH} \\
&\qquad + 10 \cdot (5,0) \cdot \mathsf{CH} + 10 \cdot (4,2) \cdot \mathsf{CH} + 10 \cdot (3,4) \cdot \mathsf{CH} + \\
&\quad 10 \cdot (2,6) \cdot \mathsf{CH} + 10 \cdot (1,8) \cdot \mathsf{CH} + 10 \cdot (0,10) \cdot \mathsf{CH} + \mathsf{empty} \cdot 0\Big)(a).
\end{aligned}
$$

Figure 3.2 Calculation of the behaviour of the process CH.

For an arbitrary set A of actions, consider the simple polynomial functor $\Sigma_A : \mathbf{Sets} \to \mathbf{Sets}$ given by

$$\Sigma_A(X) = 1 + (A \times X) + (X \times X).$$

An algebra $\Sigma_A(X) \to X$ for this functor thus consists of three operations, which we call $0 : 1 \to X$ for null-process, $\cdot : A \times X \to X$ for prefixing and $+ : X \times X \to X$ for sum. We have seen that the final coalgebra $Z_A \xrightarrow{\cong} \mathcal{P}_{\text{fin}}(Z_A)^A$ carries a Σ_A-algebra structure which we shall write as $\xi_A : \Sigma_A(Z_A) \to Z_A$. It is given by the structure described earlier (before Example 3.5.1). Thus we have a bialgebra of processes:

$$\Sigma_A(Z_A) \xrightarrow{\quad \xi_A \quad} Z_A \xrightarrow[\cong]{\quad \zeta_A \quad} \mathcal{P}_{\text{fin}}(Z_A)^A.$$

The free Σ_A-algebra on a set V consists of the terms built up from the elements from V as variables. An interesting algebra is given by the free Σ_A-algebra on the final coalgebra Z_A. It consists of terms built out of processes, as studied in [407] under the phrase 'processes as terms'. Here we shall write P_A for the initial Σ_A-algebra, i.e. the free algebra on the empty set. The set P_A contains the 'closed' process terms, without free variables. They are built up from 0 and $a \in A$. We shall write this algebra as $\alpha : \Sigma_A(P_A) \xrightarrow{\cong} P_A$. It is not hard to see that the set P_A of process terms also carries a bialgebra structure:

$$\Sigma_A(P_A) \xrightarrow[\cong]{\quad \alpha \quad} P_A \xrightarrow{\quad \beta \quad} \mathcal{P}_{\text{fin}}(P_A)^A.$$

The coalgebra β is defined by induction, following the transition rules (3.6) and (3.7). For each $a \in A$,

$$\begin{aligned} \beta(0)(a) &= \emptyset \\ \beta(b \cdot s)(a) &= \{s \mid b = a\} \\ \beta(s + t)(a) &= \beta(s)(a) \cup \beta(t)(a). \end{aligned}$$

These definitions form a very simple example of a structural operations semantics (SOS): operational behaviour defined by induction on the structure of terms.

The next result shows that the denotational semantics given by initiality and operational semantics given by finality for process terms coincide.

Proposition 3.5.2 *In the above situation we obtain two maps $P_A \to Z_A$, one by initiality and one by finality:*

$$
\begin{array}{ccc}
\Sigma_A(P_A) & \overset{\Sigma_A(\mathrm{int}_\alpha)}{\dashrightarrow} & \Sigma_A(Z_A) \\
\alpha \big\downarrow \cong & & \big\downarrow \xi \\
P_A & \underset{\mathrm{int}_\alpha}{\dashrightarrow} & Z_A
\end{array}
\qquad
\begin{array}{ccc}
\mathcal{P}_{fin}(P_A)^A & \overset{\mathcal{P}_{fin}(\mathrm{beh}_\beta)^A}{\dashrightarrow} & \mathcal{P}_{fin}(Z_A)^A \\
\beta \big\uparrow & & \cong \big\uparrow \zeta \\
P_A & \underset{\mathrm{beh}_\beta}{\dashrightarrow} & Z_A
\end{array}
$$

These two maps are equal, so that $\mathrm{int}_\alpha = \mathrm{beh}_\beta \colon P_A \to Z_A$ is a 'map of bialgebras' commuting with both the algebra and coalgebra structures. This proves in particular that the behavioural semantics beh_β of processes is compositional: it commutes with the term-forming operations.

Proof By induction on the structure of a term $s \in P_A$ we prove that $\mathrm{beh}_\beta(s) = \mathrm{int}_\alpha(s)$, or equivalently, $\zeta(\mathrm{beh}_\beta(s))(a) = \zeta(\mathrm{int}_\alpha(s))(a)$, for all $a \in A$:

$$
\begin{array}{rcl}
\zeta(\mathrm{beh}_\beta(0))(a) & = & \mathcal{P}_{\mathrm{fin}}(\mathrm{beh}_\beta)^A(\beta(0))(a) \\
& = & \mathcal{P}_{\mathrm{fin}}(\mathrm{beh}_\beta)(\beta(0)(a)) \\
& = & \mathcal{P}_{\mathrm{fin}}(\mathrm{beh}_\beta)(\emptyset) \\
& = & \emptyset \\
& = & \zeta(0)(a) \\
& = & \zeta(\mathrm{int}_\alpha(0))(a) \\
\zeta(\mathrm{beh}_\beta(b \cdot s))(a) & = & \mathcal{P}_{\mathrm{fin}}(\mathrm{beh}_\beta)(\beta(b \cdot s)(a)) \\
& = & \mathcal{P}_{\mathrm{fin}}(\mathrm{beh}_\beta)(\{s \mid b = a\}) \\
& = & \{\mathrm{beh}_\beta(s) \mid b = a\}) \\
& \overset{(\mathrm{IH})}{=} & \{\mathrm{int}_\alpha(s) \mid b = a\}) \\
& = & \zeta(b \cdot \mathrm{int}_\alpha(s))(a) \\
& = & \zeta(\mathrm{int}_\alpha(b \cdot s))(a) \\
\zeta(\mathrm{beh}_\beta(s + t))(a) & = & \mathcal{P}_{\mathrm{fin}}(\mathrm{beh}_\beta)(\beta(s + t)(a)) \\
& = & \mathcal{P}_{\mathrm{fin}}(\mathrm{beh}_\beta)(\beta(s)(a) \cup \beta(t)(a)) \\
& = & \mathcal{P}_{\mathrm{fin}}(\mathrm{beh}_\beta)(\beta(s)(a)) \cup \mathcal{P}_{\mathrm{fin}}(\mathrm{beh}_\beta)(\beta(t)(a)) \\
& = & \zeta(\mathrm{beh}_\beta(s))(a) \cup \zeta(\mathrm{beh}_\beta(t))(a) \\
& \overset{(\mathrm{IH})}{=} & \zeta(\mathrm{int}_\alpha(s))(a) \cup \zeta(\mathrm{int}_\alpha(t))(a) \\
& = & \zeta(\mathrm{int}_\alpha(s) + \mathrm{int}_\alpha(t))(a) \\
& = & \zeta(\mathrm{int}_\alpha(s + t))(a).
\end{array}
$$

❑

Proposition 3.5.3 *Still in the above situation, the bisimilarity relation $\underline{\leftrightarrow}$ on the set P_A of process terms is a congruence.*

Proof Consider the following diagram, where we abbreviate $f = \text{beh}_\beta = \text{int}_\alpha$:

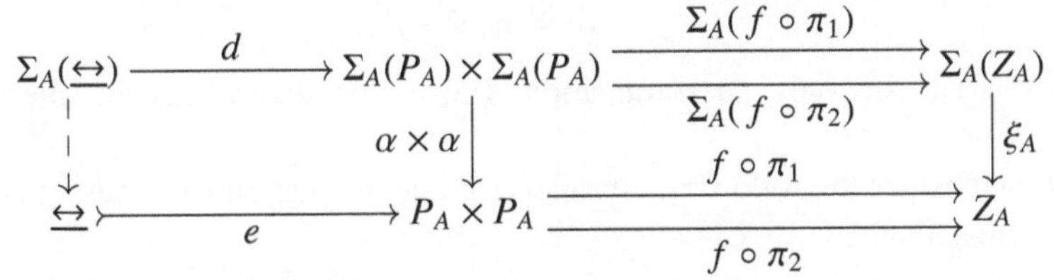

The map e is the equaliser of $f \circ \pi_1$ and $f \circ \pi_2$, using Theorem 3.4.1. The map d is the pair $\langle \Sigma_A(\pi_1 \circ e), \Sigma(\pi_2 \circ e)\rangle$. We show that $(\alpha \times \alpha) \circ d$ equalises $f \circ \pi_1, f \circ \pi_2$. The dashed arrow then exists by the universal property of the equaliser e, making $\underline{\leftrightarrow}$ a congruence (see Theorem 3.3.5). Thus, what remains is

$$\begin{aligned} & f \circ \pi_1 \circ (\alpha \times \alpha) \circ d \\ &= \text{int}_\alpha \circ \alpha \circ \pi_1 \circ d \\ &= \xi \circ \Sigma_A(\text{int}_\alpha) \circ \Sigma_A(\pi_1 \circ e) \\ &= \xi \circ \Sigma_A(\text{int}_\alpha) \circ \Sigma_A(\pi_2 \circ e) \qquad \text{since } e \text{ is equaliser and } \text{beh}_\beta = \text{int}_\alpha \\ &= \text{int}_\alpha \circ \alpha \circ \pi_2 \circ d \\ &= f \circ \pi_2 \circ (\alpha \times \alpha) \circ d. \end{aligned}$$

□

This result shows that $s \;\underline{\leftrightarrow}\; s'$ and $t \;\underline{\leftrightarrow}\; t'$ implies $a \cdot s \;\underline{\leftrightarrow}\; a \cdot s'$ and $s + t \;\underline{\leftrightarrow}\; s' + t'$. Such congruence results are fundamental in process algebra, because they show that the algebraic operations for process formation preserve indistinguishability of behaviour. Later, in Section 5.5, this topic will be studied in greater generality (following [452, 451, 58, 298]). A more abstract view on the structure we find on the final coalgebra Z_A is elaborated in [207, 202, 204], where it is shown that 'outer' structure that can be formulated on coalgebras is mimicked by 'inner' structure on the final coalgebra.

The toy example (3.5) of this section illustrates the close connection between final coalgebras and process languages. This connection is further elaborated in [430] (and [428]) where locally final coalgebras of polynomial functors are described syntactically.

Exercises

3.5.1 Complete the definition of the coalgebra $\mathsf{ch}\colon S \to \mathcal{P}_{\text{fin}}(S)^E$ in the beginning of this section.

3.5.2 Prove that $(Z_A, +, 0)$ is indeed a commutative monoid, as claimed above.

3.5.3 One can consider each action $a \in A$ as a process

$$\widehat{a} \stackrel{\text{def}}{=} a \cdot 0 \in Z_A.$$

It can do only an a-transition. Prove that this yields an injection $A \hookrightarrow Z_A$.

3.5.4 Consider the following alternative process equation for a euro change machine.

$$\begin{aligned} \mathsf{CH}' &= 5 \cdot \big((2,1) \cdot \mathsf{CH}' + (1,3) \cdot \mathsf{CH}' + (0,5) \cdot \mathsf{CH}'\big) \\ &\quad + 10 \cdot \big((5,0) \cdot \mathsf{CH}' + (4,2) \cdot \mathsf{CH}' + (3,4) \cdot \mathsf{CH}' + \\ &\qquad\qquad (2,6) \cdot \mathsf{CH}' + (1,8) \cdot \mathsf{CH}' + (0,10) \cdot \mathsf{CH}'\big) \\ &\quad + \mathsf{empty}. \end{aligned}$$

Examine the difference between CH in (3.5) and this CH′, for instance by describing a suitable transition system which forms a model of this equation.

Are CH and CH′ bisimilar?

4

Logic, Lifting and Finality

The previous three chapters have introduced some basic elements of the theory of coalgebras, focusing on coalgebraic system descriptions, homomorphisms, behaviour, finality and bisimilarity. So far, only relatively simple coalgebras have been used, for inductively defined classes of polynomial functors, on the category **Sets** of sets and functions. This chapter will go beyond these polynomial functors and will consider other examples. But more important, it will follow a different, more systematic approach, relying not on the way functors are constructed but on the properties they satisfy – and work from there. Inevitably, this chapter will technically be more challenging, requiring more categorical maturity from the reader.

The chapter starts with a concrete description of two new functors, namely the multiset and distribution functors, written as $\mathcal{M}$ and $\mathcal{D}$, respectively. As we shall see, on the one hand, from an abstract point of view, they are much like powerset $\mathcal{P}$, but on the other hand they capture different kinds of computation: $\mathcal{D}$ is used for probabilistic computation and $\mathcal{M}$ for resource-sensitive computation.

Subsequently, Sections 4.2–4.5 will take a systematic look at relation lifting – used in the previous chapter to define bisimulation relations. Relation lifting will be described as a certain logical operation, which will be developed on the basis of a moderate amount of categorical logic, in terms of so-called factorisation systems. This will give rise to the notion of 'logical bisimulation' in Section 4.5. It is compared with several alternative formulations. For weak pullback-preserving functors on **Sets** these different formulations coincide. With this theory in place Section 4.6 concentrates on the existence of final coalgebras. Recall that earlier we skipped the proof of Theorem 2.3.9, claiming the existence of final coalgebras for finite Kripke polynomial functors. Here we present general existence results, for 'bounded' endofunctors on **Sets**. Finally, Section 4.7 contains another characterisation of simple polynomial functors in terms of size and preservation properties. It also contains a characterisation of more general 'analytical' functors, which includes for instance the multiset functor $\mathcal{M}$.

4.1 Multiset and Distribution Functors

A set is a collection of elements. Such an element, if it occurs in the set, occurs only once. This sounds completely trivial. But one can imagine situations in which multiple occurrences of the same element can be relevant. For instance, in a list it is quite natural that one and the same element occurs multiple times – but at different places in the list. A **multiset** – or what computer scientists usually call a **bag** – is a 'set' in which an element x may occur multiple times. One can write this for instance as $2x$, $10x$, $1x$, or even $0x$, where the n in nx describes that x occurs n times.

Thus, one can distinguish different operators for collecting elements according to whether the order of occurrence of elements matters or whether multiplicities of elements are relevant. The table in Figure 4.1 describes some of these collection types.

An important question is how to count occurrences of elements in a multiset. The obvious choice is to use natural numbers nx, like above. Instead of writing x, we like to write $|x\rangle$ in $n|x\rangle$. This 'ket' notation is meaningless syntactic sugar that is used to distinguis the element $x \in X$ from its occurrence in such counting expressions. It turns out to be convenient to allow also negative occurrences $-2|x\rangle$, describing for instance a 'lack' or 'deficit' of two elements x. But also one may want to allow $\frac{1}{2}|x\rangle$, so that the number in front of x may be interpreted as the probability of having x in a set. This is precisely what happens in the distribution functor $\mathcal{D}$; see below.

It thus makes sense to allow a quite general form of counting elements. We shall use an arbitrary commutative monoid $M = (M, +, 0)$ for this purpose and write multiple occurrences of an element x as $m|x\rangle$ for $m \in M$.

Operator	Order Relevant	Multiplicities Relevant
List $(-)^\star$	Yes	Yes
Powerset $\mathcal{P}$	No	No
Multiset $\mathcal{M}$	No	Yes
Distribution $\mathcal{D}$	No	Yes

Figure 4.1 Various collection types described as functors.

The expression $0|x\rangle$ is then used for non-occurrence of x. Later on, in Section 5.1 we shall see that it makes sense to assume more than an additive monoid structure on multiplicities, namely also multiplication (giving a semiring). But for now a monoid structure suffices.

The operation of forming (finite) multisets of a given set is functorial. The resulting functor will be called the multiset functor and written as $\mathcal{M}$. It is called the 'monoidal exponentiation functor' in [428] (see also [185, 179]).

Definition 4.1.1 For a commutative monoid $M = (M, +, 0)$ we define the **multiset functor** $\mathcal{M}_M$: **Sets** $\rightarrow$ **Sets** on a set X as

$$\mathcal{M}_M(X) = \{\varphi\colon X \rightarrow M \mid \mathrm{supp}(\varphi) \text{ is finite}\},$$

where $\mathrm{supp}(\varphi) = \{x \in X \mid \varphi(x) \neq 0\}$ is called the support of φ.

On a function $f\colon X \rightarrow Y$ one has $\mathcal{M}_M(f)\colon \mathcal{M}_M(X) \rightarrow \mathcal{M}_M(Y)$ via

$$\mathcal{M}_M(f)(\varphi)(y) = \sum_{x\in f^{-1}(y)} \varphi(x) = \sum\{\varphi(x) \mid x \in \mathrm{supp}(\varphi) \text{ with } f(x) = y\}.$$

This definition requires some explanation. A multiset $\varphi \in \mathcal{M}_M(X)$ is a function $\varphi\colon X \rightarrow M$ which is non-zero on only finitely many elements, say $\varphi(x_1) \neq 0, \ldots, \varphi(x_n) \neq 0$. The support of φ is then the subset $\mathrm{supp}(\varphi) = \{x_1, \ldots, x_n\} \subseteq X$ of those elements that occur in the multiset, with certain (non-zero) multiplicities. The multiplicity of $x_i \in \mathrm{supp}(\varphi)$ is $\varphi(x_i) \in M$. One conveniently writes such a multiset as formal sum $m_1|x_1\rangle + \cdots + m_n|x_n\rangle$, where $m_i = \varphi(x_i) \in M$ is the multiplicity of the element x_i. By convention, the + in these formal sums is commutative and associative; $m|x\rangle + m'|x\rangle$ is the same as $(m + m')|x\rangle$, and $m|x\rangle + 0|y\rangle$ is $m|x\rangle$.

With this formal sum notation we can write the action of the functor $\mathcal{M}_M$ on functions as

$$\mathcal{M}_M(f)\big(m_1|x_1\rangle + \cdots + m_n|x_n\rangle\big) = m_1|f(x_1)\rangle + \cdots + m_n|f(x_n)\rangle.$$

This works by the above-mentioned conventions about formal sums. Preservation of composition holds simply by:

$$\begin{aligned}\mathcal{M}_M(g)(\mathcal{M}_M(f)(\textstyle\sum_i m_i|x_i\rangle)) &= \mathcal{M}_M(g)(\textstyle\sum_i m_i|f(x_i)\rangle) \\ &= \textstyle\sum_i m_i|g(f(x_i))\rangle \\ &= \mathcal{M}_M(g \circ f)(\textstyle\sum_i m_i|x_i\rangle).\end{aligned}$$

Thus, $\mathcal{M}_M(X)$ contains finite multisets, with elements from the set X and multiplicities in the monoid M. We restrict ourselves to *finite* multiset, to make sure that the sums $\sum$ in the definition of $\mathcal{M}_M(f)$ in Definition 4.1.1 exist.

Before describing examples, we mention some obvious properties.

Lemma 4.1.2 *The empty multiset and the join of multisets make the sets $\mathcal{M}_M(X)$ from Definition 4.1.1 commutative monoids. Moreover, the functions $\mathcal{M}_M(f)\colon \mathcal{M}_M(X) \to \mathcal{M}_M(Y)$ preserve this monoid structure.*

Further, for the initial (empty) set 0, for the final (singleton) set 1, and for a finite set V, there are isomorphisms:

$$\mathcal{M}_M(0) \cong 1 \qquad \mathcal{M}_M(1) \cong M \qquad \mathcal{M}_M(V) \cong M^V$$

(which are all isomorphisms of monoids).

Proof The monoid structure on $\mathcal{M}_M(X)$ can be described pointwise:

$$\varphi + \psi = \lambda x \in X.\, \varphi(x) + \psi(x) \quad \text{with zero} \quad \lambda x \in X.\, 0.$$

Alternatively, it may be described in terms of the formal sums: the operations represent the join of multisets and the empty multiset. The isomorphisms are obvious. □

Example 4.1.3 For some specific examples of monoids M we describe the multiset $\mathcal{M}_M(X)$ in some more detail. The characterisations we give as free structures can be checked per case but turn out to be instances of a more general result; see Example 5.4.3.2 later on.

1. We start with $M = \mathbb{N}$, the commutative monoid of natural numbers with addition $(0, +)$. The set $\mathcal{M}_\mathbb{N}(X)$ contains 'traditional' multisets (or bags), with natural numbers as multiplicities. This set $\mathcal{M}_\mathbb{N}(X)$ is the free commutative monoid on the set X.
2. For $M = \mathbb{Z}$ we can additionally have negative occurrences of elements in $\mathcal{M}_\mathbb{Z}(X)$. This $\mathcal{M}_\mathbb{Z}(X)$ is the free commutative (Abelian) group on X.
3. For the two-element monoid $M = 2 = \{0, 1\}$, with join $\vee$ (logical or) and unit 0 as monoid structure, there are no multiplicities except 0 and 1. Hence multisets over 2 are finite subsets: $\mathcal{M}_2(X) \cong \mathcal{P}_{\text{fin}}(X)$.
4. For $M = \mathbb{R}$ we get real numbers as multiplicities, and $\mathcal{M}_\mathbb{R}(X)$ is free vector space over $\mathbb{R}$ on X. Scalar multiplication is given by $r \bullet \varphi = \lambda x.\, r \cdot \varphi(x)$, where $\cdot$ is multiplication of real numbers.

The general description of multisets, over an arbitrary monoid, thus covers various mathematical structures. Next we illustrate how repeated multisets are of interest, namely in order to describe (multivariate) polynomials.

Example 4.1.4 Above we have used formal sums $m_1|x_1\rangle + \cdots + m_n|x_n\rangle$ as convenient notation for multisets. For monoids and groups one sometimes uses additive notation $(0, +)$ and sometimes multiplicative notation $(1, \cdot)$. Similarly, one may choose to use multiplicative notation for multisets, as in

$$x_1^{m_1} \cdots x_n^{m_n}.$$

Now let's consider the repeated multiset functor application

$$\mathcal{M}_M\big(\mathcal{M}_{\mathbb{N}}(X)\big), \tag{4.1}$$

where M is an arbitrary monoid. An element $\Phi \in \mathcal{M}_M(\mathcal{M}_{\mathbb{N}}(X))$ is described as a formal sum of multisets:

$$\Phi = \textstyle\sum_i m_i|\varphi_i\rangle \qquad \text{where} \qquad \varphi_i \in \mathcal{M}_{\mathbb{N}}(X).$$

Now it is convenient to use multiplicative notation for the inner multisets φ_i, so that we get:

$$\Phi = \textstyle\sum_i m_i|\, x_{i1}^{n_{i1}} \cdots x_{ik_i}^{n_{ik_i}}\rangle \qquad \text{for} \qquad k_i, n_{ij} \in \mathbb{N}.$$

Thus, elements of the double-multiset (4.1) are polynomials (as in algebra), with coefficients m_i from M. They are so-called multivariate polynomials, involving multiple variables x_{ij}, taken from a set of variables X. The univariate polynomials, with only one variable – say x – are obtained as special case of (4.1), namely by taking the singleton set $X = 1$. The element of $\mathcal{M}_M(\mathcal{M}_{\mathbb{N}}(1)) \cong \mathcal{M}_M(\mathbb{N})$ are formal sums $\sum_i m_i|n_i\rangle$, commonly written as $\sum_i m_i x^{n_i}$, where x is a chosen variable.

This concludes, for the time being, our investigation of multiset functors. In Section 5.1 we shall return to them and see that they carry a monad structure, provided the set of multiplicities is not only a monoid but also a semiring (with additionally multiplication, see Definition 5.1.4 and Lemma 5.1.5). In the remainder of this section we investigate the distribution functor $\mathcal{D}$, which involves probabilities as multiplicities.

Definition 4.1.5 The (discrete probability) **distribution functor** $\mathcal{D}$: **Sets** $\to$ **Sets** is defined as

$$\mathcal{D}(X) = \{\varphi\colon X \to [0,1] \mid \text{supp}(\varphi) \text{ is finite and } \textstyle\sum_x \varphi(x) = 1\},$$

where $[0,1] \subseteq \mathbb{R}$ is the unit interval of real numbers, and $\text{supp}(\varphi) \subseteq X$ is the subset of $x \in X$ where $\varphi(x) \neq 0$, as in Definition 4.1.1. For a function $f\colon X \to Y$ the map $\mathcal{D}(f)\colon \mathcal{D}(X) \to \mathcal{D}(Y)$ is defined as for multisets, namely by

$$\mathcal{D}(f)(\varphi)(y) = \sum_{x \in f^{-1}(y)} \varphi(x) = \sum \{\varphi(x) \mid x \in \text{supp}(\varphi) \text{ with } f(x) = y\}.$$

A discrete probability distribution $\varphi \in \mathcal{D}(X)$ may be identified with a formal convex sum $r_1|x_1\rangle + \cdots + r_n|x_n\rangle$, where $\mathrm{supp}(\varphi) = \{x_1, \ldots, x_n\}$ and $r_i = \varphi(x_i) \in [0,1]$ is the probability associated with the element $x_i \in X$. These probabilities are required to add up to one: $\sum_i r_i = 1$. With this formal sum notation we can again describe the functor applied to a function f succinctly as $\mathcal{D}(f)(\sum_i r_i|x_i\rangle) = \sum_i r_i|f(x_i)\rangle$, as for multiset functors. This shows that the map $\mathcal{D}(f)$ is well defined, in the sense $\sum_y \mathcal{D}(f)(\varphi)(y) = 1$.

For distributions $\sum_i r_i|x_i\rangle \in \mathcal{D}(X)$ the probabilities $r_i \in [0,1]$ add up to 1. In some situations one wishes to be more flexible and allow $\sum_i r_i \leq 1$. Such 'sub' probability distributions give rise to a 'sub' probability functor $\mathcal{D}_{\leq 1}\colon \mathbf{Sets} \to \mathbf{Sets}$. These sets $\mathcal{D}_{\leq 1}(X)$ have more structure than sets $\mathcal{D}(X)$ of (proper) probabilities. For instance, they carry an action $[0,1] \times \mathcal{D}_{\leq 1}(X) \to \mathcal{D}_{\leq 1}(X)$. And if one allows non-finite support, as in Exercise 4.1.8, then subdistributions even form a dcpo (used in [206]).

Multiset functors $\mathcal{M}_M$ are parametrised by a monoid M. In contrast, distributions in $\mathcal{D}(X)$ take their values in the (fixed) set $[0,1]$ of probabilities. It is also possible to replace $[0,1]$ by a parameterised structure, namely a so-called effect monoid; see [250]. Such effect monoids are used in the (generalised) probability theory developed in the context of quantum mechanics (see e.g. [396, 95]). However, such generality is not needed here.

In analogy with Lemma 4.1.2 we have the following results for distribution functors.

Lemma 4.1.6 *For the distribution functor $\mathcal{D}$ there are the following isomorphisms*:

$$\mathcal{D}(0) \cong 0, \qquad \mathcal{D}(1) \cong 1, \qquad \mathcal{D}(2) \cong [0,1]. \qquad \square$$

We saw that sets $\mathcal{M}_M(X)$ are commutative monoids. Sets $\mathcal{D}(X)$ also carry algebraic structure in the form of convex sums; see [247]. Further investigation of the categorical structure of distribution functors will be postponed until Section 5.1 on monads. We conclude by briefly describing coalgebras of the functors $\mathcal{M}_M$ and $\mathcal{D}$ introduced in this section.

Coalgebras $c\colon X \to \mathcal{D}(X)$ of the distribution functor map a state x to a probability distribution $c(X) \in \mathcal{D}(X)$ over successor states. Such a distribution may be written as $c(x) = r_1|x_1\rangle + \cdots + r_n|x_n\rangle$, where $\sum_i r_i = 1$. A transition $x \to x_i$ takes place with probability $r_i \in [0,1]$. Sometimes this written as $x \xrightarrow{r_i} x_i$. Such a probabilistic transition system $X \to \mathcal{D}(X)$ is also known as Markov chain. This basic form may be extended in various ways; see Figure 4.2.

Here is a very simple example of a Markov chain. Assume an abstract political landscape where only lefties (L) and righties (R) are distinguished.

It appears that with each cycle (of one year, say) 80% of lefties remain lefties and 20% become righties. The righties are more stable: 90% of them remains loyal, and 10% become lefties. This may be written as a Markov chain, or probabilistic automaton, with two states L and R:

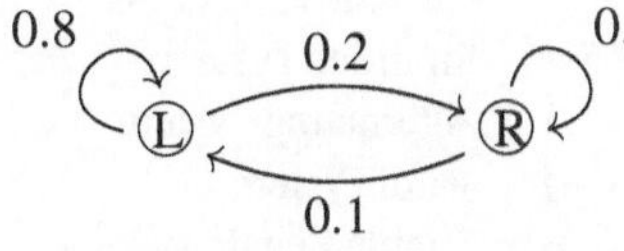

(4.2)

This Markov chain can equivalently be described as a coalgebra $c\colon \{L, R\} \to \mathcal{D}(\{L, R\})$ of the distribution functor. The function c maps each state to a distribution, written as formal convex sum:

$$c(L) = 0.8|L\rangle + 0.2|R\rangle, \qquad c(R) = 0.1|L\rangle + 0.9|R\rangle. \tag{4.3}$$

Conventionally, such a system is described via a transition matrix

$$\begin{pmatrix} 0.8 & 0.1 \\ 0.2 & 0.9 \end{pmatrix} \tag{4.4}$$

and the analysis of the system proceeds via an analysis of the matrix (see [293, 227] for more information).

So far we have seen non-deterministic systems as coalgebras of the powerset functor $\mathcal{P}$ and probabilistic systems as coalgebras of the distribution functor $\mathcal{D}$. Many systems display both non-deterministic and probabilistic behaviour, in many combinations. Via a description in terms of coalgebras the differences can be seen clearly. Figure 4.2 gives an overview of the various systems that have been studied in the literature. The table is copied from [59, 434], to which we refer for more information. Here we focus on *discrete* probabilistic systems, in contrast to *continuous* ones, taking measurable spaces as state spaces. Such continuous systems can also be described as coalgebras, namely of the 'Giry' functor (or monad); see [157] and [369, 124, 252] for more information.

Coalgebras $c\colon X \to \mathcal{M}_M(X)$ of a multiset functor are known as multigraphs [115]. They can be understood more generally in terms of resources or costs $m \in M$ associated with a transition, as in $x \xrightarrow{m} x'$ when $c(x)(x') = m$. This is characteristic of a weighted automaton; see [421, 125, 90, 76].

4.1.1 Mappings between Collection Functors

In Figure 4.1 we have seen various functors, such as list $(-)^\star$, powerset $\mathcal{P}$, multiset $\mathcal{M}$ and distribution $\mathcal{D}$ that collect elements in a certain way.

Functor F	Name for $X \to F(X)$	Reference
$\mathcal{D}$	Markov chain	
$\mathcal{P}(A \times -) \cong \mathcal{P}^A$	Labelled transition system	
$(1 + \mathcal{D}(-))^A$	Reactive system	[328, 159]
$1 + \mathcal{D}(A \times -)$	Generative systems	[159]
$1 + (A \times -) + \mathcal{D}(-)$	Stratified system	[159]
$\mathcal{D}(-) + \mathcal{P}(A \times -)$	Alternating system	[197]
$\mathcal{D}(A \times -) + \mathcal{P}(A \times -)$	Vardi system	[456]
$\mathcal{P}(A \times \mathcal{D}(-))$	Simple Segala system	[426, 425]
$\mathcal{P}\mathcal{D}(A \times -)$	Segala system	[426, 425]
$\mathcal{D}\mathcal{P}(A \times -)$	Bundle system	[118]
$\mathcal{P}\mathcal{D}\mathcal{P}(A \times -)$	Pnueli–Zuck system	[391]
$\mathcal{P}\mathcal{D}\mathcal{P}(A \times (-) + (-))$	Most general systems	

Figure 4.2 An overview of (discrete) probabilistic system types. From [59, 434].

An obvious question that comes up is whether we can relate these functors via suitable mappings. This can be done via natural transformations. Recall from Definition 2.5.4 that they are mappings between functors which work in a uniform manner. Below it will be shown that naturality is quite a subtle matter in this setting.

Consider first the probability distribution functor $\mathcal{D}$: **Sets** $\to$ **Sets**. Each distribution $\varphi \in \mathcal{D}(X)$ is a function $\varphi\colon X \to [0, 1]$ with finite support $\mathrm{supp}(\varphi) \subseteq X$, given by $\mathrm{supp}(\varphi) = \{x \in X \mid \varphi(x) \neq 0\}$ and probabilities adding up to 1: $\sum_x \varphi(x) = 1$. Thus, we have $\mathrm{supp}(\varphi) \in \mathcal{P}(X)$, or even $\mathrm{supp}(\varphi) \in \mathcal{P}_{\mathrm{fin}}(X)$. In order to emphasise that we take supports of distributions on X we label the support $\mathrm{supp}(\varphi)$ with this set X, as in $\mathrm{supp}_X(\varphi)$. Thus we have a collection of mappings:

$$\left(\mathcal{D}(X) \xrightarrow{\ \mathrm{supp}_X\ } \mathcal{P}(X)\right)_{X \in \mathbf{Sets}}.$$

This collection of functions is 'natural in X': for each function $f\colon X \to Y$ we have a commuting square:

$$\begin{array}{c} X \\ {\scriptstyle f}\big\downarrow \\ Y \end{array} \qquad\qquad \begin{array}{ccc} \mathcal{D}(X) & \xrightarrow{\ \mathrm{supp}_X\ } & \mathcal{P}(X) \\ {\scriptstyle \mathcal{D}(f)}\big\downarrow & & \big\downarrow{\scriptstyle \mathcal{P}(f)} \\ \mathcal{D}(Y) & \xrightarrow[\ \mathrm{supp}_Y\ ]{} & \mathcal{P}(Y) \end{array}$$

We check in detail that this naturality diagram commutes. For a distribution $\varphi = \sum_{1\le i\le n} r_i|x_i\rangle$ in $\mathcal{D}(X)$, with non-zero probabilities $r_i \in (0,1]$, going east to south in the rectangle yields:

$$(\mathcal{P}(f) \circ \mathrm{supp}_X)(\varphi) = \mathcal{P}(f)(\{x_1, \ldots, x_n\}) = \{f(x_1), \ldots, f(x_n)\}.$$

Notice that this result is a set, in which elements $f(x_i) = f(x_j)$ for $i \neq j$ are not distinguished. Similarly, going south to east in the rectangle above gives

$$(\mathrm{supp}_Y \circ \mathcal{D}(f))(\varphi) = \mathrm{supp}_Y(\textstyle\sum_{1\le i\le n} r_i|f(x_i)\rangle) = \{f(x_1), \ldots, f(x_n)\}.$$

A subtle point is that when $f(x_i) = f(x_j) = y$, say, then this element occurs as $(r_i + r_j)y$ in the distribution $\mathcal{D}(f)(\varphi)$. This y appears in the resulting support $\mathrm{supp}_Y(\mathcal{D}(f)(\varphi))$ because $r_i + r_j \neq 0$. This is obvious for (non-zero) probabilities $r_i, r_j \in [0,1]$ but needs to be required explicitly for multiset functors.

Call a monoid M **zerosumfree** or **positive** if $x + y = 0$ in M implies $x = y = 0$. For a positive commutative monoid M, the multiset functor $\mathcal{M}_M$ comes with a natural transformation $\mathrm{supp}\colon \mathcal{M}_M \Rightarrow \mathcal{P}_{\mathrm{fin}}$.

If we take the non-positive monoid $M = \mathbb{Z}$ we can show that the support maps $\mathrm{supp}_X\colon \mathcal{M}_{\mathbb{Z}}(X) \to \mathcal{P}_{\mathrm{fin}}(X)$ are *not* natural. Take as example $A = \{a, b\}$ with multiset $\varphi = 1|a\rangle + -1|b\rangle \in \mathcal{M}_{\mathbb{Z}}(A)$. The unique function $!\colon A \to 1 = \{*\}$ then provides a counter example to naturality. On the one hand we have

$$(\mathcal{P}_{\mathrm{fin}}(!) \circ \mathrm{supp}_A)(\varphi) = \mathcal{P}_{\mathrm{fin}}(!)(\{a, b\}) = \{!(a), !(b)\} = \{*\}.$$

But by applying $\mathcal{M}_{\mathbb{Z}}$ first and then support we get a different result:

$$\begin{aligned}(\mathrm{supp}_1 \circ \mathcal{M}_{\mathbb{Z}}(!))(\varphi) &= \mathrm{supp}_1(1|\,!(a)\rangle + (-1)|\,!(b)\rangle) \\ &= \mathrm{supp}_1(1|*\rangle + (-1)|*\rangle) = \mathrm{sup}_1(0|*\rangle) = \emptyset.\end{aligned}$$

This phenomenon that certain occurrences cancel each other out is called interference.

In Section 2.5 we have already seen that turning a list into the set of its elements yields a natural transformation $(-)^\star \Rightarrow \mathcal{P}_{\mathrm{fin}}$ and that there are no 'natural' mappings in the other direction, turning finite sets into lists (by choosing some order). Here we show that the obvious mappings $\upsilon_X\colon \mathcal{P}_{\mathrm{fin}}(X) \to \mathcal{D}(X)$, choosing the uniform distribution on a finite subset, are also *not* natural. To be more specific, this mapping υ_X is given by:

$$\upsilon_X(\{x_1, \ldots, x_n\}) = \tfrac{1}{n}|x_1\rangle + \cdots + \tfrac{1}{n}|x_n\rangle.$$

Take for instance $A = \{a, b, c\}$ and $2 = \{\top, \bot\}$ with function $f : A \to 2$ given by $f(a) = f(b) = \top$ and $f(c) = \bot$. Then, for the subset $\{a, b, c\} \in \mathcal{P}_{\text{fin}}(A)$ we have on the one hand:

$$\begin{aligned}(\mathcal{D}(f) \circ \upsilon_A)(\{a, b, c\}) &= \mathcal{D}(f)(\tfrac{1}{3}|a\rangle + \tfrac{1}{3}|b\rangle + \tfrac{1}{3}|c\rangle) \\ &= \tfrac{1}{3}|f(a)\rangle + \tfrac{1}{3}|f(b)\rangle + \tfrac{1}{3}|f(c)\rangle \\ &= \tfrac{1}{3}|\top\rangle + \tfrac{1}{3}|\top\rangle + \tfrac{1}{3}|\bot\rangle \\ &= \tfrac{2}{3}|\top\rangle + \tfrac{1}{3}|\bot\rangle.\end{aligned}$$

On the other hand:

$$\begin{aligned}(\upsilon_2 \circ \mathcal{P}_{\text{fin}}(f))(\{a, b, c\}) &= \upsilon_2(\{f(a), f(b), f(c)\}) \\ &= \upsilon_2(\top, \bot\}) \\ &= \tfrac{1}{2}|\top\rangle + \tfrac{1}{2}|\bot\rangle.\end{aligned}$$

We conclude with a diagram describing some natural transformations between collection functors.

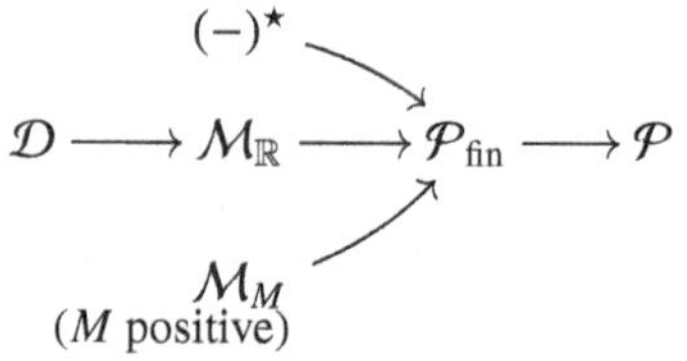

(4.5)

The map $\mathcal{D} \to \mathcal{M}_{\mathbb{R}}$ has not been discussed explicitly but is the obvious inclusion, forgetting that probabilities add up to 1.

Exercises

4.1.1 Verify that the construction $\mathcal{M}_M(X)$ from Definition 4.1.1 is functorial not only in X, but also in M. Explicitly, for a homomorphism of monoids $g : M \to L$ there is a natural transformation $\mathcal{M}_M \Rightarrow \mathcal{M}_L$.

4.1.2 1. Prove that the multiset functor sends coproducts to products, in the sense that there is an isomorphism:

$$\mathcal{M}_M(X) \times \mathcal{M}_M(Y) \cong \mathcal{M}_M(X + Y).$$

2. Prove that this isomorphism is natural in X and Y.
3. Recall from Exercise 2.1.6 that the product $M \times L$ of (commutative) monoids M, L is at the same time a coproduct (a biproduct). Check that the isomorphism in point (1) is the cotuple

$$\mathcal{M}_M(X) \times \mathcal{M}_M(Y) \xrightarrow{[\mathcal{M}_M(\kappa_1), \mathcal{M}_M(\kappa_2)]} \mathcal{M}_M(X + Y)$$

in the category of commutative monoids.

4.1.3 Interpret, as in Example 4.1.4, what double-multisets in $\mathcal{M}_M(\mathcal{M}_{\mathbb{Z}}(X))$ are. They are usually called (multivariate) *Laurent* polynomials.

4.1.4 The Dirac distribution $\eta_X : X \to \mathcal{D}(X)$ is

$$\eta(x) = 1|x\rangle = \lambda y.\,\text{if } y = x \text{ then } 1 \text{ else } 0.$$

Prove that it forms a natural transformation $\eta : \mathrm{id} \Rightarrow \mathcal{D}$.

4.1.5 Prove that functions $X \to \mathcal{D}(Y)$ and $Y \to \mathcal{D}(Z)$ can be composed to a function $X \to \mathcal{D}(Z)$, in an associative manner, with the Diract distribution as neutral element. Conclude that Markov chains $X \to \mathcal{D}(X)$ on a set X form a monoid – in analogy with the situation for powersets in Exercise 1.1.2.

4.1.6 Consider the political Markov chain (4.2), as coalgebra (4.3).

1. Describe the coalgebra composed with itself, using composition from the previous exercise.
2. Check that this corresponds to matrix composition, for the corresponding transition matrix.

4.1.7 Calculate the eigenvector of the matrix (4.4) and show that it can be described as point $1 \to \mathcal{D}(\{L, R\})$, forming a fixed point of the coalgebra $c : \{L, R\} \to \mathcal{D}(\{L, R\})$ from (4.3), using composition from Exercise 4.1.5.

4.1.8 The distribution functor $\mathcal{D}$ as introduced in Definition 4.1.5 involves distributions $\varphi : X \to [0, 1]$ with *finite* support only. This finiteness restriction is not strictly needed. One can define

$$\mathcal{D}^\infty(X) = \{\varphi : X \to [0, 1] \mid \textstyle\sum_{x \in X} \varphi(x) \text{ exists and equals } 1\}.$$

1. Prove that in this case the support $\mathrm{supp}(\varphi) = \{x \in X \mid \varphi \neq 0\}$ is necessarily a countable set.
2. Check that $\mathcal{D}^\infty$ is a functor **Sets** $\to$ **Sets**.

(This generalisation involves the existence of sums over non-finite sets; they exist in the special case of the unit interval $[0, 1]$. This infinite generalisation is less natural for multiset functors $\mathcal{M}_M$.)

4.2 Weak Pullbacks

If one starts to develop the theory of coalgebras in full categorical generality, for endofunctors $F : \mathbb{C} \to \mathbb{C}$ on an arbitrary category $\mathbb{C}$ (and not just on **Sets**), so-called weak-pullback-preserving functors become relevant, for at least two reasons:

1. For these weak-pullback-preserving functors Aczel–Mendler bisimilarity coincides with equality on the final coalgebra – as in Theorem 3.3.2.
2. For such functors there is a well-behaved categorical version of relation lifting, giving us another way to define bisimulations, namely as coalgebras of these liftings.

The current section concentrates on this first point, and the subsequent Sections 4.3–4.5 on the second one. Hence we start by describing what (weak) pullbacks are.

Definition 4.2.1 Let $\mathbb{C}$ be an arbitrary category.

1. For two morphisms $X \xrightarrow{f} Z \xleftarrow{g} Y$ in $\mathbb{C}$ with a common codomain, a **pullback** is a commuting square as on the left:

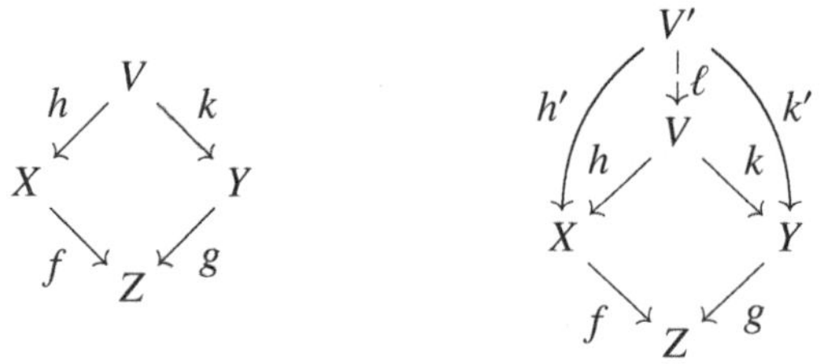

which is universal in the sense that for an arbitrary span $X \xleftarrow{h'} V' \xrightarrow{k'} Y$ with $f \circ h' = g \circ k'$ there is a unique 'mediating' map $\ell \colon V' \to V$ with $h \circ \ell = h'$ and $k \circ \ell = k'$, as on the right. In diagrams a pullback is often indicated via a small angle, as

$$\begin{array}{ccc} V & \xrightarrow{h} & X \\ {\scriptstyle k}\downarrow & \lrcorner & \downarrow{\scriptstyle f} \\ Y & \xrightarrow[g]{} & Z \end{array}$$

A **weak pullback** is like a pullback except that only existence and not unique existence of the mediating map ℓ is required.
2. A functor $F \colon \mathbb{C} \to \mathbb{D}$ is called **(weak) pullback preserving** if it maps (weak) pullback squares to (weak) pullback squares; that is, if F applied a (weak) pullback square in the category $\mathbb{C}$ forms a (weak) pullback square in $\mathbb{D}$.

The pullback of two maps $X \xrightarrow{f} Z \xleftarrow{g} Y$ in the category of sets can be described by the set

$$\mathrm{Eq}(f, g) = \{(x, y) \in X \times Y \mid f(x) = g(y)\},$$

which we used earlier in Lemma 3.2.5. It comes with obvious projections p_i to X and Y as in

$$\begin{array}{ccc}
\{(x, y) \in X \times Y \mid f(x) = g(y)\} & \xrightarrow{\quad p_2 \quad} & Y \\
{\scriptstyle p_1}\downarrow \lrcorner & & \downarrow{\scriptstyle g} \\
X & \xrightarrow[\quad f \quad]{} & Z
\end{array}$$

Also, inverse images of predicates and relations can be described via pullbacks, as in

$$\begin{array}{ccc}
h^{-1}(P) & \longrightarrow & P \\
\downarrow \lrcorner & & \downarrow \\
X & \xrightarrow[\quad h \quad]{} & Y
\end{array}
\qquad\qquad
\begin{array}{ccc}
(h \times k)^{-1}(R) & \longrightarrow & R \\
\downarrow \lrcorner & & \downarrow \\
X \times U & \xrightarrow[\quad h \times k \quad]{} & Y \times V
\end{array}$$

It is a general fact that a pullback of a mono yields a mono; see Exercise 4.2.2 below.

A pullback is a generalisation of a cartesian product of objects (as in Definition 2.1.1): in the presence of a final object $1 \in \mathbb{C}$, the product of two objects $X, Y \in \mathbb{C}$ is the same as the pullback:

$$\begin{array}{ccc}
X \times Y & \xrightarrow{\pi_2} & Y \\
{\scriptstyle \pi_1}\downarrow \lrcorner & & \downarrow{\scriptstyle !} \\
X & \xrightarrow[!]{} & 1
\end{array} \tag{4.6}$$

Similarly, the following result exploits that being a monomorphism can be expressed via pullbacks.

Lemma 4.2.2 *An arbitrary map m in a category $\mathbb{C}$ is a monomorphism if and only if the square*

$$\begin{array}{ccc}
\bullet & = & \bullet \\
\| & & \downarrow{\scriptstyle m} \\
\bullet & \xrightarrow[m]{} & \bullet
\end{array}$$

is a (weak) pullback.

A (weak) pullback-preserving functor $F: \mathbb{C} \to \mathbb{D}$ thus preserves monomorphisms: if m is mono, then so is $F(m)$. ❑

Remark 4.2.3 In category theory it is quite common that structures, such as products and coproducts in Definitions 2.1.1 and 2.1.3, are described via the 'unique existence' of a certain map. The version with only existence, not necessarily unique existence, is typically called 'weak'. For instance, one can have a *weak final object* Z, such that for each object X there is a map $X \to Z$. Each non-empty set is a weak final object in the category **Sets**. There is a third notion in between, denoted by 'semi', involving 'natural existence'; see [208, 226]. For instance, an object Z is semi-final if for each object X there is a map $s_X: X \to Z$, and for each $f: X \to Y$ one has $s_Y \circ f = s_X$.

Even though 'weak' versions are not standard, weak pullbacks are quite natural in the theory of coalgebras. In particular, preservation of weak pullbacks is a common (and useful) property. It can often be characterised in different ways, as for instance in Proposition 4.2.9 below or in Theorem 4.4.6 later on. Also it plays a crucial role in the distinction between polynomial and analytical functors in Section 4.7.

As mentioned in the introduction to this section, the following result is one of the reasons why weak-pullback preservation is important. It generalises Theorem 3.3.2.

Theorem 4.2.4 *Assume a category $\mathbb{C}$ with pullbacks, and a weak-pullback-preserving functor $F: \mathbb{C} \to \mathbb{C}$ with a final coalgebra $Z \xrightarrow{\cong} F(Z)$. Let $c: X \to F(X)$ and $d: Y \to F(Y)$ be coalgebras. The pullback relation on X, Y in*

$$\begin{array}{ccc} \mathrm{Eq}(\mathrm{beh}_c, \mathrm{beh}_d) & \longrightarrow & Z \\ \big\downarrow \lrcorner & & \big\downarrow \langle \mathrm{id}, \mathrm{id} \rangle \\ X \times Y & \xrightarrow[\mathrm{beh}_c \times \mathrm{beh}_d]{} & Z \times Z \end{array} \qquad (4.7)$$

is then the greatest Aczel–Mendler bisimulation on coalgebras c, d.

Proof First, if a relation $\langle r_1, r_2 \rangle: R \rightarrowtail X \times Y$ is an Aczel–Mendler bisimulation, that is, if R carries a coalgebra $R \to F(R)$ in $\mathbb{C}$ making the legs r_i homomorphisms in

$$\begin{array}{ccccc} F(X) & \xleftarrow{F(r_1)} & F(R) & \xrightarrow{F(r_2)} & F(Y) \\ {\scriptstyle c}\big\uparrow & & \big\uparrow & & \big\uparrow{\scriptstyle d} \\ X & \xleftarrow[r_1]{} & R & \xrightarrow[r_2]{} & Y \end{array}$$

then by finality $\mathrm{beh}_c \circ r_1 = \mathrm{beh}_d \circ r_2 \colon R \longrightarrow Z$. Hence there is a factorisation through the equality relation on Z, as in

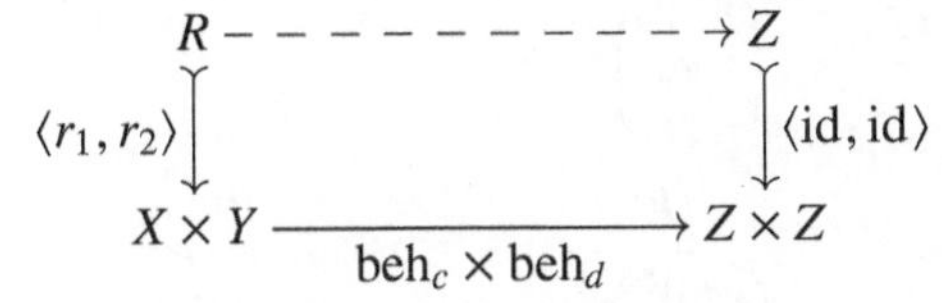

This may be read as: pairs in R are equal when mapped to the final coalgebra. It means that R factors through the pullback $\mathrm{Eq}(\mathrm{beh}_c, \mathrm{beh}_d)$ from (4.7).

We are done if we show that the pullback relation $\mathrm{Eq}(\mathrm{beh}_c, \mathrm{beh}_d) \rightarrowtail X \times Y$ from (4.7) is an Aczel–Mendler bisimulation. Here we use that F preserves weak pullbacks (and Exercise 4.2.3) to obtain a coalgebra structure in

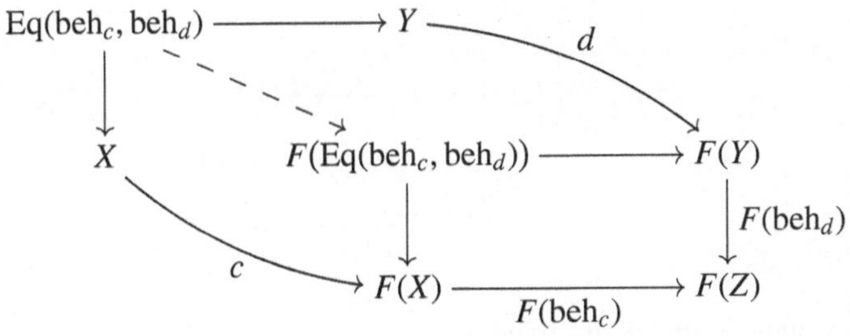

By construction, the two maps $X \leftarrow \mathrm{Eq}(\mathrm{beh}_c, \mathrm{beh}_d) \rightarrow Y$ are maps of coalgebras. ❑

We add another general result, this time involving preservation of *ordinary* (proper) pullbacks. Recall that a category of coalgebras inherits colimits (coproducts and coequalisers) from the underlying category (see Proposition 2.1.5 and Exercise 2.1.14). The situation for the dual notion of limit – including products, pullbacks and equalisers – is different. In general, they need not exist in categories of coalgebras, even if they exist in the underlying category – see also Section 6.3. The next result shows that pullbacks of coalgebras exist if the functor preserves ordinary pullbacks. As we shall see subsequently, in Proposition 4.2.6, this applies to simple polynomial functors.

Proposition 4.2.5 *Assume a functor $F \colon \mathbb{C} \to \mathbb{C}$ that preserves (ordinary) pullbacks. If the category $\mathbb{C}$ has pullbacks, then so has the category of coalgebras* **CoAlg**(F).

Proof Assume we have span of coalgebras as on the left below. Then we take the pullback in $\mathbb{C}$ of the underlying maps f, g, as on the right:

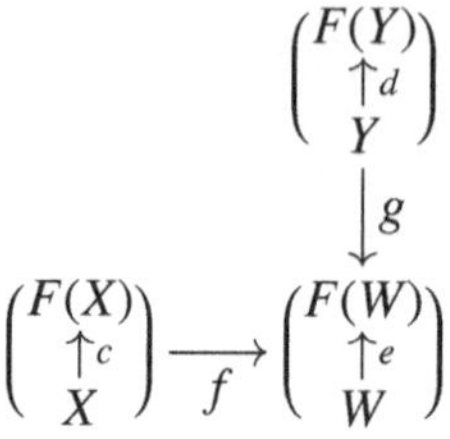

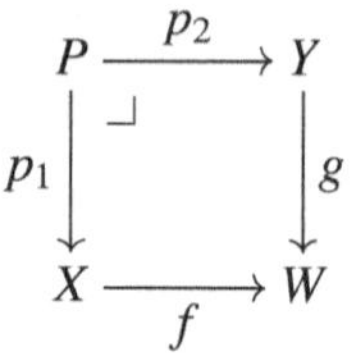

We need a coalgebra structure on the object P. Here we use that F preserves pullbacks and obtain this structure in

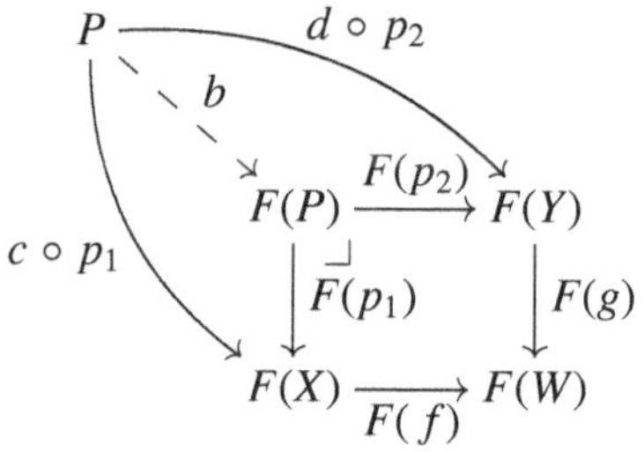

The outer maps in this diagram commute:

$$
\begin{aligned}
F(f) \circ c \circ p_1 &= e \circ f \circ p_1 && \text{since } f \text{ is a map of coalgebras} \\
&= e \circ g \circ p_2 && \text{since the pullback square in } \mathbb{C} \text{ commutes} \\
&= F(g) \circ d \circ p_2 && \text{because } g \text{ is a map of coalgebras too.}
\end{aligned}
$$

Hence the dashed map (coalgebra) $b \colon P \to F(P)$ exists. Thus we have constructed a commuting square of coalgebra maps:

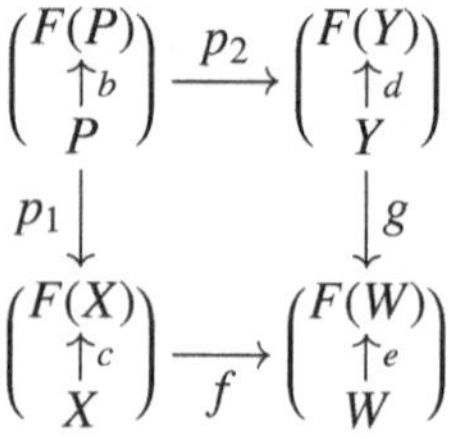

In order to obtain this square, preservation of *weak* pullbacks would have been enough. But in order to show that it is a pullback in the category of coalgebras

CoAlg(F) we need that the functor F preserves proper pullbacks. Suppose we have two coalgebra homomorphisms

$$\begin{pmatrix} F(U) \\ \uparrow a \\ U \end{pmatrix} \xrightarrow{\;h\;} \begin{pmatrix} F(X) \\ \uparrow c \\ X \end{pmatrix} \qquad \text{and} \qquad \begin{pmatrix} F(U) \\ \uparrow a \\ U \end{pmatrix} \xrightarrow{\;k\;} \begin{pmatrix} F(Y) \\ \uparrow d \\ Y \end{pmatrix}$$

with $f \circ h = g \circ k$. Then there is a unique mediating map $\ell \colon U \to P$ in $\mathbb{C}$ with $p_1 \circ \ell = h$ and $p_2 \circ \ell = k$. What remains is showing that ℓ is a homomorphism of coalgebras. Here we use that F preserves pullbacks. There are two parallel maps in

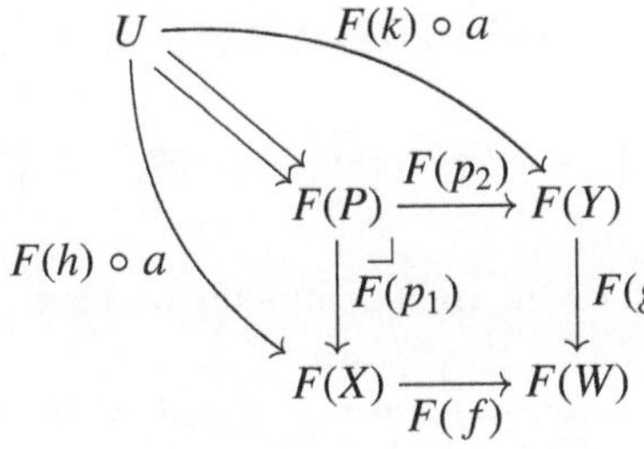

namely $b \circ \ell$ and $F(\ell) \circ a$. Thus, by uniqueness, $b \circ \ell = F(\ell) \circ a$, making ℓ a homomorphism of coalgebras. Uniqueness of ℓ in **CoAlg**(F) is left to the reader. ❑

The remainder of this section concentrates on preservation of (weak) pullbacks for specific functors. It is not hard to see that preservation of ordinary pullbacks implies preservation of weak pullbacks. This is left as exercise below. In Section 4.7 we shall see that preservation of (countably indexed) weak pullbacks and pullbacks is a key property of simple polynomial and analytical functors.

Proposition 4.2.6 *1. Every exponent polynomial functor* **Sets** $\to$ **Sets** *preserves pullbacks (and also weak ones).*

2. The powerset functors $\mathcal{P}$ and $\mathcal{P}_{fin}$ preserve weak pullbacks.

In particular, Kripke polynomial functors, finite or not, preserve weak pullbacks.

Proof 1. Trivially, identity and constant functors preserve pullbacks, so we look only at (set-indexed) coproducts and exponents. So assume for an index set I we have pullbacks in **Sets** as on the left below:

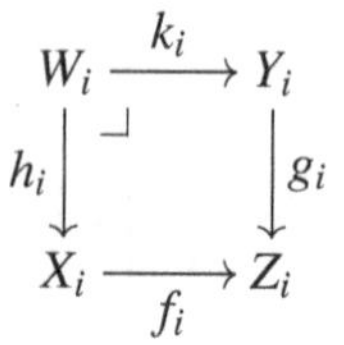

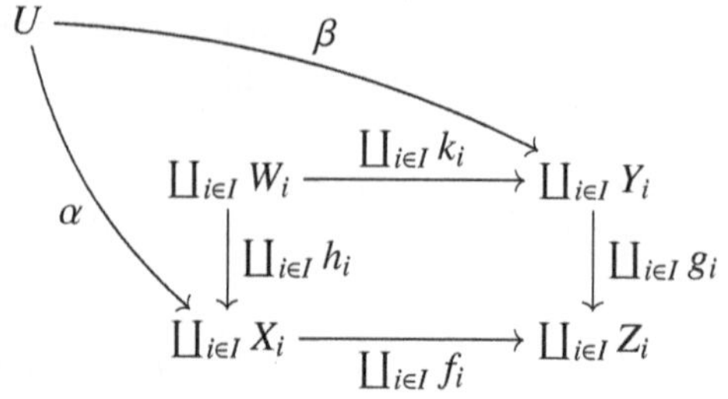

We have to show that the square on the right is again a pullback. So assume we have a set U with maps $\alpha\colon U \to \coprod_{i\in I} X_i$ and $\beta\colon U \to \coprod_{i\in I} Y_i$, as indicated, satisfying $(\coprod_{i\in I} f_i) \circ \alpha = (\coprod_{i\in I} g_i) \circ \beta$. We can decompose each of the maps α, β into two parts, $\alpha = \langle \alpha_1, \alpha_2 \rangle$ and $\beta = \langle \beta_1, \beta_2 \rangle$, where

$$\alpha_1, \beta_1 \colon U \longrightarrow I, \qquad \text{and for each } u \in U, \qquad \begin{cases} \alpha_2(u) \in X_{\alpha_1(u)} \\ \beta_2(u) \in Y_{\beta_1(u)}\,. \end{cases}$$

Because the diagram on the right commutes we have

$$\alpha_1 = \beta_1 \qquad \text{and for each } u \in U, \quad f_{\alpha_1(u)}(\alpha_2(u)) = g_{\alpha_1(u)}(\beta_2(u)).$$

Since the diagram on the left is a pullback there is for each $u \in U$ a unique element $\gamma(u) \in W_{\alpha_1(u)}$ with $h_{\alpha_1(u)}(\gamma(u)) = \alpha_2(u)$ and $k_{\alpha_1(u)}(\gamma(u)) = \beta_2(u)$. Thus we get a map $\langle \alpha_1, \gamma \rangle\colon U \to \coprod_{i\in I} W_i$ on the right. It is the unique mediating map.

For exponents assume again a pullback on the left in

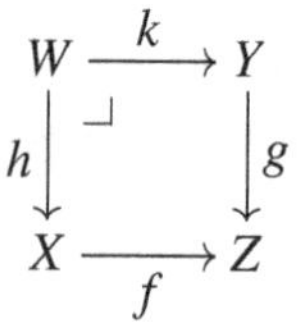

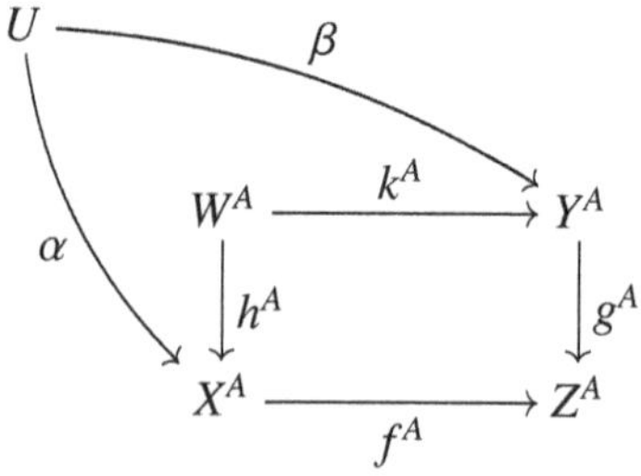

We need to find a mediating map on the right. This is straightforward by a pointwise construction: for each $u \in U$ and $a \in A$ we have

$$f(\alpha(u)(a)) = (f^A \circ \alpha)(u)(a) = (g^A \circ \beta)(u)(a) = g(\beta(u)(a)).$$

Hence there is a unique element $\gamma(u)(a) \in W$ with $h(\gamma(u)(a)) = \alpha(u)(a)$ and $k(\gamma(u)(a)) = \beta(u)(a)$. Thus we get a function $\gamma\colon U \to W^A$, as required.

2. We first consider the powerset functor, assuming a weak pullback as on the left below:

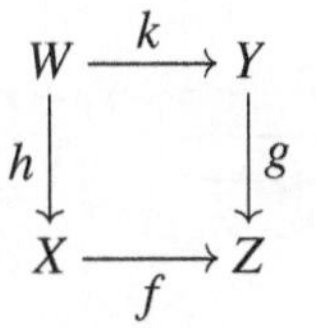

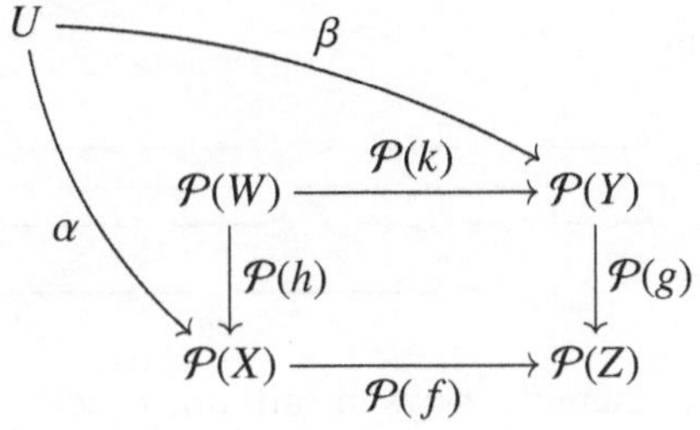

For the diagram on the right we assume $\mathcal{P}(f) \circ \alpha = \mathcal{P}(g) \circ \beta$. This means for each $u \in U$,

$$\{f(x) \mid x \in \alpha(u)\} = \{g(y) \mid y \in \beta(u)\}.$$

Thus, for each $x \in \alpha(u)$ there is an $y_x \in \beta(u)$ with $f(x) = g(y_x)$, and thus there is a $w_x \in W$ with $h(w_x) = x$ and $k(w_x) = y_x$. Similarly, for each $y \in \beta(u)$ there is an $x_y \in \alpha(u)$ with $f(x_y) = g(y)$ and so a $w_y \in W$ with $h(w_y) = x_y$ and $k(w_y) = y$. Now take $V(u) = \{w_x \mid x \in \alpha(u)\} \cup \{w_y \mid y \in \beta(u)\}$. Then

$$\begin{aligned} \mathcal{P}(h)(V(u)) &= \{h(w_x) \mid x \in \alpha(u)\} \cup \{h(w_y) \mid y \in \beta(u)\} \\ &= \{x \mid x \in \alpha(u)\} \cup \{x_y \mid y \in \beta(u)\} \\ &= \alpha(u), \end{aligned}$$

and similarly $\mathcal{P}(k)(V(u)) = \beta(u)$. ❑

The multiset functors $\mathcal{M}_M$ from Definition 4.1.1 do *not* preserve weak pullbacks in general. In [185] it is shown that this holds if and only if the commutative monoid M is both positive – so that $m + m' = 0$ implies $m = m' = 0$ – and a so-called refinement monoid (introduced in [123]).

Definition 4.2.7 A commutative monoid $M = (M, 0, +)$ is called a **refinement monoid** if for each equation

$$r_1 + r_2 = c_1 + c_2$$

there is a 2×2 'matrix' m_{ij} with

$$\begin{cases} r_i = m_{i1} + m_{i2} \\ c_j = m_{1j} + m_{2j} \end{cases} \qquad \text{as depicted in} \qquad \begin{array}{cc|c} m_{11} & m_{12} & r_1 \\ m_{21} & m_{22} & r_2 \\ \hline c_1 & c_2 & \end{array}.$$

The natural numbers $\mathbb{N}$ with addition are obviously a positive monoid. They are also a refinement monoid: assume $r_1 + r_2 = c_1 + c_2$ for $r_i, c_i \in \mathbb{N}$. The numbers m_{ij} that we need to find are written below the relevant segment of $\mathbb{N}$ in

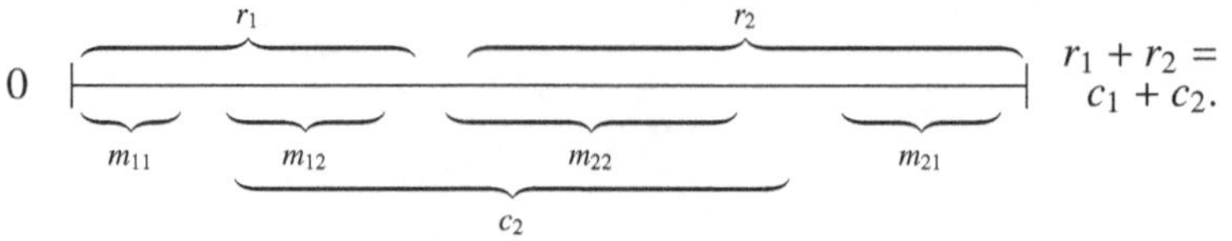

Clearly there are ways to shift this c_2 left or right, in such a way that the m_{ij} are fixed.

It is not hard to see that each abelian group is a refinement monoid: if $r_1 + r_2 = c_1 + c_2$, then we can take for example as matrix

$$\begin{array}{ccc} r_1 - c_1 & c_1 & r_1 \\ 2c_1 - r_1 & c_2 - c_1 & r_2 \\ \hline c_1 & c_2 & \end{array}$$

using (for the second row) that $(2c_1 - r_1) + (c_2 - c_1) = c_1 + c_2 - r_1 = r_1 + r_2 - r_1 = r_2$.

Lemma 4.2.8 (From [123]; also in [185]) *A commutative monoid is a positive refinement monoid if and only if the refinement property holds for each pair of numbers $(n, k) \in \mathbb{N}^2$. That is, if*

$$r_1 + \cdots + r_n = c_1 + \cdots + c_k,$$

then there is $n \times k$ matrix $(m_{ij})_{i \le n, j \le k}$ with

$$r_i = \textstyle\sum_j m_{ij} \qquad \textit{and} \qquad c_j = \textstyle\sum_i m_{ij}.$$

Proof For pairs $(0, k)$ and $(n, 0)$ positivity is used. For pairs of the form $(1, k)$ and $(n, 1)$ the result trivially holds. For the other cases it suffices to see how $(2, 3)$-refinement follows from $(2, 2)$-refinement. So assume we have $r_1 + r_2 = c_1 + c_2 + c_3$. We obtain the following successive refinements:

$$\begin{array}{ccc} m_{11} & m_{12} & r_1 \\ m_{21} & m_{22} & r_2 \\ \hline c_1 & c_2 + c_3 & \end{array} \quad \begin{array}{ccc} m'_{11} & m'_{12} & m_{12} \\ m'_{21} & m'_{22} & m_{22} \\ \hline c_2 & c_3 & \end{array} \quad \begin{array}{cccc} m_{11} & m'_{11} & m'_{12} & r_1 \\ m_{21} & m'_{21} & m'_{22} & r_2 \\ \hline c_1 & c_2 & c_3 & \end{array} .$$

□

Now we can address preservation of weak pullbacks for the multiset functors $\mathcal{M}_M$: **Sets** → **Sets** from Definition 4.1.1.

Proposition 4.2.9 (From [185]) *A multiset functor $\mathcal{M}_M$ preserves weak pullbacks if and only if the commutative monoid M is both positive and a refinement monoid.*

Proof First, assume $\mathcal{M}_M$ preserves weak pullbacks. We show that M is a refinement monoid. So assume we have $r_1 + r_2 = c_1 + c_2$. For the set $2 = \{0, 1\}$ we consider the pullback in **Sets** on the left, and the resulting weak pullback on the right:

$$\begin{array}{ccc} 2 \times 2 & \xrightarrow{\pi_2} & 2 \\ {\scriptstyle \pi_1}\downarrow & \lrcorner & \downarrow{\scriptstyle !} \\ 2 & \xrightarrow[!]{} & 1 \end{array} \qquad \begin{array}{ccc} \mathcal{M}_M(2 \times 2) & \xrightarrow{\mathcal{M}_M(\pi_2)} & \mathcal{M}_M(2) \\ {\scriptstyle \mathcal{M}_M(\pi_1)}\downarrow & & \downarrow{\scriptstyle \mathcal{M}_M(!)} \\ \mathcal{M}_M(2) & \xrightarrow[\mathcal{M}_M(!)]{} & \mathcal{M}_M(1) \cong M \end{array}$$

We use the numbers $r_i, c_i \in M$ in multisets $\varphi, \psi \in \mathcal{M}_M(2)$, namely via $\varphi(i) = r_i$ and $\psi(i) = c_i$, for $i \in 2 = \{0, 1\}$. The equation $r_1 + r_2 = c_1 + c_2$ yields $\mathcal{M}_M(!)(\varphi) = \mathcal{M}_M(!)(\psi)$. The weak pullback property gives a multiset $\chi \in \mathcal{M}_M(2{\times}2)$ with $\mathcal{M}_M(\pi_1)(\chi) = \varphi$ and $\mathcal{M}_M(\pi_2)(\chi) = \psi$. Writing $\chi(i, j) = m_{ij}$ we get the required matrix, since, for instance

$$r_1 = \varphi(0) = \mathcal{M}_M(\pi_1)(\chi)(0) = \textstyle\sum_j \chi(0, j) = m_{00} + m_{01}.$$

Next we show that M is positive. If we have $m, m' \in M$ with $m + m' = 0$ we consider the pullback in **Sets** on the left, and the resulting weak pullback on the right:

$$\begin{array}{ccc} 0 & \xrightarrow{\text{id}} & 0 \\ {\scriptstyle !}\downarrow & \lrcorner & \downarrow{\scriptstyle !} \\ 2 & \xrightarrow[!]{} & 1 \end{array} \qquad \begin{array}{ccc} \mathcal{M}_M(0) & \xrightarrow{\text{id}} & \mathcal{M}_M(0) \cong 1 \\ {\scriptstyle \mathcal{M}_M(!)}\downarrow & & \downarrow{\scriptstyle \mathcal{M}_M(!) = 0} \\ \mathcal{M}_M(2) & \xrightarrow[\mathcal{M}_M(!)]{} & \mathcal{M}_M(1) \cong M \end{array}$$

The pair m, m' gives rise to the multiset $\varphi \in \mathcal{M}_M(2)$ given by $\varphi(0) = m$ and $\varphi(1) = m'$. It satisfies $\mathcal{M}_M(!)(\varphi) = m + m' = 0 = \mathcal{M}_M(!)(0)$, for the empty multiset $0 \in \mathcal{M}_M(0)$. The weak pullback property now yields an element in $\mathcal{M}_M(0)$, necessarily the empty multiset 0, with $\mathcal{M}_M(!)(0) = \varphi$. Thus $\varphi = 0$ and so $m = m' = 0$.

Conversely, assume M is a positive refinement monoid. If we have a weak pullback as on the left, we need to show that the square on the right is also a weak pullback:

$$\begin{array}{ccc} W & \xrightarrow{k} & Y \\ {\scriptstyle h}\downarrow & & \downarrow{\scriptstyle g} \\ X & \xrightarrow[f]{} & Z \end{array} \qquad\qquad \begin{array}{ccc} \mathcal{M}_M(W) & \xrightarrow{\mathcal{M}_M(k)} & \mathcal{M}_M(Y) \\ {\scriptstyle \mathcal{M}_M(h)}\downarrow & & \downarrow{\scriptstyle \mathcal{M}_M(g)} \\ \mathcal{M}_M(X) & \xrightarrow[\mathcal{M}_M(f)]{} & \mathcal{M}_M(Z) \end{array}$$

So suppose we have multisets $\varphi \in \mathcal{M}_M(X)$ and $\psi \in \mathcal{M}_M(Y)$ with $\mathcal{M}_M(f)(\varphi) = \mathcal{M}_M(g)(\psi)$. This means

$$\sum_{x \in f^{-1}(z)} \varphi(x) = \sum_{y \in g^{-1}(z)} \psi(y)$$

for each $z \in Z$. We can limit these sums to the finite subset $Z' \subseteq Z$, given by

$$Z' = \{f(x) \mid x \in \mathrm{supp}(\varphi)\} \cup \{g(y) \mid y \in \mathrm{supp}(\psi)\}.$$

For each $z \in Z'$ we now have $\sum_{x \in f^{-1}(z)} \varphi(x) = \sum_{y \in g^{-1}(z)} \psi(y)$. This has the shape of a general refinement problem, as in Lemma 4.2.8. Thus there is a matrix (m^z_{xy}), where $x \in f^{-1}(z) \cap \mathrm{supp}(\varphi)$ and $y \in g^{-1}(z) \cap \mathrm{supp}(\psi)$ with

$$\varphi(x) = \textstyle\sum_y m^z_{xy} \quad \text{and} \quad \psi(y) = \textstyle\sum_x m^z_{xy}.$$

For each pair $x \in f^{-1}(z) \cap \mathrm{supp}(\varphi), y \in g^{-1}(z) \cap \mathrm{supp}(\psi)$ we use the weak pullback on the left and choose an element $w^z_{xy} \in W$ with $h(w^z_{xy}) = x$ and $k(w^z_{xy}) = y$. This yields a multiset $\chi_z = \sum_{x,y} m^z_{xy} w^z_{xy} \in \mathcal{M}_M(W)$.

By doing this for each $z \in Z'$ we can form $\chi = \sum_{z \in Z'} \chi_z \in \mathcal{M}_M(W)$. Notice that if $z_1 \neq z_2$, then $\mathrm{supp}(\chi_{z_1}) \cap \mathrm{supp}(\chi_{z_2}) = \emptyset$. Indeed, if $w \in \mathrm{supp}(\chi_{z_1}) \cup \mathrm{supp}(\chi_{z_2})$, then $z_1 = (f \circ h)(w) = z_2$, which is impossible. Finally,

$$\begin{aligned} &\mathcal{M}_M(h)(\chi) \\ &= \textstyle\sum_{z \in Z'} \mathcal{M}_M(h)(\chi_z) \\ &= \textstyle\sum_{z \in Z'} \sum\{m^z_{xy} | h(w^z_{xy})\rangle \mid x \in f^{-1}(z) \cap \mathrm{supp}(\varphi), y \in g^{-1}(z) \cap \mathrm{supp}(\psi)\} \\ &= \textstyle\sum_{x \in \mathrm{supp}(\varphi)} \sum\{m^z_{xy} | h(w^z_{xy})\rangle \mid z = f(x), y \in g^{-1}(z) \cap \mathrm{supp}(\psi)\} \\ &= \textstyle\sum_{x \in \mathrm{supp}(\varphi)} \varphi(x) | x\rangle \\ &= \varphi. \end{aligned}$$

Similarly one gets $\mathcal{M}_M(k)(\chi) = \psi$. ❑

For the distribution functor the situation is simpler.

Proposition 4.2.10 *The distribution functor* $\mathcal{D}$: **Sets** $\to$ **Sets** *from Definition 4.1.5 preserves weak pullbacks.*

Proof This result can be obtained in several ways. One can mimic the proof for the multiset functor $\mathcal{M}_M$, after observing that the unit interval $[0, 1]$ of probabilities is a 'partial commutative monoid' that is positive and satisfies the refinement property (in a suitably generalised sense). This approach is followed (implicitly) in [359]. Alternatively, one can follow the appendix of [459] which contains a proof using the max-flow min-cut theorem from graph theory. ❑

The 'continuous' analogue of the (discrete) distribution functor $\mathcal{D}$ on **Sets** is the 'Giry' functor $\mathcal{G}$ on the category of measurable spaces; see e.g. [157, 117, 273, 369, 124, 252]. In [458] it is shown that this functor $\mathcal{G}$ does not preserve weak pullbacks. Another such example occurs in Exercise 4.2.11 below.

Exercises

4.2.1 Describe the graph Graph(f) of a function via a pullback in **Sets**.

4.2.2 Consider in an arbitrary category a pullback

$$\begin{array}{ccc} \bullet & \longrightarrow & \bullet \\ {\scriptstyle m'}\big\downarrow & \lrcorner & \big\downarrow{\scriptstyle m} \\ \bullet & \longrightarrow & \bullet \end{array}$$

Prove that if m is a mono, then so is m'.

4.2.3 Verify that taking a pullback on the left is the same as taking a pullback on the right:

$$\begin{array}{ccc} P & \xrightarrow{p_2} & Y \\ {\scriptstyle p_1}\big\downarrow & \lrcorner & \big\downarrow{\scriptstyle g} \\ X & \xrightarrow[f]{} & Z \end{array} \qquad\qquad \begin{array}{ccc} P & \longrightarrow & Z \\ {\scriptstyle \langle p_1, p_2\rangle}\big\downarrow & & \big\downarrow{\scriptstyle \langle \mathrm{id}, \mathrm{id}\rangle} \\ X \times Y & \xrightarrow[f \times g]{} & Z \times Z \end{array}$$

4.2.4 Let $\mathbb{C}$ be a category with pullbacks. Prove that if a functor $F\colon \mathbb{C} \to \mathbb{C}$ preserves pullbacks, then it also preserves weak pullbacks.

4.2.5 Let F be a weak-pullback-preserving functor $\mathbb{C} \to \mathbb{C}$.

1. Prove that F preserves (ordinary) pullbacks of monos: if the diagram below on the left is a pullback, then so is the one on the right:

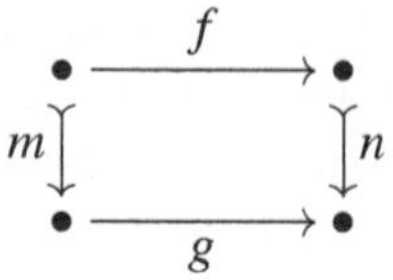

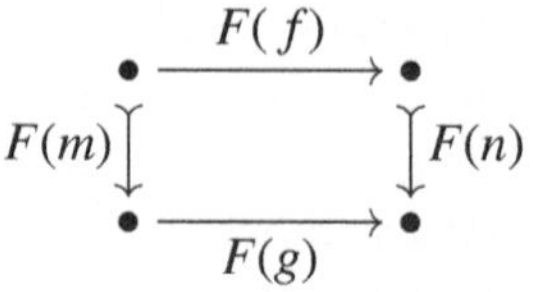

This result can be written as $F(g^{-1}(n)) \xrightarrow{\cong} F(g)^{-1}(F(n))$.

2. Conclude that F also preserves intersections $\wedge$ of monos, given as diagonal in a pullback square:

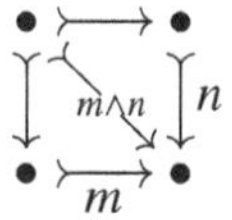

4.2.6 The following two results are known as the pullback lemmas. Prove them yourself.

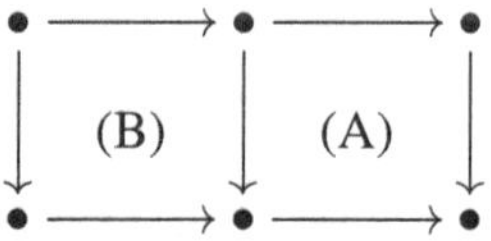

1. If (A) and (B) are pullback squares, then the outer rectangle is also a pullback square.
2. If the outer rectangle and (A) are pullback squares, then (B) is a pullback square as well.

4.2.7 Let $F\colon \mathbb{C} \to \mathbb{C}$ be an endofunctor on a category $\mathbb{C}$ with pullbacks. Prove that the category **Alg**(F) of algebras also has pullbacks, constructed as in $\mathbb{C}$.

4.2.8 Assume a category $\mathbb{C}$ with pullbacks. Prove that each slice category $\mathbb{C}/I$ (see Exercise 1.4.3) then has

1. products (and trivially also a final object)
2. pullbacks.

4.2.9 Prove, using the axiom of choice, that in the category **Sets** a diagram

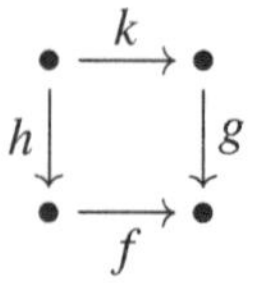

is a *weak* pullback iff $\mathrm{Eq}(f, g) \subseteq \mathrm{Im}(\langle h, k\rangle)$.

4.2.10 Consider the following pullback in **Sets**:

$$\begin{array}{ccc} \{0,1,2,3,4,5\} = 6 & \xrightarrow{\;k\;} & \{u,v,w\} \\ {\scriptstyle h}\downarrow & \lrcorner & \downarrow{\scriptstyle g} \\ \{a,b,c,d\} & \xrightarrow[\;f\;]{} & 2 = \{0,1\} \end{array}$$

where

$$\begin{cases} f(a) = f(b) = 0, f(c) = f(d) = 1 \\ g(u) = 0, g(v) = g(w) = 1 \\ h(0) = a, h(1) = b, h(2) = h(3) = c, h(4) = h(5) = d \\ k(0) = k(1) = u, k(2) = k(4) = v, k(3) = k(5) = w. \end{cases}$$

Let multisets $\varphi \in \mathcal{M}_{\mathbb{N}}(\{a,b,c,d\})$ and $\psi \in \mathcal{M}_{\mathbb{N}}(\{u,v,w\})$ be given by

$$\varphi = 2|a\rangle + 3|b\rangle + 4|c\rangle + 5|d\rangle, \qquad \psi = 5|u\rangle + 3|v\rangle + 6|w\rangle.$$

1. Check that $\mathcal{M}_{\mathbb{N}}(f)(\varphi) = \mathcal{M}_{\mathbb{N}}(g)(\psi)$.
2. Find a multiset $\chi \in \mathcal{M}_{\mathbb{N}}(6)$ with $\mathcal{M}_{\mathbb{N}}(h)(\chi) = \varphi$ and $\mathcal{M}_{\mathbb{N}}(k)(\chi) = \psi$.
3. There are four such χ; describe them all.

4.2.11 ([411]) Consider the neighbourhood functor $\mathcal{N}\colon$ **Sets** $\to$ **Sets** from Exercise 2.2.7, given by the contravariant powerset functor composed with itself. That is, $\mathcal{N}(X) = \mathcal{P}(\mathcal{P}(X))$, and for $f\colon X \to Y$ we have $\mathcal{N}(f)\colon \mathcal{P}(\mathcal{P}(X)) \to \mathcal{P}(\mathcal{P}(Y))$ by

$$\mathcal{N}(f)(A \subseteq \mathcal{P}(X)) = (f^{-1})^{-1}(A) = \{V \in \mathcal{P}(Y) \mid f^{-1}(V) \in A\}.$$

Later on, in Example 5.1.3.7, we shall recognise $\mathcal{N} = 2^{(2^{(-)})}$ as an example of a continuation monad.

Prove that $\mathcal{N}$ does *not* preserve weak pullbacks. *Hint*: Use the pullback of the following functions f, g (as in [411]):

$$\begin{cases} X = \{x_1, x_2, x_3\} \\ Z = \{z_1, z_2\} \end{cases} \qquad \begin{array}{rcl} X & \xrightarrow{\;f\;} & Z \\ x_1, x_2 & \longmapsto & y_1 \\ x_3 & \longmapsto & y_2 \end{array} \qquad \begin{array}{rcl} X & \xrightarrow{\;g\;} & Z \\ x_1 & \longmapsto & y_1 \\ x_2, x_3 & \longmapsto & y_2\,. \end{array}$$

4.3 Predicates and Relations

Recall from Section 3.1 that the notion of bisimulation (and hence of bisimilarity) is introduced via the lifting of a polynomial functor $F\colon$ **Sets** $\to$ **Sets** to an endofunctor $\mathrm{Rel}(F)\colon$ **Rel** $\to$ **Rel** on a category of relations. Bisimulations

R are then coalgebras $R \to \mathrm{Rel}(F)(R)$ of this functor, and congruences are $\mathrm{Rel}(F)$-algebras. In the next few sections we generalise this situation from **Sets** to more general categories $\mathbb{C}$. But first we need to better understand what predicates and relations are, in an arbitrary category $\mathbb{C}$. That is the topic of the present section. We shall describe these predicates and relations via 'logical' structure in $\mathbb{C}$, expressed in the form of a suitable factorisation system. As we shall see in Example 4.3.8 this factorisation structure corresponds to a predicate logic with finite conjunctions $\top, \wedge$, existential quantification $\exists$ and comprehension $\{x \colon \sigma \mid \varphi\}$.

First we need to establish some basic notions and terminology – some of which has already been used (implicitly). We start by ordering monomorphisms. Given two monos $m \colon U \rightarrowtail X$ and $n \colon V \rightarrowtail X$ one writes $m \leq n$ if there is a necessarily unique, dashed map φ in

$$\begin{array}{ccc} U & \overset{\varphi}{\dashrightarrow} & V \\ & {}_{m}\searrow \quad \swarrow_{n} & \\ & X & \end{array} \qquad \text{with} \qquad n \circ \varphi = m.$$

This order $\leq$ is reflexive ($m \leq m$) and transitive ($m \leq n$ and $n \leq k$ implies $m \leq k$) and is thus a preorder. If we write $m \cong n$ for $m \leq n$ and $n \leq m$, then $\cong$ is an equivalence relation. An equivalence class $[m] = \{n \mid n \cong m\}$ of such monos is called a **subobject**. These subobjects are seen as predicates on the object X. They are partially ordered, via $\leq$ as described above.

In practice one often writes m for the corresponding equivalence class $[m]$. Thus, we often say things like: consider a subobject $m \colon U \rightarrowtail X$ with …

In the category **Sets**, monos $U \rightarrowtail X$ are injections and subobjects $U \rightarrowtail X$ are subsets $U \subseteq X$. Thus, a relation $R \subseteq X \times Y$ is a subobject $R \rightarrowtail X \times Y$ in **Sets**. More generally, in a category $\mathbb{C}$, a relation is a subobject $R \rightarrowtail X \times Y$. Such relations carry a partial order $\leq$, as subobjects of $X \times Y$.

In practical situations it is sometimes more appropriate to consider certain subsets of monos as predicates. This will be illustrated in the case of directed complete partial orders (dcpos).

Example 4.3.1 Recall the category **Dcpo** of directed complete partial orders from Example 1.4.2.3d. These orders play an important role in the semantics of many programming constructs (see e.g. [188, 356, 34]). Here we take a closer look at predicates in **Dcpo**.

For a dcpo D a subset $U \subseteq D$ is called **admissible** if, with the order inherited from D, it is closed under directed joins – and thus a sub-dcpo. These admissible subsets look like subobjects (equivalence classes of monos) in **Dcpo**, but they are not. They correspond to a special class of monos, namely those monos that reflect the order. Thus, an admissible subset $U \subseteq D$ can be identified with an equivalence class of maps $m: E \to D$ in **Dcpo** that reflect the order: $m(x) \leq m(x')$ in D iff $x \leq x'$ in E. Such a map m is automatically injective: $m(x) = m(x')$ implies $m(x) \leq m(x')$ and $m(x') \leq m(x)$, and thus $x \leq x'$ and $x' \leq x$, so $x = x'$.

The appropriate notion to capture such situations where one wants to consider only particular subsets of monos is a factorisation system; see [56, 36, 146]. In general, such a factorisation system is given by two collections of morphisms, $\mathfrak{M}$ for 'abstract monos' and $\mathfrak{E}$ for 'abstract epis', satisfying certain properties. In the present 'logical' context we add three special requirements, in points (4)–(6) below, to the standard properties; so we will speak of a 'logical factorisation system'.

Definition 4.3.2 In an arbitrary category $\mathbb{C}$, a **logical factorisation system** is given by two collections of maps $\mathfrak{M}$ and $\mathfrak{E}$ satisfying the following properties:

1. Both $\mathfrak{M}$ and $\mathfrak{E}$ contain all isomorphisms from $\mathbb{C}$ and are closed under composition.
2. Each map $f: X \to Y$ in $\mathbb{C}$ can be factored as a map $\mathfrak{e}(f) \in \mathfrak{E}$ followed by a map $\mathfrak{m}(f) \in \mathfrak{M}$, as in

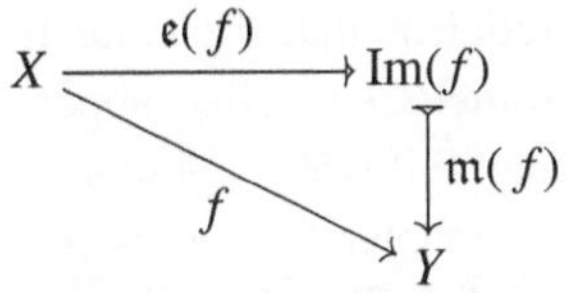

   In such diagrams we standardly write special arrows $\rightarrowtail$ for maps from $\mathfrak{M}$ and $\twoheadrightarrow$ for maps from $\mathfrak{E}$. Maps in $\mathfrak{M}$ and $\mathfrak{E}$ are sometimes called abstract monos and abstract epis, respectively.
3. The diagonal-fill-in property holds: in a commuting square as indicated below, there is a unique diagonal map as indicated, making the two triangles commute:

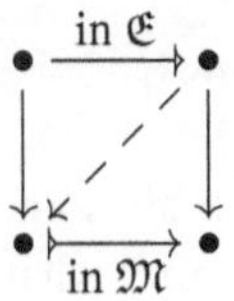

4. All maps in $\mathfrak{M}$ are monos.

5. The collection $\mathfrak{M}$ is closed under pullback: if $m \in \mathfrak{M}$, then the pullback $f^{-1}(m)$ along an arbitrary map f in $\mathbb{C}$ exists, and $f^{-1}(m) \in \mathfrak{M}$, as in

$$\begin{array}{ccc} f^{-1}(V) & \longrightarrow & V \\ {\scriptstyle f^{-1}(n)}\big\downarrow & \lrcorner & \big\downarrow{\scriptstyle n} \\ X & \xrightarrow[f]{} & Y \end{array} \tag{4.8}$$

Notice that we have overloaded the notation f^{-1} by writing both $f^{-1}(V)$ and $f^{-1}(n)$. This is often convenient.

6. For each $m \in \mathfrak{M}$ and $e \in \mathfrak{E}$ we have $m^{-1}(e) \in \mathfrak{E}$; this pullback exists by the previous point.

The standard example of a logical factorisation system is given by $\mathfrak{M}$ = (*injections*) and $\mathfrak{E}$ = (*surjections*) in **Sets**. But there are many other examples, for instance with the abstract monos $\mathfrak{M}$ given by admissible subsets on dcpos (Example 4.3.1), closed subsets of metric or topological spaces, or linearly closed subsets of vector spaces. The maps in $\mathfrak{E}$ are those whose 'obvious' factorisation has an isomorphism as monic part. Examples will be discussed explicitly in Example 4.3.7 below.

It can be shown that the collections $\mathfrak{M}$ and $\mathfrak{E}$ determine each other; see Exercise 4.3.5.

The above definition requires that the collection $\mathfrak{M}$ of abstract monos is closed under pullbacks. Sometimes it also happens that $\mathfrak{E}$ is closed under pullback, but such stability may fail (for a counter-example involving admissible subsets of dcpos, see [113, chapter 1, exercise 7]). It does hold in **Sets**, where surjections are indeed closed under pullback.

We now define categories of predicates and relations with respect to a logical factorisation system.

Definition 4.3.3 For a category $\mathbb{C}$ with a logical factorisation system $(\mathfrak{M}, \mathfrak{E})$ we define:

- the category $\mathrm{Pred}(\mathbb{C})$ of predicates in $\mathbb{C}$
- the category $\mathrm{Rel}(\mathbb{C})$ of relations in $\mathbb{C}$ – provided $\mathbb{C}$ has products $\times$.

Notice that in these notations $\mathrm{Pred}(\mathbb{C})$ and $\mathrm{Rel}(\mathbb{C})$ we leave $(\mathfrak{M}, \mathfrak{E})$ implicit. Usually it is clear for a given category what the relevant factorisation system is.

Objects of the category $\mathrm{Pred}(\mathbb{C})$ are subobjects/predicates $(m\colon U \rightarrowtail X)$ of maps $m \in \mathfrak{M}$. Morphisms from $(m\colon U \rightarrowtail X)$ to $(n\colon V \rightarrowtail Y)$ in $\mathrm{Pred}(\mathbb{C})$ are maps $f\colon X \to Y$ in $\mathbb{C}$ for which there is a necessarily unique dashed map as on the left below:

$$\begin{array}{ccc} U & \dashrightarrow & V \\ {\scriptstyle m}\downarrow & & \downarrow{\scriptstyle n} \\ X & \xrightarrow{\;f\;} & Y \end{array} \qquad\qquad \begin{array}{ccc} R & \dashrightarrow & S \\ {\scriptstyle \langle r_1, r_2\rangle}\downarrow & & \downarrow{\scriptstyle \langle s_1, s_2\rangle} \\ X_1 \times X_2 & \xrightarrow{\;f_1 \times f_2\;} & Y_1 \times Y_2 \end{array}$$

Intuitively, this says $\forall x \in X.\, U(x) \Rightarrow V(f(x))$.

The category $\mathrm{Rel}(\mathbb{C})$ has relations/subobjects $\langle r_1, r_2\rangle\colon R \rightarrowtail X_1 \times X_2$ as objects, where the tuple $\langle r_1, r_2\rangle$ is in $\mathfrak{M}$. Morphisms in $\mathrm{Rel}(\mathbb{C})$ from the relation $\langle r_1, r_2\rangle\colon R \rightarrowtail X_1 \times X_2$ to $\langle s_1, s_2\rangle\colon S \rightarrowtail Y_1 \times Y_2$ are pairs of morphisms $f_1\colon X_1 \to Y_1$, $f_2\colon X_2 \to Y_2$ in $\mathbb{C}$ for which there is a necessarily unique morphism $R \to S$ making the diagram on the right commute. It says $R(x_1, x_2) \Rightarrow S(f_1(x_1), f_2(x_2))$.

For an object $X \in \mathbb{C}$ we sometimes write $\mathrm{Pred}(X)$ for the partial order of subobjects $U \rightarrowtail X$ of an object X, coming from $\mathfrak{M}$; it may be considered as the subcategory $\mathrm{Pred}(X) \hookrightarrow \mathrm{Pred}(\mathbb{C})$ with morphisms given by the identity map in $\mathbb{C}$. Similarly, we write $\mathrm{Rel}(X_1, X_2) \hookrightarrow \mathrm{Rel}(\mathbb{C})$ for the subcategory of relations $R \rightarrowtail X_1 \times X_2$ with pairs of identities as morphisms. Thus $\mathrm{Rel}(X_1, X_2) = \mathrm{Pred}(X_1 \times X_2)$.

Applying this construction for $\mathbb{C} = \mathbf{Sets}$ yields the category $\mathrm{Rel}(\mathbf{Sets})$, which is the category **Rel** as described earlier in Definition 3.2.3.

Lemma 4.3.4 *Let $\mathbb{C}$ be a category with a logical factorisation system $(\mathfrak{M}, \mathfrak{E})$. There are obvious forgetful functors that map predicates and relations to their carriers:*

$$\begin{array}{c} \mathrm{Pred}(\mathbb{C}) \\ \downarrow \\ \mathbb{C} \end{array} \quad \begin{array}{c} U \rightarrowtail X \\ \downarrow \\ X \end{array} \qquad\qquad \begin{array}{c} \mathrm{Rel}(\mathbb{C}) \\ \downarrow \\ \mathbb{C} \times \mathbb{C} \end{array} \quad \begin{array}{c} R \rightarrowtail X_1 \times X_2 \\ \downarrow \\ (X_1, X_2) \end{array}$$

1. *Each map $f\colon X \to Y$ in $\mathbb{C}$ gives rise to a pullback functor $f^{-1}\colon \mathrm{Pred}(Y) \to \mathrm{Pred}(X)$, using diagram (4.8). Similarly, each pair of maps $f_1\colon X_1 \to Y_1$, $f_2\colon X_2 \to Y_2$ gives rise to a pullback functor $(f_1 \times f_2)^{-1}\colon \mathrm{Rel}(Y_1, Y_2) \to \mathrm{Rel}(X_1, X_2)$. In this way we get two functors:*

$$\mathbb{C}^{\mathrm{op}} \xrightarrow{\;\mathrm{Pred}(-)\;} \mathbf{PoSets} \qquad \textit{and} \qquad (\mathbb{C} \times \mathbb{C})^{\mathrm{op}} \xrightarrow{\;\mathrm{Rel}(-)\;} \mathbf{PoSets}.$$

Such functors are also known as indexed categories; see e.g. *[239].*

2. *These posets* Pred(X) *and* Rel(X_1, X_2) *have finite meets* $\top, \wedge$, *given by the identity predicate and pullbacks of predicates*

$$\top_X = \begin{pmatrix} X \\ \downarrow \mathrm{id} \\ X \end{pmatrix} \qquad \begin{array}{ccc} P & \rightarrowtail & V \\ \downarrow & \searrow^{m \wedge n} & \downarrow n \\ U & \underset{m}{\rightarrowtail} & X \end{array}$$

and similarly for relations. These meets are preserved by pullback functors f^{-1} *so the above indexed categories can be restricted to*

$$\mathbb{C}^{\mathrm{op}} \xrightarrow{\mathrm{Pred}(-)} \mathbf{MSL} \qquad \text{and} \qquad (\mathbb{C} \times \mathbb{C})^{\mathrm{op}} \xrightarrow{\mathrm{Rel}(-)} \mathbf{MSL}$$

where **MSL** *is the category of meet semilattices (and monotone functions preserving* $\top, \wedge$ *as homomorphisms).*

3. *The mapping* $X \mapsto \top_X$ *yields a 'truth' functor* $\top \colon \mathbb{C} \to \mathrm{Pred}(\mathbb{C})$ *that is a right adjoint to the forgetful functor in*

$$\begin{array}{c} \mathrm{Pred}(\mathbb{C}) \\ \downarrow \dashv \uparrow \top \\ \mathbb{C} \end{array}$$

4. *This truth functor* $\top \colon \mathbb{C} \to \mathrm{Pred}(\mathbb{C})$ *itself also has a right adjoint, mapping a predicate* $(U \rightarrowtail X)$ *to its domain* U. *This functor provides* ***comprehension*** *and will thus be written as* $\{-\}$ *in*

$$\begin{array}{c} \mathrm{Pred}(\mathbb{C}) \\ \downarrow \dashv \uparrow \top \dashv \{-\} \\ \mathbb{C} \end{array}$$

Since subobjects are equivalence classes, this domain functor $\{-\}$ *involves a choice of objects, from an isomorphism class.*

Proof 1. We first have to check that the pullback operations f^{-1} preserve the order between predicates. This is obvious. Further, we have $\mathrm{id}^{-1} = \mathrm{id}$ and also $(g \circ f)^{-1} = f^{-1} \circ g^{-1}$; the latter holds by the pullback lemma; see Exercise 4.2.6. Notice that we get equalities (instead of isomorphisms) because predicates are subobjects and thus equivalence classes of maps.

2. The identity predicate $\top_X = (\mathrm{id} \colon X \rightarrowtail X)$ is obviously the top element in the poset Pred(X). Also, the above square defining $m \wedge n \colon P \rightarrowtail X$ satisfies $k \leq m$ and $k \leq n$ iff $k \leq m \wedge n$ in Pred(X). The top element (identity map) is clearly preserved under pullback. The same holds for the meet $m \wedge n$, but this requires some low-level reasoning with pullbacks (using Exercise

4.2.6 again). In a similar manner finite meets for relations exist; they are preserved by pullback.

3. We check the adjunction *forgetful* $\dashv$ *truth*. For a predicate $U \rightarrowtail X$ and an object $Y \in \mathbb{C}$ this involves a bijective correspondence:

$$\frac{(U \overset{m}{\rightarrowtail} X) \longrightarrow (Y \overset{\text{id}}{\rightarrowtail} Y) \quad \text{in Pred}(\mathbb{C})}{X \longrightarrow Y \quad \text{in } \mathbb{C}} \qquad \text{namely} \qquad \frac{\begin{array}{ccc} U & \dashrightarrow & Y \\ {\scriptstyle m}\downarrow & & \downarrow{\scriptstyle \text{id}} \\ X & \xrightarrow{f} & Y \end{array}}{X \xrightarrow[f]{} Y}.$$

Clearly, the dashed map is uniquely determined as $f \circ m$.

4. We now prove the adjunction *truth* $\dashv$ *comprehension*. For an object $X \in \mathbb{C}$ and a predicate $n \colon V \rightarrowtail Y$ we have to prove bijective correspondences:

$$\frac{(X \overset{\text{id}}{\rightarrowtail} X) \longrightarrow (V \overset{n}{\rightarrowtail} Y) \quad \text{in Pred}(\mathbb{C})}{X \longrightarrow V \quad \text{in } \mathbb{C}} \qquad \text{namely} \qquad \frac{\begin{array}{ccc} X & \overset{f}{\dashrightarrow} & V \\ {\scriptstyle \text{id}}\downarrow & & \downarrow{\scriptstyle n} \\ X & \xrightarrow{n \circ f} & Y \end{array}}{X \xrightarrow[f]{} V}.$$

Clearly the maps above and under the double lines determine each other. ❑

We continue the investigation of the logical structure provided by factorisation systems. The next result shows that the existential quantifier $\exists$, written categorically as $\coprod$, exists for the predicates and relations defined in terms of a logical factorisation structure.

Proposition 4.3.5 *Let $\mathbb{C}$ be a category with a logical factorisation system $(\mathfrak{M}, \mathfrak{E})$.*

1. *Each map $f \colon X \to Y$ and each pair of maps $f_1 \colon X_1 \to Y_1$, $f_2 \colon X_2 \to Y_2$ in $\mathbb{C}$ give rise to functors (monotone functions) between posets:*

$$\text{Pred}(X) \xrightarrow{\coprod_f} \text{Pred}(Y) \qquad\qquad \text{Rel}(X_1, X_2) \xrightarrow{\coprod_{f_1 \times f_2}} \text{Rel}(Y_1, Y_2).$$

They are defined via factorisations: $\coprod_f(m) = \mathfrak{m}(f \circ m)$ and $\coprod_{f_1 \times f_2}(r) = \mathfrak{m}((f_1 \times f_2) \circ r)$ in

$$\begin{array}{ccc} U & \longrightarrow & \coprod_f(U) \\ {\scriptstyle m}\downarrow & & \downarrow{\scriptstyle \coprod_f(m)} \\ X & \xrightarrow[f]{} & Y \end{array} \qquad\qquad \begin{array}{ccc} R & \longrightarrow & \coprod_{f_1 \times f_2}(R) \\ {\scriptstyle r = \langle r_1, r_2 \rangle}\downarrow & & \downarrow{\scriptstyle \coprod_{f_1 \times f_2}(r)} \\ X_1 \times X_2 & \xrightarrow[f_1 \times f_2]{} & Y_1 \times Y_2 \end{array}$$

Here we have used the same overloading for $\coprod_f$ *that we used for* f^{-1} *in (4.8), namely: we apply* $\coprod_f$ *both to maps and to their domain objects.*

2. *These functors* $\coprod$ *are left (Galois) adjoints to the pullback functors from Lemma 4.3.4:*

$$\mathrm{Pred}(X) \underset{f^{-1}}{\overset{\coprod_f}{\rightleftarrows}} \mathrm{Pred}(Y) \qquad \mathrm{Rel}(X_1, X_2) \underset{(f_1 \times f_2)^{-1}}{\overset{\coprod_{f_1 \times f_2}}{\rightleftarrows}} \mathrm{Rel}(Y_1, Y_2).$$

They satisfy

$$\coprod\nolimits_{\mathrm{id}_X} = \mathrm{id}_{\mathrm{Pred}(X)} \quad \textit{and} \quad \coprod\nolimits_{g \circ f} = \coprod\nolimits_g \circ \coprod\nolimits_f . \tag{4.9}$$

3. *For each object* $X \in \mathbb{C}$ *the equality relation* $\mathrm{Eq}(X) \in \mathrm{Rel}(X, X)$ *is defined by factoring the diagonal* $\Delta = \langle \mathrm{id}, \mathrm{id} \rangle : X \rightarrowtail X \times X$, *in*

$$\begin{aligned} \mathrm{Eq}(X) &= \coprod\nolimits_{\langle \mathrm{id},\mathrm{id} \rangle}(\top) \\ &= \mathfrak{m}(\langle \mathrm{id}, \mathrm{id} \rangle) \end{aligned} \qquad \text{i.e.} \qquad \begin{array}{ccc} X & \twoheadrightarrow & \mathrm{Eq}(X) \\ & \searrow^{\langle \mathrm{id}, \mathrm{id} \rangle} & \downarrow \mathfrak{m}(\langle \mathrm{id}, \mathrm{id} \rangle) = \mathrm{Eq}(X). \\ & & X \times X \end{array}$$

Here we deliberately overload the notation $\mathrm{Eq}(X)$. *This equality forms a functor in*

$$\begin{array}{ccc} & & \mathrm{Rel}(\mathbb{C}) \\ & \nearrow^{\mathrm{Eq}(-)} & \downarrow \\ \mathbb{C} & \xrightarrow[\langle \mathrm{id}_{\mathbb{C}}, \mathrm{id}_{\mathbb{C}} \rangle]{} & \mathbb{C} \times \mathbb{C} \end{array}$$

If diagonals $\Delta = \langle \mathrm{id}, \mathrm{id} \rangle : X \rightarrowtail X \times X$ *are in the set* $\mathfrak{M}$ *of abstract monos, then* $\mathrm{Eq}(X)$ *is equal to this diagonal, and 'internal' equality in our predicate logic and 'external' equality coincide, in the sense that for two parallel maps* $f, g : Y \to X$ *one has*

$$\begin{aligned} f, g \textit{ are internally equal} &\overset{def}{\Longleftrightarrow} \top \leq \mathrm{Eq}(f, g) = \langle f, g \rangle^{-1}(\mathrm{Eq}(X)) \\ &\Longleftrightarrow f = g \\ &\overset{def}{\Longleftrightarrow} f, g \textit{ are externally equal.} \end{aligned}$$

(The direction $(\Longleftarrow)$ *of the equivalence in the middle always holds; only for* $(\Longrightarrow)$ *one needs that diagonals are in* $\mathfrak{M}$.*)*

4. *The inequalities* $\coprod_f(f^{-1}(n) \wedge m) \leq n \wedge \coprod_f(m)$ *are isomorphisms, because the collection* $\mathfrak{E}$ *is closed under pullback along* $m \in \mathfrak{M}$ *– see Definition 4.3.2.6.*

Having an equality like in the last point is usually referred to as the 'Frobenius' condition. Logically, it corresponds to the equivalence of $\exists x.(\varphi \wedge \psi)$ and $\varphi \wedge (\exists x.\psi)$ if the variable x does not occur freely in the formula φ.

Proof We do the proofs for predicates since they subsume relations.

1. Assume $m \leq m'$ in $\mathrm{Pred}(X)$, say via a map φ, then we get $\coprod_f(m) \leq \coprod_f(m')$ in $\mathrm{Pred}(Y)$ via the diagonal-fill-in property:

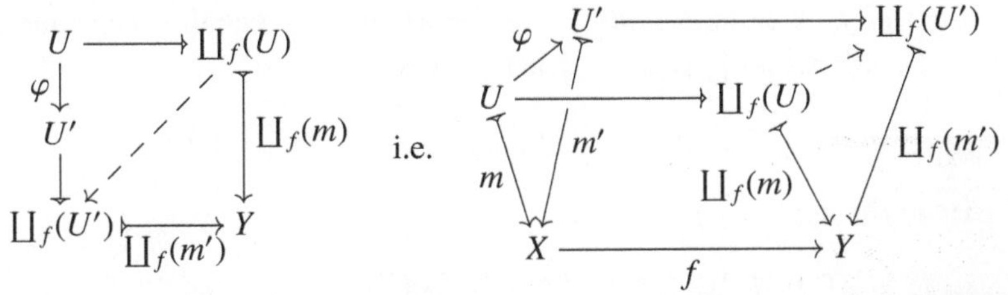

2. For predicates $m\colon U \rightarrowtail X$ and $n\colon V \rightarrowtail Y$ we have to produce a bijective correspondence:

$$\frac{\coprod_f(m) \leq n}{\overline{m \leq f^{-1}(n)}}$$ that is,

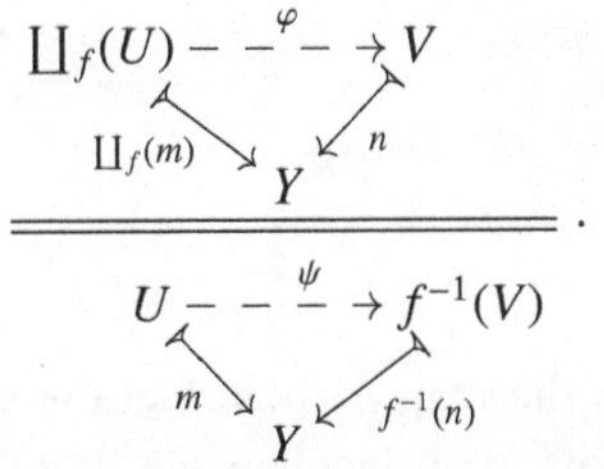

This works as follows. Given φ as indicated, we obtain ψ on the left below. The converse is sketched on the right:

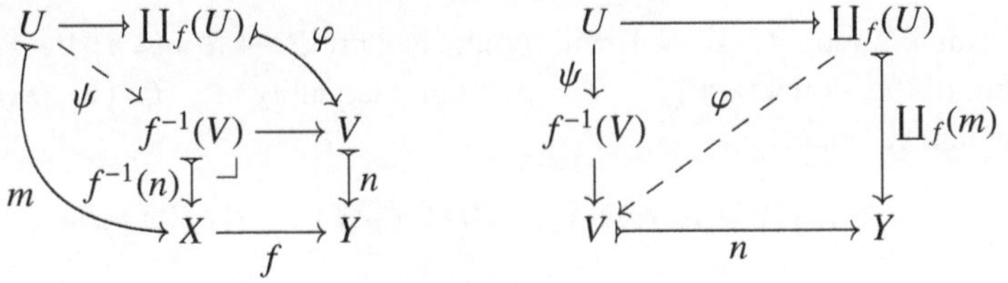

The functoriality equations (4.9) for $\coprod$ follow from the functoriality equations for pullback, via Galois connection (adjunction) $\coprod_f \dashv f^{-1}$.

3. For a map $f\colon X \to Y$ in $\mathbb{C}$ we have to show that the pair (f, f) is a morphism $\mathrm{Eq}(X) \to \mathrm{Eq}(Y)$ in $\mathrm{Rel}(\mathbb{C})$. This follows from diagonal-fill-in:

$$\begin{array}{ccc} X & \longrightarrow & \mathrm{Eq}(X) \\ {\scriptstyle f}\downarrow & & \downarrow \\ Y & \swarrow & X\times X \\ \downarrow & & \downarrow{\scriptstyle f\times f} \\ \mathrm{Eq}(Y) & \rightarrowtail & Y\times Y \end{array}$$

The outer rectangle commutes because $\langle \mathrm{id}, \mathrm{id}\rangle \circ f = (f \times f) \circ \langle \mathrm{id}, \mathrm{id}\rangle$.

Clearly, external equality $f = g$ implies internal equality, since the unit of the adjunction $\coprod_{\langle \mathrm{id},\mathrm{id}\rangle} \dashv \langle \mathrm{id}, \mathrm{id}\rangle^{-1}$ gives

$$\top \le \langle \mathrm{id}, \mathrm{id}\rangle^{-1} \coprod_{\langle \mathrm{id},\mathrm{id}\rangle}(\top) = \langle \mathrm{id}, \mathrm{id}\rangle^{-1}\mathrm{Eq}(X).$$

Hence by applying f^{-1} we get:

$$\begin{aligned} \top = f^{-1}(\top) &\le f^{-1}\langle \mathrm{id}, \mathrm{id}\rangle^{-1}\mathrm{Eq}(X) \\ &= (\langle \mathrm{id}, \mathrm{id}\rangle \circ f)^{-1}\mathrm{Eq}(X) = \langle f, f\rangle^{-1}\mathrm{Eq}(X). \end{aligned}$$

For the other direction, assume the diagonal $\Delta = \langle \mathrm{id}, \mathrm{id}\rangle$ is in $\mathfrak{M}$. Equality $\mathrm{Eq}(X)$ is then equal to this diagonal, and so $\langle f, g\rangle^{-1}(\mathrm{Eq}(X))$ is the pullback

$$\begin{array}{ccc} E & \xrightarrow{\;p_2\;} & X \\ {\scriptstyle p_1}\downarrow & \lrcorner & \downarrow{\scriptstyle \langle \mathrm{id}, \mathrm{id}\rangle = \mathrm{Eq}(X)} \\ Y & \xrightarrow[\langle f, g\rangle]{} & X\times X \end{array}$$

where the map p_1 is in fact the equaliser of f, g. Internal equality of f, g amounts to an inequality $\top \le \langle f, g\rangle^{-1}(\mathrm{Eq}(X)) = p_1$. It expresses that the map p_1 is an isomorphism, and so

$$\begin{aligned} f = \pi_1 \circ \langle f, g\rangle &= \pi_1 \circ \langle \mathrm{id}, \mathrm{id}\rangle \circ p_2 \circ p_1^{-1} \\ &= \pi_2 \circ \langle \mathrm{id}, \mathrm{id}\rangle \circ p_2 \circ p_1^{-1} = \pi_2 \circ \langle f, g\rangle = g. \end{aligned}$$

4. Assume a map $f\colon X \to Y$ with predicates $m\colon U \rightarrowtail X$ and $n\colon V \rightarrowtail Y$. The unit of the adjunction $\coprod_f \dashv f^{-1}$ gives an inequality $m \le f^{-1}(\coprod_f(m))$, from which we obtain

$$f^{-1}(n) \wedge m \le f^{-1}(n) \wedge f^{-1}(\textstyle\coprod_f(m)) = f^{-1}(n \wedge \coprod_f(m)).$$

The adjunction $\coprod_f \dashv f^{-1}$ now yields the required inequality $\coprod_f(f^{-1}(n) \wedge m) \le n \wedge \coprod_f(m)$. The second part of the statement requires more work and uses requirement (6) in Definition 4.3.2, which says that $\mathfrak{E}$ is closed under pullback along each $k \in \mathfrak{M}$. Consider the left side of the following diagram:

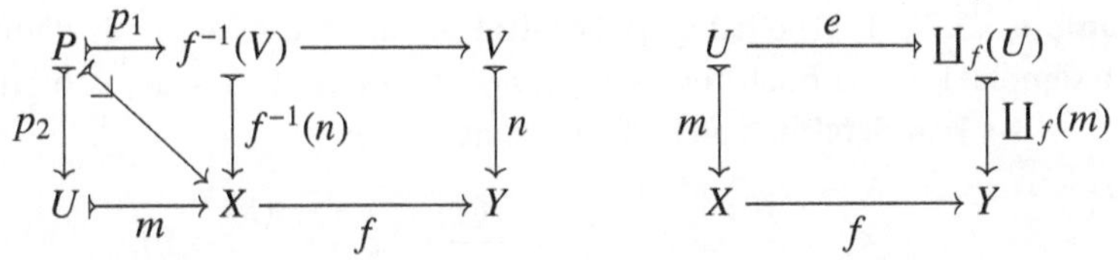

The diagonal is the conjunction $f^{-1}(n) \wedge m$. In order to get $\coprod_f(f^{-1}(n) \wedge m)$ we need to factorise the composite $f \circ (f^{-1}(n) \wedge m) = f \circ m \circ p_2$. This factorisation appears if we consider the other conjunct $n \wedge \coprod_f(m)$, arising as the diagonal on the right below:

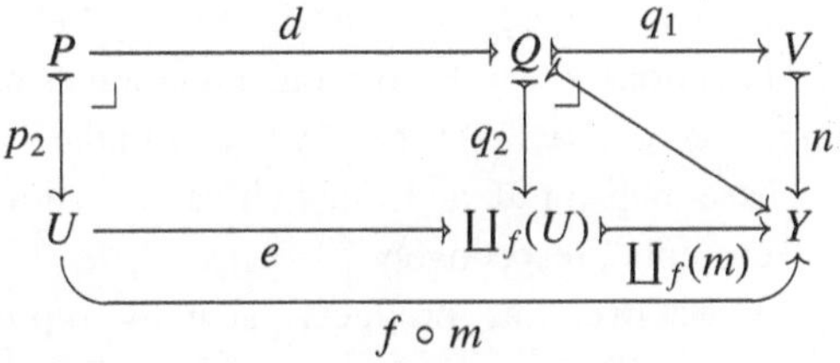

The outer rectangle is a pullback because $f \circ m = \coprod_f(m) \circ e$, as described above. Thus, by the pullback lemma (see Exercise 4.2.6), the rectangle on the left is a pullback. But then the map d arises as pullback of $e \in \mathfrak{E}$ along $q_2 \in \mathfrak{M}$. Hence, by assumption, $d \in \mathfrak{E}$. But this means that $\coprod_f(f^{-1}(n) \wedge m)$, which is by definition the $\mathfrak{M}$-part of $f \circ m \circ p_2$, is the diagonal map $n \wedge \coprod_f(m)$. Thus we are done. □

Remark 4.3.6 The third point of Proposition 4.3.5 deals with equality relations $\mathrm{Eq}(X)$, as $\mathfrak{M}$-part of the diagonal $\Delta = \langle \mathrm{id}, \mathrm{id} \rangle \colon X \rightarrowtail X \times X$. This diagonal Δ need not be in $\mathfrak{M}$. Its presence in $\mathfrak{M}$ is a non-trivial property, and therefore we have not made it part of the requirements for a 'logical factorisation system' in Definition 4.3.2. Recall, for instance, that one formulation of the Hausdorff property for topological spaces says: the diagonal relation Δ is closed. Closed subsets of topological spaces may indeed be used as abstract monos for a factorisation system. An alternative example will be elaborated in Example 4.3.7.1 below: in the category **SetsRel** of sets and relations diagonals Δ are not abstract monos.

The requirement that diagonals are in $\mathfrak{M}$ is equivalent to the requirement that all maps in $\mathfrak{E}$ are epis. This is left as an exercise below.

But if diagonals are in $\mathfrak{M}$ the resulting logic in the category satisfies the special property that internal and external equality coincide. This is for instance the case in every topos whose logic is described via *all* monos/subobjects (see e.g. [346]).

Example 4.3.7 1. The category **SetsRel** of sets and relations is defined in Example 1.4.2.4. Each relation $\langle r_1, r_2\rangle\colon R \hookrightarrow X \times Y$, seen as morphism $R\colon X \to Y$ in **SetsRel**, can be factored as

$$\big(X \xrightarrow{\;R\;} Y\big) = \big(X \xrightarrow{\;\mathfrak{e}(R)\;} Y' \xrightarrow{\;\mathfrak{m}(R)\;} Y\big),$$

where $Y' = \{y \mid \exists x.\, R(x,y)\}$ is the image of $r_2\colon R \to Y$ and

$$\mathfrak{e}(R) = \left(\begin{array}{ccc} & R & \\ {\scriptstyle r_1}\swarrow & & \searrow \\ X & & Y' \end{array}\right), \qquad \mathfrak{m}(R) = \left(\begin{array}{ccc} & Y' & \\ /\!/ & & \searrow \\ Y' & & Y \end{array}\right).$$

A factorisation system $(\mathfrak{M}, \mathfrak{E})$ exists on the category **SetsRel**: the collection $\mathfrak{M}$ consists of injections $Y' \rightarrowtail Y$, forming a relation as on the right above. And $\mathfrak{E}$ contains relations $\langle r_1, r_2\rangle\colon R \hookrightarrow X \times Y$ with the right leg r_2 forming a surjection. These maps in $\mathfrak{M}$ and $\mathfrak{E}$ are characterised in [220] as 'dagger kernels' and 'zero-epis', respectively.

The category **SetsRel** has products (actually biproducts) given by coproducts + on sets. The diagonal (relation) $\Delta = \langle \mathrm{id}, \mathrm{id}\rangle\colon Y \rightarrowtail Y + Y$ in **SetsRel** is given by the subset:

$$\Delta = \{(y, \kappa_1 y) \mid y \in Y\} \cup \{(y, \kappa_2 y) \mid y \in Y\} \subseteq Y \times (Y + Y).$$

Thus, in span form it can be written as

$$\Delta = \left(\begin{array}{ccc} & Y+Y & \\ {\scriptstyle [\mathrm{id},\mathrm{id}]}\swarrow & & \searrow\!\!\!\!= \\ Y & & Y+Y \end{array}\right)$$

with image given by the equality relation on $Y + Y$, obtained as

$$\mathrm{Eq}(Y) = \mathfrak{m}(\Delta) = \left(\begin{array}{ccc} & Y+Y & \\ /\!/ & & \backslash\!\backslash \\ Y+Y & & Y+Y \end{array}\right).$$

For parallel relations $R, S : X \to Y$ in **SetsRel** we then have

$$\langle R, S\rangle^{-1}(\mathrm{Eq}(Y)) = \{x \in X \mid \forall y, y'.\, R(x,y) \wedge S(x,y') \Rightarrow y = y'\}.$$

Hence internal equality $\top \leq \langle R, S\rangle^{-1}(\mathrm{Eq}(Y))$ is in **SetsRel** not the same as external equality.

2. Recall the category **Vect** of vector spaces (over the real numbers $\mathbb{R}$ or some other field). It carries a factorisation system $(\mathfrak{M}, \mathfrak{E})$ where

$$\mathfrak{M} = (\textit{injective linear maps}) \quad \text{and} \quad \mathfrak{E} = (\textit{surjective linear maps}).$$

The subobjects associated with $\mathfrak{M}$ are linearly closed subspaces. The product (actually biproduct) for two vector spaces is given by the product of the underlying sets, with coordinatewise structure. Hence the diagonal $\Delta = \langle \mathrm{id}, \mathrm{id} \rangle \colon V \rightarrowtail V \times V$ is the usual (set-theoretic) diagonal. Since these diagonals are in $\mathfrak{M}$, internal and external equality coincide in **Vect**.

3. We write **Hilb** for the category of Hilbert spaces (over the real or possibly also over the complex numbers). Morphisms are linear maps which are continuous (or equivalently bounded). The category **Hilb** also carries a factorisation system, where maps in $\mathfrak{M}$ correspond to subspaces which are both linearly and metrically closed. Also in this case diagonals are in $\mathfrak{M}$. More details can be found in [220].

In this book we use factorisation systems primarily for logical purposes, namely for predicate and relation lifting and for existential quantification. Factorisation systems are also used in coalgebra for minimal realisations, as in Exercise 2.5.14; see [77, 307] for more information.

We close this section with a special example of a category with a logical factorisation system. It is obtained from an ordinary logical calculus, by suitably organising the syntactic material into a category. It illustrates what kind of logical structure is relevant in this setting. Categories such as these are described as Lindenbaum–Tarski (or term model) constructions in [239]. These categories may also be described via an initiality property, but that goes beyond the current setting.

Example 4.3.8 We sketch a (multi-sorted, typed) logic, whose types, terms and formulas are given by the following BNF syntax:

$$\begin{array}{rl} \textbf{Types} & \sigma := B \mid 1 \mid \sigma \times \sigma \mid \{x \colon \sigma \mid \varphi\} \\ \textbf{Terms} & M := x \mid f(M, \ldots, M) \mid \pi_1 M \mid \pi_2 M \mid \langle M, M \rangle \\ \textbf{Formulas} & \varphi := P \mid \top \mid \varphi \wedge \varphi \mid M =_\sigma M \mid \exists x \colon \sigma. \varphi. \end{array}$$

A type is thus either a primitive type B, a singleton (or unit) type 1, a product type $\sigma \times \sigma'$ or a comprehension type $\{x \colon \sigma \mid \varphi\}$ for a formula φ. A term is either a variable x, a function application $f(M_1, \ldots, M_n)$ for a function symbol f with appropriate arity and type, a projection term $\pi_i M$ or a tuple $\langle M, M' \rangle$. Finally, a formula is one of the following: an atomic predicate P; a truth formula $\top$; a conjunction $\varphi \wedge \varphi'$; an equation $M =_\sigma M'$ for two terms M, M' both of type σ; or a (typed) existential formula $\exists x \colon \sigma. \varphi$. Here we assume that the function symbols f and atomic predicates P are somehow given, via an appropriate signature. Notice, by the way, that negation is not part of this logical language.

A proper account of our predicate logic now lists the typing rules for sequents of the form $x_1 \colon \sigma_1, \ldots, x_n \colon \sigma_n \vdash M \colon \tau$, expressing that term M, (possibly) containing typed variables $x_1, \ldots, x_n$, has type τ, and also the deduction rules for logical sequents written as $x_1 : \sigma_1, \ldots, x_n : \sigma_n \mid \varphi_1, \ldots, \varphi_n \vdash \psi$. Sometimes we write a term M as $M(x_1, \ldots, x_n)$ to make the variables occurring in M explicit (and similarly for formulas). We write $[M/x]$ for the (postfix) operation of substituting the term M for all (free) occurrences of the variable x (where M and x have the same type). We assume the reader is reasonably familiar with these rules; refer to [239] for further information. The exception we make is for the rules of the comprehension type (see below), because they are not so standard.

But first we show that we may restrict ourselves to terms and formulas containing at most one variable. Suppose a term $M(x, y)$ has two variables $x : \sigma, y : \tau$. Then we may replace x, y by a single variable $z : \sigma \times \tau$ in a product type. This z is placed in M via substitution: $M[\pi_1 z/x, \pi_2/y]$. This can of course be repeated for multiple variables. Similarly, we may replace entailments $\varphi_1, \cdots, \varphi_n \vdash \psi$ by an entailment $\varphi_1 \wedge \cdots \wedge \varphi_n \vdash \psi$ between only two formulas (using $\top$ as antecedent if $n = 0$). If we keep track of the variable involved we write such sequents as $x : \sigma \mid \varphi \vdash \psi$.

Now we can be more explicit about comprehension types. If we have a formula $\varphi(x)$, involving a variable $x : \sigma$, we can form the type $\{x : \sigma \mid \varphi\}$. It comes with introduction and elimination rules, involving tags i for 'in' and o for 'out',

$$\frac{y : \tau \vdash M : \sigma \qquad y : \tau \mid \top \vdash \varphi[M/x]}{y : \tau \vdash \mathsf{i}_\varphi(M) : \{x : \sigma \mid \varphi\}}$$

$$\frac{y : \tau \vdash N : \{x : \sigma \mid \varphi\}}{y : \tau \vdash \mathsf{o}_\varphi(N) : \sigma} \qquad \frac{x : \sigma \mid \varphi, \psi \vdash \chi}{x' : \{x : \sigma \mid \varphi\} \mid \psi[\mathsf{o}_\varphi(x')/x] \vdash \chi[\mathsf{o}_\varphi(x')/x]}$$

with associated conversions $\mathsf{o}_\varphi(\mathsf{i}_\varphi(M)) = M$ and $\mathsf{i}_\varphi(\mathsf{o}_\varphi(N)) = N$. In the sequel we omit these subscripts φ for convenience.

We now form a category $\mathbb{L}$, for logic, with

objects types σ

morphisms $\sigma \to \tau$ which are equivalence classes $[M]$ of terms $x : \vdash M : \tau$. Two terms M, M' are equivalent when one can deduce $x : \sigma \mid \top \vdash M =_\sigma M'$. Via these equivalence classes we force 'internal' (provable) equality and 'external' equality (of maps) to coincide.

The identity map $\sigma \to \sigma$ is $[x_\sigma]$, where $x_\sigma : \sigma$ is a particular (chosen) variable of type σ. The composition of maps $[M(x)] : \sigma \to \tau$ and $[N(y)] : \tau \to \rho$ is given by the term $x : \sigma \vdash N[M(x)/y] : \rho$ obtained by substitution.

This category $\mathbb{L}$ has finite (categorical) products, via the type-theoretic products $1, \sigma \times \tau$.

The category $\mathbb{L}$ also carries a logical factorisation system $(\mathfrak{M}, \mathfrak{E})$. The set $\mathfrak{M}$ contains the 'o-maps', given by a formula φ in

$$u \colon \{x \colon \sigma \mid \varphi\} \vdash \mathsf{o}(u) \colon \sigma.$$

(Actually, we take in $\mathfrak{M}$ all these o-maps composed with isomorphisms.)

These o-maps are clearly monic: if $\mathsf{o}(N) = \mathsf{o}(N')$, then $N = \mathsf{i}(\mathsf{o}(N)) = \mathsf{i}(\mathsf{o}(N')) = N'$. Also, they are closed under composition. This can be seen in the following diagram:

$$\begin{array}{ccc}
\{y \colon \{x \colon \sigma \mid \varphi\} \mid \psi\} & \xrightarrow{[\mathsf{o}]} & \{x \colon \sigma \mid \varphi\} \\
{\scriptstyle [P]}\downarrow{\scriptstyle \cong} & & \downarrow{\scriptstyle [\mathsf{o}]} \\
\{x \colon \sigma \mid \varphi \wedge \psi[\mathsf{i}(x)/y]\} & \xrightarrow[{[\mathsf{o}]}]{} & \sigma
\end{array}$$

The term $P(v)$ is $\mathsf{i}(\mathsf{o}(\mathsf{o}(v)))$ and is obtained in the following derivation:

$$\begin{array}{c}
\begin{array}{cc}
\begin{array}{c} x \colon \sigma \mid \varphi, \top \vdash \varphi \\ \hline y \colon \{x \colon \sigma \mid \varphi\} \mid \top \vdash \varphi[\mathsf{o}(y)/x] \end{array}
&
\begin{array}{c} y \colon \{x \colon \sigma \mid \varphi\} \mid \psi, \top \vdash \psi \\ \hline v \colon \{y \colon \{x \colon \sigma \mid \varphi\} \mid \psi\} \mid \top \vdash \psi[\mathsf{o}(v)/y] \end{array}
\end{array} \\ \hline
v \colon \{y \colon \{x \colon \sigma \mid \varphi\} \mid \psi\} \mid \top \vdash \varphi[\mathsf{o}(\mathsf{o}(v))/x] \wedge \psi[\mathsf{o}(v)/y] \\
= (\varphi \wedge \psi[\mathsf{i}(x)/y])[\mathsf{o}(\mathsf{o}(v))/x] \\ \hline
v \colon \{y \colon \{x \colon \sigma \mid \varphi\} \mid \psi\} \vdash P(v) \stackrel{\text{def}}{=} \mathsf{i}(\mathsf{o}(\mathsf{o}(v))) \colon \{x \colon \sigma \mid \varphi \wedge \psi[\mathsf{i}(x)/y]\}
\end{array}$$

Notice that the i and os in $P(v) = \mathsf{i}(\mathsf{o}(\mathsf{o}(v)))$ are associated with different formulas. In a similar manner one obtains the inverse term P^{-1} of P as $P^{-1}(u) = \mathsf{i}(\mathsf{i}(\mathsf{o}(u)))$.

The set $\mathfrak{M}$ is also closed under pullback, via substitution in formulas:

$$\begin{array}{ccc}
\{y \colon \tau \mid \varphi[M(y)/x]\} & \xrightarrow{[\mathsf{i}(M[\mathsf{o}(v)/y])]} & \{x \colon \sigma \mid \varphi\} \\
{\scriptstyle [\mathsf{o}]}\downarrow & & \downarrow{\scriptstyle [\mathsf{o}]} \\
\tau & \xrightarrow[{[M(y)]}]{} & \sigma
\end{array}$$

The map on top is well typed since

$$\begin{array}{c}
y \colon \tau \mid \varphi[M(y)/x], \top \vdash \varphi[M(y)/x] \\ \hline
v \colon \{y \colon \tau \mid \varphi[M(y)/x]\} \mid \top \vdash \varphi[M(y)/x][\mathsf{o}(v)/y] = \varphi[M[\mathsf{o}(v)/y]/x]. \\ \hline
v \colon \{y \colon \tau \mid \varphi[M(y)/x]\} \vdash \mathsf{i}(M[\mathsf{o}(v)/y]) \colon \{x \colon \sigma \mid \varphi\}
\end{array}$$

We leave it to the reader to check that it forms a pullback in $\mathbb{L}$.

The set $\mathfrak{M}$ also contains the equality relations via the following isomorphism:

$$\begin{array}{ccc} \sigma & \xrightarrow[\cong]{[\mathsf{i}(\langle x, x\rangle)]} & \{z\colon \sigma \times \sigma \,|\, \pi_1 z =_\sigma \pi_2 z\} \\ & {\scriptstyle \langle \mathrm{id}, \mathrm{id}\rangle = [\langle x, x\rangle]} \searrow \quad \swarrow {\scriptstyle [\mathsf{o}]} & \\ & \sigma \times \sigma & \end{array}$$

The inverse in this diagram is given by $[\pi_1 \mathsf{o}(v)] = [\pi_2 \mathsf{o}(v)]$, where the variable v has type $\{z\colon \sigma \times \sigma \,|\, \pi_1 z =_\sigma \pi_2 z\}$.

We define $\mathfrak{E}$ so that it contains those maps of the form $[M]\colon \sigma \to \{y\colon \tau \,|\, \psi\}$ that satisfy $y\colon \tau \mid \psi \vdash \exists x\colon \sigma.\, \mathsf{o}(M(x)) = y$. We leave it to the reader to show that these maps are closed under composition and also closed under pullback along o-maps (from $\mathfrak{M}$).

We come to factorisation. For an arbitrary map $[M]\colon \sigma \to \tau$ in $\mathbb{L}$ we can consider the following predicate on the codomain type τ:

$$\mathrm{Im}([M]) = (y\colon \tau \vdash \exists x\colon \sigma.\, M(x) = y).$$

Thus we can factorise the map $[M]$ as

$$\begin{array}{ccl} \sigma & \xrightarrow{[\mathsf{i}(M)]} & \{\mathrm{Im}([M])\} = \{y\colon \tau \,|\, \exists x\colon \sigma.\, M(x) = y\} \\ & {\scriptstyle [M]} \searrow & \downarrow {\scriptstyle [\mathsf{o}]} \\ & & \tau \end{array}$$

Finally, we check the diagonal-fill-in condition. Assume we have a commuting square:

$$\begin{array}{ccc} \sigma & \xrightarrow{[M]} & \{y\colon \tau \,|\, \psi\} \\ {\scriptstyle [P]} \downarrow & & \downarrow {\scriptstyle [Q]} \\ \{z\colon \rho \,|\, \varphi\} & \xrightarrow[{[\mathsf{o}]}]{} & \rho \end{array} \tag{4.10}$$

Commutation says $\mathsf{o}(P) =_\rho Q[M/y']$, where $y'\colon \{y\colon \tau \,|\, \psi\}$. It may be clear that the only choice as diagonal $\{y\colon \tau \,|\, \psi\} \to \{z\colon \rho \,|\, \varphi\}$ is the term $\mathsf{i}(Q)$, since

$$\begin{aligned} \mathsf{o}(\mathsf{i}(Q)) &= Q \\ \mathsf{i}(Q)[M/y'] &= \mathsf{i}(Q[M/y']) \\ &= \mathsf{i}(\mathsf{o}(P)) \\ &= P. \end{aligned}$$

The challenge is to show that the term $\mathsf{i}(Q)$ is appropriately typed. This is achieved via the derivation in Figure 4.3.

$$
\dfrac{y' : \{y \colon \tau \,|\, \psi\} \vdash \mathsf{o}(y') \colon \tau \qquad y \colon \tau \,|\, \psi, \top \vdash \exists x \colon \sigma.\, \mathsf{o}(M) = y}{\dfrac{y' : \{y \colon \tau \,|\, \psi\} \,|\, \top \vdash \exists x \colon \sigma.\, \mathsf{o}(M) = \mathsf{o}(y')}{y' : \{y \colon \tau \,|\, \psi\} \,|\, \top \vdash \exists x \colon \sigma.\, M = y'}}
$$

$$
\dfrac{x \colon \sigma \vdash P \colon \{z \colon \rho \,|\, \varphi\} \qquad \dfrac{z \colon \rho \,|\, \varphi, \top \,|\, \varphi}{z' : \{z \colon \rho \,|\, \varphi\} \,|\, \top \vdash \varphi[\mathsf{o}(z')/z]}}{x \colon \sigma \,|\, \top \vdash \varphi[\mathsf{o}(P)/z]}
$$

$$
\dfrac{\dfrac{\dfrac{x \colon \sigma \,|\, \top \vdash \varphi[\mathsf{o}(P)/z] \qquad \dfrac{x \colon \sigma \,|\, \top \vdash \mathsf{o}(P) = Q[M/y'] \qquad \dfrac{x \colon \sigma \vdash M \colon \{y \colon \tau \,|\, \psi\} \qquad y', y'' : \{y \colon \tau \,|\, \psi\} \,|\, \varphi[Q[y''/y']/z], y'' = y' \vdash \varphi[Q/z]}{x \colon \sigma, y' : \{y \colon \tau \,|\, \psi\} \,|\, \varphi[Q[M/y']/z], M = y' \vdash \varphi[Q/z]}}{x \colon \sigma, y' : \{y \colon \tau \,|\, \psi\} \,|\, \varphi[\mathsf{o}(P)/z], M = y' \vdash \varphi[Q/z]}}{\dfrac{x \colon \sigma, y' : \{y \colon \tau \,|\, \psi\} \,|\, M = y' \vdash \varphi[Q/z]}{y' : \{y \colon \tau \,|\, \psi\} \,|\, \exists x \colon \sigma.\, M = y' \vdash \varphi[Q/z]}} \qquad y' : \{y \colon \tau \,|\, \psi\} \,|\, \top \vdash \exists x \colon \sigma.\, M = y'}{\dfrac{y' : \{y \colon \tau \,|\, \psi\} \,|\, \top \vdash \varphi[Q/z]}{y' : \{y \colon \tau \,|\, \psi\} \vdash \mathsf{i}(Q) \colon \{z \colon \rho \,|\, \varphi\}}}
$$

Figure 4.3 Well-typedness of the diagonal map for the rectangle (4.10).

Exercises

4.3.1 Let $(\mathfrak{M}, \mathfrak{E})$ be a logical factorisation system.

1. Show that a map $f \in \mathfrak{M} \cap \mathfrak{E}$ is an isomorphism.
2. Prove that if we can factor a map g both as $g = m \circ e$ and as $g = m' \circ e'$, where $m, m' \in \mathfrak{M}$ and $e, e' \in \mathfrak{E}$, then there is a unique isomorphism φ with $m' \circ \varphi = m$ and $\varphi \circ e = e'$.
3. Show for $m \in \mathfrak{M}$ and $e \in \mathfrak{E}$ that $\mathfrak{m}(m \circ f) = m \circ \mathfrak{m}(f)$ and $\mathfrak{e}(f \circ e) = \mathfrak{m}(f) \circ e$, where $\mathfrak{m}(-)$ and $\mathfrak{e}(-)$ take the $\mathfrak{M}$-part and $\mathfrak{E}$-part as in Definition 4.3.2.2.

4.3.2 Let $F\colon \mathbb{C} \to \mathbb{C}$ be an endofunctor on a category $\mathbb{C}$ with a logical factorisation system $(\mathfrak{M}, \mathfrak{E})$.

1. Assume that F preserves abstract epis, i.e. $e \in \mathfrak{E} \Rightarrow F(e) \in \mathfrak{E}$. Prove that the category **Alg**(F) of algebras also carries a logical factorisation system. Use that pullbacks in **Alg**(F) are constructed as in $\mathbb{C}$; see Exercise 4.2.7.
2. Check that every endofunctor F: **Sets** $\to$ **Sets** satisfies this assumption, i.e. preserves surjections – if the axiom of choice holds. *Hint*: Recall that the axiom of choice can be formulated as: each surjection has a section; see Section 2.1.

4.3.3 Define the category $\mathrm{EnRel}(\mathbb{C})$ of endorelations in a category $\mathbb{C}$ (with a logical factorisation system) via the following pullback of functors:

$$\begin{array}{ccc} \mathrm{EnRel}(\mathbb{C}) & \longrightarrow & \mathrm{Rel}(\mathbb{C}) \\ \big\downarrow \lrcorner & & \big\downarrow \\ \mathbb{C} & \xrightarrow[\langle \mathrm{id}_{\mathbb{C}}, \mathrm{id}_{\mathbb{C}} \rangle]{} & \mathbb{C} \times \mathbb{C} \end{array}$$

1. Describe this category $\mathrm{EnRel}(\mathbb{C})$ in detail.
2. Demonstrate that taking equality relations on an object forms a functor $\mathrm{Eq}(-)\colon \mathbb{C} \to \mathrm{EnRel}(\mathbb{C})$.

4.3.4 Let $\mathbb{C}$ be a category with a logical factorisation system and finite coproducts $(0, +)$.

1. Show that the image of the unique map $!\colon 0 \to X$ is the least element $\perp$ in the poset $\mathrm{Pred}(X)$ of predicates on X.
2. Similarly, show that the join $m \vee n$ in $\mathrm{Pred}(X)$ of predicates $m\colon U \rightarrowtail X$ and $n\colon V \rightarrowtail Y$ is the image of the cotuple $[m, n]\colon U + V \to X$.

4.3.5 Two morphisms f, g in an arbitrary category $\mathbb{C}$ may be called orthogonal, written as $f \perp g$, if in each commuting square as below there is a unique diagonal making everything in sight commute:

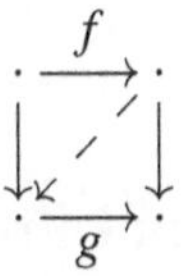

The diagonal-fill-in property for a factorisation system $(\mathfrak{M}, \mathfrak{E})$ in Definition 4.3.2 thus says that $e \perp m$ for each $m \in \mathfrak{M}$ and $e \in \mathfrak{E}$.

Now assume that a category $\mathbb{C}$ is equipped with a factorisation system $(\mathfrak{M}, \mathfrak{E})$, not necessarily 'logical'. This means that only properties (1)–(3) in Definition 4.3.2 hold.

1. Prove that $f \in \mathfrak{E}$ if and only if $f \perp m$ for all $m \in \mathfrak{M}$.
2. Similarly, prove that $g \in \mathfrak{M}$ if and only if $e \perp g$ for all $e \in \mathfrak{E}$.
3. Prove $e, d \circ e \in \mathfrak{E} \Rightarrow d \in \mathfrak{E}$.
4. Similarly (or dually), prove $m, m \circ n \in \mathfrak{M} \Rightarrow n \in \mathfrak{M}$.
5. Prove $m, n \in \mathfrak{M} \Rightarrow m \times n \in \mathfrak{M}$, assuming products exist in $\mathbb{C}$.
6. Show that diagonals $\Delta = \langle \mathrm{id}, \mathrm{id} \rangle$ are in $\mathfrak{M}$ if and only if all maps in $\mathfrak{E}$ are epis.

4.3.6 Prove the converse of Proposition 4.3.5.4: if $\coprod_f(f^{-1}(n) \wedge m) = n \wedge \coprod_f(m)$ holds for all appropriate f, m, n, then $\mathfrak{E}$ is closed under pullback along maps $m \in \mathfrak{M}$.

4.3.7 Let $(\mathfrak{M}, \mathfrak{E})$ be a factorisation system on a category $\mathbb{C}$ with finite products $1, \times$. Prove that the category of predicates $\mathrm{Pred}(\mathbb{C})$ also has finite products, via the following constructions:

1. The identity $(1 \rightarrowtail 1)$ on the final object $1 \in \mathbb{C}$ is final in $\mathrm{Pred}(\mathbb{C})$.
2. The product of predicates $(m\colon U \rightarrowtail X)$ and $(n\colon V \rightarrowtail Y)$ is the conjunction of the pullbacks $\pi_1^{-1}(m) \wedge \pi_2^{-1}(n)$, as predicate on $X \times Y$.

Show that also the category of relations $\mathrm{Rel}(\mathbb{C})$ has finite products.

4.3.8 Let $(\mathfrak{M}, \mathfrak{E})$ be a logical factorisation system on a category $\mathbb{C}$ with pullbacks. Prove that $\mathfrak{E}$ is closed under pullbacks along arbitrary maps if and only if the so-called Beck–Chevalley condition holds: for a pullback as on the left, the inequality on the right is an isomorphism:

$$\begin{array}{ccc} X & \xrightarrow{h} & Y \\ {\scriptstyle f}\downarrow & \lrcorner & \downarrow{\scriptstyle k} \\ Z & \xrightarrow[g]{} & W \end{array} \qquad\qquad \coprod_f h^{-1}(m) \leq g^{-1} \coprod_k(m)$$

4.4 Relation Lifting, Categorically

The previous section introduced predicates and relations in a category via a factorisation system, corresponding to conjunctions $\top, \wedge$, equality, existential quantification $\exists$ and comprehension $\{-\}$. In this section we use such factorisation systems to describe relation lifting for an arbitrary functor – not just for a polynomial one, as in Section 3.1. Predicate lifting with respect to such a factorisation system will be described later, in Section 6.1.2.

Definition 4.4.1 Let $\mathbb{C}$ be a category with a logical factorisation system $(\mathfrak{M}, \mathfrak{E})$ and an arbitrary endofunctor $F\colon \mathbb{C} \to \mathbb{C}$. Then we define a functor $\mathrm{Rel}(F)\colon \mathrm{Rel}(\mathbb{C}) \to \mathrm{Rel}(\mathbb{C})$ in the following way.

- Given a relation $r = \langle r_1, r_2\rangle\colon R \rightarrowtail X_1 \times X_2$ in $\mathrm{Rel}(\mathbb{C})$ we introduce a new relation $\mathrm{Rel}(F)(r)\colon \mathrm{Rel}(F)(R) \rightarrowtail F(X_1) \times F(X_2)$ via the following factorisation, describing the lifted relation as the right-hand-side leg of the triangle:

$$\begin{array}{ccc} F(R) & \xrightarrow{\qquad\qquad} & \mathrm{Rel}(F)(R) \\ {\scriptstyle \langle F(r_1), F(r_2)\rangle}\searrow & & \swarrow{\scriptstyle \mathrm{Rel}(F)(r)} \\ & F(X_1) \times F(X_2) & \end{array}$$

- Assume a morphism $(f_1, f_2)\colon R \to S$ in $\mathrm{Rel}(\mathbb{C})$, as in

$$\begin{array}{ccc} R & \overset{\varphi}{-\,-\,-\,\to} & S \\ {\scriptstyle r = \langle r_1, r_2\rangle}\downarrow & & \downarrow{\scriptstyle s = \langle s_1, s_2\rangle} \\ X_1 \times X_2 & \xrightarrow{f_1 \times f_2} & Y_1 \times Y_2 \end{array}$$

The pair of maps $(F(f_1), F(f_2))$ forms a morphism $\mathrm{Rel}(F)(r) \to \mathrm{Rel}(F)(s)$ in $\mathrm{Rel}(\mathbb{C})$ by the diagonal-fill-in property from Definition 4.3.2.3:

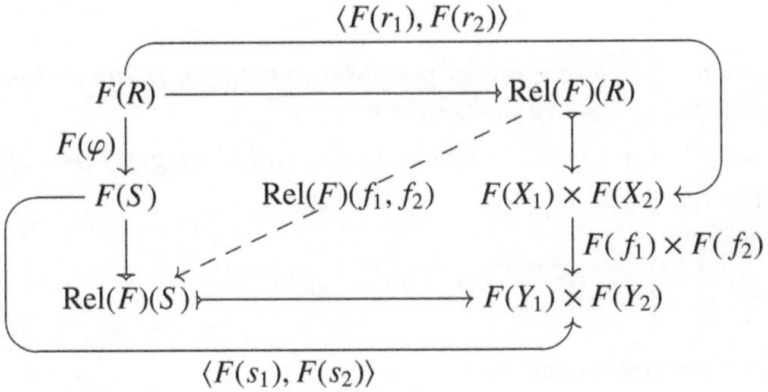

By uniqueness of such diagonals one verifies that $\mathrm{Rel}(F)(-)$ preserves identities and composition.

This definition of relation lifting generalises the situation in Lemma 3.3.1 for Kripke polynomial functors on **Sets**. We first establish some general properties of this relation lifting, much as in Section 3.2. But before we can do so we need to describe composition of relations.

For two relations $\langle r_1, r_2\rangle \colon R \rightarrowtail X \times Y$ and $\langle s_1, s_2\rangle \colon S \rightarrowtail Y \times Z$ we define their relational composition $S \circ R \rightarrowtail X \times Z$ via pullback and image: first form the object P by pullback in

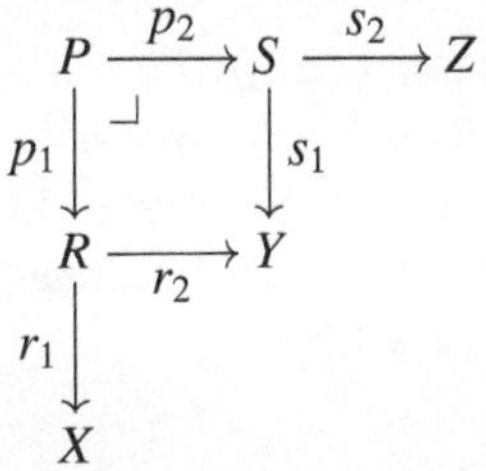

and take the image

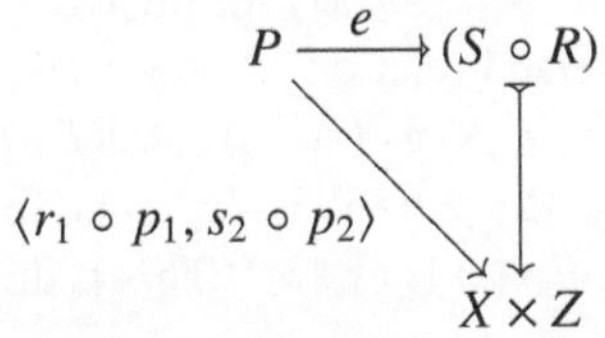

(4.11)

It is possible to turn objects with such relations between them into a category, say $\mathbb{C}$-**Rel** as in **SetsRel**, but this requires additional properties of logical factorisation systems (namely, diagonals are in $\mathfrak{M}$ and $\mathfrak{E}$ is closed under pullbacks, as in Exercise 4.3.8). See [295] for details.

Proposition 4.4.2 *Relation lifting as defined above forms a functor in a commuting diagram:*

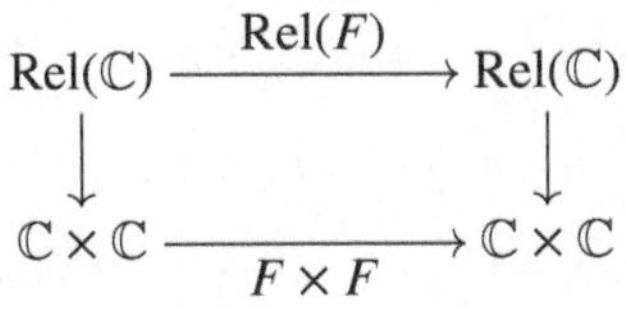

Then

1. *The functor* Rel(F) *preserves the order* $\leq$ *between relations (on the same objects):* $R \leq S \implies \mathrm{Rel}(F)(R) \leq \mathrm{Rel}(F)(()S)$.
2. *This* Rel(F) *also preserves reversal (also called daggers)* $(-)^{\dagger}$ *of relations, where*

$$\big(R \overset{\langle r_1, r_2\rangle}{\rightarrowtail} X \times Y\big)^{\dagger} = \big(R \overset{\langle r_2, r_1\rangle}{\rightarrowtail} Y \times X\big).$$

Moreover, there are inequalities:

3. $\mathrm{Eq}(F(X)) \leq \mathrm{Rel}(F)(\mathrm{Eq}(X))$ *and* $\mathrm{Eq}(F(X)) = \mathrm{Rel}(F)(\mathrm{Eq}(X))$ *in case either:*
   - *Diagonals are in* $\mathfrak{M}$, *or*
   - *F preserves abstract epis,* i.e. $e \in \mathfrak{E} \implies F(e) \in \mathfrak{E}$.
4. $\mathrm{Rel}(F)(S \circ R) \leq \mathrm{Rel}(F)(R) \circ \mathrm{Rel}(F)(S)$, *if F preserves abstract epis.*

Proof 1. Let $R \rightarrowtail X_1 \times X_2$ and $S \rightarrowtail X_1 \times X_2$ be two relations on the same objects. Then $R \leq S$ means that the pair of identities $(\mathrm{id}_{X_1}, \mathrm{id}_{X_2})$ is a morphism $R \to S$ in Rel($\mathbb{C}$). By applying Rel(F) we get a map of the form $(\mathrm{id}_{F(X_1)}, \mathrm{id}_{F(X_2)})\colon \mathrm{Rel}(F)(R) \to \mathrm{Rel}(F)(S)$ in Rel($\mathbb{C}$). This gives the required inequality $\mathrm{Rel}(F)(R) \leq \mathrm{Rel}(F)(S)$.

2. We write $\gamma = \langle \pi_2, \pi_1\rangle$ for the twist map. For a relation $\langle r_1, r_2\rangle\colon R \rightarrowtail X \times Y$, the reversed relation $R^{\dagger}$ is $\gamma \circ \langle r_1, r_2\rangle = \langle r_2, r_1\rangle\colon R \rightarrowtail Y \times X$. We write the image of $F(R)$ as $\langle s_1, s_2\rangle\colon \mathrm{Rel}(F)(R) \rightarrowtail F(X_1) \times F(X_2)$. The reversed lifted relation $\mathrm{Rel}(F)(R)^{\dagger}$ is $\langle s_2, s_1\rangle$. Thus we need to prove that the image of $\langle F(r_2), F(r_1)\rangle$ is $\langle s_2, s_1\rangle$. This is done via Exercise 4.3.1.3:

$$\begin{aligned} \mathrm{m}(\langle F(r_2), F(r_1)\rangle) &= \mathrm{m}(\gamma \circ \langle F(r_1), F(r_2)\rangle) \\ &= \gamma \circ \mathrm{m}(\langle F(r_1), F(r_2)\rangle) \quad \text{since } \gamma \text{ is an iso, thus in } \mathfrak{M} \\ &= \gamma \circ \langle s_1, s_2\rangle \\ &= \langle s_2, s_1\rangle. \end{aligned}$$

3. The equality relation Eq(X) on an object X is given by the image of the diagonal, below on the left, giving rise to the relation lifting on the right:

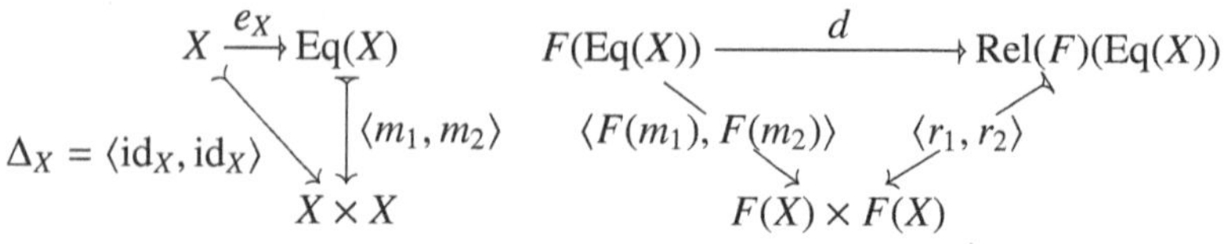

The inequality $\mathrm{Eq}(F(X)) \leq \mathrm{Rel}(F)(\mathrm{Eq}(X))$ is obtained via diagonal-fill-in:

$$\begin{array}{ccc}
F(X) & \xrightarrow{\;e_{F(X)}\;} & \mathrm{Eq}(F(X)) \\
{\scriptstyle F(e_X)}\downarrow & & \\
F(\mathrm{Eq}(X)) & \swarrow\!{\scriptstyle (\text{dashed})} & \downarrow \\
{\scriptstyle d}\downarrow & & \\
\mathrm{Rel}(F)(\mathrm{Eq}(X)) & \underset{\langle r_1, r_2\rangle}{\rightarrowtail} & F(X) \times F(X)
\end{array}$$

In case $F(e_X) \in \mathfrak{E}$ we have two factorisations of the diagonal $F(X) \rightarrowtail F(X) \times F(X)$, making the dashed map an isomorphism.

If diagonals $\Delta_X = \langle \mathrm{id}_X, \mathrm{id}_X\rangle \colon X \rightarrowtail X \times X$ are in $\mathfrak{M}$, then the equality relation $\mathrm{Eq}(X)$ on X is this diagonal Δ_X, and its lifting $\mathrm{Rel}(F)(\mathrm{Eq}(X))$ is the image of the tuple $\langle F(\mathrm{id}_X), F(\mathrm{id}_X)\rangle = \langle \mathrm{id}_{F(X)}, \mathrm{id}_{F(X)}\rangle \colon F(X) \rightarrowtail F(X) \times F(X)$. This image is the diagonal $\Delta_{F(X)}$ itself, which is the equality relation $\mathrm{Eq}(F(X))$ on $F(X)$.

4. Let $\langle r_1, r_2\rangle \colon R \rightarrowtail X \times Y$ and $\langle s_1, s_2\rangle \colon S \rightarrowtail Y \times Z$ be two relations with composition $S \circ R$ as described in (4.11). We write their liftings as two tuples $\langle r_1', r_2'\rangle \colon \mathrm{Rel}(F)(R) \rightarrowtail F(X) \times F(Y)$ and $\langle s_1', s_2'\rangle \colon \mathrm{Rel}(F)(S) \rightarrowtail F(Y) \times F(Z)$with composition:

$$\begin{array}{ccccc}
Q & \xrightarrow{\;q_2\;} & \mathrm{Rel}(F)(S) & \xrightarrow{\;s_2'\;} & F(Z) \\
{\scriptstyle q_1}\downarrow & \lrcorner & \downarrow{\scriptstyle s_1'} & & \\
\mathrm{Rel}(F)(R) & \underset{r_2'}{\longrightarrow} & F(Y) & & \\
{\scriptstyle r_1'}\downarrow & & & & \\
F(X) & & & &
\end{array}
\qquad \text{and image} \qquad
\begin{array}{ccc}
Q & \xrightarrow{\;d\;} & \mathrm{Rel}(F)(S) \circ \mathrm{Rel}(F)(R) \\
 & \searrow{\scriptstyle \langle r_1' \circ q_1,\, s_2' \circ q_2\rangle} & \downarrow \\
 & & F(X) \times F(Z)
\end{array}$$

Consider the map φ obtained in

$$\begin{array}{ccccc}
F(P) & \xrightarrow{\;F(p_2)\;} & F(S) & \longrightarrow & \\
{\scriptstyle F(p_1)}\downarrow & \searrow{\scriptstyle \varphi}\ (\text{dashed}) & & & \downarrow \\
F(R) & & Q & \xrightarrow{\;q_2\;} & \mathrm{Rel}(F)(S) \\
 & \searrow & {\scriptstyle q_1}\downarrow & \lrcorner & \downarrow{\scriptstyle s_1'} \\
 & & \mathrm{Rel}(F)(R) & \underset{r_2'}{\longrightarrow} & F(Y)
\end{array}$$

Since $F(e) \in \mathfrak{E}$, by assumption, we obtain the required inequality $\leq$ as diagonal in

$$\begin{array}{ccccc}
F(P) & \xrightarrow{F(e)} & F(S \circ R) & \rightarrowtail & \mathrm{Rel}(F)(S \circ R) \\
\downarrow{\scriptstyle\varphi} & & & & \downarrow \\
Q & & & & \\
\downarrow{\scriptstyle d} & & & & \\
\mathrm{Rel}(F)(S) \circ \mathrm{Rel}(F)(S) & & \rightarrowtail & & F(X) \times F(Z)
\end{array}$$

(with a dashed diagonal arrow from $\mathrm{Rel}(F)(S \circ R)$ to $\mathrm{Rel}(F)(S) \circ \mathrm{Rel}(F)(S)$) ❑

Stronger preservation results for lifted functors $\mathrm{Rel}(F)$ can be obtained if we assume that all abstract epis are split epis, i.e. that $\mathfrak{E} \subseteq$ *SplitEpis*. This assumption in **Sets** is equivalent to the axiom of choice; see text before Lemma 2.1.7. Also in the category **Vect** of vector spaces surjective (linear) maps are split: if $f: V \to W$ in **Vect** is surjective, then there are isomorphisms $V \cong \ker(f) \oplus \mathrm{Im}(f)$ and $\mathrm{Im}(f) \cong W$. The resulting map

$$W \xrightarrow{\cong} \mathrm{Im}(f) \xrightarrow{\kappa_2} \ker(f) \oplus \mathrm{Im}(f) \xrightarrow{\cong} V$$

is a section of f.

Proposition 4.4.3 *Let* $(\mathfrak{M}, \mathfrak{E})$ *be a logical factorisation system on a category* $\mathbb{C}$ *where* $\mathfrak{E} \subseteq$ *SplitEpis. In this case the lifting* $\mathrm{Rel}(F): \mathrm{Rel}(\mathbb{C}) \to \mathrm{Rel}(\mathbb{C})$ *of a functor* $F: \mathbb{C} \to \mathbb{C}$ *preserves reversals* $(-)^\dagger$*, equality and coproducts* $\coprod$.

Moreover, if F preserves weak pullbacks, then $\mathrm{Rel}(F)$ *preserves composition of relations and inverse images:*

$$\begin{aligned}
\mathrm{Rel}(F)(S \circ R) &= \mathrm{Rel}(F)(S) \circ \mathrm{Rel}(F)(R) \\
\mathrm{Rel}(F)((f_1 \times f_2)^{-1}(S)) &= (f(f_1) \times F(f_2))^{-1}(\mathrm{Rel}(F)(S)).
\end{aligned}$$

In particular relation lifting preserves graph relations: $\mathrm{Rel}(F)(\mathrm{Graph}(f)) = \mathrm{Graph}(F(f))$*, since* $\mathrm{Graph}(f) = (f \times \mathrm{id})^{-1}(\mathrm{Eq}(Y)) \rightarrowtail X \times Y$ *for* $f: X \to Y$.

Proof Split epis are 'absolute': they are preserved by any functor F; see Lemma 2.1.7. As a result, equality and coproducts are preserved; see Proposition 4.4.2 and Exercise 4.4.3. Hence we concentrate on the second part of the proposition and assume that the functor F preserves weak pullbacks. We shall write sections t_R in a factorisation:

$$\begin{array}{ccc}
F(R) & \overset{t_R}{\underset{e_R}{\rightleftarrows}} & \mathrm{Rel}(F)(R) \\
& {\scriptstyle \langle F(r_1), F(r_2)\rangle} \searrow & \downarrow{\scriptstyle \langle r_1', r_2' \rangle} \\
& & F(X) \times F(Y)
\end{array} \qquad \text{where} \qquad e_R \circ t_R = \mathrm{id}.$$

We then get $\langle F(r_1), F(r_2)\rangle \circ t_R = \langle r_1', r_2'\rangle$, and thus $F(r_i) \circ t_R = r_i'$, because e_R is a (split) epi:

$$\langle F(r_1), F(r_2)\rangle \circ t_R \circ e_R = \langle r_1', r_2'\rangle \circ e_R \circ t_R \circ e_R = \langle r_1', r_2'\rangle \circ e_R.$$

We first show that Rel(F) preserves composition of relations. We use pullbacks P and Q as in the proof of Proposition 4.4.2. Because F preserves weak pullbacks we obtain a map φ in

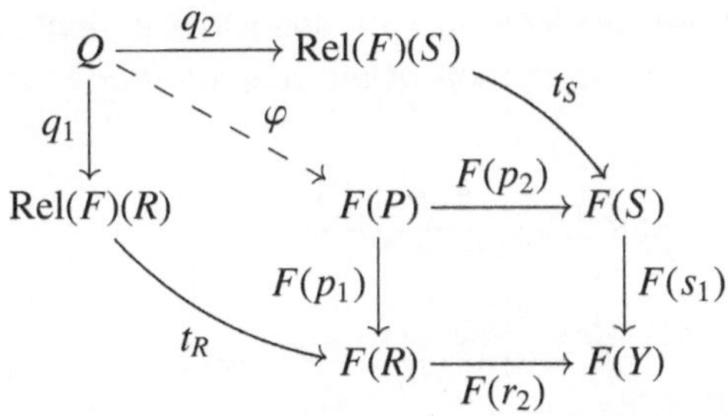

Thus we obtain a diagonal:

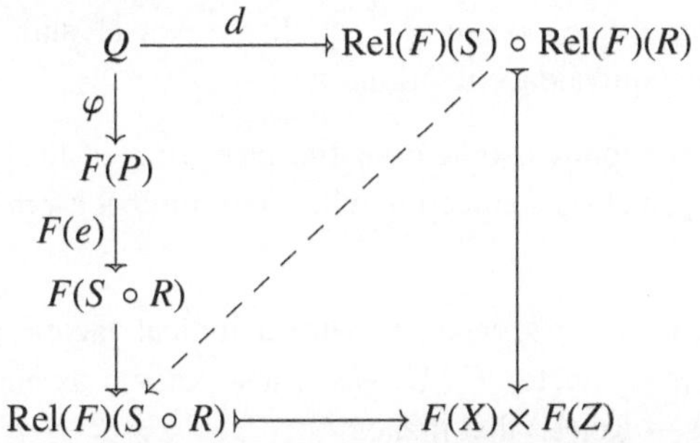

Preservation of inverse images is obtained as follows. Consider the pullback $(f(f_1) \times F(f_2))^{-1}(\mathrm{Rel}(F)(S))$ written simply as $\bullet$ in

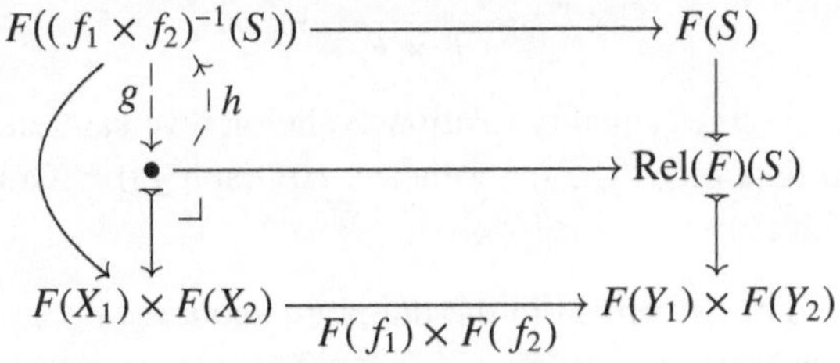

The map g is obtained because the lower square is a pullback. And h arises because F preserves weak pullbacks (and the map $F(S) \twoheadrightarrow \mathrm{Rel}(F)(S)$ is a split epi). Then $g \circ h = \mathrm{id}$, because the lower square is a (proper) pullback, so that

g is split epi. From Exercise 4.4.2 we can conclude that we have an appropriate factorisation giving $\bullet \cong \mathrm{Rel}(F)((f_1 \times f_2)^{-1}(S))$, as required. ❑

Corollary 4.4.4 *Let F be a weak-pullback-preserving endofunctor on a category $\mathbb{C}$ with a logical factorisation system $(\mathfrak{M}, \mathfrak{E})$ satisfying $\mathfrak{E} \subseteq$ SplitEpis. Then*

$$R \text{ is an equivalence relation} \Longrightarrow \mathrm{Rel}(F)(R) \text{ is an equivalence relation.}$$

Writing $\mathrm{EqRel}(\mathbb{C})$ *for the category of equivalence relations* $R \rightarrowtail X \times X$ *in* $\mathbb{C}$*, we get an obvious restriction of the relation-lifting functor* $\mathrm{Rel}(F)$ *to* $\mathrm{EqRel}(F)$ *in*

$$\begin{array}{ccc} \mathrm{EqRel}(\mathbb{C}) & \xrightarrow{\mathrm{EqRel}(F)} & \mathrm{EqRel}(\mathbb{C}) \\ \downarrow & & \downarrow \\ \mathbb{C} & \xrightarrow[F]{} & \mathbb{C} \end{array}$$

Proof The fact that $R \rightarrowtail X \times X$ is an equivalence relation can be expressed via three inequalities $\Delta = \mathrm{Eq}(X) \leq R$, $R^{\dagger} \leq R$ and $R \circ R \leq R$. By the previous result relation lifting $\mathrm{Rel}(F)$ preserves equality, reversal and composition, making $\mathrm{Rel}(F)(R)$ an equivalence relation. ❑

The next definition captures some essential properties of the lifted functor $\mathrm{Rel}(F)$. Subsequently, a close connection with weak-pullback-preserving functors is established.

Definition 4.4.5 Let $\mathbb{C}$ be a category with a logical factorisation system $(\mathfrak{M}, \mathfrak{E})$. A **relator** for a functor $F\colon \mathbb{C} \to \mathbb{C}$, also known as an F-relator, is a functor $H\colon \mathrm{Rel}(\mathbb{C}) \to \mathrm{Rel}(\mathbb{C})$ that makes

$$\begin{array}{ccc} \mathrm{Rel}(\mathbb{C}) & \xrightarrow{H} & \mathrm{Rel}(\mathbb{C}) \\ \downarrow & & \downarrow \\ \mathbb{C} \times \mathbb{C} & \xrightarrow[F \times F]{} & \mathbb{C} \times \mathbb{C} \end{array}$$

commute and preserves equality relations, relation reversal, relation composition and graph relations. (The latter means $H(\mathrm{Graph}(f)) = \mathrm{Graph}(F(f))$ and thus links H and F.)

There is some disagreement in the literature about the precise definition of an F-relator (see for instance [446, 409]), but the most reasonable requirements seem to be precisely those that yield the equivalence with weak-pullback-preservation by F in the next result. Often these relators are defined with respect to a category with relations as morphisms, such as the category **SetsRel** in Example 1.4.2.4. But the category Rel($\mathbb{C}$) with relations as objects (that we use) seems more natural in this context, for instance, because it contains bisimulations as coalgebras (see below). The situation is compared, in the set-theoretic case, more explicitly in Corollary 5.2.8 later on.

The essence of the following result comes from [450] and [94]. It applies in particular for $\mathbb{C}$ = **Sets**. A generalisation in enriched categories may be found in [70] (using preservation of exact squares instead of preservation of weak pullbacks).

Theorem 4.4.6 *Let $F\colon \mathbb{C} \to \mathbb{C}$ be a functor, where the category $\mathbb{C}$ carries a logical factorisation system $(\mathfrak{M}, \mathfrak{E})$ with $\mathfrak{E} \subseteq$ SplitEpis. Then F has a relator if and only if F preserves weak pullbacks.*

Moreover, this relator, if it exists, is uniquely determined.

Proof Proposition 4.4.3 tells us that Rel(F) is a relator if F preserves weak pullbacks. Conversely, assume H is an F-relator. We use make extensive use of the equations in Exercise 4.4.4 for showing that F preserves weak pullbacks. So assume we have a weak pullback on the left below, and a pair of maps a, b making the outer diagram on the right commute:

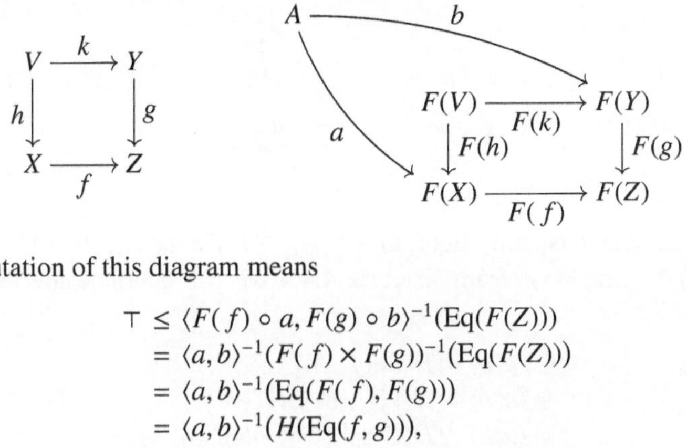

Commutation of this diagram means

$$
\begin{aligned}
\top &\leq \langle F(f) \circ a, F(g) \circ b\rangle^{-1}(\mathrm{Eq}(F(Z))) \\
&= \langle a, b\rangle^{-1}(F(f) \times F(g))^{-1}(\mathrm{Eq}(F(Z))) \\
&= \langle a, b\rangle^{-1}(\mathrm{Eq}(F(f), F(g))) \\
&= \langle a, b\rangle^{-1}(H(\mathrm{Eq}(f, g))),
\end{aligned}
\qquad (*)
$$

where the latter equation holds because in Exercise 4.4.4 equality is formulated in terms of graphs, composition and reversal, namely as $\mathrm{Eq}(f, g) = \mathrm{Graph}(g)^{\dagger} \circ \mathrm{Graph}(f)$. The two graph relations involved are obtained via pullbacks in

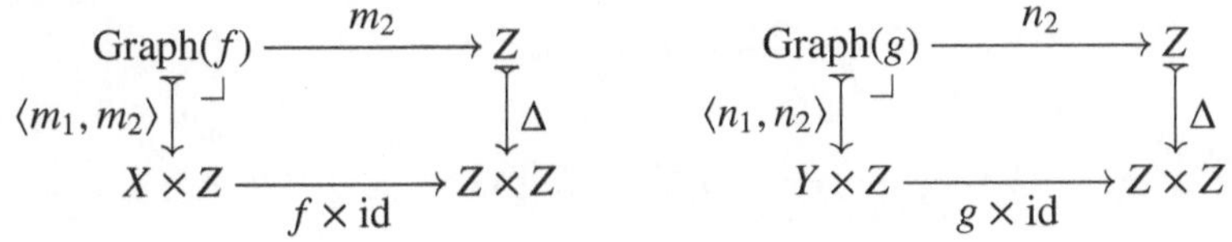

And the relation composition $\mathrm{Graph}(g)^\dagger \circ \mathrm{Graph}(f) = \mathrm{Eq}(f,g)$ results from the following pullback and image:

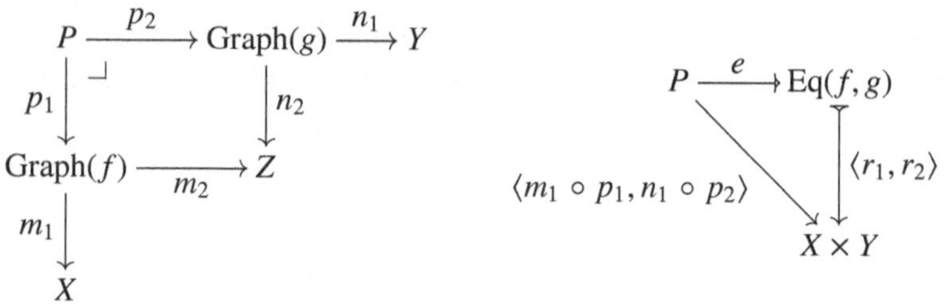

We then have

$$f \circ m_1 \circ p_1 = m_2 \circ p_1 = n_2 \circ p_2 = g \circ n_1 \circ p_2.$$

This line of reasoning started with a weak pullback. It yields a map $c\colon P \to V$, not necessarily unique, with $h \circ c = m_1 \circ p_1$ and $k \circ c = n_1 \circ p_2$. The image factorisation on the left below then gives a diagonal on the right:

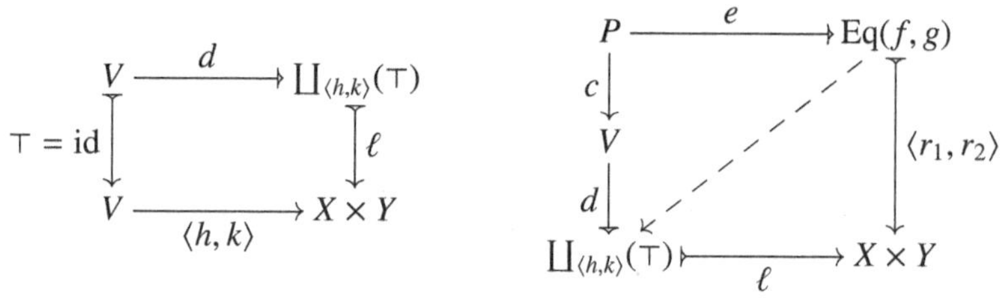

Thus we have an inequality $\mathrm{Eq}(f,g) \leq \coprod_{\langle h,k\rangle}(\top)$. Via the equation $\coprod_{\langle h,k\rangle}(\top) = \mathrm{Graph}(k) \circ \mathrm{Graph}(h)^\dagger$ from Exercise 4.4.4 we can continue the reasoning from (∗) in

$$\begin{aligned}
\top &\leq \langle a,b\rangle^{-1}(H(\mathrm{Eq}(f,g))) \\
&\leq \langle a,b\rangle^{-1}(H(\textstyle\coprod_{\langle h,k\rangle}(\top))) \\
&= \langle a,b\rangle^{-1}(H(\mathrm{Graph}(k) \circ \mathrm{Graph}(h)^\dagger)) \\
&= \langle a,b\rangle^{-1}(\mathrm{Graph}(F(k)) \circ \mathrm{Graph}(F(h))^\dagger) \\
&= \langle a,b\rangle^{-1}(\textstyle\coprod_{\langle F(h),F(k)\rangle}(\top)).
\end{aligned}$$

This inequality can be described diagrammatically as

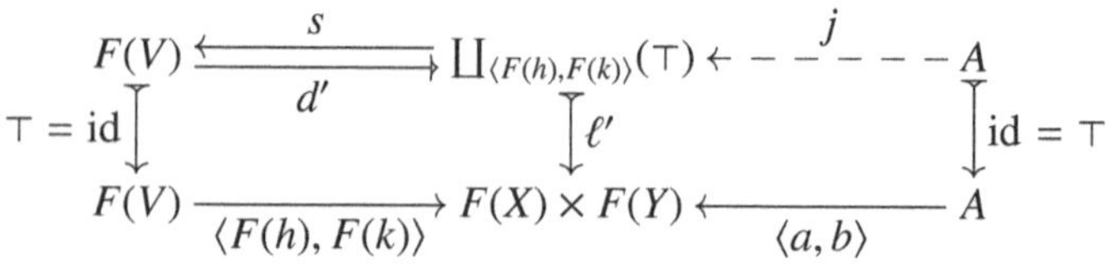

where s satisfies $d' \circ s = \mathrm{id}$, using that d' is a split epi. The resulting map $s \circ j\colon A \to F(V)$ is the mediating map that proves that F preserves weak pullbacks:

$$\langle F(h), F(k)\rangle \circ s \circ j = \ell' \circ d' \circ s \circ j = \ell' \circ j = \langle a, b\rangle.$$

Finally, via the equations in Exercise 4.4.4 and the preservation properties of relators we can prove uniqueness: one gets $H = \mathrm{Rel}(F)$ from

$$\begin{aligned} H(R) &= H(\mathrm{Graph}(r_2) \circ \mathrm{Graph}(r_1)^\dagger) \\ &= H(\mathrm{Graph}(r_2)) \circ H(\mathrm{Graph}(r_1)^\dagger) \\ &= H(\mathrm{Graph}(r_2)) \circ H(\mathrm{Graph}(r_1))^\dagger \\ &= \mathrm{Graph}(F(r_2)) \circ \mathrm{Graph}(F(r_1))^\dagger \\ &= \mathrm{Rel}(F)(\mathrm{Graph}(r_2)) \circ \mathrm{Rel}(F)(\mathrm{Graph}(r_1))^\dagger \\ &= \cdots = \mathrm{Rel}(F)(R). \end{aligned}$$

❑

Exercises

4.4.1 Verify that composition of relations as defined categorically in (4.11) is in the set-theoretic case, with classes of maps $\mathfrak{M} = $ *(injections)* and $\mathfrak{E} = $ *(surjections)*, the same as ordinary composition of relations.

4.4.2 Prove that split epis are orthogonal to all monos (where orthogonality is defined in Exercise 4.3.5).

Conclude that $\mathfrak{E} \subseteq$ *SplitEpis*, for a logical factorisation system $(\mathfrak{M}, \mathfrak{E})$, implies $\mathfrak{E} =$ *SplitEpis*.

4.4.3 Use functoriality of relation lifting to obtain:

1. $\coprod_{F(f)\times F(g)}(\mathrm{Rel}(F)(R)) \le \mathrm{Rel}(F)(\coprod_{f\times g}(R))$.
2. $\mathrm{Rel}(F)((f\times g)^{-1}(R)) \le (F(f)\times F(g))^{-1}(\mathrm{Rel}(F)(R))$.

Prove that the inequality $\le$ in (1) is an equality $=$ if the functor F preserves abstract epis.

4.4.4 Let $\mathbb{C}$ be a category with a logical factorisation system $(\mathfrak{M}, \mathfrak{E})$. Show that graph relations, composition and reversal are fundamental, in the sense that:

1. $\mathrm{Eq}(f, g) \stackrel{\mathrm{def}}{=} \langle f, g\rangle^{-1}(\mathrm{Eq}(Z)) = \mathrm{Graph}(g)^\dagger \circ \mathrm{Graph}(f)$, for $f\colon X \to Z$ and $g\colon Y \to Z$.

And if $\mathfrak{E} \subseteq$ *SplitEpis*, show that

2. $R = \mathrm{Graph}(r_2) \circ \mathrm{Graph}(r_1)^\dagger$, for $\langle r_1, r_2\rangle\colon R \rightarrowtail X \times Y$.

3. $\mathrm{Im}(\langle h, k\rangle) \stackrel{\text{def}}{=} \coprod_{\langle h,k\rangle}(\top) = \mathrm{Graph}(k) \circ \mathrm{Graph}(h)^{\dagger}$, for $h\colon V \to X$ and $k\colon V \to Y$.

4.4.5 Let F be an endofunctor on a category $\mathbb{C}$. Use the pullback in Exercise 4.3.3 to define a lifting $\mathrm{EnRel}(F)\colon \mathrm{EnRel}(\mathbb{C}) \to \mathrm{EnRel}(\mathbb{C})$ of F to an endofunctor on endorelations in $\mathbb{C}$. Describe in detail how this functor works.

4.4.6 Let $\mathbb{C}$ be a category with a logical factorisation system $(\mathfrak{M}, \mathfrak{E})$.

1. Show that a natural transformation $\sigma\colon F \Rightarrow G$ gives rise to a 'lifted' natural transformation $\mathrm{Rel}(\sigma)\colon \mathrm{Rel}(F) \Rightarrow \mathrm{Rel}(G)$ in

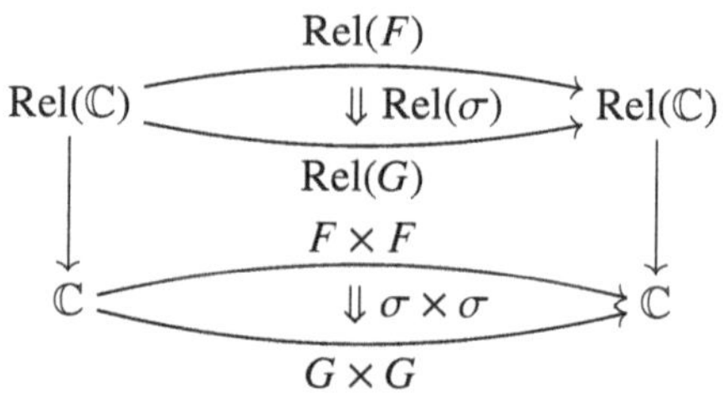

2. Prove that relation lifting is functorial, in the sense that it preserves identity natural transformations $\mathrm{Rel}(\mathrm{id}_F) = \mathrm{id}_{\mathrm{Rel}(F)}$ and composition of natural transformations: $\mathrm{Rel}(\tau \circ \sigma) = \mathrm{Rel}(\sigma) \circ \mathrm{Rel}(\tau)$.
3. Show that for two arbitrary endofunctors $F, G\colon \mathbb{C} \to \mathbb{C}$, there is a natural transformation $\mathrm{Rel}(FG) \Rightarrow \mathrm{Rel}(F)\mathrm{Rel}(G)$.

4.5 Logical Bisimulations

In the preceding sections we have first developed a general form of categorical logic using factorisation systems, and subsequently used this logic to introduce liftings $\mathrm{Rel}(F)$ of functors F to relations. This enables us to describe F-bisimulations as $\mathrm{Rel}(F)$-coalgebras, as in Lemma 3.2.4, but at a much more general level – not only for polynomial functors, not only on **Sets**.

Definition 4.5.1 Consider a functor $F\colon \mathbb{C} \to \mathbb{C}$ on a category $\mathbb{C}$ with a logical factorisation system, together with the resulting lifting $\mathrm{Rel}(F)\colon \mathrm{Rel}(\mathbb{C}) \to \mathrm{Rel}(\mathbb{C})$ as described in Definition 4.4.1.

In this setting a **logical F-bisimulation** is a $\mathrm{Rel}(F)$-coalgebra. It thus consists of a relation $R \rightarrowtail X \times Y$ with a pair of morphisms (coalgebras) $c\colon X \to F(X)$, $d\colon Y \to F(Y)$ in $\mathbb{C}$ forming a morphism in the category $\mathrm{Rel}(\mathbb{C})$:

$$\begin{array}{ccc} R & \dashrightarrow & \mathrm{Rel}(F)(R) \\ \downarrow & & \downarrow \\ X \times Y & \xrightarrow[c \times d]{} & F(X) \times F(Y) \end{array}$$

For the record we add that a **logical F-congruence** is a $\mathrm{Rel}(F)$-algebra.

In particular, a coalgebra of the functor $\mathrm{EnRel}(F)\colon \mathrm{EnRel}(\mathbb{C}) \to \mathrm{EnRel}(\mathbb{C})$ from Exercise 4.4.5 is a bisimulation on a single coalgebra. It is an endorelation itself.

The notion of $\mathrm{Rel}(F)$-coalgebra thus already contains the two underlying coalgebras. It is more common that these two coalgebras c, d are already given and that a bisimulation is a relation R on their state spaces so that the pair (c, d) is a morphism $R \to \mathrm{Rel}(F)(R)$ in $\mathrm{Rel}(\mathbb{C})$. This is just a matter of presentation.

As is to be expected, equality relations are bisimulations. This can be formulated more abstractly as a lifting property, using the category $\mathrm{EnRel}(\mathbb{C})$ of endorelations from Exercise 4.3.3.

Lemma 4.5.2 *Let $F\colon \mathbb{C} \to \mathbb{C}$ be an endofunctor on a category $\mathbb{C}$ with a logical factorisation system. For each coalgebra $c\colon X \to F(X)$ the equality relation $\mathrm{Eq}(X) \rightarrowtail X \times X$ on its state is a bisimulation.*

As a result there is a lifting of the equality functor in

$$\begin{array}{ccc} \mathbf{CoAlg}(F) & \xrightarrow{\mathrm{Eq}(-)} & \mathbf{CoAlg}(\mathrm{EnRel}(F)) \\ \downarrow & & \downarrow \\ \mathbb{C} & \xrightarrow[\mathrm{Eq}(-)]{} & \mathrm{EnRel}(\mathbb{C}) \\ \circlearrowleft F & & \circlearrowleft \mathrm{EnRel}(F) \end{array}$$

This lifting sends

$$\left(X \xrightarrow{c} F(X) \right) \longmapsto \left(\begin{array}{ccc} \mathrm{Eq}(X) & \longrightarrow & \mathrm{Eq}(F(X)) \leq \mathrm{Rel}(F)(\mathrm{Eq}(X)) \\ \downarrow & & \searrow \quad \swarrow \\ X \times X & \xrightarrow{c \times c} & F(X) \times F(X) \end{array} \right)$$

where we use that the functor $\mathrm{Rel}(F)$ and $\mathrm{EnRel}(F)$ coincide on endorelations. ❑

A similar lifting for algebras is more complicated, because in general, there is only an inequality $\mathrm{Eq}(F(X)) \leq \mathrm{Rel}(F)(\mathrm{Eq}(X))$; see Proposition 4.4.2.3. But under additional assumptions guaranteeing $\mathrm{Eq}(F(X)) = \mathrm{Rel}(F)(\mathrm{Eq}(X))$ there is also a lifted functor $\mathrm{Eq}(-)\colon \mathbf{Alg}(F) \to \mathbf{Alg}(\mathrm{EnRel}(F))$. These additional assumptions are for instance that diagonals are abstract monos, or F preserves abstract epis.

In a next step we wish to compare notions of bisimulation (as in [436]):

- The above logical one involving a coalgebra $R \to \mathrm{Rel}(F)(R)$ in $\mathrm{Rel}(\mathbb{C})$;
- The span-based formulation – with the Aczel–Mendler notion involving a coalgebra $R \to F(R)$ in $\mathbb{C}$ as special case;
- The cospan-based formulation, also known as behavioural equivalence.

Earlier, in Theorems 3.3.2 and 3.3.3 it was shown that these notions coincide in the more restricted context of polynomial functors on **Sets**. In the present general setting they diverge – but they still coincide in **Sets**, for weak-pullback-preserving functors.

Theorem 4.5.3 *In the setting of Definition 4.5.1, with a logical factorisation system* $(\mathfrak{M}, \mathfrak{E})$ *on a category* $\mathbb{C}$*, there are the following implication arrows between notions of bisimulation:*

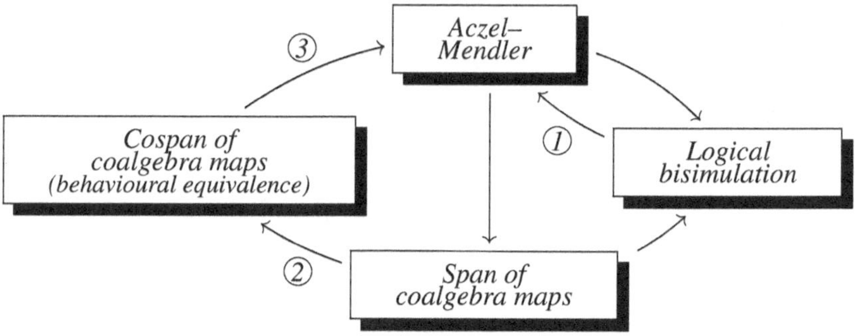

With additional side-conditions:

① *Abstract epis are split,* i.e. $\mathfrak{E} \subseteq$ *SplitEpis.*
② *The category* $\mathbb{C}$ *has pushouts.*
③ *The functor F preserves weak pullbacks and diagonals* $\Delta = \langle \mathrm{id}, \mathrm{id}\rangle$ *are in* $\mathfrak{M}$ *– or equivalently,* $\mathfrak{E} \subseteq$ *Epis; see Exercise 4.3.5.6.*

Proof Clearly, an Aczel–Mendler bisimulation $R \to F(R)$ on two coalgebras $c\colon X \to F(X)$ and $d\colon Y \to F(Y)$ forms a span of coalgebra maps $X \leftarrow R \to Y$. This implication is the (vertical) down arrow in the middle.

For the other unlabelled (unconditional) implication starting from an Aczel–Mendler bisimulation, write the relation as $\langle r_1, r_2\rangle\colon R \rightarrowtail X \times Y$, carrying a coalgebra $R \to F(R)$ as on the left below. It gives rise to a $\mathrm{Rel}(F)$-coalgebra on the right via the factorisation that defines $\mathrm{Rel}(F)(R)$:

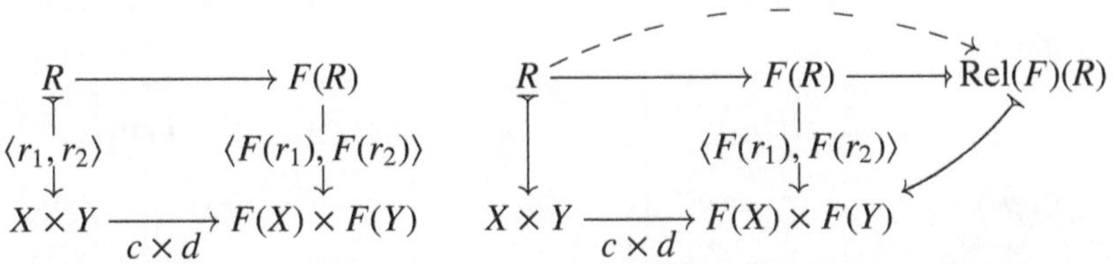

(Here we assume that the relation R is already in $\mathfrak{M}$. If not, we have to consider it as a proper span and factorise it first; see below.)

In the other direction, for the implication with label/condition ①, assume a coalgebra $(c, d)\colon R \to \mathrm{Rel}(F)(R)$ in $\mathrm{Rel}(\mathbb{C})$ as in the outer diagram on the right above. If the abstract epi $F(R) \twoheadrightarrow \mathrm{Rel}(F)(R)$ is split, we obtain a map $R \to \mathrm{Rel}(F)(R) \to F(R)$ yielding a commuting diagram as on the left.

Next assume a general span of coalgebra maps:

$$\begin{pmatrix} F(X) \\ \uparrow c \\ X \end{pmatrix} \xleftarrow{\;f\;} \begin{pmatrix} F(W) \\ \uparrow b \\ W \end{pmatrix} \xrightarrow{\;g\;} \begin{pmatrix} F(Y) \\ \uparrow d \\ Y \end{pmatrix}.$$

Clearly if the category $\mathbb{C}$ has pushouts, then so has the category $\mathbf{CoAlg}(F)$. The pushout of this diagram forms a cospan of coalgebras. This establishes the implication ②.

For the span above we consider the factorisations of $\langle f, g\rangle$ and $\langle F(r_1), F(r_2)\rangle$ in the diagram below. They make the image $\mathrm{Im}(\langle f, g\rangle)$ a logical bisimulation via diagonal-fill-in:

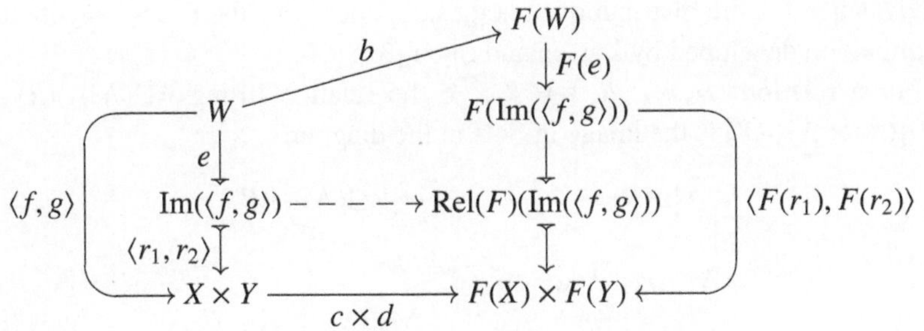

Finally, for the implication with condition ③ assume we have a cospan of coalgebra maps:

$$\begin{pmatrix} F(X) \\ \uparrow c \\ X \end{pmatrix} \xrightarrow{\;f\;} \begin{pmatrix} F(W) \\ \uparrow b \\ W \end{pmatrix} \xleftarrow{\;g\;} \begin{pmatrix} F(Y) \\ \uparrow d \\ Y \end{pmatrix}.$$

We claim that the pullback below on the left gives rise to a coalgebra map on the right:

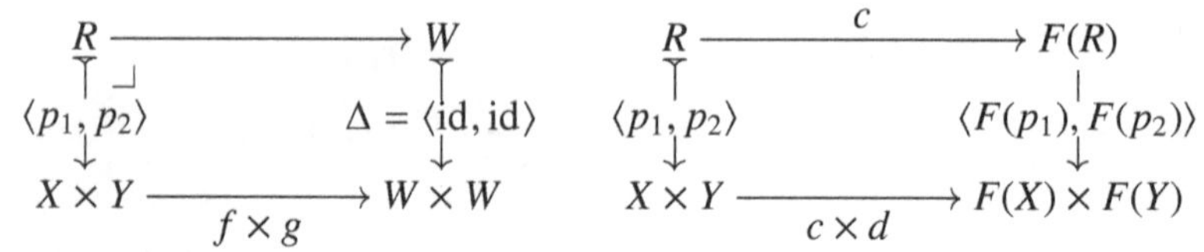

By construction the pair $X \overset{p_1}{\longleftarrow} R \overset{p_2}{\longrightarrow} Y$ is the pullback of f, g. Hence because F preserves weak pullbacks there is a (not necessarily unique) map $c\colon R \to F(R)$ in

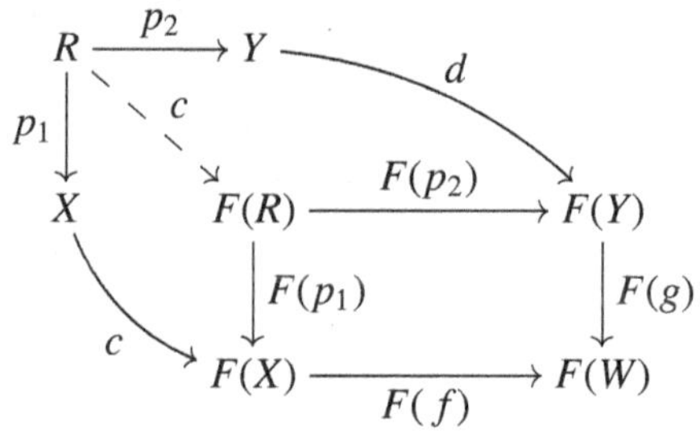

It is precisely the map we seek. ❑

After this analysis of the abstract situation we become more concrete and characterise (logical) bisimulation for the two endofunctors from Section 4.1, namely multiset $\mathcal{M}_M$ (for a commutative monoid M) and distribution $\mathcal{D}$, both on **Sets**. We follow [459] where it was first shown that bisimulation equivalence for the distribution functor coincides with the (non-coalgebraic) formulation developed by Larsen and Skou [328].

For a relation $\langle r_1, r_2\rangle\colon R \hookrightarrow X \times Y$ the relation lifting $\mathrm{Rel}(\mathcal{M}_M)(R) \subseteq \mathcal{M}_M(X) \times \mathcal{M}_M(Y)$ is the image in **Sets** in the diagram

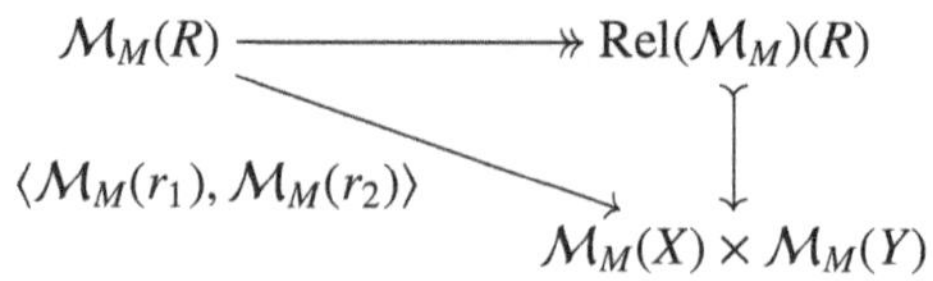

We can describe this image concretely as

$$\begin{aligned}\mathrm{Rel}(\mathcal{M}_M)(R) &= \{(\varphi, \psi) \in \mathcal{M}_M(X) \times \mathcal{M}_M(Y) \mid \exists \chi \in \mathcal{M}_M(R).\\ &\qquad\qquad \mathcal{M}_M(r_1)(\chi) = \varphi \wedge \mathcal{M}_M(r_2)(\chi) = \psi\}\\ &= \{(\mathcal{M}_M(r_1)(\chi), \mathcal{M}_M(r_2)(\chi)) \mid \chi \in \mathcal{M}_M(R)\}\\ &= \{(\textstyle\sum_i m_i|x_i\rangle, \sum_i m_i|y_i\rangle) \mid \sum_i m_i|x_i, y_i\rangle \in \mathcal{M}_M(R)\}.\end{aligned}$$

Thus, this relation R is a (logical) bisimulation for two $\mathcal{M}_M$-coalgebras $c\colon X \to \mathcal{M}_M(X)$ and $d\colon Y \to \mathcal{M}_M(Y)$ if for each pair $(x,y) \in R$ there is multiset $\sum_i m_i|x_i, y_i\rangle$ over R with $c(x) = \sum_i m_i|x_i\rangle$ and $d(x) = \sum_i m_i|y_i\rangle$.

It is not hard to see that bisimulations for the distribution functor $\mathcal{D}$ take precisely the same form, except that the multiplicities m_i must be in the unit interval $[0,1]$ and add up to 1. For the distribution functor there is an alternative description of bisimulation equivalences (i.e. for relations that are at the same time bisimulations and equivalence relations).

Proposition 4.5.4 (From [459]) *Let $c, d\colon X \to \mathcal{D}(X)$ be two coalgebras of the distribution functor, with the same state space X. An equivalence relation $R \subseteq X \times X$ is then a logical bisimulation for $\mathcal{D}$-coalgebras c, d if and only if R is a 'probabilistic bisimulation' (as defined in [328]): for all $x, y \in X$,*

$$R(x,y) \Longrightarrow c(x)[Q] = d(y)[Q], \qquad \textit{for each } R\textit{-equivalence class } Q \subseteq X$$

where for $\varphi \in \mathcal{D}(X)$ and $U \subseteq X$ we write $\varphi[U] = \sum_{x \in U} \varphi(x)$.

Proof First, assume R is a bisimulation equivalence with $R(x,y)$. As described above, there is then a formal distribution $\chi = \sum_i r_i|x_i, y_i\rangle \in \mathcal{D}(X \times X)$ with $R(x_i, y_i)$ for each i, and $c(x) = \sum_i r_i|x_i\rangle$ and $d(y) = \sum_i r_i|y_i\rangle$. Now let $Q \subseteq X$ be an R-equivalence class. Then $x_i \in Q$ iff $y_i \in Q$, since $R(x_i, y_i)$, and thus:

$$c(x)[Q] = \textstyle\sum_{x_i \in Q} r_i = \sum_{y_i \in Q} r_i = d(y)[Q].$$

Conversely, assume R is a probabilistic bisimulation with $R(x,y)$. We write $c(x) = \sum_i r_i|x_i\rangle$ and $d(y) = \sum_j s_j|y_j\rangle$. For each x_i and y_j in this sum, for which $R(x_i, y_j)$ holds, there is an equivalence class:

$$Q_{i,j} \stackrel{\text{def}}{=} [x_i]_R = [y_j]_R.$$

By assumption, $c(x)[Q_{i,j}] = d(y)[Q_{i,j}]$. These sums, say $t_{i,j} \in [0,1]$, are non-zero because by definition $x_i \in \mathrm{supp}(c(x))$ and $y_j \in \mathrm{supp}(d(y))$. We now define $\chi \in \mathcal{D}(X \times X)$ by

$$\chi(u,v) = \begin{cases} \dfrac{c(x)(x_i) \cdot d(y)(y_j)}{t_{i,j}} & \text{if } (u,v) = (x_i, y_j) \text{ and } R(x_i, y_j) \\ 0 & \text{otherwise.} \end{cases}$$

We then have for $x_i \in \mathrm{supp}(c(x))$

$$\begin{aligned}
\mathcal{D}(\pi_1)(\chi)(x_i) &= \sum_{j, R(x_i, y_j)} \chi(x_i, y_j) \\
&= \sum_{j, R(x_i, y_j)} \frac{c(x)(x_i) \cdot d(y)(y_j)}{t_{i,j}} \\
&= c(x)(x_i) \cdot \frac{\sum_{j, R(x_i, y_j)} d(y)(y_j)}{t_{i,j}} \\
&= c(x)(x_i).
\end{aligned}$$

Similarly, $\mathcal{D}(\pi_2)(\chi) = d(y)$. Finally, the probabilities in χ add up to 1 since

$$\begin{aligned}
\sum_{i, j, R(x_i, y_j)} \chi(x_i, y_j) &= \sum_i \sum_{j, R(x_i, y_j)} \chi(x_i, y_j) \\
&= \sum_i c(x)(x_i) \qquad \text{as just shown} \\
&= 1.
\end{aligned}$$

□

4.5.1 Logical Formulations of Induction and Coinduction

Earlier, in Theorem 3.1.4 we have stated that the familiar induction principle for initial algebras can be formulated in 'binary' form as: every congruence on the initial algebra is reflexive (i.e. contains the equality relation). In the present setting we can formulate this induction principle in far more general logical form, as preservation properties.

For the validity of these logical formulations we need the assumption that relation lifting preserves equality: $\mathrm{Eq}(F(X)) = \mathrm{Rel}(F)(\mathrm{Eq}(X))$. Recall from Proposition 4.4.2.3 that in general only the inequality $\leq$ holds. This property is used in the algebraic case to guarantee that equality lifts to a functor $\mathrm{Eq}(-)\colon \mathbf{Alg}(F) \to \mathbf{Alg}(\mathrm{EnRel}(F))$, analogously to Lemma 4.5.2, turning equality relations on carriers of algebras into logical congruences.

But first we have a quite general result.

Theorem 4.5.5 *Let $F\colon \mathbb{C} \to \mathbb{C}$ be an endofunctor, on a category $\mathbb{C}$ with a logical factorisation system, which has an initial algebra $\alpha\colon F(A) \xrightarrow{\cong} A$. Then each logical congruence is reflexive.*

More precisely, suppose we have two arbitrary algebras $a\colon F(X) \to X$ and $b\colon F(Y) \to Y$ and a relation $\langle r_1, r_2\rangle\colon R \rightarrowtail X \times Y$. Assume this R is a logical congruence, in the sense that the pair (a, b) forms an algebra $\mathrm{Rel}(F)(R) \to R$

in the category Rel($\mathbb{C}$) *of relations. Then there is a map* $\mathrm{Eq}(A) \to R$ *in* Rel($\mathbb{C}$), *namely:*

$$\begin{array}{ccc} \mathrm{Eq}(A) & \dashrightarrow & R \\ \downarrow & & \downarrow{\scriptstyle \langle r_1, r_2\rangle} \\ A \times A & \xrightarrow[\mathrm{int}_a \times \mathrm{int}_b]{} & X \times Y \end{array}$$

where $\mathrm{int}_a : A \to X$ *and* $\mathrm{int}_b : A \to Y$ *are the algebra homomorphisms obtained by initiality.*

Proof Exercise 4.5.1 says that the fact that R is a logical congruence may be described via a (necessarily unique) algebra structure $c\colon F(R) \to R$ with $\langle r_1, r_2\rangle \circ c = (a \times b) \circ \langle F(r_1), F(r_2)\rangle$, as in the rectangle on the right, below. It yields an algebra map $\mathrm{int}_c : A \to R$ on the left, in

$$\begin{array}{ccccc} F(A) & \overset{F(\mathrm{int}_c)}{\dashrightarrow} & F(R) & \xrightarrow{\langle F(r_1), F(r_2)\rangle} & F(X) \times F(Y) \\ {\scriptstyle \cong}\downarrow & & {\scriptstyle c}\downarrow & & \downarrow{\scriptstyle a \times b} \\ A & \underset{\mathrm{int}_c}{\dashrightarrow} & R & \underset{\langle r_1, r_2\rangle}{\rightarrowtail} & X \times Y \end{array}$$

By uniqueness we then get $r_1 \circ \mathrm{int}_c = \mathrm{int}_a$ and $r_2 \circ \mathrm{int}_c = \mathrm{int}_b$ in

$$\begin{array}{ccccc} & & \overset{\mathrm{int}_a}{\longrightarrow} & & X \\ A & \xrightarrow{\mathrm{int}_c} & R & \overset{r_1}{\nearrow} \underset{r_2}{\searrow} & \\ & & \underset{\mathrm{int}_b}{\longrightarrow} & & Y \end{array}$$

But then we obtain the map of relations $\mathrm{Eq}(A) \to R$ via diagonal-fill-in:

$$\begin{array}{ccc} A & \longrightarrow & \mathrm{Eq}(A) \\ & & \downarrow \\ {\scriptstyle \mathrm{int}_c}\downarrow & \swarrow & A \times A \\ & & \downarrow{\scriptstyle \mathrm{int}_a \times \mathrm{int}_b} \\ R & \underset{\langle r_1, r_2\rangle}{\rightarrowtail} & X \times Y \end{array}$$

□

If we restrict ourselves to endorelations (on the same object), then we can formulate this binary induction principle more abstractly as a preservation property (as in [218]).

Corollary 4.5.6 *Assuming relation lifting preserves equality, the lifted equality functor* $\mathrm{Eq}(-)\colon \mathbf{Alg}(F) \to \mathbf{Alg}(\mathrm{EnRel}(F))$ *preserves initial objects.*

Thus: if $F(A) \xrightarrow{\cong} A$ *is initial in the category* **Alg**(F), *then the logical congruence* $\mathrm{Eq}(A) \rightarrowtail A \times A$ *is initial in* **Alg**$(\mathrm{EnRel}(F))$, i.e. *is the initial logical congruence.*

Proof Let $b\colon \mathrm{EnRel}(F)(R) \to R$ be an $\mathrm{EnRel}(F)$–algebra, given by an F-algebra $b\colon F(X) \to X$ and a logical congruence relation $\langle r_1, r_2\rangle\colon R \rightarrowtail X \times X$. Theorem 4.5.5 gives the unique map of relations $\mathrm{int}_b\colon \mathrm{Eq}(A) \to R$ in the category $\mathrm{EnRel}(\mathbb{C})$ of endorelations. This makes $\mathrm{Eq}(A)$ initial in **Alg**$(\mathrm{EnRel}(F))$. Hence the functor $\mathrm{Eq}(-)\colon$ **Alg**$(F) \to$ **Alg**$(\mathrm{EnRel}(F))$ preserves initial objects. ❑

The formulation '$\mathrm{Eq}(-)\colon$ **Alg**$(F) \to$ **Alg**$(\mathrm{EnRel}(F))$ preserves initial objects' is used as a definition in [218]; it expresses that the logic involved satisfies the induction principle. The above result says that under mild assumptions (relation lifting preserves equality) the logic given by a logical factorisation system indeed satisfies the induction principle. In [218] logics are described more generally in terms of fibrations. Then it is shown that the crucial structure for this result is comprehension $\{-\}$. It is built into the kind of logics we consider here; see Section 4.3.

A bit more concretely, suppose we have a relation $R \subseteq A^\star \times A^\star$ on the initial algebra $A^\star$ of lists over a set A. Assume $R(\mathsf{nil}, \mathsf{nil})$ and $R(\sigma, \sigma') \Rightarrow R(\mathsf{cons}(a, \sigma), \mathsf{cons}(a, \sigma'))$ hold. These two assumptions express that R is a logical congruence, for the (initial) algebra $[\mathsf{nil}, \mathsf{cons}]\colon 1 + A \times A^\star \xrightarrow{\cong} A^\star$. The previous corollary then says that R must be reflexive, i.e. that $R(\sigma, \sigma)$ holds for all $\sigma \in A^\star$.

We turn to a similar logical formulation of coinduction. We can say, still following the approach of [218], that the coinduction principle is satisfied if the equality functor $\mathrm{Eq}(-)\colon$ **CoAlg**$(F) \to$ **CoAlg**$(\mathrm{EnRel}(F))$ from Lemma 4.5.2 preserves final objects. This is not automatically the case. The crucial structures we now need are quotients (instead of comprehension). We briefly explain how this works.

Definition 4.5.7 Let $\mathbb{C}$ be a category with a logical factorisation system. We say that it admits **quotients** if the equality functor $\mathrm{Eq}(-)\colon \mathbb{C} \to \mathrm{EnRel}(\mathbb{C})$ has a left adjoint, typically written as Q.

Intuitively, the above functor Q sends an endorelation $R \rightarrowtail X \times X$ to the quotient $X/\overline{R}$, where $\overline{R}$ is the least equivalence relation containing R. Exercise 4.5.5 will describe some conditions guaranteeing the existence of such quotients Q.

Theorem 4.5.8 *Assume a logical factorisation system with quotients on a category* $\mathbb{C}$, *and an endofunctor* $F\colon \mathbb{C} \to \mathbb{C}$ *whose relation lifting* $\mathrm{Rel}(F)$

preserves equality. Then the coinduction principle holds: the equality functor $\mathrm{Eq}(-)\colon \mathbf{CoAlg}(F) \to \mathbf{CoAlg}(\mathrm{EnRel}(F))$ *from Lemma 4.5.2 preserves final objects.*

Proof Let $\zeta\colon Z \xrightarrow{\cong} F(Z)$ be a final F-coalgebra. We have to prove that equality $\mathrm{Eq}(Z)$ on its carrier is the final logical bisimulation. So let $R \rightarrowtail X \times X$ be an arbitrary logical bisimulation on a coalgebra $c\colon X \to F(X)$. We have to produce a unique map of $\mathrm{Rel}(F)$-coalgebras:

$$\big(R \xrightarrow{c} \mathrm{Rel}(F)(R)\big) \longrightarrow \big(\mathrm{Eq}(Z) \xrightarrow{\zeta} \mathrm{Rel}(F)(\mathrm{Eq}(Z))\big).$$

Since such a map is by definition also a map $c \to \zeta$ in the category $\mathbf{CoAlg}(F)$ it can only be the unique map $\mathrm{beh}_c\colon X \to Z$ to the final coalgebra. Hence our task is reduced to showing that beh_c is a map of relations $R \to \mathrm{Eq}(Z)$. But since $\mathrm{Eq}(-)$ is right adjoint to quotients Q we need to find a map $Q(R) \to Z$. It arises by finality as soon as the object $Q(R)$ carries an F-coalgebra structure $Q(R) \to F(Q(R))$. Again we use the adjunction $Q \dashv \mathrm{Eq}(-)$ to obtain such a coalgebra map. It suffices to have a map of relations $R \to \mathrm{Eq}(F(Q(R)))$. The latter is obtained as from the unit $\eta\colon R \to \mathrm{Eq}(Q(R))$ of the adjunction, using that relation lifting preserves equality:

$$R \xrightarrow{c} \mathrm{Rel}(F)(R) \xrightarrow{\mathrm{Rel}(F)(\eta)} \mathrm{Rel}(F)(\mathrm{Eq}(Q(R))) = \mathrm{Eq}(F(Q(R))). \quad \square$$

Exercises

4.5.1 Let F be an endofunctor on a category $\mathbb{C}$ with a logical factorisation system. Assume algebras $a\colon F(X) \to X$ and $b\colon F(Y) \to Y$ and a relation $\langle r_1, r_2\rangle\colon R \rightarrowtail X \times Y$. Prove that the pair (a, b) is a $\mathrm{Rel}(F)$-algebra $\mathrm{Rel}(F)(R) \to R$ in $\mathrm{Rel}(\mathbb{C})$ – making R a logical congruence – if and only if the object $R \in \mathbb{C}$ carries an F-algebra $c\colon F(R) \to R$ making the r_i algebra homomorphisms in

$$\begin{array}{ccccc} F(X) & \xleftarrow{F(r_1)} & F(R) & \xrightarrow{F(r_2)} & F(Y) \\ {\scriptstyle a}\downarrow & & {\scriptstyle c}\downarrow & & \downarrow{\scriptstyle b} \\ X & \xleftarrow[r_1]{} & R & \xrightarrow[r_2]{} & X \end{array}$$

Check that this algebra c, if it exists, is unique.

4.5.2 Generalise Lemma 4.5.2 in the following manner. Assume an endofunctor $F\colon \mathbb{C} \to \mathbb{C}$ on a category $\mathbb{C}$ with a logical factorisation system.

Consider two coalgebra homomorphisms f, g with the same domain. Prove that the image $\mathrm{Im}(\langle f, g\rangle) = \coprod_{\langle f,g\rangle}(\top)$ is a logical bisimulation.

4.5.3 Suppose we have two coalgebras $X \xrightarrow{c} F(X), Y \xrightarrow{d} F(Y)$ of an endofunctor $F\colon \mathbb{C} \to \mathbb{C}$ on a category $\mathbb{C}$ with a logical factorisation system.

1. Prove that a relation $R \rightarrowtail X \times Y$ is a logical bisimulation for c, d if and only if $\coprod_{c\times d}(R) \leq \mathrm{Rel}(F)(R)$.
2. Assume that posets of relations have arbitrary joins $\bigvee$. Prove that logical bisimulations are closed under $\bigvee$, in the sense that if each R_i is a logical bisimulation, then so $\bigvee_i R_i$. *Hint*: Use that $\coprod_{c\times d}$, as left adjoint, preserves joins.

This shows that bisimilarity $\underline{\leftrightarrow}$, as join of all bisimulations, is a bisimulation itself.

4.5.4 Use Exercise 4.4.3.1 to prove that for coalgebra homomorphisms f, g one has: if R is a bisimulation, then so is $\coprod_{f\times g}(R)$.

4.5.5 Consider a category $\mathbb{C}$ with a logical factorisation system $(\mathfrak{M}, \mathfrak{E})$ with diagonals $\Delta = \langle \mathrm{id}, \mathrm{id}\rangle$ contained in $\mathfrak{M}$. Prove that if $\mathbb{C}$ has coequalisers, then its logic admits quotients – in the sense of Definition 4.5.7. *Hint*: Define the functor $Q\colon \mathrm{EnRel}(\mathbb{C}) \to \mathbb{C}$ via the coequaliser of the two legs of a relation.

4.5.6 Assume a logical factorisation system with quotients on a category $\mathbb{C}$, and an endofunctor $F\colon \mathbb{C} \to \mathbb{C}$ whose relation lifting $\mathrm{Rel}(F)$ preserves equality. Prove that a bisimulation $R \rightarrowtail X \times X$ on a coalgebra $c\colon X \to F(X)$ yields a quotient coalgebra $c/R\colon Q(R) \to F(Q(R))$ and a map of coalgebras:

$$\begin{array}{ccc} F(X) & \xrightarrow{\quad F(q) \quad} & F(Q(R)) \\ {\scriptstyle c}\uparrow & & \uparrow{\scriptstyle c/R} \\ X & \xrightarrow[\quad q \quad]{} & Q(R) \end{array}$$

This construction makes explicit what is used in the proof of Theorem 4.5.8; it generalises Theorem 3.3.4.1.

4.6 Existence of Final Coalgebras

Final coalgebras have already been used at various places in this text. They have been described explicitly for a number of special functors, such as for functors $(-)^A \times B$ for deterministic automata in Proposition 2.3.5. Often it is interesting to see what the elements of such final coalgebras are, but in actually using final coalgebras their universal property (i.e. finality) is most relevant; see for instance Section 3.5. Hence what is most important to know is whether or not a final coalgebra exists. Theorem 2.3.9 has mentioned, without proof, that a final coalgebra exists for each finite Kripke polynomial functor. It is the main aim in this section to prove a general result about the existence of final coalgebras in **Sets**, which implies the earlier mentioned Theorem 2.3.9. This general result says: bounded endofunctors on **Sets** have final coalgebras.

In this section we consider only final coalgebras for endofunctors on sets. There are more general results, applying to other categories. For instance, [469] shows that any accessible endofunctor on a locally presentable category admits a final coalgebra. Such results go beyond this introductory text. There is an extensive literature on final coalgebras [433, 12, 55, 291, 15, 18, 24, 469, 411, 176, 404, 170] that can be consulted for further information. The last two references [404, 170] describe a 'logical' construction of final coalgebras, via modal formulas (known as canonical models, in modal logic).

The section starts with a generalisation of the familiar construction of obtaining least fixed points of continuous endofunctions on directed complete posets (dcpo). It serves as a suitable introduction to the topic.

First we recall the basic fixed-point constructions for directed complete partial orders (dcpos). Assume D is a dcpo with a least element $\bot \in D$, and $f: D \to D$ is a continuous function – that is, an endomap in **Dcpo**, so f is not required to preserve $\bot$. The least fixed point $\mu f \in D$ can then be defined as a join of an ascending ω-chain of elements in D:

$$\bot \leq f(\bot) \leq f^2(\bot) \leq f^3(\bot) \leq \cdots \leq \mu f = \bigvee_{n \in \mathbb{N}} f^n(\bot),$$

where $f^0(x) = x$ and $f^{n+1}(x) = f(f^n(x))$. By continuity one obtains $f(\mu f) = \bigvee_n f(f^n(\bot)) = \bigvee_n f^{n+1}(\bot) = \mu f$. It is easy to see that μf is the least fixed point, or, better, the least *pre-fixed point*: if $f(x) \leq x$, then $\mu(f) \leq x$. By induction one obtains $f^n(\bot) \leq x$, and thus $\mu f = \bigvee_n f^n(\bot) \leq x$.

Aside: the Greek letter ω is often used in mathematical logic for the set $\mathbb{N}$ of natural numbers (considered as ordinal). It is standard in this context, in expressions such as ω-chain and ω-limit.

Since each poset is a category and a monotone function between posets is a functor, we can see $f: D \to D$ as a functor. The element μf is then the initial algebra. This construction can be generalised to categories (as in [433, lemma 2]), once the relevant notions have been suitably extended, both for algebras and for coalgebras.

Proposition 4.6.1 *Let $F\colon \mathbb{C} \to \mathbb{C}$ be an arbitrary endofunctor.*

1. *Suppose the following ω-chain starting from the initial object $0 \in \mathbb{C}$ has a colimit $A \in \mathbb{C}$:*

$$0 \xrightarrow{\;!\;} F(0) \xrightarrow{F(!)} F^2(0) \xrightarrow{F^2(!)} F^3(0) \longrightarrow \cdots \longrightarrow A. \tag{4.12}$$

If the functor F is co-continuous, in the sense that it preserves colimits of ω-chains, then there is an initial algebra structure $F(A) \xrightarrow{\cong} A$.

2. *Dually, assume there is a limit $Z \in \mathbb{C}$ of the chain starting at the final object $1 \in \mathbb{C}$.*

$$1 \xleftarrow{\;!\;} F(1) \xleftarrow{F(!)} F^2(1) \xleftarrow{F^2(!)} F^3(1) \longleftarrow \cdots \longleftarrow Z. \tag{4.13}$$

If F is continuous (preserves ω-limits), then we get a final coalgebra $Z \xrightarrow{\cong} F(Z)$ E.

Proof The two statements are each other's duals and we choose to prove only the second one; Exercise 4.6.1 elaborates on the first point. We shall make explicit what it means when an object X is a limit of a diagram:

$$X_0 \xleftarrow{f_1} X_1 \xleftarrow{f_2} X_2 \xleftarrow{f_3} X_3 \xleftarrow{f_4} \cdots \longleftarrow X.$$

It requires the presence of a 'universal cone': a collection of arrows $(X \xrightarrow{\zeta_n} X_n)_{n \in \mathbb{N}}$ satisfying $f_{n+1} \circ \zeta_{n+1} = \zeta_n$, with the following universal property. For each cone, given by an object $Y \in \mathbb{C}$ with arrows $g_n\colon Y \to X_n$ such that $f_{n+1} \circ g_{n+1} = g_n$, there is a unique map $h\colon Y \to Z$ with $\zeta_n \circ h = g_n$, for each $n \in \mathbb{N}$. In a diagram:

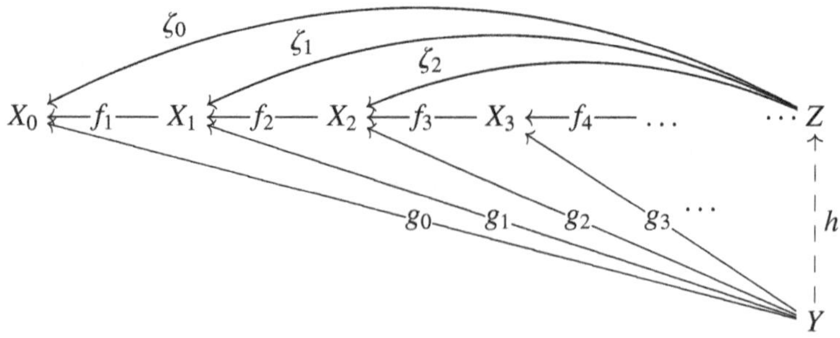

We now return to the situation in the proposition. Assume a limit (4.13) with maps $\zeta_n\colon Z \to F^n(1)$ satisfying $F^n(!) \circ \zeta_{n+1} = \zeta_n$. Applying F to the chain (4.13) and its limit Z yields another chain $(F(\zeta_n)\colon F(Z) \to F^{n+1}(1))$ with limit $F(Z)$. Using the latter's universal property yields an isomorphism

$\zeta\colon Z \xrightarrow{\cong} F(Z)$ with $F(\zeta_n) \circ \zeta = \zeta_{n+1}$. It is a final coalgebra, since for an arbitrary coalgebra $c\colon Y \to F(Y)$ we can form a collection of maps $c_n\colon Y \longrightarrow F^n(1)$ via

$$\begin{aligned}
c_0 &= (Y \xrightarrow{!} 1)\\
c_1 &= (Y \xrightarrow{c} F(Y) \xrightarrow{F(!)} F(1)) = F(c_0) \circ c\\
c_2 &= (Y \xrightarrow{c} F(Y) \xrightarrow{F(c)} F^2(1) \xrightarrow{F^2(!)} F^2(1)) = F(c_1) \circ c\\
&\vdots\\
c_{n+1} &= F(c_n) \circ c.
\end{aligned}$$

The maps c_n commute with the arrows in the chain (4.13), as is easily seen by induction. This yields a unique map $h\colon Y \to Z$ with $\zeta_n \circ h = c_n$. It forms a homomorphism of coalgebras, i.e. satisfies $\zeta \circ h = F(h) \circ c$, by uniqueness of maps $Y \to F(Z)$ to the limit $F(Z)$:

$$\begin{aligned}
F(\zeta_n) \circ \zeta \circ h &= \zeta_{n+1} \circ h\\
&= c_{n+1}\\
&= F(c_n) \circ c\\
&= F(\zeta_n) \circ F(h) \circ c.
\end{aligned}$$

❑

In order to make this ω-limit construction more concrete we consider some examples in **Sets**. The following result is then useful.

Lemma 4.6.2 *1. In* **Sets** *limits of ω-chains exist and are computed as follows. For a chain $(X_{n+1} \xrightarrow{f_n} X_n)_{n\in\mathbb{N}}$ the limit Z is a subset of the infinite product $\prod_{n\in\mathbb{N}} X_n$ given by*

$$Z = \{(x_0, x_1, x_2, \ldots) \mid \forall n \in \mathbb{N}.\, x_n \in X_n \wedge f_n(x_{n+1}) = x_n\}.$$

2. Each exponent polynomial functor F: **Sets** $\to$ **Sets** *– without powerset – is continuous.*

Proof 1. The maps $\zeta_n\colon Z \to X_n$ are the nth projections. The universal property is easily established: given a set Y with functions $g_n\colon Y \to X_n$ satisfying $f_{n+1} \circ g_{n+1} = g_n$, the unique map $h\colon Y \to Z$ with $\zeta_n \circ h = g_n$ is given by the ω-tuple $h(y) = (g_0(y), g_1(y), g_2(y), \ldots)$.

2. By induction on the structure of F, using that products, coproducts and (constant) exponents preserve the relevant constructions. ❑

Corollary 4.6.3 *Each exponent polynomial functor F:* **Sets** $\to$ **Sets** *has a final coalgebra, which can be computed as the limit of an ω-chain as in (4.13).* ❑

We show how to actually (re)calculate one such final coalgebra.

Example 4.6.4 In Corollary 2.3.6.2 we have seen that the final coalgebra for a (simple polynomial) functor $F(X) = X^A \times 2$ can be described as the set $2^{A^\star} = \mathcal{P}(A^\star) = \mathcal{L}(A)$ of languages with alphabet A. Here we shall reconstruct this coalgebra as the limit of an ω-chain.

Therefore we start by investigating what the chain (4.13) looks like for this functor F:

$$\begin{aligned} F^0(1) &= 1 \cong \mathcal{P}(0) \\ F^1(1) &= 1^A \times 2 \cong 1 \times 2 \cong 2 \cong \mathcal{P}(1) \\ F^2(1) &\cong 2^A \times 2 \cong 2^{A+1} \cong \mathcal{P}(1 + A) \\ F^3(1) &\cong (2^{A+1})^A \times 2 \cong 2^{A\times(A+1)} \times 2 \cong 2^{A^2+A+1} \cong \mathcal{P}(1 + A + A^2) \quad \text{etc.} \end{aligned}$$

One sees that

$$F^n(1) \cong \mathcal{P}(\textstyle\coprod_{i=0}^{n-1} A^i).$$

The maps $F^n(!)\colon F^{n+1}(1) \to F^n(1)$ in (4.13) are given by the inverse image κ_n^{-1} of the obvious coprojection function $\kappa_n\colon 1 + A + \cdots + A^{n-1} \longrightarrow 1 + A + \cdots + A^{n-1} + A^n$. An element $U \in Z$ of the limit Z as described in Lemma 4.6.2.1 consists of elements $U_n \subseteq 1{+}A{+}\cdots{+}A^{n-1} \longrightarrow 1{+}A{+}\cdots{+}A^{n-1}$, with the requirement that $\kappa_n^{-1}(U_{n+1}) = U_n$. The latter means that these $U_{n+1} \subseteq 1 + A + \cdots + A^{n-1} \longrightarrow 1 + A + \cdots + A^{n-1} + A^n$ can be identified with the set A^n of words of length n. Together they form a set of words, or language, $U \subseteq A^\star$, as in the original description in Corollary 2.3.6.2.

What we have done so far applies only to (co-)continuous endofunctors. But, for instance, the finite powerset is not continuous. Hence we shall need more powerful techniques to cover a larger class of functors. This is done via the notion of a *bounded* functor. It goes back to [291] and occurs regularly in the theory of coalgebras; see for instance [411, 187]. Here we shall use it for the description of final coalgebras. We first introduce the standard formulation, and then immediately introduce an equivalent alternative that is easier to work with in the current setting.

Definition 4.6.5 A functor $F\colon$ **Sets** $\to$ **Sets** is called **bounded** if there is a set M such that for each coalgebra $X \to F(X)$ and state $x \in X$ there is a subcoalgebra on $S \hookrightarrow X$ with $x \in S$, where S is strictly smaller than M, i.e. $|S| < |M|$.

We use the notation $|X|$ for the cardinality of a set X. The notion of subcoalgebra $S \hookrightarrow X$ will be investigated more systematically in the next chapter in terms of invariants. Here the meaning should be obvious, namely a coalgebra $S \to F(S)$ making the inclusion $S \hookrightarrow X$ a homomorphism of coalgebras.

Example 4.6.6 Here is an example of a bounded functor that will play an important role. Deterministic automata are described as coalgebras of the functor $D(X) = X^A \times B$, for suitable sets A, B. This functor is bounded by the set $\mathcal{P}(A^\star)$. Indeed, given an arbitrary D-coalgebra $\langle \delta, \epsilon \rangle : X \to X^A \times B$ with a state $x \in X$ we can take as subset $S \hookrightarrow X$ the set of successor states of x, given by

$$S = \{\delta^*(x, \alpha) \mid \alpha \in A^\star\},$$

where the iterated transition function δ^* is introduced in (2.22). Clearly, $x \in S$ for $\alpha = \langle\rangle$. Also, S is closed under transitions and thus carries a subcoalgebra structure. Finally, $|S| \leq |A^\star| < |\mathcal{P}(A^\star|$.

The following alternative description of bounded functors is a combination of results from [449, 187, 28].

Proposition 4.6.7 *For a functor F:* **Sets** $\to$ **Sets** *the following three statements are equivalent.*

1. *F is bounded.*
2. *F is* ***accessible****: there is a set M such that for each set X,*

$$F(X) = \bigcup\{F(U) \mid U \subseteq X \quad \text{and} \quad |U| < |M|\}.$$

3. *There are sets A, B with a natural transformation:*

$$(-)^A \times B \stackrel{\sigma}{\Longrightarrow} F$$

where for each set $X \neq \emptyset$ the component $\sigma_X : X^A \times B \to F(X)$ is surjective.

The equation in (2) is intuitively clear but a bit sloppy, since we have omitted inclusion functions $i: U \hookrightarrow X$, which turn elements $u \in F(U)$ into elements $F(i)(u) \in F(X)$.

A functor that is bounded by the set $\mathbb{N}$ of natural numbers is called ω-**accessible** or **finitary**. These ω-accessible/finitary functors are thus entirely determined by their behaviour on finite sets. They preserve ω-colimits (see Exercise 4.6.7), and thus have initial algebras. As we shall see, they also have final coalgebras.

Proof (1) $\Rightarrow$ (2) Assume that F is bounded, say via the set M (as in Definition 4.6.5). This same set can be used in the description of accessibility. The inclusion ($\supseteq$) in (2) clearly holds, so we concentrate on ($\subseteq$). If $X = \emptyset$ the result is obvious, so we may assume an element $x_0 \in X$. Let $w \in F(X)$; it yields a constant coalgebra $c = \lambda x \in X.\, w \colon X \to F(X)$. Since F is bounded by assumption, for the element $x_0 \in X$ there is a subcoalgebra $c_S \colon S \to F(S)$ of c, on a subset $i \colon S \hookrightarrow X$ with $|S| < |M|$, and an element $y_0 \in S$ with $i(y_0) = x_0$. We claim that $v = c_S(y_0) \in \bigcup\{F(U) \mid U \subseteq X \text{ and } |U| < |M|\}$ is the required element that is mapped to $w \in F(X)$:

$$\begin{aligned} F(i)(v) &= F(i)(c_S(y_0)) \\ &= c(i(y_0)) \qquad \text{since } c_S \text{ is a subcoalgebra} \\ &= w. \end{aligned}$$

(2) ⇒ (3) Assume that F is accessible, say via the set M. We take $A = M$ and $B = F(M)$, and define $\sigma_X \colon X^M \times F(M) \to F(X)$ as $\sigma_X(f, b) = F(f)(b)$. It is easy to see that σ is natural, so we concentrate on showing that σ_X is surjective for $X \neq \emptyset$.

Let $w \in F(X)$. By accessibility of F there is a subset $i \colon U \hookrightarrow X$ with $|U| < |M|$ and an element $v \in F(U)$ with $w = F(i)(v)$. Since $|U| < |M|$ there is, by definition of the cardinal order, an injection $j \colon U \rightarrowtail M$. We distinguish two cases.

- $U = \emptyset$. In that case $i \colon U \rightarrowtail X$ is the unique map $!_X \colon \emptyset \to X$, since $\emptyset$ is initial in **Sets**. Thus $w = F(!_X)(v) \in F(X)$. We take $b = F(!_M)(v) \in F(M)$ and $f = \lambda m \in M.\, x_0 \colon M \to X$, where $x_0 \in X$ is an arbitrary element. This pair $(f, b) \in X^M \times F(M)$ is mapped by σ_X to the element $w \in F(X)$ that we started from:

$$\begin{aligned} \sigma_X(f, b) = F(f)(b) = F(f)(F(!_M)(v)) &= F(f \circ\, !_M)(v) \\ &= F(!_X)(v) = w. \end{aligned}$$

- $U \neq \emptyset$. Since $j \colon U \rightarrowtail M$ is injective there is a map $k \colon M \twoheadrightarrow S$ in the reverse direction with $k \circ j = \mathrm{id}$ – as noted at the end of Section 2.1. We now take $f = i \circ k \colon M \to X$ and $b = F(j)(v) \in F(M)$. Then

$$\begin{aligned} \sigma_X(f, b) = F(f)(b) = F(i \circ k)(F(j)(v)) &= F(i \circ k \circ j)(v) \\ &= F(i)(v) = w. \end{aligned}$$

(3) ⇒ (1) This implication is easy using the previous example and Exercise 4.6.8. ❑

Lemma 4.6.8 *Each finite Kripke polynomial functor is bounded.*

Proof We shall use the third formulation from Proposition 4.6.7 and show by induction on the structure of a finite Kripke polynomial functor F that there are sets A, B with a suitable natural transformation $\sigma \colon (-)^A \times B \Rightarrow F$. We leave details to the interested reader.

- If F is the identity functor, we simply take $A = 1$ and $B = 0$.
- If F is the constant functor $X \mapsto C$, we take $A = 0$ and $B = C$.
- In case F is a product $F_1 \times F_2$ for which we have suitable natural transformations $\sigma_i \colon (-)^{A_i} \times B_i \Rightarrow F_i$, for $i = 1, 2$, we take $A = A_1 + A_2$ and $B = B_1 \times B_2$ and define $\sigma_X \colon X^A \times B \to F(X)$ by

$$\sigma_X(f, (b_1, b_2)) = \langle \sigma_1(\lambda a \in A_1.\, f(\kappa_1 a), b_1),\ \sigma_2(\lambda a \in A_2.\, f(\kappa_2 a), b_2)\rangle.$$

It is clearly natural, and surjective for $X \neq \emptyset$.
- Similarly, if $F = \coprod_{i \in I} F_i$ with $\sigma_i \colon (-)^{A_i} \times B_i \Rightarrow F_i$, we take $A = \coprod_{i \in I} A_i$ and $B = \coprod_{i \in I} B_i$ and define $\sigma_X \colon X^A \times B \to F(X)$ by

$$\sigma_X(f, \kappa_j b) = \kappa_j \sigma_j(\lambda a \in A_j.\, f(\kappa_j a), b).$$

- Next consider $F = G^C$ and assume we have a suitable natural transformation $\tau \colon (-)^A \times B \Rightarrow G$. We then define $\sigma_X \colon X^{(C \times A)} \times B^C \to G(X)^C$ as

$$\sigma_X(f, g)(c) = \tau_X(\lambda a \in A.\, f(c, a), g(c)).$$

Proving that σ_X is surjective, for $X \neq \emptyset$, involves the axiom of choice.
- Finally, assume $F = \mathcal{P}_{\text{fin}} G$, where G already comes with a natural transformation $\tau \colon (-)^A \times B \Rightarrow G$. Then we can define $\sigma_X \colon X^{(\mathbb{N} \times A)} \times B^\star \to \mathcal{P}_{\text{fin}}(GX)$ as the n-element set:

$$\begin{aligned}&\sigma_X(f, \langle b_1, \ldots, b_n\rangle)\\ &\quad = \{\tau_X(\lambda a \in A.\, f(1, a), b_1), \ldots, \tau_X(\lambda a \in A.\, f(n, a), b_n)\}.\end{aligned}$$

□

Theorem 4.6.9 *Each bounded functor* **Sets** $\to$ **Sets** *has a final coalgebra. In particular:*

- *Each finite Kripke polynomial functor, such as* $\mathcal{P}_{fin}(A \times -)$*, and*
- *Each finitary functor, such as multiset* $\mathcal{M}_M$ *or distribution* $\mathcal{D}$*,*

has a final coalgebra.

The existence of final coalgebras of finitary functors occurs already in [55].

Proof Let F be this bounded functor. The third, 'surjectivity' formulation of Proposition 4.6.7 yields a natural transformation $\sigma \colon (-)^A \times B \Rightarrow F$, for suitable sets A and B, with surjective components σ_X for $X \neq \emptyset$. Recall from Proposition 2.3.5 that the functor $(-)^A \times B$ has carrier $Z = B^{A^\star}$ of the final coalgebra $\zeta \colon Z \xrightarrow{\cong} Z^A \times B$. We define an F-coalgebra $\xi = \sigma_Z \circ \zeta \colon Z \to F(Z)$. We claim that it is 'weakly' final: for each F-coalgebra $c \colon X \to F(X)$ there is a (not necessarily unique) homomorphism of F-coalgebras $f \colon X \to Z$.

If X is the empty (initial) set $\emptyset$, there is obviously such a homomorphism $f\colon X \to Z$. Otherwise, we know that $\sigma_X\colon X^A \times B \to F(X)$ is surjective, and thus, using the axiom of choice, has a section $s\colon F(X) \to X^A \times B$ with $\sigma_X \circ s = \mathrm{id}_{F(X)}$. The coalgebra $s \circ c\colon X \to X^A \times B$ yields a homomorphism $f\colon X \to Z$ of $(-)^A \times B$-coalgebra by finality. It is then also a homomorphism of F-coalgebras in

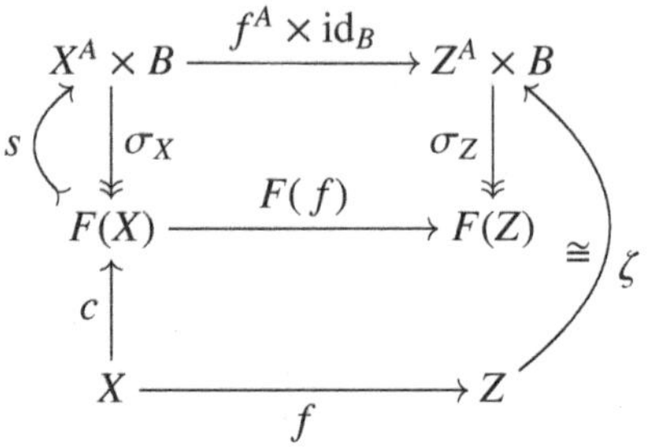

Explicitly:

$$\begin{aligned} F(f) \circ c &= F(f) \circ \sigma_X \circ s \circ c && \text{because } \sigma_X \circ s = \mathrm{id} \\ &= \sigma_Z \circ (f^A \times \mathrm{id}_B) \circ s \circ c && \text{by naturality of } \sigma \\ &= \sigma_Z \circ \zeta \circ f && \text{since } f \text{ is a homomorphism} \\ &= \xi \circ f. \end{aligned}$$

We now force the weakly final coalgebra $\xi\colon Z \to F(Z)$ to be truly final. The general theory of bisimulation from Section 4.4 will be used, for the standard logical factorisation system on **Sets**, with its quotients. Bisimilarity $\underline{\leftrightarrow}$ is a join of bisimulations, and thus a bisimulation itself by Exercise 4.5.3. Hence we can form a quotient coalgebra $\xi/\underline{\leftrightarrow}\colon W \to F(W)$ on $W = Q(\underline{\leftrightarrow}) = Z/\underline{\leftrightarrow}$ by Exercise 4.5.6. This coalgebra $\xi/\underline{\leftrightarrow}$ is final: for each coalgebra $c\colon X \to F(X)$ there is a homomorphism $X \to W$, namely the composition of the map $X \to Z$ obtained by weak finality and the quotient map $q\colon Z \to W$. This is the only one, since if we have two such homomorphisms $f, g\colon X \to W$, then the image $\mathrm{Im}(\langle f, g\rangle) \rightarrowtail W \times W$ is a bisimulation by Exercise 4.5.2. Hence $\mathrm{Im}(\langle f, g\rangle) \leq \underline{\leftrightarrow}$, so that $f = g$. ❑

Weakly final (co)algebras, as used in this proof, may also be constructed in (second order) polymorphic type theory; see [191, 471]. Under suitable parametricity conditions, these constructions yield proper final coalgebras [199, 385, 49, 149].

In the end one may ask if there is a link between the final coalgebra constructions based on limits of chains and on boundedness. It is provided in [470], via transfinite induction, going beyond ω: for instance, for a finitary

(ω-accessible) functor F the final coalgebra can be reached in $\omega + \omega$ steps as the limit of the chain $F^{\omega+m}(1) = F^m(f^\omega(1))$, where $F^\omega(1)$ is the limit of the chain $F^n(1)$.

Exercises

4.6.1 1. Spell out the notions of colimit of ω-chain and of co-continuity.

2. Check that the colimit of an ω-chain $X_0 \xrightarrow{f_0} X_1 \xrightarrow{f_1} X_2 \cdots$ in **Sets** can be described as the quotient of the disjoint union

$$\textstyle\coprod_{n\in\mathbb{N}} X_n/\sim\; = \{(n, x) \mid n \in \mathbb{N} \wedge x \in X_n\}/\sim$$

where

$$(n, x) \sim (m, y) \iff \exists p \geq n, m.\, f_{np}(x) = f_{mp}(y),$$

with $f_{qp} = f_{p-1} \circ f_{p-2} \circ \cdots \circ f_q \colon X_q \to X_p$ for $q \leq p$.

3. Prove Proposition 4.6.1.1 in detail: the initial algebra of a co-continuous functor $F\colon \mathbb{C} \to \mathbb{C}$ on a category $\mathbb{C}$ with initial object $0 \in \mathbb{C}$ can be obtained as colimit A of the ω-chain

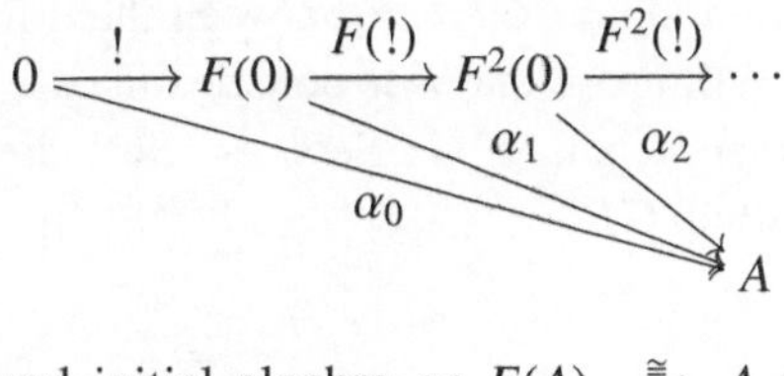

with the induced initial algebra $\alpha\colon F(A) \xrightarrow{\cong} A$ satisfying $\alpha \circ F(\alpha_n) = \alpha_{n+1}$.

4.6.2 Recall from Example 2.4.4 that the lift functor $\mathcal{L} = 1 + (-)$ on **Sets** has the natural numbers $\mathbb{N}$ as initial algebra. If we consider the functor $1 + (-)$ not on **Sets** but on **PoSets**, there are two obvious ways to add an element, namely at the bottom or at the top.

1. Check that $\mathbb{N}$ is the colimit of the chain (4.12) for the lift functor on **Sets**.

2. Write $\mathcal{L}_\perp$ for the functor which adds a bottom element $\perp$ to a poset X; prove that the natural numbers with the usual order $\leq$ form an initial algebra of the functor $\mathcal{L}_\perp\colon$ **PoSets** $\to$ **PoSets**, for instance via the chain construction (4.12).

3. Now write $\mathcal{L}_\top\colon$ **PoSets** $\to$ **PoSets** for the functor that adds an element $\top$ as top element. Check that the initial $\mathcal{L}_\top$-algebra is $(\mathbb{N}, \geq)$ – which has 0 as top element.

4.6.3 Prove that a left adjoint preserves colimits of ω-chains and, dually, that a right adjoint preserves limits of such chains.

4.6.4 Let $\alpha\colon F(A) \xrightarrow{\cong} A$ be an initial algebra constructed as colimit (4.12) for a functor F that preserves monomorphisms — like any weak-pullback-preserving, and hence any Kripke polynomial functor; see Lemma 4.2.2. Assume F also has a final coalgebra $\zeta\colon Z \xrightarrow{\cong} F(Z)$, and let $\iota\colon A \to Z$ be the unique (algebra and coalgebra) homomorphism with $\zeta \circ \iota = F(\iota) \circ \alpha^{-1}$. Prove that ι is injective. *Hint*: Define suitable $\zeta_n\colon Z \to F^n(1)$ and use that $F^n(!)\colon F^n(0) \to F^n(1)$ is mono.

4.6.5 Check that the list $(-)^\star$, multiset $\mathcal{M}_M$, distribution $\mathcal{D}$ and finite powerset $\mathcal{P}_{\text{fin}}$ functors are finitary (ω-accessible), but the ordinary powerset functor $\mathcal{P}$ is not.

4.6.6 Check that Lemma 4.6.8 specialises to: every simple polynomial functor is finitary. *Hint*: The easiest way is to use Proposition 2.2.3.

4.6.7 Prove that each finitary functor $F\colon \mathbf{Sets} \to \mathbf{Sets}$ preserves colimits of ω-chains (described explicitly in Exercise 4.6.1).

Conclude that such a functor has both an initial algebra, by Proposition 4.6.1, and a final coalgebra, by Theorem 4.6.9.

4.6.8 (See e.g. [187]) Let $F, G\colon \mathbf{Sets} \to \mathbf{Sets}$ be functors with a natural transformation $\sigma\colon G \Rightarrow F$ between them for which σ_X surjective for each $X \neq \emptyset$. Prove that F is bounded in case G is.

4.6.9 Show that if both $F, G\colon \mathbf{Sets} \to \mathbf{Sets}$ are bounded, then so is the composition GF.

4.7 Polynomial and Analytical Functors

This section will present another characterisation of simple polynomial endofunctors (on **Sets**) and also of the related class of analytical functors (introduced in [285]). The characterisation involves properties that we have seen earlier in this chapter – notably finitariness and (weak) pullbacks. Recall from Proposition 2.2.3 that simple polynomial functors are of the form $F(X) = \coprod_{i\in I} X^{\#i}$, where $\#\colon I \to \mathbb{N}$ is an 'arity'. Analytical functors are similar and have the form $F(X) = \coprod_{i\in I} X^{\#i}/G_i$ involving an additional quotient (see below for details). The main result of this section (Theorem 4.7.8) says that a functor is simple polynomial if and only if it is finitary and preserves (countable) pullbacks. Similarly, a functor is analytical if and only if it is finitary and preserves (countable) *weak* pullbacks.

These characterisation results will not be used elsewhere in this book but provide background theory about endofunctors. These results go back to [285] and have been digested and reformulated several times, as in [200] and [31], but see also [134]. The present section contains another minor re-digest, leaning heavily on [200]. Although the main result is relatively easy to formulate, its proof requires quite a bit of work.

First of all we need to generalise the notion of (weak) pullback, as introduced in Section 4.2. There, the *binary* (weak) pullback is defined for two maps $f_1 : X_1 \to Y$ and $f_2 : X_2 \to Y$ with common codomain. Here we generalise it to an arbitrary number of maps $f_i : X_i \to Y$, for indices i in an arbitrary index set I. In fact, we need only I to be countable, so we restrict ourselves to index set $I = \mathbb{N}$. The formulation for arbitrary sets I is then an obvious generalisation.

So assume we have a countable collection $(f_i : X_i \to Y)_{i \in \mathbb{N}}$ with common codomain (in an arbitrary category). The pullback of this collection is given by an object P together with maps $p_i : P \to X_i$ such that $f_i \circ p_i = f_j \circ p_j$, for all $i, j \in \mathbb{N}$, as in

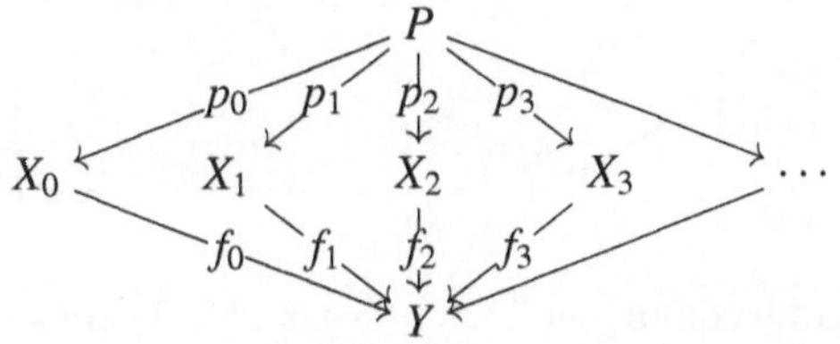

This diagram is a (countable) pullback if it is universal in the obvious way: for each object Q with maps $g_i : Q \to X_i$ satisfying $f_i \circ g_i = f_j \circ g_j$, there is a unique map $g : Q \to P$ satisfying $p_i \circ g = g_i$. It is a *weak* pullback if such a g exists, without the uniqueness requirement. A functor F preserves such a (weak) pullback if the maps $F(p_1) : F(P) \to F(X_i)$ are a (weak) pullback of the collection $(f(f_i) : F(X_i) \to F(Y))_{i \in \mathbb{N}}$. These preservation properties play a crucial role in Theorem 4.7.8 below.

The present proof of this section's main result, Theorem 4.7.8, exploits the idea, as in [31], that each functor F can be written as coproduct $F = \coprod_i G_i$ of affine functors G_i (preserving the terminal object: $G_i(1) \cong 1$). This observation goes back to [449] and will be described first. We will use it only for the category **Sets**, but the construction involved can be performed more generally and so we describe it first in a separate lemma. This construction is important and may be understood as a form of reindexing/substitution – whence the $(-)^{-1}$ notation. This construction is used for instance in [276, 274] to define the notion of 'shapely' functor; see Proposition 4.7.9.

Lemma 4.7.1 *Let $\mathbb{C}$ be a category with finite limits (binary pullbacks and terminal object $1 \in \mathbb{C}$). Let $F : \mathbb{C} \to \mathbb{C}$ be an arbitrary endofunctor, with a map $u : A \to F(1)$ in $\mathbb{C}$. Form for each object $X \in \mathbb{C}$ the pullback*

$$\begin{array}{ccc} u^{-1}(F)(X) & \xrightarrow{\ \sigma_X\ } & F(X) \\ {\scriptstyle \pi_X}\big\downarrow & \lrcorner & \big\downarrow{\scriptstyle F(!_X)} \\ A & \xrightarrow[\ u\ ]{} & F(1) \end{array} \tag{4.14}$$

1. *The mapping $X \mapsto u^{-1}(F)(X)$ extends to a functor $u^{-1}(F)\colon \mathbb{C} \to \mathbb{C}$, with a natural transformation $\sigma\colon u^{-1}(F) \Rightarrow F$.*
2. *For $X = 1$ the map $\pi_1\colon u^{-1}(F)(1) \to A$ is an isomorphism.*

Proof 1. For $f\colon X \to Y$ in $\mathbb{C}$ define $u^{-1}(F)(f)$ in

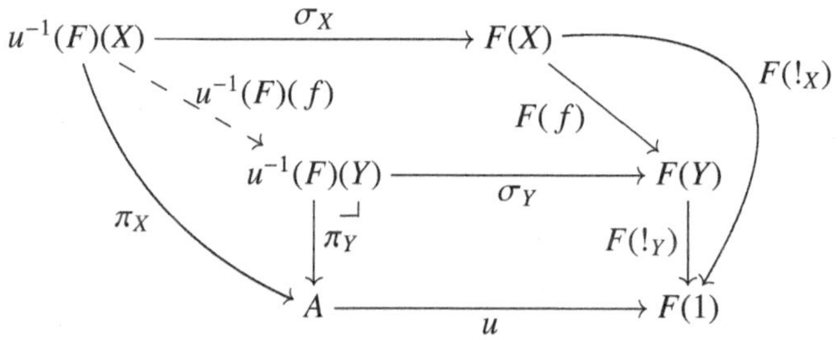

The outer diagram commutes since $!_Y \circ f = \,!_X$. By construction this yields a natural transformation $\sigma\colon u^{-1}(F) \Rightarrow F$ that commutes with the πs.

2. For $X = 1$ we have $F(!_1) = \mathrm{id}\colon F(1) \to F(1)$ as vertical map on the right-hand side in (4.14). Hence its pullback $\pi_1\colon u^{-1}(F)(1) \to A$ is an isomorphism. ❑

Proposition 4.7.2 *Consider the previous lemma for $\mathbb{C} = \mathbf{Sets}$ and $A = 1$.*

1. *For each functor $F\colon \mathbf{Sets} \to \mathbf{Sets}$ one obtains a natural isomorphism*

$$\left(\coprod_{u \in F(1)} u^{-1}(F) \right) \xRightarrow{\ \cong\ } F$$

describing the functor F as coproduct of affine functors $u^{-1}(F)$ – where we use that there is an isomorphism $u^{-1}(F)(1) \cong 1$ by Lemma 4.7.1.2.

2. *If F preserves (weak) pullbacks, then so does each $u^{-1}(F)$.*
3. *If F is finitary, then so is $u^{-1}(F)$.*

Proof 1. In the previous lemma we found a natural transformation of the form $\sigma\colon u^{-1}(F) \Rightarrow F$. We left the dependence on the map u implicit. But if we make it explicit by writing σ^u instead of σ, then the above map

$\coprod_{u \in F(1)} u^{-1}(F) \Rightarrow F$ is the cotuple $[\sigma^u]_{u \in F(1)}$. It is an isomorphism since for each $w \in F(X)$ we get $u = F(!)(w) \in F(1)$ with $w' \in u^{-1}(F)(X)$ obtained via the following pullback:

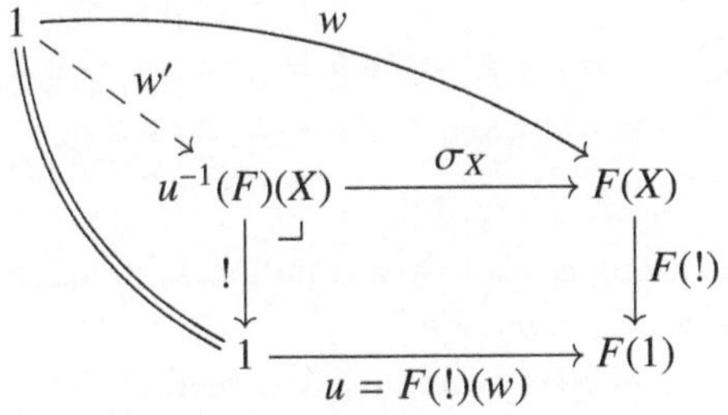

This shows surjectivity. Injectivity is obvious by the uniqueness of maps into the pullback $u^{-1}(F)(X)$.

2. Fix $u \in F(1)$; for convenience we abbreviate $F' = u^{-1}(F)$. Assume a non-empty collection of maps $f_i \colon X_i \to Y$ in $\mathbb{C}$ with weak pullback $p_i \colon P \to X_i$. If we have map $g_i \colon Q \to F'(X_i)$ with $F'(f_i) \circ g_i = F'(f_j) \circ g_j$, then we get maps $\sigma_{X_i} \circ g_i \colon Q \to F(X_i)$ with

$$\begin{aligned} F(f_i) \circ \sigma_{X_i} \circ g_i &= \sigma_Y \circ F'(f_i) \circ g_i \\ &= \sigma_Y \circ F'(f_j) \circ g_j \;=\; F(f_j) \circ \sigma_{X_j} \circ g_j. \end{aligned}$$

Since F preserves weak pullbacks this yields a map $g \colon Q \to F(P)$ with $F(p_i) \circ g = \sigma_{X_i} \circ g_i$. Then we obtain a unique map $h \colon Q \to F'(P)$ in

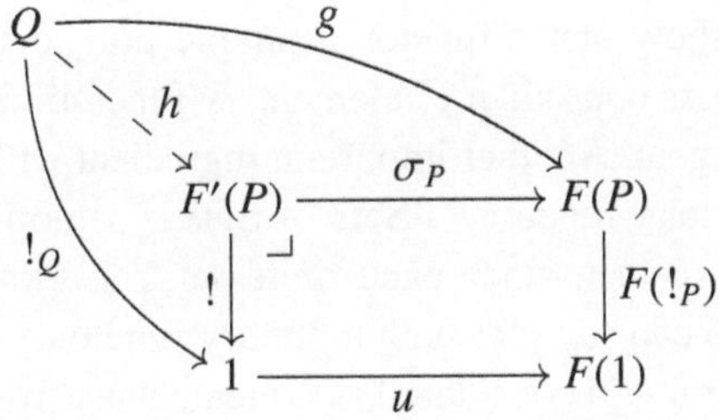

In order to see that the outer diagram commutes, we pick an arbitrary index i_0; it exists because we assumed our collection of maps f_i is non-empty. Then

$$\begin{aligned} F(!_P) \circ g &= F(!_{X_{i_0}}) \circ F(p_{i_0}) \circ g \\ &= F(!_{X_{i_0}}) \circ \sigma_{X_{i_0}} \circ g_{i_0} \\ &= u \circ \,!_{F(X_{i_0})} \circ g_{i_0} \\ &= u \circ \,!_Q. \end{aligned}$$

The resulting map $h\colon Q \to F'(P)$ satisfies $F'(p_i) \circ h = g_i$ by uniqueness of mediating maps for the pullback defining $F'(X) = u^{-1}(F)$:

$$\begin{aligned} ! \circ F'(p_i) \circ h &= \,! \\ &= \,! \circ g_i \\ \sigma_{X_i} \circ F'(p_i) \circ h &= F(p_i) \circ \sigma_P \circ h \\ &= F(p_i) \circ g \\ &= \sigma_{X_i} \circ g_i. \end{aligned}$$

In case P is a proper (non-weak) pullback, uniqueness of h is obtained from uniqueness of $g = \sigma_P \circ h$.

3. Assume $F\colon$ **Sets** $\to$ **Sets** is finitary. We need to show that $F' = u^{-1}(F)$ is finitary too. So assume an element $w \in F'(X)$. Then $\sigma_X(w) \in F(X)$. Hence there is a finite subset $\varphi\colon Y \hookrightarrow X$ with $v \in F(Y)$ such that $\sigma_X(w) = F(\varphi)(v)$. We obtain $v' \in F'(Y)$ via the pullback defining $F'(Y) = u^{-1}(f)(Y)$:

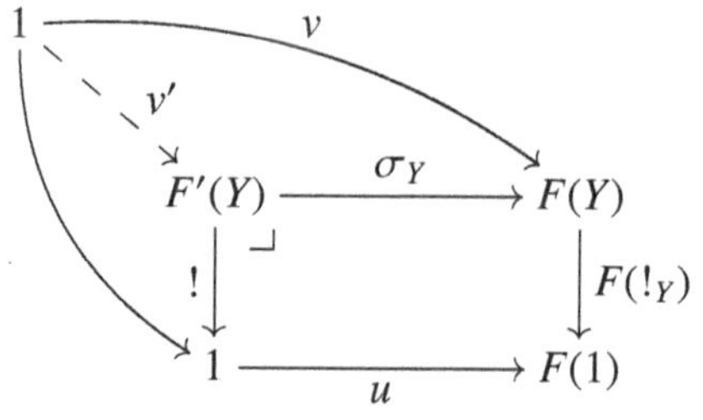

By uniqueness one obtains $F'(\varphi)(v') = w$, as required. ❑

If we wish to show that a functor is simple polynomial or analytical, we need to describe it as coproduct of elementary functors. The previous result is important for that goal. Another important ingredient will be described next.

Recall that a finitary functor $F\colon$ **Sets** $\to$ **Sets** is determined by its behaviour on finite subsets. In fact, since each finite set Y is isomorphic to a natural number $Y \cong n$, we can say that such a finitary functor F is determined by the outcomes $F(n)$, for $n \in \mathbb{N}$ considered as n-element set. We shall make this a bit more precise via the operations

$$\begin{array}{ccc} F(n) \times X^n & \xrightarrow{\quad \mathrm{ap}_n \quad} & F(X) \\ (u, t) & \longmapsto & F(t)(u). \end{array} \tag{4.15}$$

The n-tuple $t \in X^n$ is identified with a function $t\colon n \to X$, to which the functor F is applied. The following result again relates a functor to a coproduct of simpler functors.

Lemma 4.7.3 *For an arbitrary functor $F\colon \mathbf{Sets} \to \mathbf{Sets}$, the cotuple of the* ap_n *maps in (4.15) yields a natural transformation:*

$$\Big(\coprod_{n\in\mathbb{N}} F(n)\times(-)^n\Big) \xRightarrow{\ \mathrm{ap}\,=\,[\mathrm{ap}_n]_{n\in\mathbb{N}}\ } F \tag{4.16}$$

all of whose components $\coprod_n F(n)\times X^n \Rightarrow F(X)$ are surjective if and only if F is finitary.

This functor $\coprod_{n\in\mathbb{N}} F(n)\times(-)^n$ on the left-hand side is isomorphic to an arity functor $F_{\#_F}$, where the arity $\#_F$ associated with F is defined as $\#_F = \pi_1\colon (\coprod_{n\in\mathbb{N}} F(n)) \to \mathbb{N}$.

Proof First, let F be finitary and u be an element of $F(X)$. Hence there is a finite subset $i\colon Y \hookrightarrow X$ and element $v \in F(Y)$ with $F(i)(v) = u$. Let $n = |Y|$ be the size of Y and choose an isomorphism $j\colon n \xrightarrow{\cong} Y$. Take $t = i \circ j\colon n \to X$ and $w = F(j^{-1})(v) \in F(n)$. Then

$$\mathrm{ap}_n(w,t) = F(t)(w) = (F(i\circ j)\circ F(j^{-1}))(v) = F(i)(v) = u.$$

Conversely, assume the map $\coprod_n F(n)\times X^n \Rightarrow F(X)$ is surjective for each set X, and let $u \in F(X)$. Then there is an $n \in \mathbb{N}$ and $(v,t) \in F(n)\times X^n$ with $\mathrm{ap}_n(v,t) = F(t)(v) = u$. Consider the image factorisation (in **Sets**)

$$\begin{array}{ccl} n & \overset{e}{\twoheadrightarrow} & Y = \{t(i) \in X \mid i \in n\}. \\ & {\scriptstyle t}\searrow & \ \downarrow{\scriptstyle m} \\ & & \ X \end{array}$$

Now take $w = F(e)(v) \in F(Y)$. It satisfies $F(m)(w) = F(m\circ e)(v) = F(t)(v) = u$.

The arity functor $F_{\#_F}$ for the arity $\#_F$ is given by $F_{\#}(X) = \coprod_{n\in\mathbb{N},u\in F(n)} X^n$. This is obviously isomorphic to $\coprod_{n\in\mathbb{N}} F(n)\times X^n$, as used above. ☐

A finitary functor F is thus a quotient, via the ap-map (4.16), of a simple polynomial functor $\coprod_n F(n)\times(-)^n$. We shall see that further (preservation) conditions on the functor F allow us to say more about this map ap. To this end the following category of elements is useful.

Definition 4.7.4 For a functor $F\colon \mathbf{Sets} \to \mathbf{Sets}$ let $\mathbf{Elt}(F)$ be the category of 'elements' of F, defined in the following manner. Objects of $\mathbf{Elt}(F)$ are pairs $X \in \mathbf{Sets}$, $u \in F(X)$. A morphism $(u \in F(X)) \to (v \in F(Y))$ is a map $f\colon X \to Y$ in **Sets** satisfying $F(f)(u) = v$. Composition and identities are inherited from **Sets**. There is thus an obvious forgetful functor $\mathbf{Elt}(F) \to \mathbf{Sets}$.

This category **Elt**(F) of elements is relevant in the current setting since an equation $\mathrm{ap}_n(u, t) = v \in F(X)$, for $u \in F(n)$ and $t \in X^n$, means that t is a morphism $t\colon (u \in F(n)) \to (v \in F(X))$ in **Elt**(F). We are interested in getting appropriately minimal versions of such maps.

Lemma 4.7.5 *Let $F\colon$ **Sets** $\to$ **Sets** be a finitary functor.*

1. *For each $w \in F(X)$,*
   - *there is a map $(w' \in F(n)) \to (w \in F(X))$ in* **Elt**(F)*, where $n \in \mathbb{N}$,*
   - *such that for each $f\colon (v \in F(Y)) \to (w' \in F(n))$ in* **Elt**(F)*, the function $f\colon Y \to n$ is surjective.*
2. *If the functor F preserves countable weak pullbacks, then the previous point can be strengthened to: for each $w \in F(X)$,*
   - *there is a map $(w' \in F(n)) \to (w \in F(X))$ in* **Elt**(F)*,*
   - *such that each $f\colon (v \in F(Y) \to (w' \in F(n))$ is a split epi in* **Elt**(F)*.*

   If we write:

$$F^{\circ}(n) = \{w \in F(n) \mid \textit{each } (v \in F(Y)) \to (w \in F(n)) \textit{ is a split epi}\},$$

   then this last result says: for each $w \in F(X)$ there is a map $(w' \in F^{\circ}(n)) \to (w \in F(X))$.

Proof 1. Let $w \in F(X)$. Because the component at X of the map ap in (4.16) is surjective, the set $\{(n, u, t) \mid n \in \mathbb{N}, u \in F(n), t \in X^n \text{ with } \mathrm{ap}_n(u, t) = w\}$ is non-empty. Among all these elements we take the one with the least number n and call it (n, w', t). Thus $\mathrm{ap}_n(w', t) = w$, so that $t\colon (w' \in F(n)) \to (w \in F(X))$ in **Elt**(F); additionally, for each $m \in \mathbb{N}$ with $\mathrm{ap}_m(u, s) = w$ we have $n \leq m$. Next, assume a map $f\colon (v \in F(Y) \to (w' \in F(n))$ in **Elt**(F). Factorise $f\colon Y \to n$ in **Sets** as $f = i \circ f'$ where $f'\colon Y \twoheadrightarrow m$ and $i\colon m \rightarrowtail n$ (so that $m \leq n$). We then have $v' = F(f')(v) \in F(m)$ and $t' = t \circ i \in X^m$ satisfying

$$\mathrm{ap}_m(v', t') = F(t')(v') = F(t \circ i \circ f')(v) = F(t \circ f)(v) = F(t)(w') = w.$$

But then $n \leq m$ and thus $m = n$. Hence $i\colon m \rightarrowtail n$ is an isomorphism, and $f = i \circ f'$ is surjective.

2. Assume towards a contradiction that for each $w \in F(X)$ and for all $(w' \in F(n)) \to (w \in F(X))$ there is a map $f\colon (v \in F(Y)) \to (w' \in F(n))$ that is not a split epi. We proceed in a number of steps.

   a. By (1) we do have a map $t_1\colon (w_1 \in F(n_1)) \to (w \in F(X))$ such that for each $f\colon (v \in F(Y)) \to (w_1 \in F(n_1))$ the function f is surjective. But, by assumption, there is a map $f_1 : (v_1 \in F(Y_1)) \to (w_1 \in F(n_1))$ which is not a split epi.

b. We now apply the assumption to $v_1 \in F(Y_1)$. It yields, as before, a map $t_2 : (w_2 \in F(n_2)) \to (v_1 \in F(Y_1))$ as in (1), together with $f_2 : (v_2 \in F(Y_2)) \to (w_2 \in F(n_2))$ that is not a split epi.

c. Continuing in this manner we obtain a chain of maps in **Elt**(F):

$$\begin{array}{ccccccc} (w \in F(X)) & & (v_1 \in F(Y_1)) & & (v_2 \in F(Y_2)) & & \cdots \\ t_1 \uparrow & \overset{f_1}{\swarrow} & t_2 \uparrow & \overset{f_2}{\swarrow} & t_3 \uparrow & \swarrow & \\ (w_1 \in F(n_1)) & & (w_2 \in F(n_2)) & & (w_3 \in F(n_3)) & & \end{array} \qquad (4.17)$$

d. By (1), for each of the resulting maps $f_i \circ t_{i+1} : (w_{i+1} \in F(n_{i+1})) \to (w_i \in F(n_i))$ in **Elt**(F) the underlying function $f_i \circ t_{i+1} : n_{i+1} \to n_i$ is surjective. Hence $n_{i+1} \geq n_i$. We also have $n_{i+1} \neq n_i$: if $n_{i+1} = n_i$, then the map $f_i \circ t_{i+1}$ is an isomorphism, say with inverse $s : n_i \to n_{i+1}$. As a result:

- s is a map $(w_i \in F(n_i)) \to (w_{i+1} \in F(n_{i+1}))$ in **Elt**(F) since $F(s)(w_i) = (F(s) \circ F(f_i \circ t_{i+1}))(w_{i+1}) = w_{i+1}$.
- $t_{i+1} \circ s : (w_i \in F(n_i)) \to (v_i \in F(Y_i))$ is a splitting for f_i, since $f_i \circ t_{i+1} \circ s = \mathrm{id}$.

Thus, an equality $n_{i+1} = n_i$ makes f_i a split epi in **Elt**(F) – which we know is not the case. Hence $n_{i+1} > n_i$.

e. Now write g_n for the resulting maps $g_i : (w_i \in F(n_i)) \to (w \in F(X))$, obtained as chain of ts and fs in (4.17). We take the (countable) pullback $P = \prod_X n_i$ of these $g_i : n_i \to X$ in **Sets**, with projections $p_i : P \to n_i$ satisfying $g_i \circ p_i = g_j \circ p_j$. Since F preserves weak pullbacks and $F(g_i)(w_i) = w$, there is an element $u \in F(P)$ with $F(p_i)(u) = w_i$. Hence $p_i : (u \in F(P)) \to (w_i \in F(n_i))$ in **Elt**(F).

f. Since F is finitary and $u \in F(P)$ we get a map $t : (u' \in F(k)) \to (u \in F(P))$, for some $k \in \mathbb{N}$. Recall that we have an ascending sequence $n_1 < n_2 < n_3 < \ldots$, so there is an index i with $k < n_i$. At the same time we have a map

$$(u' \in F(k)) \xrightarrow{\;t\;} (u \in F(P)) \xrightarrow{\;p_i\;} (w_i \in F(n_i))$$

whose underlying function $p_i \circ t : k \to n_i$ must be surjective – by construction of the objects $(w_i \in F(n_i))$, using (1). But surjectivity of this map implies $k \geq n_i$, which is impossible.

Hence our original assumption is wrong. □

If we apply these results to affine functors, things are beginning to fall into place.

Proposition 4.7.6 *Let $F\colon \mathbf{Sets} \to \mathbf{Sets}$ be a countable weak-pullback-preserving functor which is finitary and affine.*

1. *There is a unique $n \in \mathbb{N}$ for which $\mathrm{ap}_n\colon F^\circ(n) \times (-)^n \Rightarrow F$ is surjective.*
2. *This yields a (natural) isomorphism*

$$\left(\coprod_{u \in F^\circ(n)} X^n / \sim_u\right) \xrightarrow{\;\cong\;} F(X),$$

where $\sim_u$ is the equivalence relation on X^n given by

$$t \sim_u s \iff \exists \varphi\colon n \xrightarrow{\cong} n.\, t \circ \varphi = s \quad \text{and} \quad F(\varphi)(u) = u.$$

3. *In case F preserves (proper) pullbacks, we get an isomorphism*

$$\left(\coprod_{u \in F^\circ(n)} X^n\right) \xrightarrow{\;\cong\;} F(X),$$

making F a simple polynomial functor.

Proof 1. Assume we have two elements $u \in F(X)$ and $v \in F(Y)$. We pick two maps $t\colon (u' \in F^\circ(n)) \to (u \in F(X))$ and $s\colon (v' \in F^\circ(m)) \to (v \in F(Y))$ with the properties of Lemma 4.7.5.2. The aim is to show $n = m$. The product $n \times m$ is a pullback over 1; applying F yields a weak pullback, as on the left below. Because F is affine we get $F(t)(u') = F(s)(v')$; hence there is an element $w \in F(n \times m)$ with $F(\pi_1)(w) = u'$ and $F(\pi_2)(w) = v'$. The maps π_i are then split epis, by Lemma 4.7.5.2, so we get a diagram in $\mathbf{Elt}(F)$ as on the right, with r_i splitting π_i:

$$\begin{array}{ccc} F(n \times m) & \xrightarrow{F(\pi_2)} & F(m) \\ {\scriptstyle F(\pi_1)}\downarrow & & \downarrow{\scriptstyle F(s)} \\ F(n) & \xrightarrow[F(t)]{} & F(1) \cong 1 \end{array} \qquad\qquad \begin{array}{ccc} (w \in F(n \times m)) & \underset{\pi_2}{\twoheadrightarrow} \; \overset{r_2}{\longleftarrow} & (v' \in F(m)) \\ {\scriptstyle r_1}\uparrow\;\twoheaddownarrow{\scriptstyle \pi_1} & & \downarrow{\scriptstyle s} \\ (u' \in F(n)) & \xrightarrow[t]{} & (* \in F(1)) \end{array}$$

Since $v' \in F^\circ(m)$ and $u' \in F^\circ(n)$ we obtain that the two resulting diagonal maps

$$(u' \in F(n)) \xrightarrow{r_1} (w \in F(n \times m)) \xrightarrow{\pi_2} (v' \in F(m))$$
$$(v' \in F(m)) \xrightarrow{r_2} (w \in F(n \times m)) \xrightarrow{\pi_1} (u' \in F(n))$$

are both split epis. Hence $n \leq m$ and $m \leq n$, and so $n = m$.

2. For $u \in F^\circ(n)$ we need to show

$$\mathrm{ap}_n(u, t) = \mathrm{ap}_n(u, s) \iff t \sim_u s.$$

The implication ($\Leftarrow$) is trivial: if $t \circ \varphi = s$ and $F(\varphi)(u) = u$, then

$$\begin{aligned} \mathrm{ap}_n(u, t) &= F(t)(u) = F(t)(F(\varphi)(u)) = F(t \circ \varphi)(u) \\ &= F(s)(u) = \mathrm{ap}_n(u, s). \end{aligned}$$

For the direction ($\Rightarrow$), assume $\mathrm{ap}_n(u, t) = \mathrm{ap}_n(u, s)$ in $F(X)$. We form the pullback

$$\begin{array}{ccc} n \times_X n & \xrightarrow{p_2} & n \\ {\scriptstyle p_1}\downarrow & \lrcorner & \downarrow{\scriptstyle s} \\ n & \xrightarrow[t]{} & X \end{array}$$

Applying F yields a weak pullback, and thus an element $w \in F(n \times_X n)$ with maps $p_1, p_2 \colon (w \in F(n \times_X n)) \to (u \in F(n))$ in $\mathbf{Elt}(F)$. These maps p_i are both split epis, since $u \in F^\circ(n)$, say with splittings r_i. Then $\varphi = p_1 \circ r_2 \colon n \to n$ is a split epi, and thus an isomorphism. It satisfies

$$\begin{aligned} t \circ \varphi = t \circ p_1 \circ r_2 &= s \circ p_2 \circ r_2 = s \\ F(\varphi)(u) = F(p_1 \circ r_2)(u) &= F(p_1)(w) = u. \end{aligned}$$

3. Assume now that F preserves pullbacks. Since F is affine, i.e. $F(1) \cong 1$, F preserves all finite limits, and in particular products and equalisers. We first show that there is at most one map $(u \in F^\circ(n)) \to (v \in F(X))$. If we have two of them, say t, s, we can form the equaliser in **Sets**:

$$E \overset{e}{\rightarrowtail} n \underset{s}{\overset{t}{\rightrightarrows}} X.$$

This equaliser is preserved by F. Since $F(t)(u) = v = F(s)(u)$, there is a unique element $w \in F(E)$ with $F(e)(w) = u$. This map $e \colon (w \in F(E)) \to (u \in F(n))$ is then a split epi, and thus an isomorphism. Hence $t = s$.

With this observation we see that $\mathrm{ap}_n(u, t) = \mathrm{ap}_n(u, s)$ implies $s = t$. Thus the equivalence relation $\sim_u$ on X^n used in (2) is the equality relation. ❑

The equivalence relation $\sim_u$ used in point (2) involves a set of isomorphisms

$$G_u = \{\varphi \colon n \xrightarrow{\cong} n \mid F(\varphi)(u) = u\} \tag{4.18}$$

that is a subgroup of the symmetric group S_n of permutations on n. It induces an action:

$$G_u \times X^n \longrightarrow X^n \quad \text{by} \quad (\varphi, t) \longmapsto t \circ \varphi.$$

The set of equivalence classes $X^n / \sim_u$ can then also be described as the set of orbits

$$X^n/G_u = \{[t] \mid t \in X^n\} \qquad \text{with} \qquad [t] = \{t \circ \varphi \mid \varphi \in G_u\}.$$

This is used below.

Definition 4.7.7 ([285]) A functor F: **Sets** → **Sets** is called **analytical** if it can be written (via a natural isomorphism) as

$$F \cong \left(\coprod_{i \in I} X^{\#i}/G_i \right),$$

where $\#: I \to \mathbb{N}$ is an arity and $G_i \subseteq S_{\#i}$ is a subgroup of the symmetric group of permutations on $\#i \in \mathbb{N}$.

An example of an analytical functor is the multiset (or bag) functor $\mathcal{M}_{\mathbb{N}}$, since one can write

$$\mathcal{M}_{\mathbb{N}}(X) = 1 + X + X^2/S_2 + X^3/S_3 + \cdots$$

since each finite multiset, say with n elements in total, can be identified with an n-tuple in X^n up-to-the-order, that is, with an element of the quotient X^n/S_n that destroys the order. Intriguingly this formulation of $\mathcal{M}_{\mathbb{N}}(X)$ looks very similar to the Taylor expansion of the exponential function:

$$e^x = 1 + x + \tfrac{x^2}{2!} + \tfrac{x^3}{3!} + \cdots.$$

(Recall that the order of the symmetric group S_n is $n!$.) In [285] the concept of derivative of a functor (see also [4]) is defined, in such a way that $\mathcal{M}_{\mathbb{N}}$ is its own derivative; see Exercise 4.7.3 for a glimpse.

We finally come to the main characterisation result. The second point dealing with analytical functors comes from [285]. The first point is in a sense a restriction of the second and can be found in [200] (where simple polynomial functors are called normal, after [156]).

Theorem 4.7.8 *Let F:* **Sets** → **Sets** *be an arbitrary functor.*

1. *F is simple polynomial if and only if it is finitary and preserves countable pullbacks.*
2. *F is analytical if and only if it is finitary and preserves countable weak pullbacks.*

Proof 1. We already know that a simple polynomial functor is finitary (Exercise 4.6.6) and preserves binary pullbacks (Proposition 4.2.6); it is not hard to see that the latter generalises to preservation of arbitrary pullbacks.

For the converse, assume F is finitary and preserves countable pullbacks. In order to see that F is simple polynomial we have to combine two results.

1. First, use Proposition 4.7.2 to write $F \cong \coprod_{u \in F(1)} u^{-1}(F)$ as coproduct of affine functors $u^{-1}(F)$, each of which is finitary and preserves countable pullbacks.
2. Next, use Proposition 4.7.6 to write these affine functors as $u^{-1}(F) \cong \coprod_{v \in I_u} X^{n_u}$, where $I_u = u^{-1}(F)^{\circ}(n_u)$ and $n_u \in \mathbb{N}$ is the unique number for which $I_u \times X^{n_u} \to u^{-1}(F)(X)$ is surjective.

Hence by taking $I = \coprod_{u \in F(1)} I_u$ and $\#(u, v) = n_u$ we obtain $F \cong \coprod_{i \in I} X^{\#i}$, making F a simple polynomial functor.

2. Following the same steps one shows that a finitary functor preserving countable *weak* pullbacks is of the form $\coprod_{i \in I} X^{\#i}/G_i$, where the subgroup G_i of the symmetric group on $\#i \in \mathbb{N}$ arises as in (4.18).

In the reverse direction, an analytical functor $F = \coprod_{i \in I} X^{\#i}/G_i$ is obviously finitary. It is not hard to see that it preserves countable weak pullbacks. This is left to the interested reader. ❑

We conclude this section with yet another characterisation of simple polynomial functors, namely as the 'shapely' functors from [276, 274] (see also [4, theorem 8.3]). For this result we need to know that a natural transformation is called **(weak) cartesian** if all its naturality squares are (weak) pullbacks. This terminology will also be used in the exercises below.

Proposition 4.7.9 *A functor $F\colon$ **Sets** $\to$ **Sets** is simple polynomial if and only if it is **shapely**: it preserves binary pullbacks and comes with a cartesian natural transformation $F \Rightarrow (-)^{\star}$ to the list functor.*

Proof Assume F is polynomial: $F(X) = \coprod_{i \in I} X^{\#i}$ via an arity $\#\colon I \to \mathbb{N}$. We already know that it preserves pullbacks. Each tuple $t \in X^{\#i}$ can be seen as an $\#i$ element list $t = \langle t_0, t_1, \ldots, t_{\#i-1}\rangle \in X^{\star}$. Thus we obtain a natural transformation $\sigma\colon F \Rightarrow (-)^{\star}$ as cotuple. We check that its naturality squares form pullbacks. So assume we have a function $f\colon X \to Y$ and a naturality square

$$
\begin{array}{ccc}
F(X) & \xrightarrow{\quad \sigma_X \quad} & X^{\star} \\
{\scriptstyle F(f)}\big\downarrow & & \big\downarrow{\scriptstyle f^{\star}} \\
F(Y) & \xrightarrow[\quad \sigma_Y \quad]{} & Y^{\star}
\end{array}
$$

If we have an element $v \in F(Y)$ and a list $\alpha = \langle x_1, \ldots, x_n \rangle \in X^\star$ with $\sigma_Y(v) = f^\star(\langle x_1, \ldots, x_n \rangle) = \langle f(x_1), \ldots, f(x_n) \rangle$, then $v \in F(Y) = \coprod_{i \in I} Y^{\#i}$ must be of the form $\langle i, \lambda k \leq n.\, f(x_k) \rangle$, where $\#i = n$. Hence there is precisely one mediating element $u \in F(X)$, namely $u = \langle i, \lambda k \leq n.\, x_k \rangle$.

Conversely, assume that F preserves pullbacks and that we have a cartesian natural transformation $\sigma\colon F \Rightarrow (-)^\star$. We take $I = F(1)$ with arity:

$$\# \stackrel{\text{def}}{=} (I = F(1) \xrightarrow{\sigma_1} 1^\star = \mathbb{N}).$$

Since σ is cartesian, the following naturality square is a pullback:

$$\begin{array}{ccc} F(X) & \xrightarrow{\sigma_X} & X^\star \\ {\scriptstyle F(!)}\big\downarrow & \lrcorner & \big\downarrow{\scriptstyle !^\star = \mathsf{length}} \\ I = F(1) & \xrightarrow[\sigma_1 = \#]{} & 1^\star = \mathbb{N} \end{array}$$

This means that $F(X)$ is the set of pairs $i \in I$ and $t \in X^\star$ with $\mathsf{length}(t) = \#i$. Hence such an element is a pair $i \in I, t \in X^{\#i}$. Thus $F(X) = \coprod_{i \in I} X^{\#i}$, making F simple polynomial. □

In the end we recall the description of a 'container' or 'dependent polynomial functor' from Exercise 2.2.6, as a functor of the form $F(X) = \coprod_{i \in I} X^{A_i}$, for an indexed collection $(A_i)_{i \in I}$ of not necessarily finite sets A_i. These containers are more general than simple polynomial functors. They capture the idea that many data types are given by a template that determines how data are stored. Their theory is developed, from a programming perspective, in [3, 2]. In particular, the idea of keeping track of a specific position within such datatypes (as one-hole contexts) can be formalised via derivatives of such functors; see [4]. Such a derivative of a functor goes back to [285]; it is described in Exercise 4.7.3 below, for simple polynomial functors.

Exercises

4.7.1 Assume two arities $\#\colon I \to \mathbb{N}$ and $\#\colon J \to \mathbb{N}$. Prove that there are bijective correspondences:

1.

$$\frac{\coprod_{i \in I}(-)^{\#i} \stackrel{\sigma}{\Longrightarrow} \coprod_{j \in J}(-)^{\#j}}{\overline{\overline{I \xrightarrow[f]{} J \ \text{with} \ \#f(i) \xrightarrow[\varphi_i]{} \#i}}}.$$

2. This correspondence restricts to

$$\frac{\text{(weak) cartesian } \coprod_{i\in I}(-)^{\#i} \overset{\sigma}{\Longrightarrow} \coprod_{j\in J}(-)^{\#j}}{I \xrightarrow[f]{} J \text{ with } \#f(i) \xrightarrow[\varphi_i]{\cong} \#i} .$$

4.7.2 For a functor $F\colon \mathbf{Sets} \to \mathbf{Sets}$ and a set X consider the coequaliser

$$\Big(\coprod_{n,m\in\mathbb{N}} F(m)\times n^m \times X^n\Big) \underset{d_2}{\overset{d_1}{\rightrightarrows}} \Big(\coprod_{n\in\mathbb{N}} F(n)\times X^n\Big) \overset{c}{\twoheadrightarrow} \widetilde{F}(X).$$

The two maps d_1, d_2 are given by $(u, t, s) \mapsto (F(t)(u), s)$ and $(u, t, s) \mapsto (u, s \circ t)$.

1. In this situation, describe a natural transformation $\widetilde{F} \Rightarrow F$, via the map $\mathrm{ap}\colon \coprod_{n\in\mathbb{N}} F(n)\times X^n \to F(X)$ from (4.16).
2. Show that this $\widetilde{F} \Rightarrow F$ consists of monos if F preserves (binary) weak pullbacks.
3. Show also that it consists of epis if F is finitary.

4.7.3 (From [285]) For an arbitrary functor $F\colon \mathbb{C} \to \mathbb{C}$ define the **derivative**, if it exists, to the functor $F'\colon \mathbb{C} \to \mathbb{C}$ with a universal weak cartesian natural transformation $\rho\colon F' \times \mathrm{id}_{\mathbb{C}} \Rightarrow F$. Universality means that for an arbitrary functor G with a weak cartesian $\tau\colon G \times \mathrm{id}_{\mathbb{C}} \Rightarrow F$ there is a unique weak cartesian τ' making the following diagram commute:

$$\begin{array}{ccc} F' \times \mathrm{id}_{\mathbb{C}} & \xrightarrow{\quad\rho\quad} & F \\ {\scriptstyle \tau' \times \mathrm{id}} \uparrow & \nearrow_{\tau} & \\ G \times \mathrm{id}_{\mathbb{C}} & & \end{array}$$

Prove that for a simple polynomial functor $F(X) = \coprod_{i\in I} X^{\#i}$ the derivative is

$$F'(X) = \coprod_{i\in I, \#i>0} \#i \times X^{\#i-1}.$$

Hint: Use Exercise 4.7.1.

5

Monads, Comonads and Distributive Laws

Monads and comonads – the duals of monads – form one of the basic notions in category theory, like adjunction. A monad is a special kind of endofunctor with some additional structure (unit and multiplication), a bit like for a monoid. Various computationally relevant functors are actually monads: lift, list, powerset, multiset, distribution. Associated with a (co)monad two categories are of interest.

- The *Kleisli* category captures the computations associated with a (co)monad. Coalgebras can be described within such Kleisli categories, namely as endomaps, and the Kleisli structure can be used for sequential composition of coalgebras. A prominent application involves final coalgebra inside such Kleisli categories. It gives a systematic description of so-called trace semantics; see Section 5.3.
- The *Eilenberg–Moore* category contains (co)algebras for a (co)monad as objects. They are the mathematical structures associated with the (co)monad. Where (co)algebras of a *functor* describe only operations, (co)algebras of a *(co)monad* additionally capture constraints, in the form of equations or other assertions. Many standard mathematical structures, such as monoids, vector spaces and complete lattices, are algebras of a monad. A systematic, uniform description of such structures is convenient, since it gives many results at once, such as the existence of limit and colimits.

Most of the material in this chapter is standard (basic) category theory. It is presented here with a special focus on the context of coalgebras as dynamical state-based systems. Section 5.1 introduces the basic definitions, together with the main examples. Kleisli categories will be discussed in Section 5.2, with an emphasis on the lifting of functors to such categories. These liftings play an important role in Section 5.3, on trace semantics. Section 5.4 describes Eilenberg–Moore categories, together with their basic properties. Section 5.5

concludes this chapter with bialgebras, combining both algebras (structure) and coalgebras (behaviour), for the operational semantics of programming languages.

The topic of algebras and coalgebras in combination with assertions is postponed until Chapter 6. Assertions require the notion of invariant, in the coalgebraic case. Once this concept is in place, (co)algebraic specifications can be introduced, as descriptions of (co)algebras satisfying certain logical properties.

5.1 Monads and Comonads: Definition and Examples

Monads are special endofunctors with some additional structure. Our prime example is the powerset functor $\mathcal{P}$ which comes equipped with a singleton map $\{-\}\colon X \to \mathcal{P}(X)$ and a (big) union $\bigcup\colon \mathcal{P}^2(X) \to \mathcal{P}(X)$. These operations are natural in X and satisfy some basic equations. This is axiomatised in the notion of monad; see Definition 5.1.1 below. It turns out that the 'collection' functors (list, powerset, multiset, distribution) from Figure 4.1 all have such monad structure.

As we shall see, monads have two distinct roles.

1. **Monads capturing types of computations**. In this sense monads are similar to the endofunctors F whose coalgebras $X \to F(X)$ we have been studying so far: a distinction is made between values/states in X and computations in $F(X)$, returning values in X. But the additional structure that monads have gives rise to the basic operation of sequential composition. We have already seen such composition, for instance in Exercise 1.1.2, without identifying the underlying monad structure. This composition is captured in what is called the Kleisli category associated with a monad. It will be the main focus of the next section.
2. **Monads capturing algebraic theories**. Monads also form an abstraction that describes the essentials of an algebraic theory, given by operations and equations between them. Here one associates a different category with a monad, namely its category of so-called Eilenberg–Moore algebras. These categories will be introduced in Section 5.4, but the aspect of monads as algebraic theories will be postponed to Section 6.7.

A combination of these two roles is achieved in the theory of algebraic effects [387, 388].

Historically, the algebraic view on monads preceded the computational view. This other, computational view, using Kleisli categories, has been introduced by Moggi [357]. It has been widely adopted in functional programming – notably in the programming language Haskell – in order to deal with special kinds of computational effects (see e.g. [284], and for combinations with initial algebras/final coalgebras, see [370, 44]). In fact, in this second computational approach to monads an alternative formulation of the notion of monad has emerged, in terms of so-called Kleisli extensions. Here we stick to the standard formulation in the next definition and formulate the Kleisli version separately (in Proposition 5.2.3 in the next section).

From the computational perspective monads are thus used for structuring the *outputs* of computations. At the end of this section we briefly discuss comonads, which can be used to structure *inputs* of computations. As an aside, a combined structuring can be achieved via so-called arrows, generalising both monads and comonads. They will not be discussed here, but the interested reader is referred to [229, 372, 219, 255].

This section will introduce the notion of (co)monad and focus mostly on examples – of which there are plenty, contributing to the relevance of the notion. The next section will define the associated concept of a Kleisli category.

We are now ready to see define a monad. It is an endofunctor, typically written as T, with additional structure given by two natural transformations that play the role of 'singleton' and 'flattening'. This will be illustrated in the subsequent examples.

Definition 5.1.1 A **monad** on an arbitrary category $\mathbb{C}$ consists of an endofunctor $T\colon \mathbb{C} \to \mathbb{C}$ together with two natural transformations: a **unit** $\eta\colon \mathrm{id}_{\mathbb{C}} \Rightarrow T$ and **multiplication** $\mu\colon T^2 \Rightarrow T$. These are required to make the following diagrams commute, for $X \in \mathbb{C}$:

$$
\begin{array}{ccccc}
T(X) & \xrightarrow{\eta_{T(X)}} & T^2(X) & \xleftarrow{T(\eta_X)} & T(X) \\
& \searrow\!\!= & \downarrow{\scriptstyle \mu_X} & =\!\!\swarrow & \\
& & T(X) & &
\end{array}
\qquad
\begin{array}{ccc}
T^3(X) & \xrightarrow{\mu_{T(X)}} & T^2(X) \\
{\scriptstyle T(\mu_X)}\downarrow & & \downarrow{\scriptstyle \mu_X} \\
T^2 & \xrightarrow[\mu_X]{} & T(X)
\end{array}
$$

Often we simply write T for a monad (T, η, μ) since we standardly use η and μ for the unit and multiplication involved.

Before seeing examples of monads we immediately illustrate their usefulness in the theory of coalgebras: they introduce (sequential) composition of coalgebras. This composition will be presented in slightly more general form later on, via Kleisli categories.

Lemma 5.1.2 *For a monad T, coalgebras $X \to T(X)$ form a monoid – where X is a fixed object.*

Proof For two coalgebras $c, d\colon X \to T(X)$ we define a 'c then d' coalgebra $d \circledcirc c\colon X \to T(X)$ as composite:

$$d \circledcirc c \stackrel{\text{def}}{=} \Big(X \xrightarrow{\;c\;} T(X) \xrightarrow{\;T(d)\;} T^2(X) \xrightarrow{\;\mu_X\;} T(X)\Big).$$

The associated neutral element 'skip' is the unit $\eta_X\colon X \to T(X)$ of the monad. The fact that $(\eta, \circledcirc)$ forms a monoid follows almost immediately from the monad equations. ❑

The notion of monad, like most of the concepts in category theory, is best understood via concrete examples. In order to see the common structure it may help to fill in some of the missing verifications of the monad requirements in the series of examples below.

Example 5.1.3 1. As mentioned in the beginning, the powerset functor $\mathcal{P}$ is a monad with unit and multiplication given by singleton and union:

$$\begin{array}{ccc} X \xrightarrow{\;\eta_X\;} \mathcal{P}(X) & \qquad & \mathcal{P}(\mathcal{P}(X)) \xrightarrow{\;\mu_X\;} \mathcal{P}(X) \\ x \longmapsto \{x\} & & A \longmapsto \bigcup A = \{x \in X \mid \exists U \in A.\, x \in U\}. \end{array}$$

The first two of the monad equations are easy: for $V \in \mathcal{P}(X)$,

$$\begin{aligned} (\mu_X \circ \eta_{\mathcal{P}(X)})(V) &= \bigcup\{V\} = V \\ (\mu_X \circ \mathcal{P}(\eta_X))(V) &= \bigcup\{\{x\} \mid x \in V\} = V. \end{aligned}$$

The μ-equation requires more care. For $\mathcal{A} \in \mathcal{P}^3(X) = \mathcal{P}(\mathcal{P}(\mathcal{P}(X)))$ one has

$$\begin{aligned} (\mu_X \circ \mathcal{P}(\mu_X))(\mathcal{A}) &= \bigcup\{\bigcup B \mid B \in \mathcal{A}\} \\ &= \{x \in X \mid \exists B \in \mathcal{A}.\, \exists U \in B.\, x \in U\} \\ &= \{x \in X \mid \exists U \in \bigcup \mathcal{A}.\, x \in U\} \\ &= \bigcup\bigcup \mathcal{A} \\ &= (\mu_X \circ \mu_{\mathcal{P}(X)})(\mathcal{A}). \end{aligned}$$

Also the non-empty powerset $\mathcal{P}^{\neq\emptyset}$ is a monad, and so are their finitary versions $\mathcal{P}_{\text{fin}}$ and $\mathcal{P}^{\neq\emptyset}_{\text{fin}}$ (taking only *finite* subsets).

We recall that coalgebras $X \to \mathcal{P}(X)$ of this powerset monad $\mathcal{P}$ – or of non-empty/finite variations thereof – are unlabelled transition systems, as special cases of non-deterministic automata; see Section 2.2.4. Composition $\circledcirc$ as in Lemma 5.1.2 corresponds to relational composition: for coalgebras $c, d\colon X \to \mathcal{P}(X)$ we have

$$(d \circledcirc c)(x) = \bigcup\{d(x') \mid x' \in c(x)\} = \{x'' \mid \exists x'.\, x \xrightarrow{c} x' \text{ and } x' \xrightarrow{d} x''\}.$$

2. The non-empty powerset takes subsets with *at least* one element. The 'lift' or 'maybe' functor $\mathcal{L}(X) = 1 + X$ adds a base point to a set X; it may be understood as a 'collection' functor that takes subsets of X with *at most* one element (also known as subsingletons). Lift $\mathcal{L} = 1 + (-)\colon \mathbf{Sets} \to \mathbf{Sets}$ is a monad with unit and multiplication:

$$X \xrightarrow{\eta = \kappa_2} 1 + X \qquad\qquad 1 + (1 + X) \xrightarrow{\mu = [\kappa_1, \mathrm{id}]} 1 + X.$$

This lift monad may be defined on categories other than **Sets**; see Exercise 5.1.1 (or [303]). Coalgebras $X \to \mathcal{L}(X)$ of the lift monad are partial (endo)functions on X; their sequential composition $\circledcirc$ from Lemma 5.1.2 is the usual composition of partial functions.

3. Recall the list functor $(-)^\star\colon \mathbf{Sets} \to \mathbf{Sets}$ that sends a set X to the set $X^\star = \{\langle x_1, \ldots, x_n\rangle \mid x_i \in X\}$ of finite sequences of elements of X. It forms a monad via:

$$\begin{array}{rclcrcl} X & \xrightarrow{\eta} & X^\star & \qquad & X^{\star\star} & \xrightarrow{\mu} & X^\star \\ x & \longmapsto & \langle x\rangle & & \langle \vec{x_1}, \ldots, \vec{x_n}\rangle & \longmapsto & \vec{x_1}\cdots\vec{x_n}. \end{array}$$

The multiplication μ flattens a list of lists to a single list, by removing inner brackets.

4. The distribution functor $\mathcal{D}$ from Definition 4.1.5 forms a monad with singleton 'Dirac' distribution as unit, and matrix multiplication as (monad) multiplication.

$$\begin{array}{rclcrcl} X & \xrightarrow{\eta} & \mathcal{D}(X) & \qquad & \mathcal{D}(\mathcal{D}(X)) & \xrightarrow{\mu} & \mathcal{D}(X) \\ x & \longmapsto & 1|x\rangle = \lambda y.\begin{cases} 1 & \text{if } y = x \\ 0 & \text{if } y \neq x \end{cases} & & \Psi & \longmapsto & \lambda y.\ \sum_{\varphi \in \mathcal{D}(X)} \Psi(\varphi)\cdot\varphi(y). \end{array}$$

This unit was already described as a Dirac distribution in Exercise 4.1.4. The multiplication applied to a multiset of multisets $\Psi = \sum_i r_i|\varphi_i\rangle$ yields a distribution $\mu(\Psi) \in \mathcal{D}(X)$ that assigns to $y \in X$ the probability $\mu(\Psi)(y) = \sum_i r_i \cdot \varphi_i(y)$. For instance, if we have distributions

$$\varphi = \tfrac{1}{2}|x\rangle + \tfrac{1}{2}|y\rangle \quad \text{and} \quad \psi = \tfrac{1}{3}|y\rangle + \tfrac{2}{3}|z\rangle,$$

then

$$\begin{aligned} \mu(\tfrac{3}{4}|\varphi\rangle + \tfrac{1}{4}|\psi\rangle) &= (\tfrac{3}{4}\cdot\tfrac{1}{2})|x\rangle + (\tfrac{3}{4}\cdot\tfrac{1}{2})|y\rangle + (\tfrac{1}{4}\cdot\tfrac{1}{3})|y\rangle + (\tfrac{1}{4}\cdot\tfrac{2}{3})|z\rangle \\ &= \tfrac{3}{8}|x\rangle + \tfrac{11}{24}|y\rangle + \tfrac{1}{6}|z\rangle. \end{aligned}$$

As discussed in Section 4.1, coalgebras of the distribution monad are Markov chains. Their composition, occurring already in Exercise 4.1.5, corresponds to standard composition of such chains – usually given by matrix multiplication; see Exercise 5.1.3.

It is not hard to see that subdistributions – with sum of probabilities at most 1, instead of equal to 1 – also form a monad, written as $\mathcal{D}_{\leq 1}$.

5. If M is a monoid, say with multiplicative structure $(1, \cdot)$, then the functor $M \times (-)\colon \mathbf{Sets} \to \mathbf{Sets}$ is a monad. The multiplication map $\eta\colon X \to M \times X$ is $\eta(x) = (1, x)$ and multiplication $\mu\colon M \times (M \times X) \to M \times X$ is simply $\mu(m_1, (m_2, x)) = (m_1 \cdot m_2, x)$. The monad laws follow directly from the monoid laws.

   There are close connections between monoids and monads; see for instance the adjunction in Exercise 5.2.17. On a more abstract level one can describe monads $\mathbb{C}$ as 'monoids in the category of endofunctors' on $\mathbb{C}$; see [344, VII.3] for more information.

6. The above first few examples of monads all involve collections (subsets, lists, distributions) of some sort. The remaining examples are also relevant in computing but are of a different kind. We start with the state monad. It involves a fixed set S, elements of which are seen as states that are passed around in a computation. The state monad $\mathcal{S}\colon \mathbf{Sets} \to \mathbf{Sets}$ is given by

$$\mathcal{S}(X) = (S \times X)^S.$$

   It comes equipped with unit and multiplication operations:

$$\begin{array}{ccc} X \xrightarrow{\;\eta\;} (S \times X)^S & \qquad & \big(S \times (S \times X)^S\big)^S \xrightarrow{\quad\mu\quad} (S \times X)^S \\ x \longmapsto \lambda s \in S.\, \langle s, x\rangle & & \langle h_1, h_2\rangle \longmapsto \lambda s \in S.\, h_2(s)(h_1(s)). \end{array}$$

7. Our next example, the continuation monad, is also motivated by programming semantics. It starts from a fixed set C and takes the 'double dual' of a set, where C is used as dualising object (see [300] for a more general setting). This is the pure form of the monad, which we shall describe first; some variations will be discussed subsequently. To start, we define a functor $\mathcal{C}\colon \mathbf{Sets} \to \mathbf{Sets}$ by

$$\mathcal{C}(X) = C^{(C^X)} \quad \text{and} \quad \mathcal{C}(X \xrightarrow{f} Y) = \lambda h \in C^{(C^X)}.\, \lambda g \in C^Y.\, h(g \circ f).$$

   Earlier, in Exercise 2.2.7, we have seen the neighbourhood functor as special case $\mathcal{N}(X) = 2^{(2^X)}$ for $C = 2$.

   This functor $\mathcal{C}$ forms a monad via

$$\begin{array}{ccc} X \xrightarrow{\;\eta\;} C^{(C^X)} & \qquad & C^{\left(C^{\left(C^{(C^X)}\right)}\right)} \xrightarrow{\quad\mu\quad} C^{(C^X)} \\ x \longmapsto \lambda g \in C^X.\, g(x) & & H \longmapsto \lambda g \in C^X.\, H(\lambda k \in C^{(C^X)}.\, k(g)). \end{array}$$

It requires a bit of elementary bookkeeping to check that these η and μ are natural and satisfy the three monad equations.

Coalgebras $X \to C^{(C^X)}$ of the continuation monad capture an indirect way of computing a result in C. The computation involves an explicit argument function $X \to C$ that is usually called a continuation. This may be useful to get a better handle on intermediate values and argument evaluation; see e.g. [460] for more information.

The monad C describes what may be called the 'pure' version of the continuation monad. There are many variations which restrict the kind of functions involved (and form submonads). This will be illustrated via a several examples.

- For an arbitrary set X an ultrafilter on X is a subset $\mathcal{F} \subseteq \mathcal{P}(X)$ which is a filter ($\mathcal{F}$ is closed under finite intersections and is upclosed) satisfying $\emptyset \notin \mathcal{F}$ and for each $U \in \mathcal{P}(X)$ either $U \in \mathcal{F}$ or $\neg U \in \mathcal{F}$. Such an ultrafilter can be identified with a morphism of Boolean algebras $\mathcal{P}(X) \to \{0,1\}$. The set of ultrafilters on X may thus be defined as a hom-set in the category **BA** of Boolean algebras:

$$\begin{aligned} \mathcal{UF}(X) &= \mathbf{BA}(\mathcal{P}(X), \{0,1\}) \\ &= \mathbf{BA}(\{0,1\}^X, \{0,1\}) \\ &= \{f \in \{0,1\}^{(\{0,1\}^X)} \mid f \text{ is a morphism of Boolean algebras}\}. \end{aligned}$$

  This is a subset of $C(X) = C^{(C^X)}$, with set $C = \{0,1\}$. In this case $\mathcal{UF}$ is still a monad (see e.g. [280] for more information).

- Next we take the unit interval $[0,1] \subseteq \mathbb{R}$ as constant C. It is a complete lattice and hence certainly a dcpo. Also, for each set X, the function space $[0,1]^X$ is a dcpo. In [292] the following monad is defined:

$$\mathcal{G}(X) = \{f \in [0,1]^{([0,1]^X)} \mid f \text{ is continuous and sublinear}\}.$$

  Sublinearity means that $f(r \cdot g) = r \cdot f(g)$ and $f(g_1 + g_2) \leq f(g_1) + f(g_2)$, if $g_1(x) + g_2(x) \leq 1$ for all $x \in X$. This monad $\mathcal{G}$ is used for a semantics and logic of a 'probabilistic-nondeterministic' programming language in [292].

- A variation on the previous point is the expectation monad given by:

$$\mathcal{E}(X) = \{f \in [0,1]^{([0,1]^X)} \mid f \text{ is a map of effect modules}\}.$$

  These (effect module map) requirements are slightly different and amount to $f(r \cdot g) = r \cdot f(g)$, $f(\lambda x.\, 1) = 1$ and $f(g_1 + g_2) = f(g_1) + f(g_2)$, if $g_1(x) + g_2(x) \leq 1$ for all $x \in X$. More information can be found in [257, 254, 259, 258].

8. In [264] a monad $\mathcal{J}$: **Sets** $\to$ **Sets** is introduced for the semantics of Java programs that combines several of the monads mentioned earlier. We recall that statements (and expressions) in Java:
   - can fail to terminate, typically because of an infinite ('for' or 'while') loop; this is modelled via the lift monad $\mathcal{L} = 1 + (-)$
   - can terminate normally, producing a successor state (for statements), and a successor state together with a result (for an expression) or
   - can terminate exceptionally, yielding an exception, say of type E, together with a successor state.

   When we put these ingredients together we obtain a monad:

   $$\mathcal{J}(X) = \Big(1 + (S \times X) + (S \times E)\Big)^S.$$

   The unit $\eta\colon X \to \mathcal{J}(X)$ is given by $\eta(x) = \lambda s \in S.\, \kappa_2\langle s, x\rangle$, as for the state monad. The second coprojection κ_2, used for the middle option, corresponds to normal/successful termination. The multiplication operation $\mu\colon \mathcal{J}^2(X) \to \mathcal{J}(X)$ is

   $$\mu(h)(s) = \begin{cases} \kappa_1 * & \text{if } h(s) = \kappa_1 * \\ k(s') & \text{if } h(s) = \kappa_2(s', k) \text{ where } k \in \mathcal{J}(X) \\ \kappa_3(s, e) & \text{if } h(s) = \kappa_3(s, e). \end{cases}$$

   Thus, in the first case (non-termination) and the third case (abrupt termination) the outcome of h is returned, and no subsequent evaluation takes place. This is what happens in Java. But exceptions in Java have more substructure, which is ignored here. A more complete formalisation occurs in [240].

   Producing new monads from existing ones, as we have done here, may be described more systematically in terms of so-called monad transformers; see [336, 65, 233].
9. Let I, O be two arbitrary but fixed sets, used as sets of 'inputs' and 'outputs'. Consider for each set X the initial algebra $F^*(X)$ of the functor

   $$Y \longmapsto X + F(Y), \qquad \text{where} \qquad F(Y) = Y^I + (O \times Y),$$

   with initial algebra map

   $$X + F^*(X)^I + (O \times F^*(X)) \xrightarrow{\quad\cong\quad} F^*(X).$$

   The mapping $X \mapsto F^*(X)$ is then a monad. Later on, with Proposition 5.1.8, we shall recognise this as an instance of a more general construction of free monads on a functor. The above special case is called the I/O monad; see [357, 65]. See also [171] for similar store/stack/tape monads.

10. In point (5.1.3) we have seen the state monad $X \mapsto (S \times X)^S$, for a fixed set of states. There are associated predicate transformer monads on **Sets** (see [253]), which come in two flavours, namely after Dijkstra [439] and after Hoare [364, 440]. The Dijkstra monad $\mathfrak{D}$ on **Sets** is given by

$$\mathfrak{D}(X) = \mathcal{P}(S)^{\mathcal{P}(S\times X)} \quad \text{and} \quad \mathfrak{D}(X \xrightarrow{f} Y) = \lambda h.\, \lambda Q.\, h((\mathrm{id}_S \times f)^{-1}(Q)).$$

Such functions $\mathcal{P}(S \times X) \to \mathcal{P}(S)$ in $\mathfrak{D}(X)$ are predicate transformers that send a postcondition predicate $Q \subseteq S \times X$ on the state S and the output type X, to a precondition $P \subseteq S$. The associated unit and multiplication are

$$\begin{array}{ccc} X \xrightarrow{\;\eta\;} \mathcal{P}(S)^{\mathcal{P}(S\times X)} & \qquad & \mathcal{P}(S)^{\mathcal{P}(S\times \mathcal{P}(S)^{\mathcal{P}(S\times X)})} \xrightarrow{\quad\mu\quad} \mathcal{P}(S)^{\mathcal{P}(S\times X)} \\ x \mapsto \lambda P.\, \{s \mid \langle s, x\rangle \in P\} & & \Phi \longmapsto \lambda P.\, \Phi(\{\langle s, h\rangle \mid s \in h(P)\}). \end{array}$$

Notice that this unit $\eta(x)\colon \mathcal{P}(S \times X) \to \mathcal{P}(S)$ is the inverse image $(\eta^S(x))^{-1}$ of the unit of the state monad $\mathcal{S}$ from point (5.1.3). In fact, there is a 'map of monads' from state to Dijkstra; see Exercise 5.1.10.

The Hoare monad $\mathcal{H}$ on **Sets** involves dependent sums and products:

$$\mathcal{H}(X) = \coprod_{Pre\subseteq S} \; \coprod_{Post\subseteq S\times X\times S} \; \prod_{s\in Pre} (s \times \mathrm{id} \times \mathrm{id})^{-1}(Post).$$

An element of $\mathcal{H}(X)$ is thus a 'Hoare triple' given by (1) a precondition predicate $Pre \subseteq S$ on the state space S; (2) a postcondition relation $Post \subseteq S \times X \times S$ on an initial state, an output element in X and a final state; (3) a dependent function h which sends a state s satisfying the precondition Pre to a pair of elements $h(s) = (x, s') \in X \times S$ such that $Post(s, x, s')$ – i.e. such that $h(s) \in (s \times \mathrm{id} \times \mathrm{id})^{-1}(Post)$.

For a function $f\colon X \to Y$ we obtain $\mathcal{H}(f)\colon \mathcal{H}(X) \to \mathcal{H}(Y)$ given by

$$\mathcal{H}(f)(P, Q, h) = \langle P, \{(s, f(x), s') \mid (s, x, s') \in Q\}, (f \times \mathrm{id}) \circ h\rangle.$$

It is not hard to see that this triple is well defined.

The unit $\eta\colon X \to \mathcal{H}(X)$ of this Hoare monad is given by

$$\eta(x) = \langle \top, \{(s, x, s) \mid s \in S\}, \lambda s \in S.\, \langle x, s\rangle\rangle.$$

Its precondition is the truth predicate $\top \subseteq S$; its postcondition relates the input value x with all identical initial and final states; and its operation is essentially the unit $\eta^S(x)$ of the state monad from point (5.1.3).

The multiplication operation $\mu\colon \mathcal{H}^2(X) \to \mathcal{H}(X)$ is basically the same as multiplication for the state monad, except that the relevant pre- and post-conditions have to be take into a account. Thus μ acts on a triple $\langle P, Q, h\rangle \in \mathcal{H}^2(X)$, where $P \subseteq S$, $Q \subseteq S \times \mathcal{H}(X) \times S$ and

$h(s) \in (s \times \mathrm{id} \times \mathrm{id})^{-1}(Q)$, for each $s \in P$. We will produce a triple $\mu(P, Q, h) \in \mathcal{H}(X)$, where

$$\begin{aligned}\mu(P, Q, h) = \langle\, & \{s \mid P(s) \wedge \forall g, s'.\, Q(s, g, s') \Rightarrow (\pi_1 g)(s')\}, \\ & \{(s, x, s'') \mid \exists g, s'.\, Q(s, g, s') \wedge (\pi_2 g)(s', x, s'')\}, \\ & \lambda s.\, \mathsf{let}\ (g, s') = h(s)\ \mathsf{in}\ (\pi_3 g)(s') \,\rangle.\end{aligned}$$

To see that μ is well defined, write $\mu(P, Q, h) = (P', Q', h')$ and assume $s \in P'$. Then $s \in P$ and $Q(s, g, s') \Rightarrow (\pi_1 g)(s')$. The first conjunct yields $h(s) \in \{(g, s') \mid (s, g, s') \in Q\}$. So write $h(s) = (g, s')$, where $(s, g, s') \in Q$. Let's write $g \in \mathcal{H}(X)$ as $g = (U, V, k)$. Then $U(s') = (\pi_1 g)(s')$ holds by assumption. Hence $k(s') \in \{(x, s'') \mid (s', x, s'') \in V\}$. Now we write $k(s') = (x, s'')$, where $V(s', x, s'') = (\pi_2 g)(s', x, s'')$. Hence we are done.

We leave the verifications that $\mathcal{H}$ is a monad to the interested reader. Programs represented as maps $X \to \mathcal{H}(Y)$ automatically incorporate appropriate pre- and post-conditions via the dependent types involved. This is the essence of the programming language Ynot; see [364].

One may expect that the multiset functor $\mathcal{M}_M$ from Definition 4.1.1 also forms a monad (like distribution above). This is indeed the case if one assumes that the multiplicities in M used for counting elements not only form an (additive) commutative monoid, but also carry a multiplication operation. In short, multiplicities must form a semiring, that is, a 'ring without additive inverses', sometimes also called a 'rig'. For convenience, we define it explicitly.

Definition 5.1.4 A **semiring** is a set S carrying:

- a commutative monoid structure, written additively as $(0, +)$
- another monoid structure, written multiplicatively as $(1, \cdot)$

in such a way that multiplication distributes over addition:

$$\begin{aligned} 0 \cdot z &= 0 & \qquad z \cdot 0 &= 0 \\ (x + y) \cdot z &= x \cdot z + y \cdot y & \qquad z \cdot (x + y) &= z \cdot x + z \cdot y. \end{aligned}$$

The semiring is called commutative if its multiplication operation $\cdot$ is commutative.

A morphism of semirings $f \colon S \to R$ is a function between the underlying sets which is both additively and a multiplicatively a homomorphism of monoids. We write **SRng** for the category of semirings and their homomorphisms. The full subcategory of commutative semirings is then written as **CSRng** $\hookrightarrow$ **SRng**.

Recall that a ring is a semiring with additive inverses, so the additive monoid involved is an (Abelian) group. The natural numbers $\mathbb{N}$ form a semiring, but not a ring. Similarly, the non-negative natural numbers $\mathbb{R}_{\geq 0} = \{r \in \mathbb{R} \mid r \geq 0\}$ form a semiring but not a ring.

Lemma 5.1.5 *If S is a semiring, then the multiset functor $\mathcal{M}_S : \mathbf{Sets} \to \mathbf{Sets}$, taking multiplicities in S, is a monad, with unit and multiplication as for the distribution monad:*

$$\eta(x) = 1|x\rangle \qquad \textit{and} \qquad \mu(\textstyle\sum_i s_i|\varphi_i\rangle)(x) = \sum_i s_i \cdot \varphi_i(x),$$

where $\varphi_i \in \mathcal{M}_S(X) = \{\psi \colon X \to S \mid \mathrm{supp}(\psi)$ is finite$\}$. ❑

The proof involves some routine calculations and is left to the reader. Basically, the multiplication μ is given by matrix multiplication; see Exercise 5.1.3 below.

We turn to some more categorical aspects of monads. The following basic result (see e.g. [344, VI 1]) shows that adjunctions form a source of monads.

Lemma 5.1.6 *For each adjunction $F \dashv G$ the endofunctor GF is a monad, as depicted in*

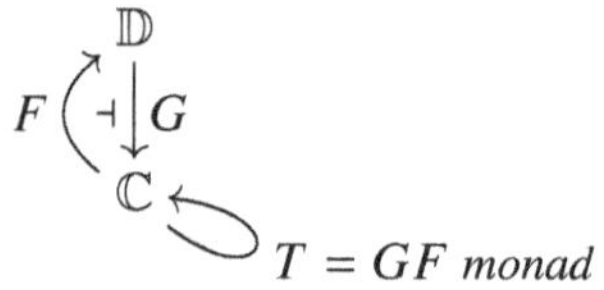

Proof Write the unit and counit of the adjunction $F \dashv G$ as $\eta\colon \mathrm{id}_{\mathbb{C}} \Rightarrow GF$ and $\varepsilon\colon FG \Rightarrow \mathrm{id}_{\mathbb{D}}$. They satisfy the triangular identities $G(\varepsilon_Y) \circ \eta_{GY} = \mathrm{id}_{GY}$ and $\varepsilon_{FX} \circ F(\eta_X) = \mathrm{id}_{FX}$ described in Exercise 2.5.7. We write $T = GF\colon \mathbb{C} \to \mathbb{C}$. The unit η of the adjunction forms the unit $\eta\colon \mathrm{id}_{\mathbb{C}} \Rightarrow T$ for the monad T. The multiplication $\mu\colon T^2 \Rightarrow T$ is defined with component at $X \in \mathbb{C}$:

$$TT(X) = GFGF(X) \xrightarrow{\;\mu_X \stackrel{\text{def}}{=} G(\varepsilon_{FX})\;} GF(X) = T(X). \tag{5.1}$$

It is easy to see that this is a natural transformation satisfying the monad requirements from Definition 5.1.1. For instance,

$$\begin{aligned} \mu_X \circ T(\eta_X) &= G(\varepsilon_{FX}) \circ GF(\eta_X) \\ &= G(\varepsilon_{FX} \circ F(\eta_X)) \\ &= G(\mathrm{id}_{FX}) \\ &= \mathrm{id}_{TX}. \end{aligned}$$

❑

Of course, in category theory one does not introduce a notion without saying what the associated morphism is.

Definition 5.1.7 Let $T, S : \mathbb{C} \to \mathbb{C}$ be two monads on the same category. A **map of monads** $\sigma\colon T \Rightarrow S$ is a natural transformation that commutes with the respective units and multiplications, as in

$$\begin{array}{ccc} X & = & X \\ {\scriptstyle \eta_X}\downarrow & & \downarrow{\scriptstyle \eta_X} \\ T(X) & \xrightarrow[\sigma_X]{} & S(X) \end{array} \qquad \begin{array}{ccccc} T^2(X) & \xrightarrow{\sigma_{TX}} & S(T(X)) & \xrightarrow{S(\sigma_X)} & S^2(X) \\ {\scriptstyle \mu_X}\downarrow & & & & \downarrow{\scriptstyle \mu_X} \\ T(X) & & \xrightarrow[\sigma_X]{} & & S(X) \end{array}$$

In this way one obtains a category **Mnd**($\mathbb{C}$) of monads on $\mathbb{C}$ with maps between them.

Maps of monads arise naturally; see for instance Exercises 5.1.7 and 5.1.8 describing the functoriality of some monad constructions from Example 5.1.3. Also, the earlier examples (4.5) of natural transformations between collection functors are all maps of monads.

We also use maps of monads in the following result. It describes how every endofunctor can be turned into a monad, in a free manner, assuming that certain initial algebras exist.

Proposition 5.1.8 *Let $F\colon \mathbb{C} \to \mathbb{C}$ be an endofunctor on a category $\mathbb{C}$ with coproducts $+$. Assume for each object X, the functor $X + F(-)\colon \mathbb{C} \to \mathbb{C}$ has an initial algebra, written as*

$$X + F(F^*(X)) \xrightarrow[\cong]{\alpha_X} F^*(X).$$

The mapping $X \mapsto F^*(X)$ forms a monad, that is the free one on the functor F, via a universal natural transformation $\theta\colon F \Rightarrow F^*$.

Proof We first show that $X \mapsto F^*(X)$ is functorial. For a map $f\colon X \to Y$ in $\mathbb{C}$ write $F^*(f)\colon F^*(X) \to F^*(Y)$ for the unique algebra homomorphism obtained by initiality in

$$\begin{array}{ccc} X + F(F^*(X)) & \xrightarrow{\mathrm{id} + F(F^*(f))} & X + F(F^*(Y)) \\ & & \downarrow{\scriptstyle f + \mathrm{id}} \\ {\scriptstyle \alpha_X}\downarrow{\scriptstyle \cong} & & Y + F(F^*(Y)) \\ & & {\scriptstyle \cong}\downarrow{\scriptstyle \alpha_Y} \\ F^*(X) & \xrightarrow[F^*(f)]{} & F^*(Y) \end{array} \tag{5.2}$$

It is easy to see that F^* preserves identities and composition. We get a unit natural transformation $\eta\colon \mathrm{id} \Rightarrow F^*$ with components

$$\eta_X \stackrel{\text{def}}{=} \Big(X \xrightarrow{\;\kappa_1\;} X + F(F^*(X)) \xrightarrow[\cong]{\;\alpha_X\;} F^*(X)\Big).$$

It is natural since for $f\colon X \to Y$,

$$\begin{aligned} F^*(f) \circ \eta_X &= \alpha_Y \circ (f + \mathrm{id}) \circ (\mathrm{id} + F(F^*(f))) \circ \alpha_X^{-1} \circ \alpha_X \circ \kappa_1 \\ &= \alpha_Y \circ \kappa_1 \circ f \;=\; \eta_Y \circ f. \end{aligned}$$

The multiplication map $\mu\colon F^*F^* \Rightarrow F^*$ arises as a unique map in

$$\begin{array}{ccc} F^*(X) + F(F^*F^*(X)) & \xrightarrow{\;\mathrm{id} + F(\mu_X)\;} & F^*(X) + F(F^*(X)) \\ {\scriptstyle \alpha_{F^*(X)}}\big\downarrow{\scriptstyle \cong} & & \big\downarrow{\scriptstyle [\mathrm{id},\, \alpha_X \circ \kappa_2]} \\ F^*F^*(X) & \xrightarrow[\;\mu_X\;]{} & F^*(X) \end{array}$$

The first of the monad equations is easy:

$$\begin{aligned} \mu_X \circ \eta_{F^*(X)} &= \mu_X \circ \alpha_{F^*(X)} \circ \kappa_1 \\ &= [\mathrm{id}, \alpha_X \circ \kappa_2] \circ (\mathrm{id} + F(\mu_X)) \circ \kappa_1 \\ &= [\mathrm{id}, \alpha_X \circ \kappa_2] \circ \kappa_1 \\ &= \mathrm{id}. \end{aligned}$$

The other two equations can be obtained via (the uniqueness part of) initiality – which we leave to the reader. We continue with the universal property. There is a natural transformation $\theta\colon F \Rightarrow F^*$ with components

$$\theta_X \stackrel{\text{def}}{=} \Big(F(X) \xrightarrow{\;F(\eta_X)\;} F(F^*(X)) \xrightarrow{\;\kappa_2\;} X + F(F^*(X)) \xrightarrow[\cong]{\;\alpha_X\;} F^*(X)\Big).$$

If $T = (T, \eta^T, \mu^T)$ is an arbitrary monad with a natural transformation $\sigma\colon F \Rightarrow T$, then there is a unique map of monads $\overline{\sigma}\colon F^* \Rightarrow T$ with $\overline{\sigma} \circ \theta = \sigma$. This $\overline{\sigma}$ is obtained by initiality in

$$\begin{array}{ccc} X + F(F^*(X)) & \xrightarrow{\;\mathrm{id} + F(\overline{\sigma}_X)\;} & X + F(T(X)) \\ {\scriptstyle \alpha_X}\big\downarrow{\scriptstyle \cong} & & \big\downarrow{\scriptstyle [\eta^T_X,\, \mu^T_X \circ \sigma_{T(X)}]} \\ F^*(X) & \xrightarrow[\;\overline{\sigma}_X\;]{} & T(X) \end{array}$$

Then indeed

$$\begin{aligned}
\overline{\sigma}_X \circ \theta_X &= \overline{\sigma}_X \circ \alpha_X \circ \kappa_2 \circ F(\eta_X) \\
&= [\eta^T_X, \mu^T_X \circ \sigma_{T(X)}] \circ (\mathrm{id} + F(\overline{\sigma}_X)) \circ \kappa_2 \circ F(\eta_X) \\
&= \mu^T_X \circ \sigma_{T(X)} \circ F(\overline{\sigma}_X)) \circ F(\eta_X) \\
&= \mu^T_X \circ T(\overline{\sigma}_X) \circ T(\eta_X) \circ \sigma_X \\
&= \mu^T_X \circ T(\overline{\sigma}_X \circ \alpha_X \circ \kappa_1) \circ \sigma_X \\
&= \mu^T_X \circ T([\eta^T_X, \mu^T_X \circ \sigma_{T(X)}] \circ (\mathrm{id} + F(\overline{\sigma}_X)) \circ \kappa_1) \circ \sigma_X \\
&= \mu^T_X \circ T(\eta^T_X) \circ \sigma_X \\
&= \sigma_X.
\end{aligned}$$

Uniqueness of $\overline{\sigma}$ is left to the interested reader. □

Free monads will be used to give a systematic description of terms in algebraic specification; see Sections 6.6 and 6.7. There we shall see further properties of such free monads (on **Sets**), such as: F^* is finitary (or weak pullback preserving) if F is; see Lemma 6.6.4 and Exercise 6.7.4.

5.1.1 Comonads

A comonad is the dual of a monad, in the sense that a comonad on category $\mathbb{C}$ is a monad on the opposite category $\mathbb{C}^{\mathrm{op}}$. Basically, we can now use what we have already learned about monads. However, it is convenient to make a few things explicit.

Definition 5.1.9 A **comonad** on a category $\mathbb{C}$ is given by a functor $S : \mathbb{C} \to \mathbb{C}$ together with two natural transformations, namely a **counit** $\varepsilon\colon S \Rightarrow \mathrm{id}$ and a **comultiplication** $\delta\colon S \Rightarrow S^2$ making for each object $X \in \mathbb{C}$ the following diagrams commute:

$$\begin{array}{ccccc}
 & & S(X) & & \\
 & \swarrow\!\!= & \downarrow{\scriptstyle \delta_X} & =\!\!\searrow & \\
S(X) & \xleftarrow[S(\varepsilon_X)]{} & S^2(X) & \xrightarrow[\varepsilon_{S(X)}]{} & S(X)
\end{array}
\qquad
\begin{array}{ccc}
S(X) & \xrightarrow{\delta_X} & S^2(X) \\
{\scriptstyle \delta_X}\downarrow & & \downarrow{\scriptstyle S(\delta_X)} \\
S^2(X) & \xrightarrow[\delta_{S(X)}]{} & S^3(X)
\end{array}$$

A **map of comonads** is a natural transformation $\sigma\colon S \Rightarrow S'$ that commutes appropriately with the counits ε and comultiplications δ.

In Lemma 5.1.6 we have seen that an adjunction $F \dashv G$ gives rise to a monad GF. It is not hard to see that the (other) composite FG is a comonad. Still, comonads are not so common as monads. We conclude by listing some examples. They can all be understood as providing structure to the context of computations (as in [454]). Further examples can be obtained as cofree comonads on a functor, via final coalgebras (see Exercise 5.1.15). In Section 6.8 we will see how to obtain comonads from coalgebraic specifications.

Example 5.1.10 Probably the clearest example of a comonad is given by streams, i.e. by the functor $X \mapsto X^{\mathbb{N}}$ sending a set X to the set $X^{\mathbb{N}}$ of infinite sequences of its elements:

$$\begin{array}{ccccc} X & \xleftarrow{\quad\varepsilon\quad} & X^{\mathbb{N}} & \xrightarrow{\quad\delta\quad} & (X^{\mathbb{N}})^{\mathbb{N}} \\ \alpha(0) & \longleftarrow\!\shortmid & \alpha & \longmapsto & \lambda n.\,\lambda m.\,\alpha(n+m). \end{array}$$

The counit ε thus selects the head of the stream. The comultiplication δ maps a stream to a stream of streams, where the ith stream of $\delta(\alpha)$ is the substream $\alpha(i), \alpha(i+1), \alpha(i+2), \ldots$ of α starting at i. It is not hard to see that this forms a comonad.

One way of understanding this stream comonad $X^{\mathbb{N}}$ is as providing infinitely many copies of X as input for a computation. This is also the idea of the exponential ! in linear logic [155] – which is standardly modelled via a comonad; see e.g. [424, 64]. There are some variations on the stream comonad, with additional structure keeping track of a position in the sequence. We shall describe them in concrete form, for both discrete time and real time inputs. Exercise 5.1.11 describes some of these comonads in more abstract form.

The diagram below presents the discrete time comonads, together with comonad maps between them:

$$\begin{array}{ccccc} X^{\star} \times X & \xleftarrow{\quad\text{no future}\quad} & X^{\mathbb{N}} \times \mathbb{N} & \xrightarrow{\quad\text{no past}\quad} & X^{\mathbb{N}} \\ (\langle \alpha(0), \ldots, \alpha(n-1) \rangle, \alpha(n)) & \longleftarrow\!\shortmid & (\alpha, n) & \longmapsto & \lambda m.\,\alpha(n+m). \end{array} \tag{5.3}$$

The comonad structure for $X^{\mathbb{N}} \times \mathbb{N}$ in the middle is

$$\begin{array}{ccccc} X & \xleftarrow{\quad\varepsilon\quad} & X^{\mathbb{N}} \times \mathbb{N} & \xrightarrow{\quad\delta\quad} & (X^{\mathbb{N}} \times \mathbb{N})^{\mathbb{N}} \times \mathbb{N} \\ \alpha(n) & \longleftarrow\!\shortmid & (\alpha, n) & \longmapsto & (\lambda m.\,(\alpha, m), n). \end{array}$$

The intuition for a pair $(\alpha, n) \in X^{\mathbb{N}} \times \mathbb{N}$ is that the number n represents the present stage in the stream $\alpha = \langle \alpha(0), \alpha(1), \ldots, \alpha(n-1), \alpha(n), \alpha(n+1), \ldots \rangle$, where everything before n is past input, and everything after n is future input.

The comonad structure for $X^{\star} \times X$ on the left in (5.3) is

$$\begin{array}{ccccc} X & \xleftarrow{\ \varepsilon\ } & X^{\star} \times X & \xrightarrow{\quad\delta\quad} & (X^{\star} \times X)^{\star} \times X \\ x & \longleftarrow\!\shortmid & (\alpha, x) & \longmapsto & (\langle (\langle\rangle, x_1), \ldots, (\langle x_1, \ldots, x_{n-1} \rangle, x_n) \rangle,\ (\alpha, x)), \end{array}$$

where $\alpha = \langle x_1, \ldots, x_n \rangle$. The comultiplication δ thus turns a pair (α, x) into another pair $(\beta, (\alpha, x))$, where β is a list of pairs $(\alpha_{<i}, x_i)$, in which x_i is the ith element in α and $\alpha_{<i}$ is the sublist of α of elements up to i.

The two arrows in the diagram (5.3) are homomorphisms of comonads, commuting with the relevant comonad/context structure.

The real-time analogue uses the (semiring of) non-negative real numbers $\mathbb{R}_{\geq 0} = \{r \in \mathbb{R} \mid r \geq 0\}$ to represent time. One obtains a diagram like (5.3):

$$\Big(\coprod_{t \in \mathbb{R}_{\geq 0}} X^{[0,t)} \Big) \times X \longleftarrow X^{\mathbb{R}_{\geq 0}} \times \mathbb{R}_{\geq 0} \longrightarrow X^{\mathbb{R}_{\geq 0}}. \qquad (5.4)$$

The leftmost comonad may also be described as: $\coprod_{t \in \mathbb{R}_{\geq 0}} X^{[0,t]}$.

The notion of 'arrow' developed in [229] forms a generalisation of both monads and comonads; see also [394, 255, 337].

Exercises

5.1.1 Let $\mathbb{C}$ be a category with coproducts $+$. Fix an arbitrary object $A \in \mathbb{C}$ and prove that the functor $X \mapsto A + X$ forms a monad on $\mathbb{C}$.

5.1.2 Describe sequential composition $\circledcirc$ for Java programs, modelled as coalgebras $X \to \mathcal{J}(X)$, using the Java monad $\mathcal{J}$ described in Example 5.1.3.8.

5.1.3 Let S be a semiring and X a finite set, say $X = \{x_1, \ldots, x_n\}$.

1. Check that coalgebras $c \colon X \to \mathcal{M}_S(X)$ can be identified with $n \times n$ matrices, with entries in S; write M_c for the matrix corresponding to c.
2. Prove that $M_{d \circledcirc c} = M_c M_d$, where the left-hand side uses sequential composition $\circledcirc$ for coalgebras (from Lemma 5.1.2) and the right-hand side involves matrix composition.
3. Check that the same works for finite Markov chains, as coalgebras $X \to \mathcal{D}(X)$, where X is finite.

5.1.4 Fix a set C and consider the contravariant functor $X \mapsto C^X$. Prove that there is an adjunction

$$\mathbf{Sets}^{\mathrm{op}} \underset{C^{(-)}}{\overset{C^{(-)}}{\rightleftarrows}} \mathbf{Sets} \quad (\top)$$

Check that the monad induced by this adjunction on **Sets**, as in Lemma 5.1.6, is the continuation monad from Example 5.1.3.7.

5.1.5 Fix a set C and consider the continuation monad $C(X) = C^{(C^X)}$ from Example 5.1.3.7. Assume the set C carries an operation $f: C^n \to C$, for some $n \in \mathbb{N}$. Prove that it gives rise to a natural transformation:

$$(C(X))^n \Longrightarrow C(X).$$

5.1.6 Recall from Section 2.2.6 that Turing machines can be modelled as coalgebras $X \to T(n \cdot X)^n$, where n is the number of (control) states.

1. ([249]) Show that the mapping $X \mapsto T(n \cdot X)^n$ is a monad if T is a monad.
2. Describe the two-step composite $c \circledcirc c$ for the Turing-machine-as-coalgebra example monad c described in (2.27) and (2.28).

5.1.7 Show that sending a monoid M to the monad $M \times (-)$, as in Example 5.1.3.5, yields a functor **Mon** $\to$ **Mnd**(**Sets**). (This functor has a right adjoint; see Exercise 5.2.17.)

5.1.8 Prove that mapping a semiring S to the corresponding multiset monad $\mathcal{M}_S$ yields a functor **SRng** $\to$ **Mnd**(**Sets**).

5.1.9 Prove that the support natural transformation $\mathcal{D} \Rightarrow \mathcal{P}$ from (4.5) is a map of monads.

5.1.10 Prove that there is a map of monads $\mathcal{S} \Rightarrow \mathfrak{D}$, from the state monad $\mathcal{S}$ to the Dijkstra monad $\mathfrak{D}$ (see Examples 5.1.3.6 and 5.1.3.10), given by inverse image

$$\mathcal{S}(X) = (S \times X)^S \xrightarrow{\ h \,\mapsto\, h^{-1}\ } \mathcal{P}(S)^{\mathcal{P}(S \times X)} = \mathfrak{D}(X).$$

It sends a program to the corresponding weakest precondition operation.

5.1.11 Show that for each monoid M and set E the following functors are comonads on **Sets**:

1. $X \mapsto X^M$
2. $X \mapsto X^E \times E$.

Show also that there is a map of comonads $(-)^M \times M \Rightarrow (-)^M$, as used twice in Example 5.1.10.

5.1.12 Prove that the arrows $X^{\mathbb{N}} \times \mathbb{N} \to X^\star \times X$ in Diagram (5.3) form a map of comonads.

5.1.13 Define the comonad structure on $(\coprod_{t \in \mathbb{R}_{\geq 0}} (-)^{[0,t)}) \times X$ and the comonad map $(-)^{\mathbb{R}_{\geq 0}} \times \mathbb{R}_{\geq 0} \Rightarrow (\coprod_{t \in \mathbb{R}_{\geq 0}} (-)^{[0,t)}) \times X$ in Diagram (5.4).

5.1.14 Let T be a monad on a category $\mathbb{C}$, and let $\mathbb{A}, \mathbb{B}$ be arbitrary categories. Prove that pre- and post-composition with T yields monads $(-) \circ T \colon \mathbb{A}^{\mathbb{C}} \to \mathbb{A}^{\mathbb{C}}$ and $T \circ (-) \colon \mathbb{C}^{\mathbb{B}} \to \mathbb{C}^{\mathbb{B}}$ on categories of functors $\mathbb{C} \to \mathbb{A}$ and $\mathbb{B} \to \mathbb{C}$.

5.1.15 Prove the dual of Proposition 5.1.8: the cofree comonad on a functor $F \colon \mathbb{C} \to \mathbb{C}$ can be constructed as the mapping $F^{\infty} \colon \mathbb{C} \to \mathbb{C}$ that sends $X \in \mathbb{C}$ to the carrier of the final coalgebra of the functor $X \times F(-)$, assuming this final coalgebra exists. Check that:

1. This comonad arises from the cofree coalgebra construction in Proposition 2.5.3.
2. The stream comonad $X \mapsto X^{\mathbb{N}}$ is the cofree comonad on the identity functor.

5.2 Kleisli Categories and Distributive Laws

This section shows how to associate with a monad T a category of 'computations of type T'. This category is called the Kleisli category of T and is written as $\mathcal{K}\ell(T)$. Composition of morphisms in Kleisli categories generalises composition $\circledcirc$ of coalgebras from Lemma 5.1.2. Dually, for comonads S there is a also a Kleisli category, for which we use the same notation: $\mathcal{K}\ell(S)$. In general, it will be clear from the context whether this is a Kleisli category of a monad or of a comonad.

The basics of the theory of Kleisli categories are described in this section. Included is the notion of distributive law that will be used to lift functors to Kleisli categories. Such liftings play a crucial role in the description of trace semantics for coalgebras in the next section.

Definition 5.2.1 Let T be a monad on a category $\mathbb{C}$. The **Kleisli** category $\mathcal{K}\ell(T)$ of T has the same objects as $\mathbb{C}$, but morphisms $X \to Y$ in $\mathcal{K}\ell(T)$ are maps $X \to T(Y)$ in $\mathbb{C}$.

The identity map $X \to X$ in $\mathcal{K}\ell(T)$ is the unit $\eta_X \colon X \to T(X)$. Composition of $f \colon X \to Y$ and $g \colon Y \to Z$ in $\mathcal{K}\ell(T)$, that is, of $f \colon X \to T(Y)$ and $g \colon Y \to T(Z)$ in $\mathbb{C}$, is written as $\circledcirc$ and defined as in Lemma 5.1.2:

$$g \circledcirc f \stackrel{\text{def}}{=} \Big(X \xrightarrow{\;f\;} T(Y) \xrightarrow{\;T(g)\;} T^2(Z) \xrightarrow{\;\mu_Z\;} T(Z)\Big).$$

Having a separate notation for composition in Kleisli categories may help to distinguish composition in $\mathbb{C}$ from composition in $\mathcal{K}\ell(T)$. Kleisli maps $X \to Y$ may be seen as generalised coalgebras $X \to T(Y)$, with different domain and codomain. The composition monoid described in Lemma 5.1.2 is the monoid of endomorphisms $X \to X$ in the Kleisli category $\mathcal{K}\ell(T)$.

The Kleisli construction captures many well-known categories. For instance, the Kleisli category $\mathcal{K}\ell(\mathcal{P})$ of the powerset monad is the category **SetsRel** of sets with relations between them – using the correspondence (2.16). The Kleisli category $\mathcal{K}\ell(\mathcal{L})$ of the lift monad is the category of sets and partial functions. The Kleisli category $\mathcal{K}\ell(\mathcal{D})$ of the distribution monad is the category of sets with Markov chains or 'stochastic relations' between them.

There are some general observations about Kleisli categories.

Proposition 5.2.2 *Let T be a monad on an arbitrary category $\mathbb{C}$.*

1. *There is an adjunction*

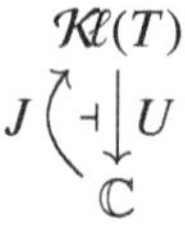

where $\begin{cases} U(Y) = T(Y) \\ U(f) = \mu \circ T(f) \end{cases}$ *and* $\begin{cases} J(X) = X \\ J(f) = \eta \circ f. \end{cases}$

2. *If the monad T comes as $T = HL$ from an adjunction $L \dashv H$, as in Lemma 5.1.6, then there is a 'comparison' functor K in*

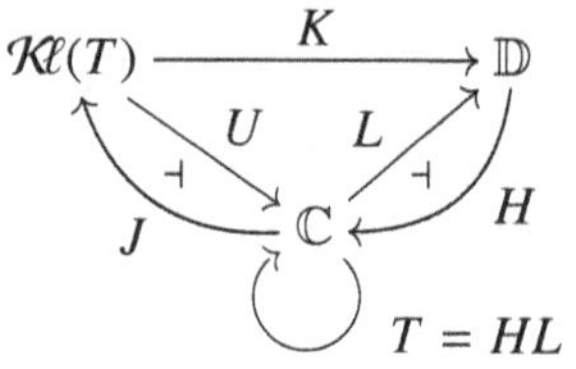

where $\begin{cases} K(Y) = L(Y) \\ K(f) = \varepsilon \circ L(f). \end{cases}$

This functor comparison functor K satisfies $K \circ J = L$ and $H \circ K = U$.

3. *The Kleisli category inherits coproducts and coequalisers from $\mathbb{C}$ – if any – and $J\colon \mathbb{C} \to \mathcal{K}\ell(T)$ preserves them.*

Proof 1. It is not hard to see that J and U as defined above are functors, using some care with respect to the order of compositions. For instance

$$\begin{aligned} U(g \circledcirc f) &= \mu \circ T(\mu \circ T(g) \circ f) \\ &= \mu \circ \mu \circ T^2(g) \circ T(f) && \text{via the } \mu\text{-law from Definition 5.1.1} \\ &= \mu \circ T(g) \circ \mu \circ T(f) && \text{by naturality of } \mu \\ &= U(g) \circ U(f) \\ J(g \circ f) &= \eta \circ g \circ f \\ &= T(g) \circ \eta \circ f \\ &= \mu \circ T(\eta \circ g) \circ \eta \circ f \\ &= (\eta \circ g) \circledcirc (\eta \circ f) \\ &= J(g) \circledcirc J(f). \end{aligned}$$

The adjunction $J \dashv U$ follows directly by unravelling the structure:

$$\frac{J(X) = X \xrightarrow{\;f\;} Y \qquad \text{in } \mathcal{K}\ell(T)}{X \xrightarrow[f]{} T(Y) = U(Y) \qquad \text{in } \mathbb{C}.}$$

2. Let $\eta\colon \mathrm{id} \Rightarrow HL = T$ and $\varepsilon\colon LH \Rightarrow \mathrm{id}$ be the unit and counit of the adjunction $L \dashv H$. For a morphism $f\colon X \to Y$ in $\mathcal{K}\ell(T)$, that is for $f\colon X \to HLY$ in $\mathbb{C}$, the transpose $K(f) = \varepsilon_{LY} \circ L(f)\colon LX \to LY$ yields a map in $\mathbb{D}$. This K preserves identities and composition, since

$$\begin{aligned} K(\mathrm{id}_X) &= K(\eta_X) \\ &= \varepsilon_{LX} \circ L(\eta_X) \\ &= \mathrm{id}_{LX} && \text{by the triangular identities} \\ K(g \circledcirc f) &= K(\mu \circ T(g) \circ f) \\ &= \varepsilon_{LZ} \circ L(H(\varepsilon_{LZ}) \circ HL(g) \circ f) && \text{by (5.1)} \\ &= \varepsilon_{LZ} \circ L(g) \circ \varepsilon_{LY} \circ L(f) && \text{by naturality} \\ &= K(g) \circ K(f). \end{aligned}$$

3. We shall do the (simple) case of binary coproducts $+$ in $\mathbb{C}$. They also form coproducts in $\mathcal{K}\ell(T)$ with coprojections $J(\kappa_i) = \eta \circ \kappa_i\colon X_i \to X_1 + X_2$. For two maps $f_i\colon X_i \to Y$ in $\mathcal{K}\ell(T)$, that is for $f_i\colon X_i \to T(Y)$ in $\mathbb{C}$, their cotuple $[f_1, f_2]\colon X_1 + X_2 \to T(Y)$ in $\mathbb{C}$ forms the cotuple in $\mathcal{K}\ell(T)$. Alternatively, one can use the bijective correspondences

$$\cfrac{\cfrac{X + Y \longrightarrow Z}{X + Y \longrightarrow T(Z)}}{\cfrac{X \longrightarrow T(Z)}{X \longrightarrow Z} \qquad \cfrac{Y \longrightarrow T(Z)}{Y \longrightarrow Z}}$$

where the arrows at the top and bottom are in $\mathcal{K}\ell(T)$. □

This Kleisli construction inspires an alternative formulation of the notion of monad that is popular in the (functional) programming community. It avoids functoriality and naturality and speaks only about a type constructor with operations satisfying some equations.

Proposition 5.2.3 *Let $\mathbb{C}$ be an arbitrary category with*

- *a mapping $\mathbb{C} \ni X \longmapsto T(X) \in \mathbb{C}$, on objects only*
- *a map $\eta_X\colon X \to T(X)$, for each $X \in \mathbb{C}$.*

The following are then equivalent:

1. *The mapping $X \mapsto T(X)$ is functorial, η is a natural transformation, and there is a natural transformation $\mu\colon T^2 \Rightarrow T$ making (T, η, μ) a monad.*
2. *There is a 'Kleisli extension' operation $(-)^\$$ sending*

$$X \xrightarrow{\;f\;} T(Y) \qquad \textit{to} \qquad T(X) \xrightarrow{\;f^\$\;} T(Y) \tag{5.5}$$

in such a way that the following equations hold:

$$f^\$ \circ \eta = f \qquad \eta^\$ = \mathrm{id} \qquad g^\$ \circ f^\$ = (g^\$ \circ f)^\$.$$

Proof First assume we have a monad (T, η, μ). Then define extension $(-)^\$$ as

$$f^\$ \stackrel{\text{def}}{=} \Big(T(X) \xrightarrow{\;T(f)\;} T^2(Y) \xrightarrow{\;\mu_Y\;} T(Y)\Big).$$

It is not hard to check that it satisfies the above three equations.

Conversely, assume a Kleisli extension operation $(-)^\$$ as in (5.5). We first turn the mapping $X \mapsto T(X)$ into a functor by defining for a map $f\colon X \to Y$ a new map $T(f) = (\eta_Y \circ f)^\$\colon T(X) \to T(Y)$. Then $T(\mathrm{id}) = \mathrm{id}$ obviously holds, and

$$\begin{aligned} T(g) \circ T(f) &= (\eta \circ g)^\$ \circ (\eta \circ f)^\$ \\ &= \big((\eta \circ g)^\$ \circ \eta \circ f\big)^\$ \\ &= (\eta \circ g \circ f)^\$ \\ &= T(g \circ f). \end{aligned}$$

This also makes the maps $\eta_X\colon X \to T(X)$ natural: for $f\colon X \to Y$,

$$T(f) \circ \eta_X = (\eta_Y \circ f)^\$ \circ \eta_X = \eta_Y \circ f.$$

The multiplication $\mu_X\colon T^2(X) \to T(X)$ can then be defined as $\mu_X = (\mathrm{id}_{TX})^\$$. Remaining details are left to the interested reader. ❑

With this Kleisli extension $(-)^\$$ one can define composition $g \circledcirc f$ in a Kleisli category as $g^\$ \circ f$.

For the record we mention that the **Kleisli category** category $\mathcal{K}\ell(S)$ of a comonad S has the same objects as $\mathbb{C}$ and its morphisms $X \to Y$ are maps $S(X) \to Y$ in $\mathbb{C}$. The obvious functor $\mathcal{K}\ell(S) \to \mathbb{C}$ from the Kleisli category of a comonad has a right adjoint. Further, Kleisli categories of comonads inherit limits (products and equalisers) from the underlying category.

We come to the next big topic of this section.

Definition 5.2.4 Let $T\colon \mathbb{C} \to \mathbb{C}$ be a monad and $F\colon \mathbb{C} \to \mathbb{C}$ an ordinary endofunctor, both on the same category $\mathbb{C}$. A **distributive law** or also a **$\mathcal{K}\ell$-law** of F over T is a natural transformation $\lambda\colon FT \Rightarrow TF$ that is compatible with T's monad structure:

$$\begin{array}{ccc} F(X) & = & F(X) \\ {\scriptstyle F(\eta_X)}\downarrow & & \downarrow{\scriptstyle \eta_{F(X)}} \\ FT(X) & \xrightarrow[\lambda_X]{} & TF(X) \end{array} \qquad \begin{array}{ccccc} FT^2(X) & \xrightarrow{\lambda_{T(X)}} & TFT(X) & \xrightarrow{T(\lambda_X)} & T^2F(X) \\ {\scriptstyle F(\mu_X)}\downarrow & & & & \downarrow{\scriptstyle \mu_{F(X)}} \\ FT(X) & & \xrightarrow[\lambda_X]{} & & TF(X) \end{array} \tag{5.6}$$

Distributive laws $FT \Rightarrow TF$ as described here correspond to liftings to Kleisli categories $\mathcal{K}\ell(T)$. That is why we also refer to these distributive laws as $\mathcal{K}\ell$-laws. Such liftings play an important role in the next section. Later on, we shall also see distributive laws $TG \Rightarrow GT$ that correspond to liftings to Eilenberg–Moore categories $\mathcal{EM}(T)$. We shall call such distributive laws $\mathcal{EM}$-laws.

Proposition 5.2.5 *Let $T\colon \mathbb{C} \to \mathbb{C}$ be a monad and $F\colon \mathbb{C} \to \mathbb{C}$ a functor. Then there is a bijective correspondence between*

1. *a distributive $\mathcal{K}\ell$-law $\lambda\colon FT \Rightarrow TF$*
2. *a lifting of $F\colon \mathbb{C} \to \mathbb{C}$ to a functor $\mathcal{K}\ell(F)\colon \mathcal{K}\ell(T) \to \mathcal{K}\ell(T)$ in a commuting diagram:*

$$\begin{array}{ccc} \mathcal{K}\ell(T) & \xrightarrow{\mathcal{K}\ell(F)} & \mathcal{K}\ell(T) \\ {\scriptstyle J}\uparrow & & \uparrow{\scriptstyle J} \\ \mathbb{C} & \xrightarrow[F]{} & \mathbb{C} \end{array} \tag{5.7}$$

The notation $\mathcal{K}\ell(F)$ for this lifting is convenient, but a bit dangerous: not only does it leave the dependence on λ implicit – so $\mathcal{K}\ell_\lambda(F)$ would be better – but it may also give the wrong impression that the Kleisli construction as such is functorial. Bearing that caveat in mind, we continue to write $\mathcal{K}\ell(F)$ for the lifting.

Proof Assuming a distributive law $\lambda\colon FT \Rightarrow TF$ we can define the functor $\mathcal{K}\ell(F)\colon \mathcal{K}\ell(T) \to \mathcal{K}\ell(T)$ as

$$\mathcal{K}\ell(F)(X) = F(X) \qquad \mathcal{K}\ell(F)(X \xrightarrow{f} Y) = (F(X) \xrightarrow{F(f)} FT(Y) \xrightarrow{\lambda_Y} TF(Y)).$$

The above two requirements in (5.6) for λ precisely say that $\mathcal{K}\ell(F)$ is a functor. For instance,

$$\begin{aligned}
\mathcal{K}\ell(F)(g \circledcirc f) &= \lambda \circ F(\mu \circ T(g) \circ f) \\
&= \mu \circ T(\lambda) \circ \lambda \circ FT(g) \circ F(f) \\
&= \mu \circ T(\lambda) \circ TF(g) \circ \lambda \circ F(f) \\
&= \mu \circ T(\mathcal{K}\ell(F)(g)) \circ \mathcal{K}\ell(f) \\
&= \mathcal{K}\ell(F)(g) \circledcirc \mathcal{K}\ell(F)(f).
\end{aligned}$$

Conversely, assume there is a functor $G\colon \mathcal{K}\ell(T) \to \mathcal{K}\ell(T)$ in a commuting square as in (5.7). Then, on objects, $G(X) = F(X)$. Further, for a map $f\colon X \to T(Y)$ in $\mathbb{C}$ we get $G(f)\colon F(X) \to TF(Y)$ in $\mathbb{C}$. This suggests how to define a distributive law: the identity map $\mathrm{id}_{TX}\colon T(X) \to T(X)$ in $\mathbb{C}$ forms a map $T(X) \to X$ in $\mathcal{K}\ell(T)$, so that we can define $\lambda_X = G(\mathrm{id}_{TX})\colon FT(X) \to TF(X)$ in $\mathbb{C}$. We check that these λs form a natural transformation, again using commutation of the diagram: for an arbitrary $f\colon X \to Y$ in $\mathbb{C}$ we have $G(\eta \circ f) = \eta \circ F(f)$. Then we get in $\mathbb{C}$:

$$\begin{aligned}
TF(f) \circ \lambda_X &= TF(f) \circ G(\mathrm{id}_{TX}) \\
&= \mu \circ T(\eta \circ F(f)) \circ G(\mathrm{id}_{TX}) \\
&= G(\eta \circ f) \circledcirc G(\mathrm{id}_{TX}) \\
&= G((\eta \circ f) \circledcirc \mathrm{id}_{TX}) \\
&= G(\mu \circ T(\eta \circ f) \circ \mathrm{id}_{TX}) \\
&= GT(f) \\
&= G(\mu \circ T(\mathrm{id}_{TY}) \circ \eta \circ Tf) \\
&= G(\mathrm{id}_{TY} \circledcirc (\eta \circ T(f))) \\
&= G(\mathrm{id}_{TY}) \circledcirc G(\eta \circ T(f)) \\
&= \mu \circ T(\lambda_Y) \circ \eta \circ FT(f) \\
&= \mu \circ \eta \circ \lambda_Y \circ FT(f) \\
&= \lambda_Y \circ FT(f) \\
&= \lambda_Y \circ FT(f).
\end{aligned}$$

Similarly one verifies that λ interacts appropriately with η and μ. □

The commuting diagram (5.7) in this result yields an identity as isomorphism in diagram (2.31) in Theorem 2.5.9, needed to lift an adjunction to categories of algebras. Thus we obtain the following result as immediate consequence. At this stage it may look like a formality, but it turns out to be very useful in the next section.

Corollary 5.2.6 *Let F be an endofunctor and T a monad on a category $\mathbb{C}$, together with a distributive $\mathcal{K}\ell$-law $FT \Rightarrow TF$ of F over T. The adjunction $J \dashv U$ between the Kleisli category $\mathcal{K}\ell(T)$ and $\mathbb{C}$ lifts to an adjunction* $\mathbf{Alg}(J) \dashv \mathbf{Alg}(U)$ *by Theorem 2.5.9:*

$$\begin{array}{ccc} & \mathbf{Alg}(J) & \\ \mathbf{Alg}(F) & \underset{\longleftarrow}{\overset{\longrightarrow}{\bot}} & \mathbf{Alg}(\mathcal{K}\ell(F)) \\ & \mathbf{Alg}(U) & \\ \downarrow & & \downarrow \\ & J & \\ F \circlearrowright \mathbb{C} & \underset{\longleftarrow}{\overset{\longrightarrow}{\bot}} & \mathcal{K}\ell(T) \circlearrowleft \mathcal{K}\ell(F) \\ & U & \end{array}$$

□

Explicitly, the functor $\mathbf{Alg}(J)\colon \mathbf{Alg}(F) \to \mathbf{Alg}(\mathcal{K}\ell(F))$ at the top of this diagram maps an algebra $a\colon F(X) \to X$ to the composite:

$$\mathcal{K}\ell(F)(X) = \Big(F(X) \xrightarrow{\;a\;} X \xrightarrow{\;\eta\;} T(X)\Big).$$

It is an algebra $\mathcal{K}\ell(F)(X) \to X$ in the Kleisli category $\mathcal{K}\ell(T)$. Conversely, the functor $\mathbf{Alg}(U)\colon \mathbf{Alg}(\mathcal{K}\ell(F)) \to \mathbf{Alg}(F)$ sends an algebra $b\colon F(Y) \to T(Y)$ to the F-algebra:

$$FT(Y) \xrightarrow{\;\lambda\;} TF(Y) \xrightarrow{\;T(b)\;} T^2(Y) \xrightarrow{\;\mu\;} T(Y).$$

We now concentrate on distributive $\mathcal{K}\ell$-laws where the monad involved is powerset $\mathcal{P}$. The next result from [244] derives a distributive law with respect to the powerset monad from preservation of weak pullbacks. It is described as ∇, following [457], because of its role in Moss' coalgebraic modal logic, with its ∇ operator; see Section 6.5.2. This ∇ should not be confused with the codiagonal notation $\nabla = [\mathrm{id}, \mathrm{id}]\colon X + X \to X$.

Lemma 5.2.7 *For each weak-pullback-preserving functor $F\colon$ **Sets** $\to$ **Sets** there is a distributive $\mathcal{K}\ell$-law $\nabla\colon F\mathcal{P} \Rightarrow \mathcal{P}F$ obtained by applying relation lifting to the reverse inhabitation relation $\ni \subseteq \mathcal{P}(X) \times X$.*

This ∇ may be used to describe relation lifting: if, for an arbitrary relation $R \subseteq X \times Y$ we write $\widehat{R}\colon X \to \mathcal{P}(Y)$ for the corresponding function, as in (2.16), then the following triangle commutes:

$$\begin{array}{ccc} & \overset{F(\widehat{R})}{\nearrow} & F\mathcal{P}(Y) \\ F(X) & & \downarrow \nabla \\ & \underset{\widehat{\mathrm{Rel}(F)(R)}}{\searrow} & \mathcal{P}F(Y) \end{array} \qquad (5.8)$$

At this stage we use relation lifting as described in Definition 4.4.1, for arbitrary endofunctors on **Sets**, using the standard logical factorisation system of injections and surjections.

Proof By applying relation lifting $\mathrm{Rel}(F)$ to $\ni_X \subseteq \mathcal{P}(X) \times X$ we obtain a relation $\mathrm{Rel}(F)(\ni_X) \subseteq F(\mathcal{P}(X)) \times F(X)$. It can be reformulated, via (2.16), as

$$\begin{array}{rcl} F(\mathcal{P}(X)) & \xrightarrow{\quad \nabla_X \quad} & \mathcal{P}(F(X)) \\ u & \longmapsto & \{v \in F(X) \mid (u,v) \in \mathrm{Rel}(F)(\ni_X)\}. \end{array} \tag{5.9}$$

In order to show that the ∇s behave appropriately we need the basic properties of relation lifting from Proposition 4.4.3 – which all apply because epis are split in **Sets** – together with basic connections between inhabitation and the unit $\eta = \{-\}$ and multiplication $\mu = \bigcup$ of the powerset monad (see Example 5.1.3.1).

a. $(\mathcal{P}(f) \times \mathrm{id})^{-1}(\ni_Y) = \coprod_{\mathrm{id}\times f}(\ni_X)$
b. $(\eta_X \times \mathrm{id})^{-1}(\ni_X) = \mathrm{Eq}(X)$
c. $(\mu_X \times \mathrm{id})^{-1}(\ni_X) = \ni_X \circ \ni_{\mathcal{P}(X)}$.

The proof of these three properties is left as an exercise for the reader.

For naturality of ∇ we use that relation lifting preserves inverse and direct images. For $f\colon X \to Y$ and $u \in F(\mathcal{P}(X))$,

$$\begin{array}{rcl} (\nabla_Y \circ F\mathcal{P}(f))(u) & = & \{w \in F(Y) \mid (F\mathcal{P}(f)(u), w) \in \mathrm{Rel}(F)(\ni_Y)\} \\ & = & \{w \mid (u,w) \in (F(\mathcal{P}(f)) \times \mathrm{id})^{-1}(\mathrm{Rel}(F)(\ni_Y))\} \\ & = & \{w \mid (u,w) \in \mathrm{Rel}(F)((\mathcal{P}(f) \times \mathrm{id})^{-1}(\ni_Y))\} \\ & \stackrel{(a)}{=} & \{w \mid (u,w) \in \mathrm{Rel}(F)(\coprod_{\mathrm{id}\times f}(\ni_X))\} \\ & = & \{w \mid (u,w) \in \coprod_{F(\mathrm{id})\times F(f)}(\mathrm{Rel}(F)(\ni_X))\} \\ & = & \{w \mid \exists (u',v).\, u' = u \wedge F(f)(v) = w \wedge (u',v) \in \mathrm{Rel}(F)(\ni_X)\} \\ & = & \{F(f)(v) \mid (u,v) \in \mathrm{Rel}(F)(\ni_X)\} \\ & = & (\mathcal{P}F(f) \circ \nabla_X)(u). \end{array}$$

This ∇ also interacts appropriately with the unit $\eta = \{-\}$ and multiplication $\mu = \bigcup$ of the powerset monad, because relation lifting preserves equality and composition, and because of the above points (b) and (c):

$$\begin{array}{l} (\nabla_X \circ F(\eta_X))(u) \\ \quad = \{v \mid (F(\eta)(u), v) \in \mathrm{Rel}(F)(\ni_X)\} \\ \quad = \{v \mid (u,v) \in (F(\eta) \times F(\mathrm{id}))^{-1}(\mathrm{Rel}(F)(\ni_X))\} \\ \quad = \{v \mid (u,v) \in \mathrm{Rel}(F)((\eta \times \mathrm{id})^{-1}(\ni_X))\} \\ \quad \stackrel{(b)}{=} \{v \mid (u,v) \in \mathrm{Rel}(F)(\mathrm{Eq}(X))\} \\ \quad = \{v \mid (u,v) \in \mathrm{Eq}(F(X))\} \\ \quad = \{u\} \\ \quad = \eta_{F(X)}(u) \end{array}$$

$$
\begin{aligned}
&(\mu_{F(X)} \circ \mathcal{P}(\nabla_X) \circ \nabla_{\mathcal{P}(X)})(u) \\
&\quad = \bigcup \{\nabla_X(v) \mid (u, v) \in \mathrm{Rel}(F)(\ni_{\mathcal{P}(X)})\} \\
&\quad = \{w \mid \exists v.\, (v, w) \in \mathrm{Rel}(F)(\ni_X) \wedge (u, v) \in \mathrm{Rel}(F)(\ni_{\mathcal{P}(X)})\} \\
&\quad = \{w \mid (u, w) \in \mathrm{Rel}(F)(\ni_X) \circ \mathrm{Rel}(F)(\ni_{\mathcal{P}(X)})\} \\
&\quad = \{w \mid (u, w) \in \mathrm{Rel}(F)(\ni_X \circ \ni_{\mathcal{P}(X)})\} \\
&\quad \overset{(c)}{=} \{w \mid (u, w) \in \mathrm{Rel}(F)((\mu_X \times \mathrm{id})^{-1}(\ni_X))\} \\
&\quad = \{w \mid (u, w) \in (F(\mu_X) \times F(\mathrm{id}))^{-1}(\mathrm{Rel}(F)(\ni_X))\} \\
&\quad = \{w \mid (F(\mu_X)(u), w) \in \mathrm{Rel}(F)(\ni_X)\} \\
&\quad = (\nabla_X \circ F(\mu_X))(u).
\end{aligned}
$$

Commutation of the triangle (5.8) follows from an easy calculation. For $u \in F(X)$ and $v \in F(Y)$ we have:

$$
\begin{aligned}
v \in (\nabla \circ F(\widehat{R}))(u) &\iff (F(\widehat{R})(u), v) \in \mathrm{Rel}(F)(\ni) \\
&\iff (u, v) \in (F(\widehat{R}) \times \mathrm{id})^{-1}(\mathrm{Rel}(F)(\ni)) \\
&\iff (u, v) \in \mathrm{Rel}(F)((\widehat{R} \times \mathrm{id})^{-1}(\ni)) \\
&\iff (u, v) \in \mathrm{Rel}(F)(R) \\
&\iff v \in \widehat{\mathrm{Rel}(F)(R)}(u).
\end{aligned}
$$ ❑

For a weak-pullback-preserving functor on **Sets** there are thus *two* liftings to two categories of relations: once with relations as morphisms and once as objects. Since they are easily confused, we describe them explicitly.

Corollary 5.2.8 *A weak-pullback-preserving functor F:* **Sets** $\to$ **Sets** *has two liftings:*

1. *a 'computational' one, to* **SetsRel** $\to$ **SetsRel** *via Proposition 5.2.5, using the distributive $\mathcal{K}\ell$-law $\nabla\colon F\mathcal{P} \Rightarrow \mathcal{P}F$ from the previous lemma and the fact that the category* **SetsRel** *of sets and relations* as morphisms *is the Kleisli category $\mathcal{K}\ell(\mathcal{P})$ of the powerset monad*
2. *a 'logical' one, to a relator* **Rel** $\to$ **Rel** *via Theorem 4.4.6, where* **Rel** = Rel(**Sets**) *is the category with relations* as objects *and morphisms preserving relation inclusion.*

The action $(X \xrightarrow{R} \mathcal{P}(Y)) \longmapsto (F(X) \xrightarrow{\nabla \circ F(R)} \mathcal{P}(F(Y)))$ of the functor in point 1 on morphisms is the same as the action $(R \rightarrowtail X \times Y) \longmapsto (\mathrm{Rel}(F)(R) \rightarrowtail F(X) \times F(Y))$ of the functor in point 2 on objects, via the triangle (5.8). ❑

We close this section by extending the notion of strength from functors to monads, together with the associated property of commutativity for monads. Commutativity allows us to define some more distributive laws and to define a tensor on Kleisli categories.

In Exercise 2.5.4 we have already briefly seen strength for functors on **Sets**. In general, for functors $F \colon \mathbb{C} \to \mathbb{C}$ on a category $\mathbb{C}$ with finite products, a functor is called **strong** if it comes with a **strength** natural transformation st with components

$$F(X) \times Y \xrightarrow{\ \mathrm{st}_{X,Y}\ } F(X \times Y)$$

that commute appropriately with trivial projections and associativity isomorphisms. Here are the diagrams, starting with naturality:

$$\begin{array}{ccc} F(X) \times Y & \xrightarrow{\ \mathrm{st}\ } & F(X \times Y) \\ {\scriptstyle F(f) \times g}\big\downarrow & & \big\downarrow{\scriptstyle F(f \times g)} \\ F(Z) \times W & \xrightarrow{\ \mathrm{st}\ } & F(Z \times W) \end{array} \qquad \begin{array}{ccc} F(X) \times 1 & \xrightarrow{\ \mathrm{st}\ } & F(X \times 1) \\ & {\scriptstyle \pi_1}\searrow^{\cong} & \cong\big\downarrow{\scriptstyle F(\pi_1)} \\ & & F(X) \end{array}$$

$$\begin{array}{ccccc} (F(X) \times Y) \times Z & \xrightarrow{\ \mathrm{st} \times \mathrm{id}\ } & F(X \times Y) \times Z & \xrightarrow{\ \mathrm{st}\ } & F((X \times Y) \times Z) \\ \cong\big\downarrow & & & & \big\downarrow\cong \\ F(X) \times (Y \times Z) & & \xrightarrow[\ \mathrm{st}\ ]{} & & F(X \times (Y \times Z)) \end{array} \tag{5.10}$$

As an aside: cartesian products are not really needed in order to define strength; monoidal (tensor) products suffice, since we do not really need projections or diagonals.

A monad T is called **strong** if it has a strength $\mathrm{st} \colon T(X) \times Y \to T(X \times Y)$, as above, that commutes with the unit and multiplication of the monad:

$$\begin{array}{ccc} X \times Y & = & X \times Y \\ {\scriptstyle \eta \times \mathrm{id}}\big\downarrow & & \big\downarrow{\scriptstyle \eta} \\ T(X) \times Y & \xrightarrow[\mathrm{st}]{} & T(X \times Y) \end{array} \qquad \begin{array}{ccccc} T^2(X) \times Y & \xrightarrow{\mathrm{st}} & T(T(X) \times Y) & \xrightarrow{T(\mathrm{st})} & T^2(X \times Y) \\ {\scriptstyle \mu \times \mathrm{id}}\big\downarrow & & & & \big\downarrow{\scriptstyle \mu} \\ T(X) \times Y & & \xrightarrow[\ \mathrm{st}\ ]{} & & T(X \times Y) \end{array} \tag{5.11}$$

Given such a strength $\mathrm{st} \colon F(X) \times Y \to F(X \times Y)$ one can also form a 'swapped' strength in the second argument, written as

$$\mathrm{st}' \stackrel{\text{def}}{=} \Big(X \times F(Y) \xrightarrow[\cong]{\mathrm{swap}} F(Y) \times X \xrightarrow{\mathrm{st}} F(Y \times X) \xrightarrow[\cong]{F(\mathrm{swap})} F(X \times Y) \Big) \tag{5.12}$$

where, of course, $\mathrm{swap} = \langle \pi_2, \pi_1 \rangle$.

Given both st and st′ maps for a monad T, there are two ways of going $T(X) \times T(Y) \to T(X \times Y)$, namely via first st and then st′ or the other way around, in

$$\begin{array}{ccccc} & & T(X \times T(Y)) \xrightarrow{T(\mathrm{st}')} T^2(X \times Y) & & \\ & \overset{\mathrm{st}}{\nearrow} & & \overset{\mu}{\searrow} & \\ T(X) \times T(Y) & & & & T(X \times Y) \\ & \underset{\mathrm{st}'}{\searrow} & & \underset{\mu}{\nearrow} & \\ & & T(T(X) \times Y) \xrightarrow[T(\mathrm{st})]{} T^2(X \times Y) & & \end{array} \qquad (5.13)$$

The monad T is called **commutative** if this diagram commutes. We wrap up the foregoing in a definition.

Definition 5.2.9 Let $\mathbb{C}$ be a category with finite products.

1. A functor $F\colon \mathbb{C} \to \mathbb{C}$ is called **strong** if there is strength natural transformation st satisfying (5.10).
2. A monad $T\colon \mathbb{C} \to \mathbb{C}$ is **strong** if it comes equipped with a strength map satisfying both (5.10) and (5.11).
3. A strong monad is called **commutative** if its strength makes diagram (5.13) commute. In that case we sometimes write $\mathrm{dst}\colon T(X) \times T(Y) \to T(X \times Y)$ for the resulting 'double strength' map.
4. If we have two strong monads T and S, then a map of monads $\sigma\colon T \Rightarrow S$ is called a **strong map** if it commutes with strengths, as in

$$\begin{array}{ccc} T(X) \times Y & \xrightarrow{\sigma_X \times \mathrm{id}} & S(X) \times Y \\ {\scriptstyle \mathrm{st}^T} \big\downarrow & & \big\downarrow {\scriptstyle \mathrm{st}^S} \\ T(X \times Y) & \xrightarrow[\sigma_{X \times Y}]{} & S(X \times Y) \end{array} \qquad (5.14)$$

This σ then also commutes with st′ and dst.

We shall write $\mathbf{CMnd}(\mathbb{C}) \hookrightarrow \mathbf{StMnd}(\mathbb{C}) \hookrightarrow \mathbf{Mnd}(\mathbb{C})$ for the subcategories of commutative and strong monads, with strong monad maps between them.

Commutativity of monads can also be formulated by saying that the monad is monoidal (see e.g. [299]), but this requires the machinery of monoidal categories which is out of scope. Alternative formulations using exponents occur in Exercise 5.2.16.

In Exercise 2.5.4 we have seen that every endofunctor on **Sets** is strong, via a uniform strength. The same holds for monads – and also for comonads; see Exercise 5.2.14.

Lemma 5.2.10 **StMnd(Sets)** = **Mnd(Sets)**. *That is, each monad on* **Sets**, *and each monad map between them, is automatically strong.*

Proof Recall from Exercise 2.5.4 the definition of $\mathrm{st}\colon T(X) \times Y \to T(X \times Y)$:

$$\mathrm{st}(u, y) = T(\lambda x \in X.\, (x, y))(u).$$

It is not hard to see that it commutes with T's unit η and multiplication μ, using their naturality. For instance,

$$\begin{aligned}(\mathrm{st}\circ(\eta\times\mathrm{id}))(z,y) &= T(\lambda x\in X.\,(x,y))(\eta(z))\\ &= \big(T(\lambda x\in X.\,(x,y))\circ\eta\big)(z)\\ &= \big(\eta\circ(\lambda x\in X.\,(x,y))\big)(z)\\ &= \eta(z,y).\end{aligned}$$

And for a map of monads $\sigma\colon T \Rightarrow S$ the relevant diagram (5.14) automatically commutes, by naturality:

$$\begin{aligned}(\mathrm{st}^T\circ(\sigma\times\mathrm{id}))(u,y) = \mathrm{st}^T(\sigma(u),y) &= T(\lambda x\in X.\,(x,y))(\sigma(u))\\ &= \sigma(S(\lambda x\in X.\,(x,y))(u))\\ &= (\sigma\circ\mathrm{st}^S)(u,y).\end{aligned}$$ ❑

The next result shows that commutativity of monads is a reasonable notion.

Lemma 5.2.11 *1. A monoid M is commutative if and only if the associated monad $M\times(-)$:* **Sets** $\to$ **Sets** *is commutative.*
2. A semiring S is commutative if and only if the multiset monad $\mathcal{M}_S$: **Sets** $\to$ **Sets** *is commutative.*

Proof By the previous lemma both monads are strong. For the monad $M\times(-)$ the strength map st: $(M\times X)\times Y \to M\times(X\times Y)$ is associativity: $\mathrm{st}((m,x),y) = (m,(x,y))$. The swapped strength map $\mathrm{st}'\colon X\times(M\times Y)\to M\times(X\times Y)$ is similar: $\mathrm{st}'(x,(m,y)) = (m,(x,y))$. The upper and lower path in (5.13) are

$$(M\times X)\times(M\times Y) \xrightarrow{((m,x),(k,y))\,\mapsto\,(m\cdot k,(x,y))} M\times(X\times Y)$$
$$(M\times X)\times(M\times Y) \xrightarrow[((m,x),(k,y))\,\mapsto\,(k\cdot m,(x,y))]{} M\times(X\times Y)$$

Hence they are equal if and only if M is commutative.

The strength st: $\mathcal{M}_S(X)\times Y\to\mathcal{M}_S(X\times Y)$ and the swapped strength st$'$ for the multiset monad are given by

$$\mathrm{st}(\textstyle\sum_i s_i|x_i\rangle, y) = \sum_i s_i|x_i,y\rangle \qquad \mathrm{st}'(x,\sum_j t_j|y_j\rangle) = \sum_j t_j|x,y_j\rangle.$$

The resulting two paths in (5.13) are

$$\mathcal{M}_S(X)\times\mathcal{M}_S(Y) \xrightarrow{(\sum_i s_i|x_i\rangle,\,\sum_j t_j|y_j\rangle)\,\mapsto\,\sum_{i,j} s_i\cdot t_j|x_i,y_j\rangle} \mathcal{M}_S(X\times Y)$$
$$\mathcal{M}_S(X)\times\mathcal{M}_S(Y) \xrightarrow[(\sum_i s_i|x_i\rangle,\,\sum_j t_j|y_j\rangle)\,\mapsto\,\sum_{i,j} t_j\cdot s_i|x_i,y_j\rangle]{} \mathcal{M}_S(X\times Y)$$

Thus, commutativity of the (multiplication of the) semiring S implies commutativity of the monad $\mathcal{M}_S$. For the converse, it suffices to choose $X = Y = 1$ and use that $\mathcal{M}_S(1) \cong S$. ❑

Commutativity of monads is not only a reasonable but also a useful notion because it gives us distributive laws, and thus liftings to Kleisli categories. The following result comes from [206] and defines a distributive law of a simple polynomial functor – with identity, constants, products $\times$ and coproducts $\coprod$ only (see Definition 2.2.1) – over an arbitrary commutative monad.

Lemma 5.2.12 *Let T: **Sets** $\to$ **Sets** be a commutative monad, and F: **Sets** $\to$ **Sets** a simple polynomial functor. Then there is a distributive $\mathcal{K}\ell$-law $\lambda\colon FT \Rightarrow TF$.*

Proof The construction of the distributive law λ proceeds by induction on the structure of the functor F.

- If F is the identity functor, then λ is the identity natural transformation $T \Rightarrow T$.
- If F is a constant functor, say $X \mapsto A$, then λ is the unit map $\eta_A\colon A \to T(A)$.
- If $F = F_1 \times F_2$, we use induction and assume distributive laws $\lambda_i\colon F_iT \Rightarrow TF_i$ for $i \in \{1, 2\}$. Then we can form a new distributive law, using that T is commutative:

$$F_1T(X) \times F_2T(X) \xrightarrow{\lambda_1 \times \lambda_2} TF_1(X) \times TF_2(X) \xrightarrow{\text{dst}} T(F_1(X) \times F_2(X)).$$

- If F is a coproduct $\coprod_{i\in I} F_i$ then we may assume laws $\lambda_i\colon F_iT \Rightarrow TF_i$ for $i \in I$, and define

$$\textstyle\coprod_{i\in I} F_iT(X) \xrightarrow{[T(\kappa_i) \circ \lambda_i]_{i\in I}} T(\coprod_{i\in I} F_i(X)).$$

It is straightforward to check that these λs are natural and compatible with the monad structure. ❑

We conclude by remarking that the Kleisli category of a commutative monad carries a tensor $\otimes$. Formally, this is expressed in terms of a symmetric monoidal structure (see [344]). Here we describe it more concretely.

Proposition 5.2.13 *Let $\mathbb{C}$ be a category with finite products and $T\colon \mathbb{C} \to \mathbb{C}$ a commutative monoid. The cartesian product then becomes a functor $\times\colon \mathcal{K}\ell(T) \times \mathcal{K}\ell(T) \to \mathcal{K}\ell(T)$.*

Proof For $f\colon X \to Y$ and $g\colon Z \to W$ in $\mathcal{K}\ell(T)$ we have to define a map $X \times Z \to Y \times W$. Using double strength we take

$$X \times Z \xrightarrow{f \times g} T(Y) \times T(W) \xrightarrow{\text{dst}} T(Y \times W).$$

Via the properties of Exercise 5.2.15 it easy to see that identities and composition are preserved. □

Exercises

5.2.1 Show that a map of monads $\sigma\colon T \Rightarrow S$ induces a functor $\mathcal{K}\ell(T) \to \mathcal{K}\ell(S)$.

5.2.2 Let $T_1, T_2\colon \mathbb{C} \to \mathbb{C}$ be two monads, with unit and multiplication η_i, μ_i and Kleisli extension $(-)^{\$_i}$. Prove that a natural transformation $\sigma\colon T_1 \Rightarrow T_2$ is a map of monads (i.e. commutes with the η_i, μ_i) if and only if σ commutes with the η_i (that is, $\sigma \circ \eta_1 = \eta_2$) and σ commutes with Kleisli extension: for each map $f\colon X \to T_1(Y)$ the following diagram commutes:

$$\begin{array}{ccc} T_1(Y) & \xrightarrow{\sigma_Y} & T_2(Y) \\ {\scriptstyle f^{\$_1}}\uparrow & & \uparrow{\scriptstyle (\sigma_Y \circ f)^{\$_2}} \\ T_1(X) & \xrightarrow[\sigma_X]{} & T_2(X) \end{array}$$

5.2.3 Prove the three properties (a)–(c) in the proof of Lemma 5.2.7.

5.2.4 Prove in detail that if a category $\mathbb{C}$ has coequalisers, then so has the Kleisli category $\mathcal{K}\ell(T)$ for a monad T on $\mathbb{C}$ – as claimed in Proposition 5.2.2.

5.2.5 Let T be a monad on a category $\mathbb{C}$ with an initial object $0 \in \mathbb{C}$. Prove that the following two statements are equivalent.

1. The object $T(0) \in \mathbb{C}$ is final.
2. The object $0 \in \mathcal{K}\ell(T)$ is a zero object; i.e. 0 is both initial and final in $\mathcal{K}\ell(T)$.

5.2.6 Prove that there is a bijective correspondence

$$\frac{X \longrightarrow Y \qquad \text{in } \mathcal{K}\ell(\mathcal{P}) = \textbf{SetsRel}}{\overline{\mathcal{P}(Y) \longrightarrow \mathcal{P}(X) \qquad \text{in } \textbf{CL}_\wedge}}$$

where $\textbf{CL}_\wedge$ is the category of complete lattices and arbitrary meet preserving functions. (This is the heart of the correspondence between (non-deterministic) state transformers $X \to Y$ and predicate transformers $\mathcal{P}(Y) \to \mathcal{P}(X)$ that forms the basis for the weakest precondition condition calculus in program verification; see [122, 341, 254].)

5.2.7 Recall that non-deterministic automata can be turned into deterministic ones by changing the state space. If we consider the state transition map only, this can be described as transforming a coalgebra $X \to \mathcal{P}(X)^A$ into a coalgebra $\mathcal{P}(X) \to \mathcal{P}(X)^A$.

1. Describe this transformation concretely.
2. Describe this transformation via (pointwise) Kleisli extension.
3. Show that it leads to a functor $\mathbf{CoAlg}(\mathcal{P}(-)^A) \to \mathbf{CoAlg}((-)^A)$.

5.2.8 In Definition 5.2.4 we have seen a distributive $\mathcal{K}\ell$-law $FT \Rightarrow TF$ of a functor F over a monad T. There is also a notion of distributive law when F is a monad too (going back to [61]). In that case the natural transformation should commute with all units and multiplications.

1. Write down the relevant diagrams.
2. Let $ST \Rightarrow TS$ be such a distributive law, where S and T are monads, and prove that the composite TS is then also a monad.
3. Prove that the units yield maps of monads $S \Rightarrow TS$ and $T \Rightarrow TS$.
4. Check that the lifted functor $\mathcal{K}\ell(S)\colon \mathcal{K}\ell(T) \to \mathcal{K}\ell(T)$ is again a monad.
5. Show that the Kleisli category $\mathcal{K}\ell(\mathcal{K}\ell(S))$ of this lifted monad is the same as the Kleisli category $\mathcal{K}\ell(TS)$ of the composite monad.

We consider only the last two points. As unit $\overline{\eta}_X\colon X \to \mathcal{K}\ell(S)(X)$ in $\mathcal{K}\ell(T)$ we take $\overline{\eta}_X = \eta^T_{SX} \circ \eta^S_X\colon X \to TS(X)$. As multiplication $\overline{\mu}_X\colon \mathcal{K}\ell(S)^2(X) \to \mathcal{K}\ell(S)(X)$ we take $\overline{\mu}_X = \eta^T_{SX} \circ \mu^S_X\colon S^2(X) \to TS(X)$. Hence they are obtained via the inclusion functor $F\colon \mathbb{C} \to \mathcal{K}\ell(T)$, so all relevant monad equations hold because S is a monad.

5.2.9 Let $T\colon \mathbb{C} \to \mathbb{C}$ be an arbitrary monad on a category $\mathbb{C}$ with coproducts. Fix an object $A \in \mathbb{C}$, recall from Exercise 5.1.1 that $A + (-)$ is a monad on $\mathbb{C}$ and prove that the maps

$$A + T(X) \xrightarrow{\lambda_X = [T(\kappa_1) \circ \eta_A, T(\kappa_2)]} T(A + X)$$

form a $\mathcal{K}\ell$-law of monads (as defined in the previous exercise). Conclude that $T(A + (-))$ is also a monad on $\mathbb{C}$.

In the special case where A is the singleton (final) set 1, and T is the distribution on **Sets**, check that the monad $\mathcal{D}(1 + (-))$ is the subdistribution $\mathcal{D}_{\leq 1}$.

5.2.10 Consider the mapping $T \mapsto \mathcal{E}(T) \stackrel{\text{def}}{=} T(A + (-))$ from the previous exercise.

1. Prove that it gives rise to a functor $\mathcal{E}\colon \mathbf{Mnd}(\mathbb{C}) \to \mathbf{Mnd}(\mathbb{C})$.

2. Show that $\mathcal{E}$ is even a monad, on the category $\mathbf{Mnd}(\mathbb{C})$ of monads on $\mathbb{C}$.

(This mapping $\mathcal{E}$ is often called the exception monad transformer. It is investigated more systematically in [335] together with associated operations for handling exceptions in programs. The proof that $\mathcal{E}$ is a monad requires a certain level of categorical fearlessness but is in essence not difficult.)

5.2.11 Let T be a strong monad. Check that strength gives for each object Y a $\mathcal{K}\ell$-law of the functor $(-) \times Y$ over T.

5.2.12 ([233, proposition 1]) Let F be an endofunctor and $T = (T, \eta, \mu)$ be a monad, both on the same category. Show that the maps $\lambda_X = \eta_{FT(X)} \circ F(\mu_X) \colon FTT(X) \to FT(X) \to TFT(X)$ form a distributive $\mathcal{K}\ell$-law of the functor FT over the monad T.

5.2.13 Show that the powerset $\mathcal{P}$ and distribution $\mathcal{D}$ monad are commutative, with double strengths:

$$\begin{array}{ccc} \mathcal{P}(X) \times \mathcal{P}(Y) & \xrightarrow{\text{dst}} & \mathcal{P}(X \times Y) \\ (U, V) & \longmapsto & U \times V \end{array} \qquad \begin{array}{ccc} \mathcal{D}(X) \times \mathcal{D}(Y) & \xrightarrow{\text{dst}} & \mathcal{D}(X \times Y) \\ (\varphi, \psi) & \longmapsto & \lambda(x, y).\, \varphi(x) \cdot \psi(x). \end{array}$$

Check that the list monad $(-)^\star$ is *not* commutative.

5.2.14 Prove the following analogue of Lemma 5.2.10 for comonads: for each comonad $S : \mathbf{Sets} \to \mathbf{Sets}$, the associated strength map st makes the following diagrams commute:

$$\begin{array}{ccc} S(X) \times Y & \xrightarrow{\text{st}} & S(X \times Y) \\ {\scriptstyle \varepsilon \times \text{id}} \downarrow & & \downarrow {\scriptstyle \varepsilon} \\ X \times Y & = & X \times Y \end{array}$$

$$\begin{array}{ccccc} S(X) \times Y & & \xrightarrow{\text{st}} & & S(X \times Y) \\ {\scriptstyle \delta \times \text{id}} \downarrow & & & & \downarrow {\scriptstyle \delta} \\ S^2(X) \times Y & \xrightarrow[\text{st}]{} & S(S(X) \times Y) & \xrightarrow[S(\text{st})]{} & S^2(X \times Y) \end{array}$$

5.2.15 Show that the double strength map dst of a commutative monad T is a natural transformation and makes the following diagrams commute:

$$\begin{array}{ccc} X \times Y & = & X \times Y \\ {\scriptstyle \eta \times \eta} \downarrow & & \downarrow {\scriptstyle \eta} \\ T(X) \times T(Y) & \xrightarrow[\text{dst}]{} & T(X \times Y) \end{array}$$

$$\begin{array}{ccccc} T^2(X)\times T^2(Y) & \xrightarrow{\text{dst}} & T(T(X)\times T(Y)) & \xrightarrow{T(\text{dst})} & T^2(X\times Y) \\ {\scriptstyle \mu\times\mu}\downarrow & & & & \downarrow{\scriptstyle \mu} \\ T(X)\times T(Y) & & \xrightarrow[\text{dst}]{\hspace{6em}} & & T(X\times Y) \end{array}$$

5.2.16 This exercise involves an alternative formulation of strength st, both for functors and for monads, using exponents. Let $\mathbb{C}$ be a cartesian closed category with an endofunctor $F\colon \mathbb{C}\to\mathbb{C}$.

1. Prove that F is strong if and only if there are maps $r_{X,Y}\colon F(X^Y)\to F(X)^Y$ which are natural in X and Y and make the following three diagrams commute:

$$\begin{array}{ccc} F(X_1^{Y_2}) & \xrightarrow{\;r\;} & F(X_1)^{Y_2} \\ {\scriptstyle F(f^g)}\downarrow & & \downarrow{\scriptstyle F(f)^g} \\ F(X_2^{Y_1}) & \xrightarrow[\;r\;]{} & F(X_2)^{Y_1} \end{array}$$

where $f^g = \Lambda(f\circ \text{ev}\circ(\text{id}\times g))$, and

$$\begin{array}{ccc} F(X) & = & F(X) \\ {\scriptstyle F(\Lambda(\pi_1))}\downarrow{\scriptstyle\cong} & & {\scriptstyle\cong}\downarrow{\scriptstyle\Lambda(\pi_1)} \\ F(X^1) & \xrightarrow[r]{} & F(X)^1 \end{array} \qquad \begin{array}{ccccc} F((X^Y)^Z) & \xrightarrow{r} & F(X^Y)^Z & \xrightarrow{r^Z} & (F(X)^Y)^Z \\ {\scriptstyle\cong}\downarrow & & & & \downarrow{\scriptstyle\cong} \\ F(X^{Z\times Y}) & & \xrightarrow[r]{\hspace{4em}} & & F(X)^{Z\times Y} \end{array}$$

where the isomorphisms on the right are $\Lambda(\text{ev}\circ(\text{ev}\times\text{id})\circ\alpha)$, with associativity isomorphism $\alpha = \langle(\text{id}\times\pi_1), \pi_2\circ\pi_2\rangle\colon A\times(B\times C)\xrightarrow{\cong}(A\times B)\times C$.

2. Prove also that a monad $T\colon\mathbb{C}\to\mathbb{C}$ is strong if and only if there are such $r\colon T(X^Y)\to T(X)^Y$ satisfying

$$\begin{array}{ccc} X^Y & = & X^Y \\ {\scriptstyle\eta}\downarrow & & \downarrow{\scriptstyle\eta^Y} \\ T(X^Y) & \xrightarrow[r]{} & T(X)^Y \end{array} \qquad \begin{array}{ccccc} T^2(X^Y) & \xrightarrow{T(r)} & T(T(X)^Y) & \xrightarrow{r} & T^2(X)^Y \\ {\scriptstyle\mu}\downarrow & & & & \downarrow{\scriptstyle\mu^Y} \\ T(X^Y) & & \xrightarrow[r]{\hspace{4em}} & & T(X)^Y \end{array}$$

5.2.17 Show that there are adjunctions between (commutative) monoids and (commutative) monads:

$$\begin{array}{ccc} & \mathbf{Mon} & \\ M\mapsto M\times(-) & \dashv & T\mapsto T(1) \\ & \mathbf{Mnd(Sets)} & \end{array} \qquad \begin{array}{ccc} & \mathbf{CMon} & \\ M\mapsto M\times(-) & \dashv & T\mapsto T(1) \\ & \mathbf{CMnd(Sets)} & \end{array}$$

The adjunction on the left comes from [466] where it is described in the more general setting of monoidal categories. See also [111] for more such adjunctions, also involving the multiset monad. *Hint*: Define a monoid structure on $T(1)$ via strength, as $T(1) \times T(1) \to T(1 \times T(1)) \xrightarrow{\cong} T^2(1) \to T(1)$.

5.3 Trace Semantics via Finality in Kleisli Categories

This section describes how finite traces of observables of computations can be described via finality, not in the familiar category **Sets** but in the Kleisli category $\mathcal{K}\ell(T)$ of a monad T on **Sets**. We shall be considering coalgebras of the form $X \to TF(X)$, where T is our monad capturing the kind of computation involved (non-deterministic, probabilistic, etc.) and F is a functor that describes the branching structure. The main examples that will be elaborated here involve non-deterministic and probabilistic automata and context-free grammars. The use of Kleisli categories for trace semantics goes back to [395, 244]. (The latter reference [244] deals not with *finite* trace semantics – as in this section – but with the infinite version. It involves only weak finality; i.e. it lacks uniqueness of the trace map.)

We begin with an illustration that uses the powerset monad $\mathcal{P}$ on **Sets**. Recall that its Kleisli category $\mathcal{K}\ell(\mathcal{P})$ is the category **SetsRel** of sets and relations between them, since a relation $R \subseteq X \times Y$ may equivalently be described as a characteristic function $R: X \to \mathcal{P}(Y)$. We shall often switch back-and-forth between these two notations. For convenience we recall that the identity map $X \to X$ in **SetsRel** is the equality relation $\mathrm{Eq}(X) \subseteq X \times X$. And (Kleisli) composition of $R: X \to Y$ and $S: Y \to Z$ is the usual relational composition: $S \circledcirc R = \{(x, z) \in X \times Z \mid \exists y \in Y. R(x, y) \wedge S(y, z)\}$. There is an obvious 'graph' functor **Sets** $\to$ **SetsRel** as already described in Example 1.4.4.4. It maps a set to itself, and a function $f: X \to Y$ to its graph relation $\mathrm{Graph}(f) = \{(x, y) \in X \times Y \mid f(x) = y\}$. This graph functor is the standard map $J: \mathbb{C} \to \mathcal{K}\ell(T)$ associated with a Kleisli category; see Proposition 5.2.2.

Let us now fix a set L of 'labels' and write F for the 'list' functor **Sets** $\to$ **Sets** given by $X \mapsto 1 + (L \times X)$. We shall consider transition systems of the form $c: X \to \mathcal{P}F(X) = \mathcal{P}(1 + (L \times X))$. As noticed in Exercise 2.2.3, such transition systems correspond to non-deterministic automata $X \to \mathcal{P}(X)^L \times 2$. Here we prefer the formulation with the powerset on the outside, because it allows us to describe such automata as maps $X \to F(X)$ in the Kleisli category $\mathcal{K}\ell(\mathcal{P}) =$ **SetsRel**.

These transition systems $X \to \mathcal{P}(1 + (L \times X))$ describe besides usual transitions $x \xrightarrow{a} x'$, as shorthand for $(a, x') \in c(x)$, also terminating transitions $x \to *$ for $* \in c(x)$. A *trace* for such a coalgebra starting in a state $x \in X$ consists of a list of elements $\langle a_1, \ldots, a_n \rangle \in L^\star$ for which there is a list of states $x_0, \ldots, x_n \in X$ where:

- $x_0 = x$
- $x_i \xrightarrow{a_i} x_{i+1}$ for all $i < n$
- $x_n \to *$.

Thus, traces are sequences of labels appearing in terminating sequences of computations.

One can then define a function $\text{trace}_c \colon X \to \mathcal{P}(L^\star)$ that maps a state to the set of traces starting in that state. We shall show how to define this function by finality. It turns out not to be a coincidence that the set $L^\star$ of lists is the initial algebra of the functor F.

First we need to move from sets to relations as morphisms. We shall lift the functor $F \colon \mathbf{Sets} \to \mathbf{Sets}$ to $\mathcal{K}\ell(F) \colon \mathbf{SetsRel} \to \mathbf{SetsRel}$. Essentially, this lifting to Kleisli categories happens as in Proposition 5.2.5 –using the distributive law from Lemma 5.2.7 –but at this moment we wish to proceed more concretely. The lifting $\mathcal{K}\ell(F) \colon \mathbf{SetsRel} \to \mathbf{SetsRel}$ maps an object (or set) X to $F(X)$, and a morphism $R \colon X \to Y$ in $\mathbf{SetsRel}$ to the morphism $\text{Rel}(F)(R) \colon F(X) \to F(Y)$ obtained by relation lifting; see Corollary 5.2.8. Explicitly, according to Definition 3.1.1, the lifted relation $\text{Rel}(F)(R) \subseteq F(X) \times F(Y)$ is described by

$$\text{Rel}(F)(R) = \{(*, *)\} \cup \{\langle (a, x), (a, x') \rangle \mid a \in L \text{ and } R(x, x')\}.$$

We now have three important observations.

1. The first one is simple: a transition system, or $\mathcal{P}F$-coalgebra, $c \colon X \to \mathcal{P}F(X)$ as considered above, is an $\mathcal{K}\ell(F)$-coalgebra $X \to \mathcal{K}\ell(F)(X)$ in the category $\mathbf{SetsRel}$, for the lifted functor $\mathcal{K}\ell(F) \colon \mathbf{SetsRel} \to \mathbf{SetsRel}$.
2. Further, the above trace map $\text{trace}_c \colon X \to \mathcal{P}(L^\star)$, considered as a homomorphism $X \to L^\star$ in $\mathbf{SetsRel}$, is a homomorphism of $\mathcal{K}\ell(F)$-coalgebras in $\mathbf{SetsRel}$:

$$\begin{array}{ccc} F(X) & \xrightarrow{\text{Rel}(F)(\text{trace}_c)} & F(L^\star) \\ {\scriptstyle c}\uparrow & & \cong\uparrow{\scriptstyle \text{Graph}(\alpha^{-1})} \\ X & \xrightarrow[\text{trace}_c]{} & L^\star \end{array} \tag{5.15}$$

where $\alpha = [\mathsf{nil}, \mathsf{cons}] \colon F(L^\star) \xrightarrow{\cong} L^\star$ is the initial algebra map, and the graph functor $\text{Graph}(-) \colon \mathbf{Sets} \to \mathbf{SetsRel}$ is the canonical map $J \colon \mathbf{Sets} \to$

$\mathcal{K}\ell(\mathcal{P})$ described in Proposition 5.2.2. Commutation of the diagram (5.15) amounts to

$$\sigma \in \mathrm{trace}_c(x) \iff \exists u \in c(x).\, \langle u, \alpha^{-1}(\sigma)\rangle \in \mathrm{Rel}(F)(\mathrm{trace}_c).$$

This may be split in two cases, depending on whether the list $\sigma \in L^\star$ is empty or not:

$$\begin{aligned}
\mathsf{nil} \in \mathrm{trace}_c(x) &\iff \exists u \in c(x).\, \langle u, *\rangle \in \mathrm{Rel}(F)(\mathrm{trace}_c) \\
&\iff * \in c(x) \\
&\iff x \to * \\
\mathsf{cons}(a,\sigma) \in \mathrm{trace}_c(x) &\iff \exists u \in c(x).\, \langle u, (a,\sigma)\rangle \in \mathrm{Rel}(F)(\mathrm{trace}_c) \\
&\iff \exists x' \in X.\, (a, x') \in c(x) \wedge \langle x', \sigma\rangle \in \mathrm{trace}_c \\
&\iff \exists x' \in X.\, x \xrightarrow{a} x' \wedge \sigma \in \mathrm{trace}_c(x').
\end{aligned}$$

This shows that the trace map trace_c in (5.15) indeed makes the diagram commute.

3. Finally, this trace map trace_c is obtained by coinduction since the graph of the initial F-algebra $\mathrm{Graph}(\alpha^{-1})\colon L^\star \xrightarrow{\cong} F(L^\star)$ is a final $\mathcal{K}\ell(F)$-coalgebra in **SetsRel**.

   We have already seen that there is a homomorphism trace_c for an arbitrary $\mathcal{K}\ell(F)$-coalgebra c in **SetsRel** $= \mathcal{K}\ell(\mathcal{P})$. Here we check that it is unique: if a relation $S \subseteq X \times L^\star$ also satisfies $\mathrm{Graph}(\alpha^{-1}) \circledcirc S = \mathrm{Rel}(F)(S) \circledcirc c$, then $S = \mathrm{trace}_c$. Recall that we write $\circledcirc$ for composition in the Kleisli category $\mathcal{K}\ell(\mathcal{P})$, which we identify with relational composition. Uniqueness holds, since for $x \in X$ one obtains $\sigma \in S(x) \Leftrightarrow \sigma \in \mathrm{trace}_c(x)$ by induction on $\sigma \in L^\star$:

$$\begin{aligned}
\mathsf{nil} \in S(x) &\iff \exists \tau \in L^\star.\, * = \alpha^{-1}(\tau) \wedge \tau \in S(x) \\
&\iff \exists \tau \in L^\star.\, * \in \mathrm{Graph}(\alpha^{-1})(\tau) \wedge \tau \in S(x) \\
&\iff (x, *) \in \mathrm{Graph}(\alpha^{-1}) \circledcirc S \\
&\iff (x, *) \in \mathrm{Rel}(F)(S) \circledcirc c \\
&\iff \exists u \in c(x).\, * \in \mathrm{Rel}(F)(S)(u) \\
&\iff * \in c(x) \\
&\iff \mathsf{nil} \in \mathrm{trace}_c(x) \\
\mathsf{cons}(a,\sigma) \in S(x) &\iff \exists \tau \in L^\star.\, (a,\sigma) = \alpha^{-1}(\tau) \wedge \tau \in S(x) \\
&\iff \exists \tau \in L^\star.\, (a,\sigma) \in \mathrm{Graph}(\alpha^{-1})(\tau) \wedge \tau \in S(x) \\
&\iff (x, (a,\sigma)) \in \mathrm{Graph}(\alpha^{-1}) \circledcirc S \\
&\iff (x, (a,\sigma)) \in \mathrm{Rel}(F)(S) \circledcirc c \\
&\iff \exists u \in c(x).\, (a,\sigma) \in \mathrm{Rel}(F)(S)(u) \\
&\iff \exists x' \in X.\, \sigma \in S(x') \wedge (a, x') \in c(x) \\
&\overset{\text{(IH)}}{\iff} \exists x' \in X.\, \sigma \in \mathrm{trace}_c(x') \wedge x \xrightarrow{a} x' \\
&\iff \mathsf{cons}(a,\sigma) \in \mathrm{trace}_c(x).
\end{aligned}$$

In the next result we shall describe this situation for the powerset monad in greater generality, following [203]. Later on we shall describe the situation in still more general form, for other monads, as in [206]. But we prefer to stick to the powerset monad first, because it involves a very common situation and because it has a snappy proof that avoids domain-theoretic machinery.

Theorem 5.3.1 *Let* $F\colon \mathbf{Sets} \to \mathbf{Sets}$ *be a weak-pullback-preserving functor with an initial algebra* $\alpha\colon F(A) \xrightarrow{\cong} A$. *Lemma 5.2.7 then yields a distributive* $\mathcal{K}\ell$*-law* $F\mathcal{P} \Rightarrow \mathcal{P}F$ *and thus a lifting of* F *to an endofunctor* $\mathcal{K}\ell(F)$ *on the Kleisli category* $\mathcal{K}\ell(\mathcal{P}) = \mathbf{SetsRel}$.

The map $\mathrm{Graph}(\alpha^{-1})\colon A \xrightarrow{\cong} \mathcal{K}\ell(F)(A)$ *is then a final coalgebra in* $\mathbf{SetsRel}$, *for the lifted functor* $\mathcal{K}\ell(F)\colon \mathbf{SetsRel} \to \mathbf{SetsRel}$.

Proof Our first observation is that reversal of relations, sending $R \subseteq X \times Y$ to $R^{\dagger} \subseteq Y \times X$, makes the (Kleisli) category $\mathbf{SetsRel} = \mathcal{K}\ell(\mathcal{P})$ self-dual: $\mathbf{SetsRel} \cong \mathbf{SetsRel}^{\mathrm{op}}$. When we combine this duality with the lifting of Corollary 5.2.6 we obtain the following situation:

$$\begin{array}{ccccc}
\mathbf{Alg}(F) & \underset{\mathbf{Alg}(U)}{\overset{\mathbf{Alg}(J)}{\rightleftarrows}}\;\bot & \mathbf{Alg}(\mathcal{K}\ell(F)) & \cong & \mathbf{Alg}(\mathcal{K}\ell(F)^{\mathrm{op}}) = \mathbf{CoAlg}(\mathcal{K}\ell(F))^{\mathrm{op}} \\
\downarrow & & \downarrow & & \downarrow \\
\mathbf{Sets} & \underset{U}{\overset{J}{\rightleftarrows}}\;\bot & \mathbf{SetsRel} & \cong & \mathbf{SetsRel}^{\mathrm{op}} \\
\circlearrowleft F & & \circlearrowleft \mathcal{K}\ell(F) & & \circlearrowleft \mathcal{K}\ell(F)^{\mathrm{op}}
\end{array}$$

First we consider the bottom row. By composing the adjunction $J \dashv U$ from Proposition 5.2.2 with the isomorphism $\mathbf{SetsRel} \cong \mathbf{SetsRel}^{\mathrm{op}}$ we obtain an adjunction between the categories $\mathbf{Sets}$ and $\mathbf{SetsRel}^{\mathrm{op}}$, say $J^{\dagger} \dashv U^{\dagger}$, where $J^{\dagger}\colon \mathbf{Sets} \to \mathbf{SetsRel}^{\mathrm{op}}$ is given by $J^{\dagger}(X) = X$ and $J^{\dagger}(f\colon X \to Y) = \mathrm{Graph}(f)^{\dagger} = \{(y, x) \mid f(x) = y\} \subseteq Y \times X$. In the reverse direction we have $U^{\dagger}(Y) = \mathcal{P}(Y)$ and $U^{\dagger}(R\colon X \to Y) = \lambda U \in \mathcal{P}(Y).\,\{x \in X \mid \exists y \in U.\, R(x, y)\}$.

The resulting lifted adjunction in the top row, between the categories of algebra $\mathbf{Alg}(F)$ and $\mathbf{Alg}(\mathcal{K}\ell(F)^{\mathrm{op}}) = \mathbf{CoAlg}(\mathcal{K}\ell(F))^{\mathrm{op}}$, say $\mathbf{Alg}(J^{\dagger}) \dashv \mathbf{Alg}(U^{\dagger})$, allows us to finish the argument, since left adjoints preserve initial objects. Hence $\mathbf{Alg}(J^{\dagger})$ sends the initial F-algebra $F(A) \xrightarrow{\cong} A$ to the initial object in $\mathbf{Alg}(\mathcal{K}\ell(F)^{\mathrm{op}})$, i.e. to the final object in $\mathbf{CoAlg}(\mathcal{K}\ell(F))$, i.e. to the final coalgebra of the lifted functor $\mathcal{K}\ell(F)\colon \mathbf{SetsRel} \to \mathbf{SetsRel}$. ❑

This theorem captures traces for transition systems as described in the beginning of this section. The main application of [203] is the parsed language associated with a context-free grammar; see Example 5.3.2.

Example 5.3.2 Recall from Section 2.2.5 that a context-free grammar (CFG) is described coalgebraically as a function $g\colon X \to \mathcal{P}((X+L)^\star)$ where X is the state space of nonterminals, and L is the alphabet of terminals (or tokens). Such a CFG is thus a coalgebra in **SetsRel** of the lifting of the functor $F(X) = (X+L)^\star$ on **Sets**. Let us write its initial F-algebra as

$$(L^\vartriangle + L)^\star \xrightarrow[\cong]{\alpha} L^\vartriangle .$$

Notice that $F(X) = (X+L)^\star = \coprod_{n\in\mathbb{N}}(X+L)^n = \coprod_{\sigma\in(1+L)^\star} X^{\|\sigma\|}$ where $\|\sigma\| \in \mathbb{N}$ is the number of $* \in 1$ occurring in $\sigma \in (1+L)^\star$. Hence an F-algebra $F(X) \to X$ involves an n-ary operation $X^n \to X$ for each $\sigma \in (1+L)^\star$ with $n = \|\sigma\|$. An example of such a σ is $\langle \mathsf{if}, *, \mathsf{then}, *, \mathsf{else}, *, \mathsf{fi}\rangle$. It may be understood as a ternary operation $X^3 \to X$. The initial algebra $L^\vartriangle$ then consists of terms built with such operations. Hence it contains all structured or parsed words over L (according to the grammar g).

Theorem 5.3.1 gives for each CFG $g\colon X \to \mathcal{P}((X+L)^\star)$ a unique homomorphism $\mathrm{trace}_g\colon X \to \mathcal{P}(L^\vartriangle)$. It maps a nonterminal $x \in X$ to the collection of parsed words that can be produced from x. Coalgebraic trace semantics was first used to describe the language of a context-free grammar in [203]; see also [464]. The situation is elaborated further in Exercise 5.3.2.

At this stage we wish to generalise the result of Theorem 5.3.1 to monads other than powerset $\mathcal{P}$. The theorem itself says that an initial algebra $F(A) \xrightarrow{\cong} A$ yields a final coalgebra $A \xrightarrow{\cong} F(A) = \mathcal{K}\ell(F)(A)$ of the lifting $\mathcal{K}\ell(F)$ of the functor F to a Kleisli category. One aspect that we ignored is that the initial algebra $F(A) \xrightarrow{\cong} A$ also yields an initial algebra in the Kleisli category, since the canonical functor $J\colon \mathbf{Sets} \to \mathcal{K}\ell(\mathcal{P})$ is a left adjoint (see Proposition 5.2.2) and thus preserves colimits of ω-chains (Exercise 4.6.3). Thus A carries both an initial algebra and a final coalgebra structure in the Kleisli category. Such coincidence has been a topic of extensive study; see notably [433, 144, 145, 132]. Here we concentrate on the essentials and refer to these original sources for further information.

There is one more concept that we need. A category $\mathbb{C}$ is called **dcpo-enriched** if each hom-set $\mathbb{C}(X, Y)$ of map $X \to Y$ in $\mathbb{C}$ forms a directed complete partial order (dcpo) and the dcpo structure is preserved under composition: both pre- and post-composition $h \circ (-)$ and $(-) \circ g$ are maps in the category **Dcpo**. Again, the Kleisli category $\mathcal{K}\ell(\mathcal{P})$ is an example: for a collection of maps $f_i\colon X \to \mathcal{P}(Y)$ we can form the pointwise join $\bigvee_i f_i\colon X \to \mathcal{P}(Y)$, namely $(\bigvee_i f_i)(x) = \bigcup_i f_i(x)$. It is not hard to see that $h \circ (\bigvee_i f_i) = \bigvee_i (h \circ f_i)$ and similarly $(\bigvee_i f_i) \circ g = \bigvee_i (f_i \circ g)$.

In such a dcpo-enriched category the hom-sets $\mathbb{C}(X, Y)$ carry a partial order $\leq$ for which directed joins exist. An arbitrary map $f: X \to Y$ is called an **embedding** if there is a 'projection' map $f^p: Y \to X$ in the other direction with

$$f^p \circ f = \mathrm{id}_X \quad \text{and} \quad f \circ f^p \leq \mathrm{id}_Y.$$

This map f^p, if it exists, is uniquely determined. As a consequence, $(g \circ f)^p = f^p \circ g^p$. The pair (f, f^p) is sometimes called an embedding-projection pair. More formally, we have an adjunction (Galois connection) with an isomorphism as unit (also known as coreflection).

The heart of the matter is the following limit-colimit coincidence result from [433].

Proposition 5.3.3 *Let $\mathbb{C}$ be a dcpo-enriched category. Assume an ω-chain*

$$X_0 \xrightarrow{f_0} X_1 \xrightarrow{f_1} X_2 \xrightarrow{f_2} \cdots$$

with colimit $A \in \mathbb{C}$. If the maps f_i are embeddings, then the colimit A is also a limit in $\mathbb{C}$, namely the limit of the ω-chain of associated projections $f_i^p: X_{i+1} \to X_i$. This will be proven via the following intermediate results.

1. *The coprojection maps $\kappa_n: X_n \to A$ associated with the colimit A are embeddings, and their projections $\pi_n = \kappa_n^p: A \to X_n$ form a cone,* i.e. *satisfy $f_n^p \circ \pi_{n+1} = \pi_n$.*
2. *These (co)projections satisfy $\bigvee_{n\in\mathbb{N}} \kappa_n \circ \pi_n = \mathrm{id}_A$.*
3. *The projections $\pi_n: A \to X_n$ are jointly monic: if $h, h': Y \to A$ satisfy $\pi_n \circ h = \pi_n \circ h'$ for all $n \in \mathbb{N}$, then $h = h'$.*
4. *For a cone $g_n: Y \to X_n$, where $f_n^p \circ g_{n+1} = g_n$, there is a unique mediating map $g: Y \to A$, given by $g = \bigvee_n \kappa_n \circ g_n$.*

Proof 1. For each $n \in \mathbb{N}$ we first show that the object X_n forms a cone: for $m \in \mathbb{N}$ there is a map $f_{mn}: X_m \to X_n$, namely:

$$f_{mn} = \begin{cases} f_{n-1} \circ \cdots \circ f_m \; : \; X_m \to X_{m+1} \to \cdots \to X_n & \text{if } m \leq n \\ f_n^p \circ \cdots \circ f_{m-1}^p \; : \; X_m \to X_{m-1} \to \cdots \to X_n & \text{if } m > n. \end{cases}$$

These maps f_{mn} commute with the maps $f_i: X_i \to X_{i+1}$ in the chain: $f_{(m+1)n} \circ f_m = f_{mn}$, and thus form a cone. Since A is the colimit, there is a unique map $\pi_n: A \to X_n$ with $\pi_n \circ \kappa_m = f_{mn}$. In particular, for $m = n$ we get $\pi_n \circ \kappa_n = f_{nn} = \mathrm{id}$. We postpone the proof that $\kappa_n \circ \pi_n \leq \mathrm{id}_A$ for a moment.

2. The maps $\kappa_n \circ \pi_n: A \to A$ form an ascending chain, since $\kappa_n \circ \pi_n = \kappa_{n+1} \circ f_n \circ f_n^p \circ \pi_{n+1} \leq \kappa_{n+1} \circ \pi_{n+1}$. Hence their join exists, which we abbreviate as $f = \bigvee_n \kappa_n \circ \pi_n: A \to A$. Then for each $i \in \mathbb{N}$,

$$\begin{aligned} f \circ \kappa_i &= \bigvee_{n\in\mathbb{N}} \kappa_n \circ \pi_n \circ \kappa_i \\ &= \bigvee_{n\geq i} \kappa_n \circ \pi_n \circ \kappa_i \\ &= \bigvee_{n\geq i} \kappa_n \circ \pi_n \circ \kappa_n \circ f_{in} \\ &= \bigvee_{n\geq i} \kappa_n \circ f_{in} \\ &= \bigvee_{n\geq i} \kappa_i \\ &= \kappa_i. \end{aligned}$$

Hence $f \colon A \to A$ must be the identity, by uniqueness of mediating maps out of colimits.

At this stage we see

$$\kappa_i \circ \pi_i \leq \bigvee_{n\in\mathbb{N}} \kappa_n \circ \pi_n = \mathrm{id}_A.$$

Now we can conclude that $\kappa_i \colon X_i \to A$ is an embedding, with associated projection $\kappa_i^p = \pi_i$. Further, these projections π_i form a cone for the ω-chain $f_n^p : X_{n+1} \to X_n$ since $f_n^p \circ \pi_{n+1} = f_n^p \circ \kappa_{n+1}^p = (\kappa_{n+1} \circ f_n)^p = \kappa_n^p = \pi_n$.

3. Assume $h, h' : Y \to A$ satisfy $\pi_n \circ h = \pi_n \circ h'$, for all $n \in \mathbb{N}$. Then by (2),

$$\begin{aligned} h = \mathrm{id}_A \circ h = \big(\bigvee_n \kappa_n \circ \pi_n\big) \circ h &= \bigvee_n \kappa_n \circ \pi_n \circ h \\ &= \bigvee_n \kappa_n \circ \pi_n \circ h' \\ &= \big(\bigvee_n \kappa_n \circ \pi_n\big) \circ h' = h'. \end{aligned}$$

4. Assume $g_n \colon Y \to X_n$ satisfying $f_n^p \circ g_{n+1} = g_n$. The maps $\kappa_n \circ g_n \colon Y \to A$ then form an ascending chain, since $\kappa_n \circ g_n = \kappa_{n+1} \circ f_n \circ f_n^p \circ g_{n+1} \leq \kappa_{n+1} \circ g_{n+1}$. Hence their join $g = \bigvee_n \kappa_n \circ g_n \colon Y \to A$ exists. It is a mediating map since

$$\begin{aligned} \pi_i \circ g &= \bigvee_{n\in\mathbb{N}} \pi_i \circ \kappa_n \circ g_n \\ &= \bigvee_{n>i} \pi_i \circ \kappa_n \circ g_n \\ &= \bigvee_{n>i} f_{ni} \circ \pi_n \circ \kappa_n \circ g_n \\ &= \bigvee_{n>i} f_{ni} \circ g_n \\ &= \bigvee_{n>i} g_i \\ &= g_i. \end{aligned}$$

Moreover, this g is unique by the previous point. ❑

We now wish to obtain a fixed point $F(A) \xrightarrow{\cong} A$ which is an initial algebra and its inverse $A \xrightarrow{\cong} F(A)$ is a final coalgebra. The idea is to combine the previous result about limit-colimit coincidence with the chain-based approach from Proposition 4.6.1, where one uses a colimit of $F^n(0)$ to obtain the initial algebra and the limit of $F^n(1)$ for the final coalgebra. But if we are in a situation where we have a zero object 0 – which is both initial and final – then we obtain the same chain with coinciding limit and colimit.

Our next assumption is thus that the category at hand has a zero object, commonly written as 0. The singleton monoid is an example of a zero object in the category of monoids; see Exercise 2.1.6. Interestingly, the empty set 0 is also a zero object in the Kleisli category $\mathcal{K}\ell(\mathcal{P}) = \mathbf{SetsRel}$ of the powerset monad: for each set X there is a unique arrow $0 \to \mathcal{P}(X)$ in **Sets**, because 0 is initial in **Sets**, and there is a unique arrow $X \to \mathcal{P}(0)$, since $\mathcal{P}(0) = 1$ is final in **Sets**.

Given such a zero object 0 in a category $\mathbb{C}$, there is for each pair of objects $X, Y \in \mathbb{C}$ an arrow $X \to Y$, which we write as

$$\perp_{X,Y} = (X \xrightarrow{\;!\;} 0 \xrightarrow{\;!\;} Y), \tag{5.16}$$

using that 0 is both initial and final. Such a map $\perp_{X,Y}$ is sometimes called a zero map; it is typical in an algebraic setting, for instance in a category of groups or of vector or Hilbert spaces. The subscripts X, Y are often omitted. Notice that $\perp \circ f = \perp = g \circ \perp$, for all maps f, g.

We use Propositions 5.3.3 and 4.6.1 to generalise Theorem 5.3.1 to more general monads, as in [206].

Theorem 5.3.4 *Assume we have the following data on a category* $\mathbb{C}$.

- *Both an initial and a final object* $0, 1 \in \mathbb{C}$.
- *A monad* $T\colon \mathbb{C} \to \mathbb{C}$ *for which*
  - *the Kleisli category* $\mathcal{K}\ell(T)$ *is dcpo-enriched*
  - *the map* $T(0) \to 1$ *is an isomorphism, so that* $0 \in \mathcal{K}\ell(T)$ *is a zero object (see Exercise 5.2.5)*
  - *the resulting zero maps (5.16) in* $\mathcal{K}\ell(T)$ *are least elements in the hom-set dcpos.*
- *A functor* $F\colon \mathbb{C} \to \mathbb{C}$ *for which*
  - *there is a distributive* $\mathcal{K}\ell$*-law* $FT \Rightarrow TF$*, or equivalently by Proposition 5.2.5, a lifting* $\mathcal{K}\ell(F)\colon \mathcal{K}\ell(T) \to \mathcal{K}\ell(T)$*;*
  - *this lifting* $\mathcal{K}\ell(F)\colon \mathcal{K}\ell(T) \to \mathcal{K}\ell(T)$ *is 'locally monotone':* $f \leq g$ *implies* $\mathcal{K}\ell(T)(f) \leq \mathcal{K}\ell(T)(g)$*;*
  - *there is an initial algebra* $\alpha\colon F(A) \xrightarrow{\cong} A$ *in* $\mathbb{C}$*, obtained as colimit of the chain* $F^n(0)$ *as in (4.12).*

The map $J(\alpha^{-1})\colon A \xrightarrow{\cong} F(A) = \mathcal{K}\ell(F)(A)$ *is then a final coalgebra in the Kleisli category* $\mathcal{K}\ell(T)$*, for the lifted functor* $\mathcal{K}\ell(F)\colon \mathcal{K}\ell(T) \to \mathcal{K}\ell(T)$.

Proof By assumption, the initial algebra structure $\alpha\colon F(A) \xrightarrow{\cong} A$ is obtained on the colimit A of the ω-chain $F^n(0)$, as in (4.12). The functor $J\colon \mathbb{C} \to \mathcal{K}\ell(T)$ is a left adjoint, and thus preserves this colimit; see Exercise 4.6.3. Hence in the Kleisli category $\mathcal{K}\ell(T)$ we obtain $A = J(A)$ as colimit of the chain $J(F^n(0)) = \mathcal{K}\ell(F)^n(J(0)) = \mathcal{K}\ell(F)^n(0) = F^n(0)$. This ω-chain in $\mathcal{K}\ell(T)$ thus starts from the zero object 0 and can be described explicitly as

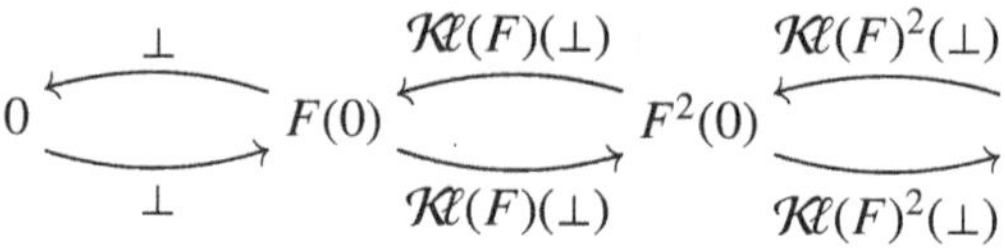

(5.17)

The two maps written as $\perp$ are the unique map $0 \to F(0)$ obtained by initiality and the unique map $F(0) \to 0$ obtained by finality, as in (5.16). The maps $\mathcal{K}\ell(F)^n(\perp)\colon F^n(0) \to F^{n+1}(0)$ are embeddings, since 0 is zero object (on the left below) and $\perp$ is least (on the right):

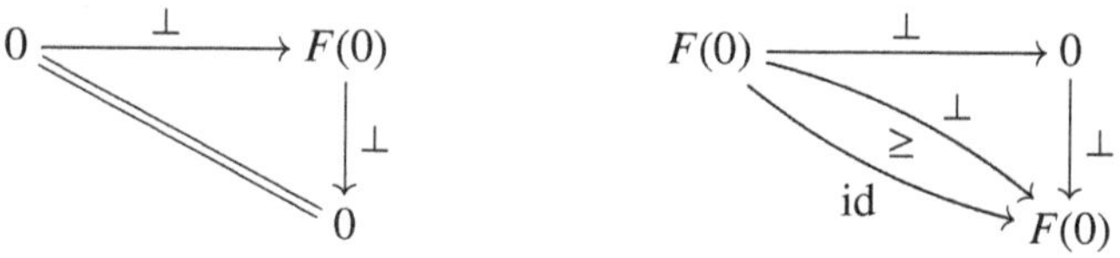

After applying $\mathcal{K}\ell(F)^n$ we obtain a chain of embeddings, by local monotonicity.

Thus, Proposition 5.3.3 applies and the colimit $A \in \mathcal{K}\ell(T)$ of the ω-chain of embeddings $\mathcal{K}\ell(F)^n(\perp)\colon F^n(0) \to F^{n+1}(0)$ in (5.17) is also the limit of the chain of associated projections $\mathcal{K}\ell(F)^n(\perp)\colon F^{n+1}(0) \to F^n(0)$. Hence it carries a final coalgebra structure $J(\alpha^{-1})\colon A \xrightarrow{\cong} F(A)$.

The finality of $J(\alpha^{-1})\colon A \xrightarrow{\cong} F(A)$ in $\mathcal{K}\ell(T)$ can also be proven directly, using the points (1)–(4) listed in Proposition 5.3.3. For a coalgebra $c\colon X \to TF(X)$ in $\mathbb{C}$, that is, for a coalgebra $X \to \mathcal{K}\ell(F)(X)$ in $\mathcal{K}\ell(T)$ we obtain a map $\mathrm{trace}_c\colon X \to A$ in $\mathcal{K}\ell(T)$ as join:

$$\mathrm{trace}_c = \bigvee_{n\in\mathbb{N}} \kappa_n \circledcirc c_n \text{ where } \begin{cases} c_0 &= \perp\colon X \to 0 = F^0(0) \\ c_{n+1} &= \mathcal{K}\ell(F)(c_n) \circledcirc c\colon X \to F(X) \to F^{n+1}(0) \end{cases}$$

and where the maps $\kappa_n\colon F^n(0) \to A$ form the colimit cone, so that $J(\alpha) \circledcirc \mathcal{K}\ell(F)(\kappa_n) = \kappa_{n+1}$. This gives $\mathcal{K}\ell(F)(\pi_n) \circledcirc J(\alpha^{-1}) = \pi_{n+1}$ by taking the associated projections $(-)^p$. This trace map satisfies $\pi_n \circledcirc \mathrm{trace}_c = c_n$, by construction; see Proposition 5.3.3.4. We then get a commuting diagram in $\mathcal{K}\ell(T)$:

$$\begin{array}{ccc} F(X) & \xrightarrow{\mathcal{K}\ell(F)(\mathrm{trace}_c)} & F(A) \\ c\uparrow & & \cong\uparrow J(\alpha^{-1}) \\ X & \xrightarrow[\mathrm{trace}_c]{} & A \end{array}$$

Commutation is obtained by using that the projections $\pi_n = \kappa_n^p \colon A \to F^n(0)$ are jointly monic (see Proposition 5.3.3.3):

$$\begin{aligned} \pi_n \circledcirc J(\alpha) \circledcirc \mathcal{K}\ell(F)(\mathrm{trace}_c) \circledcirc c &= \mathcal{K}\ell(F)(\pi_{n-1}) \circledcirc \mathcal{K}\ell(F)(\mathrm{trace}_c) \circledcirc c \\ &= \mathcal{K}\ell(F)(\pi_{n-1} \circledcirc \mathrm{trace}_c) \circledcirc c \\ &= \mathcal{K}\ell(F)(c_{n-1}) \circledcirc c \\ &= c_n \\ &= \pi_n \circledcirc \mathrm{trace}_c. \end{aligned}$$

In a similar way one obtains uniqueness: assume a map $g \colon X \to A$ in $\mathcal{K}\ell(T)$ satisfies $\mathcal{K}\ell(F)(g) \circledcirc c = J(\alpha^{-1}) \circledcirc g$. The equation $g = \mathrm{trace}_c$ follows if $\pi_n \circledcirc g = \pi_n \circledcirc \mathrm{trace}_c$ holds for each n. This is shown by induction on n, using that $\pi_n \circledcirc \mathrm{trace}_c = c_n$. The case $n = 0$ is trivial, and

$$\begin{aligned} \pi_{n+1} \circledcirc g &= \pi_{n+1} \circledcirc J(\alpha) \circledcirc \mathcal{K}\ell(F)(g) \circledcirc c \\ &= \mathcal{K}\ell(F)(\pi_n) \circledcirc \mathcal{K}\ell(F)(g) \circledcirc c \\ &= \mathcal{K}\ell(F)(\pi_n \circledcirc g) \circledcirc c \\ &\overset{\text{(IH)}}{=} \mathcal{K}\ell(F)(c_n) \circledcirc c \\ &= c_{n+1}. \end{aligned}$$

□

This result requires that the lifted functor $\mathcal{K}\ell(F)$ is only locally monotone, instead of locally continuous. The latter is a more common assumption in this setting; it leads to a slightly simpler proof (see Exercise 5.3.4).

Example 5.3.5 The powerset monad $\mathcal{P}$ satisfies the assumptions of Theorem 5.3.4. Another example is the lift monad $\mathcal{L} = 1 + (-)$ on **Sets**, where maps $f, g \colon X \to \mathcal{L}(Y)$ are ordered via the so-called flat order: $f \leq g$ iff $f(x) = y \in Y$ implies $g(x) = y$. A more interesting example is obtained via the infinite subdistribution monad $\mathcal{D}^{\infty}_{\leq 1}$ on **Sets**. It involves distributions $\varphi \in \mathcal{D}^{\infty}_{\leq 1}(X)$ given by functions $\varphi \colon X \to [0,1]$ satisfying $\sum_x \varphi(x) \leq 1$. Notice that there is no finiteness requirement for the support (but supports have at countably many elements, as in Exercise 4.1.8). Clearly, $\mathcal{D}^{\infty}_{\leq 1}(0) \cong 1$. The resulting zero maps $\bot \colon X \to \mathcal{D}^{\infty}_{\leq 1}(Y)$ in the Kleisli category are given by $\bot(x)(y) = 0$. Subdistributions $\varphi, \psi \in \mathcal{D}^{\infty}_{\leq 1}(X)$ can be ordered pointwise: $\varphi \leq \psi$ iff $\varphi(x) \leq \psi(x)$ for all $x \in X$. This yields a dcpo, with $\bot$ as least element.

As functor $F \colon$ **Sets** $\to$ **Sets** we take $F(X) = 1+(L\times X)$, as before, with initial algebra $[\mathsf{nil}, \mathsf{cons}] \colon F(L^\star) \xrightarrow{\cong} L^\star$ given by lists of labels. There is a distributive law $\lambda\colon F(\mathcal{D}^{\infty}_{\leq 1}(X)) \to \mathcal{D}^{\infty}_{\leq 1}(F(X))$ by Lemma 5.2.12, since the monad $\mathcal{D}^{\infty}_{\leq 1}$ is commutative. Explicitly, this λ is given by

$$\lambda(*)(u) = \begin{cases} 1 & \text{if } u = * \\ 0 & \text{otherwise} \end{cases} \quad \text{and} \quad \lambda(a,\varphi)(u) = \begin{cases} \varphi(x) & \text{if } u = (a,x) \\ 0 & \text{otherwise.} \end{cases}$$

The resulting lifting $\mathcal{K}\ell(F)\colon \mathcal{K}\ell(\mathcal{D}^{\infty}_{\leq 1}) \to \mathcal{K}\ell(\mathcal{D}^{\infty}_{\leq 1})$ sends a map $f\colon X \to \mathcal{D}^{\infty}_{\leq 1}(Y)$ to the map $\mathcal{K}\ell(F)(\backslash, f)\colon 1 + L \times X \to \mathcal{D}^{\infty}_{\leq 1}(1 + L \times Y)$ given by

$$\begin{aligned} \mathcal{K}\ell(f)(*)(u) &= \begin{cases} 1 & \text{if } u = * \\ 0 & \text{otherwise} \end{cases} \\ \mathcal{K}\ell(F)(f)(a,x)(u) &= \begin{cases} f(x)(y) & \text{if } u = (a,y) \\ 0 & \text{otherwise.} \end{cases} \end{aligned}$$

Theorem 5.3.4 now says that we have a final coalgebra $\eta \circ [\mathsf{nil}, \mathsf{cons}]^{-1} \colon L^{\star} \xrightarrow{\cong} F(L^{\star})$ in $\mathcal{K}\ell(\mathcal{D}^{\infty}_{\leq 1})$. This will be illustrated via an example.

Assume our state space X consists of an urn containing black, white and red balls. Thus X can be described as the set of multisets $\mathcal{M}_{\mathbb{N}}(3) = \{\varphi\colon 3 \to \mathbb{N} \mid \mathrm{supp}(\varphi) \text{ is finite}\} \cong \mathbb{N}^3$, where $3 = \{\mathsf{B}, \mathsf{W}, \mathsf{R}\}$ is the three-element set of balls. A typical state $\varphi \in \mathcal{M}_{\mathbb{N}}(3)$ has the multiset form $\varphi = n_B|\mathsf{B}\rangle + n_W|\mathsf{W}\rangle + n_R|\mathsf{R}\rangle$, where $n_B = \varphi(\mathsf{B}), n_W = \varphi(\mathsf{W}), n_R = \varphi(\mathsf{R})$ describe the number of balls of each colour. We write $\|\varphi\| = \varphi(\mathsf{B}) + \varphi(\mathsf{W}) + \varphi(\mathsf{R})$ for the total number of balls, and $\varphi - \mathsf{B} \in \mathcal{M}_{\mathbb{N}}(3)$ for the multiset with one black ball removed (if any). Formally,

$$(\varphi - \mathsf{B})(\mathsf{B}) = \begin{cases} \varphi(\mathsf{B}) - 1 & \text{if } \varphi(\mathsf{B}) > 0 \\ 0 & \text{otherwise} \end{cases} \qquad \begin{aligned} (\varphi - \mathsf{B})(\mathsf{W}) &= \varphi(\mathsf{W}) \\ (\varphi - \mathsf{B})(\mathsf{R}) &= \varphi(\mathsf{R}). \end{aligned}$$

The multisets $\varphi - \mathsf{W}$ and $\varphi - \mathsf{R}$ are defined similarly.

Taking out a ball constitutes a transition. The probability that you take a certain colour is determined by the relative multiplicities: the balls are taken blindly; they cannot be chosen. If a black or white ball is taken out, a next ball may be picked, but the process stops with a red ball. All this is formalised via a coalgebra:

$$\begin{array}{ccc} X & \xrightarrow{\qquad\qquad c \qquad\qquad} & \mathcal{D}^{\infty}_{\leq 1}(1 + 3 \times X) \\ \varphi & \longmapsto & \dfrac{\varphi(\mathsf{B})}{\|\varphi\|}|\mathsf{B}, \varphi - \mathsf{B}\rangle + \dfrac{\varphi(\mathsf{W})}{\|\varphi\|}|\mathsf{W}, \varphi - \mathsf{W}\rangle + \dfrac{\varphi(\mathsf{R})}{\|\varphi\|}|*\rangle \end{array}$$

where $\varphi \in X = \mathcal{M}_{\mathbb{N}}(3)$ is a multiset of coloured balls, representing the contents of the urn. Notice that in order to describe this coalgebra we could have used the monad $\mathcal{D}$ of ordinary distributions instead of $\mathcal{D}^{\infty}_{\leq 1}$. However, a finite system may have infinitely many traces, which can be modelled only via $\mathcal{D}^{\infty}_{\leq 1}$.

Theorem 5.3.4 now gives a coalgebra map $\text{trace}_c : X \to \mathcal{D}^{\infty}_{\leq 1}(3^{\star})$ in the Kleisli category $\mathcal{K}\ell(\mathcal{D}^{\infty}_{\leq 1})$. For a filled urn $\varphi \in X$, this map $\text{trace}_c(\varphi)$ gives a probability distribution over sequences of (consecutive) balls, described as elements of the set $3^{\star} = \{\mathsf{B}, \mathsf{W}, \mathsf{R}\}^{\star}$. For a specific sequence σ of (white and black) balls $\text{trace}_c(\varphi)(\sigma)$ gives the probability of consecutively taking the balls in σ, until a red ball appears. The finality diagram in $\mathcal{K}\ell(\mathcal{D}^{\infty}_{\leq 1})$ precisely captures how to compute these probabilities. This diagram in $\mathcal{K}\ell(\mathcal{D}^{\infty}_{\leq 1})$ is

$$
\begin{array}{ccc}
1 + 3 \times X & \xrightarrow{\mathcal{K}\ell(F)(\text{trace}_c)} & 1 + 3 \times 3^{\star} \\
{\scriptstyle c}\uparrow & & \uparrow {\scriptstyle \cong\ J([\mathsf{nil}, \mathsf{cons}]^{-1})} \\
X & \xrightarrow[\text{trace}_c]{} & 3^{\star}
\end{array}
$$

Commutation says:

$$
\begin{aligned}
\text{trace}_c(\varphi)(\sigma) &= (\mathcal{K}\ell(F)(\text{trace}_c) \circledcirc J(\alpha) \circledcirc c)(\varphi)(\sigma) \\
&= \sum_{u \in 1+3\times X} \; \sum_{v \in 1+3\times 3^{\star}} c(\varphi)(u) \cdot \mathcal{K}\ell(F)(\text{trace}_c)(u)(v) \cdot J(\alpha)(v)(\sigma) \\
&= \begin{cases} c(\varphi)(*) & \text{if } \sigma = \mathsf{nil} \\ c(\varphi)(C, \varphi - C) \cdot \text{trace}_c(\varphi - C)(\sigma') & \text{if } \sigma = \mathsf{cons}(C, \sigma') \end{cases} \\
&= \begin{cases} \frac{\varphi(\mathsf{R})}{\|\varphi\|} & \text{if } \sigma = \mathsf{nil} \\ \frac{\varphi(\mathsf{B})}{\|\varphi\|} \cdot \text{trace}_c(\varphi - \mathsf{B})(\sigma') & \text{if } \sigma = \mathsf{cons}(\mathsf{B}, \sigma') \\ \frac{\varphi(\mathsf{W})}{\|\varphi\|} \cdot \text{trace}_c(\varphi - \mathsf{W})(\sigma') & \text{if } \sigma = \mathsf{cons}(\mathsf{W}, \sigma') \\ 0 & \text{if } \sigma = \mathsf{cons}(\mathsf{R}, \sigma'). \end{cases}
\end{aligned}
$$

Thus, for instance, if we start with a state/urn $\varphi = 3|\mathsf{B}\rangle + 2|\mathsf{W}\rangle + 1|\mathsf{R}\rangle$, then the probability of a trace $\langle \mathsf{B}, \mathsf{W}, \mathsf{B}, \mathsf{B} \rangle$ can be calculated as

$$
\begin{aligned}
&\text{trace}_c(3|\mathsf{B}\rangle + 2|\mathsf{W}\rangle + 1|\mathsf{R}\rangle)(\langle \mathsf{B}, \mathsf{W}, \mathsf{B}, \mathsf{B} \rangle) \\
&= \tfrac{3}{6} \cdot \text{trace}_c(2|\mathsf{B}\rangle + 2|\mathsf{W}\rangle + 1|\mathsf{R}\rangle)(\langle \mathsf{W}, \mathsf{B}, \mathsf{B} \rangle) \\
&= \tfrac{3}{6} \cdot \tfrac{2}{5} \cdot \text{trace}_c(2|\mathsf{B}\rangle + 1|\mathsf{W}\rangle + 1|\mathsf{R}\rangle)(\langle \mathsf{B}, \mathsf{B} \rangle) \\
&= \tfrac{3}{6} \cdot \tfrac{2}{5} \cdot \tfrac{2}{4} \cdot \text{trace}_c(1|\mathsf{B}\rangle + 1|\mathsf{W}\rangle + 1|\mathsf{R}\rangle)(\langle \mathsf{B} \rangle) \\
&= \tfrac{3}{6} \cdot \tfrac{2}{5} \cdot \tfrac{2}{4} \cdot \tfrac{1}{3} \cdot \text{trace}_c(1|\mathsf{W}\rangle + 1|\mathsf{R}\rangle)(\langle\rangle) \\
&= \tfrac{3}{6} \cdot \tfrac{2}{5} \cdot \tfrac{2}{4} \cdot \tfrac{1}{3} \cdot \tfrac{1}{2} \\
&= \tfrac{1}{60}.
\end{aligned}
$$

Example 5.3.6 Our next illustration again uses the subdistribution monad $\mathcal{D}^{\infty}_{\leq 1}$. It shows how an unfair coin can be simulated by two fair ones. This is based on [310, 311]. Consider the coalgebra $c : 2 \to \mathcal{D}^{\infty}_{\leq 1}(\{\mathsf{H}, \mathsf{T}\} + 2)$, with two-element state space $2 = \{0, 1\}$, described in the following diagram:

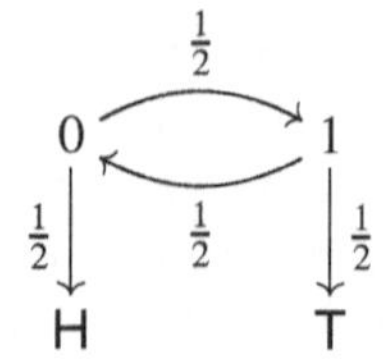

Explicitly, this coalgebra c can be described on the state space 2 via convex sums:

$$c(0) = \tfrac{1}{2}|\mathsf{H}\rangle + \tfrac{1}{2}|1\rangle \qquad \text{and} \qquad c(1) = \tfrac{1}{2}|\mathsf{T}\rangle + \tfrac{1}{2}|0\rangle.$$

The coalgebra is of the form $2 \to \mathcal{D}^{\infty}_{\leq 1}(F(2))$ for the functor $F(X) = \{\mathsf{H}, \mathsf{T}\} + X$. The initial F-algebra is $\mathbb{N} \times \{\mathsf{H}, \mathsf{T}\}$, with structure map:

$$\{\mathsf{H}, \mathsf{T}\} + \mathbb{N} \times \{\mathsf{H}, \mathsf{T}\} \xrightarrow[\cong]{\alpha} \mathbb{N} \times \{\mathsf{H}, \mathsf{T}\} \quad \text{given by} \quad \begin{cases} \alpha(\mathsf{A}) = (0, \mathsf{A}) \\ \alpha(n, \mathsf{A}) = (n+1, \mathsf{A}), \end{cases}$$

where $\mathsf{A} \in \{\mathsf{H}, \mathsf{T}\}$.

The theory described above now yields a trace map by finality in the Kleisli category. It is of the form

$$2 \xrightarrow{\;\text{trace}_c\;} \mathcal{D}^{\infty}_{\leq 1}(\mathbb{N} \times \{\mathsf{H}, \mathsf{T}\}),$$

given by the infinite convex sums

$$\begin{aligned} \text{trace}_c(0) &= \tfrac{1}{2}|0, \mathsf{H}\rangle + \tfrac{1}{4}|1, \mathsf{T}\rangle + \tfrac{1}{8}|2, \mathsf{H}\rangle + \tfrac{1}{16}|3, \mathsf{T}\rangle + \cdots \\ &= \sum_{n \in \mathbb{N}} \frac{1}{2^{2n+1}}|2n, \mathsf{H}\rangle + \frac{1}{2^{2n+2}}|2n+1, \mathsf{T}\rangle \\ \text{trace}_c(1) &= \tfrac{1}{2}|0, \mathsf{T}\rangle + \tfrac{1}{4}|1, \mathsf{H}\rangle + \tfrac{1}{8}|2, \mathsf{T}\rangle + \tfrac{1}{16}|3, \mathsf{H}\rangle + \cdots \\ &= \sum_{n \in \mathbb{N}} \frac{1}{2^{2n+1}}|2n, \mathsf{T}\rangle + \frac{1}{2^{2n+2}}|2n+1, \mathsf{H}\rangle. \end{aligned}$$

A summand like $\tfrac{1}{8}|2, \mathsf{H}\rangle$ in $\text{trace}_c(0)$ represents the probability of $\tfrac{1}{8}$ of getting outcome H via two transitions, starting in state 0.

We can add all the summands with the same coin together, by applying the second projection $\pi_2 \colon \mathbb{N} \times \{\mathsf{H}, \mathsf{T}\} \to \{\mathsf{H}, \mathsf{T}\}$ to these sums. The resulting composite map

$$\text{unfair-coin} = \Big(2 \xrightarrow{\;\text{trace}_c\;} \mathcal{D}^{\infty}_{\leq 1}(\mathbb{N} \times \{\mathsf{H}, \mathsf{T}\}) \xrightarrow{\;\mathcal{D}^{\infty}_{\leq 1}(\pi_2)\;} \mathcal{D}^{\infty}_{\leq 1}(\{\mathsf{H}, \mathsf{T}\})\Big)$$

is given by

$$\text{unfair-coin}(0) = \left(\sum_{n\in\mathbb{N}} \frac{1}{2^{2n+1}}\right)|\mathsf{H}\rangle + \left(\sum_{n\in\mathbb{N}} \frac{1}{2^{2n+2}}\right)|\mathsf{T}\rangle = \tfrac{2}{3}|\mathsf{H}\rangle + \tfrac{1}{3}|\mathsf{T}\rangle$$
$$\text{unfair-coin}(1) = \left(\sum_{n\in\mathbb{N}} \frac{1}{2^{2n+1}}\right)|\mathsf{T}\rangle + \left(\sum_{n\in\mathbb{N}} \frac{1}{2^{2n+2}}\right)|\mathsf{H}\rangle = \tfrac{2}{3}|\mathsf{T}\rangle + \tfrac{1}{3}|\mathsf{H}\rangle.$$

For these last steps we use the familiar equation $\sum_n a^n = \frac{1}{1-a}$ if $|a| < 1$, for instance in

$$\textstyle\sum_n \frac{1}{2^{2n+1}} = \frac{1}{2}\sum_n \left(\frac{1}{2}\right)^{2n} = \frac{1}{2}\sum_n \left(\frac{1}{4}\right)^n = \frac{1}{2}\cdot\frac{1}{1-\frac{1}{4}} = \frac{1}{2}\cdot\frac{4}{3} = \frac{2}{3}.$$

The initial algebra $\mathbb{N}\times\{\mathsf{H},\mathsf{T}\}$ can be described more abstractly as a copower $\mathbb{N}\cdot\{\mathsf{H},\mathsf{T}\}$. The second projection π_2 used above then becomes a codiagonal $\nabla\colon \mathbb{N}\cdot\{\mathsf{H},\mathsf{T}\} \to \{\mathsf{H},\mathsf{T}\}$. This description is used in [248], where a 'monoidal trace operator' in a Kleisli category is constructed from coalgebraic traces. This monoidal trace can thus be used to compute the outcome of the unfair coin directly. Alternatively, as in [311] one can use recursive coalgebras; see [93, 21].

Finally, we briefly return to the transition system example that we used in the beginning of this section.

Remark 5.3.7 For the special case of transition systems $X \to \mathcal{P}F(X) = \mathcal{P}(1+(L\times X))$ involving the powerset monad, there are two alternative ways to obtain its trace semantics: one via determinisation and one via logical interpretation. We briefly review these constructions.

1. As already noted –for instance in Exercise 2.2.3 – a transition system of the form $X \to \mathcal{P}(1 + (L \times X))$ can also be viewed as a non-deterministic automaton $\langle\delta,\varepsilon\rangle\colon X \to \mathcal{P}(X)^L\times 2$. As such it may be 'determinised' into an automaton $\langle\delta',\varepsilon'\rangle\colon \mathcal{P}(X) \to \mathcal{P}(X)^L \times 2$ via the standard definitions:

$$\delta'(U)(a) = \bigcup\{\delta(x)(a) \mid x \in U\} \qquad \varepsilon'(U) = 1 \iff \exists x \in U.\, \varepsilon(x) = 1.$$

   See also [427] and Exercise 5.2.7. Since $\mathcal{P}(L^\star)$ is the final deterministic automaton (see Corollary 2.3.6.2), we obtain a unique coalgebra homomorphism $h\colon \mathcal{P}(X) \to \mathcal{P}(L^\star)$. The trace map $X \to \mathcal{P}(L^\star)$ is then $h \circ \{-\}$. This approach is elaborated in Example 5.4.12.1, and in [429, 272].
2. The logically oriented approach uses the fact that the set $\mathcal{P}(X)$ of predicates on the state space of our non-deterministic automaton $\langle\delta,\varepsilon\rangle\colon X\to\mathcal{P}(X)^L\times 2$ carries elementary logical structure, in the form of an algebra for the functor

$F = 1 + (L \times -)$. Indeed, there is an algebra $[\varepsilon, \neg \bigcirc \neg] \colon F(\mathcal{P}(X)) \to \mathcal{P}(X)$ given by $\varepsilon \colon 1 \to \mathcal{P}(X)$ as a subset of final states and by a labelled (strong) nexttime $\neg\bigcirc\neg \colon L \times \mathcal{P}(X) \to \mathcal{P}(X)$ operator. It maps a pair $(a, P) \in L \times \mathcal{P}(X)$ to the predicate $\neg \bigcirc \neg(a, P)$, commonly written as $\neg \bigcirc_a \neg(P)$, given by

$$\neg \bigcirc_a \neg(P) = \{x \in X \mid \exists x' \in X.\, x \xrightarrow{a} x' \wedge P(x')\}.$$

By initiality of $L^\star$ we then obtain an algebra homomorphism $\ell \colon L^\star \to \mathcal{P}(X)$. The adjunction $\mathbf{Sets}^{\mathrm{op}} \leftrightarrows \mathbf{Sets}$ from Exercise 2.5.2 now yields the trace map $X \to \mathcal{P}(L^\star)$. This second approach fits in the 'testing' framework of [377] and uses coalgebraic modal logic; see Section 6.5 for a more general account.

The trace semantics that we have described in this section induces a new type of equivalence on states (of suitable coalgebras), namely trace equivalence, given by the relation $\{(x, y) \mid \mathrm{trace}(x) = \mathrm{trace}(y)\}$. Exercise 5.3.5 below shows that bisimilarity implies trace equivalence, but not the other way around.

Exercises

5.3.1 Suppose we have a monad $T \colon \mathbb{C} \to \mathbb{C}$ that satisfies $T(0) \cong 0$ and $T(1) \cong 1$. Use Exercise 5.2.9 to transform T into a monad $T' = T(1 + (-))$ that satisfies $T'(0) \cong 1$, as needed in Theorem 5.3.4. (This transformation trick applies for instance to the ordinary distribution monad $\mathcal{D}$; a non-trivial dcpo structure on these discrete distributions in $\mathcal{D}(X)$ is described in [106].)

5.3.2 Consider the alphabet $\{a, b\}$ with two non-terminals V, W and productions

$$V \longrightarrow a \cdot V \qquad V \longrightarrow W \qquad W \longrightarrow b \cdot W \cdot a \qquad W \longrightarrow \langle\rangle.$$

1. Describe this CFG as a coalgebra $X \to \mathcal{P}(FX)$ with state space $X = \{V, W\}$ and functor $F = ((-) + \{a, b\})^\star$.
2. Check that the unparsed language generated by the symbol V is the set $\{a^n b^m a^m \mid n, m \in \mathbb{N}\}$.
3. Consider the parsed language map $\mathrm{trace}_g \colon X \to \{a, b\}^\triangle$ from Example 5.3.2 and show that a term $t \in \mathrm{trace}_g(V)$ can be drawn as a tree:

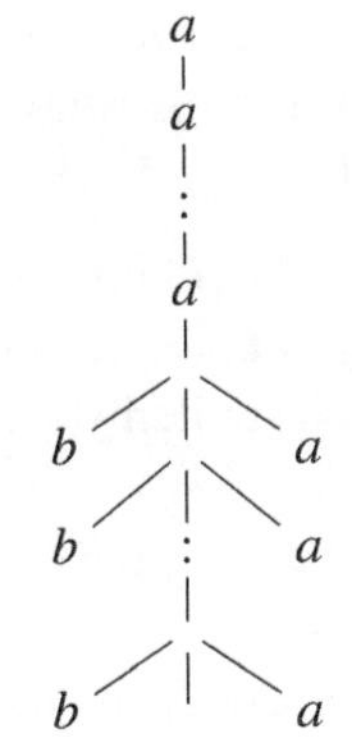

5.3.3 1. Define a function $L^{\triangle} \to L^{\star}$ that maps parsed words to 'flat' words by initiality. Prove that this function is a split epi.

2. Prove that the assignment $L \mapsto L^{\triangle}$ is functorial and that the mapping $L^{\triangle} \twoheadrightarrow L^{\star}$ from (1) forms a natural transformation.

3. Let $L^{\wedge}$ be the final coalgebra of the functor $X \mapsto (X + L)^{\star}$ from Example 5.3.2. It consists of both the finite and infinite parsed words. Define a map $L^{\triangle} \to L^{\wedge}$ and show that it is injective.

(This gives the following fundamental span of languages $L^{\star} \twoheadleftarrow L^{\triangle} \rightarrowtail L^{\wedge}$. The three operations involved $L \mapsto L^{\star}, L^{\triangle}, L^{\wedge}$ are all monads and the mappings between them preserve the monad structure; see [203].)

5.3.4 Consider the situation as described in Theorem 5.3.4, but with the stronger assumption that the lifted functor $\mathcal{K}\ell(F)\colon \mathcal{K}\ell(T) \to \mathcal{K}\ell(T)$ is *locally continuous*. This means that directed joins in hom-sets are preserved: $\mathcal{K}\ell(F)(\bigvee_i f_i) = \bigvee_i \mathcal{K}\ell(F)(f_i)$. Check that under this assumption the proof of Theorem 5.3.4 can be simplified in the following manner. For an initial F-algebra $\alpha\colon F(A) \xrightarrow{\cong} A$ and a coalgebra $c\colon X \to \mathcal{K}\ell(F)(X)$, consider the following two operators on Kleisli hom-sets:

$$\begin{array}{rclcrcl} \mathcal{K}\ell(T)(A,A) & \xrightarrow{\Phi} & \mathcal{K}\ell(T)(A,A) & \qquad & \mathcal{K}\ell(T)(X,A) & \xrightarrow{\Psi} & \mathcal{K}\ell(T)(X,A) \\ g & \longmapsto & \mathcal{K}\ell(F)(g) \circledcirc J(\alpha) \circledcirc J(\alpha^{-1}) & & h & \longmapsto & \mathcal{K}\ell(F)(h) \circledcirc J(\alpha) \circledcirc c. \end{array}$$

1. Prove that both Φ and Ψ are continuous.
2. Prove that the least fixed point $g = \mathrm{fix}(\Phi) = \bigvee_{n \in \mathbb{N}} \Phi^n(\bot)$ is a map of algebras $J(\alpha) \to J(\alpha)$. Conclude $g = \mathrm{id}$ from Proposition 5.2.6.
3. Check that the least fixed point $h = \mathrm{fix}(\Psi) = \bigvee_{n \in \mathbb{N}} \Psi^n(\bot)$ is a map of coalgebras $c \to J(\alpha^{-1})$.
4. Prove that this h is the unique such map. *Hint*: Use that if h' is also such a coalgebra map, then $\Phi^n(\bot) \circledcirc h' = \Psi^n(\bot)$.

5.3.5 Let $c\colon X \to \mathcal{P}F(X)$ and $d\colon Y \to \mathcal{P}F(Y)$ be two coalgebras, where $F\colon \textbf{Sets} \to \textbf{Sets}$ preserves weak pullbacks and has an initial algebra $F(A) \xrightarrow{\cong} A$ (as in Theorem 5.3.1).

1. Let $f\colon X \to Y$ be a homomorphisms (of $\mathcal{P}F$-coalgebras). Prove that $\mathrm{trace}_d \circledcirc f = \mathrm{trace}_c\colon X \to A$ in $\mathcal{K}\ell(\mathcal{P})$.
2. Conclude that bisimilarity is included in trace equivalence:

$$x \;{}_c\underline{\leftrightarrow}_d\; y \Longrightarrow \mathrm{trace}_c(x) = \mathrm{trace}_d(y).$$

3. Check that the reverse implication ($\Longleftarrow$) in (2) does not hold, for instance via the following two pictures:

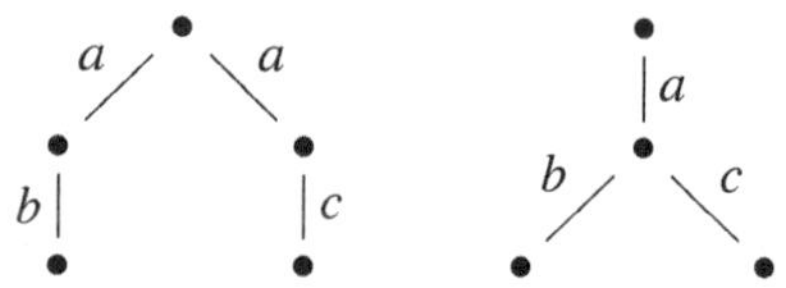

5.4 Eilenberg–Moore Categories and Distributive Laws

This section introduces the category $\mathcal{EM}(T)$, called the Eilenberg–Moore category, for a monad or comonad T. Its objects are algebras or coalgebras of T, but with some additional requirements. These requirements make them different from coalgebras of a functor.

Definition 5.4.1 Given a monad $T = (T, \eta, \mu)$ on a category $\mathbb{C}$ (see Definition 5.1.1), one defines an **algebra** of this monad as an algebra $\alpha\colon T(X) \to X$ of the functor T satisfying two additional requirements, expressed via the diagrams

$$\begin{array}{ccc} X & \xrightarrow{\eta_X} & T(X) \\ & \searrow{=} & \downarrow{\alpha} \\ & & X \end{array} \qquad\qquad \begin{array}{ccc} T^2(X) & \xrightarrow{\mu_X} & T(X) \\ {\scriptstyle T(\alpha)}\downarrow & & \downarrow{\scriptstyle \alpha} \\ T(X) & \xrightarrow[\alpha]{} & X \end{array} \tag{5.18}$$

We write $\mathcal{EM}(T) = \mathcal{EM}(T, \eta, \mu)$ for the category with such monad-algebras as objects. The morphisms are the same as homomorphisms of functor-algebras: a morphism from $\alpha\colon T(X) \to X$ to $\beta\colon T(Y) \to Y$ is a map $f\colon X \to Y$ in $\mathbb{C}$ with $\beta \circ T(f) = f \circ \alpha$.

Similarly, for a comonad $S = (S, \varepsilon, \delta)$ a **coalgebra** is a coalgebra $\alpha\colon X \to S(X)$ of the functor S satisfying

$$\begin{array}{ccc} S(X) & \xrightarrow{\varepsilon_X} & X \\ {\scriptstyle\alpha}\uparrow & \nearrow\!\!\!\!= & \\ X & & \end{array} \qquad \begin{array}{ccc} S(X) & \xrightarrow{\delta_X} & S^2(X) \\ {\scriptstyle\alpha}\uparrow & & \uparrow{\scriptstyle S(\alpha)} \\ X & \xrightarrow[\alpha]{} & S(X) \end{array} \tag{5.19}$$

We write $\mathcal{EM}(S) = \mathcal{EM}(S, \varepsilon, \delta)$ for this category of comonad coalgebras, with homomorphisms given by ordinary coalgebra homomorphisms.

There is ample room for confusion at this stage. First of all, the same notation $\mathcal{EM}(T)$ is used for both a category of algebras, if T is a monad, and for a category of coalgebras, if T is a comonad. The context should make clear which option is intended. What might help a bit is that we generally use the letter T for a monad and S for a comonad. Notice, by the way, that for Kleisli categories we have the same overloading of notation.

Next, 'algebra' or 'coalgebra' may be used both for a functor – in which case it is just a map of a certain form – and for a (co)monad; in the latter case such a map should satisfy additional equations, as described in (5.18) and (5.19) above. In order to emphasise what is intended we sometimes explicitly speak of a *functor algebra* or of a *monad algebra/Eilenberg–Moore algebra* – and similarly, we can have a *functor coalgebra* and a *comonad coalgebra/Eilenberg–Moore coalgebra*.

Recall that we use the notation **Alg**(F) and **CoAlg**(F) for the categories of algebras and coalgebras of a functor F. They are related to categories $\mathcal{EM}(-)$ of Eilenberg–Moore algebras and coalgebras via two full and faithful (horizontal) functors in commuting triangles:

$$\begin{array}{ccc} \mathcal{EM}(T, \eta, \mu) & \longrightarrow & \mathbf{Alg}(T) \\ & \searrow \quad \swarrow & \\ & \mathbb{C} & \end{array} \qquad \begin{array}{ccc} \mathcal{EM}(S, \varepsilon, \delta) & \longrightarrow & \mathbf{CoAlg}(S) \\ & \searrow \quad \swarrow & \\ & \mathbb{C} & \end{array}$$

Here, T is a monad and S is a comonad. In writing **Alg**(T) and **CoAlg**(S) we consider them as ordinary functors, forgetting about their (co)unit and (co)multiplication.

Lemma 5.4.2 *Let T be a monad and S a comonad on a category $\mathbb{C}$. The obvious forgetful functors have left and right adjoints*

$$\begin{array}{c} \mathcal{EM}(T) \\ \mathcal{F}\left(\dashv\downarrow\right. \\ \mathbb{C} \end{array} \qquad\qquad \begin{array}{c} \mathcal{EM}(S) \\ \left.\downarrow\dashv\right)\mathcal{G} \\ \mathbb{C} \end{array}$$

given by multiplication and comultiplication:

$$\mathcal{F}(X) = \begin{pmatrix} T^2(X) \\ \downarrow{\scriptstyle \mu_X} \\ T(X) \end{pmatrix} \qquad \textit{and} \qquad \mathcal{G}(Y) = \begin{pmatrix} S^2(Y) \\ \uparrow{\scriptstyle \delta_Y} \\ S(Y) \end{pmatrix}.$$

These adjunctions give rise to a comonad on the category $\mathcal{EM}(T)$ of algebras, and to a monad on the category $\mathcal{EM}(S)$ of coalgebras; see Exercise 5.4.8.

Also the Kleisli categories can be embedded full and faithfully in Eilenberg–Moore categories, via commuting triangles:

$$\begin{array}{ccc} \mathcal{K}\ell(T) & \longrightarrow & \mathcal{EM}(T) \\ & \searrow \;\; \swarrow & \\ & \mathbb{C} & \end{array} \qquad\qquad \begin{array}{ccc} \mathcal{K}\ell(S) & \longrightarrow & \mathcal{EM}(S) \\ & \searrow \;\; \swarrow & \\ & \mathbb{C} & \end{array}$$

Recall that maps in Kleisli categories are used to represent computations. Through the full and faithful embeddings into Eilenberg–Moore categories these computations can also be studied, between free algebras or cofree coalgebras, in richer universes, with more categorical structure; see Lemma 5.4.5 and Proposition 5.4.6 below.

Proof We shall sketch the proof for the comonad case. The monad case is dual. For an arbitrary comonad coalgebra $\alpha\colon X \to S(X)$ we have to produce a bijective correspondence:

$$\frac{\begin{pmatrix} S(X) \\ \alpha\uparrow \\ X \end{pmatrix} \xrightarrow{\quad f \quad} \begin{pmatrix} S^2(Y) \\ \uparrow{\scriptstyle \delta_Y} \\ S(Y) \end{pmatrix}}{X \xrightarrow[\quad g \quad]{} Y}.$$

It is given as follows.

- For a map of coalgebras $f\colon X \to S(Y)$ take $\overline{f} = \varepsilon \circ f\colon X \to Y$.

- For an ordinary map $g\colon X \to Y$ in $\mathbb{C}$ take $\overline{g} = S(g) \circ \alpha\colon X \to S(Y)$. It is a map of coalgebras since

$$\begin{aligned} S(\overline{g}) \circ \alpha &= S^2(g) \circ S(\alpha) \circ \alpha \\ &= S^2(g) \circ \delta \circ \alpha && \text{see (5.19)} \\ &= \delta \circ S(g) \circ \alpha && \text{by naturality of } \delta \\ &= \delta \circ \overline{g}. \end{aligned}$$

These transformations are each other's inverse – i.e. $\overline{\overline{f}} = f$ and $\overline{\overline{g}} = g$ – via the coalgebra laws (5.19).

The functor $\mathcal{K}\ell(S) \to \mathcal{EM}(S)$ is on objects $Y \mapsto \mathcal{G}(Y) = (\delta_Y\colon S(Y) \to S^2(Y))$. A Kleisli map $f\colon Y \to Z$, where $f\colon S(Y) \to Z$ in $\mathbb{C}$ yields $S(f) \circ \delta\colon S(Y) \to S(Z)$. We skip the proof that this is a map of coalgebras and show that each coalgebra map $g\colon \mathcal{G}(Y) \to \mathcal{G}(Z)$ is of this form. Take $f = \varepsilon \circ g\colon S(Y) \to Z$. Then

$$S(f) \circ \delta = S(\varepsilon) \circ S(g) \circ \delta = S(\varepsilon) \circ \delta \circ g = g. \qquad \square$$

We turn to examples of Eilenberg–Moore categories. It may happen that quite a few details have to be checked to verify that a certain category is (isomorphic to) an Eilenberg–Moore category. We sketch only the essentials of the various constructions. In Theorem 6.7.11 we describe a general way to obtain Eilenberg–Moore categories via specifications (algebras or coalgebras with assertions). In general, one calls a category **algebraic** or **monadic** if it is isomorphic to a category of Eilenberg–Moore algebras.

Example 5.4.3 1. For a monoid $M = (M, 1, \cdot)$ there is an associated monad $M \times (-)\colon$ **Sets** $\to$ **Sets**; see Example 5.1.3.5 and Exercise 5.1.7. Recall that the unit is $\eta(x) = (1, x)$ and the multiplication $\mu(m, k, x) = (m \cdot k, x)$. An Eilenberg–Moore algebra $\alpha\colon M \times X \to X$ is given by a set X with this operation α. It corresponds to a **monoid action**, since the equations (5.18) amount to

$$\alpha(1, x) = x \qquad \alpha(m, \alpha(k, x)) = \alpha(m \cdot k, x).$$

Usually such an action is written as a dot, as in a scalar multiplication:

$$1 \bullet x = x \qquad m \bullet (k \bullet x) = (m \cdot k) \bullet x,$$

where $m \bullet x = \alpha(m, x)$. It is easy to see that homomorphisms of algebras, that is, maps in $\mathcal{EM}(M \times -)$, correspond to scalar-product-preserving functions f, satisfying $f(m \bullet x) = m \bullet f(x)$.

2. For a semiring S we have the multiset monad $\mathcal{M}_S$; see Lemma 5.1.5. An algebra $\alpha\colon \mathcal{M}_S(X) \to X$ forms a clever encoding of both a scalar

multiplication $\bullet$ and a commutative monoid structure $(0, +)$ on X. This structure can be extracted from α as follows:

$$0 = \alpha(0) \qquad x + y = \alpha(1|x\rangle + 1|y\rangle) \qquad s \bullet x = \alpha(s|x\rangle).$$

The element $0 \in \mathcal{M}_S(X)$ on the right in the first equation is the empty multiset, and $1|x\rangle = \eta(x) \in \mathcal{M}_S(X)$ is the singleton multiset.

These operations turn X into a **module** over the semiring S. As illustration, we check the zero law for the monoid structure. It involves translating the required equality into one of the algebra laws (5.18):

$$\begin{aligned} x + 0 = \alpha(1|x\rangle + 1|0\rangle) &\overset{(5.18)}{=} \alpha(1|\alpha(1|x\rangle)\rangle + 1|\alpha(0)\rangle) \\ &= \alpha(\mathcal{M}_S(\alpha)(1|\,1|x\rangle\rangle + 1|0\rangle)) \\ &= \alpha(\mathcal{M}_S(\alpha)(1|\,1|x\rangle\rangle + 1|0|0\rangle\rangle)) \\ &\overset{(5.18)}{=} \alpha(\mu(1|\,1|x\rangle\rangle + 1|0|0\rangle\rangle)) \\ &= \alpha((1 \cdot 1)|x\rangle + (1 \cdot 0)|0\rangle) = \alpha(1|x\rangle) = x. \end{aligned}$$

Conversely, if Y is a module over S, then we can define an algebra structure $\beta\colon \mathcal{M}_S(Y) \to Y$ by

$$\beta(s_1|y_1\rangle + \cdots + s_n|y_n\rangle) = s_1 \bullet y_1 + \cdots + s_n \bullet y_n.$$

Notice that the sum on the left is a formal sum (multiset) in $\mathcal{M}_S(Y)$, whereas the sum on the right is an actual sum, in the module Y. It is not hard to see that algebra maps correspond to module maps, preserving scalar multiplication and sums. Thus we have

$$\mathcal{EM}(\mathcal{M}_S) \cong \mathbf{Mod}_S,$$

where $\mathbf{Mod}_S$ is the category of modules over the semiring S.

Some special cases are worth mentioning:

$\mathbf{Mod}_{\mathbb{N}} \cong \mathbf{CMon}$	the category of commutative monoids
$\mathbf{Mod}_{\mathbb{Z}} \cong \mathbf{Ab}$	the category of commutative/Abelian groups
$\mathbf{Mod}_K \cong \mathbf{Vect}_K$	the category of vector spaces over a field K.

We see that monads and their algebras give a uniform description of these mathematical structures.

The distribution monad $\mathcal{D}$ is similar to the multiset monad, but its algebras are quite different. They are **convex sets**, where each formal *convex* combination $\sum_i r_i|x_i\rangle$, with $r_i \in [0, 1]$ satisfying $\sum_i r_i = 1$, has an actual sum; see [438, 441, 247].

3. Algebras $\alpha\colon \mathcal{P}(X) \to X$ of the powerset monad encode complete lattice structure on the set X. For an arbitrary subset $U \subseteq X$ one defines

$$\bigvee U = \alpha(U) \in X.$$

This is the join with respect to an order on X defined as $x \leq y$ iff $y = \bigvee\{x, y\} = \alpha(\{x, y\})$. This is a partial order. Reflexivity $x \leq x$ holds because $\alpha(\{x, x\}) = \alpha(\{x\}) = x$, since the singleton map is the unit of the powerset monad; see Example 5.1.3.1. Antisymmetry is trivial, and for transitivity one proceeds as follows. If $x \leq y$ and $y \leq z$, that is $\alpha(\{x, y\}) = y$ and $\alpha(\{y, z\}) = z$, then $x \leq z$ since

$$\begin{aligned}
\alpha(\{x, z\}) &= \alpha(\{\alpha(\{x\}), \alpha(\{y, z\})) \\
&= \alpha(\mathcal{P}(\alpha)(\{\{x\}, \{y, z\}\})) \\
&\overset{(5.18)}{=} \alpha(\bigcup\{\{x\}, \{y, z\}\}) \\
&= \alpha(\{x, y, z\}) \\
&\overset{(5.18)}{=} \alpha(\bigcup\{\{x, y\}, \{z\}\}) \\
&= \alpha(\mathcal{P}(\alpha)(\{\{x, y\}, \{z\}\})) \\
&= \alpha(\{\alpha(\{x, y\}), \alpha(\{z\})) \\
&= \alpha(\{y, z\}) \\
&= z.
\end{aligned}$$

Similarly one can prove $\forall x \in U. x \leq y$ iff $\bigvee U \leq y$.

Conversely, each complete lattice L forms a $\mathcal{P}$-algebra $\mathcal{P}(L) \to L$ via $U \mapsto \bigvee U$. Algebra maps correspond to $\bigvee$-preserving (also known as linear) maps. In this way we obtain isomorphisms

$$\mathcal{EM}(\mathcal{P}) \cong \mathbf{CL} \qquad \text{and} \qquad \mathcal{EM}(\mathcal{P}_{\text{fin}}) \cong \mathbf{JSL},$$

where **CL** is the category of complete lattices and linear maps, and **JSL** is the category of join semi-lattices and join-preserving maps. In such join semilattices one has only finite joins $(\bot, \vee)$. The situation for meet semilattices is described in Exercise 5.4.6.

4. On the category **PoSets** one can define the ideal monad Idl. An ideal in a poset $X = (X, \leq)$ is a directed downset $I \subseteq X$; thus, I is non-empty and satisfies:

   - if $x \leq y \in I$, then $x \in I$
   - if $y, y' \in I$, then there is an element $x \in I$ with $y \leq x$ and $y' \leq x$.

   We write Idl(X) for the poset of ideals in X, ordered by inclusion. The mapping $X \mapsto \text{Idl}(X)$ forms a monad on **PoSets**, with unit $\eta\colon X \to \text{Idl}(X)$ given by $\eta(x) = \downarrow x = \{y \in X \mid y \leq x\}$. Union is multiplication. As in the previous point, algebras $\text{Idl}(X) \to X$ for this monad are supremum maps. Since they are defined for directed (down)sets only, they turn the set X into a directed complete partial order (dcpo). Thus,

   $$\mathcal{EM}(\text{Idl}) \cong \mathbf{Dcpo}, \qquad \text{the category of dcpos and continuous maps.}$$

5. Most of our examples of comonad coalgebras appear in the next chapter, as coalgebras satisfying certain assertions. We shall consider one example here. A coalgebra $X \to X^{\mathbb{N}}$ of the stream comonad from Example 5.1.10 can be identified with an endomap $f\colon X \to X$. The coalgebra is then given by $c(x) = \langle x, f(x), f^2(x), f^3(x), \ldots\rangle$. The reason is that such a coalgebra can be identified with a map of monoids $c'\colon \mathbb{N} \to X^X$. Hence it is determined by the function $f = c'(1)\colon X \to X$.

   More generally, for a monoid M we know that the functor $M \times (-)$ is a monad, by Example 5.1.3.5, and that $(-)^M$ is a comonad, by Exercise 5.1.11. In that case it can be shown that there is a bijective correspondence

$$\frac{\dfrac{\text{coalgebras } X \longrightarrow X^M}{\text{algebras } M \times X \longrightarrow X}}{\text{monoid maps } M \longrightarrow X^X}$$

We continue with some basic categorical facts. Eilenberg–Moore categories inherit limits (for algebras) and colimits (for coalgebras) from the underlying categories. This is the same as for functor (co)algebras.

Lemma 5.4.4 *The Eilenberg–Moore category $\mathcal{EM}(T)$ for a monad $T\colon \mathbb{C} \to \mathbb{C}$ has the same kind of limits as the underlying category $\mathbb{C}$. These limits are constructed elementwise, as in*

$$\begin{pmatrix} T(X) \\ \alpha\big\downarrow \\ X \end{pmatrix} \times \begin{pmatrix} T(Y) \\ \big\downarrow\beta \\ Y \end{pmatrix} = \begin{pmatrix} T(X \times Y) \\ \big\downarrow \langle \alpha \circ T(\pi_1), \beta \circ T(\pi_2)\rangle \\ X \times Y \end{pmatrix}.$$

The same holds for colimits in $\mathcal{EM}(S)$, for a comonad S. □

The following useful result due to Linton depends on some special properties of the category **Sets**. For a proof we refer to [56, §9.3, prop. 4]. It means that a category $\mathcal{EM}(T)$ of algebras for a monad T on **Sets** always has both limits and colimits.

Lemma 5.4.5 *For a monad T on* **Sets**, *the category $\mathcal{EM}(T)$ of algebras is cocomplete (and complete).* □

There are some more pleasant properties of such Eilenberg–Moore categories of monads on **Sets**. For instance, they always carry a logical factorisation system (see Exercise 5.4.3), given by injective and surjective algebra homomorphisms.

The structure of categories of Eilenberg–Moore coalgebras is investigated systematically in [282, 283].

In [111] it is shown that Eilenberg–Moore $\mathcal{EM}(T)$ and Kleisli categories $\mathcal{K}\ell(T)$ have biproducts if and only if the monad T is 'additive', i.e. maps coproducts to products, as in Exercise 2.1.10 for the (finite) powerset monad $\mathcal{P}$ and in Exercise 4.1.2 for the multiset monads $\mathcal{M}_S$.

We mention a related result, also without proof, for monads on **Sets**. It describes monoidal closed structure (see [344]) in categories of algebras, as in Proposition 5.2.13 for Kleisli categories. It explains why categories of abelian groups or vector spaces have tensors ⊗ and associated function spaces. The result can be generalised to other categories, provided suitable coequalisers of algebras exist.

Proposition 5.4.6 (From [302, 301]) *Let T be a commutative monad on* **Sets**. *The associated Eilenberg–Moore category $\mathcal{EM}(T)$ is then symmetric monoidal closed. The free functor F:* **Sets** $\to \mathcal{EM}(T)$ *sends cartesian products $(1, \times)$ to monoidal products $(I, \otimes)$, in the sense that $F(1) \cong I$ is the tensor unit, and $F(X \times Y) \cong F(X) \otimes F(Y)$.* □

Let $T(X) \xrightarrow{\alpha} X$, $T(Y) \xrightarrow{\beta} Y$ and $T(Z) \xrightarrow{\gamma} Z$ be algebras. The tensors ⊗ for algebras are constructed in such a way that there is a bijective correspondence:

$$\frac{\text{homomorphisms of algebras } \alpha \otimes \beta \longrightarrow \gamma}{\text{bihomomorphisms } X \times Y \longrightarrow Z}.$$

Intuitively, a bihomomorphism $f \colon X \times Y \to Z$ is a homomorphism in each coordinate separately: $f(x, -) \colon Y \to Z$ and $f(-, y) \colon X \to Z$ are algebra maps, for each $x \in X$ and $y \in Y$. More formally this is described via a commuting diagram, involving the 'double strength' map dst of a commutative monad (see Definition 5.2.9):

$$\begin{array}{ccccc} T(X) \times T(Y) & \xrightarrow{\text{dst}} & T(X \times Y) & \xrightarrow{T(f)} & T(Z) \\ {\scriptstyle \alpha \times \beta}\downarrow & & & & \downarrow{\scriptstyle \gamma} \\ X \times Y & & \xrightarrow[f]{\qquad\qquad} & & Z \end{array}$$

Under suitable circumstances one can construct for a functor F the free monad or the cofree comonad on F. The Eilenberg–Moore categories of these (co)free extensions turn out to be the same as the categories of (co)algebras of the original functor F. This is the content of the next result.

Proposition 5.4.7 *Let $F \colon \mathbb{C} \to \mathbb{C}$ be an arbitrary functor. Recall the free monad $F^* \colon \mathbb{C} \to \mathbb{C}$ on F from Proposition 5.1.8, and the cofree comonad the* *cofree comonad $F^\infty\colon \mathbb{C} \to \mathbb{C}$ on F from Exercise 5.1.15, described via the initial algebras and final coalgebras:*

$$X + F(F^*(X)) \xrightarrow[\cong]{\alpha_X} F^*(X) \qquad \textit{and} \qquad F^\infty(X) \xrightarrow[\cong]{\zeta_X} X \times F(F^\infty(X)).$$

Assuming these constructions exist in $\mathbb{C}$, there are isomorphisms of categories of monad (co)algebras and functor (co)algebras:

$$\mathcal{EM}(F^*) \cong \mathbf{Alg}(F) \quad \textit{and} \quad \mathcal{EM}(F^\infty) \cong \mathbf{CoAlg}(F).$$

Proof We do the proof in the coalgebra case only. We first list the relevant structure. The counit $\varepsilon_X\colon F^\infty(X) \to X$ is given by $\varepsilon_X = \pi_1 \circ \zeta_X$. The comultiplication $\delta_X\colon F^\infty(X) \to F^\infty F^\infty(X)$ is obtained by finality in

$$\begin{array}{ccc}
F^\infty(X) \times F(F^\infty(X)) & \xrightarrow{\mathrm{id} \times F(\delta_X)} & F^\infty(X) \times F(F^\infty F^\infty(X)) \\
{\scriptstyle \langle \mathrm{id}, \pi_2 \circ \zeta_X \rangle}\uparrow & & \cong\uparrow{\scriptstyle \zeta_{F^\infty(X)}} \\
F^\infty(X) & \xrightarrow[\delta_X]{} & F^\infty F^\infty(X)
\end{array}$$

There is a universal natural transformation $\theta\colon F^\infty \Rightarrow F$ by

$$\theta_X = \Big(F^\infty(X) \xrightarrow[\cong]{\zeta_X} X \times F(F^\infty(X)) \xrightarrow{\pi_2} F(F^\infty(X)) \xrightarrow{F(\varepsilon_X)} F(X)\Big).$$

We now define two functors:

$$\mathcal{EM}(F^\infty) \underset{K}{\overset{L}{\rightleftarrows}} \mathbf{CoAlg}(F).$$

The functor L is easy: it takes a comonad coalgebra $\beta\colon X \to F^\infty(X)$ and sends it to $\theta_X \circ \beta\colon X \to F(X)$. Naturality of θ makes L a functor.

In the other direction, the functor K sends a functor coalgebra $c\colon X \to F(X)$ to the map $K(c)$ defined by finality in

$$\begin{array}{ccc}
X \times F(X) & \xrightarrow{\mathrm{id} \times F(K(c))} & X \times F(F^\infty(X)) \\
{\scriptstyle \langle \mathrm{id}, c \rangle}\uparrow & & \cong\uparrow{\scriptstyle \zeta_X} \\
X & \xrightarrow[K(c)]{} & F^\infty(X)
\end{array}$$

Then

$$\begin{aligned}
\varepsilon_X \circ K(c) = \pi_1 \circ \zeta_X \circ K(c) &= \pi_1 \circ (\mathrm{id} \times F(K(c))) \circ \langle \mathrm{id}, c \rangle \\
&= \pi_1 \circ \langle \mathrm{id}, c \rangle = \mathrm{id}.
\end{aligned}$$

Using uniqueness of maps to final coalgebras one can prove that $\delta \circ K(c) = F^\infty(K(c)) \circ K(c)$. Hence $K(c)\colon X \to F^\infty(X)$ is a comonad coalgebra. Also by uniqueness one can show that maps of functor coalgebras are also maps of monad coalgebras.

We have $LK = \mathrm{id}$ since:

$$\begin{aligned} LK(c) = \theta_X \circ K(c) &= F(\varepsilon_X) \circ \pi_2 \circ \zeta_X \circ K(c) \\ &= F(\varepsilon_X) \circ \pi_2 \circ (\mathrm{id} \times F(K(c))) \circ \langle \mathrm{id}, c\rangle \\ &= F(\varepsilon_X) \circ F(K(c)) \circ \pi_2 \circ \langle \mathrm{id}, c\rangle \\ &= F(\mathrm{id}) \circ c \\ &= c. \end{aligned}$$

We get $KL(\beta) = \beta$ because $\beta\colon X \to F^\infty(X)$ is a map of coalgebras in the diagram defining $KL(\beta)$, namely:

$$\begin{array}{ccc} X \times F(X) & \xrightarrow{\ \mathrm{id} \times F(\beta)\ } & X \times F(F^\infty(X)) \\ {\scriptstyle \langle \mathrm{id}, L(\beta)\rangle}\uparrow & & \uparrow{\scriptstyle \cong\ \zeta_X} \\ X & \xrightarrow[\ \beta\ ]{} & F^\infty(X) \end{array}$$

This diagram commutes since

$$\begin{aligned} (\mathrm{id} \times F(\beta)) \circ \langle \mathrm{id}, L(\beta)\rangle &= \langle \mathrm{id}, F(\beta) \circ \theta \circ \beta\rangle \\ &= \langle \mathrm{id}, \theta \circ F^*(\beta) \circ \beta\rangle \\ &= \langle \mathrm{id}, \theta \circ \delta \circ \beta\rangle \\ &= \langle \mathrm{id}, F(\varepsilon) \circ \pi_2 \circ \zeta \circ \delta \circ \beta\rangle \\ &= \langle \mathrm{id}, F(\varepsilon) \circ \pi_2 \circ (\mathrm{id} \times F(\delta)) \circ \langle \mathrm{id}, \pi_2 \circ \zeta\rangle \circ \beta\rangle \\ &= \langle \varepsilon \circ \beta, F(\varepsilon) \circ F(\delta) \circ \pi_2 \circ \langle \mathrm{id}, \pi_2 \circ \zeta\rangle \circ \beta\rangle \\ &= \langle \pi_1 \circ \zeta \circ \beta, \pi_2 \circ \zeta \circ \beta\rangle \\ &= \zeta \circ \beta. \end{aligned}$$

□

In Definition 5.2.4 we have seen distributive '$\mathcal{K}\ell$' laws $FT \Rightarrow TF$ that correspond to liftings $\mathcal{K}\ell(F)\colon \mathcal{K}\ell(T) \to \mathcal{K}\ell(T)$ of the functor $F\colon \mathbb{C} \to \mathbb{C}$ to Kleisli categories of a monad T. There is an analogous result for Eilenberg–Moore categories.

Definition 5.4.8 Let $T\colon \mathbb{C} \to \mathbb{C}$ be a monad and $G\colon \mathbb{C} \to \mathbb{C}$ an ordinary functor. A **distributive law** or an **$\mathcal{EM}$-law** of T over G is a natural transformation $\rho\colon TG \Rightarrow GT$ that commutes appropriately with the unit and multiplication of the monad T, as in

$$\begin{array}{ccc} G(X) & = & G(X) \\ {\scriptstyle \eta_{GX}}\downarrow & & \downarrow{\scriptstyle G(\eta_X)} \\ TG(X) & \xrightarrow[\rho_X]{} & GT(X) \end{array} \qquad \begin{array}{ccccc} T^2G(X) & \xrightarrow{T(\rho_X)} & TGT(X) & \xrightarrow{\rho_{TX}} & GT^2(X) \\ {\scriptstyle \mu_{GX}}\downarrow & & & & \downarrow{\scriptstyle G(\mu_X)} \\ TG(X) & & \xrightarrow[\rho_X]{\qquad\qquad} & & GT(X) \end{array} \tag{5.20}$$

We now show that these $\mathcal{EM}$-laws correspond to liftings to Eilenberg–Moore categories. We repeat the lifting result for Kleisli categories from Proposition 5.2.5 in order to get a clear picture of the whole situation (following [272]). A more abstract account in terms of 2-categories may be found in [333].

Proposition 5.4.9 ('Laws and liftings') *Let T be a monad and F, G be endofunctors on the same category $\mathbb{C}$, as above. There are bijective correspondences between $\mathcal{K}\ell/\mathcal{EM}$-laws and liftings of F to $\mathcal{K}\ell/\mathcal{EM}$-categories, in*

$$\begin{array}{c} \mathcal{K}\ell\text{-law } FT \stackrel{\lambda}{\Longrightarrow} TF \\ \hline\hline \begin{array}{ccc} \mathcal{K}\ell(T) & \xrightarrow{L} & \mathcal{K}\ell(T) \\ \downarrow & & \downarrow \\ \mathbb{C} & \xrightarrow{F} & \mathbb{C} \end{array} \end{array} \qquad\qquad \begin{array}{c} \mathcal{EM}\text{-law } TG \stackrel{\rho}{\Longrightarrow} GT \\ \hline\hline \begin{array}{ccc} \mathcal{EM}(T) & \xrightarrow{R} & \mathcal{EM}(T) \\ \downarrow & & \downarrow \\ \mathbb{C} & \xrightarrow{G} & \mathbb{C} \end{array} \end{array}$$

We standardly write $\mathcal{K}\ell(F) = L$ and $\mathcal{EM}(G) = R$ for these liftings, if confusion is unlikely. Thus we leave the laws λ and ρ implicit.

Proof The proof of first part about $\mathcal{K}\ell$-liftings has already been given in Proposition 5.2.5, so we concentrate on the second part. Thus, assume we have an $\mathcal{EM}$-law $\rho\colon TG \Rightarrow GT$. It gives rise to a functor $R\colon \mathcal{EM}(T) \to \mathcal{EM}(T)$ by

$$\begin{pmatrix} T(X) \\ \downarrow\alpha \\ X \end{pmatrix} \longmapsto \begin{pmatrix} T(X) \\ \downarrow G(\alpha)\circ\rho \\ X \end{pmatrix} \qquad \text{and} \qquad f \longmapsto G(f).$$

Equations (5.20) guarantee that this yields a new T-algebra.

In the reverse direction, let $R\colon \mathcal{EM}(T) \to \mathcal{EM}(T)$ be a lifting of G. Applying R to the multiplication μ_X yields an algebra $R(\mu_X)\colon TGT(X) \to GT(X)$. We then define $\rho_X = R(\mu_X) \circ TG(\eta_X)\colon TG(X) \to GT(X)$. Remaining details are left to the reader. ❑

The previous section illustrated how distributive $\mathcal{K}\ell$-laws are used to obtain final coalgebras in Kleisli categories. This requires non-trivial side-conditions, such as enrichment in dcpos; see Theorem 5.3.4. For $\mathcal{EM}$-laws the situation is much easier; see below. Instances of this result have been studied in [427]; see also [51].

Proposition 5.4.10 *Let T be a monad and G an endofunctor on a category $\mathbb{C}$, with an $\mathcal{EM}$-law $\rho\colon TG \Rightarrow GT$ between them. If G has a final coalgebra $\zeta\colon Z \xrightarrow{\cong} G(Z)$ in $\mathbb{C}$, then Z carries an Eilenberg–Moore algebra structure obtained by finality, as in*

$$\begin{array}{ccc} GT(Z) & \xrightarrow{\;G(\beta)\;} & G(Z) \\ {\scriptstyle \rho \circ T(\zeta)}\uparrow & & \uparrow{\scriptstyle \zeta}\;\cong \\ T(Z) & \xrightarrow[\;\beta\;]{} & Z \end{array} \tag{5.21}$$

The map ζ then forms a map of algebras as below, which is the final coalgebra for the lifted functor $\mathcal{EM}(G)\colon \mathcal{EM}(T) \to \mathcal{EM}(T)$:

$$\begin{pmatrix} T(Z) \\ \downarrow\beta \\ Z \end{pmatrix} \xrightarrow[\cong]{\;\zeta\;} \mathcal{EM}(G)\begin{pmatrix} T(Z) \\ \downarrow\beta \\ Z \end{pmatrix} = \begin{pmatrix} TG(Z) \\ \downarrow G(\beta)\circ\rho \\ G(Z) \end{pmatrix}.$$

Proof We leave it to the reader to verify that the map β defined in (5.21) is a T-algebra. By construction of β, the map ζ is a homomorphism of algebras $\beta \to \mathcal{EM}(G)(\beta)$. Suppose for an arbitrary algebra $\gamma\colon T(Y) \to Y$ we have a $\mathcal{EM}(G)$-coalgebra $c\colon \gamma \to \mathcal{EM}(G)(\gamma)$. Then $c\colon Y \to G(Y)$ satisfies $G(\gamma) \circ \rho \circ T(c) = c \circ \gamma$. By finality in **CoAlg**(G) there is a unique map $f\colon Y \to Z$ with $\zeta \circ f = G(f) \circ c$. This f is then the unique map $\gamma \to \beta$ in $\mathcal{EM}(T)$. ❑

In Section 5.5 we shall see that this result describes not only a final coalgebra in a category of algebras but a final bialgebra.

These $\mathcal{EM}$-laws are used for 'state extension'. This involves turning a coalgebra of the form $X \to GT(X)$ into a coalgebra in the category of Eilenberg–Moore algebras, with the free algebra $T(X)$ as state space. This state extension is functorial, in the following manner.

Lemma 5.4.11 *Given an $\mathcal{EM}$-law of the form $TG \Rightarrow GT$, the free algebra functor $\mathcal{F}\colon \mathbb{C} \to \mathcal{EM}(T)$ can be lifted, as in*

$$\begin{array}{ccc} \mathbf{CoAlg}(GT) & \xrightarrow{\;\mathcal{F}_{\mathcal{EM}}\;} & \mathbf{CoAlg}(\mathcal{EM}(G)) \\ \downarrow & & \downarrow \\ G \circlearrowright \mathbb{C} \circlearrowleft T & \xrightarrow{\;\mathcal{F}\;} & \mathcal{EM}(T) \circlearrowleft \mathcal{EM}(G) \end{array}$$

The functor $\mathcal{F}_{\mathcal{EM}}\colon$ **CoAlg**$(GT) \to$ **CoAlg**$(\mathcal{EM}(G))$ gives an abstract description of what is called the generalised powerset construction in [429]. A similar functor exists for Kleisli categories; see Exercise 5.4.13 below.

Proof Let $\rho\colon TG \Rightarrow GT$ be an $\mathcal{EM}$-law. The functor $\mathcal{F}_{\mathcal{EM}}\colon \mathbf{CoAlg}(GT) \to \mathbf{CoAlg}(\mathcal{EM}(G))$ is defined by

$$\mathcal{F}_{\mathcal{EM}}\big(X \xrightarrow{c} GT(X)\big) = \big(T(X) \xrightarrow{T(c)} TGT(X) \xrightarrow{\rho_{T(X)}} GT^2(X) \xrightarrow{G(\mu)} GT(X)\big). \quad (5.22)$$

This $\mathcal{F}_{\mathcal{EM}}(c)$ is a well-defined coalgebra $\mu_X \to \mathcal{EM}(G)(\mu_X)$ on the free algebra $\mathcal{F}(X) = \mu_X$ since it is a map of algebras:

$$\begin{aligned}
\mathcal{EM}(G)(\mu_X) \circ T(\mathcal{F}_{\mathcal{EM}}(c)) &= G(\mu) \circ \rho \circ TG(\mu) \circ T(\rho) \circ T^2(c) \\
&= G(\mu) \circ GT(\mu) \circ \rho \circ T(\rho) \circ T^2(c) \\
&= G(\mu) \circ G(\mu) \circ \rho \circ T(\rho) \circ T^2(c) \\
&= G(\mu) \circ \rho \circ \mu \circ T^2(c) \\
&= G(\mu) \circ \rho \circ T(c) \circ \mu \\
&= \mathcal{F}_{\mathcal{EM}}(c) \circ \mu.
\end{aligned}$$

On morphisms one simply has $\mathcal{F}_{\mathcal{EM}}(f) = T(f)$.

Further, if $f\colon X \to Y$ is a map of GT-algebras, from $c\colon X \to GTX$ to $d\colon Y \to GTY$, then $\mathcal{F}_{\mathcal{EM}}(c)(f) = T(f)$ is obviously a map of algebras $\mu_X \to \mu_Y$ and also a map of $\mathcal{EM}(G)$-coalgebras:

$$\begin{aligned}
\mathcal{EM}(G)(\mathcal{F}_{\mathcal{EM}}(f)) \circ \mathcal{F}_{\mathcal{EM}}(c) &= GT(f) \circ G(\mu) \circ \rho \circ T(c) \\
&= G(\mu) \circ GT^2(f) \circ \rho \circ T(c) \\
&= G(\mu) \circ \rho \circ TGT(f) \circ T(c) \\
&= G(\mu) \circ \rho \circ T(d) \circ T(f) \\
&= \mathcal{F}_{\mathcal{EM}}(d) \circ \mathcal{F}_{\mathcal{EM}}(f).
\end{aligned}$$

❑

Example 5.4.12 1. A non-deterministic automaton can be described as a coalgebra $\langle \delta, \epsilon \rangle\colon X \to (\mathcal{P}X)^A \times 2$, which is of the form $X \to GT(X)$, where G is the functor $(-)^A \times 2$ and T is the powerset monad $\mathcal{P}$ on **Sets**. Since $2 = \{0, 1\}$ is the (carrier of the) free algebra $\mathcal{P}(1)$ on the singleton set $1 = \{*\}$, there is an $\mathcal{EM}$-law $TG \Rightarrow GT$ by Exercise 5.4.4. It is $\rho = \langle \rho_1, \rho_2 \rangle\colon \mathcal{P}(2 \times X^A) \to \mathcal{P}(X)^A \times 2$, given by

$$\begin{cases} x \in \rho_1(U)(a) \iff \exists \langle b, h \rangle \in U.\, h(a) = x \\ \quad \rho_1(U) = 1 \iff \exists h \in X^A.\, \langle 1, h \rangle \in U. \end{cases}$$

Lemma 5.4.11 now yields a coalgebra $\mathcal{F}_{\mathcal{EM}}(\langle \delta, \epsilon \rangle) = \langle \delta_{\mathcal{EM}}, \epsilon_{\mathcal{EM}} \rangle\colon \mathcal{P}(X) \to \mathcal{P}(X)^A \times 2$ in the category $\mathcal{EM}(\mathcal{P}) \cong \mathbf{CL}$ of complete lattices; see Example 5.4.3.3. This coalgebra is given by

$$\begin{cases} \delta_{\mathcal{EM}}(U)(a) = \bigcup_{x \in U} \delta(x)(a) \\ \epsilon_{\mathcal{EM}}(U) = 1 \Leftrightarrow \exists x \in U.\, \epsilon(x) = 1. \end{cases}$$

By Proposition 5.4.10 the final G-coalgebra $2^{A^\star} = \mathcal{P}(A^\star)$ of languages is also final for the lifted functor $\mathcal{EM}(G)$ on $\mathbf{CL} \cong \mathcal{EM}(\mathcal{P})$. Hence we get a map $\mathcal{P}(X) \to \mathcal{P}(A^\star)$ of $\mathcal{EM}(G)$-coalgebras by finality. Applying this map to the singleton set $\{x\} \in \mathcal{P}(X)$ yields the set of words that are accepted in the state $x \in X$. This yields the trace semantics for a non-deterministic automaton $X \to \mathcal{P}(X)^A \times 2$, via determinisation in the Eilenberg–Moore category. This was first described in [427, 429] and elaborated in terms of $\mathcal{EM}$-laws in [271, 272].

2. Let S be a semiring and $\mathcal{M}_S$ the associated multiset monad on **Sets**, with as Eilenberg–Moore category $\mathcal{EM}(\mathcal{M}_S) = \mathbf{Mod}_S$ the category of modules over S; see Example 5.4.3.2. Consider for a set A of labels the functor $G(X) = X^A \times S$ on **Sets**, with final coalgebra $S^{A^\star}$; see Proposition 2.3.5. We claim that this functor G can be lifted to $\mathcal{EM}(G)$ on $\mathbf{Mod}_S$ via the $\mathcal{EM}$-law $\mathcal{M}_S G \Rightarrow G\mathcal{M}_S$ of the form

$$\begin{array}{ccc} \mathcal{M}_S(X^A \times S) & \longrightarrow & \mathcal{M}_S(X)^A \times S \\ \sum_i s_i | f_i, t_i \rangle & \longmapsto & \langle \lambda a.\ \sum_i s_i | f_i(a) \rangle,\ \sum_i s_i \cdot t_i \rangle \end{array}$$

The final coalgebra $S^{A^\star}$ in **Sets** is then also final in the Eilenberg–Moore category $\mathbf{Mod}_S$, with obvious (pointwise) module structure; see Proposition 5.4.10. It is final for the lifted functor $(-)^A \times S$ on $\mathbf{Mod}_S$.

Trace semantics for a weighted automaton $c\colon X \to \mathcal{M}_S(X)^A \times S$ is now obtained as follows. The codomain $\mathcal{M}_S(X)^A \times S$ is a module over S. Hence we can freely extend the coalgebra c to a map $\overline{c}\colon \mathcal{M}_S(X) \to \mathcal{M}_S(X)^A \times S$ in $\mathbf{Mod}_S$. This $\overline{c}$ is a coalgebra of the lifted endofunctor $(-)^A \times S$ on $\mathbf{Mod}_S$. Hence there is unique map of coalgebras $\mathrm{beh}_{\overline{c}}\colon \mathcal{M}_S(X) \to S^{A^\star}$. By pre-composition with the unit we get the trace map $X \to S^{A^\star}$.

3. Recall from Figure 4.2 that a 'simple Segala' system is a coalgebra of the form $X \to \mathcal{P}(A \times \mathcal{D}(X))$, combining non-deterministic and probabilistic computation. It can be turned into a non-deterministic transition system $\mathcal{D}(X) \to \mathcal{P}(A \times \mathcal{D}(X))$ with distributions as states. As before this is done via a $\mathcal{EM}$-law, namely of the form $\mathcal{DP}(A \times -) \Rightarrow \mathcal{P}(A \times -)\mathcal{D}$, of the monad $\mathcal{D}$ over the functor $\mathcal{P}(A \times -)$.

In [272] it is shown that such $\mathcal{EM}$-laws exist more generally, as soon as we have a map of monads. More concretely, each map of monads $\sigma\colon T \Rightarrow S$ induces an $\mathcal{EM}$-law

$$TS(A \times -) \stackrel{\rho}{\Longrightarrow} S(A \times T(-))$$

of the *monad* T over the *functor* $S(A \times -)$. The components of ρ are given by

$$\rho_X = \Big(TS(A \times X) \xrightarrow{\sigma} S^2(A \times X) \xrightarrow{S^2(\mathrm{id} \times \eta)} S^2(A \times TX) \xrightarrow{\mu} S(A \times TX)\Big).$$

Verifications are left to the interested reader. For simple Segala systems we can apply this general construction with $T = \mathcal{D}$ and $S = \mathcal{P}$, and the support map of monads $\mathcal{D} \Rightarrow \mathcal{P}$ from Exercise 5.1.9.

We conclude with another result about the free monads F^* on a functor F from Proposition 5.1.8.

Lemma 5.4.13 *For two endofunctors $F, G\colon \mathbb{C} \to \mathbb{C}$, there is a bijective correspondence*

$$\frac{\text{distributive } \mathcal{EM}\text{-laws } \; F^*G \overset{\rho}{\Longrightarrow} GF^*}{\text{natural transformations } \; FG \underset{\tau}{\Longrightarrow} GF^*}$$

where F^* is the free monad from Proposition 5.1.8.

Proof The correspondence is given as follows.

- Starting from an $\mathcal{EM}$-law $\rho\colon F^*G \Rightarrow GF^*$ we take $\overline{\rho}_X = \rho_X \circ \theta_{GX}\colon FG(X) \to F^*G(X) \to GF^*(X)$, where $\theta\colon F \Rightarrow F^*$ is the universal map.
- Conversely, given a natural transformation $\tau\colon FG \Rightarrow GF^*$ we obtain a map $\overline{\tau}_X\colon F^*G(X) \to GF^*(X)$ by initiality in

$$\begin{array}{ccc}
G(X) + F(F^*G(X)) & \xrightarrow{\text{id} + F(\overline{\tau}_X)} & G(X) + F(GF^*(X)) \\
{\scriptstyle \alpha_{G(X)}} \downarrow \cong & & \downarrow {\scriptstyle [G(\eta), G(\mu) \circ \tau]} \\
F^*G(X) & \xrightarrow[\overline{\tau}_X]{} & GF^*(X)
\end{array} \qquad (5.23)$$

  We leave it to the reader to verify that this map $\overline{\tau}$ is natural and commutes with the F^*'s unit and multiplication, as required in (5.20).

Proving that $\overline{\overline{\rho}} = \rho$ and $\overline{\overline{\tau}} = \tau$ is elementary. ❑

Exercises

5.4.1 Prove that the product of Eilenberg–Moore algebras, as described in Lemma 5.4.4, forms again an Eilenberg–Moore algebra. Formulate and prove a dual result for coalgebras.

5.4.2 Let $\alpha\colon T(X) \to X$ be an algebra of a monad T. Prove that α is a coequaliser in the category $\mathcal{EM}(T)$ of algebras, in a diagram:

$$\begin{pmatrix} T^3(X) \\ \downarrow\mu \\ T^2(X) \end{pmatrix} \overset{T(\alpha)}{\underset{\mu}{\rightrightarrows}} \begin{pmatrix} T^2(X) \\ \downarrow\mu \\ T(X) \end{pmatrix} \xrightarrow{\ \alpha\ }\!\!\!\!\!\twoheadrightarrow \begin{pmatrix} T(X) \\ \downarrow\alpha \\ X \end{pmatrix}.$$

5.4.3 Let T be a monad on a category $\mathbb{C}$ with a logical factorisation system $(\mathfrak{M}, \mathfrak{E})$. Prove, as in Exercise 4.3.2, that if T preserves abstract epis, $\mathcal{EM}(T)$ also carries a logical factorisation system. Check that this is the case for monads on **Sets**, using the axiom of choice; see text before Lemma 2.1.7.

5.4.4 Let $T\colon \mathbb{C} \to \mathbb{C}$ be an arbitrary monad. Show that for the various functors $G\colon \mathbb{C} \to \mathbb{C}$ described below an $\mathcal{EM}$-law $TG \Rightarrow GT$ exists.

1. If G is the identity functor $\mathbb{C} \to \mathbb{C}$.
2. If $G = B$, the constant functor mapping $X \mapsto B$, where B carries an algebra $\beta\colon T(B) \to B$.
3. If $G = G_1 \times G_2$ and there are $\mathcal{EM}$-laws $\rho_i\colon TG_i \Rightarrow G_iT$.
4. If $G = H^A$, provided: $\mathbb{C}$ is cartesian closed, there is an $\mathcal{EM}$-law $TH \Rightarrow HT$, and T is a strong monad.
5. If $G = G_1 \circ G_2$ and there are $\mathcal{EM}$-laws $\rho_i\colon TG_i \Rightarrow G_iT$.
6. Assuming that T preserves coproducts we also have: if $G = \coprod_i G_i$ and there are $\mathcal{EM}$-laws $\rho_i\colon TG_i \Rightarrow G_iT$.

5.4.5 For each set X, the function space $[0,1]^X$ is the set of fuzzy predicates.

1. Check that $[0,1]^X$ is a convex set.
2. Define an algebra $\mathcal{D}([0,1]^X) \to [0,1]^X$ of the distribution monad $\mathcal{D}$.

5.4.6 Let **MSL** be the category of meet semilattices.

1. Prove that the forgetful functor **MSL** $\to$ **Sets** has a left adjoint $X \mapsto \mathcal{P}_{\text{fin}}(X)^{\text{op}}$, where $\mathcal{P}_{\text{fin}}(X)^{\text{op}}$ is the poset of finite subsets of X, with reverse inclusion $U \supseteq V$ as order.
2. Describe the unit and multiplication of the monad $\mathcal{P}_{\text{fin}}^{\text{op}}\colon$ **Sets** $\to$ **Sets** induced by this adjunction.
3. Prove **MSL** $\cong \mathcal{EM}(\mathcal{P}_{\text{fin}}^{\text{op}})$.

5.4.7 1. Prove in detail that taking ideals leads to a monad $\text{Idl}\colon$ **PoSets** $\to$ **PoSets**, as claimed in Example 5.4.3.4, and also that $\mathcal{EM}(\text{Idl}) \cong$ **Dcpo**.

2. Define along the same lines a downset monad Dwn on **PoSets**, which sends a poset to the set of all its downclosed subsets, ordered by inclusion.
3. Prove that $\mathcal{EM}(\mathrm{Dwn}) \cong \mathbf{CL}$.

5.4.8
1. For an arbitrary monad T on a category $\mathbb{C}$, the free algebra adjunction $\mathcal{EM}(T) \leftrightarrows \mathbb{C}$ induces a comonad $\widehat{T}\colon \mathcal{EM}(T) \to \mathcal{EM}(T)$. Describe the counit and comultiplication of this comonad in detail.
2. Write $\widehat{\mathrm{Idl}}\colon \mathbf{Dcpo} \to \mathbf{Dcpo}$ for the comonad induced in this way by $\mathbf{Dcpo} \leftrightarrows \mathbf{PoSets}$; see the previous exercise. Prove that the category $\mathcal{EM}(\widehat{\mathrm{Idl}})$ of Eilenberg–Moore coalgebras of this comonad is the category of *continuous* posets.
3. Write $\mathcal{M} = \mathcal{M}_{\mathbb{R}}$ for the multiset monad over the real numbers $\mathbb{R}$, with vector spaces over $\mathbb{R}$ as algebras; see Example 5.4.3.2. Prove that a coalgebra of the induced comonad $\widehat{\mathcal{M}}$ on a vector space corresponds to a basis for this space.

(For a more systematic study of coalgebras of the induced comonad $\widehat{T}$, see [251] and the references therein.)

5.4.9 Consider the distributive $\mathcal{K}\ell$-law $\nabla\colon (-)^{\star}\mathcal{P} \Rightarrow \mathcal{P}(-)^{\star}$ described in Lemma 5.2.7, for the list functor $(-)^{\star}$.

1. Prove that this ∇ is in fact a distributive law between two monads, as in Exercise 5.2.8.
2. As a result, there is a language monad $X \mapsto \mathcal{P}(X^{\star})$; describe its unit and multiplication in detail.
3. Verify that the Eilenberg–Moore algebras of this monad $\mathcal{P}((-)^{\star})$ are complete lattices with a monoid structure where multiplication preserves joins in both arguments separately. They are known as unital quantales (see [401]) or as Kleene algebras with arbitrary joins.

5.4.10 For a semiring S consider the category $\mathbf{Mod}_S = \mathcal{EM}(\mathcal{M}_S)$ of modules over S, as in Example 5.4.3.2.

1. Show that the (covariant) powerset functor can be lifted to modules as in

$$\begin{array}{ccc} \mathbf{Mod}_S & \xrightarrow{\quad \mathcal{P} \quad} & \mathbf{Mod}_S \\ \downarrow & & \downarrow \\ \mathbf{Sets} & \xrightarrow{\quad \mathcal{P} \quad} & \mathbf{Sets} \end{array}$$

and similarly for the finite powerset functor $\mathcal{P}_{\mathrm{fin}}$.

2. Check that $\mathcal{P}\colon \mathbf{Mod}_S \to \mathbf{Mod}_S$ is also a monad –with unit and multiplication as on **Sets**.
3. Describe the $\mathcal{EM}$-law $\mathcal{M}_S\mathcal{P} \Rightarrow \mathcal{P}\mathcal{M}_S$ corresponding to this lifting $\mathcal{P}\colon \mathbf{Mod}_S \to \mathbf{Mod}_S$, as in Proposition 5.4.9.

5.4.11 Let $T\colon \mathbf{Sets} \to \mathbf{Sets}$ be a monad, and $\alpha\colon T(A) \to A$ an arbitrary (but fixed) Eilenberg–Moore algebra. Show that there is an adjunction

$$\mathbf{Sets}^{\mathrm{op}} \underset{Hom(-,\alpha)}{\overset{A^{(-)}}{\underset{\longleftarrow}{\overset{\longrightarrow}{\top}}}} \mathcal{EM}(T).$$

The canonical choice is to take $A = T(1)$, the free monad on the singleton set 1. Elaborate on what the resulting adjoint functors are in the cases $\mathbf{Sets}^{\mathrm{op}} \leftrightarrows \mathbf{JSL}$ and $\mathbf{Sets}^{\mathrm{op}} \leftrightarrows \mathbf{Mod}_S$, involving the finite powerset functor $\mathcal{P}_{\mathrm{fin}}$ and the multiset functor $\mathcal{M}_S$. This result plays a role in the abstract description of modalities; see [201, 221].

5.4.12 (From [51, 272]) Let $\lambda\colon FT \Rightarrow TF$ be a $\mathcal{K}\ell$-law and $\rho\colon TG \Rightarrow GT$ an $\mathcal{EM}$-law. Show that the standard adjunctions $\mathbb{C} \rightleftarrows \mathcal{K}\ell(T)$ from Proposition 5.2.2 and $\mathbb{C} \rightleftarrows \mathcal{EM}(T)$ from Lemma 5.4.2 lift to adjunctions between categories of, respectively, (functor) algebras and coalgebras, as described below:

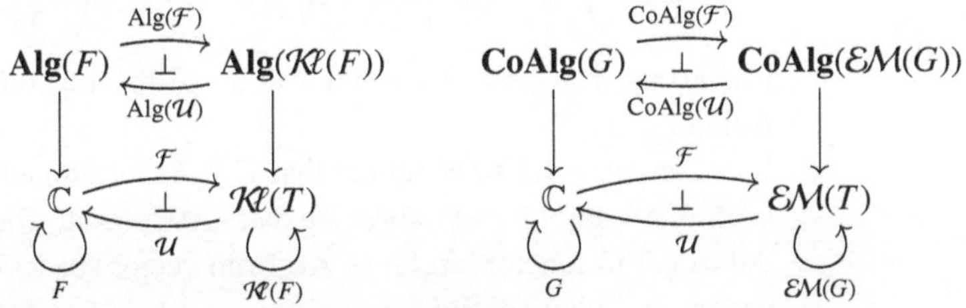

Hint: Use Theorem 2.5.9 and its dual.

5.4.13 Prove, in analogy with Lemma 5.4.11, that in presence of a $\mathcal{K}\ell$-law $FT \Rightarrow TF$ the free functor $\mathcal{F}\colon \mathbb{C} \to \mathcal{K}\ell(T)$ can be lifted, as in

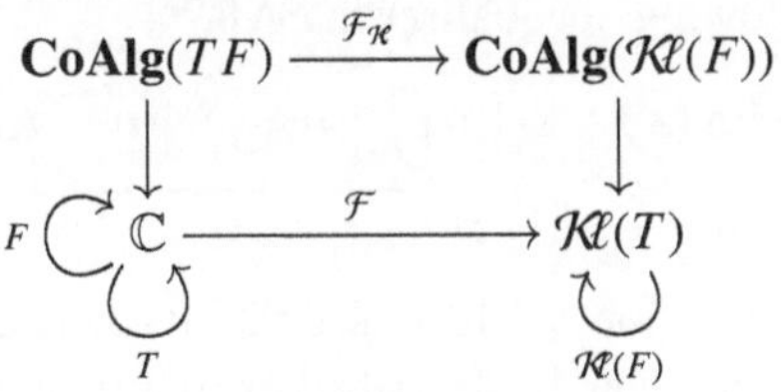

5.4.14 Recall that for two sets A, B, the deterministic automaton functor $G(X) = X^A \times B$ has final coalgebra $B^{A^\star}$; see Proposition 2.3.5. Let

T: **Sets** $\to$ **Sets** be an arbitrary monad, which is automatically strong by Lemma 5.2.10. Let $\beta\colon T(B) \to B$ be an algebra.

1. Give an explicit description of the $\mathcal{EM}$-law $T(X^A \times B) \to T(X)^A \times B$, following Exercise 5.4.4. The 'exponent' version r of strength, from Exercise 5.2.16, is convenient here.
2. Describe the resulting lifted functor $\mathcal{EM}(G)\colon \mathcal{EM}(T) \to \mathcal{EM}(T)$ explicitly.
3. Prove that the T-algebra (5.21) induced on the final G-coalgebra $B^{A^\star}$ is given by

$$T(B^{A^\star}) \xrightarrow{\;r\;} T(B)^{A^\star} \xrightarrow{\;\beta^{A^\star}\;} B^{A^\star}.$$

5.4.15 Let $\sigma\colon T \Rightarrow S$ be a map of monads.

1. Show that it induces a functor $(-) \circ \sigma\colon \mathcal{EM}(S) \to \mathcal{EM}(T)$ that commutes with the forgetful functors.
2. Assume that the category $\mathcal{EM}(S)$ has coequalisers. Consider the mapping that sends a T-algebra $\alpha\colon T(X) \to X$ to the following coequaliser in $\mathcal{EM}(S)$:

$$\begin{pmatrix} S^2T(X) \\ \downarrow\mu \\ ST(X) \end{pmatrix} \overset{\mu\circ S(\sigma)}{\underset{S(\alpha)}{\rightrightarrows}} \begin{pmatrix} S^2(X) \\ \downarrow\mu \\ S(X) \end{pmatrix} \xrightarrow{\;c\;} \begin{pmatrix} S(X_\sigma) \\ \downarrow\alpha_\sigma \\ X_\sigma \end{pmatrix}.$$

Prove that $\alpha \mapsto \alpha_\sigma$ is the left adjoint to the functor $(-) \circ \sigma$ from (1).

3. Use Lemma 5.4.5 to conclude that if T, S are monads on **Sets**, such a left adjoint $(-)_\sigma$ always exists. As a result, for instance, the forgetful functor $\mathbf{Vect}_{\mathbb{R}} \to \mathbf{Ab}$ from vector spaces to abelian groups has a left adjoint.
4. Assume, as before, that the category $\mathcal{EM}(S)$ of S-algebras has coequalisers. Prove that the coproduct of two Eilenberg–Moore algebras $\alpha\colon S(X) \to X$ and $\beta\colon S(Y) \to Y$ is given by the following coequaliser in $\mathcal{EM}(S)$:

$$\begin{pmatrix} S^2(S(X)+S(Y)) \\ \downarrow\mu \\ S(S(X)+S(Y)) \end{pmatrix} \overset{\mu\circ S([S(\kappa_1),S(\kappa_2)])}{\underset{S(\alpha+\beta)}{\rightrightarrows}} \begin{pmatrix} S^2(X+Y) \\ \downarrow\mu \\ S(X+Y) \end{pmatrix} \twoheadrightarrow \begin{pmatrix} S(U) \\ \downarrow \\ U \end{pmatrix}.$$

5.4.16 Use Proposition 5.4.13 to see that there is an $\mathcal{EM}$-law $F^*F \Rightarrow FF^*$. Check that it can be described explicitly as

$$F^*F(X) \xrightarrow[\cong]{\alpha^{-1}_{F(X)}} F(X) + F(F^*F(X)) \xrightarrow{[F(\eta_X),\, F(\mu_X \circ F^*(\theta_X))]} FF^*(X)$$

where $\theta\colon F \Rightarrow F^*$ is the universal map as in the proof of Proposition 5.1.8.

5.4.17 Assuming the relevant coproducts + and free monads $(-)^*$ exist, define a natural isomorphism:

$$(F+G)^* \cong F^* + G^*,$$

where the sum + on the right-hand side denotes the coproduct in the category of monads. (In [233, theorem 4] it is shown that the coproduct of monads $F^* + T$, where T is a monad, is the composite $T(FT)^*$.)

5.4.18 Let $\mathbb{C}$ be a category with a logical factorisation system $(\mathfrak{M}, \mathfrak{E})$, and let $F\colon \mathbb{C} \to \mathbb{C}$ be a functor with free monad $F^*\colon \mathbb{C} \to \mathbb{C}$.

1. Deduce from Lemma 4.4.6 that there is a natural transformation $\mathrm{Rel}(F) \Rightarrow \mathrm{Rel}(F^*)$ and that the relation lifting functor $\mathrm{Rel}(F^*)\colon \mathrm{Rel}(\mathbb{C}) \to \mathrm{Rel}(\mathbb{C})$ is a monad; conclude that there is a map of monads $\mathrm{Rel}(F)^* \Rightarrow \mathrm{Rel}(F^*)$.
2. Use Exercise 4.5.1 and Proposition 5.4.7 to show that F-congruences are the same as F^*-congruences. More formally, show that there is an isomorphism between categories of algebras:

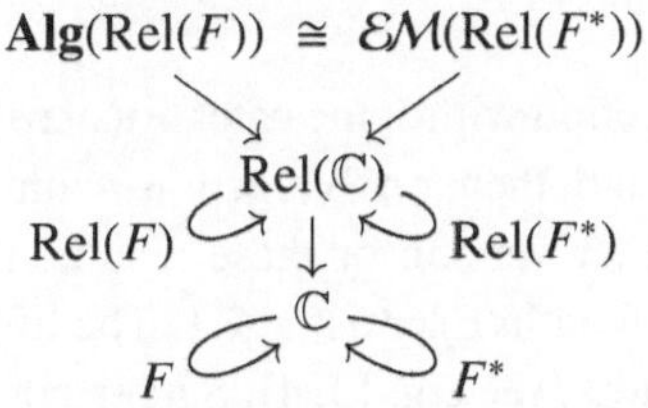

5.4.19 This exercise describes an analogon of Proposition 5.2.2 for Eilenberg–Moore categories. Let T be a monad on an arbitrary category $\mathbb{C}$, and let functor $L\colon \mathbb{C} \to \mathbb{D}$ have a right adjoint H, so that $T = HL$ is the induced monad. Define a 'comparison' functor $K\colon \mathbb{D} \to \mathcal{EM}(T)$ in

$$\begin{array}{ccc} \mathbb{D} & \xrightarrow{\;K\;} & \mathcal{EM}(T) \\ & {\scriptstyle H}\searrow \quad \swarrow & \\ & \mathbb{C} & \end{array}$$

1. Check that the functor $\mathcal{K}\ell(T) \to \mathcal{EM}(T)$ in Lemma 5.4.2 arises in this manner.
2. What is the dual statement for comonads?

5.4.20 (See [334, proposition 26]) Let $T\colon \mathbb{C} \to \mathbb{C}$ be an arbitrary strong monad on a cartesian (or monoidal) closed category $\mathbb{C}$. For an object $C \in \mathbb{C}$ there is the continuation monad $X \mapsto C^{(C^X)}$ from Example 5.1.3.7. Prove that there is a bijective correspondence between monad maps $T \Rightarrow C^{(C^{(-)})}$ commuting with strength, and Eilenberg–Moore algebras $T(C) \to C$.

5.5 Bialgebras and Operational Semantics

We continue our investigation of distributive laws. The new perspective that will be explored is that an $\mathcal{EM}$-law $TG \Rightarrow GT$ induces algebraic structure on the final G-coalgebra, of the kind we have seen before on streams and on processes, namely in Sections 1.2 and 3.5.2. The connection between distributive laws and specification formats for algebraic operations was first made in [452] (see also [451]), with preliminary ideas stemming from [407]. This connection forms one of the main achievements of the discipline of coalgebras. Traditionally in process algebras these formats are described as 'structured operational semantics' (SOS) rules, following [383]. A subtle matter is which rule formats correspond precisely to which rules. This is investigated for instance in [58, 334]. An overview of this material may be found in [298].

We start this section with some categorical results, first on liftings of monads and comonads, and then on bialgebras with respect to a distributive law. A slightly smoother version of these results, involving comonads instead of functors, is described in Exercise 5.5.1. The liftings may also be described for 'copointed' functors; see e.g. [334]. Such a copointed functor F comes with a counit (or copoint) $\varepsilon\colon F \Rightarrow \mathrm{id}$. This notion sits in between ordinary functors and comonads.

Proposition 5.5.1 *Let T be a monad and F, G be two endofunctors, all on the same category $\mathbb{C}$. Distributive $\mathcal{K}\ell$- and $\mathcal{EM}$-laws give rise to liftings of the monad T to new monads* Alg(T) *and* CoAlg(T) *in*

$$\frac{\mathcal{K}\ell\text{-law}\ \ FT \stackrel{\lambda}{\Longrightarrow} TF}{\begin{array}{ccc}\mathbf{Alg}(F) & \xrightarrow{\mathrm{Alg}(T)} & \mathbf{Alg}(F)\\ \downarrow & & \downarrow\\ \mathbb{C} & \xrightarrow{\quad T\quad} & \mathbb{C}\end{array}} \qquad \frac{\mathcal{EM}\text{-law}\ \ TG \stackrel{\rho}{\Longrightarrow} GT}{\begin{array}{ccc}\mathbf{CoAlg}(G) & \xrightarrow{\mathrm{CoAlg}(T)} & \mathbf{CoAlg}(G)\\ \downarrow & & \downarrow\\ \mathbb{C} & \xrightarrow{\quad T\quad} & \mathbb{C}\end{array}}.$$

Explicitly, these liftings are given by

$$\begin{pmatrix} F(X)\\ \downarrow a\\ X\end{pmatrix} \xmapsto{\ \mathrm{Alg}(T)\ } \begin{pmatrix} FT(X)\\ \downarrow T(a)\circ\lambda\\ T(X)\end{pmatrix} \quad \textit{and} \quad \begin{pmatrix} G(X)\\ \uparrow c\\ X\end{pmatrix} \xmapsto{\ \mathrm{CoAlg}(T)\ } \begin{pmatrix} GT(X)\\ \uparrow \rho\circ T(c)\\ T(X)\end{pmatrix}.$$

On morphisms both functors are given by $f \mapsto T(f)$. The unit and multiplication of T are also unit and multiplication of Alg(T) *and* CoAlg(T), *in the appropriate categories.*

Notice that the rules in this proposition have only a single line, indicating a passage from top to bottom, and not a bidirectional correspondence as in Proposition 5.4.9.

Proof It is easy to see that Alg(T) and CoAlg(T) are functors. The functor Alg(T) is a monad, because the unit and multiplication η, μ of T also form maps of F-algebras:

$$\begin{pmatrix} F(X) \\ \downarrow a \\ X \end{pmatrix} \xrightarrow{\ \eta\ } \mathrm{Alg}(T)\begin{pmatrix} F(X) \\ \downarrow a \\ X \end{pmatrix} \xleftarrow{\ \mu\ } \mathrm{Alg}(T)^2\begin{pmatrix} F(X) \\ \downarrow a \\ X \end{pmatrix}.$$

This is checked via the properties of $\mathcal{KL}$-laws from (5.6):

$$\begin{aligned}
\mathrm{Alg}(T)(a) \circ F(\eta) &= T(a) \circ \lambda \circ F(\eta) \\
&= T(a) \circ T(\eta) \\
&= \mathrm{id} \\
\mathrm{Alg}(T)(a) \circ F(\mu) &= T(a) \circ \lambda \circ F(\mu) \\
&= T(a) \circ \mu \circ T(\lambda) \circ \lambda \\
&= \mu \circ T(T(a) \circ \lambda) \circ \lambda \\
&= \mu \circ \mathrm{Alg}(T)^2(a).
\end{aligned}$$

Similarly one shows that CoAlg(T) is a monad. □

In the remainder of this section we concentrate on $\mathcal{EM}$-laws. First we define bialgebras for such laws. They are combinations of algebras and coalgebras which interact appropriately via the distributive law. This turns out to be a useful notion. Intuitively one can think of the algebra as describing some programming language, and of the coalgebra as its operational semantics, in the form of transition steps that the various program constructs can make.

Definition 5.5.2 Let $\rho\colon TG \Rightarrow GT$ be an $\mathcal{EM}$-law, for a monad T and an endofunctor G. We write **BiAlg**$(TG \stackrel{\rho}{\Rightarrow} GT)$ for the category of ρ-bialgebras. Its objects are algebra-coalgebra pairs $T(X) \stackrel{\alpha}{\to} X \stackrel{c}{\to} G(X)$ on the same carrier, where α is an Eilenberg–Moore algebra, and the following diagram commutes:

$$\begin{array}{ccccc}
T(X) & \xrightarrow{\ \alpha\ } & X & \xrightarrow{\ c\ } & G(X) \\
{\scriptstyle T(c)}\downarrow & & & & \uparrow{\scriptstyle G(\alpha)} \\
TG(X) & & \xrightarrow[\ \rho_X\ ]{} & & GT(X)
\end{array} \qquad (5.24)$$

A map of bialgebras, from $T(X) \overset{\alpha}{\to} X \overset{c}{\to} G(X)$ to $T(Y) \overset{\beta}{\to} Y \overset{d}{\to} G(Y)$ is a map $f\colon X \to Y$ in the underlying category which is a map of both algebras and coalgebras, as in

$$\begin{array}{ccc} T(X) & \xrightarrow{T(f)} & T(Y) \\ {\scriptstyle\alpha}\downarrow & & \downarrow{\scriptstyle\beta} \\ X & \xrightarrow{f} & Y \\ {\scriptstyle c}\downarrow & & \downarrow{\scriptstyle d} \\ G(X) & \xrightarrow[G(f)]{} & G(Y) \end{array}$$

Thus, by construction, there are two forgetful functors:

$$\mathcal{EM}(T) \longleftarrow \mathbf{BiAlg}(TG \overset{\rho}{\Rightarrow} GT) \longrightarrow \mathbf{CoAlg}(G).$$

The next result comes from [452] and forms the basis for categorical operational semantics.

Theorem 5.5.3 *For an $\mathcal{EM}$-law $\rho\colon TG \Rightarrow GT$ consider the associated liftings $\mathcal{EM}(G)$ from Proposition 5.4.9 and* CoAlg(T) *from Proposition 5.5.1 in*

$$\begin{array}{ccccc} \mathcal{EM}(T) & \longleftarrow & \mathbf{BiAlg}(TG \overset{\rho}{\Rightarrow} GT) & \longrightarrow & \mathbf{CoAlg}(G). \\ \circlearrowleft & & & & \circlearrowright \\ \mathcal{EM}(G) & & & & \mathrm{CoAlg}(T) \end{array}$$

Then there are isomorphism and adjoints as in

$$\begin{array}{ccccc} \mathbf{CoAlg}(\mathcal{EM}(G)) & \cong & \mathbf{BiAlg}(TG \overset{\rho}{\Rightarrow} GT) & \cong & \mathcal{EM}(\mathrm{CoAlg}(T)). \\ \downarrow \dashv \uparrow \mathcal{G} & & & & \mathcal{F} \uparrow \dashv \downarrow \\ \mathcal{EM}(T) & & & & \mathbf{CoAlg}(G) \end{array} \tag{5.25}$$

The left adjoint $\mathcal{F}$ is the lifting of the free algebra functor $\mathcal{F}\colon \mathbb{C} \to \mathcal{EM}(T)$ from Lemma 5.4.2. The right adjoint $\mathcal{G}$ exists, if we assume the existence of cofree G-coalgebras.

Proof Diagram (5.24) can be read at the same time as

- $c\colon X \to G(X)$ is a map of T-algebras:

$$\begin{pmatrix} T(X) \\ \downarrow\alpha \\ X \end{pmatrix} \xrightarrow{\quad c \quad} \mathcal{EM}(G)\begin{pmatrix} T(X) \\ \downarrow\alpha \\ X \end{pmatrix} = \begin{pmatrix} TG(X) \\ \downarrow G(\alpha)\circ\rho \\ G(X) \end{pmatrix}.$$

- $\alpha\colon T(X) \to X$ is a map of G-coalgebras:

$$\begin{pmatrix} GT(X) \\ \uparrow{\scriptstyle \rho\circ T(c)} \\ T(X)\end{pmatrix} = \mathrm{CoAlg}(T)\begin{pmatrix} G(X)\\ \uparrow{\scriptstyle c}\\ X\end{pmatrix} \xrightarrow{\;\alpha\;} \begin{pmatrix} G(X)\\ \uparrow{\scriptstyle c}\\ X\end{pmatrix}.$$

This is the essence of the isomorphisms in (5.25).

We also need to check that $f\colon X \to Y$ is a map of bialgebras

$$\big(T(X) \overset{\alpha}{\to} X \overset{c}{\to} G(X)\big) \longrightarrow \big(T(Y) \overset{\beta}{\to} Y \overset{d}{\to} G(Y)\big)$$

if and only if

- f is both a map of T-algebras $\alpha \to \beta$ and a map of $\mathcal{EM}(G)$-coalgebras $c \to d$
- f is both a map of G-coalgebras $c \to d$ and a map of CoAlg(T)-algebras $\alpha \to \beta$.

This is easy. We turn to the left adjoint $\mathcal{F}\colon \mathbf{CoAlg}(G) \to \mathbf{BiAlg}(TG \overset{\rho}{\Rightarrow} GT)$. Let $c\colon X \to G(X)$ be a coalgebra. We take the free algebra $\mu\colon T^2(X) \to T(X)$ and change c into a G-coalgebra $\mathbf{CoAlg}(T)(c) = \rho \circ T(c)\colon T(X) \to TG(X) \to GT(X)$. Together they form a bialgebra, since the next rectangle commutes:

$$\begin{array}{ccccc}
T^2(X) & \xrightarrow{\;\mu\;} & T(X) & \xrightarrow{\;\rho\,\circ\, T(c)\;} & GT(X) \\
{\scriptstyle T(\rho\,\circ\, T(c))}\downarrow & & & & \uparrow{\scriptstyle G(\mu)} \\
TGT(X) & & \xrightarrow[\;\rho\;]{} & & GT^2(X)
\end{array}$$

Since

$$G(\mu) \circ \rho \circ T(\rho \circ T(c)) \overset{(5.20)}{=} \rho \circ \mu \circ T^2(c) = \rho \circ T(c) \circ \mu.$$

Moreover, we have an adjoint correspondence between f and h in

$$\begin{array}{ccc}
T^2(X) & \xrightarrow{\;T(f)\;} & T(Y) \\
{\scriptstyle \mu}\downarrow & & \downarrow{\scriptstyle \beta} \\
T(X) & \xrightarrow{\;f\;} & Y \\
{\scriptstyle \rho\circ T(c)}\downarrow & & \downarrow{\scriptstyle d} \\
GT(X) & \xrightarrow{\;G(f)\;} & G(Y) \\
\hline\hline
G(X) & \xrightarrow{\;G(h)\;} & G(Y) \\
{\scriptstyle c}\uparrow & & \uparrow{\scriptstyle d} \\
X & \xrightarrow{\;h\;} & Y
\end{array}.$$

This the usual free algebra adjoint correspondence, given by $\overline{f} = f \circ \eta$ and $\overline{h} = \beta \circ T(h)$.

Next we assume the existence of cofree coalgebras, in the form of a right adjoint $\mathcal{G}\colon \mathbb{C} \to \mathbf{CoAlg}(G)$ to the forgetful functor. We show how it can be adapted to a functor $\mathcal{G}\colon \mathcal{EM}(T) \to \mathbf{BiAlg}(TG \overset{\rho}{\Rightarrow} GT)$. For an arbitrary object $Y \in \mathbb{C}$, we write the cofree coalgebra as $\zeta_Y\colon \mathcal{G}(Y) \to G(\mathcal{G}(Y))$, with counit $\varepsilon\colon \mathcal{G}(Y) \to Y$.

For an Eilenberg–Moore algebra $\beta\colon T(Y) \to Y$ consider the composite map $\beta \circ T(\varepsilon)\colon T(\mathcal{G}(Y)) \to T(Y) \to Y$. Its carrier can be equipped with a G-coalgebra $\rho \circ T(\zeta_Y)$ and yields a unique coalgebra map $\widehat{\beta}\colon T(\mathcal{G}(Y)) \to \mathcal{G}(Y)$ with $\varepsilon_Y \circ \widehat{\beta} = \beta \circ T(\varepsilon)$ in

$$\begin{array}{ccc}
GT\mathcal{G}(Y) & \xrightarrow{\;G(\widehat{\beta})\;} & G\mathcal{G}(Y) \\
{\scriptstyle\rho}\uparrow & & \\
TG\mathcal{G}(Y) & & \uparrow{\scriptstyle\zeta_Y} \\
{\scriptstyle T(\zeta_Y)}\uparrow & & \\
T\mathcal{G}(Y) & \xrightarrow[\;\widehat{\beta}\;]{} & \mathcal{G}(Y)
\end{array} \qquad (5.26)$$

We leave it to the reader to check that $\widehat{\beta}$ is an Eilenberg–Moore algebra. By construction, the pair $(\widehat{\beta}, \zeta_Y)$ is a bialgebra. Further, for an arbitrary bialgebra (α, c) there is a bijective correspondence between f and h in

$$\begin{array}{ccc}
T(X) & \xrightarrow{\;T(f)\;} & T\mathcal{G}(Y) \\
{\scriptstyle\alpha}\downarrow & & \downarrow{\scriptstyle\widehat{\beta}} \\
X & \xrightarrow{\;f\;} & \mathcal{G}(Y) \\
{\scriptstyle c}\downarrow & & \downarrow{\scriptstyle\zeta_Y} \\
G(X) & \xrightarrow{\;G(f)\;} & G\mathcal{G}(Y) \\
\hline\hline
T(X) & \xrightarrow{\;T(h)\;} & T(Y) \\
{\scriptstyle\alpha}\downarrow & & \downarrow{\scriptstyle\beta} \\
X & \xrightarrow{\;h\;} & Y
\end{array} .$$

This correspondence arises as follows.

- Given a map of bialgebras $f\colon X \to \mathcal{G}(Y)$ we take $\overline{f} = \varepsilon \circ f\colon X \to Y$. It forms a map of algebras:

$$\beta \circ T(\overline{f}) = \beta \circ T(\varepsilon) \circ T(f) = \varepsilon \circ \widehat{\beta} \circ T(f) = \varepsilon \circ f \circ \alpha = \overline{f} \circ \alpha.$$

- In the other direction, let $h\colon X \to Y$ be a map of algebras. By cofreeness we obtain a unique map of coalgebras $\overline{h}\colon X \to \mathcal{G}(Y)$, with $\zeta_Y \circ \overline{h} = G(\overline{h}) \circ c$

and $\varepsilon \circ \overline{h} = h$. This $\overline{h}$ is a map of bialgebras. The missing equation $\widehat{\beta} \circ T(\overline{h}) = \overline{h} \circ \alpha$ can be obtained as follows. Both sides of the latter equation form a map of coalgebras in

$$\begin{array}{ccc} GT(X) & \longrightarrow & G\mathcal{G}(Y) \\ {\scriptstyle \rho \circ T(c)}\uparrow & & \uparrow{\scriptstyle \zeta_Y} \\ T(X) & \longrightarrow & \mathcal{G}(Y) \end{array}$$

And both sides are equal when post-composed with the counit ε:

$$\varepsilon \circ \widehat{\beta} \circ T(\overline{h}) = \beta \circ T(\varepsilon) \circ T(\overline{h}) = \beta \circ T(h) = h \circ \alpha = \varepsilon \circ \overline{h} \circ \alpha.$$

By construction we have $\overline{\overline{h}} = \varepsilon \circ \overline{h} = h$. And $\overline{\overline{f}} = f$ holds by uniqueness. ❑

Corollary 5.5.4 *Let $\rho\colon TG \Rightarrow GT$ be an $\mathcal{EM}$-law for a monad T and an endofunctor G on a category $\mathbb{C}$.*

1. *If $\mathbb{C}$ has an initial object 0, then the category* **BiAlg**$(TG \stackrel{\rho}{\Rightarrow} GT)$ *of bialgebras has an initial object, namely*

$$T^2(0) \xrightarrow{\ \mu\ } T(0) \xrightarrow{\ \rho \circ T(!)\ } GT(0).$$

It is given by the initial T-algebra $T^2(0) \to T(0)$ and the initial G-coalgebra $0 \to G(0)$.

2. *If the functor G has a final coalgebras $Z \xrightarrow[\cong]{\zeta} G(Z)$, then there is also a final bialgebra*

$$T(Z) \xrightarrow{\ \beta\ } Z \xrightarrow[\cong]{\ \zeta\ } G(Z),$$

where the Eilenberg–Moore algebra $\beta\colon T(Z) \to Z$ is obtained by finality from the coalgebra $\rho \circ T(\zeta)\colon T(Z) \to GT(Z)$, as in (5.21).

Proof The left adjoint $\mathcal{F}$ in (5.25) preserves initial objects, so applying it to the initial G-algebra $!\colon 0 \to G(0)$ yields the initial bialgebra.

Similarly, a final G-coalgebra $\zeta\colon Z \stackrel{\cong}{\to} G(Z)$ is the cofree coalgebra $\mathcal{G}(1)$ at the final object $1 \in \mathbb{C}$. Hence the final T-algebra $!\colon T(1) \to 1$ yields an algebra $T(Z) \to Z$ as in (5.26) –or, equivalently, as in (5.21). ❑

Theorem 5.5.5 *Assume the presence of both the initial and final bialgebra as described in Corollary 5.5.4.*

1. *The algebraic semantics obtained by initiality and the coalgebraic semantics by finality coincide with the unique bialgebra map*

$$\begin{array}{ccc} T^2(0) & \dashrightarrow & T(Z) \\ {\scriptstyle\mu}\downarrow & & \downarrow{\scriptstyle\beta} \\ T(0) & \overset{\text{int = beh}}{\dashrightarrow} & Z \\ {\scriptstyle\rho\,\circ\, T(!)}\downarrow & & \cong\downarrow{\scriptstyle\zeta} \\ GT(0) & \dashrightarrow & G(Z) \end{array} \tag{5.27}$$

2. *As a result, bisimilarity/observational equivalence on $T(0)$ is a congruence.*

Proof 1. The algebra morphism $\text{int}\colon T(0) \to Z$ is obtained by initiality of the algebra $\mu\colon T^2(0) \to T(0)$ in $\mathcal{EM}(T)$. We wish to show that it is also a coalgebra map, i.e. that it satisfies $\zeta \circ \text{int} = G(\text{int}) \circ \rho \circ T(!)$ in (5.27). Then, by uniqueness, int = beh. Define

$$\beta' = \Big(TG(Z) \xrightarrow[\cong]{T(\zeta^{-1})} T(Z) \xrightarrow{\ \beta\ } Z \xrightarrow[\cong]{\ \zeta\ } G(Z)\Big).$$

It is not hard to see that β' is an Eilenberg–Moore algebra. Hence we are done, by initiality, if both $\zeta \circ \text{int}$ and $G(\text{int}) \circ \rho \circ T(!)$ are algebra maps $\mu \to \beta'$ in a situation:

$$\begin{array}{ccc} T^2(0) & \rightrightarrows & TG(Z) \\ {\scriptstyle\mu}\downarrow & & \downarrow{\scriptstyle\beta'} \\ T(0) & \underset{G(\text{int})\,\circ\,\rho\,\circ\, T(!)}{\overset{\zeta\,\circ\,\text{int}}{\rightrightarrows}} & G(Z) \end{array}$$

This is easy:

$$\begin{aligned}
&\beta' \circ T(\zeta \circ \text{int}) \\
&\quad= \zeta \circ \beta \circ T(\zeta^{-1}) \circ T(\zeta) \circ T(\text{int}) \\
&\quad= \zeta \circ \beta \circ T(\text{int}) \\
&\quad= \zeta \circ \text{int} \circ \mu \\
&\beta' \circ T(G(\text{int}) \circ \rho \circ T(!)) \\
&\quad= \zeta \circ \beta \circ T(\zeta^{-1}) \circ TG(\text{int}) \circ T(\rho) \circ T^2(!) \\
&\quad\overset{(5.21)}{=} G(\beta) \circ \rho \circ T(\zeta) \circ T(\zeta^{-1}) \circ TG(\text{int}) \circ T(\rho) \circ T^2(!) \\
&\quad= G(\beta) \circ \rho \circ TG(\text{int}) \circ T(\rho) \circ T^2(!) \\
&\quad= G(\beta) \circ GT(\text{int}) \circ \rho \circ T(\rho) \circ T^2(!) \\
&\quad= G(\text{int}) \circ G(\mu) \circ \rho \circ T(\rho) \circ T^2(!) \\
&\quad\overset{(5.20)}{=} G(\text{int}) \circ \rho \circ \mu \circ T^2(!) \\
&\quad= G(\text{int}) \circ \rho \circ T(!) \circ \mu.
\end{aligned}$$

2. Consider the following diagram, where the arrow e is an equaliser, describing the kernel of the (co)algebra map $\mathrm{int} = \mathrm{beh}\colon T(0) \to Z$ to the final coalgebra:

$$\begin{array}{ccccc}
T(\underline{\leftrightarrow}) & \xrightarrow{\langle T(\pi_1 \circ e), T(\pi_2 \circ e)\rangle} & T^2(0) \times T^2(0) & \overset{T(\mathrm{int}) \circ \pi_1}{\underset{T(\mathrm{int}) \circ \pi_2}{\rightrightarrows}} & T(Z) \\
{\scriptstyle \gamma}\downarrow & & \downarrow{\scriptstyle \mu \times \mu} & & \downarrow{\scriptstyle \beta} \\
\underline{\leftrightarrow} & \stackrel{e}{\rightarrowtail} & T(0) \times T(0) & \overset{\mathrm{int} \circ \pi_1}{\underset{\mathrm{int} \circ \pi_2}{\rightrightarrows}} & Z
\end{array}$$

The existence of the map γ follows from an easy diagram chase. It is an Eilenberg–Moore algebra and shows that bisimilarity $\underline{\leftrightarrow}$ on the initial algebra $T(0)$ is a congruence. ❑

In the remainder of this section we show how distributive laws can be used to define operations on final coalgebras, in the form of an Eilenberg–Moore algebra $T(Z) \to Z$ as above. Such an algebra exists as soon as we have an $\mathcal{EM}$-law $TG \Rightarrow GT$. In fact, this law forms a specification format for the algebra. Here we shall give not the most *general* format but the most *common* format. Such laws are called GSOS rules, because they correspond to the 'general structured operational semantics' specification format introduced in [74]. Here we do not go into the syntactic formulation of these rules, but only consider their categorical counterparts. For more information about the connection, see [58, 298].

Definition 5.5.6 Let T be a monad and G an endofunctor on a category $\mathbb{C}$. A **GSOS-law** is an $\mathcal{EM}$-law of the form

$$T(G \times \mathrm{id}) \stackrel{\sigma}{\Longrightarrow} (G \times \mathrm{id})T \qquad \text{satisfying} \qquad \pi_2 \circ \sigma = T(\pi_2).$$

For such a law we define a category $\mathbf{BiAlg}(T(G \times \mathrm{id}) \stackrel{\rho}{\Rightarrow} (G \times \mathrm{id})T)$ whose objects are bialgebras given by a pair of an Eilenberg–Moore algebra $\alpha\colon T(X) \to X$ and a coalgebra $c\colon X \to G(X)$ making the following diagram commute:

$$\begin{array}{ccccc}
T(X) & \xrightarrow{\alpha} & X & \xrightarrow{c} & G(X) \\
{\scriptstyle T(\langle c, \mathrm{id}\rangle)}\downarrow & & & & \uparrow{\scriptstyle G(\alpha)} \\
T(G(X) \times X) & \xrightarrow[\sigma_X]{} & GT(X) \times T(X) & \xrightarrow[\pi_1]{} & GT(X)
\end{array}$$

Morphisms of bialgebras are, as before, pairs of algebra maps and coalgebra maps.

A GSOS-law is thus a distributive law of the monad T over the functor $G \times \mathrm{id}\colon \mathbb{C} \to \mathbb{C}$, given by $X \mapsto G(X) \times X$. The formulation used here (following [246]) generalises the original one from [452] for a free monad F^*.

Lemma 5.5.7 *For two endofunctors $F, G\colon \mathbb{C} \to \mathbb{C}$, there is a bijective correspondence:*

$$\frac{\text{GSOS-laws } F^*(G \times \mathrm{id}) \stackrel{\sigma}{\Longrightarrow} (G \times \mathrm{id})F^*}{\text{natural transformations } F(G \times \mathrm{id}) \underset{\tau}{\Longrightarrow} GF^*}\,.$$

Thus this GSOS format involves maps $F(G(X) \times X) \to GF^*(X)$. These are more useful than the maps $FG(X) \to GF^*(X)$ from Lemma 5.4.13. The additional occurrence of the X on the input side increases the expressive power, as illustrated below. In view of this result we shall also call natural transformations $\tau\colon F(G \times \mathrm{id}) \Rightarrow GF^*$ GSOS-laws.

Proof We have the following correspondences:

$$\begin{array}{c}
\text{nat. transf. } F(G \times \mathrm{id}) \stackrel{\tau}{\Longrightarrow} GF^* \\
\hline\hline
\text{nat. transf. } F(G \times \mathrm{id}) \stackrel{\tau}{\Longrightarrow} (G \times \mathrm{id})F^* \text{ with } \pi_2 \circ \tau = \theta \circ F(\pi_2) \\
\hline\hline
\mathcal{EM}\text{-laws } F^*(G \times \mathrm{id}) \underset{\sigma}{\Longrightarrow} (G \times \mathrm{id})F^* \text{ with } \pi_2 \circ \sigma = F^*(\pi_2)
\end{array}$$

The first correspondence is easy and the second one is a minor adaptation of Lemma 5.4.13. ❑

For GSOS-laws we do not elaborate analogues of Theorem 5.5.3 and Corollary 5.5.4, but concentrate on the essentials for what is needed below.

Theorem 5.5.8 *Let $\sigma\colon T(G \times \mathrm{id}) \Rightarrow (G \times \mathrm{id})T$ be a GSOS-law. If the underlying category has an initial object 0, then the initial bialgebra exists with carrier $T(0)$ and algebra-coalgebra structure:*

$$T^2(0) \xrightarrow{\mu} T(0) \xrightarrow{T(!)} T(G(0) \times 0) \xrightarrow{\sigma_0} GT(0) \times T(0) \xrightarrow{\pi_1} GT(0).$$

If there is a final coalgebra $\zeta\colon Z \xrightarrow{\cong} G(Z)$, then there is also a final bialgebra with carrier Z, and structure maps

$$T(Z) \xrightarrow{\beta} Z \underset{\cong}{\xrightarrow{\zeta}} G(Z),$$

where the Eilenberg–Moore algebra structure $\beta\colon T(Z) \to Z$ is obtained by finality in

$$\begin{array}{ccc} GT(Z) & \xrightarrow{\quad G(\beta) \quad} & G(Z) \\ {\scriptstyle \pi_1 \circ \sigma \circ T(\langle \zeta, \mathrm{id} \rangle)} \uparrow & & \uparrow {\scriptstyle \cong\, \zeta} \\ T(Z) & \xrightarrow[\quad \beta \quad]{} & Z \end{array} \tag{5.28}$$

The analogue of Theorem 5.5.5 holds in this situation: the map $T(0) \to Z$ obtained by initiality is the same as the map $T(0) \to Z$ by finality and is the unique map of bialgebras $T(0) \to Z$. And also: bisimilarity $\leftrightarrows$ on $T(0)$ is a congruence.

Proof Let $T(X) \xrightarrow{\alpha} X \xrightarrow{c} G(X)$ be an arbitrary bialgebra. Consider $f = \alpha \circ T(!)\colon T(0) \to T(X) \to X$. This f is a map of bialgebras. Similarly, the unique coalgebra map $X \to Z$ is a map of bialgebras. ❑

Suppose we have a final coalgebra $\zeta\colon Z \to G(Z)$ and we wish to define function interpretations $f_i\colon Z^{\#i} \to Z$ on Z, for some arity $\#\colon I \to \mathbb{N}$. That is, we wish to define a (functor) algebra $F_\#(Z) \to Z$, where $F_\#$ is the arity functor $F_\#(X) = \coprod_{i \in I} X^{\#i}$ associated with #. Equivalently, by Proposition 5.4.7, we can define a monad algebra $F_\#^*(Z) \to Z$. The associated free monad $F_\#^*$ is characterised as the initial algebra

$$X + \coprod_{i \in I} F_\#^*(X)^{\#i} \xrightarrow[\cong]{\quad \alpha_X \quad} F_\#^*(X);$$

see Proposition 5.1.8. The unit of the free monad $F_\#^*$ is then $\eta = \alpha \circ \kappa_1\colon X \to F_\#^*(X)$.

Theorem 5.5.8 says that in order to define such an interpretation $F_\#^*(Z) \to Z$ it suffices to define a natural transformation $\tau\colon F_\#(G \times \mathrm{id}) \Rightarrow (G \times \mathrm{id})F_\#^*$. This means that for each $i \in I$ we need (natural) maps

$$(G(X) \times X)^{\#i} \cong G(X)^{\#i} \times X^{\#i} \xrightarrow{\quad \tau_i \quad} GF_\#^*(X),$$

one for each operation f_i. This τ_i describes the relevant behaviour associated with the operation f_i: it sends the appropriate number of behaviours and states $(u_k, x_k) \in G(X) \times X$, for k below the arity $\#i \in \mathbb{N}$ of f_i, to a behaviour over terms, in $GF_\#^*(X)$.

Now assume we have such τ_i as described above. Together they correspond to a cotuple natural transformation $\tau = [\tau_i]_{i \in I}\colon F_\#(G \times \mathrm{id}) \Rightarrow GF_\#^*$, which corresponds by Lemma 5.5.7 to a GSOS-law $\overline{\tau}\colon F^*_\#(G \times \mathrm{id}) \Rightarrow (G \times \mathrm{id})F^*_\#$. By Theorem 5.5.8 we have a final bialgebra,

$$F^*_\#(Z) \xrightarrow{\ \beta\ } Z \xrightarrow[\cong]{\ \zeta\ } G(Z),$$

where the map β is obtained by finality as in (5.28). The interpretation $f_i\colon Z^n \to Z$ of each operation f_i with arity $\#i \in \mathbb{N}$ can then be described as the composite:

$$Z^{\#i} \xrightarrow{\ \eta^{\#i}\ } F^*_\#(Z)^{\#i} \xrightarrow{\ \kappa_i\ } \textstyle\coprod_{i \in I} F^*_\#(Z)^{\#i} \xrightarrow{\ \alpha \circ \kappa_2\ } F^*_\#(Z) \xrightarrow{\ \beta\ } Z.$$

We claim that these maps $f_i\colon Z^{\#i} \to Z$ are determined by the following equation:

$$\zeta \circ f_i = G(\beta \circ \mu) \circ \tau_i \circ \langle G(\eta) \circ \zeta, \eta\rangle^{\#i}. \qquad (5.29)$$

It follows from a diagram chase:

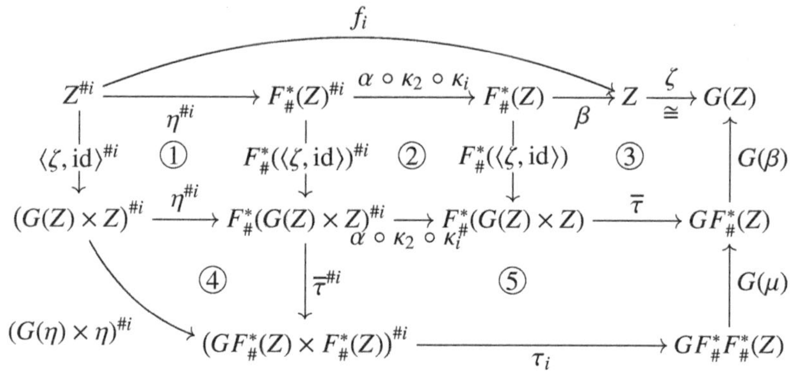

We list why the various numbered subdiagrams commute:

① by naturality of η
② by definition of $F^*_\#$ on maps; see (5.2)
③ because the pair (β, ζ) is a bialgebra, by construction of β in (5.28)
④ and ⑤ by definition of $\overline{\tau}$ in (5.23), using that $\eta = \alpha \circ \kappa_1$.

Bisimilarity on the set $F^*_\#(0)$ of closed terms is then a congruence.

Example 5.5.9 We briefly review several ways of employing the above uniform approach to obtain algebraic operations on final coalgebras and coalgebraic structure on closed terms.

1. Recall from Section 1.2 the final coalgebra of finite and infinite sequences,

$$A^\infty \xrightarrow[\cong]{\ \mathsf{next}\ } 1 + (A \times A^\infty),$$

for the functor $G(X) = 1 + A \times X$ on **Sets**. Suppose we wish to define a sequential composition operation $\mathsf{comp}\colon A^\infty \times A^\infty \to A^\infty$ on such sequences. We expect the following rules for composition $\mathsf{comp}(s,t)$ of two sequences $s, t \in A^\infty$:

$$\frac{s \not\rightarrow \quad t \not\rightarrow}{\mathsf{comp}(s,t) \not\rightarrow} \qquad \frac{s \not\rightarrow \qquad t \xrightarrow{a} t'}{\mathsf{comp}(s,t) \xrightarrow{a} \mathsf{comp}(s,t')} \qquad \frac{s \xrightarrow{a} s'}{\mathsf{comp}(s,t) \xrightarrow{a} \mathsf{comp}(s',t)}$$

where we recall from (1.5) that $s \not\rightarrow$ means $\mathsf{next}(s) = *$ and $s \xrightarrow{a} s'$ means $\mathsf{next}(s) = (a, s')$. Hence $\mathsf{comp}(s,t)$ is s if s is infinite; otherwise it is $s \cdot t$, where $\cdot$ is prefixing.

The arity functor for this single, binary composition operation is $F(X) = X \times X$. The associated free monad on F is written as F^*, as in Proposition 5.1.8, and determined as the initial algebra

$$X + F^*(X) \times F^*(X) \xrightarrow[\cong]{\alpha_X} F^*(X).$$

We now define a GSOS-law $\tau\colon F(G \times \mathrm{id}) \Rightarrow GF^*$ for sequential composition as follows:

$$\begin{array}{rcl}
\big((1 + A \times X) \times X\big) \times \big((1 + A \times X) \times X\big) & \xrightarrow{\quad\tau_X\quad} & 1 + A \times F^*(X). \\
(\langle *, x\rangle,\ \langle *, y\rangle) & \longmapsto & * \\
(\langle *, x\rangle,\ \langle\langle b, y'\rangle, y\rangle) & \longmapsto & \langle b, \alpha(\eta(x), \eta(y'))\rangle \\
(\langle\langle a, x'\rangle, x\rangle,\ \langle u, y\rangle) & \longmapsto & \langle a, \alpha(\eta(x'), \eta(y))\rangle.
\end{array}$$

The three cases of this GSOS-law clearly resemble the above three rules for comp.

As above we now get an algebra $\beta\colon F^*(A^\infty) \to A^\infty$, via (5.28). The actual function $\mathsf{comp}\colon A^\infty \times A^\infty \to A^\infty$ is the composite:

$$A^\infty \times A^\infty \xrightarrow{\eta \times \eta} F^*(A^\infty) \times F^*(A^\infty) \xrightarrow{\alpha \circ \kappa_2} F^*(A^\infty) \xrightarrow{\beta} A^\infty.$$

It satisfies, as in (5.29),

$$\mathsf{next} \circ \mathsf{comp} = G(\beta \circ \mu) \circ \tau \circ (\langle G(\eta) \circ \mathsf{next}, \eta\rangle \times \langle G(\eta) \circ \mathsf{next}, \eta\rangle).$$

It allows use to check that if $\mathsf{next}(s) = \mathsf{next}(t) = *$, then

$$\begin{aligned}
&\mathsf{next}(\mathsf{comp}(s,t)) \\
&= \big(G(\beta \circ \mu) \circ \tau \circ (\langle G(\eta) \circ \mathsf{next}, \eta\rangle \times \langle G(\eta) \circ \mathsf{next}, \eta\rangle)\big)(s,t) \\
&= \big(G(\beta \circ \mu) \circ \tau\big)(\langle G(\eta)(\mathsf{next}(s)), \eta(s)\rangle,\ \langle G(\eta)(\mathsf{next}(t)), \eta(t)\rangle) \\
&= \big(G(\beta \circ \mu) \circ \tau\big)(\langle G(\eta)(*), \eta(s)\rangle,\ \langle G(\eta)(*), \eta(t)\rangle) \\
&= G(\beta \circ \mu)(\tau(\langle *, \eta(s)\rangle,\ \langle *, \eta(t)\rangle)) \\
&= G(\beta \circ \mu)(*) \\
&= *.
\end{aligned}$$

In a similar manner one can prove

$$\mathsf{next}(\mathsf{comp}(s,t)) = \begin{cases} (b, \mathsf{comp}(s,t')) & \text{if } \mathsf{next}(s) = *,\ \mathsf{next}(t) = (b,t') \\ (a, \mathsf{comp}(s',t)) & \text{if } \mathsf{next}(s) = (a,s'). \end{cases}$$

Hence we have obtained a sequential composition function, described informally via the above three rules, and defined precisely via the GSOS-law/natural transformation τ.

2. In Section 3.5.2 we have seen a final coalgebra of processes

$$Z_A \xrightarrow[\cong]{\zeta_A} \mathcal{P}_{\text{fin}}(Z_A)^A$$

for the functor $G(X) = \mathcal{P}_{\text{fin}}(X)^A$ on **Sets**, where A is a (fixed) set of labels. A simple process language was defined via an arity functor

$$\Sigma_A(X) = 1 + (A \times X) + (X \times X)$$

with initial algebra $\xi_A \colon \Sigma_A(P_A) \xrightarrow{\cong} P_A$. Alternatively, we can describe this set P_A of process terms as initial monad algebra $\Sigma_A^*(0)$. Here we shall describe the situation in terms of bialgebras and distributive laws.

To start, there is an $\mathcal{EM}$-law of the form $\Sigma_A \mathcal{P}_{\text{fin}}(-)^A \Rightarrow \mathcal{P}_{\text{fin}}(-)^A \Sigma_A^*$, via Lemma 5.4.13, namely,

$$1 + (A \times \mathcal{P}_{\text{fin}}(X)^A) + (\mathcal{P}_{\text{fin}}(X)^A \times \mathcal{P}_{\text{fin}}(X)^A) \xrightarrow{\ \tau_X\ } \mathcal{P}_{\text{fin}}(\Sigma_A^*(X))^A$$

given by the following three clauses:

$$\begin{cases} * \longmapsto \lambda b \in A.\, \emptyset \\ (a,h) \longmapsto \lambda b \in A.\, \begin{cases} h(b) & \text{if } a = b \\ \emptyset \end{cases} \\ (h,k) \longmapsto \lambda b \in A.\, h(b) \cup k(b). \end{cases}$$

Corollary 5.5.4 now tells us that there is a unique map $P_A \to Z_A$ that is both a map of algebras and a map of coalgebras. This map that sends a process term to its interpretation as a process is thus compositional. Moreover, by Theorem 5.5.5 bisimilarity on P_A is a congruence. Earlier in this book, in Propositions 3.5.2 and 3.5.3, we had to prove these results by hand.

3. In [246] it is described how regular languages and automata can be understood in terms of bialgebras and distributive laws. The arity functor for regular languages is:

$$\mathcal{R}(X) = \underbrace{1}_{\text{zero}} + \underbrace{1}_{\text{one}} + \underbrace{(X \times X)}_{\text{sum } +} + \underbrace{(X \times X)}_{\text{product } \cdot} + \underbrace{X}_{\text{star } (-)^*} .$$

For an arbitrary set A, the free monad $\mathcal{R}^*(A)$ captures the regular expressions with elements $a \in A$ as atomic expressions. There is the familiar interpretation $\mathcal{R}^*(A) \to \mathcal{P}(A^\star)$ of regular expressions as languages. This set of languages $\mathcal{P}(A^\star)$ is the final coalgebra of the deterministic automaton functor $G(X) = X^A \times 2$; see Corollary 2.3.6.2. This interpretation $\mathcal{R}^*(A) \to \mathcal{P}(A^\star)$ can then be described as a bialgebra for the GSOS-law:

$$\begin{array}{rcl}
\mathcal{R}(X^A \times 2 \times X) & \longrightarrow & \mathcal{R}^*(X)^A \times 2 \\
0 & \stackrel{\text{zero}}{\longmapsto} & (\lambda a \in A.\, 0, 0) \\
1 & \stackrel{\text{one}}{\longmapsto} & (\lambda a \in A.\, 0, 1) \\
\langle (h_1, b_1, x_1), (h_2, b_2, x_2) \rangle & \stackrel{\text{plus}}{\longmapsto} & (\lambda a \in A.\, h_1(a) + h_2(a), b_1 \vee b_2) \\
\langle (h_1, b_1, x_1), (h_2, b_2, x_2) \rangle & \stackrel{\text{product}}{\longmapsto} & (\lambda a \in A.\, h_1(a) \cdot x_2 + b_1 \cdot h_2(a), b_1 \wedge b_2) \\
(h, b, x) & \stackrel{\text{star}}{\longmapsto} & (\lambda a \in A.\, h(a) \cdot x^*, 1).
\end{array}$$

More details can be found in [246].

We conclude this section with a strengthened form of coinduction for distributive $\mathcal{EM}$-laws. We shall refer to this generalised version as '$\mathcal{EM}$-coinduction'. An even stronger 'corecursive' version that builds on it is described in Exercise 5.5.4.

Proposition 5.5.10 *Assume we have*

- *a functor $G\colon \mathbb{C} \to \mathbb{C}$ with a final coalgebra $\zeta\colon Z \xrightarrow{\cong} G(Z)$*
- *a monad $T\colon \mathbb{C} \to \mathbb{C}$, on the same category, with an $\mathcal{EM}$-law $\rho\colon TG \Rightarrow GT$.*

Then the following '$\mathcal{EM}$-coinduction' principle holds: for each map $e\colon X \to GT(X)$ there is a unique map $s\colon X \to Z$ making the following diagram commute:

$$\begin{array}{ccc} GT(X) & \xrightarrow{\;GT(s)\;} & GT(Z) \\ & & \big\downarrow G(\beta) \\ e \big\uparrow & & G(Z) \\ & & \cong \big\uparrow \zeta \\ X & \xrightarrow[\;s\;]{} & Z \end{array}$$

where $\beta\colon T(Z) \to Z$ is the induced Eilenberg–Moore algebra as in (5.21).

Proof There is a direct way to prove the result, and an indirect way via bialgebras. We shall present both proofs.

For the direct proof, first transform e into a G-coalgebra $\overline{e}$ as in

$$\overline{e} = \mathcal{F}_{\mathcal{EM}}(e) = \Big(T(X) \xrightarrow{\;T(e)\;} TGT(X) \xrightarrow{\;\rho_{T(X)}\;} GT^2(X) \xrightarrow{\;G(\mu_X)\;} GT(X)\Big),$$

where $\mathcal{F}_{\mathcal{EM}}\colon \mathbf{CoAlg}(GT) \to \mathbf{CoAlg}(\mathcal{EM}(G))$ is the state extension functor defined in (5.22). This coalgebra satisfies $\overline{e} \circ \eta = e$. By finality we get a unique coalgebra map $f\colon T(X) \to Z$ with $\zeta \circ f = G(f) \circ \overline{e}$. One can then show that $s = f \circ \eta\colon X \to Z$ is the required map.

$$\begin{aligned} G(\beta) \circ GT(s) \circ e &= G(\beta) \circ GT(f) \circ GT(\eta) \circ e \\ &\overset{(*)}{=} G(f) \circ G(\mu) \circ GT(\eta) \circ e \\ &= G(f) \circ e \\ &= G(f) \circ \overline{e} \circ \eta \\ &= \zeta \circ f \circ \eta \\ &= \zeta \circ s \end{aligned}$$

The marked equation holds because $\beta \circ T(f) = f \circ \mu$ by uniqueness of coalgebra maps $(\rho \circ T(\overline{e})) \to \zeta$.

The indirect proof uses the following bijective correspondence:

$$\frac{\text{maps } X \xrightarrow{\;c\;} GT(X)}{\text{bialgebras } T^2(X) \xrightarrow[\;\mu\;]{} T(X) \xrightarrow[\;d\;]{} GT(X)}. \qquad (5.30)$$

This correspondence sends e to $\overline{e}$ as defined above. In the reverse direction d is mapped to $\overline{d} = d \circ \eta$.

Corollary 5.5.4 says that the pair (β, ζ) is a final bialgebra. Hence we get a unique map of bialgebras $(\mu, \overline{e}) \to (\beta, \zeta)$. It involves a map $f\colon T(X) \to Z$

which is a map both of algebras and of coalgebras. Then $s = f \circ \eta$ is the desired map, since

$$\begin{aligned} G(\beta) \circ GT(s) \circ e &= G(\beta) \circ GT(f \circ \eta) \circ \overline{e} \circ \eta \\ &= G(f) \circ G(\mu) \circ GT(\eta) \circ \overline{e} \circ \eta \\ &= G(f) \circ \overline{e} \circ \eta \\ &= \zeta \circ f \circ \eta \\ &= \zeta \circ s. \end{aligned}$$

□

In this result we have used letters 'e' and 's' for 'equation' and 'solution'. Indeed, maps of the form $X \to GT(X)$ can be understood as abstract recursive equations, with maps $s\colon X \to Z$ as solutions in a final coalgebra. This view is elaborated in a series of papers (see e.g. [360, 11, 25, 26, 27]) formalising and extending in coalgebraic terms earlier work of especially Elgot, Bloom and Esik (see [75]). Of particular interest in this setting is the so-called free iterative monad F^∞ on a functor F, where $F^\infty(X)$ is the final coalgebra of the functor $X + F(-)$. The connections between this monad, distributive laws and $\mathcal{EM}$-coinduction are elaborated in [246].

Exercises

5.5.1 This exercise gives a more symmetric version of the situation described in the beginning of this section, dealing not with a monad and a functor but with a monad and a comonad. So let $T = (T, \eta, \mu)$ be a monad and $S = (S, \varepsilon, \delta)$ a comonad, on the same category $\mathbb{C}$. A distributive law of the monad T over the comonad S is a natural transformation $\rho\colon TS \Rightarrow ST$ which commutes both with the monad and with the comonad structure, as in

$$\begin{array}{ccc} S & = & S \\ {\scriptstyle \eta}\downarrow & & \downarrow{\scriptstyle S(\eta)} \\ TS & \xrightarrow{\rho} & ST \\ {\scriptstyle T(\varepsilon)}\downarrow & & \downarrow{\scriptstyle \varepsilon} \\ T & = & T \end{array} \qquad \begin{array}{ccccc} T^2S & \xrightarrow{T(\rho)} & TST & \xrightarrow{\rho} & ST^2 \\ {\scriptstyle \mu}\downarrow & & & & \downarrow{\scriptstyle S(\mu)} \\ TS & & \xrightarrow{\rho} & & ST \\ {\scriptstyle T(\delta)}\downarrow & & & & \downarrow{\scriptstyle \delta} \\ TS^2 & \xrightarrow{\rho} & STS & \xrightarrow{S(\rho)} & S^2T \end{array}$$

1. Define a suitable category of bialgebras $\mathbf{BiAlg}(TS \overset{\rho}{\Rightarrow} ST)$, together with two forgetful functors:

$$\mathcal{EM}(T) \longleftarrow \mathbf{BiAlg}(TS \overset{\rho}{\Rightarrow} ST) \longrightarrow \mathcal{EM}(S).$$

2. Define liftings of the monad T to a monad $\mathcal{EM}(T)$ and of the comonad S to a comonad $\mathcal{EM}(S)$ in the following diagrams:

$$\begin{array}{ccc} \mathcal{EM}(S) & \xrightarrow{\mathcal{EM}(T)} & \mathcal{EM}(S) \\ \downarrow & & \downarrow \\ \mathbb{C} & \xrightarrow[T]{} & \mathbb{C} \end{array} \qquad \begin{array}{ccc} \mathcal{EM}(T) & \xrightarrow{\mathcal{EM}(S)} & \mathcal{EM}(T) \\ \downarrow & & \downarrow \\ \mathbb{C} & \xrightarrow[S]{} & \mathbb{C} \end{array}$$

3. Strengthen the previous point by proving that there are bijective correspondences between distributive laws and liftings of (co)monads to (co)monads in

$$\frac{\text{distr. laws } TS \overset{\rho}{\Longrightarrow} ST}{\begin{array}{ccc} \mathcal{EM}(S) & \xrightarrow{L} & \mathcal{EM}(S) \\ \downarrow & & \downarrow \\ \mathbb{C} & \xrightarrow{T} & \mathbb{C} \end{array}} \qquad \frac{\text{distr. laws } TS \overset{\rho}{\Longrightarrow} ST}{\begin{array}{ccc} \mathcal{EM}(T) & \xrightarrow{R} & \mathcal{EM}(T) \\ \downarrow & & \downarrow \\ \mathbb{C} & \xrightarrow{S} & \mathbb{C} \end{array}}$$

4. Show that there are isomorphisms of categories

$$\mathcal{EM}(\mathcal{EM}(T)) \cong \mathbf{BiAlg}(TS \overset{\rho}{\Rightarrow} ST) \cong \mathcal{EM}(\mathcal{EM}(S)).$$

5. Prove that the (downwards pointing) forgetful functors have adjoints in

$$\begin{array}{ccc} & \mathbf{BiAlg}(TS \overset{\rho}{\Rightarrow} ST) & \\ \swarrow\!\nearrow \ \dashv & & \dashv \ \nwarrow\!\searrow \\ \mathcal{EM}(T) & & \mathcal{EM}(S) \end{array}$$

6. Conclude that if the category $\mathbb{C}$ has a final object 1 and a initial object 0, then, in the category of bialgebra one has both

$$\begin{cases} \text{final object:} & TS(1) \xrightarrow{\rho} ST(1) \xrightarrow{S(!)} S(1) \xrightarrow{\delta} S^2(1) \\ \text{initial object:} & T^2(0) \xrightarrow{\mu} T(0) \xrightarrow{T(!)} TS(0) \xrightarrow{\rho} ST(0). \end{cases}$$

7. Let $G\colon \mathbb{C} \to \mathbb{C}$ be an arbitrary functor, with cofree comonad G^∞. Show that each $\mathcal{EM}$-law $TG \Rightarrow GT$ gives rise to a distributive law $TG^\infty \Rightarrow G^\infty T$.

5.5.2 Recall the operation merge: $A^\infty \times A^\infty \to A^\infty$ on sequences from Section 1.2. Define a GSOS-law corresponding to the three rules describing the behaviour of merge, as for sequential composition in Example 5.5.9.1.

5.5.3 Elaborate the details of the bijective correspondence (5.30). Show that it can be used to construct a functor $\mathbf{CoAlg}(GT) \to \mathbf{BiAlg}(TG \stackrel{\rho}{\Rightarrow} GT)$.

5.5.4 Let $G\colon \mathbb{C} \to \mathbb{C}$ be a functor with final coalgebra $\zeta\colon Z \xrightarrow{\cong} G(Z)$, and let $T\colon \mathbb{C} \to \mathbb{C}$ be a monad with an $\mathcal{EM}$-law $\rho\colon TG \Rightarrow GT$.

1. Recall from Exercise 5.2.9 that $T^Z = T(Z + (-))$ is also a monad. Transform ρ into an $\mathcal{EM}$-law $\rho^Z\colon T^Z G \Rightarrow GT^Z$. (This works for any G-coalgebra on Z, not necessarily the final one.)
2. Let $\beta\colon T(Z) \to Z$ be the Eilenberg–Moore algebra induced by ρ, as in (5.21). Check that the algebra induced by ρ^Z is $\beta \circ T([\mathrm{id}, \mathrm{id}])\colon T(Z + Z) \to Z$.
3. Deduce the following '$\mathcal{EM}$-corecursion' principle: for each map $e\colon X \to T(Z + X)$ there is a unique map $s\colon X \to Z$ in

$$\begin{array}{ccc}
GT(Z+X) & \xrightarrow{\;GT([\mathrm{id}, s])\;} & GT(Z) \\
\Big\uparrow e & & \Big\downarrow G(\beta) \\
 & & G(Z) \\
 & & \cong \Big\uparrow \zeta \\
X & \xrightarrow[\;s\;]{} & Z
\end{array}$$

4. Conclude that this yields the dual of recursion, as described in Proposition 2.4.7, when T is the identity monad.
5. Is there also a strengthened form of '$\mathcal{EM}$-induction', dual to point (3), involving a comonad and a suitable distributive law?

5.5.5 Let A be a set and L a complete lattice, and consider the deterministic automaton functor $G(X) = X^A \times L$. What follows comes essentially from [58].

1. Use Exercise 5.4.4 to construct an $\mathcal{EM}$-law $\rho\colon \mathcal{P}G \Rightarrow G\mathcal{P}$, where $\mathcal{P}$ is the powerset monad.
2. Recall from Proposition 2.3.5 that the final G-coalgebra is $L^{A^\star}$. Describe the induced $\mathcal{P}$-algebra $\beta\colon \mathcal{P}(L^{A^\star}) \to L^{A^\star}$ according to (5.21).
3. Prove that via this $\mathcal{EM}$-law and the $\mathcal{EM}$-coinduction principle from Proposition 5.5.10 one can obtain trace semantics for non-deterministic automata, when L is the two-element lattice 2.

6
Invariants and Assertions

The second important notion in the logic of coalgebras, besides bisimularity, is invariance. Whereas a bisimulation is a binary relation on state spaces that is closed under transitions, an invariant is a predicate, or unary relation if you like, on a state space which is closed under transitions. This means, once an invariant holds, it will continue to hold, no matter which state transition operations are applied. That is: coalgebras maintain their invariants.

Invariants are important in the description of systems, because they often express certain implicit assumptions, such as this integer value will always be non-zero (so that dividing by the integer is safe), or the contents of this tank will never be below a given minimum value. Thus, invariants are 'safety properties', which express that something bad will never happen.

This chapter will introduce a general notion of invariant for a coalgebra, via predicate lifting. Predicate lifting is the unary analogue of relation lifting. First it will be introduced for polynomial functors on sets, but later also for more general functors, using the categorical logic introduced in Section 4.3. Various properties of invariants are established, in particular their intimate relation to subcoalgebras. An important application of invariants lies in a generic temporal logic for coalgebras, involving henceforth $\square$ and eventually $\lozenge$ operators on predicates (on a state space for a coalgebra), which will be introduced in Section 6.4. It uses $\square P$ as the greatest invariant that is contained in the predicate P. Further, invariants play a role in the construction of equalisers and products for coalgebras.

The operator $\square P$ involves closure of the predicate P under *all* operations of a coalgebra. In many situations one also likes to express closure under specific operations only. This can be done via the modal logic for coalgebras introduced in Section 6.5. The modal operators are themselves described via a functor as signature, and the meaning of the operators is given by a suitable natural transformation. In examples it is shown how to use such operations to describe coalgebras satisfying certain logical assertions.

The semantics of such assertions – in a set-theoretic context – is the topic of the second part of this chapter, starting in Section 6.6. First, the relatively familiar situation of algebras satisfying assertions is reviewed. In the style of Lawvere's functorial semantics it is shown that monad/functor algebras correspond to finite product preserving functors from a Kleisli category to **Sets**. Subsequently, Section 6.7 describes assertions as axioms for a monad (or functor). This leads to an important conceptual result, namely that functor algebras satisfying certain assertions correspond to Eilenberg–Moore algebras of an associated (quotient) monad. A dual result for coalgebras is described in Section 6.8: functor coalgebras satisfying assertions correspond to Eilenberg–Moore coalgebras of an associated (subset) comonad. This will be illustrated by actually determining such subset comonads in concrete cases. In order to obtain these results we use earlier work on congruences and invariants, in order to turn the given axioms into suitable least congruences and greatest invariants, so that quotient and subset constructions can be applied. The chapter (and book) closes with Section 6.9, illustrating coalgebras and assertions in the context of specification of classes, as in object-oriented programming languages.

6.1 Predicate Lifting

This section will introduce the technique of predicate lifting. It will be used in the next section in the definition of invariance for (co)algebras. Here we will first establish various elementary, useful properties of predicate lifting. Special attention will be paid to the left adjoint to predicate lifting, called predicate lowering.

Predicate lifting for a Kripke polynomial functor $F\colon \mathbf{Sets} \to \mathbf{Sets}$ is an operation $\mathrm{Pred}(F)$ which sends a predicate $P \subseteq X$ on a set X to a 'lifted' predicate $\mathrm{Pred}(F)(P) \subseteq F(X)$ on the result of applying the functor F to X. The idea is that P should hold on all occurrences of X inside $F(X)$, as suggested in

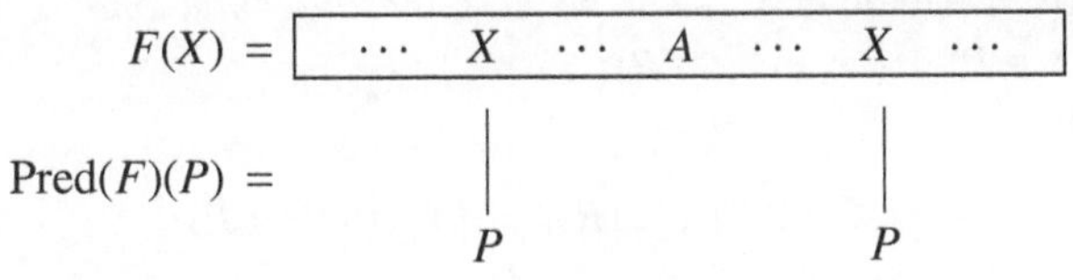

(6.1)

The formal definition proceeds by induction on the structure of the functor F, just as for relation lifting in Definition 3.1.1.

Definition 6.1.1 (Predicate lifting) Let $F\colon \mathbf{Sets} \to \mathbf{Sets}$ be a Kripke polynomial functor, and let X be an arbitrary set. The mapping $\mathrm{Pred}(F)$ which sends a predicate $P \subseteq X$ to a 'lifted' predicate $\mathrm{Pred}(F)(P) \subseteq F(X)$ is defined by induction on the structure of F, in accordance with the steps in Definition 2.2.1.

1. For the identity functor $\mathrm{id}\colon \mathbf{Sets} \to \mathbf{Sets}$ we have

$$\mathrm{Pred}(\mathrm{id})(P) = P.$$

2. For a constant functor $A\colon \mathbf{Sets} \to \mathbf{Sets}$, given by $Y \mapsto A$,

$$\mathrm{Pred}(A)(P) = \top_A = (A \subseteq A).$$

3. For a product functor,

$$\begin{aligned}&\mathrm{Pred}(F_1 \times F_2)(P)\\&= \{(u,v) \in F_1(X) \times F_2(X) \mid \mathrm{Pred}(F_1)(P)(u) \wedge \mathrm{Pred}(F_2)(P)(v)\}.\end{aligned}$$

4. For a set-indexed coproduct,

$$\mathrm{Pred}(\textstyle\coprod_{i \in I} F_i)(P) = \bigcup_{j \in I} \{\kappa_j(u) \in \coprod_{i \in I} F_i(X) \mid \mathrm{Pred}(F_j)(P)(u)\}.$$

5. For an exponent functor,

$$\mathrm{Pred}(F^A)(P) = \{f \in F(X)^A \mid \forall a \in A.\, \mathrm{Pred}(F)(P)(f(a))\}.$$

6. For powerset functor,

$$\mathrm{Pred}(\mathcal{P}(F))(P) = \{U \subseteq X \mid \forall u \in U.\, \mathrm{Pred}(F)(P)(u)\}.$$

   This same formula will be used for a *finite* powerset $\mathcal{P}_{\mathrm{fin}}(F)$.

First we show that predicate and relation lifting are closely related.

Lemma 6.1.2 *1. Relation lifting* $\mathrm{Rel}(F)$ *and predicate lifting* $\mathrm{Pred}(F)$ *for a polynomial functor* $F\colon \mathbf{Sets} \to \mathbf{Sets}$ *are related in the following way:*

$$\mathrm{Rel}(F)(\textstyle\coprod_{\Delta_X}(P)) = \coprod_{\Delta_{F(X)}}(\mathrm{Pred}(F)(P)),$$

where $\Delta_X = \langle \mathrm{id}, \mathrm{id}\rangle\colon X \to X \times X$ *is the diagonal, and so* $\coprod_{\Delta_X}(P) = \{\Delta_X(x) \mid x \in P\} = \{(x,x) \mid x \in P\}$.

2. Similarly,

$$\mathrm{Pred}(F)(\textstyle\coprod_{\pi_i}(R)) = \coprod_{\pi_i}(\mathrm{Rel}(F)(R)),$$

where $\coprod_{\pi_1}(R) = \{x_1 \mid \exists x_2.\, R(x_1,x_2)\}$ *is the domain of the relation* R*, and* $\coprod_{\pi_2}(R) = \{x_2 \mid \exists x_1.\, R(x_1,x_2)\}$ *is its codomain.*

3. *As a result, predicate lifting can be expressed in terms of relation lifting:*

$$\mathrm{Pred}(F)(P) \;=\; \coprod_{\pi_i} (\mathrm{Rel}(F)(\coprod_{\Delta}(P))),$$

for both $i = 1$ *and* $i = 2$.

Proof 1. and 2. By induction on the structure of F.
3. Since $\coprod_{\pi_i} \circ \coprod_{\Delta} = \coprod_{\pi_i \circ \Delta} = \coprod_{\mathrm{id}} = \mathrm{id}$, we use the previous point to get

$$\mathrm{Pred}(F)(P) \;=\; \coprod_{\pi_i} \coprod_{\Delta} \mathrm{Pred}(F)(P) \;=\; \coprod_{\pi_i} \mathrm{Rel}(F)(\coprod_{\Delta} P). \qquad \square$$

Despite this last result, it is useful to study predicate lifting on its own, because it has some special properties that relation lifting does not enjoy – such as preservation of intersections (see the next result), and thus existence of a left adjoint; see Section 6.1.1.

Lemma 6.1.3 *Predicate lifting* Pred(F) *with respect to a Kripke polynomial functor* F: **Sets** $\to$ **Sets** *satisfies the following properties.*

1. *It preserves arbitrary intersections: for every collection of predicates* $(P_i \subseteq X)_{i \in I}$,

$$\mathrm{Pred}(F)(\textstyle\bigcap_{i \in I} P_i) \;=\; \bigcap_{i \in I} \mathrm{Pred}(F)(P_i).$$

A special case (intersection over $I = \emptyset$*) worth mentioning is preservation of truth:*

$$\mathrm{Pred}(F)(\top_X) \;=\; \top_{F(X)}.$$

Another consequence is that predicate lifting is monotone:

$$P \subseteq Q \Longrightarrow \mathrm{Pred}(F)(P) \subseteq \mathrm{Pred}(F)(Q).$$

2. *It preserves inverse images: for a function* $f : X \to Y$ *and predicate* $Q \subseteq Y$,

$$\mathrm{Pred}(F)(f^{-1}(Q)) \;=\; F(f)^{-1}(\mathrm{Pred}(F)(Q)).$$

3. *Predicate lifting also preserves direct images: for* $f : X \to Y$ *and* $P \subseteq X$,

$$\mathrm{Pred}(F)(\coprod_f(P)) \;=\; \coprod_{F(f)}(\mathrm{Pred}(F)(P)).$$

Proof 1. and 2. By induction on the structure of F.
3. This is easily seen via the link with relation lifting:

$$\begin{array}{lll} & \mathrm{Pred}(F)(\coprod_f P) & \\ = & \coprod_{\pi_1} \mathrm{Rel}(F)(\coprod_{\Delta} \coprod_f P) & \text{by Lemma 6.1.2.3} \\ = & \coprod_{\pi_1} \mathrm{Rel}(F)(\coprod_{f \times f} \coprod_{\Delta} P) & \\ = & \coprod_{\pi_1} \coprod_{F(f) \times F(f)} \mathrm{Rel}(F)(\coprod_{\Delta} P) & \text{by Lemma 3.2.2.2} \\ = & \coprod_{F(f)} \coprod_{\pi_1} \mathrm{Rel}(F)(\coprod_{\Delta} P) & \\ = & \coprod_{F(f)}(\mathrm{Pred}(F)(P)) & \text{again by Lemma 6.1.2.3.} \end{array} \qquad \square$$

Corollary 6.1.4 *Predicate lifting* Pred(F) *may be described as a natural transformation, both with the contra- and co-variant powerset functor: by Lemma 6.1.3.2 it forms a map*

$$\mathcal{P} \xRightarrow{\text{Pred}(F)} \mathcal{P}F \qquad \textit{for the contravariant} \qquad \mathbf{Sets}^{\text{op}} \xrightarrow{\mathcal{P}} \mathbf{Sets} \tag{6.2}$$

and also, by Lemma 6.1.3.3,

$$\mathcal{P} \xRightarrow{\text{Pred}(F)} \mathcal{P}F \qquad \textit{for the covariant} \qquad \mathbf{Sets} \xrightarrow{\mathcal{P}} \mathbf{Sets}. \tag{6.3}$$

Predicate liftings are described as natural transformations (6.2) with respect to the contravariant powerset functor in [373], as starting point for temporal logics, as in Section 6.4.

Relation lifting for polynomial functors is functorial, on the category **Rel** = Rel(**Sets**) of relations; see Definition 3.2.3 and more generally Definition 4.3.3. Similarly, predicate lifting yields a functor on the category **Pred** = Pred(**Sets**) of predicates on sets. Explicitly, the category **Pred** has subsets $P \subseteq X$ as objects. A morphism $(P \subseteq X) \longrightarrow (Q \subseteq Y)$ consists of a function $f\colon X \to Y$ with $P(x) \implies Q(f(x))$ for all $x \in X$. This amounts to the existence of the necessarily unique dashed map in

$$\begin{array}{ccc} P & \dashrightarrow & Q \\ \downarrow & & \downarrow \\ X & \xrightarrow[f]{} & Y \end{array}$$

Equivalently, $P \subseteq f^{-1}(Q)$, or $\coprod_f(P) \subseteq Q$, by the correspondence (2.15). There is a forgetful functor **Pred** → **Sets** which is so obvious that it does not get its own name.

Corollary 6.1.5 *Predicate lifting for a polynomial functor* $F\colon \mathbf{Sets} \to \mathbf{Sets}$ *yields a functor* Pred(F) *in a commuting square:*

$$\begin{array}{ccc} \mathbf{Pred} & \xrightarrow{\text{Pred}(F)} & \mathbf{Pred} \\ \downarrow & & \downarrow \\ \mathbf{Sets} & \xrightarrow{F} & \mathbf{Sets} \end{array}$$

Proof Given a map $f\colon (P \subseteq X) \to (Q \subseteq Y)$ in the category **Pred** we get a map $F(f)\colon (\text{Pred}(F)(P) \subseteq F(X)) \to (\text{Pred}(F)(Q) \subseteq F(Y))$. This works as follows, using Lemma 6.1.3:

$$\begin{aligned} P \subseteq f^{-1}(Q) &\Longrightarrow \mathrm{Pred}(F)(P) \subseteq \mathrm{Pred}(F)(f^{-1}(Q)) \\ &\qquad \text{since predicate lifting is monotone} \\ &\Longrightarrow \mathrm{Pred}(F)(P) \subseteq F(f)^{-1}(\mathrm{Pred}(F)(Q)) \\ &\qquad \text{since predicate lifting commutes with substitution.} \end{aligned}$$ ❑

The next result is the analogue for predicate lifting of Lemma 3.3.1 for relation lifting. It relies on considering predicates as sets themselves and may be understood as: predicate lifting commutes with comprehension $\{-\}$; see Lemma 6.3.9. This is described in a more general logical setting in Section 6.1.2 and in [218]. But in the current set-theoretic context the connection between predicate lifting and functor application is extremely simple.

Lemma 6.1.6 *Let F be a Kripke polynomial functor. Applying predicate lifting* $\mathrm{Pred}(F)$ *to a predicate $m\colon P \rightarrowtail X$ is the same as applying the functor F itself to the domain P, in*

$$\begin{array}{ccc} \mathrm{Pred}(F)(P) & = & F(P) \\ & \searrow \quad \swarrow {\scriptstyle F(m)} & \\ & F(X) & \end{array}$$

Note that from the fact that Kripke polynomial functors preserve weak pullbacks (Proposition 4.2.6) we can already conclude that $F(m)$ is a mono; see Lemma 4.2.2.

Proof Formally, one proves for $z \in F(X)$,

$$z \in \mathrm{Pred}(F)(P) \Longleftrightarrow \exists! z' \in F(P).\, F(m)(z') = z.$$

This is obtained by induction on the structure of the polynomial functor F. ❑

6.1.1 Predicate Lowering as Liftings Left Adjoint

We continue this section with a Galois connection involving predicate lifting. In the next section we shall use predicate lifting for a nexttime operator $\bigcirc$ in a temporal logic of coalgebras. There is also a lasttime operator $\overleftarrow{\bigcirc}$ (see Section 6.4.1), for which we shall need this left adjoint (or lower Galois adjoint) to predicate lifting $\mathrm{Pred}(F)$. We shall write this left (or lower) adjoint as $\underleftarrow{\mathrm{Pred}}(F)$ and shall call it 'predicate lowering'. By Lemma 6.1.3.1, predicate lifting preserves arbitrary intersections, and thus has such a left adjoint for abstract reasons; see e.g. [348] or [280, I, theorem 4.2]. But the adjoint can also be defined concretely, by induction on the structure of the functor. This is what we shall do.

Proposition 6.1.7 (From [238, 243]) *Predicate lifting for a Kripke polynomial functor* $F\colon \mathbf{Sets} \to \mathbf{Sets}$ *forms a monotone function* $\mathrm{Pred}(F)\colon \mathcal{P}(X) \to \mathcal{P}(F(X))$ *between powerset posets. In the opposite direction there is also an operation* $\overleftarrow{\mathrm{Pred}}(F)\colon \mathcal{P}(F(X)) \to \mathcal{P}(X)$ *satisfying*

$$\overleftarrow{\mathrm{Pred}}(F)(Q) \subseteq P \iff Q \subseteq \mathrm{Pred}(F)(P).$$

Hence $\overleftarrow{\mathrm{Pred}}(F)$ *is the left adjoint of* $\mathrm{Pred}(F)$ *in a Galois connection* $\overleftarrow{\mathrm{Pred}}(F) \dashv \mathrm{Pred}(F)$.

Proof One can define $\overleftarrow{\mathrm{Pred}}(F)(Q) \subseteq X$ for $Q \subseteq F(X)$ by induction on the structure of the functor F.

1. For the identity functor $\mathrm{id}\colon \mathbf{Sets} \to \mathbf{Sets}$,

$$\overleftarrow{\mathrm{Pred}}(\mathrm{id})(Q) = Q.$$

2. For a constant functor A,

$$\overleftarrow{\mathrm{Pred}}(A)(Q) = \bot_A = (\emptyset \subseteq A).$$

3. For a product functor,

$$\begin{aligned}\overleftarrow{\mathrm{Pred}}(F_1 \times F_2)(Q) &= \overleftarrow{\mathrm{Pred}}(F_1)(\textstyle\coprod_{\pi_1}(Q)) \cup \overleftarrow{\mathrm{Pred}}(F_2)(\textstyle\coprod_{\pi_2}(Q)) \\ &= \overleftarrow{\mathrm{Pred}}(F_1)(\{u \in F_1(X) \mid \exists v \in F_2(X).\, Q(u,v)\}) \\ &\qquad \cup \overleftarrow{\mathrm{Pred}}(F_2)(\{v \in F_2(X) \mid \exists u \in F_1(X).\, Q(u,v)\}).\end{aligned}$$

4. For a set-indexed coproduct functor,

$$\begin{aligned}\overleftarrow{\mathrm{Pred}}(\textstyle\coprod_{i \in I} F_i)(Q) &= \textstyle\bigcup_{i \in I} \overleftarrow{\mathrm{Pred}}(F_i)(\kappa_i^{-1}(Q)) \\ &= \textstyle\bigcup_{i \in I} \overleftarrow{\mathrm{Pred}}(F_i)(\{u \mid Q(\kappa_i(u))\}).\end{aligned}$$

5. For an exponent functor,

$$\overleftarrow{\mathrm{Pred}}(F^A)(Q) = \overleftarrow{\mathrm{Pred}}(F)(\{f(a) \mid a \in A \text{ and } f \in F(X)^A \text{ with } Q(f)\}).$$

6. For a powerset functor,

$$\overleftarrow{\mathrm{Pred}}(\mathcal{P}(F))(Q) = \overleftarrow{\mathrm{Pred}}(F)(\textstyle\bigcup Q).$$

This same formula will be used for a *finite* powerset $\mathcal{P}_{\mathrm{fin}}(F)$. ❑

Being a left adjoint means that functions $\overleftarrow{\mathrm{Pred}}(F)$ preserve certain 'colimit' structures.

Lemma 6.1.8 *Let F be a Kripke polynomial functor. Its opposite predicate lifting operations* $\overleftarrow{\mathrm{Pred}}(F)\colon \mathcal{P}(F(X)) \to \mathcal{P}(X)$ *preserve:*

1. *Unions* $\bigcup$ *of predicates*
2. *Direct images* $\coprod$, *in the sense that for* $f\colon X \to Y$,

$$\underleftarrow{\mathrm{Pred}}(F)(\textstyle\coprod_{F(f)}(Q)) = \coprod_f \underleftarrow{\mathrm{Pred}}(F)(Q).$$

This means that $\underleftarrow{\mathrm{Pred}}(F)$ *forms a natural transformation:*

$$\mathcal{P}F \xRightarrow{\;\underleftarrow{\mathrm{Pred}}(F)\;} \mathcal{P} \qquad \text{for the covariant} \qquad \mathbf{Sets} \xrightarrow{\;\mathcal{P}\;} \mathbf{Sets}$$

as in Corollary 6.1.4.

Proof 1. This is a general property of left (Galois) adjoints, as illustrated in the beginning of Section 2.5.

2. One can use a 'composition of adjoints' argument, or reason directly with the adjunctions:

$$\begin{array}{l}
\underleftarrow{\mathrm{Pred}}(F)(\coprod_{F(f)}(Q)) \subseteq P \\
\Longleftrightarrow \coprod_{F(f)}(Q) \subseteq \mathrm{Pred}(F)(P) \\
\Longleftrightarrow Q \subseteq F(f)^{-1}\mathrm{Pred}(F)(P) = \mathrm{Pred}(F)(f^{-1}(P)) \text{ by Lemma 6.1.3.2} \\
\Longleftrightarrow \underleftarrow{\mathrm{Pred}}(F)(Q) \subseteq f^{-1}(Q) \\
\Longleftrightarrow \coprod_f \underleftarrow{\mathrm{Pred}}(F)(Q) \subseteq P.
\end{array}$$

□

This left adjoint to predicate lifting gives rise to a special kind of mapping, written as sts, from an arbitrary Kripke polynomial functor F to the powerset functor $\mathcal{P}$. The maps $\mathrm{sts}_X\colon F(X) \to \mathcal{P}(X)$ collect the states inside $F(X)$ – as suggested in Diagram (6.1). With this mapping each coalgebra can be turned into an unlabelled transition system.

Proposition 6.1.9 *For a Kripke polynomial functor* F *and an arbitrary set* X, *write* $\mathrm{sts}_X\colon F(X) \to \mathcal{P}(X)$ *for the following composite:*

$$\mathrm{sts}_X \stackrel{\text{def}}{=} \Big(F(X) \xrightarrow{\;\{-\}\;} \mathcal{P}(F(X)) \xrightarrow{\;\underleftarrow{\mathrm{Pred}}(F)\;} \mathcal{P}(X)\Big).$$

These sts *maps form a natural transformation* $F \Rightarrow \mathcal{P}$: *for each function* $f\colon X \to Y$ *the following diagram commutes:*

$$\begin{array}{ccc}
F(X) & \xrightarrow{\;\mathrm{sts}_X\;} & \mathcal{P}(X) \\
{\scriptstyle F(f)}\downarrow & & \downarrow{\scriptstyle \mathcal{P}(f) = \coprod_f} \\
F(Y) & \xrightarrow[\;\mathrm{sts}_Y\;]{} & \mathcal{P}(Y)
\end{array}$$

Proof We need only to prove naturality. For $u \in F(X)$:

$$\begin{aligned}
\textstyle\coprod_f (\mathrm{sts}_X(u)) &= \textstyle\coprod_f \underleftarrow{\mathrm{Pred}}(F)(\{u\}) \\
&= \underleftarrow{\mathrm{Pred}}(F)(\textstyle\coprod_{F(f)}\{u\}) \quad \text{by Lemma 6.1.8.2} \\
&= \underleftarrow{\mathrm{Pred}}(F)(\{F(f)(u)\}) \\
&= \mathrm{sts}_Y(F(f)(u)).
\end{aligned}$$

□

Example 6.1.10 Consider the deterministic automaton functor $(-)^A \times B$. The states contained in $(h, b) \in X^A \times B$ can be computed by following the inductive clauses in the proof of Proposition 6.1.7:

$$\begin{aligned}
\mathrm{sts}(h, b) &= \underleftarrow{\mathrm{Pred}}((-)^A \times B)(\{(h, b)\}) \\
&= \underleftarrow{\mathrm{Pred}}((-)^A)(h) \cup \underleftarrow{\mathrm{Pred}}(B)(b) \\
&= \underleftarrow{\mathrm{Pred}}(\mathrm{id})(\{h(a) \mid a \in A\}) \cup \emptyset \\
&= \{h(a) \mid a \in A\} \subseteq X.
\end{aligned}$$

By combining Propositions 6.1.9 and 2.5.5 we obtain a functor

$$\mathbf{CoAlg}(F) \xrightarrow{\ \mathrm{sts} \circ (-)\ } \mathbf{CoAlg}(\mathcal{P}) \tag{6.4}$$

from coalgebras of a polynomial functor to unlabelled transition systems. This translation will be further investigated in Section 6.4. As can be seen in Exercise 6.1.3, the translation removes much of the structure of the coalgebra. However, it makes the intuitive idea precise that states of a coalgebra can make transitions.

6.1.2 Predicate Lifting, Categorically

We have defined relation lifting concretely for polynomial functors, and more abstractly for functors on a category $\mathbb{C}$ carrying a logical factorisation system; see Definition 4.4.1. Here we shall now give a similarly abstract definition of predicate lifting. It involves the category $\mathrm{Pred}(\mathbb{C})$ of predicates with respect to such a logical factorisation system $(\mathfrak{M}, \mathfrak{E})$ on $\mathbb{C}$; see Definition 4.3.3.

Definition 6.1.11 Let $F\colon \mathbb{C} \to \mathbb{C}$ be an endofunctor on a category $\mathbb{C}$ with a logical factorisation system $(\mathfrak{M}, \mathfrak{E})$. For a predicate $(m\colon U \rightarrowtail X) \in \mathrm{Pred}(\mathbb{C})$, where $m \in \mathfrak{M}$, we define a new predicate $\mathrm{Pred}(F)(U) \rightarrowtail F(X)$ on $F(X)$ via factorisation in

$$\begin{array}{ccc}
F(U) & \xrightarrow{\ \mathfrak{e}(F(m))\ } & \mathrm{Pred}(F)(U) \\
 & {\scriptstyle F(m)} \searrow \quad \swarrow {\scriptstyle \mathfrak{m}(F(m))} & \\
 & F(X) &
\end{array}$$

This yields a functor

$$\begin{array}{ccc} \mathrm{Pred}(\mathbb{C}) & \xrightarrow{\mathrm{Pred}(F)} & \mathrm{Pred}(\mathbb{C}) \\ \downarrow & & \downarrow \\ \mathbb{C} & \xrightarrow{F} & \mathbb{C} \end{array}$$

since for a map of predicates $f\colon (m\colon U \rightarrowtail X) \to (n\colon V \rightarrowtail X)$ the map $F(f)$ is a map $\mathrm{Pred}(F)(U) \to \mathrm{Pred}(F)(V)$ via diagonal-fill-in

$$\begin{array}{ccc} F(U) & \xrightarrow{\mathfrak{e}(F(m))} & \mathrm{Pred}(F)(U) \\ \downarrow & & \downarrow \mathfrak{m}(F(m)) \\ F(V) & & F(X) \\ \downarrow & & \downarrow F(f) \\ \mathrm{Pred}(F)(V) & \xrightarrow[\mathfrak{m}(F(n))]{} & F(Y) \end{array}$$

where the vertical map $F(U) \to F(V)$ on the left is obtained by applying F to the map $U \to V$ that exists since f is a map of predicates (see in Definition 4.3.3).

Some of the properties we have seen earlier on in this section also hold for the abstract form of predicate lifting, under additional assumptions.

Proposition 6.1.12 *Let $F\colon \mathbb{C} \to \mathbb{C}$ be an endofunctor on a category $\mathbb{C}$ with a logical factorisation system $(\mathfrak{M}, \mathfrak{E})$.*

1. *Predicate lifting is monotone and preserves truth $\top$.*
2. *If diagonals $\Delta = \langle \mathrm{id}, \mathrm{id}\rangle\colon X \to X \times X$ are in $\mathfrak{M}$ – that is, if internal and external equality coincide (see Remark 4.3.6) – then $\coprod_\Delta \mathrm{Pred}(F) = \mathrm{Rel}(F) \coprod_\Delta$, as in the square on the left below:*

$$\begin{array}{ccc} \mathrm{Pred}(X) & \xrightarrow{\coprod_\Delta} & \mathrm{Rel}(X) \\ {\scriptstyle \mathrm{Pred}(F)}\downarrow & & \downarrow{\scriptstyle \mathrm{Rel}(F)} \\ \mathrm{Pred}(F(X)) & \xrightarrow[\coprod_\Delta]{} & \mathrm{Rel}(F(X)) \end{array} \qquad \begin{array}{ccc} \mathrm{Pred}(X) & \xleftarrow{\coprod_{\pi_i}} & \mathrm{Rel}(X) \\ {\scriptstyle \mathrm{Pred}(F)}\downarrow & & \downarrow{\scriptstyle \mathrm{Rel}(F)} \\ \mathrm{Pred}(F(X)) & \xleftarrow[\coprod_{\pi_i}]{} & \mathrm{Rel}(F(X)) \end{array}$$

3. *If the functor F preserves abstract epis, then the rectangle on the right also commutes, for $i \in \{1, 2\}$.*
4. *If F preserves abstract epis, predicate lifting commutes with sums (direct images) $\coprod$, as in*

$$\mathrm{Pred}(F)(\textstyle\coprod_f(U)) = \coprod_{F(f)} \mathrm{Pred}(F)(U).$$

5. *If $\mathfrak{E} \subseteq$ SplitEpis and F preserves weak pullbacks, then predicate lifting commutes with inverse images:*

$$\mathrm{Pred}(F)(f^{-1}(V)) = F(f)^{-1}(\mathrm{Pred}(F)(V)).$$

Additionally, predicate lifting preserves meets $\wedge$ of predicates.

Proof 1. This is obvious.

2. We use some properties of images $\mathfrak{m}(-)$: for a predicate $(m\colon U \rightarrowtail X)$,

$$\begin{aligned}
\mathrm{Rel}(F)(\textstyle\coprod_{\Delta}(U)) &= \mathrm{Rel}(F)(\mathfrak{m}(\Delta \circ m)) \\
&= \mathrm{Rel}(F)(\Delta \circ m) && \text{since } \Delta, m \in \mathfrak{M} \\
&= \mathrm{Rel}(F)(\langle m, m\rangle) \\
&= \mathfrak{m}(\langle F(m), F(m)\rangle) \\
&= \mathfrak{m}(\Delta \circ F(m)) \\
&= \Delta \circ \mathfrak{m}(F(m)) && \text{by Exercise 4.3.1.3} \\
&= \textstyle\coprod_{\Delta}(\mathrm{Pred}(F)(P)).
\end{aligned}$$

3. We recall the basic constructions: for a relation $\langle r_1, r_2\rangle\colon R \rightarrowtail X \times X$,

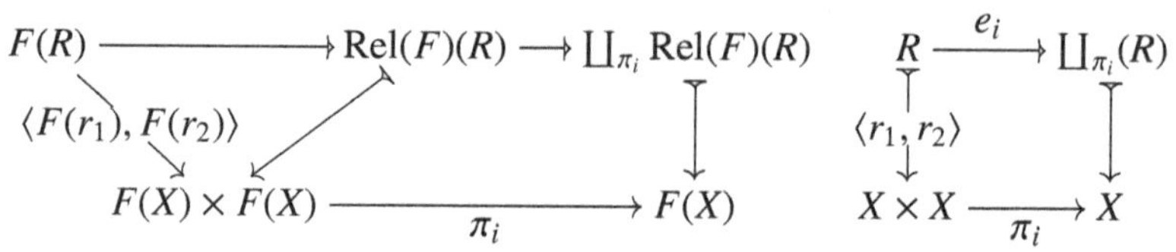

By assumption F preserves abstract epis. Thus $F(e_i) \in \mathfrak{E}$, which yields two dashed diagonals in the following rectangle – and thus the required equality of subobjects – since the south-east diagonal equals $F(r_i)$:

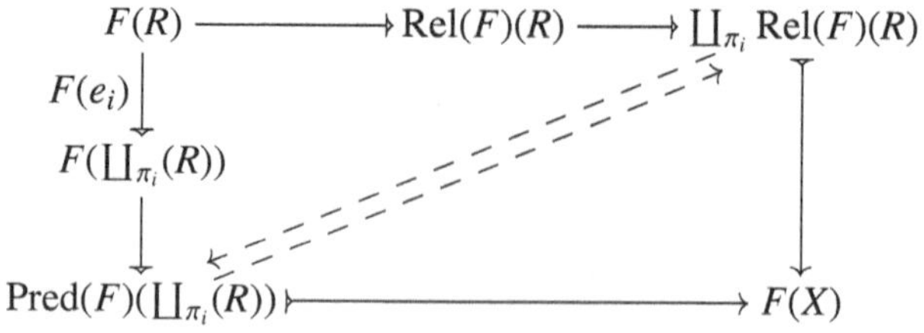

4. One first forms the sum by factorisation:

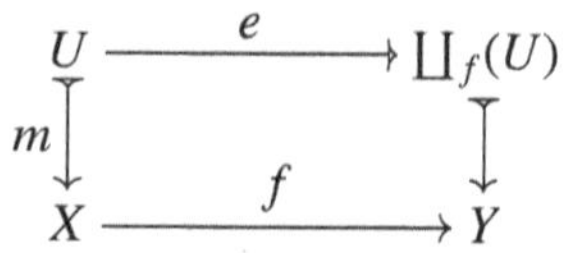

The required equality follows since $F(e) \in \mathfrak{E}$, using

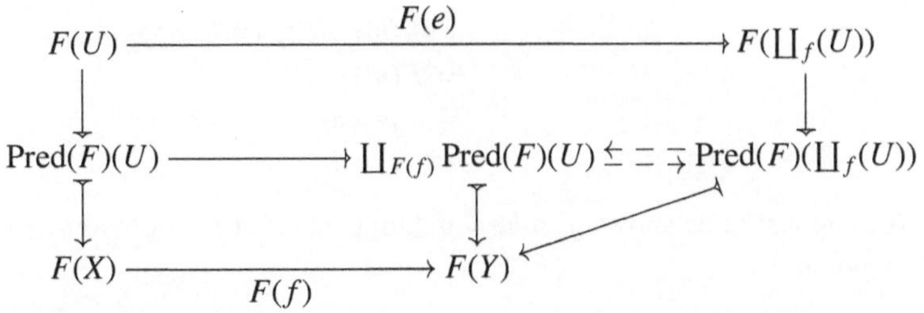

One sees that the south-east map $F(f \circ m) : F(U) \to F(Y)$ is factorised in two ways.

5. Preservation of inverse images is proven as in Proposition 4.4.3. Preservation of meets works as follows. Assume we have predicates $(m : U \rightarrowtail X)$ and $(n : V \rightarrowtail X)$ with meet given by

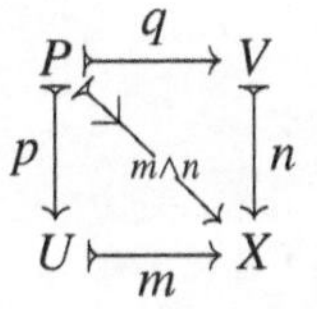

We show that $\mathrm{Pred}(F)(P)$ is the pullback of $\mathrm{Pred}(F)(U)$ and $\mathrm{Pred}(F)(V)$, using that abstract epis split, via the maps s in

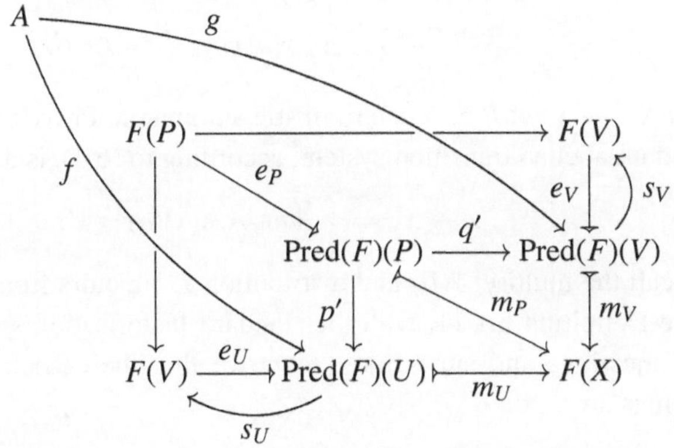

We assume that $m_U \circ f = m_V \circ g$. The two maps $f' = s_U \circ f : A \to F(U)$ and $g' = s_V \circ g : A \to F(V)$ satisfy $F(m) \circ f' = F(n) \circ g'$, since $e_U \circ s_U = \mathrm{id}$ (and similarly for V). Since F preserves weak pullbacks, there is a map $h : A \to F(P)$ with $F(p) \circ h = f'$ and $F(q) \circ h = g'$. We claim that $h' = e_P \circ h : A \to \mathrm{Pred}(F)(P)$ is the required mediating map. The equation $p' \circ h' = f$ holds, since m_U is monic:

$$\begin{aligned} m_U \circ p' \circ h' &= m_P \circ e_P \circ h \\ &= F(m \wedge n) \circ h \\ &= F(m) \circ F(p) \circ h \\ &= F(m) \circ f' \\ &= m_U \circ e_U \circ s_U \circ f \\ &= m_U \circ f. \end{aligned}$$

Analogously one shows $q' \circ h' = g$. Uniqueness of h' is obvious, since m_P is monic. ❑

Exercises

6.1.1 Show for a list functor $F^\star$ – using the description (2.17) – that for $P \subseteq X$,

$$\mathrm{Pred}(F^\star)(P) = \{\langle u_1, \ldots, u_n\rangle \in F(X)^\star \mid \forall i \leq n.\, \mathrm{Pred}(F)(P)(u_i)\}.$$

And that for $Q \subseteq Y$,

$$\begin{aligned} &\overleftarrow{\mathrm{Pred}}(F^\star)(Q) \\ &= \bigcup_{n \in \mathbb{N}} \overleftarrow{\mathrm{Pred}}(F)(\{u_i \mid i \leq n,\ \langle u_1, \ldots, u_n\rangle \in F(Y)^n,\ Q(\langle u_1, \ldots, u_n\rangle)\}). \end{aligned}$$

6.1.2 Use Lemmas 6.1.6 and 3.3.1 to check that relation lifting can also be expressed via predicate lifting. For a relation $\langle r_1, r_2\rangle\colon R \hookrightarrow X \times Y$,

$$\mathrm{Rel}(F)(R) = \coprod\nolimits_{\langle F(r_1), F(r_2)\rangle} \mathrm{Pred}(F)(R).$$

6.1.3 Let $X \xrightarrow{\langle \delta, \varepsilon\rangle} X^A \times B$ be a deterministic automaton. Prove that the associated unlabelled transition system, according to (6.4), is described by:

$$x \to x' \iff \exists a \in A.\, \delta(x)(a) = x'.$$

6.1.4 Recall the multiset $\mathcal{M}_M$ and distribution $\mathcal{D}$ functors from Section 4.1. Use Definition 6.1.11, with the standard factorisation system on **Sets** of injections and surjections, to prove that the associated predicate liftings are

$$\begin{aligned} \mathrm{Pred}(\mathcal{M}_M)(P \subseteq X) &= \{\varphi \in \mathcal{M}_M(X) \mid \forall x.\, \varphi(x) \neq 0 \Rightarrow P(x)\} \\ \mathrm{Pred}(\mathcal{D})(P \subseteq X) &= \{\varphi \in \mathcal{D}(X) \mid \forall x.\, \varphi(x) \neq 0 \Rightarrow P(x)\}. \end{aligned}$$

Hint: Recall from Propositions 4.2.9 and 4.2.10 that the functors $\mathcal{M}_M$ and $\mathcal{D}$ preserve weak pullbacks, and thus injections, by Lemma 4.2.2.

6.1.5 (From [419, proposition 16]) Let $F\colon$ **Sets** $\to$ **Sets** be an arbitrary functor. Prove that there is a bijective correspondence between natural transformations $\mathcal{P} \Rightarrow \mathcal{P}F$ as in (6.2) and subsets of $F(2)$ – where $2 = \{0, 1\}$. *Hint*: Use that predicates $P \subseteq X$ can be identified with their characteristic functions $X \to 2$.

6.1.6 Prove functoriality of predicate lifting – as in Exercise 4.4.6 for relation lifting: a natural transformation $\sigma\colon F \Rightarrow G$ gives a lifting as in

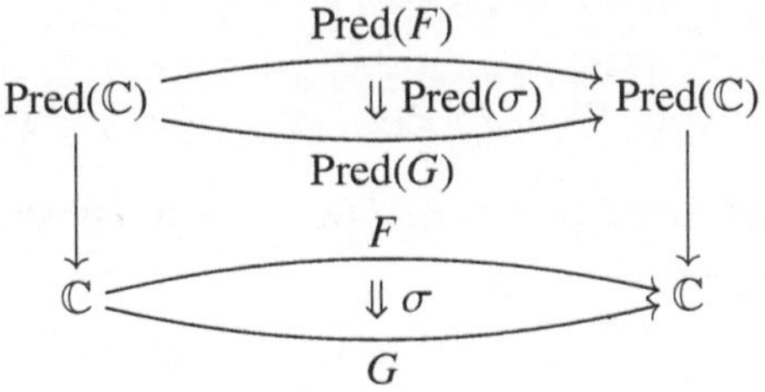

Prove also, for arbitrary functors $F, G\colon \mathbb{C} \to \mathbb{C}$, the existence of a natural transformation $\mathrm{Pred}(FG) \Rightarrow \mathrm{Pred}(F)\mathrm{Pred}(G)$.

6.2 Invariants

Invariants are predicates on the state space of a coalgebra with the special property that they are closed under transitions: once they are true, they remain true no matter which steps are taken (using the coalgebra). This section will introduce invariants via predicate lifting (from the previous section). It will first concentrate on invariants for coalgebras of polynomial functors, and later deal with more general functors. Invariants are closely related to subcoalgebras. Many of the results we describe for invariants occur in [411, section 6], but with subcoalgebra terminology, and thus with slightly different proofs.

We shall define the notion of invariant, both for coalgebras and for algebras, as (co)algebra of a predicate lifting functor $\mathrm{Pred}(F)$. In both cases an invariant is a predicate which is closed under the state transition operations. There does not seem to be an established (separate) terminology in algebra, so we simply use the phrase 'invariant' both for algebras and for coalgebras.

Definition 6.2.1 Let $F\colon \mathbf{Sets} \to \mathbf{Sets}$ be a Kripke polynomial functor, with predicate lifting functor $\mathrm{Pred}(F)\colon \mathbf{Pred} \to \mathbf{Pred}$ as in Corollary 6.1.5. Abstractly, an **invariant** is either a $\mathrm{Pred}(F)$-coalgebra $P \to \mathrm{Pred}(F)(P)$ or a $\mathrm{Pred}(F)$-algebra $\mathrm{Pred}(F)(P) \to P$, as in

More concretely, this means the following.

1. An **invariant** for a coalgebra $c\colon X \to F(X)$ is a predicate $P \subseteq X$ satisfying for all $x \in X$,

$$x \in P \Longrightarrow c(x) \in \mathrm{Pred}(F)(P).$$

   Equivalently,

$$P \subseteq c^{-1}(\mathrm{Pred}(F)(P)) \quad \text{or} \quad \coprod_c(P) \subseteq \mathrm{Pred}(F)(P).$$

2. An **invariant** for an algebra $a\colon F(X) \to X$ is a predicate $P \subseteq X$ satisfying for all $u \in F(X)$,

$$u \in \mathrm{Pred}(F)(P) \Longrightarrow a(u) \in P.$$

   That is,

$$\mathrm{Pred}(F)(P) \subseteq a^{-1}(P) \quad \text{or} \quad \coprod_a(\mathrm{Pred}(F)(P)) \subseteq P.$$

This section concentrates on invariants for coalgebras, but occasionally invariants for algebras are also considered. We first relate invariants to bisimulations. There are similar results for congruences; see Exercise 6.2.1.

Lemma 6.2.2 *Consider two coalgebras $c\colon X \to F(X)$ and $d\colon Y \to F(Y)$ of a Kripke polynomial functor F.*

1. *If $R \subseteq X \times Y$ is a bisimulation, then both its domain $\coprod_{\pi_1} R = \{x \mid \exists y.\, R(x,y)\}$ and codomain $\coprod_{\pi_2} R = \{y \mid \exists x.\, R(x,y)\}$ are invariants.*
2. *An invariant $P \subseteq X$ yields a bisimulation $\coprod_\Delta P = \{(x,x) \mid x \in P\} \subseteq X \times X$.*

Proof 1. If the relation R is a bisimulation, then the predicate $\coprod_{\pi_1} R \subseteq X$ is an invariant, since

$$\begin{aligned}\textstyle\coprod_c \coprod_{\pi_1} R &= \textstyle\coprod_{\pi_1} \coprod_{c\times d} R \\ &\subseteq \textstyle\coprod_{\pi_1} \mathrm{Rel}(F)(R) && \text{because } R \text{ is a bisimulation} \\ &= \textstyle\mathrm{Pred}(F)(\coprod_{\pi_1} R) && \text{by Lemma 6.1.2.2.}\end{aligned}$$

Similarly, $\coprod_{\pi_2} R \subseteq Y$ is an invariant for the coalgebra d.

2. Suppose now that $P \subseteq X$ is an invariant. Then

$$\begin{aligned}\textstyle\coprod_{c\times c} \coprod_\Delta P &= \textstyle\coprod_\Delta \coprod_c P \\ &\subseteq \textstyle\coprod_\Delta \mathrm{Pred}(F)(P) && \text{since } P \text{ is an invariant} \\ &= \textstyle\mathrm{Rel}(F)(\coprod_\Delta P) && \text{by Lemma 6.1.2.1.}\end{aligned}$$

❑

Example 6.2.3 We consider invariants for both deterministic and non-deterministic automata.

1. As is well known by now, a deterministic automaton $\langle\delta, \epsilon\rangle\colon X \to X^A \times B$ is a coalgebra for the functor $F = \mathrm{id}^A \times B$. Predicate lifting for this functor yields for a predicate $P \subseteq X$ a new predicate $\mathrm{Pred}(F)(P) \subseteq (X^A \times B)$, given by

$$\mathrm{Pred}(F)(P)(f, b) \iff \forall a \in A.\, P(f(a)).$$

A predicate $P \subseteq X$ is thus an invariant with respect to the coalgebra $\langle\delta, \epsilon\rangle\colon X \to X^A \times B$ if, for all $x \in X$,

$$\begin{aligned} P(x) &\Longrightarrow \mathrm{Pred}(F)(P)((\delta(x), \epsilon(x))) \\ &\iff \forall a \in A.\, P(\delta(x)(a)) \\ &\iff \forall a \in A.\, \forall x' \in X.\, x \xrightarrow{a} x' \Rightarrow P(x'). \end{aligned}$$

Thus, once a state x is in an invariant P, all its – immediate and non-immediate – successor states are also in P. Once an invariant holds, it will continue to hold.

2. A non-deterministic automaton $\langle\delta, \epsilon\rangle\colon X \to \mathcal{P}(X)^A \times B$ is a coalgebra for the functor $F = \mathcal{P}(\mathrm{id})^A \times B$. Predicate lifting for this functor sends a predicate $P \subseteq X$ to the predicate $\mathrm{Pred}(F)(P) \subseteq (\mathcal{P}(X)^A \times B)$ given by:

$$\mathrm{Pred}(F)(P)(f, b) \iff \forall a \in A.\, \forall x' \in f(a).\, P(x')\,.$$

This $P \subseteq X$ is then an invariant for the automaton $\langle\delta, \epsilon\rangle\colon X \to \mathcal{P}(X)^A \times B$ if for all $x \in X$,

$$\begin{aligned} P(x) &\Longrightarrow \mathrm{Pred}(F)(P)(\delta(x), \epsilon(x)) \\ &\iff \forall a \in A.\, \forall x' \in \delta(x)(a).\, P(x') \\ &\iff \forall a \in A.\, \forall x' \in X.\, x \xrightarrow{a} x' \Rightarrow P(x'). \end{aligned}$$

Proposition 6.2.4 *Let $X \xrightarrow{c} F(X)$ and $Y \xrightarrow{d} F(Y)$ be two coalgebras of a Kripke polynomial functor $F\colon$ **Sets** $\to$ **Sets**.*

1. *Invariants are closed under arbitrary unions and intersections: if predicates $P_i \subseteq X$ are invariants for $i \in I$, then their union $\bigcup_{i\in I} P_i$ and intersection $\bigcap_{i\in I} P_i$ are invariants.*

   In particular, falsity $\bot$ (union over $I = \emptyset$) and truth $\top$ (intersection over $I = \emptyset$) are invariants.

2. *Invariants are closed under direct and inverse images along homomorphisms: if $f\colon X \to Y$ is a homomorphism of coalgebras, and $P \subseteq X$ and $Q \subseteq Y$ are invariants, then so are $\coprod_f(P) \subseteq Y$ and $f^{-1}(Q) \subseteq X$.*

   In particular, the image $\mathrm{Im}(f) = \coprod_f(\top)$ of a coalgebra homomorphism is an invariant.

Proof 1. First we note that inverse images preserve both unions and intersections. Closure of invariants under unions then follows from monotonicity of predicate lifting: $P_i \subseteq c^{-1}(\mathrm{Pred}(F)(P_i)) \subseteq c^{-1}(\mathrm{Pred}(F)(\bigcup_{i\in I} P_i)$ for each $i \in I$, so that we may conclude $\bigcup_{i\in I} P_i \subseteq c^{-1}\mathrm{Pred}(F)(\bigcup_{i\in I} P_i)$. Similarly, closure under intersection follows because predicate lifting preserves intersections; see Lemma 6.1.3.1.

2. For preservation of direct images assume that $P \subseteq X$ is an invariant. Then

$$\begin{array}{rcll} \coprod_d \coprod_f P & = & \coprod_{F(f)} \coprod_c P & \text{because } f \text{ is a homomorphism} \\ & \subseteq & \coprod_{F(f)} \mathrm{Pred}(F)(P) & \text{since } P \text{ is an invariant} \\ & = & \mathrm{Pred}(F)(\coprod_f P) & \text{by Lemma 6.1.3.3.} \end{array}$$

Similarly, if $Q \subseteq Y$ is an invariant, then

$$\begin{array}{rcll} f^{-1}(Q) & \subseteq & f^{-1}d^{-1}(\mathrm{Pred}(F)(Q)) & \text{because } Q \text{ is an invariant} \\ & = & c^{-1}F(f)^{-1}(\mathrm{Pred}(F)(Q)) & \text{because } f \text{ is a homomorphism} \\ & = & c^{-1}(\mathrm{Pred}(F)(f^{-1}(Q)) & \text{by Lemma 6.1.3.2.} \end{array}$$

□

The next result readily follows from Lemma 6.1.6. It is the analogue of Theorem 3.3.2 and has important consequences.

Theorem 6.2.5 *Let F : **Sets** $\to$ **Sets** be a Kripke polynomial functor.*

1. *A predicate $m \colon P \rightarrowtail X$ on the state space of a coalgebra $c \colon X \to F(X)$ is an invariant if and only if $P \rightarrowtail X$ is a* ***subcoalgebra****: there is a (necessarily unique) coalgebra structure $P \to F(P)$ making $m \colon P \rightarrowtail X$ a homomorphism of coalgebras:*

$$\begin{array}{ccc} F(P) & \xrightarrow{F(m)} & F(X) \\ \uparrow & & \uparrow{\scriptstyle c} \\ P & \xrightarrow[m]{} & X \end{array}$$

Uniqueness of this coalgebra $P \to F(P)$ follows because $F(m)$ is injective by Lemma 4.2.2.

2. *Similarly, a predicate $m \colon P \rightarrowtail X$ is an invariant for an algebra $a \colon F(X) \to X$ if P carries a (necessarily unique)* ***subalgebra*** *structure $F(P) \to P$ making $m \colon P \rightarrowtail X$ a homomorphism of algebras.* □

Earlier we have seen a generic binary induction principle in Theorem 3.1.4. At this stage we can prove the familiar unary induction principle for initial algebras.

Theorem 6.2.6 (Unary induction proof principle) *An invariant on an initial algebra is always true.*

Equivalently, the truth predicate is the only invariant on an initial algebra. The proof is a generalisation of the argument we have used in Example 2.4.4 to derive induction for the natural numbers from initiality.

Proof Let $m\colon P \rightarrowtail A$ be an invariant on the initial algebra $F(A) \xrightarrow{\cong} A$. This means by the previous theorem that P itself carries a subalgebra structure $F(P) \to P$, making the square below on the right commute. This subalgebra yields a homomorphism $f\colon A \to P$ by initiality, as on the left:

$$\begin{array}{ccccc} F(A) & \xrightarrow{F(f)} & F(P) & \overset{F(m)}{\rightarrowtail} & F(A) \\ {\scriptstyle\cong}\downarrow & & \downarrow & & \downarrow{\scriptstyle\cong} \\ A & \xrightarrow[f]{} & P & \underset{m}{\rightarrowtail} & A \end{array}$$

By uniqueness we then get $m \circ f = \mathrm{id}_A$, which tells that $t \in P$, for all $t \in A$. □

Example 6.2.7 Consider the binary trees from Example 2.4.5 as algebras of the functor $F(X) = 1 + (X \times A \times X)$, with initial algebra

$$1 + (\mathsf{BinTree}(A) \times A \times \mathsf{BinTree}(A)) \xrightarrow[\cong]{[\mathsf{nil}, \mathsf{node}]} \mathsf{BinTree}(A).$$

Predicate lifting $\mathrm{Pred}(F)(P) \subseteq F(X)$ of an arbitrary predicate $P \subseteq X$ is given by

$$\mathrm{Pred}(F)(P) = \{\kappa_1(*)\} \cup \{\kappa_2(x_1, a, x_2) \mid a \in A \wedge P(x_1) \wedge P(x_2)\}.$$

Therefore, a predicate $P \subseteq \mathsf{BinTree}(A)$ on the initial algebra is an invariant if both

$$\begin{cases} P(\mathsf{nil}) \\ P(x_1) \wedge P(x_2) \Rightarrow P(\mathsf{node}(x_1, a, x_2)). \end{cases}$$

The unary induction principle then says that such a P must hold for all binary trees $t \in \mathsf{BinTree}(A)$. This may be rephrased in rule form as

$$\frac{P(\mathsf{nil}) \qquad P(x_1) \wedge P(x_2) \Rightarrow P(\mathsf{node}(x_1, a, x_2))}{P(t)}\,.$$

6.2.1 Invariants, Categorically

The description in Definition 6.2.1 of invariants as (co)algebras of a predicate lifting functor $\mathrm{Pred}(F)$ generalises immediately from polynomial functors to arbitrary functors: if the underlying category $\mathbb{C}$ carries a logical factorisation system $(\mathfrak{M}, \mathfrak{E})$, a predicate lifting functor $\mathrm{Pred}(F)\colon \mathrm{Pred}(\mathbb{C}) \to \mathrm{Pred}(\mathbb{C})$ exists as in Definition 6.1.11. An invariant is then a predicate $(m\colon U \rightarrowtail X) \in \mathrm{Pred}(\mathbb{C})$

on the carrier $X \in \mathbb{C}$ of a coalgebra $c: X \to F(X)$, or of an algebra $a: F(X) \to X$, for which there are (dashed) maps in $\mathbb{C}$:

This is the same as in Definition 6.2.1. Not all of the results that hold for invariants of (co)algebras of polynomial functors also hold in the abstract case. In particular, the tight connection between invariants of a coalgebra and subcoalgebras is lost – but it still holds in the algebraic case. We briefly discuss the main results.

Lemma 6.2.8 *Let $F: \mathbb{C} \to \mathbb{C}$ be an endofunctor on a category $\mathbb{C}$ with a logical factorisation system $(\mathfrak{M}, \mathfrak{E})$. Assume predicates $m: U \rightarrowtail X$ and $n: V \rightarrowtail Y$, on the carriers of a coalgebra $c: X \to F(X)$ and an algebra $a: F(Y) \to Y$.*

1. *If $U \rightarrowtail X$ carries a subcoalgebra $c': U \to F(U)$, then U is an invariant. The converse holds if abstract epis are split,* i.e. *if $\mathfrak{E} \subseteq$ SplitEpis.*
2. *The predicate $V \rightarrowtail Y$ carries an subalgebra if and only if it is an invariant.*

Proof 1. A subcoalgebra $c': U \to F(U)$ gives rise to a dashed map $e_U \circ c': U \to \text{Pred}(F)(U)$ by composition:

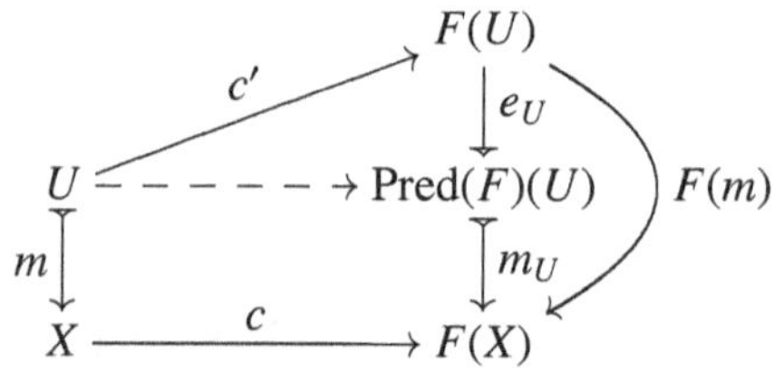

If the map $e_U \in \mathfrak{E}$ is a split epi, say via $s: \text{Pred}(F)(U) \to F(U)$, then an invariant yields a subcoalgebra, by post-composition with s.

2. For algebras an invariant $\text{Pred}(F)(V) \to V$ gives rise to a subalgebra via pre–composition with $e_V: F(V) \twoheadrightarrow \text{Pred}(F)(V)$. In the reverse direction, from a subalgebra $a': F(U) \to U$ to an invariant, we use diagonal-fill-in:

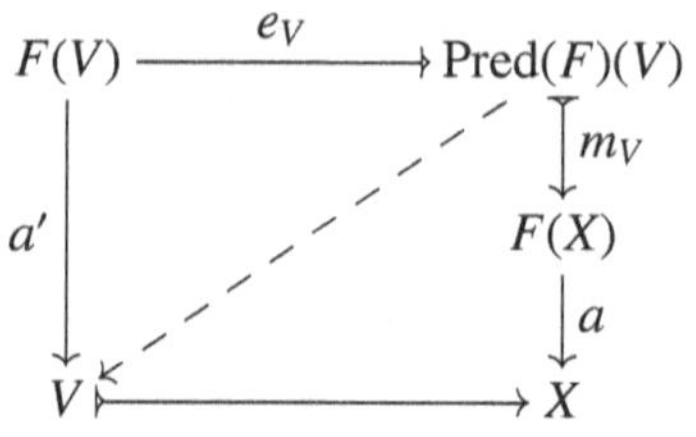

❑

Exercises

6.2.1 Let $F(X) \xrightarrow{a} X$ and $F(Y) \xrightarrow{b} Y$ be algebras of a Kripke polynomial functor F. Prove, in analogy with Lemma 6.2.2 that

1. If $R \subseteq X \times Y$ is a congruence, then both its domain $\coprod_{\pi_1} R \subseteq X$ and its codomain $\coprod_{\pi_2} R \subseteq Y$ are invariants.
2. If $P \subseteq X$ is an invariant, then $\coprod_{\Delta} P \subseteq X \times X$ is a congruence.

6.2.2 Use binary induction in Theorem 3.1.4, together with the previous exercise, to give an alternative proof of unary induction from Theorem 6.2.6.

6.2.3 Prove in the general context of Section 6.2.1 that for a coalgebra homomorphism f direct images $\coprod_f$, as in defined in Proposition 4.3.5, preserve invariants. Conclude that the image $\mathrm{Im}(f) = \coprod_f(\top) \rightarrowtail Y$ of a coalgebra homomorphism $f\colon X \to Y$ is an invariant.

6.2.4 The next result from [154] is the analogue of Exercise 3.2.7; it describes when a function is definable by coinduction. Let $Z \xrightarrow{\cong} F(Z)$ be final coalgebra of a finite Kripke polynomial functor F. Prove that an arbitrary function $f\colon X \to Z$ is defined by finality (i.e. is beh_c for some coalgebra $c\colon X \to F(X)$ on its domain X) if and only if its image $\mathrm{Im}(f) \subseteq Z$ is an invariant. *Hint*: Use the splitting of surjective functions from Lemma 2.1.7.

6.2.5 Let $S : \mathbb{C} \to \mathbb{C}$ be a comonad on a category $\mathbb{C}$ with a logical factorisation system $(\mathfrak{M}, \mathfrak{E})$.

1. Use the Exercise 6.1.6 to derive that $\mathrm{Pred}(S)$ is a comonad on the category $\mathrm{Pred}(\mathbb{C})$ via

$$\mathrm{id} = \mathrm{Pred}(\mathrm{id}) \Longleftarrow \mathrm{Pred}(S) \Longrightarrow \mathrm{Pred}(S^2) \Rightarrow \mathrm{Pred}(S)^2.$$

2. Consider a pair of maps in a commuting square:

$$\begin{array}{ccc} U & \xrightarrow{\quad \gamma' \quad} & \mathrm{Pred}(S)(U) \\ \big\downarrow & & \big\downarrow \\ X & \xrightarrow[\quad \gamma \quad]{} & S(X) \end{array}$$

Prove that if γ is an Eilenberg–Moore coalgebra for the comonad S, then the pair (γ, γ') is automatically an Eilenberg–Moore coalgebra for the comonad $\mathrm{Pred}(S)$ – and thus an invariant for γ.

3. Let $(X \xrightarrow{\gamma} S(X)) \xrightarrow{f} (Y \xrightarrow{\beta} S(Y))$ be a map of Eilenberg–Moore coalgebras. Prove, as in Exercise 6.2.3 for functor coalgebras, that

if $P \rightarrowtail X$ is an invariant, i.e. Pred(S)-coalgebra, for the Eilenberg–Moore coalgebra γ, then so is $\coprod_f(P)$ for β.

6.2.6 Let $T: \mathbb{C} \to \mathbb{C}$ now be a monad on a category $\mathbb{C}$ with a logical factorisation system $(\mathfrak{M}, \mathfrak{E})$. Assume that T preserves abstract epis, i.e. $e \in \mathfrak{E} \Rightarrow T(e) \in \mathfrak{E}$.

1. Prove, via Exercise 4.4.6, that relation lifting $\mathrm{Rel}(T): \mathrm{Rel}(\mathbb{C}) \to \mathrm{Rel}(\mathbb{C})$ is a monad. Describe its unit and multiplication explicitly.
2. Assume a commuting square:

$$\begin{array}{ccc} \mathrm{Rel}(T)(R) & \xrightarrow{\quad\gamma\quad} & R \\ \downarrow & & \downarrow \\ T(X) \times T(Y) & \xrightarrow[\alpha \times \beta]{} & X \times Y \end{array}$$

Prove that if α and β are algebras for the monad T, then the above square is automatically an Eilenberg–Moore algebra for the monad $\mathrm{Rel}(T)$ – and thus a congruence for α, β.

6.3 Greatest Invariants and Limits of Coalgebras

In the first chapter – in Definition 1.3.2 to be precise – we introduced the predicate $\Box P$, describing 'henceforth P' for a predicate P on sequences. The meaning of $(\Box P)(x)$ is that P holds in state x and for all of its successor states. Here we shall extend this same idea to arbitrary coalgebras by defining the predicate $\Box P$ in terms of greatest invariants. These greatest invariants are useful in various constructions, most important in this section, in the construction of equalisers and products for coalgebras. In the next section it will be shown that they are important in a temporal logic for coalgebras.

Definition 6.3.1 Let $c: X \to F(X)$ be a coalgebra of a Kripke polynomial functor F: **Sets** $\to$ **Sets**. For an arbitrary predicate $P \subseteq X$ on the state space of c, we define a new predicate $\Box P \subseteq X$, called **henceforth** P, as

$$(\Box P)(x) \text{ iff } \exists Q \subseteq X.\, Q \text{ is an invariant for } c \wedge Q \subseteq P \wedge Q(x),$$

that is,

$$\Box P = \bigcup \{Q \subseteq P \mid Q \text{ is an invariant}\}.$$

Since invariants are closed under union – by Proposition 6.2.4.1 – $\Box P$ is an invariant itself. Among all the invariants $Q \subseteq X$, it is the greatest one that is contained in P.

The definition of 'henceforth' resembles the definition of bisimilarity (see Definition 3.1.5). In fact, one could push the similarity by defining for an arbitrary relation R, another relation $\Box R$ as the greatest bisimilarity contained in R – so that bisimilarity $\leftrightarrow$ would appear as $\Box\top$. But there seems to be no clear use for this extra generality.

The next lemma lists some obvious properties of $\Box$. Some of these are already mentioned in Exercise 1.3.3 for sequences.

Lemma 6.3.2 *Consider the henceforth operator $\Box$ for a coalgebra $c\colon X \to F(X)$. The first three properties below express that $\Box$ is an interior operator. The fourth property says that its opens are invariants.*

1. $\Box P \subseteq P$
2. $\Box P \subseteq \Box\Box P$
3. $P \subseteq Q \Rightarrow \Box P \subseteq \Box Q$
4. *P is an invariant if and only if $P = \Box P$.*

Proof 1. This is obvious: if $\Box P(x)$, then $Q(x)$ for some invariant Q with $Q \subseteq P$. Hence $P(x)$.

2. If $\Box P(x)$, then we have an invariant Q, namely $\Box P$, with $Q(x)$ and $Q \subseteq \Box P$. Hence $\Box\Box P(x)$.
3. This is obvious.
4. The (if)-part is clear because we have already seen that $\Box P$ is an invariant. For the (only if)-part, by (1) we have only to prove $P \subseteq \Box P$, if P is an invariant. So assume $P(x)$, then we have an invariant Q, namely P, with $Q(x)$ and $Q \subseteq P$. Hence $\Box P(x)$. □

The following result gives an important structural property of greatest invariants. It may be understood abstractly as providing a form of comprehension for coalgebras, as elaborated in Section 6.3.1 below.

Proposition 6.3.3 *Consider a coalgebra $c\colon X \to F(X)$ of a Kripke polynomial functor F with an arbitrary predicate $P \subseteq X$. By Theorem 6.2.5.1 the greatest invariant $\Box P \subseteq P \subseteq X$ carries a subcoalgebra structure, say c_P, in*

$$\begin{array}{ccc} F(\Box P) & \xrightarrow{F(m)} & F(X) \\ {\scriptstyle c_P}\uparrow & & \uparrow{\scriptstyle c} \\ \Box P & \xrightarrow[m]{} & X \end{array}$$

This subcoalgebra has the following universal property: each coalgebra homomorphism $f\colon (Y \xrightarrow{d} F(Y)) \longrightarrow (X \xrightarrow{c} F(X))$ which factors through

$P \hookrightarrow X$ – i.e. *satisfies* $f(y) \in P$ *for all* $y \in Y$ – *also factors through* $\Box P$, *namely as (unique) coalgebra homomorphism* $f' \colon (Y \xrightarrow{d} F(Y)) \to (\Box P \xrightarrow{c_P} F(\Box P))$ *with* $m \circ f' = f$.

Proof The assumption that f factors through $P \subseteq X$ may be rephrased as an inclusion $\mathrm{Im}(f) = \coprod_f(\top) \subseteq P$. But since the image along a homomorphism is an invariant, (see Proposition 6.2.4.2), we get an inclusion $\mathrm{Im}(f) \subseteq \Box P$. This gives the factorisation

$$\big(Y \xrightarrow{\;f\;} X\big) = \big(Y \xrightarrow{\;f'\;} \Box P \overset{m}{\hookrightarrow} X\big).$$

We have only to show that f' is a homomorphism of coalgebras. But this follows because $F(m)$ is injective; see Lemma 4.2.2. It yields $c_P \circ f' = F(f') \circ d$ since

$$\begin{aligned} F(m) \circ c_P \circ f' &= c \circ m \circ f' \\ &= c \circ f \\ &= F(f) \circ d \\ &= F(m) \circ F(f') \circ d. \end{aligned}$$

❑

In this section we shall use greatest invariants to prove the existence of limits (equalisers and cartesian products) for coalgebras of Kripke polynomial functors. The constructions can be adapted easily to more general functors, provided the relevant structures, such as $\Box$ and cofree coalgebras, exist.

Recall from Proposition 2.1.5 and Exercise 2.1.14 that colimits (coproducts and coequalisers) of coalgebras are easy: they are constructed just like for sets. The product structure of coalgebras, however, is less trivial. First results appeared in [468], for 'bounded' endofunctors on **Sets**; see Definition 4.6.5 later on. This was generalised in [186, 283, 231] and [242] (which is followed below). We begin with equalisers, which are easy using henceforth $\Box$.

Theorem 6.3.4 (Equalisers of coalgebras) *The category* **CoAlg**(F) *of coalgebras of a Kripke polynomial functor* $F\colon$ **Sets** $\to$ **Sets** *has equalisers: for two coalgebras* $X \xrightarrow{c} F(X)$ *and* $Y \xrightarrow{d} F(Y)$ *with two homomorphisms* $f, g \colon X \to Y$ *between them, there is an equaliser diagram in* **CoAlg**(F),

$$\begin{pmatrix} F(\Box E(f,g)) \\ \uparrow \\ \Box E(f,g) \end{pmatrix} \overset{m}{\rightarrowtail} \begin{pmatrix} F(X) \\ \uparrow c \\ X \end{pmatrix} \underset{g}{\overset{f}{\rightrightarrows}} \begin{pmatrix} F(Y) \\ \uparrow d \\ Y \end{pmatrix}$$

where $E(f,g) \hookrightarrow X$ *is the equaliser* $\{x \in X \mid f(x) = g(x)\}$ *as in* **Sets**. *The greatest invariant invariant* $\Box E(f,g) \hookrightarrow E(f,g)$ *carries a subcoalgebra structure by the previous proposition.*

Proof We show that the diagram above is universal in **CoAlg**(F): for each coalgebra $e\colon Z \to F(Z)$ with homomorphism $h\colon Z \to X$ satisfying $f \circ h = g \circ h$, the map h factors through $Z \to \mathrm{E}(f, g)$ via a unique function. By Proposition 6.3.3 this h restricts to a (unique) coalgebra homomorphism $Z \to \Box\mathrm{E}(f, g)$. ❑

The next result requires a restriction to *finite* polynomial functors because the proof uses cofree coalgebras; see Proposition 2.5.3.

Theorem 6.3.5 (Products of coalgebras) *For a finite Kripke polynomial functor* $F\colon$ **Sets** $\to$ **Sets**, *the category* **CoAlg**(F) *of coalgebras has arbitrary products* $\prod$.

Proof We shall construct the product of two coalgebras $c_i\colon X_i \to F(X_i)$, for $i = 1, 2$, and leave the general case to the reader. We first form the product $X_1 \times X_2$ of the underlying sets and consider the cofree coalgebra on it; see Proposition 2.5.3. It will be written as $e\colon UG(X_1 \times X_2) \to F(UG(X_1 \times X_2))$, where $U\colon$ **CoAlg**$(F) \to$ **Sets** is the forgetful functor, and G its right adjoint. This coalgebra e comes with a universal map $\varepsilon\colon UG(X_1 \times X_2) \to X_1 \times X_2$. We write $\varepsilon_i = \pi_i \circ \varepsilon\colon UG(X_1 \times X_2) \to X_i$.

Next we form the following equaliser (in **Sets**):

$$E = \{u \in UG(X_1 \times X_2) \mid \forall i \in \{1, 2\}.\, (c_i \circ \varepsilon_i)(u) = (F(\varepsilon_i) \circ e)(u)\}.$$

Then we take its greatest invariant $\Box E \subseteq E$, as in the diagram below, describing E explicitly as equaliser:

$$\begin{array}{ccccccc} & & & & F(UG(X_1 \times X_2)) & & \\ & & & \overset{e}{\nearrow} & & \searrow^{\langle F(\varepsilon_1), F(\varepsilon_2)\rangle} & \\ \Box E & \overset{n}{\hookrightarrow} & E \overset{m}{\hookrightarrow} & UG(X_1 \times X_2) & & & F(X_1) \times F(X_2) \\ & & & \underset{\varepsilon}{\searrow} & & \nearrow_{c_1 \times c_2} & \\ & & & & X_1 \times X_2 & & \end{array} \qquad (6.5)$$

By Proposition 6.3.3, the subset $\Box E \hookrightarrow UG(X_1 \times X_2)$ carries an F-subcoalgebra structure, for which we write $c_1 \dot{\times} c_2$ in

$$\begin{array}{ccc} F(\Box E) & \overset{F(m \circ n)}{\rightarrowtail} & F(UG(X_1 \times X_2)) \\ {\scriptstyle c_1 \dot{\times} c_2}\uparrow & & \uparrow{\scriptstyle e} \\ \Box E & \underset{m \circ n}{\rightarrowtail} & UG(X_1 \times X_2) \end{array} \qquad (6.6)$$

The dot in $\dot{\times}$ is ad hoc notation used to distinguish this product of objects (coalgebras) from the product $c_1 \times c_2$ of functions, as used in the equaliser diagram above.

We claim this coalgebra $c_1 \dot{\times} c_2 \colon \Box E \to F(\Box E)$ is the product of the two coalgebras c_1 and c_2, in the category **CoAlg**(F). We thus follow the categorical description of product, from Definition 2.1.1. The two projection maps are

$$p_i \stackrel{\text{def}}{=} \Big(\Box E \xrightarrow{\;n\;} E \xrightarrow{\;m\;} UG(X_1 \times X_2) \xrightarrow{\;\varepsilon_i\;} X_i\Big).$$

We have to show that they are homomorphisms of coalgebras $c_1 \dot{\times} c_2 \to c_i$. This follows from easy calculations:

$$\begin{aligned}
F(p_i) \circ (c_1 \dot{\times} c_2) &= F(\varepsilon_i) \circ F(m \circ n) \circ (c_1 \dot{\times} c_2) \\
&= F(\varepsilon_i) \circ e \circ m \circ n && \text{see Diagram (6.6) above} \\
&= \pi_i \circ (c_1 \times c_2) \circ \varepsilon \circ m \circ n && m \text{ is an equaliser in (6.5)} \\
&= c_i \circ \pi_i \circ \varepsilon \circ m \circ n \\
&= c_i \circ p_i.
\end{aligned}$$

Next we have to construct pairs, for coalgebra homomorphisms $f_i \colon (Y \xrightarrow{d} F(Y)) \longrightarrow (X_i \xrightarrow{c_i} F(X_i))$. To start, we can form the ordinary pair $\langle f_1, f_2\rangle \colon Y \to X_1 \times X_2$ in **Sets**. By cofreeness it gives rise to unique function $g \colon Y \to UG(X_1 \times X_2)$ forming a coalgebra homomorphism $d \to e$, with $\varepsilon \circ g = \langle f_1, f_2\rangle$. This g has the following equalising property in (6.5):

$$\begin{aligned}
\langle F(\varepsilon_1), F(\varepsilon_2)\rangle \circ e \circ g &= \langle F(\pi_1 \circ \varepsilon), F(\pi_2 \circ \varepsilon)\rangle \circ F(g) \circ d \\
&\qquad \text{since } g \text{ is a coalgebra homomorphism } d \to e \\
&= \langle F(\pi_1 \circ \varepsilon \circ g) \circ d, F(\pi_2 \circ \varepsilon \circ g) \circ d\rangle \\
&= \langle F(f_1) \circ d, F(f_2) \circ d\rangle \\
&= \langle c_1 \circ f_1, c_2 \circ f_2\rangle \\
&\qquad \text{because } f_i \text{ is a coalgebra map } d \to c_i \\
&= \langle c_1 \circ \pi_1 \circ \varepsilon \circ g, c_2 \circ \pi_2 \circ \varepsilon \circ g\rangle \\
&= (c_1 \times c_2) \circ \varepsilon \circ g.
\end{aligned}$$

As a result, g factors through $m \colon E \hookrightarrow UG(X_1 \times X_2)$, say as $g = m \circ g'$. But then, by Proposition 6.3.3, g' also factors through $\Box E$. This yields the pair we seek: we write $\langle\langle f_1, f_2\rangle\rangle$ for the unique map $Y \to \Box E$ with $n \circ \langle\langle f_1, f_2\rangle\rangle = g'$.

We still have to show that this pair $\langle\langle f_1, f_2\rangle\rangle$ satisfies the required properties.

- The equations $p_i \circ \langle\langle f_1, f_2\rangle\rangle = f_i$ hold, since

$$\begin{aligned}
p_i \circ \langle\langle f_1, f_2\rangle\rangle &= \pi_i \circ \varepsilon \circ m \circ n \circ \langle\langle f_1, f_2\rangle\rangle \\
&= \pi_i \circ \varepsilon \circ m \circ g' \\
&= \pi_i \circ \varepsilon \circ g \\
&= \pi_i \circ \langle f_1, f_2\rangle \\
&= f_i.
\end{aligned}$$

- The pair $\langle\langle f_1, f_2\rangle\rangle$ is the unique homomorphism with $p_i \circ \langle\langle f_1, f_2\rangle\rangle = f_i$. Indeed, if $h: Y \to \Box E$ is also a coalgebra map $d \to (c_1 \dot{\times} c_2)$ with $p_i \circ h = f_i$, then $m \circ n \circ h$ is a coalgebra map $d \to e$ which satisfies

$$\begin{aligned} \varepsilon \circ m \circ n \circ h &= \langle \pi_1 \circ \varepsilon \circ m \circ n \circ h, \pi_2 \circ \varepsilon \circ m \circ n \circ h\rangle \\ &= \langle p_1 \circ h, p_2 \circ h\rangle \\ &= \langle f_1, f_2\rangle. \end{aligned}$$

Hence by definition of g:

$$m \circ n \circ h = g = m \circ g' = m \circ n \circ \langle\langle f_1, f_2\rangle\rangle,$$

Because both m and n are injections we get the required uniqueness: $h = \langle\langle f_1, f_2\rangle\rangle$. ❑

Since we have already seen that equalisers exist for coalgebras, we now know that all limits exist (see for instance [344, V, 2]). Proposition 2.1.5 and Exercise 2.1.14 showed that colimits also exist. Hence we can summarise the situation as follows.

Corollary 6.3.6 *The category* **CoAlg**(F) *of coalgebras of a finite Kripke polynomial functor is both complete and cocomplete.* ❑

Structure in categories of coalgebras is investigated further in [283], for endofunctors on more general categories than **Sets**. For instance, a construction of a 'subobject classifier' is given. It captures the correspondence between predicates $P \subseteq X$ and classifying maps $X \to 2$ in general categorical terms. Such subobject classifiers are an essential ingredient of a 'topos'. However, not all topos structure is present in categories of coalgebras (of functors preserving weak pullbacks): effectivity of equivalence relations may fail.

6.3.1 Greatest Invariants and Subcoalgebras, Categorically

The goal of the remainder of this section is to define in the abstract categorical setting of factorisation systems what it means to have greatest invariants $\Box$. Since in this setting invariants and subcoalgebras need not be the same – see Lemma 6.2.8 – we shall also describe greatest subcoalgebras (via comprehension). In principle, an easy direct characterisation of $\Box P$ is possible, namely as the greatest invariant $Q \leq P$. Below we shall give a more fancy description via an adjunction. This subsection is mostly an exercise in categorical formulations and is not of direct relevance in the sequel. It starts by describing the set-theoretic situation that we have dealt with so far a bit more systematically.

For an endofunctor $F:$ **Sets** $\to$ **Sets** we introduce a category **PredCoAlg**(F) of 'predicates on coalgebras'. Its objects are coalgebra-predicate pairs

$\langle X \to F(X), P \subseteq X\rangle$. And its morphisms $\langle X \to F(X), P \subseteq X\rangle \longrightarrow \langle Y \to F(Y), Q \subseteq Y\rangle$ are coalgebra homomorphisms $f\colon (X \to F(X)) \longrightarrow (Y \to F(Y))$ which are at the same time morphisms of predicates: $P \subseteq f^{-1}(Q)$, or equivalently, $\coprod_f(P) \subseteq Q$.

From this new category $\mathbf{PredCoAlg}(F)$ there are obvious forgetful functors to the categories of coalgebras and of predicates. Moreover, one can show that they form a pullback of categories:

$$\begin{array}{ccc} \mathbf{PredCoAlg}(F) & \longrightarrow & \mathbf{Pred} \\ \downarrow \lrcorner & & \downarrow \\ \mathbf{CoAlg}(F) & \longrightarrow & \mathbf{Sets} \end{array} \tag{6.7}$$

Lemma 6.3.7 *For a Kripke polynomial functor $F\colon$ **Sets** $\to$ **Sets** in the context described above, the greatest invariant operation $\Box$ yields a right adjoint in a commuting triangle:*

$$\begin{array}{ccc} \mathbf{PredCoAlg}(F) & \underset{\longleftarrow}{\overset{\Box}{\longrightarrow}}\ \top & \mathbf{CoAlg}(\mathrm{Pred}(F)) \\ & \searrow \quad \swarrow & \\ & \mathbf{Sets} & \end{array}$$

Proof Let $X \xrightarrow{c} F(X)$ and $Y \xrightarrow{d} F(Y)$ be two coalgebras with a map of coalgebras $f\colon X \to Y$ between them. Let $P \subseteq X$ be an invariant, and $Q \subseteq Y$ an ordinary predicate. The above adjunction then involves a bijective correspondence:

$$\frac{(c, P) \xrightarrow{\ f\ } (d, Q) \qquad \text{in } \mathbf{PredCoAlg}(F)}{\overline{(c, P) \xrightarrow[\ f\ ]{} (d, \Box\, Q) \qquad \text{in } \mathbf{CoAlg}(\mathrm{Pred}(F)).}}$$

Above the double lines we have $\coprod_f(P) \subseteq Q$. But since P is an invariant and $\coprod_f$ preserves invariants, this is equivalent to having $\coprod_f(P) \subseteq \Box\, Q$, like below the lines. ❑

This leads to the following obvious generalisation.

Definition 6.3.8 Let $F\colon \mathbb{C} \to \mathbb{C}$ be a functor on a category $\mathbb{C}$ with a factorisation system $(\mathfrak{M}, \mathfrak{E})$, inducing a lifting $\mathrm{Pred}(F)\colon \mathrm{Pred}(\mathbb{C}) \to \mathrm{Pred}(\mathbb{C})$ as in Definition 6.1.11. Form the category $\mathbf{PredCoAlg}(F)$ as on the left below:

$$\begin{array}{ccc} \mathbf{PredCoAlg}(F) & \longrightarrow & \mathrm{Pred}(\mathbb{C}) \\ \downarrow & \lrcorner & \downarrow \\ \mathbf{CoAlg}(F) & \longrightarrow & \mathbb{C} \end{array}$$

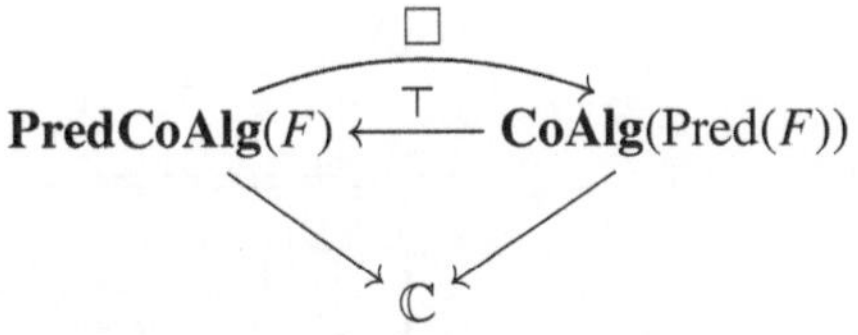

We say that the functor F admits **greatest invariants** if there is a right adjoint $\Box$ making the triangle on the right commute.

We turn to greatest subcoalgebras. Recall from Theorem 6.2.5 that they coincide with invariants in the set-theoretic case. But more generally, they require a different description, which we provide in terms of a comprehension functor $\{-\}$. As before, we first recall the set-theoretic situation. A systematic account of comprehension can be found in [239].

Consider the forgetful functor **Pred** $\to$ **Sets** that sends a predicate $(P \subseteq X)$ to its underlying set X. There is an obvious 'truth' predicate functor $\top\colon$ **Sets** $\to$ **Pred** sending a set X to the truth predicate $\top(X) = (X \subseteq X)$. It is not hard to see that $\top$ is right adjoint to the forgetful functor **Pred** $\to$ **Sets**.

In this situation there is another adjoint, namely the 'comprehension' or 'subset type' functor $\{-\}\colon$ **Pred** $\to$ **Sets**, given by $(P \subseteq X) \mapsto P$. One can prove that $\{-\}$ is right adjoint to truth $\top$, so that there is a situation:

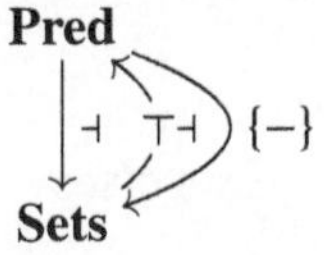

(6.8)

Lemma 6.3.9 *For a Kripke polynomial functor $F\colon$ **Sets** $\to$ **Sets**, consider the category **PredCoAlg**(F) described in (6.7).*

1. *There is a truth predicate functor $\top\colon$ **CoAlg**$(F) \to$ **PredCoAlg**(F) which is right adjoint to the forgetful functor **PredCoAlg**$(F) \to$ **CoAlg**(F).*
2. *This functor $\top$ has a right adjoint $\{-\}\colon$ **PredCoAlg**$(F) \to$ **CoAlg**(F) given by*

$$\langle X \xrightarrow{c} F(X), P \subseteq X\rangle \longmapsto (\Box P \xrightarrow{c_P} F(\Box P))$$

using the induced coalgebra c_P on the greatest invariant $\Box P$ from Proposition 6.3.3.

Proof The truth functor $\top\colon$ **CoAlg**$(F) \to$ **PredCoAlg**(F) is given by

$$(X \xrightarrow{c} F(X)) \longmapsto (c, \top) \qquad \text{and} \qquad f \longmapsto f.$$

Next let $c\colon X \to F(X)$ and $d\colon Y \to F(Y)$ be two coalgebras, with a predicate $Q \subseteq Y$. We write $d_Q\colon \Box Q \to F(\Box Q)$ for the induced coalgebra on the greatest invariant $\pi_Q\colon \Box Q \rightarrowtail Y$. We prove the comprehension adjunction:

$$\frac{\left\langle \begin{pmatrix} F(X) \\ \uparrow c \\ X \end{pmatrix}, \top \right\rangle \xrightarrow{\qquad f \qquad} \left\langle \begin{pmatrix} F(Y) \\ \uparrow d \\ Y \end{pmatrix}, Q \right\rangle}{\begin{pmatrix} F(X) \\ \uparrow c \\ X \end{pmatrix} \xrightarrow[\qquad g \qquad]{} \begin{pmatrix} F(\Box Q) \\ \uparrow d' \\ \Box Q \end{pmatrix}} \; .$$

This correspondence works as follows.

- Given a map f in the category **PredCoAlg**(F), we have $\top \subseteq f^{-1}(Q)$, so that $f(x) \in Q$, for all $x \in X$. By Proposition 6.3.3 f then restricts to a unique coalgebra homomorphism $\overline{f}\colon X \to \Box Q$ with $\pi_Q \circ \overline{f} = f$.
- Conversely, given a coalgebra homomorphism $g\colon X \to \Box Q$, we get a homomorphism $\overline{g} = \pi_Q \circ g\colon X \to Y$. By construction its image is contained in Q. ❑

It is easy to generalise this situation.

Definition 6.3.10 For a functor $F\colon \mathbb{C} \to \mathbb{C}$ on a category $\mathbb{C}$ with a factorisation system $(\mathfrak{M}, \mathfrak{E})$, consider the category **PredCoAlg**(F) described in Definition 6.3.8. There is an obvious truth functor $\top\colon$ **CoAlg**$(F) \to$ **PredCoAlg**(F). We say that the functor F admits **greatest subcoalgebras** if this truth functor $\top$ has a right adjoint $\{-\}$.

Exercises

6.3.1 Fix two sets A, B and consider the associated functor $F(X) = X^A \times B$ for deterministic automata.

1. Check that the cofree coalgebra functor $G\colon$ **Sets** $\to$ **CoAlg**(F) is given by $Y \mapsto (B \times Y)^{A^\star}$.
2. Assume two automata $\langle \delta_i, \epsilon_i \rangle\colon X_i \to X_i^A \times B$. Show that the product coalgebra, as constructed in the proof of Theorem 6.3.5, has carrier W given by the pullback of the maps to the final coalgebra:

$$\begin{array}{ccc} W & \longrightarrow & X_2 \\ \big\downarrow & \lrcorner & \big\downarrow \mathrm{beh}_{\langle \delta_2, \epsilon_2 \rangle} \\ X_1 & \xrightarrow[\mathrm{beh}_{\langle \delta_1, \epsilon_1 \rangle}]{} & B^{A^\star} \end{array}$$

   (Proposition 2.3.5 describes $B^{A^\star}$ as the final F-coalgebra.)
3. Describe the product coalgebra structure on W explicitly.
4. Explain this pullback outcome using Propositions 4.2.5 and 4.2.6.1 and the construction of products from pullbacks in Diagram (4.6).

6.3.2 Let $c\colon X \to F(X)$ be a coalgebra of a Kripke polynomial functor F. For two predicates $P, Q \subseteq X$ define a new predicate $P \mathrel{\mathsf{andthen}} Q = P \wedge \Box Q$. Prove that $\mathsf{andthen}$ forms a monoid on the poset $\mathcal{P}(X)$ of predicates on X, with truth as neutral element.

6.3.3 The next categorical result is a mild generalisation of [411, theorem 17.1]. It involves an arbitrary functor K between categories of coalgebras, instead of a special functor induced by a natural transformation, as in Proposition 2.5.5. Also the proof hint that we give leads to a slightly more elementary proof than in [411] because it avoids bisimilarity and uses an equaliser (in **Sets**) instead, much as in the proof of Theorem 6.3.5.

Consider two finite Kripke polynomial functors $F, H\colon \mathbf{Sets} \to \mathbf{Sets}$. Assume that there is a functor K between categories of coalgebras, commuting with the corresponding forgetful functors U_F and U_H, as in

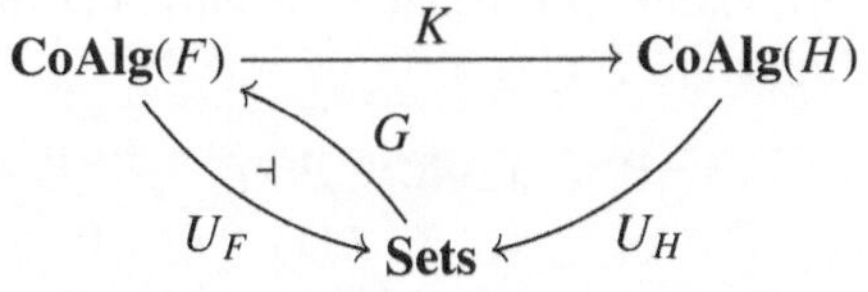

Prove that if F has cofree coalgebras, given by a right adjoint G to the forgetful functor U_F as in the diagram (and as in Proposition 2.5.3), then K has a right adjoint. *Hint*: For an arbitrary H-coalgebra $d\colon Y \to H(Y)$, first consider the cofree F-coalgebra on Y, say $e\colon U_F G(Y) \to F(U_F G(Y))$, and then form the equaliser

$$E = \{u \in U_F G(Y) \mid (K(e) \circ H(\varepsilon_Y))(u) = (d \circ \varepsilon_Y)(u)\}.$$

The greatest invariant $\Box E$ is then the carrier of the required F-coalgebra.

6.4 Temporal Logic for Coalgebras

Modal logic is a branch of logic in which the notions of necessity and possibility are investigated, via special modal operators. It has developed into a field in which other notions such as time, knowledge, program execution and provability are analysed in comparable manners; see for instance [128, 167, 228, 313, 198, 437, 73]. The use of temporal logic for reasoning about (reactive) state-based systems is advocated especially in [389, 390, 343], concentrating on temporal operators for transition systems – which may be seen as special instances of coalgebras (see Section 2.2.4). The coalgebraic approach to temporal logic extends these operators from transition systems to other coalgebras, in a uniform manner, following ideas first put forward by Moss [359], Kurz [322], Pattinson [373] and many others; see the overview papers [318, 323, 99]. This section will consider what we call *temporal* logic of coalgebras, involving logical modalities that cover all possible transitions by a particular coalgebra. Section 6.5 deals with a more refined *modal* logic, with modalities capturing specific moves to successor states. In this section we focus on (Kripke) polynomial functors on **Sets**.

We have already seen a few constructions with predicate lifting and invariants. Here we will elaborate the logical aspects and will in particular illustrate how a tailor-made temporal logic can be associated with a coalgebra, via a generic definition. This follows [243]. The exposition starts with 'forwards' temporal operators, talking about future states, and will continue with 'backwards' operators in Section 6.4.1.

The logic in this section will deal with predicates on the state spaces of coalgebras. We extend the usual boolean connectives to predicates, via pointwise definitions: for $P, Q \subseteq X$,

$$\begin{aligned} \neg P &= \{x \in X \mid \neg P(x)\} \\ P \wedge Q &= \{x \in X \mid P(x) \wedge Q(x)\} \\ P \Rightarrow Q &= \{x \in X \mid P(x) \Rightarrow Q(x)\}, \quad \text{etc.} \end{aligned}$$

In Section 1.3 we described a nexttime operator $\bigcirc$ for sequences. We start by generalising it to other coalgebras. This $\bigcirc$ will be used to construct more temporal operators.

Definition 6.4.1 Let $c\colon X \to F(X)$ be a coalgebra of a Kripke polynomial functor F. We define the **nexttime** operator $\bigcirc\colon \mathcal{P}(X) \to \mathcal{P}(X)$ as

$$\bigcirc P = c^{-1}(\mathrm{Pred}(F)(P)) = \{x \in X \mid c(x) \in \mathrm{Pred}(F)(P)\}.$$

That is, the operation $\bigcirc\colon \mathcal{P}(X) \to \mathcal{P}(X)$ is defined as the composite

$$\bigcirc = \Big(\mathcal{P}(X) \xrightarrow{\;\mathrm{Pred}(F)\;} \mathcal{P}(FX) \xrightarrow{\;c^{-1}\;} \mathcal{P}(X)\Big).$$

We understand the predicate $\bigcirc P$ as true for those states x, all of whose immediate successor states, if any, satisfy P. This will be made precise in Proposition 6.4.7 below. Notice that we leave the dependence of the operator $\bigcirc$ on the coalgebra c (and the functor) implicit. Usually, this does not lead to confusion.

Here are some obvious results.

Lemma 6.4.2 *The above nexttime operator $\bigcirc$ satisfies the following properties.*

1. *It is monotone:* $P \subseteq Q \Rightarrow \bigcirc P \subseteq \bigcirc Q$. *Hence* $\bigcirc$ *is an endofunctor* $\mathcal{P}(X) \to \mathcal{P}(X)$ *on the poset category of predicates ordered by inclusion.*
2. *It commutes with inverse images:* $\bigcirc(f^{-1}Q) = f^{-1}(\bigcirc Q)$.
3. *It has invariants as its coalgebras:* $P \subseteq X$ *is an invariant if and only if* $P \subseteq \bigcirc P$.
4. *It preserves meets (intersections) of predicates.*
5. *The greatest invariant* $\Box P$ *from Definition 6.3.1 is the cofree* $\bigcirc$*-coalgebra on P: it is the final coalgebra – or greatest fixed point – of the operator* $S \mapsto P \wedge \bigcirc S$ *on* $\mathcal{P}(X)$.

Proof We illustrate only the second and the last point. For a homomorphism of coalgebras $(X \xrightarrow{c} FX) \xrightarrow{f} (Y \xrightarrow{d} FY)$ and a predicate $Q \subseteq Y$ we have:

$$\begin{aligned} \bigcirc(f^{-1}Q) &= c^{-1}\mathrm{Pred}(F)(f^{-1}Q) \\ &= c^{-1}F(f)^{-1}\mathrm{Pred}(F)(Q) && \text{by Lemma 6.1.3.2} \\ &= f^{-1}d^{-1}\mathrm{Pred}(F)(Q) && \text{since } f \text{ is a homomorphism} \\ &= f^{-1}(\bigcirc Q). \end{aligned}$$

For the last point of the lemma, first note that the predicate $\Box P$ is a coalgebra of the functor $P \wedge \bigcirc(-)$ on $\mathcal{P}(X)$. Indeed, $\Box P \subseteq P \wedge \bigcirc(\Box P)$, because $\Box P$ is contained in P and is an invariant. Next, $\Box P$ is the greatest such coalgebra, and hence the final one: if $Q \subseteq P \wedge \bigcirc Q$, then Q is an invariant contained in P, so that $Q \subseteq \Box P$. We conclude that $\Box P$ is the cofree $\bigcirc(-)$-coalgebra. ❑

The nexttime operator $\bigcirc$ is fundamental in temporal logic. By combining it with negations, least fixed points μ and greatest fixed points ν one can define other temporal operators. For instance $\neg \bigcirc \neg$ is the so-called **strong nexttime** operator. It holds for those states for which there actually is a successor state satisfying P. Figure 6.1 shows a few standard operators.

We shall next illustrate the temporal logic of coalgebras in two examples.

Example 6.4.3 Douglas Hofstadter explains in his famous book *Gödel, Escher, Bach* [224] the object- and meta-level perspective on formal systems

Notation	Meaning	Definition
$\bigcirc P$	Nexttime P	$c^{-1}\mathrm{Pred}(F)(P)$
$\Box P$	Henceforth P	$\nu S.(P \wedge \bigcirc S)$
$\Diamond P$	Eventually P	$\neg \Box \neg P$
$P \,\mathcal{U}\, Q$	P until Q	$\mu S.(Q \vee (P \wedge \neg \bigcirc \neg S))$

Figure 6.1 Standard (forwards) temporal operators, where μ and ν are the least and greatest fixed point operators, respectively.

using a simple 'MU-puzzle'. It consists of a simple 'Post' production system (see e.g. [114, section 5.1]) or rewriting system for generating certain strings containing the symbols M, I, U. The meta-question that is considered is whether the string MU can be produced. Both this production system and this question (and also its answer) can be (re)formulated in coalgebraic terminology.

Let therefore our alphabet A be the set $\{\mathsf{M}, \mathsf{I}, \mathsf{U}\}$ of relevant symbols. We will describe the production system as an unlabelled transition system (UTS) $A^\star \to \mathcal{P}_{\text{fin}}(A^\star)$ on the set $A^\star$ of strings over this alphabet. This is given by the following transitions (from [224]), which are parametrised by strings $x, y \in A^\star$:

$$x\mathsf{I} \to x\mathsf{IU} \qquad \mathsf{M}x \to \mathsf{M}xx \qquad x\mathsf{III}y \to x\mathsf{U}y \qquad x\mathsf{UU}y \to xy.$$

Thus, the associated transition system $A^\star \to \mathcal{P}_{\text{fin}}(A^\star)$ is given by

$$\begin{aligned} w \longmapsto \{z \in A^\star \mid \exists x \in A^\star.\ & (w = x\mathsf{I} \wedge z = x\mathsf{IU}) \\ & \vee (w = \mathsf{M}x \wedge z = \mathsf{M}xx) \\ \vee\ \exists x, y \in A^\star.\ & (w = x\mathsf{III}y \wedge z = x\mathsf{U}y) \\ & \vee (w = x\mathsf{UU}y \wedge z = xy)\}. \end{aligned}$$

It is not hard to see that for each word $w \in A^\star$ this set on the right-hand side is finite.

The question considered in [224] is whether the string MU can be obtained from MI; that is, whether $\mathsf{MI} \longrightarrow^* \mathsf{MU}$. Or to put it in temporal terminology, whether the predicate 'equal to MU' eventually holds, starting from MI:

$$\begin{aligned} & \Diamond(\{x \in A^\star \mid x = \mathsf{MU}\})(\mathsf{MI}) \\ \Longleftrightarrow\ & \neg\Box(\neg\{x \in A^\star \mid x = \mathsf{MU}\})(\mathsf{MI}) \\ \Longleftrightarrow\ & \neg\exists P \text{ invariant}.\ P \subseteq \neg\{x \in A^\star \mid x = \mathsf{MU}\} \wedge P(\mathsf{MI}) \\ \Longleftrightarrow\ & \forall P \text{ invariant}.\ \neg(\forall x \in P.\ x \neq \mathsf{MU}) \wedge P(\mathsf{MI})) \\ \Longleftrightarrow\ & \forall P \text{ invariant}.\ P(\mathsf{MI}) \implies \exists x \in P.\ x = \mathsf{MU} \\ \Longleftrightarrow\ & \forall P \text{ invariant}.\ P(\mathsf{MI}) \implies P(\mathsf{MU}). \end{aligned}$$

Hofstadter [224] provides a counter-example, by producing an invariant $P \subseteq A^\star$ for which $P(\mathsf{MI})$, but not $P(\mathsf{MU})$, namely:

$$P(x) \overset{\text{def}}{\Longleftrightarrow} \text{the number of Is in } x \text{ is not a multiple of } 3.$$

This P is clearly an invariant: of the above four parametrised transitions, the first and last one do not change the number of Is; in the second transition $\mathsf{M}x \to \mathsf{M}xx$, if the number of Is in the right-hand side, i.e. in xx, is $3n$, then n must be even, so that the number of Is in x (and hence in $\mathsf{M}x$) must already be a multiple of 3; a similar argument applies to the third transition. Thus, property P is an invariant. Once we have reached this stage we have P as counter example: clearly $P(\mathsf{MI})$, but not $P(\mathsf{MU})$. Thus MU cannot be obtained from MI.

This proof is essentially the same proof that Hofstadter provides, but of course he does not use the same coalgebraic formulation and terminology. However, he does call the property P 'hereditary'.

This concludes the example. The relation we have used between $\longrightarrow^*$ and $\Diamond$ will be investigated more systematically below; see especially in Proposition 6.4.7.

Here is another, more technical, illustration.

Example 6.4.4 This example assumes some familiarity with the untyped lambda-calculus, and especially with its theory of Böhm trees; see [54, chapter 10]. It involves an operational model for head normal form reduction, consisting of a final coalgebra of certain trees. Temporal logic will be used to define an appropriate notion of 'free variable' on these trees.

We fix a set V and think of its elements as variables. We consider the polynomial functor $F\colon \mathbf{Sets} \to \mathbf{Sets}$ given by

$$F(X) = 1 + (V^\star \times V \times X^\star). \tag{6.9}$$

In this example we shall often omit the coprojections κ_i and simply write $*$ for $\kappa_1(*) \in 1 + (V^\star \times V \times X^\star)$ and $(\vec{v}, w, \vec{x})$ for $\kappa_2(\vec{v}, w, \vec{x}) \in 1 + (V^\star \times V \times X^\star)$. Also, we shall write $\zeta\colon \mathcal{B} \xrightarrow{\cong} F(\mathcal{B})$ for the final F-coalgebra – which exists by Theorem 2.3.9.

Lambda terms are obtained from variables $x \in V$, application MN of two λ-terms M, N, and abstraction $\lambda x.\, M$. The main reduction rule is $(\lambda x.\, M)N \to M[N/x]$. By an easy induction on the structure of λ-terms one then sees that an arbitrary term can be written of the form $\lambda x_1 \ldots x_n.\, yM_1 \ldots M_m$. The set Λ of λ-terms thus carries an F-coalgebra structure, given by the head-normal-form function $\mathsf{hnf}\colon \Lambda \to F(\Lambda)$ (see [54, section 8.3]): for $M \in \Lambda$,

$$\mathsf{hnf}(M) = \begin{cases} * & \text{if } M \text{ has no head normal form} \\ (\langle x_1, \ldots, x_n\rangle, y, \langle M_1, \ldots, M_m\rangle) & \text{if } M \text{ has head normal form} \\ & \lambda x_1 \ldots x_1.\, yM_1 \ldots M_m. \end{cases}$$

We can now define the Böhm tree $\mathsf{BT}(M)$ of a λ-term M via finality:

$$\begin{array}{ccc} 1 + (V^\star \times V \times \Lambda^\star) & \xrightarrow{\mathrm{id} + (\mathrm{id} \times \mathrm{id} \times \mathsf{BT}^\star)} & 1 + (V^\star \times V \times \mathcal{B}^\star) \\ {\scriptstyle \mathsf{hnf}}\uparrow & & \cong\uparrow{\scriptstyle \zeta} \\ \Lambda & \xrightarrow[\mathsf{BT}]{\qquad\qquad} & \mathcal{B} \end{array}$$

We call the elements of $\mathcal{B}$ (abstract[1]) Böhm trees. We do not really need to know what these elements look like, because we can work with the universal property of $\mathcal{B}$, namely finality. A picture may be useful. For $A \in \mathcal{B}$ we can write

$\zeta(A) = \bot$ or $\zeta(A) =$

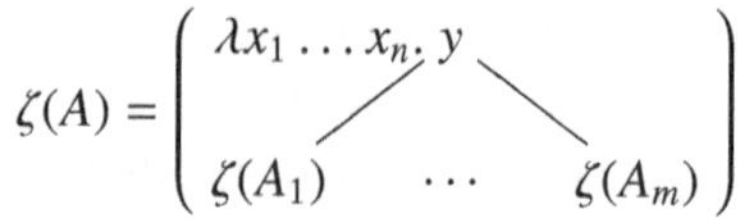

where the second picture applies when $\zeta(A) = (\langle x_1, \ldots, x_n \rangle, y, \langle A_1, \ldots, A_m \rangle)$. The '$\lambda$' is just syntactic sugar, used to suggest the analogy with the standard notation for Böhm trees [54]. The elements of $\mathcal{B}$ are thus finitely branching, possibly infinite, rooted trees, with labels of the form $\lambda x_1 \ldots x_n.\, y$, for variables $x_i, y \in V$.

Using the inverse $\zeta^{-1} \colon 1 + (\mathcal{B}^\star \times V \times \mathcal{B}^\star) \to \mathcal{B}$ of the final coalgebra we can explicitly construct Böhm trees. We give a few examples.

- Let us write $\bot_{\mathcal{B}} \in \mathcal{B}$ for $\zeta^{-1}(*)$. This the 'empty' Böhm tree.
- The Böhm tree $\lambda x.\, x$ is obtained as $\zeta^{-1}(\langle x \rangle, x, \langle\rangle)$. In a similar way one can construct various kinds of finite Böhm trees. For instance, the **S** combinator $\lambda xyz.\, xz(yz)$ is obtained as

$$\zeta^{-1}(\langle x, y, z \rangle, x, \langle \zeta^{-1}(\langle\rangle, z, \langle\rangle), \zeta^{-1}(\langle\rangle, y, \langle \zeta^{-1}(\langle\rangle, z, \langle\rangle)\rangle)\rangle).$$

  Its picture is:

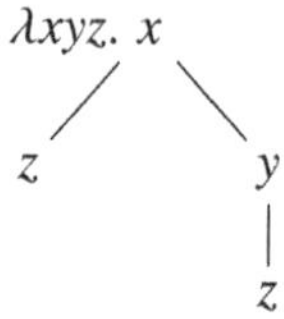

- Given an arbitrary Böhm tree $A \in \mathcal{B}$, we can define a new tree $\lambda x.\, A \in \mathcal{B}$ via λ-abstraction:

$$\lambda x.\, A = \begin{cases} \bot_{\mathcal{B}} & \text{if } \zeta(A) = * \\ \zeta^{-1}(x \cdot \vec{y}, z, \vec{B}) & \text{if } \zeta(A) = (\vec{y}, z, \vec{B}). \end{cases}$$

We proceed by using temporal logic to define free variables for Böhm trees. Such a definition is non-trivial since Böhm trees may be infinite objects. Some preliminary definitions are required. Let $x \in V$ be an arbitrary variable. It will

[1] One may have a more restricted view and call 'Böhm tree' only those elements in $\mathcal{B}$ which actually come from λ-terms, i.e. which are in the image of the function $\mathsf{BT} \colon \Lambda \to \mathcal{B}$. Then one may wish to call the elements of the whole set $\mathcal{B}$ 'abstract' Böhm trees. We shall not do so. But it is good to keep in mind that the function BT is not surjective. For example, Böhm trees coming from λ-terms can have only a finite number of free variables (as defined below), whereas elements of $\mathcal{B}$ can have arbitrarily many.

be used in the auxiliary predicates Abs_x and Hv_x on Böhm trees, which are defined as follows: for $B \in \mathcal{B}$,

$$\begin{aligned} \mathsf{Abs}_x(B) &\iff \exists x_1, \ldots, x_n.\, \exists B_1, \ldots, B_m. \\ &\qquad\qquad B = \lambda x_1 \ldots x_n.\, yB_1 \ldots B_m \text{ and } x = x_i \text{ for some } i \\ \mathsf{Hv}_x(B) &\iff \exists x_1, \ldots, x_n.\, \exists B_1, \ldots, B_m. \\ &\qquad\qquad B = \lambda x_1 \ldots x_n.\, yB_1 \ldots B_m \text{ and } x = y. \end{aligned}$$

Thus the predicate Abs_x describes the occurrence of x in the abstracted variables, and Hv_x captures that x is the head variable.

For a Böhm tree $A \in \mathcal{B}$ we can now define the set $\mathsf{FV}(A) \subseteq V$ of *free variables* in A via the until operator $\mathcal{U}$ from Figure 6.1:

$$x \in \mathsf{FV}(A) \iff (\neg\mathsf{Abs}_x\ \mathcal{U}\ (\mathsf{Hv}_x \wedge \neg\mathsf{Abs}_x))(A).$$

In words: a variable x is free in a Böhm tree A if there is a successor tree B of A in which x occurs as 'head variable', and in all successor trees of A until that tree B is reached, including B itself, x is not used in a lambda abstraction. This until formula then defines a predicate on $\mathcal{B}$, namely '$x \in \mathsf{FV}(-)$'.

There are then two crucial properties that we would like to hold for a Böhm tree A.

1. If $A = \perp_{\mathcal{B}}$, then

$$\mathsf{FV}(A) = \emptyset.$$

   This holds because if $A = \perp_{\mathcal{B}}$, then both $\mathsf{Abs}_x(A)$ and $\mathsf{Hv}_x(A)$ are false, so that the least fixed point in Figure 6.1 defining $\mathcal{U}$ at A in $x \in \mathsf{FV}(A)$ is $\mu S.\, \neg \bigcirc \neg S$. This yields the empty set.
2. If $A = \lambda x_1 \ldots x_n.\, yA_1 \ldots A_m$, then

$$\mathsf{FV}(A) = (\{y\} \cup \mathsf{FV}(A_1) \cup \cdots \cup \mathsf{FV}(A_m)) - \{x_1, \ldots, x_n\}.$$

   This result follows from the fixed point property (indicated as 'f.p.' below) defining the until operator $\mathcal{U}$ in Figure 6.1:

$$\begin{aligned} x \in \mathsf{FV}(A) &\overset{\text{def}}{\iff} [\neg\mathsf{Abs}_x\ \mathcal{U}\ (\mathsf{Hv}_x \wedge \neg\mathsf{Abs}_x)](A) \\ &\overset{\text{f.p.}}{\iff} [(\mathsf{Hv}_x \wedge \neg\mathsf{Abs}_x) \vee (\neg\mathsf{Abs}_x \wedge \neg \bigcirc \neg(x \in \mathsf{FV}(-)))](A) \\ &\iff \neg\mathsf{Abs}_x(A) \wedge (\mathsf{Hv}_x(A) \vee \neg \bigcirc \neg(x \in \mathsf{FV}(-))(A)) \\ &\iff x \notin \{x_1, \ldots, x_n\} \wedge (x = y \vee \exists j \leq m.\, x \in \mathsf{FV}(A_j)) \\ &\iff x \in (\{y\} \cup \mathsf{FV}(A_1) \cup \cdots \cup \mathsf{FV}(A_m)) - \{x_1, \ldots, x_n\}. \end{aligned}$$

This shows how temporal operators can be used to define sensible predicates on infinite structures. The generic definitions provide adequate expressive power in concrete situations. We should emphasise however that the final coalgebra $\mathcal{B}$ of Böhm trees is only an operational model of the lambda calculus and not a denotational one: for instance, it is not clear how to define an application operation $\mathcal{B} \times \mathcal{B} \to \mathcal{B}$ on our abstract Böhm trees via coinduction. Such application is defined on the Böhm model used in [54, section 18.3] via finite approximations. For more information on models of the (untyped) λ-calculus, see e.g. [54, part V], [239, section 2.5], or [136]. See also [324] for a coalgebraic approach to the infinitary λ-calculus.

Our next application of temporal logic does not involve a specific functor, as for Böhm trees above, but is generic. It involves an (unlabelled) transition relation for an arbitrary coalgebra. Before we give the definition it is useful to introduce some special notation and some associated results.

Lemma 6.4.5 *For an arbitrary set X and an element $x \in X$ we define a 'singleton' and 'non-singleton' predicate on X as*

$$\begin{aligned}(\cdot = x) &= \{y \in X \mid y = x\} = \{x\}.\\ (\cdot \neq x) &= \{y \in X \mid y \neq x\} = \neg(\cdot = x).\end{aligned}$$

Then

1. *For a predicate $P \subseteq X$,*

$$P \subseteq (\cdot \neq x) \iff \neg P(x).$$

2. *For a function $f\colon Y \to X$,*

$$f^{-1}(\cdot \neq x) = \bigcap_{y \in f^{-1}(x)} (\cdot \neq y).$$

3. *And for a Kripke polynomial functor F and a predicate $Q \subseteq F(X)$,*

$$\overleftarrow{\mathrm{Pred}}(F)(Q) = \{x \in X \mid Q \not\subseteq \mathrm{Pred}(F)(\cdot \neq x)\}$$

where $\overleftarrow{\mathrm{Pred}}(F)$ is the 'predicate lowering' left adjoint to predicate lifting $\mathrm{Pred}(F)$ *from Section 6.1.1.*

These 'non-singletons' $(\cdot \neq x)$ are also called coequations, for instance in [177, 17, 422], and pronounced as 'avoid x'. They are used for a sound and complete logical deduction calculus for coalgebras via a 'child' rule (capturing henceforth $\Box$) and a 'recolouring' rule (capturing $\boxtimes$; see Exercise 6.8.8). Here these coequations $(\cdot \neq x)$ arise naturally in a characterisation of predicate lowering in point (3).

Proof Points (1) and (2) follow immediately from the definition. For (3) we use (1) and the adjoint/Galois correspondence in

$$\begin{aligned} x \in \overleftarrow{\mathrm{Pred}}(F)(Q) &\iff \overleftarrow{\mathrm{Pred}}(F)(Q) \not\subseteq (\cdot \neq x) \\ &\iff Q \not\subseteq \mathrm{Pred}(F)(\cdot \neq x). \end{aligned}$$

□

In Proposition 6.1.9 we have seen an abstract way to turn an arbitrary coalgebra into an unlabelled transition system. Here, and later on in Theorem 6.4.9, we shall reconsider this topic from a temporal perspective.

Definition 6.4.6 Let $c\colon X \to F(X)$ be a coalgebra of a polynomial functor F. On states $x, x' \in X$ we define a transition relation via the strong nexttime operator, as

$$\begin{aligned} x \to x' &\iff x \in (\neg \bigcirc \neg)(\cdot = x') \\ &\iff x \notin \bigcirc(\cdot \neq x') \\ &\iff c(x) \notin \mathrm{Pred}(F)((\cdot \neq x')). \end{aligned}$$

This says that there is a transition $x \to x'$ if and only if there is successor state of x which is equal to x'. In this way we turn an arbitrary coalgebra into an unlabelled transition system.

We shall first investigate the properties of this new transition system $\to$, and only later in Theorem 6.4.9 show that it is actually the same as the earlier translation from coalgebras to transition systems from Section 6.1.1.

So let us first consider what we get for a coalgebra $c\colon X \to \mathcal{P}(X)$ of the powerset functor. Then the notation $x \to x'$ is standardly used for $x' \in c(x)$. This coincides with Definition 6.4.6 since

$$\begin{aligned} x \notin \bigcirc(\cdot \neq x') &\iff x \notin c^{-1}(\mathrm{Pred}(\mathcal{P})(\cdot \neq x')) \\ &\iff c(x) \notin \{a \mid a \subseteq (\cdot \neq x')\} \\ &\iff c(x) \notin \{a \mid x' \notin a\}, \qquad \text{by Lemma 6.4.5.1} \\ &\iff x' \in c(x). \end{aligned}$$

Now that we have gained some confidence in this temporal transition definition, we consider further properties. It turns out that the temporal operators can be expressed in terms of the new transition relation.

Proposition 6.4.7 *The transition relation $\to$ from Definition 6.4.6, induced by a coalgebra $X \to F(X)$, and its reflexive transitive closure $\longrightarrow^*$, satisfy the following properties.*

1. *For a predicate $P \subseteq X$,*

    a. $\bigcirc P = \{x \in X \mid \forall x'.\, x \to x' \implies P(x')\}$.
    b. $\Box P = \{x \in X \mid \forall x'.\, x \longrightarrow^* x' \implies P(x')\}$.
    c. $\Diamond P = \{x \in X \mid \exists x'.\, x \longrightarrow^* x' \wedge P(x')\}$.

This says that the temporal operators on the original coalgebra are the same as the ones on the induced unlabelled transition system.

2. *For a predicate $P \subseteq X$, the following three statements are equivalent.*

   a. *P is an invariant.*
   b. $\forall x, x' \in X.\, P(x) \wedge x \to x' \implies P(x')$.
   c. $\forall x, x' \in X.\, P(x) \wedge x \longrightarrow^{*} x' \implies P(x')$.

3. *For arbitrary states $x, x' \in X$, the following are equivalent.*

   a. $x \longrightarrow^{*} x'$.
   b. $P(x) \Rightarrow P(x')$, *for all invariants $P \subseteq X$.*
   c. $x \in \Diamond(\cdot = x')$, i.e. *eventually there is a successor state of x that is equal to x'.*

Proof 1. We reason as follows.

$$\begin{aligned} x \in \bigcirc P &\iff c(x) \in \mathrm{Pred}(F)(P) \\ &\iff \{c(x)\} \subseteq \mathrm{Pred}(F)(P) \\ &\iff \overleftarrow{\mathrm{Pred}}(F)(\{c(x)\}) \subseteq P \\ &\iff \forall x'.\, x' \in \overleftarrow{\mathrm{Pred}}(F)(\{c(x)\}) \Rightarrow P(x') \\ &\iff \forall x'.\, \{c(x)\} \not\subseteq \mathrm{Pred}(F)(\cdot \neq x') \Rightarrow P(x'), \text{ by Lemma 6.4.5.3} \\ &\iff \forall x'.\, c(x) \notin \mathrm{Pred}(F)(\cdot \neq x') \Rightarrow P(x') \\ &\iff \forall x'.\, x \to x' \Rightarrow P(x'). \end{aligned}$$

For the inclusion ($\subseteq$) of (b) we can use (a) an appropriate number of times since $\Box P \subseteq \bigcirc \Box P$ and $\Box P \subseteq P$. For ($\supseteq$) we use that the predicate $\{x \mid \forall x'.\, x \longrightarrow^{*} x' \implies P(x')\}$ contains P and is an invariant, via (a); hence it is contained in $\Box P$.

The third point (c) follows directly from (b) since $\Diamond = \neg \Box \neg$.

2. This is Immediate from (1) since P is an invariant if and only if $P \subseteq \bigcirc P$, if and only if $P \subseteq \Box P$.
3. The equivalence (b) $\Leftrightarrow$ (c) follows by unfolding the definitions. The implication (a) $\Rightarrow$ (b) follows directly from (2), but for the reverse we have to do a bit of work. Assume $P(x) \Rightarrow P(x')$ for all invariants P. In order to prove $x \longrightarrow^{*} x'$, consider the predicate Q defined by $Q(y) \Leftrightarrow x \longrightarrow^{*} y$. Clearly $Q(x)$, so $Q(x')$ follows once we have established that Q is an invariant. But this is an easy consequence using (2): if $Q(y)$, i.e. $x \longrightarrow^{*} y$, and $y \to y'$, then clearly $x \longrightarrow^{*} y'$, which is $Q(y')$. ❑

6.4.1 Backwards Reasoning

So far in this section we have concentrated on 'forwards' reasoning, by considering only operators that talk about future states. However, within the setting of coalgebras there is also a natural way to reason about previous states. This happens via predicate lowering instead of via predicate lifting, i.e. via the left adjoint $\underleftarrow{\mathrm{Pred}}(F)$ to $\mathrm{Pred}(F)$, introduced in Section 6.1.1.

It turns out that the forwards temporal operators have backwards counterparts. We shall use notation with backwards underarrows for these analogues: $\underleftarrow{\bigcirc}$, $\underleftarrow{\Box}$ and $\underleftarrow{\Diamond}$ are backwards versions of $\bigcirc$, $\Box$ and $\Diamond$.

Definition 6.4.8 For a coalgebra $c\colon X \to F(X)$ of a polynomial functor F, and a predicate $P \subseteq X$ on its carrier X, we define a new predicate **lasttime** P on X by

$$\underleftarrow{\bigcirc} P = \underleftarrow{\mathrm{Pred}}(F)(\textstyle\coprod_c P) = \underleftarrow{\mathrm{Pred}}(F)(\{c(x) \mid x \in P\}).$$

Thus

$$\underleftarrow{\bigcirc} = \Big(\mathcal{P}(X) \xrightarrow{\ \coprod_c\ } \mathcal{P}(FX) \xrightarrow{\ \underleftarrow{\mathrm{Pred}}(F)\ } \mathcal{P}(X)\Big).$$

This is the so-called **strong lasttime** operator, which holds of a state x if there is an (immediate) predecessor state of x which satisfies P. The corresponding **weak lasttime** is $\neg\underleftarrow{\bigcirc}\neg$.

One can easily define an infinite extension of $\underleftarrow{\bigcirc}$, called **earlier**:

$$\begin{aligned}\underleftarrow{\Diamond} P &= \text{`the least invariant containing } P\text{'}\\ &= \{x \in X \mid \forall Q, \text{invariant}.\ P \subseteq Q \implies Q(x)\}\\ &= \textstyle\bigcap\{Q \supseteq P \mid Q \text{ is an invariant}\}.\end{aligned}$$

This predicate $\underleftarrow{\Diamond} P$ holds of a state x if there is some (non-immediate) predecessor state of x for which P holds.

Figure 6.2 gives a brief overview of the main backwards temporal operators. In the remainder of this section we shall concentrate on the relation between the backwards temporal operators and transitions.

But first we give a result that was already announced. It states an equivalence between various (unlabelled) transition systems induced by coalgebras.

Notation	Meaning	Definition	Galois Connection
$\underleftarrow{\bigcirc} P$	lasttime P	$\underleftarrow{\mathrm{Pred}}(F)(\coprod_c P)$	$\underleftarrow{\bigcirc} \dashv \bigcirc$
$\underleftarrow{\Diamond} P$	(sometime) earlier P	$\mu S.\,(P \vee \underleftarrow{\bigcirc} S)$	$\underleftarrow{\Diamond} \dashv \Box$
$\underleftarrow{\Box} P$	(always) before P	$\neg\underleftarrow{\Diamond}\neg P$	$\Diamond \dashv \underleftarrow{\Box}$
$P\,\mathcal{S}\,Q$	P since Q	$\mu S.\,(Q \vee (P \wedge \underleftarrow{\bigcirc} S))$	

Figure 6.2 Standard (backwards) temporal operators.

Theorem 6.4.9 *Consider a coalgebra $c\colon X \to F(X)$ of a Kripke polynomial functor F. Using the lasttime operator $\overleftarrow{\bigcirc}$ one can also define an unlabelled transition system by*

$$\begin{aligned} x \to x' &\iff \text{'there is an immediate predecessor state of } x'\text{, equal to } x\text{'} \\ &\iff x' \in \overleftarrow{\bigcirc}(\cdot = x). \end{aligned}$$

This transition relation is then the same as

1. $x \notin \bigcirc(\cdot \neq x')$ *from Definition 6.4.6*
2. $x' \in \mathrm{sts}(c(x)) = \overleftarrow{\mathrm{Pred}}(F)(\{c(x)\})$, *used in the translation in (6.4).*

Proof All these forwards and backwards transition definitions are equivalent because

$$\begin{aligned} x' \in \overleftarrow{\bigcirc}(\cdot = x) &\iff x' \in \overleftarrow{\mathrm{Pred}}(F)(\textstyle\coprod_c(\cdot = x)) && \text{by Definition 6.4.8} \\ &\iff x' \in \overleftarrow{\mathrm{Pred}}(F)(\{c(x)\}) && \text{as used in (6.4)} \\ &\iff \{c(x)\} \not\subseteq \mathrm{Pred}(F)(\cdot \neq x') && \text{by Lemma 6.4.5.3} \\ &\iff x \notin \bigcirc(\cdot \neq x') && \text{as used in Definition 6.4.6.} \end{aligned}$$ ❑

Finally we mention the descriptions of the backwards temporal operators in terms of transitions, as in Proposition 6.4.7.1.

Proposition 6.4.10 *For a predicate $P \subseteq X$ on the state space of a coalgebra,*

1. $\overleftarrow{\bigcirc} P = \{x \in X \mid \exists y.\, y \to x \wedge P(y)\}$.
2. $\overleftarrow{\Diamond} P = \{x \in X \mid \exists y.\, y \longrightarrow^{*} x \wedge P(y)\}$.
3. $\overleftarrow{\Box} P = \{x \in X \mid \forall y.\, y \longrightarrow^{*} x \implies P(y)\}$.

Proof Let $c\colon X \to F(X)$ be the underlying coalgebra.

1. $$\begin{aligned} \overleftarrow{\bigcirc} P &= \overleftarrow{\mathrm{Pred}}(F)(\{c(y) \mid y \in P\}) \\ &= \overleftarrow{\mathrm{Pred}}(F)(\textstyle\bigcup_{y\in P}\{c(y)\}) \\ &= \textstyle\bigcup_{y\in P} \overleftarrow{\mathrm{Pred}}(F)(\{c(y)\}) && \text{since } \overleftarrow{\mathrm{Pred}}(F) \text{ is a left adjoint} \\ &= \{x \in X \mid \exists y \in P.\, x \in \overleftarrow{\bigcirc}(\cdot = y)\} \\ &= \{x \in X \mid \exists y.\, y \to x \wedge P(y)\}. \end{aligned}$$
2. Let us write $P' = \{x \in X \mid \exists y.\, y \longrightarrow^{*} x \wedge P(y)\}$ for the right-hand side. We have to prove that P' is the least invariant containing P.
   - Clearly $P \subseteq P'$, by taking no transition.
   - Also P' is an invariant, by Proposition 6.4.7.2: if $P'(x)$, say with $y \longrightarrow^{*} x$ where $P(y)$, and $x \to x'$, then also $y \longrightarrow^{*} x'$ and thus $P'(x')$.
   - If $Q \subseteq X$ is an invariant containing P, then $P' \subseteq Q$: if $P'(x)$, say with $y \longrightarrow^{*} x$ where $P(y)$, then $Q(y)$, and thus $Q(x)$ by Proposition 6.4.7.2.
3. Immediately from the definition $\overleftarrow{\Box} = \neg \overleftarrow{\Diamond} \neg$. ❑

Exercises

6.4.1 Consider the transition system $A^\star \to \mathcal{P}_{\text{fin}}(A^\star)$ from Example 6.4.3 and prove

$$\Box(\{x \in A^\star \mid \exists y \in A^\star .\, x = \mathsf{M}y\})(\mathsf{MI}).$$

This property is also mentioned in [224]. In words: each successor of MI starts with M.

6.4.2 Prove the following 'induction rule of temporal logic':

$$P \wedge \Box(P \Rightarrow \bigcirc P) \subseteq \Box P.$$

(Aside: using the term 'induction' for a rule that follows from a greatest fixed point property is maybe not very fortunate.)

6.4.3 Prove that for a predicate P on the state space of coalgebra of a Kripke polynomial functor,

$$\Box P = \bigcap_{n\in\mathbb{N}} \bigcirc^n P \quad \text{and} \quad \overleftarrow{\Diamond} P = \bigcup_{n\in\mathbb{N}} \overleftarrow{\bigcirc}^n P$$

(where $\bigcirc^0 P = P$, and $\bigcirc^{n+1} P = \bigcirc\bigcirc^n P$, and similarly for $\overleftarrow{\bigcirc}$).

6.4.4 Prove that

$$\begin{aligned} \Box(f^{-1}Q) &= f^{-1}(\Box Q) \\ \overleftarrow{\bigcirc}(\textstyle\coprod_f P) &= \textstyle\coprod_f(\overleftarrow{\bigcirc} P) \\ \overleftarrow{\Diamond}(\textstyle\coprod_f P) &= \textstyle\coprod_f(\overleftarrow{\Diamond} P) \end{aligned}$$

when f is a homomorphism of coalgebras.

6.4.5 Consider coalgebras $c\colon X \to \mathcal{M}_M(X)$ and $d\colon Y \to \mathcal{D}(Y)$ of the multiset and distribution functors. Use Exercise 6.1.4 to prove

$$\bigcirc(P \subseteq X) = \{x \mid \forall x'.\, c(x)(x') \neq 0 \Rightarrow P(x')\}.$$

What is $\Box(Q \subseteq Y)$?

6.4.6 Demonstrate that the least fixed point $\mu\bigcirc$ of the nexttime operator $\bigcirc\colon \mathcal{P}(X) \to \mathcal{P}(X)$ can be characterised as

$$\mu\bigcirc = \{x \in X \mid \text{there are no infinite paths } x \to x_1 \to x_2 \to \cdots\}.$$

6.4.7 Consider the transition relation $\to$ from Definition 6.4.6 and use Lemma 6.4.5.2 to prove that for a homomorphism $f\colon X \to Y$ of coalgebras,

$$f(x) \to y \iff \exists x'.\, x \to x' \wedge f(x') = y.$$

Note that this captures the functoriality of the translation from coalgebras to transition systems as in (6.4).

6.4.8 Show that each subset can be written as intersection of non-singletons (coequations): for $U \subseteq X$,

$$U = \bigcap_{x \in \neg U} (\cdot \neq x).$$

This result forms the basis for formulating a (predicate) logic for coalgebras in terms of these 'coequations' $(\cdot \neq x)$; see [177].

6.4.9 Prove – and explain in words – that

$$x \longrightarrow^{*} x' \iff x' \in \underleftarrow{\Diamond}(\cdot = x).$$

(The notation $\langle x \rangle = \{x' \mid x \longrightarrow^{*} x'\}$ and $\langle P \rangle = \{x' \mid \exists x \in P.\, x \longrightarrow^{*} x'\}$ is used in [411] for the least invariants $\underleftarrow{\Diamond}\{x\} = \underleftarrow{\Diamond}(\cdot = x)$ and $\underleftarrow{\Diamond}\, P$ containing an element x or a predicate P.)

6.4.10 Verify the Galois connections in Figure 6.2. (Such Galois connections for temporal logic are studied systematically in [288, 243].)

6.4.11 Check that $P\, \mathcal{U}\, P = P$.

6.4.12 The following is taken from [120, section 5], where it is referred to as the Whisky Problem. It is used there as a challenge in proof automation in linear temporal logic. Here it will be formulated in the temporal logic of an arbitrary coalgebra (of a Kripke polynomial functor). Consider an arbitrary set A with an endofunction $h\colon A \to A$. Let $P\colon A \to \mathcal{P}(X)$ be a parametrised predicate on the state space of a coalgebra X, satisfying for a specific $a \in A$ and $y \in X$:

1. $P(a)(y)$
2. $\forall b \in A.\, P(b)(y) \Rightarrow P(h(b))(y)$
3. $\Box(\{x \in X \mid \forall b \in A.\, P(h(b))(x) \Rightarrow \bigcirc(P(b))(x)\})(y)$.

Prove then that $\Box(P(a))(y)$ *Hint*: Use Exercise 6.4.3.

6.4.13 Describe the nexttime operator $\bigcirc$ as a natural transformation $\mathcal{P}U \Rightarrow \mathcal{P}U$, as in Corollary 6.1.4, where $U\colon \mathbf{CoAlg}(F) \to \mathbf{Sets}$ is the forgetful functor and $\mathcal{P}$ is the contravariant powerset functor. Show that it can be described as a composition of natural transformations:

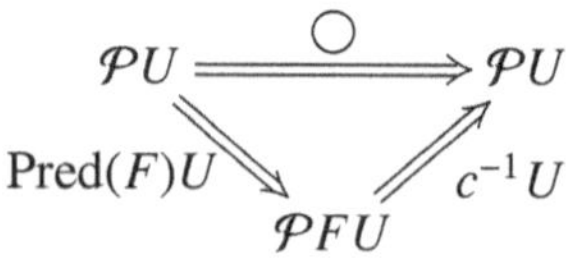

6.4.14 Prove that the 'until' and 'since' operators $\mathcal{U}, \mathcal{S}\colon \mathcal{P}(X) \times \mathcal{P}(X) \to \mathcal{P}(X)$ on the state space X of a coalgebra (see Figures 6.1 and 6.2) can be described in the following way in terms of the transition relation $\to \subseteq X \times X$ from Definition 6.4.6:

$$P\,\mathcal{U}\,Q = \{x \mid \exists n.\, \exists x_0, \ldots, x_n.\, x_0 = x \wedge (\forall i < n.\, x_i \to x_{i+1}) \wedge Q(x_n) \wedge \forall i < n.\, P(x_i)\}$$
$$P\,\mathcal{S}\,Q = \{x \mid \exists n.\, \exists x_0, \ldots, x_n.\, x_n = x \wedge (\forall i < n.\, x_i \to x_{i+1}) \wedge Q(x_0) \wedge \forall i > 0.\, P(x_i)\}.$$

6.4.15 We consider the strong nexttime operator $\neg \bigcirc \neg$ associated with a coalgebra and call a predicate P **maintainable** if $P \subseteq \neg \bigcirc \neg P$. Notice that such a predicate is a $\neg \bigcirc \neg$-coalgebra.

1. Investigate what this requirement means, for instance for a few concrete coalgebras.
2. Let us use the notation $\mathsf{EA}\,P$ for the greatest maintainable predicate contained in P. Describe $\mathsf{EA}\,P$ in terms of the transition relation $\to$ from Definition 6.4.6.
3. Similarly for $\mathsf{AE}\,P \stackrel{\text{def}}{=} \neg \mathsf{EA}\,\neg P$

(Operators like such as EA and AE are used in computation tree logic (CTL); see e.g. [128] to reason about paths in trees of computations. The interpretations we use here involve infinite paths.)

6.5 Modal Logic for Coalgebras

In the previous section we have seen a temporal logic for coalgebras based on the nexttime operator; see Definition 6.4.1. The meaning of $\bigcirc P$ is that the predicate P holds in *all* direct successor states. This operator is very useful for expressing safety and liveness properties, via the derived henceforth and eventually operators $\Box$ and $\Diamond$, expressing 'for all/some future states …'. But this temporal logic is not very useful for expressing more refined properties dealing for instance with one particular branch. Modal logics (including dynamic logics [198]) are widely used in computer science to reason about various kinds of dynamical systems. Often they are tailored to a specific domain of reasoning. As usual in coalgebra, the goal is to capture many of these variations in a common abstract framework.

Consider a simple coalgebra $c = \langle c_1, c_2 \rangle \colon X \to X \times X$ involving two transition maps $c_1, c_2 \colon X \to X$. The meaning of $\bigcirc(P)$ from temporal logic for this coalgebra is

$$\bigcirc(P) = \{x \mid c_1(x) \in P \wedge c_2(x) \in P\}.$$

Thus it contains those states x *all* of whose successors $c_1(x)$ and $c_2(x)$ satisfy P. It would be useful to have two separate logical operators, say $\bigcirc_1$ and $\bigcirc_2$, talking specifically about transition maps c_1 and c_2, respectively, as in

$$\bigcirc_1(P) = \{x \mid c_1(x) \in P\} \quad \text{and} \quad \bigcirc_2(P) = \{x \mid c_2(x) \in P\}. \tag{6.10}$$

Coalgebraic modal logic allows us to describe such operators.

To see another example/motivation, consider a coalgebra $c\colon X \to \mathcal{M}_{\mathbb{N}}(X)$ of the multiset (or bag) functor $\mathcal{M}_{\mathbb{N}}$, counting in the natural numbers $\mathbb{N}$. Thus we may write $x \xrightarrow{n} x'$ if $c(x)(x') = n \in \mathbb{N}$, expressing a transition which costs for instance n resources. For each $N \in \mathbb{N}$, a so-called graded modality $\bigcirc_N$ (see [131]) is defined as

$$\bigcirc_N(P) = \{x \mid \forall x'.\, c(x)(x') \geq N \Rightarrow P(x')\}.$$

This section describes how to obtain such operators in the research field known as 'coalgebraic modal logic'. The approach that we follow is rather concrete and 'hands-on'. The literature on the topic is extensive (see e.g. [359, 322, 404, 373, 317, 296, 315, 273] or the overview papers [318, 323, 99]), but there is a tendency to wander off into meta-theory and to omit specific examples (and what the modal logic might be useful for).

We begin with an abstract definition describing the common way that coalgebraic logics are now understood. It involves a generalised form of predicate lifting, as will be explained and illustrated subsequently. An even more abstract approach will be described later on, in Section 6.5.1.

Definition 6.5.1 For a functor $F\colon \mathbf{Sets} \to \mathbf{Sets}$, a **coalgebraic modal logic** is given by a 'modal signature functor' $L\colon \mathbf{Sets} \to \mathbf{Sets}$ and a natural transformation

$$L\mathcal{P} \overset{\delta}{\Longrightarrow} \mathcal{P}F,$$

where $\mathcal{P} = 2^{(-)}\colon \mathbf{Sets}^{\mathrm{op}} \to \mathbf{Sets}$ is the contravariant powerset functor.

Given such a $\delta\colon L\mathcal{P} \Rightarrow \mathcal{P}F$, each F-coalgebra $c\colon X \to F(X)$ yields an L-algebra on the set of predicates $\mathcal{P}(X)$, namely:

$$L(\mathcal{P}(X)) \xrightarrow{\delta_X} \mathcal{P}(F(X)) \xrightarrow{c^{-1} = \mathcal{P}(c)} \mathcal{P}(X).$$

This yields a functor $\mathbf{CoAlg}(F)^{\mathrm{op}} \to \mathbf{Alg}(L)$ in a commuting diagram:

$$\begin{array}{ccc} \mathbf{CoAlg}(F)^{\mathrm{op}} & \longrightarrow & \mathbf{Alg}(L) \\ \big\downarrow & & \big\downarrow \\ \mathbf{Sets}^{\mathrm{op}} & \xrightarrow{\quad \mathcal{P} \quad} & \mathbf{Sets} \end{array} \tag{6.11}$$

Recall from Corollary 6.1.4 that predicate lifting is described as a functor $\mathrm{Pred}(F)\colon \mathcal{P} \Rightarrow \mathcal{P}F$. In the above definition this is generalised by adding a functor L in front, yielding $L\mathcal{P} \Rightarrow \mathcal{P}F$. This L makes more flexible liftings – and thus more flexible modal operators – possible, as will be illustrated next.

A less general and more clumsy approach using 'ingredients' of functors is used in [241]. One can think of the functor L as describing the type of these ingredient operators, which becomes explicit if one considers the induced L-algebra $L(\mathcal{P}(X)) \to \mathcal{P}(X)$. For instance, for the two operators from (6.10) we use $L(Y) = Y + Y$, so that we can describe the pair of nexttime operators in (6.10) as L-algebra:

$$L(\mathcal{P}(X)) = \mathcal{P}(X) + \mathcal{P}(X) \xrightarrow{[\bigcirc_1, \bigcirc_2]} \mathcal{P}(X),$$

arising according to the pattern in the definition via a map δ in

$$L(\mathcal{P}(X)) = \mathcal{P}(X) + \mathcal{P}(X) \xrightarrow{\delta} \mathcal{P}(X \times X) \xrightarrow{\langle c_1, c_2\rangle^{-1}} \mathcal{P}(X),$$

where $\delta = [\delta_1, \delta_2] \colon L(\mathcal{P}(X)) \to \mathcal{P}(X \times X)$ is given by

$$\begin{aligned} \delta_1(P) &= \pi_1^{-1} = \mathcal{P}(\pi_1) & \delta_2(P) &= \pi_2^{-1} = \mathcal{P}(\pi_2) \\ &= \{(x_1, x_2) \in X \times X \mid P(x_1)\} & &= \{(x_1, x_2) \in X \times X \mid P(x_2)\}. \end{aligned}$$

Then indeed, $\bigcirc_i(P) = \langle c_1, c_2\rangle^{-1} \circ \delta \circ \kappa_i = c_i^{-1}$.

We turn to a more elaborate illustration.

Example 6.5.2 Suppose we wish to describe a simple bank account coalgebraically. It involves a state space X, considered as black box, accessible only via the next three balance, deposit and withdraw operations.

$\mathsf{bal} \colon X \longrightarrow \mathbb{N}$	for learning the balance of the account, which, for simplicity, is represented as a natural number
$\mathsf{dep} \colon X \times \mathbb{N} \longrightarrow X$	for depositing a certain amount of money, given as parameter, into the account
$\mathsf{wdw} \colon X \times \mathbb{N} \longrightarrow X + X$	for withdrawing a certain amount of money, given as parameter, from the account; the first/left output option of + is used for a successful withdrawal, when the balance before the withdrawal exceeds the retrieved amount. The second/right +-option is used for unsuccessful withdrawals; in that case the balance remains unchanged.

Together these maps form a coalgebra of the form

$$X \xrightarrow{\langle \mathsf{bal}, \mathsf{dep}, \mathsf{wdw}\rangle} F(X) \quad \text{for} \quad F(X) = \mathbb{N} \times X^{\mathbb{N}} \times (X + X)^{\mathbb{N}}.$$

We would like to express the above informal descriptions of the behaviour of this coalgebra in precise logical terms, using modal operators. Therefore we define predicates:

$$\begin{aligned} \mathsf{bal}\downarrow n &= \{x \in X \mid \mathsf{bal}(x) = n\} \\ [\mathsf{dep}(n)](P) &= \{x \in X \mid \mathsf{dep}(x,n) \in P\}. \end{aligned}$$

Now we can require

$$\mathsf{bal}\downarrow m \vdash [\mathsf{dep}(n)](\mathsf{bal}\downarrow(m+n)),$$

where $\vdash$ should be understood as subset inclusion $\subseteq$. Obviously this captures the intended behaviour of 'deposit'.

The modal operator for withdrawal is more subtle because of the two output options in $X+X$. Therefore we define two modal operators, one for each option:

$$\begin{aligned} [\mathsf{wdw}(n)]_1(P) &= \{x \in X \mid \forall x'.\, \mathsf{wdw}(x,n) = \kappa_1 x' \Rightarrow P(x')\} \\ [\mathsf{wdw}(n)]_2(P) &= \{x \in X \mid \forall x'.\, \mathsf{wdw}(x,n) = \kappa_2 x' \Rightarrow P(x')\}. \end{aligned}$$

One may now expect a requirement:

$$\mathsf{bal}\downarrow(m+n) \vdash [\mathsf{wdw}(n)]_1(\mathsf{bal}\downarrow m).$$

But a little thought reveals that this is too weak, since it does not enforce that a successful withdrawal yields an output in the first/left +-option. We need to use the derived operator $\langle f\rangle(P) = \neg[f](\neg P)$ so that

$$\langle\mathsf{wdw}(n)\rangle_i(P) = \neg[\mathsf{wdw}(n)]_i(\neg P) = \{x \in X \mid \exists x'.\, \mathsf{wdw}(x,n) = \kappa_i x' \wedge P(x')\}.$$

Now we can express the remaining requirements as

$$\begin{aligned} \mathsf{bal}\downarrow(m+n) &\vdash \langle\mathsf{wdw}(n)\rangle_1(\mathsf{bal}\downarrow m) \\ \mathsf{bal}\downarrow(m) &\vdash \langle\mathsf{wdw}(m+n+1)\rangle_2(\mathsf{bal}\downarrow m). \end{aligned}$$

Thus we have used four logical operators to lay down the behaviour of a bank account, via restriction of the underlying coalgebra $X \to F(X)$. Such restrictions will be studied more systematically in Section 6.8. Here we concentrate on the modal/dynamic operations. The four of them can be described jointly in the format of Definition 6.5.1, namely as 4-cotuple:

$$[\,\mathsf{bal}\downarrow(-),\ [\mathsf{dep}(-)],\ [\mathsf{wdw}_1(-)],\ [\mathsf{wdw}_1(-)]\,],$$

forming an algebra:

$$\mathbb{N} + (\mathbb{N}\times\mathcal{P}(X)) + (\mathbb{N}\times\mathcal{P}(X)) + (\mathbb{N}\times\mathcal{P}(X)) \longrightarrow \mathcal{P}(X). \qquad (6.12)$$

It is an algebra $L(\mathcal{P}(X)) \to \mathcal{P}(X)$ for the modal signature functor $L\colon \mathbf{Sets} \to \mathbf{Sets}$ given by

$$L(Y) = \mathbb{N} + (\mathbb{N} \times Y) + (\mathbb{N} \times Y) + (\mathbb{N} \times Y).$$

The algebra of operators (6.12) is the composition of two maps:

$$L(\mathcal{P}(X)) \xrightarrow{\quad \delta \quad} \mathcal{P}(F(X)) \xrightarrow{\quad c^{-1} \quad} \mathcal{P}(X),$$

where $c = \langle \mathsf{bal}, \mathsf{dep}, \mathsf{wdw} \rangle\colon X \to F(X) = \mathbb{N} \times X^{\mathbb{N}} \times (X + X)^{\mathbb{N}}$ is the coalgebra involved. The map $\delta\colon L(\mathcal{P}(X)) \to \mathcal{P}(F(X))$ is a 4-cotuple of the form $\delta = [\delta_{\mathsf{bal}}, \delta_{\mathsf{dep}}, \delta_{\mathsf{wdw},1}, \delta_{\mathsf{wdw},2}]$, where

$$\begin{aligned}
\delta_{\mathsf{bal}}(n) &= \{(m, f, g) \in F(X) \mid m = n\} \\
\delta_{\mathsf{dep}}(n, P) &= \{(m, f, g) \in F(X) \mid f(n) \in P\} \\
\delta_{\mathsf{wdw},1}(n, P) &= \{(m, f, g) \in F(X) \mid \forall x.\, g(n) = \kappa_1 x \Rightarrow P(x)\} \\
\delta_{\mathsf{wdw},2}(n, P) &= \{(m, f, g) \in F(X) \mid \forall x.\, g(n) = \kappa_2 x \Rightarrow P(x)\}.
\end{aligned}$$

It is not hard to see that this δ is a natural transformation $L\mathcal{P} \Rightarrow \mathcal{P}F$.

A next step is to see how the modal signature functor L and the natural transformation $\delta\colon L\mathcal{P} \Rightarrow \mathcal{P}F$ in Definition 6.5.1 arise. In general, this is a matter of choice. But in many cases, for instance when the functor is a polynomial, there are some canonical choices. We shall illustrate this in two steps, namely by first defining coalgebraic logics for several basic functors, and subsequently showing that coalgebraic logics can be combined via several constructions, such as composition and (co)product. A similar, but more language-oriented, modular approach is described in [100].

Definition 6.5.3 Coalgebraic logics $(L, L\mathcal{P} \stackrel{\delta}{\Rightarrow} \mathcal{P}F)$ can be defined when F is the identity/constant/powerset/multiset/distribution functor in the following manner.

1. For the identity functor $\mathrm{id}\colon \mathbf{Sets} \to \mathbf{Sets}$ we take $L = \mathrm{id}$, with identity natural transformation:

$$L(\mathcal{P}(X)) = \mathcal{P}(X) \xrightarrow{\quad \delta = \mathrm{id} \quad} \mathcal{P}(X) = \mathcal{P}(\mathrm{id}(X)).$$

2. For a constant functor $A\colon \mathbf{Sets} \to \mathbf{Sets}$, given by $X \mapsto A$, we also take $L = A$, with singleton (unit) map:

$$L(\mathcal{P}(X)) = A \xrightarrow{\quad \delta = \{-\} \quad} \mathcal{P}(A) = \mathcal{P}(A(X)).$$

3. For the (covariant) powerset functor $\mathcal{P}$ we take $L = \mathrm{id}$, with

$$L(\mathcal{P}(X)) = \mathcal{P}(X) \xrightarrow{\quad \delta \quad} \mathcal{P}(\mathcal{P}(X))$$

given by $\delta(P) = \{U \in \mathcal{P}(X) \mid U \subseteq P\}$. For the finite powerset functor $\mathcal{P}_{\mathrm{fin}}$ an analogous map $\mathcal{P}(X) \to \mathcal{P}(\mathcal{P}_{\mathrm{fin}}(X))$ is used.

4. For the distribution functor $\mathcal{D}$: **Sets** $\to$ **Sets** take $L(Y) = [0,1]_{\mathbb{Q}} \times Y$, where $[0,1]_{\mathbb{Q}} = [0,1] \cap \mathbb{Q}$ is the unit interval of rational numbers. The associated natural transformation follows the ideas of [210]:

$$L(\mathcal{P}(X)) = [0,1]_{\mathbb{Q}} \times \mathcal{P}(X) \xrightarrow{\quad \delta \quad} \mathcal{P}(\mathcal{D}X),$$

where $\delta(r, P) = \{\varphi \in \mathcal{D}(X) \mid \forall x'.\, \varphi(x)(x') \geq r \Rightarrow P(x')\}$.

5. The same coalgebraic modal logic can be used for the multiset functor $\mathcal{M}_M$, assuming the monoid carries an order. This yields the graded operators, as sketched in the beginning of this section for the multiset/bag functor $\mathcal{M}_{\mathbb{N}}$ over the natural numbers.

6. For the neighbourhood functor $\mathcal{N}(X) = 2^{(2^X)}$ from Exercise 2.2.7 one takes $L = \mathrm{id}$ with natural transformation

$$L(\mathcal{P}(X)) = \mathcal{P}(X) \xrightarrow{\quad \delta \quad} \mathcal{P}(\mathcal{N}X) = \mathcal{PPP}(X),$$

given by $\delta(P) = \{V \in \mathcal{PP}(X) \mid P \in V\}$.

Lemma 6.5.4 *Coalgebraic modal logics can be combined in the following ways.*

1. *For two functors F_1, F_2:* **Sets** $\to$ **Sets**, *each with a coalgebraic modal logic (L_i, δ_i), there is also a coalgebraic modal logic for the composite functor $F_1 \circ F_2$, namely $L_1 \circ L_2$ with natural transformation given by*

$$L_1 L_2 \mathcal{P}(X) \xrightarrow{\quad L_1(\delta_{2,X}) \quad} L_1 \mathcal{P} F_2(X) \xrightarrow{\quad \delta_{1,F_2(X)} \quad} \mathcal{P} F_1 F_2(X).$$

2. *Given an I-indexed collection of functors $(F_i)_{i\in I}$ with logics (L_i, δ_i) we define a modal signature functor $L(Y) = \coprod_{i\in I} L_i(Y)$ for the coproduct functor $\coprod_{i\in I} F_i$, with natural transformation*

$$L(\mathcal{P}(X)) = \coprod_{i\in I} L_i(\mathcal{P}(X)) \xrightarrow{\quad [\coprod_{\kappa_i} \circ\, \delta_i]_{i\in I} \quad} \mathcal{P}(\coprod_{i\in I} F_i(X)).$$

We write $\coprod_{\kappa_i} : \mathcal{P}(F_i(X)) \to \mathcal{P}(\coprod_i F_i(X))$ for the direct *image, using $\mathcal{P}$ as a covariant functor.*

3. *Similarly, for a product functor* $\prod_{i\in I} F_i$ *we use the coproduct* $\coprod_{i\in I} L_i$ *of associated modal signature functors, with*

$$L(\mathcal{P}(X)) = \coprod_{i\in I} L_i(\mathcal{P}(X)) \xrightarrow{[\pi_i^{-1} \circ \delta_i]_{i\in I}} \mathcal{P}(\prod_{i\in I} F_i(X)).$$

4. *As a special case of the previous point we make the exponent functor* F^A *explicit. It has a modal signature functor* $A \times L(-)$*, assuming a coalgebraic modal logic* (L, δ) *for* F*, with natural transformation*

$$A \times L(\mathcal{P}(X)) \longrightarrow \mathcal{P}(F(X)^A),$$

given by $(a, u) \mapsto \{f \in F(X)^A \mid f(a) \in \delta(u)\}$.

Proof One has only to check that the new δs are natural transformations; this is easy. □

As a result, each Kripke polynomial functor has a (canonical) coalgebraic modal logic.

6.5.1 Coalgebraic Modal logic, More Abstractly

In Definition 6.5.1 we have introduced a coalgebraic modal logic for a functor F as a pair (L, δ), where $\delta\colon L\mathcal{P} \Rightarrow \mathcal{P}F$. As we have seen, this approach works well for many endofunctors on **Sets**. Still, the picture is a bit too simple, for two reasons.

- Modal operators usually satisfy certain preservation properties. In particular, almost all of them preserve finite conjunctions $(\top, \wedge)$. This can be captured by restricting the modal signature functor L, from an endofunctor on **Sets** to an endofunctor on the category **MSL** of meet semilattices.
- The approach is defined for endofunctors on **Sets** and not for endofunctors on an arbitrary category. In order to take care of this additional generality we will use the more general logic described in terms of factorisation systems in Section 4.3.

We will tackle both these issues at the same time. We proceed in a somewhat informal manner; refer to the literature [323] for further details and ramifications.

Let $F\colon \mathbb{C} \to \mathbb{C}$ be an endofunctor on a category $\mathbb{C}$ with a logical factorisation system $(\mathfrak{M}, \mathfrak{E})$. Recall from Lemma 4.3.4 that it gives rise to an indexed category

$$\mathbb{C}^{\mathrm{op}} \xrightarrow{\mathrm{Pred}(-)} \mathbf{MSL}.$$

For each object $X \in \mathbb{C}$ the poset Pred(X) of predicates $(U \rightarrowtail X) \in \mathfrak{M}$ on X has finite meets $\top, \wedge$; additionally, a map $f\colon X \to Y$ in $\mathbb{C}$ yields a finite meet-preserving substitution functor $\text{Pred}(f) = f^{-1}\colon \text{Pred}(Y) \to \text{Pred}(X)$ by pullback.

In this general situation, a **coalgebraic modal logic** for $F\colon \mathbb{C} \to \mathbb{C}$ consists of a functor $L\colon \mathbf{MSL} \to \mathbf{MSL}$ together with a natural transformation $\delta\colon L\text{Pred} \Rightarrow \text{Pred}F$. Thus, δ is a natural transformation between the following two parallel functors:

$$\mathbb{C}^{\text{op}} \xrightarrow{\text{Pred}} \mathbf{MSL} \xrightarrow{L} \mathbf{MSL} \qquad \mathbb{C}^{\text{op}} \xrightarrow{F} \mathbb{C}^{\text{op}} \xrightarrow{\text{Pred}} \mathbf{MSL}$$

In Exercise 6.5.9 it is shown that all coalgebraic models described in Definition 6.5.3 live in this way in the category **MSL**, except for the neighbourhood functor $\mathcal{N}$. Also, by Exercise 6.5.8 all the constructions on coalgebraic modal logics in Lemma 6.5.4 can be performed in the category **MSL**.

A coalgebra $c\colon X \to F(X)$ in the base category $\mathbb{C}$ gives rise to an L-algebra, as before in Definition 6.5.1, but this time in the category **MSL**:

$$L(\text{Pred}(X)) \xrightarrow{\ \delta\ } \text{Pred}(F(X)) \xrightarrow{c^{-1} = \text{Pred}(c)} \text{Pred}(X).$$

This yields a lifted functor $\text{Pred}\colon \mathbf{CoAlg}(F)^{\text{op}} \to \mathbf{Alg}(L)$, in a slightly different diagram from (6.11):

$$\begin{array}{ccc} \mathbf{CoAlg}(F)^{\text{op}} & \xrightarrow{\text{Pred}} & \mathbf{Alg}(L) \\ \downarrow & & \downarrow \\ \mathbf{Sets}^{\text{op}} & \xrightarrow{\text{Pred}} & \mathbf{MSL} \end{array}$$

Next assume that the functor $L\colon \mathbf{MSL} \to \mathbf{MSL}$ has an initial algebra. We shall write it as

$$L(\mathit{Form}) \xrightarrow{\ \cong\ } \mathit{Form} \tag{6.13}$$

where '*Form*' stands for 'formulas'. This set $\mathit{Form} \in \mathbf{MSL}$ is by construction closed under finite conjunctions $(\top, \wedge)$ and comes equipped with modal operators via the above algebra $L(\mathit{Form}) \to \mathit{Form}$. By initiality we get a unique homomorphism in **MSL**:

$$\begin{array}{ccc} L(\mathit{Form}) & \xrightarrow{L([\![-]\!]_c)} & L(\mathrm{Pred}(X)) \\ {\scriptstyle\cong}\big\downarrow & & \big\downarrow{\scriptstyle \mathrm{Pred}(c)\,\circ\,\delta} \\ \mathit{Form} & \xrightarrow[{[\![-]\!]_c}]{} & \mathrm{Pred}(X) \end{array} \tag{6.14}$$

It maps a formula $\varphi \in \mathit{Form}$ to its interpretation $[\![\varphi]\!]_c \in \mathrm{Pred}(X)$, as an $\mathfrak{M}$-subobject of the state space X. This map $[\![-]\!]$ preserves finite meets and preserves the modal operators.

Remark 6.5.5 The collection $\mathit{Form} \in \mathbf{MSL}$ of logical formulas defined in (6.13) has finite meets by construction. What if we would like to have all Boolean operations on formulas? The obvious way would be to construct *Form* as an initial algebra in the category **BA** of Boolean algebras. But this approach does not work, because in general the modal operations $L(\mathit{Form}) \to \mathit{Form}$ preserve only finite meets and for instance not $\neg$ or $\vee$.

There is a neat trick around this; see e.g. [273, 323]. We use that the forgetful functor $U\colon \mathbf{BA} \to \mathbf{MSL}$ from Boolean algebras to meet semilattices has a left adjoint $F\colon \mathbf{MSL} \to \mathbf{BA}$ – which follows from Exercise 5.4.15.3. Now we consider the functor

$$L' = \Big(\mathbf{BA} \xrightarrow{\;U\;} \mathbf{MSL} \xrightarrow{\;L\;} \mathbf{MSL} \xrightarrow{\;F\;} \mathbf{BA}\Big).$$

There is a direct (adjoint) correspondence between L'- and L-algebras: for a Boolean algebra B,

$$\begin{array}{cl} FLU(B) = L'(B) \longrightarrow B & \text{in } \mathbf{BA} \\ \hline\hline L(UB) \longrightarrow UB & \text{in } \mathbf{MSL}. \end{array}$$

Thus if we now define the collection *Form′* as initial algebra of the functor L' in **BA**, then *Form′* carries all Boolean structure and has modal operators $L(\mathit{Form}') \to \mathit{Form}'$ that preserve only finite meets.

In order to proceed further we need another assumption, namely that the predicate functor $\mathrm{Pred}\colon \mathbb{C}^{\mathrm{op}} \to \mathbf{MSL}$ has a left adjoint $\mathcal{S}$. Thus we have an adjoint situation:

$$\mathbb{C}^{\mathrm{op}} \underset{\mathcal{S}}{\overset{\mathrm{Pred}}{\underset{\longleftarrow}{\overset{\longrightarrow}{\top}}}} \mathbf{MSL}. \tag{6.15}$$

Exercise 6.5.7 deals with some situations where this is the case. Such 'dual' adjunctions form the basis for many dualities; see [280], relating predicates and states, for instance in domain theory [6], probabilistic computing [308], or in quantum computing [121, 257] and see also Exercise 5.4.11.

In presence of this adjunction (6.15), two things are relevant.

- The natural transformation $\delta\colon L\mathrm{Pred} \Rightarrow \mathrm{Pred}F$, forming the modal coalgebraic logic for the functor F, bijectively corresponds to another natural transformation $\overline{\delta}$, as in

$$\frac{L\mathrm{Pred} \overset{\delta}{\Longrightarrow} \mathrm{Pred}F}{FS \underset{\overline{\delta}}{\Longrightarrow} SL} \,. \tag{6.16}$$

  Working out this correspondence is left to the interested reader, in Exercise 6.5.10.

- We can take the transpose of the interpretation map $[\![-]\!]_c\colon \mathit{Form} \to \mathrm{Pred}(X)$ from (6.14). It yields a 'theory' map $th_c : X \to S(\mathit{Form})$ that intuitively sends a state to the formulas that hold for this state. The relation containing the states for which the same formulas hold is given as kernel/equaliser $\equiv_c \rightarrowtail X \times X$ in $\mathbb{C}$:

$$\equiv_c = \mathrm{Ker}(th_c) \rightarrowtail X \times X \underset{th_c \circ \pi_2}{\overset{th_c \circ \pi_1}{\rightrightarrows}} S(\mathit{Form}). \tag{6.17}$$

An important question is how this relation $\equiv_c$ relates to the notions of indistinguishability that we have seen for coalgebras. It turns out that the behavioural equivalence (cospan) definition works best in this situation. The next result is based on [419] and also, in more categorical form, on [377, 297, 273].

Theorem 6.5.6 *Consider the situation described above, where the functor* $F\colon \mathbb{C} \to \mathbb{C}$ *has a modal coalgebraic logic* $\delta\colon L\mathrm{Pred} \Rightarrow \mathrm{Pred}F$ *with initial algebra* $L(\mathit{Form}) \xrightarrow{\cong} \mathit{Form}$*, and where there is a left adjoint* S *to the indexed category* $\mathrm{Pred}\colon \mathbb{C}^{\mathrm{op}} \to \mathbf{MSL}$ *associated with the logical factorisation system* $(\mathfrak{M}, \mathfrak{E})$ *on* $\mathbb{C}$*, as in (6.15).*

1. *Observationally equivalent states satisfy the same logical formulas: each kernel of a coalgebra map factors through the equaliser* $\equiv_c \rightarrowtail X \times X$ *in (6.17).*
2. *If the functor* F *preserves abstract monos (in* $\mathfrak{M}$*) and the transpose* $\overline{\delta}\colon FS \Rightarrow SL$ *in (6.16) consists of abstract monos, then the converse is also true: states that make the same formulas true are observationally equivalent.*

The latter property is usually called **expressivity** of the logic, or the **Hennessy–Milner** property. Originally, this property was proven in [212] but only for finitely branching transition systems. The most significant assumption in this much more general theorem for coalgebras is the $\overline{\delta}$-injectivity property of the coalgebra modal logic.

Proof 1. Let $f\colon X \to Y$ be a map of coalgebras, from $c\colon X \to F(X)$ to $d\colon Y \to F(Y)$, with kernel

$$\mathrm{Ker}(f) \xrightarrow{\langle k_1, k_2\rangle} X \times X \underset{f \circ \pi_2}{\overset{f \circ \pi_1}{\rightrightarrows}} Y.$$

By initiality we get $\mathrm{Pred}(f) \circ [\![-]\!]_d = [\![-]\!]_c$ in

$$\begin{array}{ccccc}
L(\mathit{Form}) & \longrightarrow & L(\mathrm{Pred}(Y)) & \longrightarrow & L(\mathrm{Pred}(X)) \\
\cong\downarrow & & \downarrow \mathrm{Pred}(d) \circ \delta_Y & & \downarrow \mathrm{Pred}(c) \circ \delta_X \\
\mathit{Form} & \xrightarrow{[\![-]\!]_d} & \mathrm{Pred}(Y) & \xrightarrow{\mathrm{Pred}(f)} & \mathrm{Pred}(X)
\end{array}$$

with the curved arrow $\mathit{Form} \xrightarrow{[\![-]\!]_c} \mathrm{Pred}(X)$ below.

Now we can see that $\mathrm{Ker}(f)$ factors through the equaliser $\equiv_c \rightarrowtail X \times X$ in (6.17):

$$\begin{aligned}
th_c \circ k_1 &= \mathcal{S}([\![-]\!]_c) \circ \eta \circ k_1 \\
&= \mathcal{S}(\mathrm{Pred}(f) \circ [\![-]\!]_d) \circ \mathcal{S}(\mathrm{Pred}(k_1)) \circ \eta \\
&= \mathcal{S}(\mathrm{Pred}(k_1) \circ \mathrm{Pred}(f) \circ [\![-]\!]_d) \circ \eta \\
&= \mathcal{S}(\mathrm{Pred}(f \circ k_1) \circ [\![-]\!]_d) \circ \eta \\
&= \mathcal{S}(\mathrm{Pred}(f \circ k_2) \circ [\![-]\!]_d) \circ \eta \\
&= \cdots \\
&= th_c \circ k_2.
\end{aligned}$$

2. We first observe that the transpose $\overline{\delta}$ makes the following diagram commute:

$$\begin{array}{ccccc}
F(X) & \xrightarrow{F(th_c)} & F\mathcal{S}(\mathit{Form}) & \overset{\overline{\delta}}{\rightarrowtail} & \mathcal{S}L(\mathit{Form}) \\
c\uparrow & & & & \cong\uparrow \mathcal{S}(\alpha) \\
X & & \xrightarrow{\quad th_c \quad} & & \mathcal{S}(\mathit{Form})
\end{array}$$

where we write the initial algebra map in (6.13) as $\alpha\colon L(\mathit{Form}) \xrightarrow{\cong} \mathit{Form}$. Commutation of this rectangle follows from the explicit description $\overline{\delta} = \mathcal{S}L(\eta) \circ \mathcal{S}(\delta) \circ \varepsilon$ in Exercise 6.5.10:

$$\begin{aligned}
\overline{\delta} \circ F(th_c) \circ c &= \mathcal{S}L(\eta) \circ \mathcal{S}(\delta) \circ \varepsilon \circ F(th_c) \circ c \\
&= \mathcal{S}L(\eta) \circ \mathcal{S}(\delta) \circ \mathcal{S}\mathrm{Pred}(F(th_c) \circ c) \circ \varepsilon \\
&= \mathcal{S}(\mathrm{Pred}(c) \circ \mathrm{Pred}F(th_c) \circ \delta \circ L(\eta)) \circ \varepsilon \\
&= \mathcal{S}(\mathrm{Pred}(c) \circ \delta \circ L\mathrm{Pred}(th_c) \circ L(\eta)) \circ \varepsilon \\
&= \mathcal{S}(\mathrm{Pred}(c) \circ \delta \circ L(\llbracket - \rrbracket_c)) \circ \varepsilon \\
&= \mathcal{S}(\llbracket - \rrbracket_c \circ \alpha) \circ \varepsilon \qquad \text{by (6.14)} \\
&= \mathcal{S}(\alpha) \circ \mathcal{S}(\mathrm{Pred}(th_c) \circ \eta) \circ \varepsilon \\
&= \mathcal{S}(\alpha) \circ \mathcal{S}(\eta) \circ \mathcal{S}\mathrm{Pred}(th_c) \circ \varepsilon \\
&= \mathcal{S}(\alpha) \circ \mathcal{S}(\eta) \circ \varepsilon \circ th_c \\
&= \mathcal{S}(\alpha) \circ th_c.
\end{aligned}$$

Next we take the factorisation of the theory map th_c in

$$th_c = \Big(X \xrightarrow{\;e\;} Y \xrightarrow{\;m\;} \mathcal{S}(Form)\Big).$$

Since the functor F preserves maps in $\mathfrak{M}$ a coalgebra d can be defined on the image Y via diagonal-fill-in:

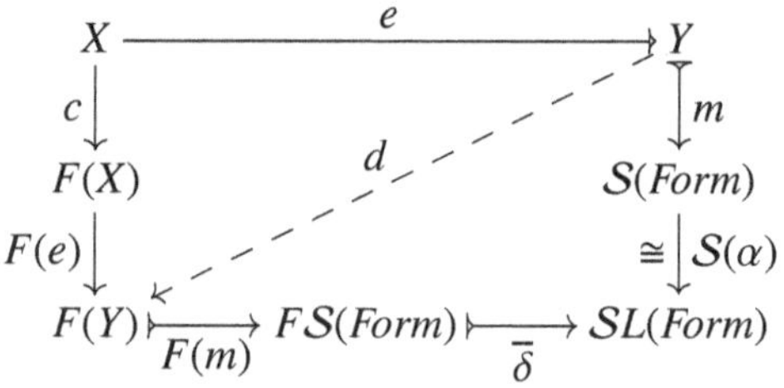

In particular, the abstract epi e is a map of coalgebras $c \to d$.

If we write $\langle r_1, r_2\rangle \colon \equiv_c \rightarrowtail X \times X$ for the equaliser map in (6.17), then, by construction:

$$m \circ e \circ r_1 \;=\; th_c \circ r_1 \;=\; th_c \circ r_2 \;=\; m \circ e \circ r_2.$$

Since m is monic, this yields $e \circ r_1 = e \circ r_2$. Thus $\equiv_c$ is contained in the kernel Ker(e) of a map of coalgebras. Hence, states related by $\equiv_c$ are behaviourally equivalent. ❑

6.5.2 Modal Logic Based on Relation Lifting

What we have described above is coalgebraic modal logic based on (an extension of) *predicate* lifting. The first form of coalgebraic modal logic, introduced by Moss [359], was based however on *relation* lifting. This lifting is applied to the set membership relation $\in$, as in Lemma 5.2.7. In fact, the distributive law $\nabla \colon F\mathcal{P} \Rightarrow \mathcal{P}F$ described there captures the essence of the logic. This ∇ is understood as a so-called logical 'cover' operator. It leads to a non-standard syntax, which we briefly illustrate.

- Consider the functor $F(X) = X \times X$, as used in the beginning of this section. It leads to an operator $\nabla\colon \mathcal{P}(X) \times \mathcal{P}(X) \to \mathcal{P}(X \times X)$, given by

$$\nabla(P, Q) = \{(x, x') \mid P(x) \wedge Q(x')\}.$$

  We see that ∇ works on multiple predicates at the same time and returns a predicate that combines the application of these predicates. (In fact, in this case we get the 'double strength' operator dst for the powerset from Exercise 5.2.13.)

- Next consider the 'bank account' functor $F(X) = \mathbb{N} \times X^{\mathbb{N}} \times (X + X)^{\mathbb{N}}$ from Example 6.5.2. The associated $\nabla\colon F(\mathcal{P}(X)) \to \mathcal{P}(F(X))$ takes the form

$$\begin{aligned}\nabla(n, P, Q) = \{(m, f, g) \in F(X) \mid m = n \wedge \forall k \in \mathbb{N}.\, f(k) \in P(k) \wedge \\ \forall x, U.\, g(k) = \kappa_i x \Rightarrow Q(k) = \kappa_i U \wedge x \in U\}.\end{aligned}$$

  With some effort one can recognise within this formulation the four logical operators $\mathsf{bal}\downarrow(-)$, $[\mathsf{dep}(-)]$, $[\mathsf{wdw}_1(-)]$, $[\mathsf{wdw}_1(-)]$ that we used in Example 6.5.2. But certainly, this ∇ is not very convenient or illuminating: just try to formulate the bank requirements using ∇.

Hence this ∇-based modal logic is mostly of theoretical interest. There are ways of translating between modal logic based on predicate lifting and this ∇-logic based on relation lifting; see [323]. In that case it is convenient to use also n-ary coalgebraic modal logics, involving maps $\delta\colon \mathcal{P}(X^n) \to \mathcal{P}(FX)$.

One advantage that is sometimes claimed for this ∇-logic is that it is generic, in the sense that the logical syntax is obtained directly from the functor and does not require a choice of (L, δ) as in Definition 6.5.1. However, this argument is hardly convincing if it leads to such a non-standard syntax. Moreover, having more choice and flexibility can be both useful and convenient – in the presence of good default choices, as offered by Definition 6.5.3 and Lemma 6.5.4. For instance, consider the multiset/bag functor $\mathcal{M}_{\mathbb{N}}$. If we wish for $n, m \in \mathbb{N}$ a modality $\bigcirc_{n,m}$ that selects the outcomes in the interval $[n, m]$, we can do so easily via the predicate-lifting-based coalgebraic modal logic, via the functor $L(Y) = \mathbb{N} \times \mathbb{N} \times Y$, with $\delta\colon L\mathcal{P} \Rightarrow \mathcal{P}\mathcal{M}_{\mathbb{N}}$ given by

$$\delta(n, m, P) = \{\varphi \in \mathcal{M}_{\mathbb{N}}(X) \mid \forall x.\, n \leq \varphi(x) \leq m \Rightarrow P(x)\}.$$

Here we conclude our brief introduction to coalgebraic modal logic. It is one of the more active subfields in coalgebra, involving much more than was covered here, such as proof theory, decidability and extensions like fixed point logic. Refer to [318] for more information and references.

Exercises

6.5.1 Check that coalgebraic temporal logic (see Section 6.4) is a special case of coalgebraic modal logic.

6.5.2 Show that a coalgebraic modal logic $\delta\colon L\mathcal{P} \Rightarrow \mathcal{P}F$ induces a functor $\mathbf{CoAlg}(F)^{\mathrm{op}} \to \mathbf{Alg}(L)$, as claimed in Definition 6.5.1, and that it makes the diagram (6.11) commute.

6.5.3 Consider in Example 6.5.2 the (senseless) action $\mathsf{wdw}(0)$ of withdrawing nothing. According to the requirements given there, does $\mathsf{wdw}(0)$ end up in the left or in the right option in $X + X$? Reformulate the requirements in such a way that $\mathsf{wdw}(0)$ is handled differently, via the other +-option.

6.5.4 Give two different implementations of the bank account coalgebra in Example 6.5.2:

1. One with the natural numbers $\mathbb{N}$ as state space, and $\mathsf{bal} = \mathrm{id}\colon \mathbb{N} \to \mathbb{N}$;
2. A 'history' model with non-empty lists $\mathbb{N}^+$ of natural numbers as states, where $\mathsf{bal} = \mathsf{last}\colon \mathbb{N}^+ \to \mathbb{N}$.

Of course, the requirements in Example 6.5.2 must hold for these implementations.

6.5.5 Recall from Proposition 2.2.3 that each simple polynomial functor F can be written as an arity functor of the form $F_\#(X) = \coprod_{i\in I} X^{\#i}$, for an arity $\#\colon I \to \mathbb{N}$. Show that the modal signature functor for this arity is $L_\#(Y) = \coprod_{i\in I}(\#i) \times X$ and describe the natural transformation $\delta_\#\colon L_\#(\mathcal{P}(X)) \to \mathcal{P}(F_\#(X))$ according to Definition 6.5.3 and Lemma 6.5.4.

6.5.6 Apply the previous exercise to the functor $F(X) = 1 + X + (E \times X)$ used in Section 1.1 for a simplified representation of statements of the Java programming language. What are the associated modal operations, and how are they interpreted in a Java context?

6.5.7
1. Show that the powerset functor $\mathcal{P}\colon \mathbf{Sets}^{\mathrm{op}} \to \mathbf{MSL}$ has a left adjoint $\mathcal{S}$ that sends a meet semilattice D to the set of filters in D: upwards closed subsets $U \subseteq D$ with $\top \in U$ and $x, y \in U \Rightarrow x \wedge y \in U$.
2. Sending a topological space X to its opens $\mathcal{O}(X)$ yields a functor $\mathbf{Sp}^{\mathrm{op}} \to \mathbf{MSL}$ from the opposite of the category $\mathbf{Sp}$ of topological spaces and continuous maps to the category $\mathbf{MSL}$. Show that it also has the filter functor $\mathcal{S}$ as left adjoint, where $\mathcal{S}(D)$ carries the smallest topology that makes the subsets of filters $\eta(a) = \{U \in \mathcal{S}(L) \mid a \in U\}$ open, for $a \in D$.

3. (See also [273]) Similarly show that sending a measurable space to its measurable subsets yields a functor **Meas** $\to$ **MSL**, with the filter functor $\mathcal{S}$ as left adjoint.

6.5.8 This exercise looks at products and coproducts in the category **MSL** of meet semilattices.

1. Check that **MSL** has products $\prod_{i\in I} D_i$ as in **Sets**, with componentwise order.
2. Show that there is an inclusion functor **MSL** $\to$ **CMon** of meet semilattices in commutative monoids and that the category **MSL** has finite biproducts $\oplus$ – just like **CMon** has; see Exercise 2.1.6.
3. For arbitrary, set-indexed products, show that the following construction works:

$$\coprod_{i\in I} D_i = \{\varphi\colon I \to \bigcup_i D_i \mid \forall i.\, \varphi(i) \in D_i \text{ and } \mathrm{supp}(\varphi) \text{ is finite}\},$$

where $\mathrm{supp}(\varphi) = \{i \mid \varphi(i) \neq \top\}$. Top and meet are defined pointwise.

4. For a set A, describe the copower $A \cdot D = \coprod_{a\in A} D$ and power $D^A = \prod_{a\in A} D$ for $D \in$ **MSL** explicitly.

6.5.9 Prove that the various functors $L\colon$ **Sets** $\to$ **Sets** introduced in Definition 6.5.3 can in fact be understood as functors $L\colon$ **MSL** $\to$ **MSL** and that the associated maps $\delta_X\colon L\mathcal{P}(X) \to \mathcal{P}(F(X))$ are maps in **MSL**, i.e. preserve finite conjunctions – except for the neighbourhood functor $\mathcal{N}$.

6.5.10 In the bijective correspondence (6.16), one defines $\overline{\delta}$ as

$$\overline{\delta} \stackrel{\text{def}}{=} \Big(F\mathcal{S} \xRightarrow{\varepsilon F\mathcal{S}} \mathcal{S}\mathrm{Pred}F\mathcal{S} \xRightarrow{\mathcal{S}\delta\mathcal{S}} \mathcal{S}L\mathrm{Pred}\mathcal{S} \xRightarrow{\mathcal{S}L\eta} \mathcal{S}L\Big).$$

Define the correspondence also in the other direction, and prove that these constructions are each other's, inverses.

6.5.11 Check that the map $\overline{\delta}\colon F\mathcal{S} \Rightarrow L\mathcal{S}$ is injective for $F_\#, L_\#, \delta_\#$ from Exercise 6.5.5, and $\mathcal{S}\colon$ **MSL** $\to$ **Sets**^{op} as in Exercise 6.5.7.1.

6.6 Algebras and Terms

At this stage we take a step back and look at the traditional way to handle logical assertions. Such assertions are predicates (or, more generally, relations) on carriers describing restrictions for algebraic (or coalgebraic) operations. This section starts with algebras and first reviews some basic constructions and definitions involving terms. The next section will look at assertions in an algebraic context. Subsequent sections will deal with the coalgebraic situation.

Traditionally in universal algebra, the material at hand is presented using terms and equations between them. Here we quickly move to a more abstract level and use (free) monads, culminating in Theorem 6.6.3, as the main result of this section. But we make a gentle start by first describing free algebras for arity functors $F_{\#}$. The elements of such a free algebra can be described as terms, built up inductively from variables and operations. This yields an explicit description of the free monad $F_{\#}^{*}$ on the functor $F_{\#}$; see Proposition 5.1.8. The term 'description' will be used to introduce equational logic for algebras.

Let $\#\colon I \to \mathbb{N}$ be an arity, as introduced in Definition 2.2.2, with associated endofunctor $F_{\#}(X) = \coprod_{i\in I} X^{\#i}$ on **Sets**. For each $i \in I$ we choose a function symbol, say f_i, and consider it with arity $\#i$. This yields a collection $(\mathsf{f}_i)_{i\in I}$. If V is a set 'of variables', we can form terms in the familiar way: we define the set $\mathcal{T}_{\#}(V)$ to be the least set satisfying the following two requirements:

- $V \subseteq \mathcal{T}_{\#}(V)$
- For each $i \in I$, if $t_1, \ldots t_{\#i} \in \mathcal{T}_{\#}(V)$, then $\mathsf{f}_i(t_1, \ldots, t_{\#i}) \in \mathcal{T}_{\#}(V)$.

The first requirement yields a map $V \to \mathcal{T}_{\#}(V)$. The second requirement provides the set of terms with an algebra structure $F_{\#}(\mathcal{T}_{\#}(V)) \to \mathcal{T}_{\#}(V)$ of the functor $F_{\#}$ associated with the arity, as in (2.18). Together these two maps yield a (cotuple) algebra structure

$$\begin{array}{rcl} V + F_{\#}(\mathcal{T}_{\#}(V)) & \longrightarrow & \mathcal{T}_{\#}(V) \\ v & \longmapsto & v \\ \langle i, (t_1, \ldots, t_{\#i})\rangle & \longmapsto & \mathsf{f}_i(t_1, \ldots, t_{\#i}). \end{array}$$

Thus we have an algebra of the functor $V + F_{\#}(-)$. It turns out that terms $\mathcal{T}_{\#}(V)$ form the initial algebra, and thus yield the free monad on the functor $F_{\#}$, following the characterisation of such monads in Proposition 5.1.8.

Proposition 6.6.1 *The set of terms $\mathcal{T}_{\#}(V)$ built from an arity $\#\colon I \to \mathbb{N}$ and a set of variables V is the free $F_{\#}$-algebra on the set V. The induced monad is the free monad $F_{\#}^{*}$ on $F_{\#}$ with $\mathbf{Alg}(F_{\#}) \cong \mathcal{EM}(F_{\#}^{*})$ by Proposition 5.4.7, as summarised in*

$$\begin{array}{c} \mathbf{Alg}(F_{\#}) \cong \mathcal{EM}(F_{\#}^{*}) \\ \mathcal{T}_{\#}\Big(\dashv\Big\downarrow U \\ \mathbf{Sets} \circlearrowleft F_{\#}^{*} = U\mathcal{T}_{\#} \end{array}$$

In essence this adjunction captures inductively defined compositional semantics, written as interpretation function $[\![-]\!]$: given an arbitrary algebra

$F_\#(X) \to X$, for each 'valuation' function $\rho\colon V \to X$ that sends variables to elements of the algebra's carrier X, there is a unique homomorphism of algebras $[\![-]\!]_\rho\colon \mathcal{T}_\#(V) \to X$ with $[\![-]\!]_\rho \circ \eta_V = \rho$, where the unit $\eta_V\colon V \to \mathcal{T}_\#(V)$ is obtained from the first bullet above.

Proof For an algebra $F_\#(X) \to X$, the interpretation map $[\![-]\!]_\rho\colon \mathcal{T}_\#(V) \to X$ extends the valuation function $\rho\colon V \to X$ from variables to terms, via the (inductive) definition

$$\begin{aligned} [\![v]\!]_\rho &= \rho(v), \qquad \text{for } v \in V \\ [\![\mathsf{f}_i(t_1, \ldots, t_{\#i})]\!]_\rho &= f_i([\![t_1]\!]_\rho, \ldots, [\![t_{\#i}]\!]_\rho), \end{aligned}$$

where the function $f_i\colon X^{\#i} \to X$ is the ith component of the algebra structure $F_\#(X) = \coprod_{j\in I} X^{\#j} \to X$. By construction $[\![-]\!]_\rho$ is a homomorphism of algebras $\mathcal{T}_\#(V) \to X$ such that $V \hookrightarrow \mathcal{T}_\#(V) \to X$ is ρ. Uniqueness is trivial. We have an adjunction because we have established a bijective correspondence:

$$\frac{V \xrightarrow{\;\rho\;} X}{\begin{pmatrix} F_\#(\mathcal{T}_\#(V)) \\ \downarrow \\ \mathcal{T}_\#(V) \end{pmatrix} \xrightarrow[\;[\![-]\!]_\rho\;]{} \begin{pmatrix} F_\#(X) \\ \downarrow \\ X \end{pmatrix}} .$$

□

From the uniqueness of the interpretation maps $[\![-]\!]$ we can easily derive the following properties:

$$\begin{cases} [\![t]\!]_\eta = t \\ h([\![t]\!]_\rho) = [\![t]\!]_{h\circ\rho}, \end{cases} \quad \text{i.e.} \quad \begin{cases} [\![-]\!]_\eta = \mathrm{id} \\ h \circ [\![-]\!]_\rho = [\![-]\!]_{h\circ\rho}, \end{cases} \tag{6.18}$$

where η is the inclusion $V \hookrightarrow \mathcal{T}_\#(V)$ and h is a homomorphism of algebras. (Implicitly, these properties already played a role in Exercise 2.5.17.)

From now on we shall use the free monad notation $F_\#^*$ instead of the terms notation $\mathcal{T}_\#$. More generally, for an arbitrary functor $F\colon \mathbf{Sets} \to \mathbf{Sets}$, we understand $F^*(V)$ as the algebra of terms with operations given by the functor F. These terms can be organised in a category, via a basic categorical construction described in the previous chapter, namely the Kleisli category $\mathcal{K}\ell(-)$. Since terms contain only finitely many variables we restrict the objects to finite sets $n = \{0, 1, \ldots, n-1\}$. We shall write $\mathcal{K}\ell_{\mathbb{N}}(-)$ when such restrictions are applied. One can understand the number/set n as a context with n variables $v_0, \ldots, v_{n-1}$.

Definition 6.6.2 For a monad T on **Sets**, we write $\mathcal{K}\ell_{\mathbb{N}}(T) \hookrightarrow \mathcal{K}\ell(T)$ for the full subcategory with $n \in \mathbb{N}$ as objects. It will be called the **finitary** Kleisli category of T. We write $\mathbf{Law}(T) = \mathcal{K}\ell_{\mathbb{N}}(T)^{\mathrm{op}}$ for the **Lawvere theory** associated with the monad T.

A **model** of this monad is a finite product preserving functor $\mathbf{Law}(T) \to \mathbf{Sets}$. We write $\mathbf{Model}(T) = [\mathbf{Law}(T), \mathbf{Sets}]_{\mathrm{fp}}$ for the category of finite product preserving functors $\mathbf{Law}(T) \to \mathbf{Sets}$, and natural transformations between them.

Lawvere theories were been introduced in [329] (see also [304, 234]) as categories with natural numbers $n \in \mathbb{N}$ as objects and finite products given by $(0, +)$. The main examples are opposites of finitary Kleisli categories, as above. Since these theories involve both the opposite $(-)^{\mathrm{op}}$ and the Kleisli construction $\mathcal{K}\ell_{\mathbb{N}}(-)$ it requires some unravelling to get a handle on their morphisms. Below we argue step by step that, nevertheless, these categories $\mathbf{Law}(T)$ form a natural way of representing terms.

- The category $\mathcal{K}\ell_{\mathbb{N}}(T)$ has coproducts, inherited from the underlying category **Sets**; see Proposition 5.2.2.3. They can simply be described as sums of natural numbers: $n + m \in \mathcal{K}\ell_{\mathbb{N}}(T)$ is the coproduct of objects $n, m \in \mathcal{K}\ell_{\mathbb{N}}(T)$, and $0 \in \mathcal{K}\ell_{\mathbb{N}}(T)$ is the initial object. As a result $(0, +)$ yield products in the opposite category $\mathcal{K}\ell_{\mathbb{N}}(T)^{\mathrm{op}} = \mathbf{Law}(T)$.
- Using that $n = 1 + \cdots + 1$ there are bijective correspondences:

$$\frac{\dfrac{\dfrac{m \longrightarrow n \quad \text{in } \mathbf{Law}(T)}{n \longrightarrow m \quad \text{in } \mathcal{K}\ell_{\mathbb{N}}(T)}}{n \longrightarrow T(m) \quad \text{in } \mathbf{Sets}}}{n \text{ `terms' } t_1, \ldots, t_n \in T(m) \text{ with } m \text{ `free variables'}}$$

  Since free variables form inputs and are usually positioned on the left (of a turnstile $\vdash$ or of an arrow), the direction of arrows in the category $\mathbf{Law}(T)$ is the most natural one for organising terms.

  Speaking of 'terms' and 'free variables' is justified for the free monad $F_\#^*$ on an arity functor $F_\#$; see Proposition 6.6.1. Here we stretch the terminology and use it for an arbitrary monad T on **Sets**.
- The coprojections $\kappa_i \colon 1 \to m$ in **Sets** yield coprojections $\eta \circ \kappa_i \colon 1 \to m$ in $\mathcal{K}\ell_{\mathbb{N}}(T)$; see point (3) in the proof of Proposition 5.2.2. In the category $\mathbf{Law}(T)$ this map $m \to 1$ is a projection, which may be understood as the ith variable v_i which is projected out of a context of m variables $v_1, \ldots, v_m$.
- Kleisli composition corresponds to substitution in terms. This can best be illustrated for a free monad $F_\#^*$ on an arity functor $F_\#$. Assume we have composable maps in the Lawvere theory $\mathbf{Law}(F_\#^*)$:

$$k \xrightarrow{\;s = \langle s_1, \ldots, s_m\rangle\;} m \qquad \text{and} \qquad m \xrightarrow{\;t = \langle t_1, \ldots, t_n\rangle\;} n.$$

That is, we have maps of cotuples:

$$n \xrightarrow{t = [t_1, \ldots, t_n]} F^*_\#(m) \qquad \text{and} \qquad m \xrightarrow{s = [s_1, \ldots, s_m]} F^*_\#(k).$$

The Kleisli composition $s \circledcirc t \colon n \to F^*_\#(k)$ is given by the n-cotuple of maps

$$t_i[s_1/v_1, \ldots, s_m/v_m] \in F^*_\#(k),$$

where we write $v_1, \ldots, v_m$ for the m variables in the terms $t_i \in F^*_\#(m)$. Thus in the Lawvere theory $\mathbf{Law}(F^*_\#)$ we have as composite

$$k \xrightarrow{t \circ s = \langle t_1[\vec{s}/\vec{v}], \ldots, t_n[\vec{s}/\vec{v}]\rangle} n.$$

These terms $t_i[\vec{s}/\vec{v}]$ are the result of substituting s_j for all occurrences of v_j in t_i (if any). They are defined by induction on the structure

$$\begin{aligned} w[\vec{s}/\vec{v}] &= \begin{cases} s_i & \text{if } w = v_i \\ w & \text{otherwise} \end{cases} \\ \mathsf{f}(r_1, \ldots, r_m)[\vec{s}/\vec{v}] &= \mathsf{f}(r_1[\vec{s}/\vec{v}], \ldots, r_m[\vec{s}/\vec{v}]), \end{aligned} \tag{6.19}$$

where f is a function symbol with arity $m = \#\mathsf{f}$.

- Weakening involves moving a term to a bigger context with additional variables (which don't occur in the term). In Kleisli categories this happens via post-composition with a coprojection $\kappa_1 : m \to m + k$, as in

$$n \xrightarrow{t} T(m) \xrightarrow{T(\kappa_1)} T(m + k).$$

That is, weakening in $\mathbf{Law}(T) = \mathcal{K}\ell_{\mathbb{N}}(T)^{\mathrm{op}}$ is described as

$$m + k \xrightarrow{\pi_1} m \xrightarrow{t} n.$$

Similarly, contraction involves replacing multiple occurrences v, v' of variables by a single variable via substitution $[w/v, w/v']$. In Kleisli categories this is done via post composition with a codiagonal $\nabla = [\mathrm{id}, \mathrm{id}]$, and in the associated Lawvere theory via a diagonal $\Delta = \langle \mathrm{id}, \mathrm{id} \rangle$.

- As mentioned, categories with natural numbers as objects and finite products given by sums $(0, +)$ are called **Lawvere theories**; see [329, 304, 386, 233, 234]. A (set-theoretic) model of such a theory is a finite product preserving functor to **Sets**. Here we consider only such models in **Sets**, but the definition of model of a Lawvere theory easily generalises to arbitrary categories with finite products. Understanding theories as categories and models as structure-preserving functors is the essence of Lawvere's functorial semantics.

(Sometimes people use an ‘opposite’ description of Lawvere theories, as categories with natural numbers as objects and finite coproducts (see [111]); in that case the finitary Kleisli categories $\mathcal{K}\ell_{\mathbb{N}}(T)$ are prime examples.)

We have described algebras of a functor or monad as models of certain operations. This model-theoretic aspect is made explicit in the next two (standard) results, connecting algebras and functorial semantics. For more general, enriched, versions, see [393]. The proof of the theorem below is quite long, even if we leave many details to the reader. A non-finitary analogue of this result is described in Exercise 6.6.5.

Theorem 6.6.3 *For a monad T : **Sets** → **Sets** there is a faithful functor from Eilenberg–Moore algebras to models:*

$$\begin{array}{ccc} \mathcal{EM}(T) & \xrightarrow{\quad\mathcal{L}\quad} & \mathbf{Model}(T) = [\mathbf{Law}(T), \mathbf{Sets}]_{\mathrm{fp}} \\ (T(X) \stackrel{\alpha}{\to} X) & \longmapsto & (n \longmapsto X^n). \end{array}$$

This $\mathcal{L}$ is an equivalence if T is finitary functor.

Each category of Eilenberg–Moore algebras (over **Sets**) can thus be embedded in a category of presheaves.

Proof On objects, the functor $\mathcal{L}\colon \mathcal{EM}(T) \to [\mathbf{Law}(T), \mathbf{Sets}]_{\mathrm{fp}}$ is described as follows. Given an Eilenberg–Moore algebra $\alpha\colon T(X) \to X$, we obtain a functor

$$\begin{array}{ccc} \mathbf{Law}(T) & \xrightarrow{\quad\mathcal{L}(X,\alpha)\quad} & \mathbf{Sets} \\ n & \longmapsto & X^n \\ (n \xrightarrow{\langle t_1,\ldots,t_m\rangle} m) & \longmapsto & (X^n \xrightarrow{\langle [\![t_1]\!],\ldots,[\![t_m]\!]\rangle} X^m). \end{array}$$

The interpretation $[\![t]\!]\colon X^n \to X$ of a term $t \in T(n)$ is obtained on $h \in X^n$ as

$$[\![t]\!](h) = \alpha(T(h)(t)) = \alpha \circ T(h) \circ t\colon\ 1 \longrightarrow T(n) \longrightarrow T(X) \longrightarrow X.$$

The identity $n \to n$ in $\mathbf{Law}(T)$ is given by the unit $\eta\colon n \to T(n)$ in $\mathcal{K}\ell_{\mathbb{N}}(T)$, consisting of n terms $\eta(i) \in T(n)$, for $i \in n$. They are interpreted as projections π_i since

$$[\![\eta(i)]\!](h) = \alpha \circ T(h) \circ \eta \circ \kappa_i = \alpha \circ \eta \circ h \circ \kappa_i = h(i) = \pi_i(h).$$

Hence $\mathcal{L}(X,\alpha)$ preserves identities. Similarly, this functor preserves composition in $\mathbf{Law}(T)$. Clearly, the functor $\mathcal{L}(X,\alpha)$ preserves products; i.e. it sends products $(0, +)$ in $\mathbf{Law}(T)$ to products in **Sets**: $\mathcal{L}(X, a)(0) = X^0 \cong 1$ and

$$\mathcal{L}(X,a)(n+m) = X^{n+m} \cong X^n \times X^m = \mathcal{L}(X,a)(n) \times \mathcal{L}(X,a)(m).$$

For a map of algebras $f\colon (T(X) \xrightarrow{\alpha} X) \to (T(Y) \xrightarrow{\beta} Y)$ one obtains a natural transformation $\mathcal{L}(f)\colon \mathcal{L}(X,\alpha) \Rightarrow \mathcal{L}(Y,\beta)$ with components

$$\mathcal{L}(X,a)(n) = X^n \xrightarrow{\quad \mathcal{L}(f)_n = f^n \quad} Y^n = \mathcal{L}(Y,b).$$

This is natural in n: for a map $t = \langle t_1, \ldots, t_m \rangle\colon n \to m$ in $\mathbf{Law}(T)$ one easily checks that there is a commuting diagram:

$$\begin{array}{rcl} X^n & \xrightarrow{\quad f^n \quad} & Y^m \\ {\scriptstyle \langle [\![t_1]\!]^\alpha, \ldots, [\![t_m]\!]^\alpha \rangle}\big\downarrow & & \big\downarrow{\scriptstyle \langle [\![t_1]\!]^\beta, \ldots, [\![t_m]\!]^\beta \rangle.} \\ X^m & \xrightarrow[\quad f^m \quad]{} & Y^m \end{array}$$

Obviously, the mapping $f \mapsto \mathcal{L}(f) = (f^n)_n$ is injective, making $\mathcal{L}$ a faithful functor.

We now assume that T is finitary. Our first goal is to show that the functor $\mathcal{L}\colon \mathcal{EM}(T) \to [\mathbf{Law}(T), \mathbf{Sets}]_{\mathrm{fp}}$ is full: if we have a natural transformation $\sigma\colon \mathcal{L}(X,\alpha) \to \mathcal{L}(Y,\beta)$, then the component at the object 1 yields a map between the two carriers:

$$X = X^1 = \mathcal{L}(X,\alpha)(1) \xrightarrow{\quad \sigma_1 \quad} \mathcal{L}(Y,\beta)(1) = Y^1 = Y.$$

We show in a moment that σ_1 is an algebra map. But first we check that $\mathcal{L}(\sigma_1) = \sigma$. For each $n \in \mathbb{N}$ and $i \in n$ we have a term $\eta(i) \in T(n)$, forming a map $n \to 1$ in $\mathbf{Law}(T)$, with interpretation $[\![\eta(i)]\!] = \pi_i$. Therefore, the following naturality square commutes:

$$\begin{array}{rcl} X^n = \mathcal{L}(X,\alpha)(n) & \xrightarrow{\quad \sigma_n \quad} & \mathcal{L}(Y,\beta)(n) = Y^n \\ {\scriptstyle \pi_i = [\![\eta(i)]\!]^\alpha}\big\downarrow & & \big\downarrow{\scriptstyle [\![\eta(i)]\!]^\beta = \pi_i} \\ X = \mathcal{L}(X,\alpha)(1) & \xrightarrow[\quad \sigma_1 \quad]{} & \mathcal{L}(Y,\beta)(1) = Y \end{array}$$

Hence $\sigma_n = \sigma_1^n = \mathcal{L}(\sigma_1)_n$.

We can now check that σ_1 is an algebra map. For an element $u \in T(X)$ we need to prove $\beta(T(\sigma_1)(u)) = \sigma_1(\alpha(u))$. Since T is finitary, we may assume $u = T(h)(t)$, for some $n \in \mathbb{N}$, $h\colon n \hookrightarrow X$ and $t \in T(n)$. Hence

$$
\begin{aligned}
\beta(T(\sigma_1)(u)) &= \beta(T(\sigma_1)(T(h)(t))) \\
&= \beta(T(\sigma_1 \circ h)(t))) \\
&= [\![t]\!]^{\beta}(\sigma_1 \circ h) \\
&= [\![t]\!]^{\beta}((\sigma_1)^n(h)) \\
&= [\![t]\!]^{\beta}(\sigma_n(h)) && \text{as just shown} \\
&= \sigma_1([\![t]\!]^{\alpha}(h)) && \text{by naturality of } \sigma \\
&= \sigma_1(\alpha(T(h)(t))) \\
&= \sigma_1(\alpha(u)).
\end{aligned}
$$

In order to show that $\mathcal{L}\colon \mathcal{EM}(T) \to [\mathbf{Law}(T), \mathbf{Sets}]_{\mathrm{fp}}$ is an equivalence, it suffices to show that it is 'essentially surjective' (see e.g. [46, prop. 7.25]): for each $M \in [\mathbf{Law}(T), \mathbf{Sets}]_{\mathrm{fp}}$ we need to find an Eilenberg–Moore algebra $\alpha\colon F(X) \to X$ such that $\mathcal{L}(X, \alpha) \cong M$.

Given a finite product preserving $M\colon \mathbf{Law}(T) \to \mathbf{Sets}$, we take $X = M(1) \in \mathbf{Sets}$ as carrier. On this carrier X an algebra structure $\alpha_M\colon T(X) \to X$ can be defined by using (again) that T is finitary. For $u \in T(X)$ there is an $n \in \mathbb{N}$ and $h\colon n \hookrightarrow X$ with $t \in T(n)$ such that $T(h)(t) = u$. This t forms a map $t\colon n \to 1$ in $\mathbf{Law}(T)$. By applying the functor M we get in $\mathbf{Sets}$

$$X^n = M(1)^n \xrightarrow[\cong]{\varphi_n^{-1}} M(n) \xrightarrow{\quad M(t) \quad} M(1) = X,$$

where the product-preservation isomorphism $\varphi_n\colon M(n) \to M(1)^n$ can be described explicitly as $\varphi_n(y)(i) = M(n \xrightarrow{\pi_i} 1)(y)$.

We define an algebra structure $\alpha_M\colon T(X) \to X$ as $\alpha_M(u) = M(t)(\varphi_n^{-1}(h))$. This outcome does not depend on the choice of n, h, t, as long as $T(h)(t) = u$.

We check that $\alpha_M \circ \eta = \mathrm{id}$. Fix $x \in X$ and consider $\eta(x) \in T(X)$. We can take $t = \eta(*) \in T(1)$ and $h = x\colon 1 \to X$ satisfying

$$T(h)(t) \;=\; T(h)(\eta(*)) \;=\; \eta(h(*)) \;=\; \eta(x).$$

Hence we get our required equality:

$$\alpha_M(\eta(x)) \;=\; \alpha_M(T(h)(t)) \;=\; M(t)(\varphi_1^{-1}(h)) \;=\; M(\eta(*))(h) \;=\; M(\mathrm{id})(x) \;=\; x.$$

Similarly one proves $\alpha_M \circ \mu = \alpha_M \circ T(\alpha_M)$.

Finally, in order to show that $\mathcal{L}(X, \alpha_M) \cong M$, we have on objects $n \in \mathbb{N}$:

$$\mathcal{L}(X, \alpha_M)(n) \;=\; X^n \;=\; M(1)^n \;\cong\; M(n).$$

On morphisms one has $[\![t]\!]^{\alpha_M} = M(t) \circ \varphi_n^{-1}\colon X^n \to X$, for $t \in F^*(n)$. ❑

The previous result is stated for monads but can be adapted to endofunctors F, via the associated free monad F^*. In order to do so we use the following result.

Lemma 6.6.4 *Let* $F\colon \mathbf{Sets} \to \mathbf{Sets}$ *be a functor for which the free monad* F^* *on* F *exists. If* F *is finitary, then so is* F^*.

Proof We use initiality of the algebra $\alpha_X\colon X + F(F^*(X)) \xrightarrow{\cong} F^*(X)$, defining the free monad F^*; see Proposition 5.4.7. Fix a set X and define the subset/predicate $i\colon P \hookrightarrow F^*(X)$ as

$$P = \{u \in F^*(X) \mid \exists(n, h, t).\, n \in \mathbb{N}, h \in X^n, t \in F^*(n) \text{ with } F^*(h)(t) = u\}.$$

Our aim is to define an algebra structure $b\colon X+F(P) \to P$ making the inclusion $i\colon P \hookrightarrow F^*(X)$ a map of algebras $b \to \alpha_X$. Then by initiality we also get a map of algebras $\mathrm{int}_b\colon F^*(X) \to W$ with $i \circ \mathrm{int}_b = \mathrm{id}$. This yields $P = F^*(X)$ and makes F^* finitary.

We define the required algebra $b = [b_1, b_2]\colon X + F(P) \to P$ in two steps.

- For $x \in X$ we define $b_1(x) = \eta_X(x) \in F^*(X)$, where $\eta_X = \alpha_X \circ \kappa_1\colon X \to F^*(X)$ is the unit of the monad F^*. This $b_1(x)$ is in the subset P via the triple $(1, x, \eta_1(*))$, where $x\colon 1 \to X$, since by naturality of η:
$$F^*(x)(\eta(*)) = (\eta \circ x)(*) = \eta(x).$$
Moreover, by construction, $i(b_1(x)) = \eta(x) = \alpha(\kappa_1 x)$.
- For an element $v \in F(P)$ we use that the functor F is finitary to get $m \in \mathbb{N}$ with $g\colon m \to P$ and $s \in F(m)$ such that $F(g)(s) = v$. For each $i \in m$ we pick a triple (n_i, h_i, t_i) with $F^*(h_i)(t_i) = g(i) \in F^*(X)$. Next we take $n = n_1 + \cdots + n_m$ and $h = [h_1, \ldots, h_m]\colon n \to X$ and $t = [F^*(\kappa_1) \circ t_1, \ldots F^*(\kappa_m) \circ t_m]\colon m \to F^*(n)$, where $\kappa_i\colon n_i \to n$ is the appropriate insertion/coprojection map. We use the universal map $\theta\colon F \Rightarrow F^*$ from the proof of Proposition 5.1.8 to get $\theta(s) \in F^*(m)$ and then $r = \alpha \circ \kappa_2 \circ F(t) \circ s\colon 1 \to F^*(n)$. This yields a new element $b_2(v) = \alpha(\kappa_2 v) \in F^*(X)$, which is in P via the triple (n, h, r), since
$$\begin{aligned} F^*(h) \circ r &= F^*(h) \circ \alpha \circ \kappa_2 \circ F(t) \circ s \\ &= \alpha \circ \kappa_2 \circ F(F^*(h)) \circ F(t) \circ s \\ &= \alpha \circ \kappa_2 \circ F([F^*(h_1) \circ t_1, \ldots, F^*(h_n) \circ t_n]) \circ s \\ &= \alpha \circ \kappa_2 \circ F(g) \circ s \\ &= b_2(v). \end{aligned}$$
❑

The following is now an easy consequence of Theorem 6.6.3.

Corollary 6.6.5 *Let* $F\colon \mathbf{Sets} \to \mathbf{Sets}$ *be a functor with free monad* F^*. *Then there is a faithful functor from functor algebras to models:*

$$\mathbf{Alg}(F) \xrightarrow{\quad \mathcal{L} \quad} \mathbf{Model}(F^*) = [\mathbf{Law}(F^*), \mathbf{Sets}]_{\mathrm{fp}}.$$

This $\mathcal{L}$ *is an equivalence if the functor* F *is finitary.*

Proof Proposition 5.4.7 describes an isomorphism of categories $\mathbf{Alg}(F) \cong \mathcal{EM}(F^*)$, which immediately gives the first part of the result. For the second part we use the previous lemma. □

With these results in place we are ready to consider assertions relating terms in the next section.

Exercises

6.6.1 Show that the action of the functor $F^*_\# = \mathcal{T}_\#$ from Proposition 6.6.1 on functions f can be described by simultaneous substitution:

$$F_\#(f)(t) = t[f(x_1)/x_1, \ldots, f(x_n)/x_n],$$

if $x_1, \ldots, x_n$ are the free variables in the term t.

6.6.2 Conclude from the fact that each term $t \in F^*_\#(V)$ contains only finitely many (free) variables that the functor/monad $F^*_\#$ is finitary.

6.6.3 Consider the functor $\mathcal{L}\colon \mathbf{Alg}(F) \to [\mathbf{Law}(F^*), \mathbf{Sets}]_{\mathrm{fp}}$ from Corollary 6.6.5. Let $a\colon F(X) \to X$ be an algebra, with associated functor $\mathcal{L}(X, a)\colon \mathbf{Law}(F^*) \to \mathbf{Sets}$. Show that the interpretation $[\![\, t \,]\!]\colon X^m \to X$ of a term $t \in F^*(m)$, considered as a map $m \to 1$ in the Lawvere theory $\mathbf{Law}(F^*)$, is obtained by initiality in

$$\begin{array}{ccc} m + F(F^*(m)) & \dashrightarrow & m + F(X^{X^m}) \\ {\scriptstyle \alpha_m}\big\downarrow{\scriptstyle \cong} & & \big\downarrow{\scriptstyle a_m} \\ F^*(m) & \underset{[\![-]\!]}{\dashrightarrow} & X^{X^m} \end{array}$$

where the algebra $a_m = [a_{m,1}, a_{m,2}]\colon m + F(X^{X^m}) \to X^{X^m}$ on the right-hand side is given by

$$\begin{cases} a_{m,1} = \lambda i \in m.\, \lambda h \in X^m.\, h(i) : m \longrightarrow X^{X^m} \\ a_{m,2} = \Big(F(X^{X^m}) \stackrel{r}{\longrightarrow} F(X)^{X^m} \stackrel{a^{X^m}}{\longrightarrow} X^{X^m}\Big). \end{cases}$$

The r map is the 'exponent version' of strength from Exercise 5.2.16.

6.6.4 The aim of this exercise is to elaborate a concrete instance of Corollary 6.6.5, stating the correspondence between algebras and models. We chose a simple (arity) functor $F(X) = A + X$, for a fixed set A.

1. Check that the free monad F^* on F is of the form $F^*(V) = \mathbb{N} \times (V + A)$. Describe the unit $\eta\colon V \to F^*(V)$, multiplication $\mu\colon F^*F^*(V) \to F^*(V)$ and universal map $\theta\colon F(V) \Rightarrow F^*(V)$ explicitly.

2. Describe morphisms $t: n \to m$ in the categories $\mathcal{K}\ell_{\mathbb{N}}(F^*)$ and $\mathbf{Law}(F^*)$ explicitly. Give concrete descriptions of identity maps and of composition.
3. Describe the two functors $\mathbf{Alg}(F) \leftrightarrows [\mathbf{Law}(F^*), \mathbf{Sets}]_{\mathrm{fp}}$ of the equivalence of Corollary 6.6.5 concretely.

6.6.5 Let $\mathbb{C}$ a category with arbitrary products.

1. Fix an object $A \in \mathbb{C}$. Write C_A: **Sets** $\to$ **Sets** for the functor given by

$$C_A(X) = \mathbb{C}(A^X, A).$$

Show that C_A is a monad, generalising the continuation monad from Example 5.1.3.7.

2. Let T be a monad on **Sets**. Prove that there is a bijective correspondence (up-to-isomorphism) between

$$\begin{array}{c} \mathcal{K}\ell(T)^{\mathrm{op}} \longrightarrow \mathbb{C} \ \text{ preserving arbitrary products} \\ \hline\hline A \in \mathbb{C} \ \text{ with a map of monads } T \Longrightarrow C_A \end{array} .$$

This gives an infinitary version of Theorem 6.6.3, after [221].

6.6.6 Show that each term $t \in F^*(V)$ gives rise to a functor in a commuting triangle:

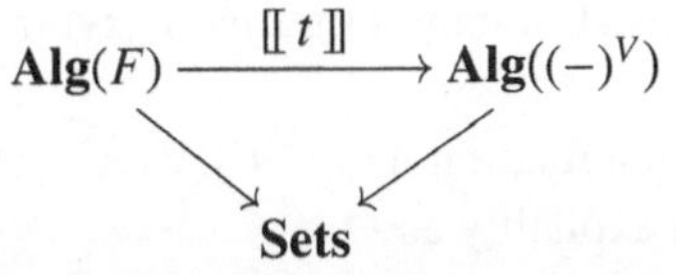

This view on terms is elaborated in [135].

6.7 Algebras and Assertions

This section covers logical assertions in an algebraic context, at first in the form of equations between terms. This material forms the basic theory of (untyped, single-sorted) algebraic specifications and may be found in many places in the literature, such as [463, 126, 465, 398, 340]. Our presentation follows the monad-based approach from the previous section, so that the similarity/duality with the coalgebraic situation in subsequent sections becomes clear. The main result of this section, Theorem 6.7.11, shows how logical assertions give rise to a quotient monad, whose Eilenberg–Moore algebras are models of the assertions. These are standard results in the theory of monads. What is new here is the systematic presentation in terms of relation lifting and quotients (of equivalence relations and of congruence equivalences).

In the end, the most important point is that operations are captured by algebras of *functors* and that operations with assertions require algebras of *monads*. This same distinction applies in the coalgebraic case.

We start with an illustration, giving a formal description of groups. Their arity function # can be seen as a map $\#\colon \{\mathsf{e}, \mathsf{m}, \mathsf{i}\} \to \mathbb{N}$, where

- e is the symbol of the unit element, with arity $\#\mathsf{e} = 0$
- m is used for multiplication, with $\#\mathsf{m} = 2$
- i is the symbol for the inverse operation, whose arity is one: $\#\mathsf{i} = 1$.

An algebra for the arity functor $F_{\#}(X) = 1 + (X + X) + X$ associated with # consists of a set A with a map $1 + (A \times A) + A \to A$, i.e. with interpretations $1 \to A$, $A \times A \to A$ and $A \to A$ of the three function symbols $\mathsf{e}, \mathsf{m}, \mathsf{i}$.

Until now we have talked only about interpretation of the function symbols and not about validity of the familiar group axioms:

$$
\begin{array}{rclcrcl}
\mathsf{m}(\mathsf{e}, v) & = & v & \qquad & \mathsf{m}(\mathsf{i}(v), v) & = & \mathsf{e} \\
\mathsf{m}(v, \mathsf{e}) & = & v & & \mathsf{m}(v, \mathsf{i}(v)) & = & \mathsf{e} \\
\multicolumn{7}{c}{\mathsf{m}(v_1, \mathsf{m}(v_2, v_3)) = \mathsf{m}(\mathsf{m}(v_1, v_2), v_3).}
\end{array} \tag{6.20}
$$

These equations consist of pairs of terms (t_1, t_2) in the free algebra $F_{\#}^{*}(V)$, for a set of variables V.

Such axioms form a relation $Ax_V \subseteq F_{\#}^{*}(V) \times F_{\#}^{*}(V)$ on the carrier of the free algebra on V, given explicitly as

$$
\begin{array}{rcl}
Ax_V & = & \{\langle \mathsf{m}(\mathsf{e}, v), v\rangle \mid v \in V\} \cup \{\langle \mathsf{m}(v, \mathsf{e}), v\rangle \mid v \in V\} \\
& & \cup\ \{\langle \mathsf{m}(\mathsf{i}(v), v), \mathsf{e}\rangle \mid v \in V\} \cup \{\langle \mathsf{m}(v, \mathsf{i}(v)), \mathsf{e}\rangle \mid v \in V\} \\
& & \cup\ \{\langle \mathsf{m}(v_1, \mathsf{m}(v_2, v_3)), \mathsf{m}(\mathsf{m}(v_1, v_2), v_3)\rangle \mid v_1, v_2, v_3 \in V\}.
\end{array} \tag{6.21}
$$

In the computer science literature such a pair $(\#, Ax)$ is usually called an algebraic specification.

Our main focus will be on models of such specifications. But we also wish to use axioms for reasoning and proving results such as uniqueness of inverses:

$$\mathsf{m}(v, w) = \mathsf{e} \Longrightarrow v = \mathsf{i}(w). \tag{6.22}$$

In order to do so we need derivation rules for equations $t_1 = t_2$. In general, assuming an arity # and a set of axioms $Ax_V \subseteq F_{\#}^{*}(V) \times F_{\#}^{*}(V)$, as in (6.20), one standardly uses the following logical rules:

$$
\begin{array}{ll}
(a) & \dfrac{Ax}{t_1 = t_2} \quad (\text{ if } (t_1, t_2) \in Ax_V; \ \text{ but see (6.24) below }) \\[2ex]
(b) & \dfrac{}{t = t} \qquad \dfrac{t_1 = t_2}{t_2 = t_1} \qquad \dfrac{t_1 = t_2 \quad t_2 = t_3}{t_2 = t_3} \\[2ex]
(c) & \dfrac{t_1 = t_1' \quad \cdots \quad t_m = t_m'}{\mathsf{f}(t_1, \ldots t_m) = \mathsf{f}(t_1', \ldots t_m')} \quad (\text{ for a function symbol } \mathsf{f} \text{ of arity } m).
\end{array}
\tag{6.23}
$$

The rules in (b) turn the equality relation = into an equivalence relation. And the rules in (c) make it a

congruence. This can be expressed as: the equality relation = is an algebra of the relation lifting functor $\mathrm{EqRel}(F_\#)\colon \mathrm{EqRel}(\mathbf{Sets}) \to \mathrm{EqRel}(\mathbf{Sets})$, restricted to equivalence relations as in Corollary 4.4.4.

In general an equation $t_1 = t_2$ is said to be derivable from a collection $Ax = (Ax_V)_V$ of relations on terms if there is a derivation tree structured by these rules with $t_1 = t_2$ as conclusion. One then often writes $Ax \vdash t_1 = t_2$. Derivable equations are also called theorems. We write $\mathrm{Th}(Ax) \subseteq F_\#^*(V) \times F_\#^*(V)$ for the relation containing precisely the equations that are derivable from Ax. It is the free congruence equivalence on Ax.

Example 6.7.1 The implication $\mathsf{m}(v, w) = \mathsf{e} \Rightarrow v = \mathsf{i}(w)$ from (6.22) has a formal derivation in the theory of groups: Figure 6.3 shows a derivation tree with the equation $v = \mathsf{i}(w)$ as conclusion, and with $\mathsf{m}(v, w) = \mathsf{e}$ as only assumption. In this tree the term $\mathsf{i}(v)$ is written as $\mathsf{i}v$ in order to save parentheses. The relation $GrAx$ refers to the group axioms from (6.20).

Remark 6.7.2 The meticulous reader may have noticed that we have cheated a bit, namely in the two rightmost occurrences of the 'axiom' rule in Figure 6.3. They use instantiations of axioms. For instance the rightmost rule involves an equation $\mathsf{m}(\mathsf{e}, \mathsf{i}w) = \mathsf{i}w$. Strictly speaking this is not an axiom but a *substitution instance* $\mathsf{m}(\mathsf{e}, v)[\mathsf{i}w/v] = v[\mathsf{i}w/v]$ of the axiom $\mathsf{m}(\mathsf{e}, v) = v$, namely with $\mathsf{i}w$ in place of v. This can be formalised as follows.

The improved 'axiom rule' (1) in (6.23) now reads

$$
(a') \quad \frac{Ax}{t_1[\vec{s}/\vec{v}] = t_2[\vec{s}/\vec{v}]} \quad (\text{for } (t_1, t_2) \in Ax) \tag{6.24}
$$

For convenience we will assume from now on that our sets of axioms are closed under substitutions, so that there is no difference between the rules (a) in (6.23) and (a') in (6.24). Basically this means that the axioms are formulated in a slightly different manner. For groups this would involve replacing the formulation used in (6.21) by

$$
\dfrac{
\dfrac{
\dfrac{\dfrac{GrAx}{\mathsf{m}(v,\mathsf{e}) = v}}{v = \mathsf{m}(v,\mathsf{e})}
\qquad
\dfrac{\dfrac{}{v = v} \quad \dfrac{\dfrac{GrAx}{\mathsf{m}(w,\mathsf{i}w) = \mathsf{e}}}{\mathsf{e} = \mathsf{m}(w,\mathsf{i}w)}}{\mathsf{m}(v,\mathsf{e}) = \mathsf{m}(v,\mathsf{m}(w,\mathsf{i}w))}
}{v = \mathsf{m}(v,\mathsf{m}(w,\mathsf{i}w))}
\qquad
\dfrac{
\dfrac{GrAx}{\mathsf{m}(v,\mathsf{m}(w,\mathsf{i}w)) = \mathsf{m}(\mathsf{m}(v,w),\mathsf{i}w)}
\qquad
\dfrac{\dfrac{\mathsf{m}(v,w) = \mathsf{e} \quad \dfrac{}{\mathsf{i}w = \mathsf{i}w}}{\mathsf{m}(\mathsf{m}(v,w),\mathsf{i}w) = \mathsf{m}(\mathsf{e},\mathsf{i}w)} \quad \dfrac{GrAx}{\mathsf{m}(\mathsf{e},\mathsf{i}w) = \mathsf{i}w}}{\mathsf{m}(\mathsf{m}(v,w),\mathsf{i}w) = \mathsf{i}w}
}{\mathsf{m}(v,\mathsf{m}(w,\mathsf{i}w)) = \mathsf{i}w}
}{v = \mathsf{i}w}
$$

Figure 6.3 Derivation of the implication (6.22) from the group axioms *GrAx*.

$$\begin{aligned} Ax_V = {} & \{\langle \mathsf{m}(\mathsf{e}, t), t\rangle \mid t \in F_\#^*(V)\} \cup \{\langle \mathsf{m}(t, \mathsf{e}), t\rangle \mid t \in F_\#^*(V)\} \\ & \cup \{\langle \mathsf{m}(\mathsf{i}(t), t), \mathsf{e}\rangle \mid t \in F_\#^*(V)\} \cup \{\langle \mathsf{m}(t, \mathsf{i}(t)), \mathsf{e}\rangle \mid t \in F_\#^*(V)\} \\ & \cup \{\langle \mathsf{m}(t_1, \mathsf{m}(t_2, t_3)), \mathsf{m}(\mathsf{m}(t_1, t_2), t_3)\rangle \mid t_1, t_2, t_3 \in F_\#^*(V)\}. \end{aligned}$$

If we assume that axioms Ax are closed under substitution, then also derivable equations are closed under substitution, in the sense that

$$Ax \vdash t_1 = t_2 \implies Ax \vdash t_1[\vec{s}/\vec{v}] = t_2[\vec{s}/\vec{v}]. \tag{6.25}$$

This is obtained by induction on the length of the derivation.

In more categorical fashion one can formulate axioms as a functor. We shall do so for the finitary Kleisli categories $\mathcal{K}\ell_{\mathbb{N}}(T)$ from Definition 6.6.2, because (terms in) axioms involve only finitely many variables. Closure of axioms under substitution is guaranteed, by construction.

Definition 6.7.3 An **axiom system** for a monad T on **Sets** is a functor $\mathcal{A}$ in a commuting triangle

$$\begin{array}{ccc} & & \mathbf{EnRel} \\ & \overset{\mathcal{A}}{\nearrow} & \downarrow \\ \mathcal{K}\ell_{\mathbb{N}}(T) & \xrightarrow[U]{} & \mathbf{Sets} \end{array} \tag{6.26}$$

where **EnRel** is the category with endorelations $R \rightarrowtail X \times X$ on a single carrier as objects, and where $U: \mathcal{K}\ell_{\mathbb{N}}(T) \to \mathbf{Sets}$ is the standard (right adjoint) functor from Proposition 5.2.2.

An axiom system for an *endofunctor* F is an axiom system for the associated free monad F^* and is thus given by a functor $\mathcal{A}: \mathcal{K}\ell_{\mathbb{N}}(F^*) \to \mathbf{EnRel}$ as above.

In a (functorial) model of the monad T, in the form of a finite product preserving functor

$$\mathbf{Law}(T) = \mathcal{K}\ell_{\mathbb{N}}(T)^{\mathrm{op}} \xrightarrow{\;M\;} \mathbf{Sets},$$

the **axioms hold** – or, equivalently, M **satisfies the axioms** – if for each parallel pair of maps

$$n \underset{t = \langle t_1, \ldots, t_m\rangle}{\overset{s = \langle s_1, \ldots, s_m\rangle}{\rightrightarrows}} m$$

in **Law**(T), one has

$$[\forall i \in m.\, (s_i, t_i) \in \mathcal{A}(n)] \implies M(s) = M(t).$$

In that case we write $M \models \mathcal{A}$. This determines a full subcategory

$$\mathbf{Model}(T, \mathcal{A}) \hookrightarrow \mathbf{Model}(T)$$

of models in which the axioms $\mathcal{A}$ hold.

Similarly, for a monad T and functor F, we have full subcategories

$$\mathcal{EM}(T, \mathcal{A}) \hookrightarrow \mathcal{EM}(T) \qquad \text{and} \qquad \mathbf{Alg}(F, \mathcal{A}) \hookrightarrow \mathbf{Alg}(F)$$

of monad and functor algebras satisfying $\mathcal{A}$. This means that the axioms $\mathcal{A}$ hold in the corresponding functorial models, obtained via Theorem 6.6.3 and Corollary 6.6.5.

The functorial description of axioms (6.26) can be unravelled as follows. For each $n \in \mathbb{N}$, considered as object $n \in \mathcal{K}\ell_{\mathbb{N}}(T)$ in the finitary Kleisli category, there is a relation $\mathcal{A}(n) \rightarrowtail T(n) \times T(n)$ on $U(n) = T(n) \in \mathbf{Sets}$, containing the pairs of terms with n variables that are equated by the axioms in $\mathcal{A}$. And for each Kleisli map $f: n \to T(m)$ there is a commuting diagram

$$\begin{array}{ccc} \mathcal{A}(n) & \dashrightarrow & \mathcal{A}(m) \\ \downarrow & & \downarrow \\ T(n) \times T(n) & \xrightarrow[f^\$ \times f^\$]{} & T(m) \times T(m) \end{array}$$

where $U(f) = f^\$ = \mu \circ T(f)$ is the Kleisli extension of f; see Proposition 5.2.3. This guarantees that the relations $\mathcal{A}(n)$ are closed under substitution. If this is too abstract, it may be helpful to elaborate on the details of this closure condition for the special case when T is a free monad $F_\#^*$ on an arity functor.

A slightly different formulation of axiom systems is given in Exercise 6.7.3.

We briefly reformulate validity of axioms in more traditional terms, along the lines of the free construction in Proposition 6.6.1 using terms.

Lemma 6.7.4 *Let # be an arity and $\mathcal{A}: \mathcal{K}\ell_{\mathbb{N}}(F_\#^*) \to$ **EnRel** an axiom system for the associated arity functor $F_\#$. For an algebra $a: F_\#(X) \to X$ the following two statements are equivalent.*

- *The axiom system $\mathcal{A}$ holds in the algebra $a: F_\#(X) \to X$.*
- *For each $n \in \mathbb{N}$ and for each pair of terms $t, s \in F_\#^*(n)$ containing at most n variables, if $(t, s) \in \mathcal{A}(n)$, then $[\![t]\!]_\rho = [\![s]\!]_\rho$ for each valuation function $\rho: n \to X$, with interpretations $[\![-]\!]_\rho$ as defined in the proof of Proposition 6.6.1.*

Diagrammatically this means that each valuation yields a map of relations, from axioms to equality:

$$\begin{array}{ccc}
\mathcal{A}(n) & \dashrightarrow & Eq(X) = X \\
\downarrow & & \downarrow \Delta = \langle \mathrm{id}, \mathrm{id} \rangle \\
F^*_\#(n) \times F^*_\#(n) & \xrightarrow[\llbracket - \rrbracket_\rho \times \llbracket - \rrbracket_\rho]{} & X \times X
\end{array}$$

Proof The functor $\mathcal{L}(X, a)\colon \mathbf{Law}(F^*_\#) \to \mathbf{Sets}$ associated in Corollary 6.6.5 with the algebra sends the terms $t, s \in F_\#(n)$ to functions $\llbracket t \rrbracket, \llbracket s \rrbracket\colon X^n \to X$, as described explicitly in Exercise 6.6.4. Validity of $\mathcal{A}$ in the model $\mathcal{L}(X, a)$ means that $\llbracket t \rrbracket = \llbracket s \rrbracket$. The resulting mapping $\rho \mapsto \llbracket t \rrbracket(\rho)$ yields the adjoint transpose in the setting of Proposition 6.6.1, since $\llbracket \eta(i) \rrbracket(\rho) = \pi_i(\rho) = \rho(i)$. Thus, the (validity) equation $\llbracket t \rrbracket = \llbracket s \rrbracket$ is equivalent to $\llbracket t \rrbracket_\rho = \llbracket t \rrbracket_\rho$ for any valuation $\rho\colon n \to X$. ❑

Example 6.7.5 For the functor $F(X) = 1 + (X \times X) + X$ capturing the group operations and the group axioms *GrAx* described in (6.20) we obtain that the category **Grp** of groups and group homomorphisms can be described as **Grp** = **Alg**(F, GrAx).

We now generalise from free monads $F^*_\#$ on arity functors to arbitrary monads T (on **Sets**) and consider congruences for such monads. Recall that such congruences correspond to algebras of a relation-lifting functor – defined here with respect to the standard set-theoretic logical factorisation system of injections and surjections. For a monad T we use Eilenberg–Moore algebras of the associated lifting $\mathrm{Rel}(T)$ as T-congruences. Recall that $\mathrm{Rel}(T)$ is a monad by Exercise 4.4.6. Using Eilenberg–Moore algebras properly generalises the situation for endofunctors F because $\mathbf{Alg}(\mathrm{Rel}(F)) \cong \mathcal{EM}(\mathrm{Rel}(F^*))$, by Exercise 5.4.18. In the remainder we restrict to liftings to endorelations; more specifically, we consider liftings $\mathrm{EqRel}(T)$ to equivalence relations, as in Corollary 4.4.4.

Lemma 6.7.6 *Let* $T\colon \mathbf{Sets} \to \mathbf{Sets}$ *be a monad, with associated monad*

$$\mathbf{EqRel} \xrightarrow{\mathrm{EqRel}(T)} \mathbf{EqRel}$$

obtained by lifting to equivalence relations. For an axiom system $\mathcal{A}\colon \mathcal{K}\ell_{\mathbb{N}}(T) \to \mathbf{EnRel}$ *we define for each* $n \in \mathbb{N}$

$$\mathrm{Th}(\mathcal{A})(n) = [\,\textit{the free } \mathrm{EqRel}(T)\textit{-algebra on } \mathcal{A}(n)\,].$$

This yields a new axiom system Th($\mathcal{A}$) *of congruence equivalences in*

$$\begin{array}{ccc} & & \mathcal{EM}(\mathrm{EqRel}(T)) \\ & \overset{\mathrm{Th}(\mathcal{A})}{\nearrow} & \downarrow \\ & & \mathbf{EqRel} \circlearrowleft \mathrm{EqRel}(T) \\ & & \downarrow \\ \mathcal{K}\ell_{\mathbb{N}}(T) & \xrightarrow[U]{} & \mathbf{Sets} \circlearrowleft T \end{array}$$

If T is finitary, then for each model $M \in$ $\mathbf{Model}(T) = [\mathbf{Law}(T), \mathbf{Sets}]_{\mathrm{fp}}$ *we have*

$$M \models \mathcal{A} \iff M \models \mathrm{Th}(\mathcal{A}).$$

Proof For each $n \in \mathbb{N}$ there is a monotone function between posets of relations:

$$\mathcal{P}(T(n) \times T(n)) \xrightarrow{\;\overline{\mathcal{A}(n)} \vee \coprod_{\mu\times\mu} \mathrm{EqRel}(T)(-)\;} \mathcal{P}(T(n) \times T(n))$$

where $\overline{\mathcal{A}(n)}$ is the least equivalence relation containing the axioms $\mathcal{A}(n) \subseteq T(n){\times}T(n)$. This function has a least fixed point, by the Knaster–Tarski fixpoint theorem (see e.g. [119, chapter 4]). By Proposition 5.1.8 this is the free algebra $\coprod_{\mu\times\mu} \mathrm{EqRel}(T)(\mathrm{Th}(\mathcal{A}(n))) \subseteq \mathrm{Th}(\mathcal{A})(n)$ on $\overline{\mathcal{A}(n)}$. The inclusion expresses that $\mathrm{Th}(\mathcal{A})(n)$ is a Rel(T)-functor algebra in

$$\begin{array}{ccc} \mathrm{EqRel}(T)(\mathrm{Th}(\mathcal{A})(n)) & \dashrightarrow & \mathrm{Th}(\mathcal{A})(n) \\ \downarrow & & \downarrow \\ T^2(n) \times T^2(n) & \xrightarrow[\mu\times\mu]{} & T(n) \times T(n) \end{array} \qquad (6.27)$$

By Exercise 6.2.6 this relation $\mathrm{Th}(\mathcal{A})(n)$ is automatically an Eilenberg–Moore algebra and thus a T-congruence.

Suppose $M \models \mathcal{A}$. We wish to prove $M \models \mathrm{Th}(\mathcal{A})$. The other direction is trivial since $\mathcal{A}(n) \subseteq \overline{\mathcal{A}(n)} \subseteq \mathrm{Th}(\mathcal{A})(n)$. For each pair of terms $t, s \in \mathcal{A}(n)$, considered as parallel maps $n \rightrightarrows 1$ in $\mathbf{Law}(T)$, we have $M(s) = M(t) \colon M(n) \to M(1)$. The relation

$$R = \{(s,t) \mid s, t \in T(n) \text{ with } M(s) = M(t)\}$$

thus contains $\mathcal{A}(n)$. This R is clearly an equivalence relation, so also $\overline{\mathcal{A}(n)} \subseteq R$. In order to show that R is also a T-congruence, we need to define an Eilenberg–Moore algebra structure $\beta \colon T(R) \to R$. First we name the inclusion explicitly as $\langle r_1, r_2\rangle \colon R \hookrightarrow T(n) \times T(n)$. Since T is finitary, an element $u \in T(R)$ can be written as $u = T(h)(v)$, for some $m \in \mathbb{N}$, $h \colon m \to R$ and $v \in T(m)$. Write

$h(i) = (s_i, t_i) \in R$. Then $M(s_i) = M(t_i)$. Moreover, these terms yield cotuples $[t_1, \ldots, t_m] = r_1 \circ h$ and $[s_1, \ldots, s_m] = r_2 \circ h$, forming functions $m \to T(n)$ and thus tuples $n \to m$ in $\mathbf{Law}(T)$. Since the model M preserves products, we get $M(r_1 \circ h) = M(r_2 \circ h)$, as functions $M(n) \to M(m) \cong M(1)^m$.

Now we can prove that the pair $\langle \mu(T(r_1)(u)), \mu(T(r_2)(u)) \rangle \in T(n) \times T(n)$ is in the relation R:

$$\begin{aligned} M(\mu(T(r_1)(u))) = M(\mu \circ T(r_1) \circ T(h) \circ v) &= M((r_1 \circ h) \circledcirc v) \\ &= M(v) \circ M(r_1 \circ h) \\ &= M(v) \circ M(r_2 \circ h) \\ &= M((r_2 \circ h) \circledcirc v) \\ &= M(\mu(T(r_2)(u))). \end{aligned}$$

We thus get a unique $\beta(u) \in R$ with $\langle r_1, r_2 \rangle(\beta(u)) = \langle \mu(T(r_1)(u)), \mu(T(r_2)(u)) \rangle$. It is easy to check that the resulting function $\beta \colon T(R) \to R$ is an Eilenberg–Moore algebra.

Thus we have that R is a congruence containing $\overline{\mathcal{A}(n)}$. Hence $\mathrm{Th}(\mathcal{A})(n) \subseteq R$. This means $M \models \mathrm{Th}(\mathcal{A})$. ❑

The next lemma is a categorical analogue of Exercise 6.7.2, saying that simultaneous substitution of equal terms yields equal results.

Lemma 6.7.7 *Let T be a monad with axiom system $\mathcal{A}$. Then*

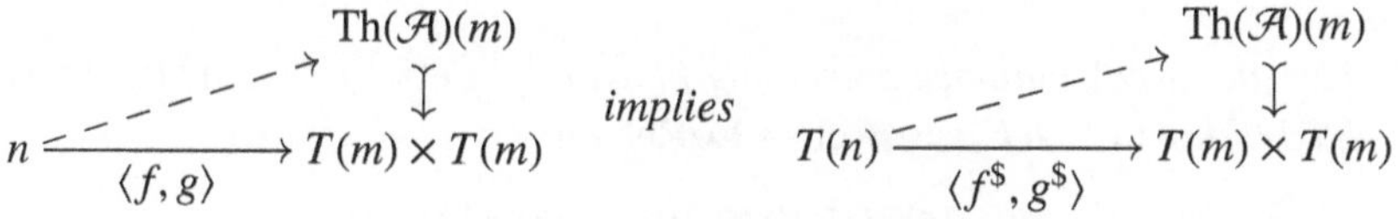

Proof We use that $f^\$ = \mu \circ T(f)$ and that the theory is a congruence. Let $h \colon n \to \mathrm{Th}(\mathcal{A})(m)$ be the dashed map in the above triangle on the left. Then

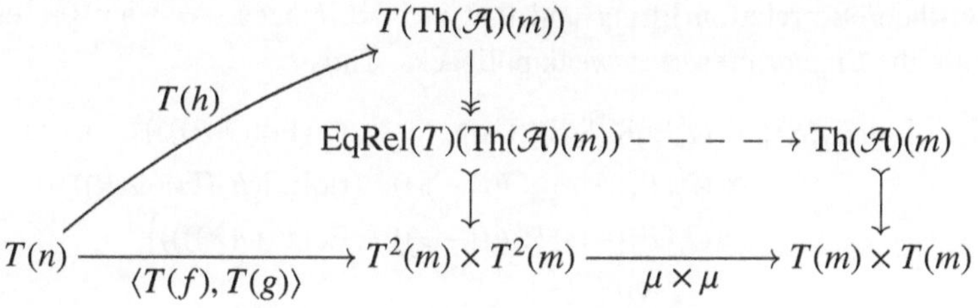

The dashed arrow in this diagram comes from (6.27). ❑

We continue with quotients in **Sets** and first collect some basic results. We use the description of quotients as left adjoint to equality from Definition 4.5.7. Such quotients exist for **Sets** by Exercise 4.5.5.

Lemma 6.7.8 *Consider the quotient functor Q, as left adjoint to equality in*

$$\begin{array}{c}\mathbf{EqRel}\\ Q\left(\dashv\right)\uparrow Eq\\ \mathbf{Sets}\end{array} \qquad \text{so that} \qquad \frac{R \longrightarrow Eq(Y)}{\overline{\overline{Q(R) \longrightarrow Y}}}.$$

This functor Q sends an equivalence relation $R \subseteq X \times X$ to the quotient $Q(R) = X/R$, with canonical (unit) map $[-]_R \colon X \to Q(R)$. This set-theoretic quotient Q satisfies the following three properties.

1. *Each equivalence relation $R \subseteq X \times X$ is the kernel of its quotient map $[-]_R$:*

$$R = \mathrm{Ker}([-]_R) = ([-]_R \times [-]_R)^{-1}(Eq(Q(R))).$$

   More concretely, this means

$$R(x, x') \iff [x]_R = [x']_R.$$

2. *Each surjection $f \colon X \twoheadrightarrow Y$ is (isomorphic to) the quotient map of its kernel* $\mathrm{Ker}(f) \subseteq X \times X$, *as in*

$$\begin{array}{ccccc}\mathrm{Ker}(f) & \rightrightarrows & X & \xrightarrow{[-]} \!\!\!\!\twoheadrightarrow & Q(\mathrm{Ker}(f))\\ & & & \searrow^{f} & \downarrow \cong\\ & & & & Y\end{array}$$

3. *For each weak pullback preserving functor $F \colon \mathbf{Sets} \to \mathbf{Sets}$, with associated lifting* $\mathrm{EqRel}(F) \colon \mathbf{EqRel} \to \mathbf{EqRel}$, *one has*

$$Q(\mathrm{EqRel}(F)(R)) \cong F(Q(R)).$$

Proof The first two points are completely standard, so we concentrate on the third one, using Proposition 4.4.3: it says that equality relations are preserved by set-theoretic relation lifting and that inverse images $(-)^{-1}$ are preserved because the functor preserves weak pullbacks. Thus

$$\begin{array}{rll} Q\big(\mathrm{EqRel}(F)(R)\big) & \cong Q\big(\mathrm{EqRel}(F)(([-]_R \times [-]_R)^{-1}(\mathrm{Eq}(Q(R))))\big) & \text{by (1)}\\ & \cong Q\Big((F([-]_R) \times F([-]_R))^{-1}(\mathrm{EqRel}(F)(\mathrm{Eq}(Q(R))))\Big) & \\ & = Q\Big((F([-]_R) \times F([-]_R))^{-1}(\mathrm{Eq}(F(Q(R))))\Big) & \\ & = Q(\mathrm{Ker}(F([-]_R))) & \\ & \cong F(Q(R)) & \text{by (2).}\end{array}$$

The final step is justified because by the axiom of choice ('each surjection is split') the functor F preserves surjections; see Lemma 2.1.7. ❑

These observations are used to show that quotients lift to algebras.

Proposition 6.7.9 *For a weak pullback preserving functor F: **Sets** $\to$ **Sets** the quotient functor Q lifts to (functor) algebras as on the left below:*

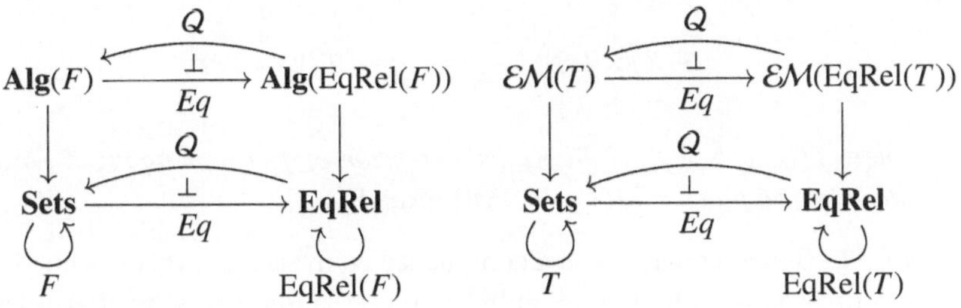

*For a weak pullback preserving monad T: **Sets** $\to$ **Sets** this lifting specialises to (monad) algebras as on the right above.*

Proof The lifted adjunction is a consequence of Theorem 2.5.9, using the isomorphism $FQ \cong Q\mathrm{EqRel}(F)$ from the previous lemma. For the restriction to monad algebras we need to show that the algebra equations remain valid. But this is obvious because the unit and multiplication of T and $\mathrm{EqRel}(T)$ are essentially the same; see Exercise 4.4.6. ❑

It is useful to extend axioms from relations on carriers $T(n)$ for $n \in \mathbb{N}$ to carriers $T(X)$ with arbitrary sets X.

Lemma 6.7.10 *Let T be a monad on **Sets** that is finitary and preserves weak pullbacks, and let $\mathcal{A}$ be an axiom system for T. For an arbitrary set X define a relation $\mathcal{A}_X \subseteq T(X) \times T(X)$ as*

$$\mathcal{A}_X = \bigcup_{n\in\mathbb{N}} \bigcup_{h\in X^n} \big\{\langle T(h)(s), T(h)(t)\rangle \;\big|\; \langle s,t\rangle \in \mathrm{Th}(\mathcal{A})(n)\big\}.$$

1. *These relations $\mathcal{A}_X$ are congruence equivalences.*
2. *For $m \in \mathbb{N}$ one has $\mathcal{A}_m = \mathrm{Th}(\mathcal{A})(m)$.*
3. *The axioms $\mathcal{A}$ hold in an arbitrary Eilenberg–Moore algebra $\alpha\colon T(X) \to X$ iff there is a map of relations*

$$\begin{array}{ccc} \mathcal{A}_X & \dashrightarrow & Eq(X) = X \\ \downarrow & & \downarrow{\scriptstyle \Delta = \langle \mathrm{id}, \mathrm{id}\rangle} \\ T(X)\times T(X) & \xrightarrow[\alpha\times\alpha]{} & X \times X \end{array}$$

4. *For each Kleisli map $f\colon X \to T(Y)$ there is a map of relations*

$$\begin{array}{ccc} \mathcal{A}_X & \dashrightarrow & \mathcal{A}_Y \\ \downarrow & & \downarrow \\ T(X) \times T(X) & \xrightarrow[f^{\$} \times f^{\$}]{} & T(Y) \times T(Y) \end{array}$$

where $f^{\$} = \mu \circ T(f)$ is the Kleisli extension of f. The mapping $X \mapsto \mathcal{A}_X$ thus yields a functor $\mathcal{K}\ell(T) \to \mathcal{EM}(\mathrm{EqRel}(T))$.

Proof 1. Showing that $\mathcal{A}_X$ is a congruence equivalence involves some low-level reasoning, where weak pullback preservation is used for transitivity. Details are left to the interested reader.

2. The inclusion $\mathrm{Th}(\mathcal{A})(m) \subseteq \mathcal{A}_m$ is obvious. For the other direction, assume $\langle u, v\rangle \in \mathcal{A}_m$, say with $u = T(h)(s), v = T(h)(t)$ for $h\colon n \to m$ and $\langle s, t\rangle \in \mathrm{Th}(\mathcal{A})(n)$. Since $T(h) = (\eta \circ h)^{\$}$, functoriality of $\mathrm{Th}(\mathcal{A})$ in Lemma 6.7.6 yields that $T(h)$ is a map of relations $\mathrm{Th}(\mathcal{A})(n) \to \mathrm{Th}(\mathcal{A})(m)$. Then $\langle u, v\rangle = \langle T(h)(s), T(h)(t)\rangle \in \mathrm{Th}(\mathcal{A})(m)$.

3. Validity of $\mathcal{A}$ in an algebra $\alpha\colon T(X) \to X$ means that for each pair $(s, t) \in \mathcal{A}(n)$ one has $[\![s]\!] = [\![t]\!]\colon X^n \to X$, where $[\![s]\!](h) = \alpha(T(h)(s))$. This precisely means for each pair $(u, v) \in \mathcal{A}_X$ one has $\alpha(u) = \alpha(v)$.

4. Assume $f\colon X \to T(Y)$ and $(u_1, u_2) \in \mathcal{A}_X$, say with $u_i = T(h)(s_i)$ for $h \in X^n$ and $(s_1, s_2) \in \mathrm{Th}(\mathcal{A})(n)$. We get $f \circ h \in T(Y)^n$. Since T is finitary we can choose for each $j \in n$ a $g_j \in Y^{m_j}$ and $r_j \in T(m_j)$ with $T(g_j)(r_j) = f(h(j))$. Put $m = m_1 + \cdots + m_n$ and $g = [g_1, \ldots, g_n]\colon m \to Y$, and $r = [T(\kappa_1) \circ r_1, \ldots, T(\kappa_n) \circ r_n]\colon n \to T(m)$, where $\kappa_i\colon m_i \rightarrowtail m$ is the appropriate insertion map. Since axioms are closed under substitution we get $(r^{\$}(s_1), r^{\$}(s_2)) \in \mathrm{Th}(\mathcal{A})(m)$. These elements prove that the pair $(f^{\$}(u_1), f^{\$}(u_2))$ is in $\mathcal{A}_Y$, since

$$\begin{aligned} f^{\$}(u_i) &= \mu \circ T(f) \circ T(h) \circ s_i \\ &= \mu \circ T([T(g_1) \circ r_1, \ldots, T(g_n) \circ r_n]) \circ s_i \\ &= \mu \circ T^2(g) \circ T([T(\kappa_1) \circ r_1, \ldots, T(\kappa_n) \circ r_n]) \circ s_i \\ &= T(g) \circ \mu \circ T(r) \circ s_i \\ &= T(g)(r^{\$}(s_i)). \end{aligned}$$

❑

We now come to the main technical result of this section, showing that axioms can be captured by quotient monads.

Theorem 6.7.11 *Let T:* **Sets** $\to$ **Sets** *be a monad that is finitary and weak pullback preserving, with an axiom system $\mathcal{A}\colon \mathcal{K}\ell_{\mathbb{N}}(T) \to$* **EnRel**. *Then there is a 'quotient' monad $T/\mathcal{A}$:* **Sets** $\to$ **Sets** *with a monad map $[-]\colon T \Rightarrow T/\mathcal{A}$, giving rise to an isomorphism $\mathcal{EM}(T/\mathcal{A}) \xrightarrow{\cong} \mathcal{EM}(T, \mathcal{A})$ in*

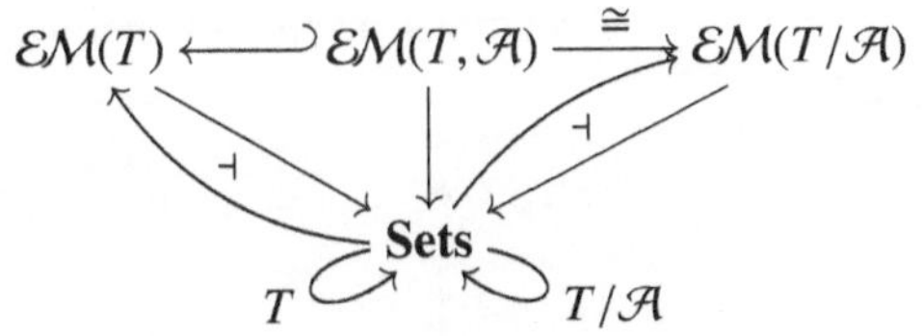

Proof For a set X consider the relation $\mathcal{A}_X \hookrightarrow T(X) \times T(X)$ from the previous lemma, as algebra of the lifted monad $\mathrm{EqRel}(T)$. The lifted quotient functor $Q\colon \mathcal{EM}(\mathrm{EqRel}(F)) \to \mathcal{EM}(T)$ from Proposition 6.7.9 yields a map of Eilenberg–Moore algebras:

$$\begin{array}{ccc} T^2(X) & \xrightarrow{T([-]_{\mathcal{A}_X})} & T(T/\mathcal{A}(X)) \\ {\scriptstyle \mu}\downarrow & & \downarrow{\scriptstyle \xi_X} \\ T(X) & \xrightarrow[{[-]_{\mathcal{A}_X}}]{} & T/\mathcal{A}(X) \stackrel{\text{def}}{=} Q(\mathcal{A}_X) = T(X)/\mathcal{A}_X \end{array}$$

The mapping $X \mapsto T/\mathcal{A}(X)$ is functorial, since $X \mapsto \mathcal{A}_X$ is functorial, so that $[-]\colon T \Rightarrow T/\mathcal{A}$ becomes a natural transformation. Moreover, by construction, the axioms $\mathcal{A}$ hold in the algebra $\xi_X\colon T(T/\mathcal{A}(X)) \to T/\mathcal{A}(X)$, via Lemma 6.7.10: consider $(u, v) \in \mathcal{A}_{T/\mathcal{A}(X)}$, say $u = T(h)(s), v = T(h)(t)$ for $h \in (T/\mathcal{A}(X))^n$ and $(s, t) \in \mathrm{Th}(\mathcal{A})(n)$. We need to show $\xi_X(u) = \xi_X(v)$. We can choose a $g \in T(X)^n$ with $h = [-] \circ g$. Define

$$s' = \mu(T(g)(s)) = g^{\$}(s) \qquad t' = \mu(T(g)(t)) = g^{\$}(t).$$

By Lemma 6.7.10.4 we get $(s', t') \in \mathcal{A}_X$. Hence $[s'] = [t'] \in T/\mathcal{A}(X)$. But now we can finish the validity argument:

$$\begin{aligned} \xi(u) = \xi \circ T(h) \circ s = \xi \circ T([-]) \circ T(g) \circ s &= [-] \circ \mu \circ T(g) \circ s \\ &= [-] \circ g^{\$} \circ s \\ &= [-] \circ g^{\$} \circ t = \cdots = \xi(v). \end{aligned}$$

Next we show that the mapping $X \mapsto \xi_X$ is left adjoint to the forgetful functor $\mathcal{EM}(T, \mathcal{A}) \to$ **Sets**. For an Eilenberg–Moore algebra $\beta\colon T(Y) \to Y$ satisfying $\mathcal{A}$ we have a bijective correspondence

$$\frac{X \xrightarrow{\ \rho\ } Y}{\hline \left(\begin{array}{c} T(T/\mathcal{A}(X)) \\ {\scriptstyle \xi_X}\downarrow \\ T/\mathcal{A}(X) \end{array}\right) \xrightarrow[{[\![-]\!]_\rho}]{} \left(\begin{array}{c} T(Y) \\ \downarrow{\scriptstyle \beta} \\ Y \end{array}\right)} . \tag{6.28}$$

Given a 'valuation' $\rho\colon X \to Y$, we obtain $\beta \circ T(\rho)\colon T(X) \to Y$, forming a map of algebras $\mu_X \to \beta$. By Lemma 6.7.10.4 the function $T(\rho) = (\eta \circ$

$\rho)^{\$}: T(X) \to T(Y)$ is a map of relations on the left below. The map on the right exists because $\beta \models \mathcal{A}$:

$$\begin{array}{ccccc} \mathcal{A}_X & \dashrightarrow & \mathcal{A}_Y & \dashrightarrow & \mathrm{Eq}(Y) = Y \\ \downarrow & & \downarrow & & \downarrow \langle \mathrm{id}, \mathrm{id} \rangle \\ T(X) \times T(X) & \xrightarrow[T(\rho) \times T(\rho)]{} & T(Y) \times T(Y) & \xrightarrow[\beta \times \beta]{} & Y \times Y \end{array}$$

Hence there is a unique map $[\![-]\!]_\rho : T/\mathcal{A}(X) \to Y$ with $[\![-]\!]_\rho \circ [-] = \beta \circ T(\rho)$.

The monad arising from this adjunction $\mathcal{EM}(T, \mathcal{A}) \rightleftarrows$ **Sets** is what we call the quotient monad $T/\mathcal{A}$: **Sets** $\to$ **Sets**. We have as unit $\eta^{T/\mathcal{A}} = [-] \circ \eta^T$ and as multiplication $\mu^{T/\mathcal{A}} = [\![-]\!]_{\mathrm{id}} : (T/\mathcal{A})^2(X) \to T/\mathcal{A}(X)$; this is the map associated via the correspondence (6.28) with the identity map on $T/\mathcal{A}(X)$ as valuation. Thus, $\mu^{T/\mathcal{A}}$ is the unique map with $\mu^{T/\mathcal{A}} \circ [-] = \xi_X$. This allows us to show that $[-]: T \Rightarrow T/\mathcal{A}$ is a map of monads (see Definition 5.1.7):

$$\mu^{T/\mathcal{A}} \circ [-] \circ T([-]) = \xi \circ T([-]) = [-] \circ \mu^T.$$

Our next step is to prove that the comparison functor $K: \mathcal{EM}(T, \mathcal{A}) \to \mathcal{EM}(T/\mathcal{A})$ – see Exercise 5.4.19 – is an isomorphism. This functor K sends an algebra $\beta: T(Y) \to Y$ satisfying $\mathcal{A}$ to the interpretation $[\![-]\!]_{\mathrm{id}} : T/\mathcal{A}(Y) \to Y$ arising from the identity $Y \to Y$ as valuation. This $K(\beta)$ is thus the unique map with $K(\beta) \circ [-] = \beta$.

In the other direction, given an Eilenberg–Moore algebra $\alpha: T/\mathcal{A}(X) \to X$ there is a map $K^{-1}(\alpha) = \alpha \circ [-]: T(X) \to X$, which is an algebra because $[-]: T \Rightarrow T/\mathcal{A}$ is a map of monads. We have $K^{-1}(\alpha) \models \mathcal{A}$ since

$$\begin{array}{ccccc} \mathcal{A}_X & \dashrightarrow & T/\mathcal{A}(X) & \dashrightarrow & X \\ \downarrow & & \downarrow \langle \mathrm{id}, \mathrm{id} \rangle & & \downarrow \langle \mathrm{id}, \mathrm{id} \rangle \\ T(X) \times T(X) & \xrightarrow[{[-] \times [-]}]{} & T/\mathcal{A}(X) \times T/\mathcal{A}(X) & \xrightarrow[\alpha \times \alpha]{} & X \times X \end{array}$$

This functor K is an isomorphism, since

$$K^{-1}(K(\beta)) = [\![-]\!]_{\mathrm{id}} \circ [-] = \beta \circ T(\mathrm{id}) = \beta.$$

The equation $K(K^{-1}(\alpha)) = \alpha$ follows directly from the definition of K. ❑

Before describing things more abstractly we consider terms from arities again, together with axioms. Quotienting terms by axioms gives another free construction, analogously to Proposition 6.6.1.

Corollary 6.7.12 *Let # be an arity with axioms $\mathcal{A}$. Then there is a finitary monad $T\colon$ **Sets** $\rightarrow$ **Sets** with a natural transformation $F_\# \Rightarrow T$, such that $F_\#$-algebras satisfying $\mathcal{A}$ correspond to Eilenberg–Moore T-algebras, as in*

$$\begin{array}{c}\mathbf{Alg}(F_\#, \mathcal{A}) \cong \mathcal{EM}(T) \\ \uparrow\dashv\downarrow \\ \mathbf{Sets} \circlearrowleft T\end{array}$$

Proof The free monad $F^*_\#$ is finitary by Exercise 6.6.2, and it preserves weak pullbacks by Exercise 6.7.4. Hence we can take T to be the quotient monad $F^*_\#/\mathcal{A}$ from Theorem 6.7.11, giving us a natural transformation $F_\# \Rightarrow F^*_\# \Rightarrow F^*_\#/\mathcal{A} = T$ and an isomorphism $\mathbf{Alg}(F_\#, \mathcal{A}) \cong \mathcal{EM}(F^*_\#, \mathcal{A}) \cong \mathcal{EM}(T)$. The monad T is finitary because $F^*_\#$ is finitary and the map $F^*_\# \Rightarrow T$ consists of surjective quotient maps $[-]$. □

The monad T in this corollary can be defined explicitly on a finite set $n \in \mathbb{N}$ as the quotient

$$T(n) = F^*_\#(n)/\mathrm{Th}(\mathcal{A})(n).$$

This holds since $\mathcal{A}_n = \mathrm{Th}(\mathcal{A})(n)$ by Lemma 6.7.10.2.

For a term $[t] \in T(n)$ and a valuation $\rho\colon n \to T(m)$ with $\rho(i) = [s_i]$ we get

$$[\![\, t \,]\!]_\rho = [\, t[s_1/v_1, \ldots, s_n/v_n] \,]_{\mathrm{Th}(\mathcal{A})}, \qquad (6.29)$$

where $v_1, \ldots, v_n$ are used as variables in t. It is used in the following basic result.

Theorem 6.7.13 (Soundness and completeness) *Consider an arity # with axioms $\mathcal{A}$. For two terms $t_1, t_2 \in F^*_\#(n)$ the following are equivalent.*

1. *$\mathcal{A} \vdash t_1 = t_2$*
2. *For each model $M \in$ **Model**$(F_\#, \mathcal{A})$, one has $M(t_1) = M(t_2)$*
3. *For each algebra $F_\#(X) \xrightarrow{a} X$ in **Alg**$(F_\#, \mathcal{A})$, one has $[\![\, t_1 \,]\!]_\rho = [\![\, t_2 \,]\!]_\rho$, for every $\rho\colon n \to X$.*

Proof The implication (1) $\Rightarrow$ (2) is soundness and follows from the implication $M \models \mathcal{A} \Rightarrow M \models \mathrm{Th}(\mathcal{A})$ from Lemma 6.7.6. The implication (2) $\Rightarrow$ (3) is based on Corollary 6.6.5 (and Exercise 6.6.4), relating algebras and models. For the implication (3) $\Rightarrow$ (1) we choose the quotient $T(n) = F^*_\#(n)/\mathrm{Th}(\mathcal{A})(n)$ from Corollary 6.7.12, with algebra structure

$$F_\#(T(n)) \xrightarrow{\ \theta\ } F^*_\#(T(n)) \xrightarrow{\ \xi_n\ } T(n),$$

as in the proof of Theorem 6.7.11. We now take the monad's unit as valuation $\eta\colon n \to T(n)$. It yields $[\![t_1]\!]_\eta = [\![t_2]\!]_\eta$, by assumption. This means $[t_1] = [t_2]$, by (6.29), and thus $(t_1, t_2) \in \mathrm{Th}(\mathcal{A})(n)$. The latter says $\mathcal{A} \vdash t_1 = t_2$, as required. □

In Corollary 6.7.12 we have seen that an arity with equations can be described by a finitary monad. We conclude this section by showing that each finitary monad is essentially given by an arity with equations.

Recall from Lemma 4.7.3 that for a functor $F\colon \mathbf{Sets} \to \mathbf{Sets}$ there is an arity $\#_F = \pi_1\colon (\coprod_{i\in\mathbb{N}} F(i)) \to \mathbb{N}$, together with a natural transformation $\mathrm{ap}\colon F_{\#_F} \Rightarrow F$. This natural transformation has components $\mathrm{ap}_n\colon F(n) \times X^n \to F(X)$ given by $\mathrm{ap}_n(u, h) = F(h)(u)$. Lemma 4.7.3 states that these components are surjective if and only if F is finitary.

If we apply these constructions to a monad $T\colon \mathbf{Sets} \to \mathbf{Sets}$ we get an arity $\#_T$ and a natural transformation $\mathrm{ap}\colon F_{\#_T} \Rightarrow T$. The free monad $F^*_{\#_T}$ on $F_{\#_T}$ thus yields a map of monads $\overline{\mathrm{ap}}\colon F^*_{\#_T} \Rightarrow T$ with $\overline{\mathrm{ap}} \circ \theta = \mathrm{ap}$; see Proposition 5.1.8. It can be described also via initiality, as in

$$\begin{array}{ccc} X + F_{\#_T}(F^*_{\#_T}(X)) & \dashrightarrow & X + F_{\#_T}(T(X)) \\ \alpha_X \downarrow \cong & & \downarrow [\eta, \mu \circ \mathrm{ap}] \\ F^*_{\#_T}(X) & \underset{\overline{\mathrm{ap}}_X}{\dashrightarrow} & T(X) \end{array} \qquad (6.30)$$

Proposition 6.7.14 *For a monad T on* **Sets** *with monad map $\overline{\mathrm{ap}}\colon F^*_{\#_T} \Rightarrow T$ from* (6.30)*, we define the* ***theory of*** *T as the axiom system* $\mathrm{Th}(T)\colon \mathcal{K}\ell_{\mathbb{N}}(F^*_{\#_T}) \to$ **EqRel** *defined as follows. For $n \in \mathbb{N}$,*

$$\mathrm{Th}(T)(n) = \mathrm{Ker}(\overline{\mathrm{ap}}_n) = \{(t_1, t_2) \in F^*_{\#_T}(n) \times F^*_{\#_T}(n) \mid \overline{\mathrm{ap}}_n(t_1) = \overline{\mathrm{ap}}_n(t_2)\}.$$

Then

1. *The relations* $\mathrm{Th}(T)(n)$ *are congruence equivalences (and hence theories).*
2. *Assuming T is finitary, the maps $\overline{\mathrm{ap}}_X\colon F^*_{\#_T}(X) \Rightarrow T(X)$ are surjective.*

Proof 1. For convenience we shall write F for $F_{\#_T}$. We first need to check that the mapping $n \mapsto \mathrm{Th}(T)(n)$ yields a functor $\mathcal{K}\ell_{\mathbb{N}}(F^*) \to$ **EqRel**. The relations $\mathrm{Th}(T)(n)$ are obviously equivalence relations. We check functoriality: for a map $f\colon n \to F^*(m)$ we need to show $\langle t_1, t_2\rangle \in \mathrm{Th}(T)(n) \Rightarrow \langle f^\$(t_1), f^\$(t_2)\rangle \in \mathrm{Th}(T)(m)$, where $f^\$ = \mu^{F^*} \circ F^*(f)$ is the Kleisli extension of f. We have

$$\begin{aligned}
\overline{\mathrm{ap}}_m(f^{\$}(t_1)) &= (\overline{\mathrm{ap}}_m \circ \mu^{F^*} \circ F^*(f))(t_1) \\
&= (\mu^T \circ T(\overline{\mathrm{ap}}_m) \circ \overline{\mathrm{ap}}_m \circ F^*(f))(t_1) \quad \overline{\mathrm{ap}} \text{ is a map of monads} \\
&= (\mu^T \circ T(\overline{\mathrm{ap}}_m) \circ T(f) \circ \overline{\mathrm{ap}}_n)(t_1) \\
&= (\mu^T \circ T(\overline{\mathrm{ap}}_m) \circ T(f) \circ \overline{\mathrm{ap}}_n)(t_2) \quad \text{since } \langle t_1, t_2\rangle \in \mathrm{Th}(T)(n) \\
&= \cdots \\
&= \overline{\mathrm{ap}}_m(f^{\$}(t_2)).
\end{aligned}$$

Next we need to check that $\mathrm{Th}(T)(n)$ is a congruence. We do so by defining an algebra $b\colon F(\mathrm{Th}(T)(n)) \to \mathrm{Th}(T)(n)$. So assume we have a triple $\langle m \in \mathbb{N}, t \in T(m), h \in \mathrm{Th}(T)(n)^m\rangle \in F(\mathrm{Th}(T)(n))$, using that F is the arity functor $F_{\#_T}$. We can consider h as a tuple $h = \langle h_1, h_2\rangle$ of maps $h_i\colon m \to F^*(n)$. Hence we get two elements:

$$\langle m \in \mathbb{N}, t \in T(m), h_i \in F^*(n)^m\rangle \in F(F^*(n)).$$

By applying the second component of the initial algebra $\alpha_n\colon n+F(F^*(n)) \xrightarrow{\cong} F^*(n)$ we can define

$$b(m,t,h) = \langle \alpha_n(\kappa_2(m,t,h_1)),\ \alpha_n(\kappa_2(m,t,h_2))\rangle.$$

Using Diagram (6.30) one can show $b(m,t,h) \in \mathrm{Th}(T)(n)$.

2. Assuming the monad T is finitary, the maps $\overline{\mathrm{ap}}\colon F^*(X) \Rightarrow T(X)$ are surjective: for $u \in T(X)$ we can find $n \in \mathbb{N}, h \in X^n$ and $t \in T(n)$ with $T(h)(t) = u$. The triple $\langle n, t, h\rangle$ is then an element of $F_{\#_T}(X)$, which we simply write as $F(X)$. Using the free extension map $\theta\colon F \Rightarrow F^*$ from Proposition 5.1.8 we get $\theta(n,t,h) \in F^*(X)$. It proves surjectivity of $\overline{\mathrm{ap}}$:

$$\begin{aligned}
\overline{\mathrm{ap}}(\theta(n,t,h)) &= \mathrm{ap}(n,t,h) \quad \text{by definition of } \overline{\mathrm{ap}} \\
&= T(h)(t) \quad \text{by definition of ap; see Lemma 4.7.3} \\
&= u.
\end{aligned}$$

□

We now obtain a standard result in the theory of monads see e.g. [344, 340]. The restriction to finitary monads can be avoided if one allows operations with arbitrary arities – and not just finite arities as we use here.

Corollary 6.7.15 *Each finitary monad T can be described via operations and equations, namely: via the arity $\#_T$ and theory* $\mathrm{Th}(T)$ *from the previous result one has*

$$T \cong F^*_{\#_T}/\mathrm{Th}(T),$$

using the quotient monad of Theorem 6.7.11.

Proof Since both $F^*_{\#_T}$ and T are finitary, it suffices to prove the isomorphism for finite sets $n \in \mathbb{N}$. We know the maps $\overline{\mathrm{ap}}\colon F^*_{\#_T}(n) \to T(n)$ are surjective. Hence, by Proposition 6.7.8.2, $T(n)$ is isomorphic to the quotient $F^*_{\#_T}(n)/\mathrm{Th}(T)(n)$, via the kernel

$$\mathrm{Ker}(\overline{\mathrm{ap}}) = \{(s,t) \in F^*_{\#_T}(n) \times F^*_{\#_T}(n) \mid \overline{\mathrm{ap}}(s) = \overline{\mathrm{ap}}(t)\} = \mathrm{Th}(T)(n). \qquad \square$$

Remark 6.7.16 A more abstract approach may be found in [334]. There, equations for a monad T are given by an endofunctor E together with two natural transformations $\tau_1, \tau_2 \colon E \Rightarrow T$. Terms $t, s \in T(X)$ are then related if they are in the image of the tuple $\langle \tau_1, \tau_2 \rangle \colon E(X) \to T(X) \times T(X)$. The free monad E^* yields two maps of monads $\overline{\tau_1}, \overline{\tau_2} \colon E^* \Rightarrow T$, as in Proposition 5.1.8. Next, a quotient monad T/E is obtained by taking the coequaliser of $\overline{\tau_1}, \overline{\tau_2}$ *in the category of monads.*

The approach followed here, especially in Theorem 6.7.11, is a bit more concrete. Moreover, the notion of equation that we use, in Definition 6.7.3, has closure under substitution built in.

Exercises

6.7.1 Let $a \colon F(X) \to X$ and $b \colon F(Y) \to Y$ be two algebras with a surjective homomorphism $X \twoheadrightarrow Y$ between them. Use (6.18) to prove

$$a \models \mathcal{A} \Longrightarrow b \models \mathcal{A}.$$

6.7.2 Let $\mathcal{A}$ be an axiom system for an arity functor $F_\#$. Prove

$$\mathcal{A} \vdash s_i = r_i \Longrightarrow \mathcal{A} \vdash t[\vec{s}/\vec{v}] = t[\vec{r}/\vec{v}].$$

6.7.3 For a monad T on **Sets** consider the category $\mathbf{EnRel}_{\mathbb{N}}(T)$ obtained by pullback in

$$\begin{array}{ccc} \mathbf{EnRel}_{\mathbb{N}}(T) & \longrightarrow & \mathbf{EnRel} \\ \downarrow \lrcorner & & \downarrow \\ \mathcal{K}\ell_{\mathbb{N}}(T) & \longrightarrow & \mathbf{Sets} \end{array}$$

1. Give an explicit description of this category $\mathbf{EnRel}_{\mathbb{N}}(T)$.
2. Check that an axiom system as introduced in Definition 6.7.3 corresponds to a section of the forgetful functor in

$$\begin{array}{c} \mathbf{EnRel}_{\mathbb{N}}(T) \\ \mathcal{A}\big(\ \big\downarrow \\ \mathcal{K}\ell_{\mathbb{N}}(T) \end{array}$$

6.7.4 Consider an arity $\# \colon I \to \mathbb{N}$ with associated arity functor $F_\#$ and free monad $F^*_\#$, sending a set V to the set of terms $F^*_\#(V) = \mathcal{T}_\#(V)$, as in Proposition 6.6.1. Prove that $F^*_\#\colon \mathbf{Sets} \to \mathbf{Sets}$ preserves weak pullbacks. *Hint*: Recall Exercise 6.6.1.

6.8 Coalgebras and Assertions

This section returns to the study of coalgebras, in particular, coalgebras with assertions. It follows the main lines of the previous two sections on algebras and assertions, using predicates instead of relations as axioms.

First we look into cofree coalgebras for arity functors. They always exist, by Proposition 2.5.3, and are built from suitable observations. The explicit construction is a bit more complicated than for free algebras.

For an arity $\#\colon I \to \mathbb{N}$ and a set C we shall consider infinite, finitely branching trees of the form

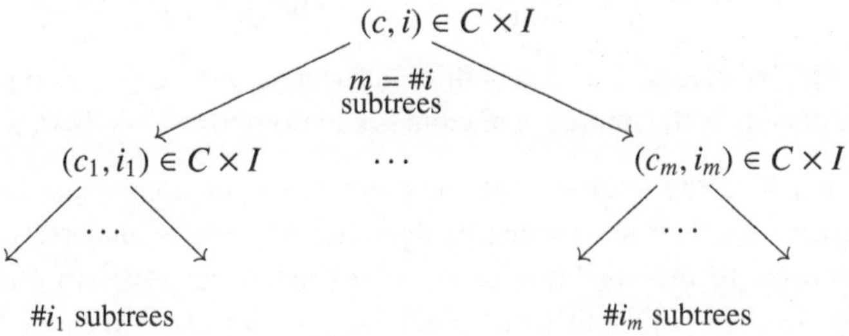

(6.31)

More formally, we describe the set of these trees as

$$O_\#(C) = \Big\{\varphi\colon \mathbb{N} \to (C \times I)^\star \;\Big|\; |\varphi(0)| = 1 \wedge \forall n.\, |\varphi(n+1)| = \#\varphi(n)\Big\}, \qquad (6.32)$$

where $|\sigma| \in \mathbb{N}$ denotes the length of a list σ, and

$$\#\langle (c_1, i_1), \ldots, (c_n, i_n)\rangle = \#i_1 + \cdots + \#i_n.$$

The elements of the set C are sometimes called colours, because they occur as labels throughout the trees in $O_\#(C)$. They give rise to the following dual version of Proposition 6.6.1; see also [29].

Proposition 6.8.1 *For an arity $\#\colon I \to \mathbb{N}$ and a set C, the above set of trees $O_\#(C)$ describes the cofree coalgebra on C for the arity functor $F_\#$. This gives rise to a right adjoint, as below, where the induced comonad is the cofree comonad $F^\infty_\#$ on the arity functor $F_\#$, with isomorphism $\mathbf{CoAlg}(F_\#) \cong \mathcal{EM}(F^\infty_\#)$ from Proposition 5.4.7:*

$$\begin{array}{c}\mathcal{EM}(F_\#^\infty) \cong \mathbf{CoAlg}(F_\#) \\ U \big\downarrow \dashv \big\uparrow O_\# \\ F_\#^\infty = UO_\# \circlearrowright \mathbf{Sets}\end{array}$$

Proof The map $\varepsilon_C \colon O_\#(C) \to C$ takes the C-element at the root of the tree, as in

$$\varepsilon_C(\varphi) = \pi_1\varphi(0) \in C.$$

The coalgebra $\zeta_C \colon O_\#(C) \to \coprod_{i\in I} O_\#(C)^{\#i}$ sends a tree φ to $i = \pi_2\varphi(0) \in I$ with its $m = \#i$ subtrees $\varphi_1, \ldots, \varphi_m$, as described in (6.31).

If we have an arbitrary coalgebra $d \colon X \to F_\#(X)$ with a function $f \colon X \to C$, there is a map of coalgebras $\mathrm{beh}_d \colon X \to O_\#(C)$ given as

$$\mathrm{beh}_d(x) = \left(\begin{array}{ccc} & (f(x), i) & \\ \swarrow & \cdots & \searrow \\ \mathrm{beh}_d(x_1) & & \mathrm{beh}_d(x_m)\end{array}\right)$$

where $d(x) = (i, \langle x_1, \ldots, x_m\rangle)$ with $m = \#i \in \mathbb{N}$. Clearly, $\varepsilon_C \circ \mathrm{beh}_d = f$. It is not hard to see that beh_d is the unique coalgebra homomorphism $X \to O_\#(C)$. ❑

The next result is the analogue of Theorem 6.6.3 for comonads, in the style of Exercise 6.6.5. It shows that coalgebras can also be understood as functorial models. In the algebraic case we restricted ourselves to *finitary* Kleisli categories $\mathcal{K}\ell_{\mathbb{N}}(T)$, with natural numbers as objects. This is natural because terms in an algebraic context typically involve only finitely many variables. In a coalgebraic context the corresponding restriction to only finitely many observable outcomes is less natural. Therefore we use ordinary Kleisli categories in the coalgebraic setting. A consequence is that we do not get exactly the same result as Theorem 6.6.3.

Theorem 6.8.2 *For a comonad S : **Sets** $\to$ **Sets** there is a full and faithful functor from Eilenberg–Moore coalgebras to models:*

$$\begin{array}{ccc}\mathcal{EM}(S)^{\mathrm{op}} & \xrightarrow{\quad\mathcal{L}\quad} & [\mathcal{K}\ell(S), \mathbf{Sets}]_{\mathrm{fp}} \\ (X \xrightarrow{\gamma} S(X)) & \longmapsto & (U \longmapsto U^X).\end{array}$$

Proof The action $\mathcal{L}(X,\gamma)(U) = U^X$ is functorial in U, since each Kleisli map $f \colon S(U) \to V$ yields a map $\mathcal{L}(X,\gamma)(f) \colon U^X \to V^X$ given by abstraction and strength as

$$\Lambda\Big(U^X \times X \xrightarrow{\ \mathrm{id}\times\gamma\ } U^X \times S(X) \xrightarrow{\ \mathrm{st}'\ } S(U^X \times X) \xrightarrow{\ S(\mathrm{ev})\ } S(U) \xrightarrow{\ f\ } V\Big),$$

where the swapped strength map st′ is as in (5.12). Explicitly,

$$\mathcal{L}(X,\gamma)(f) = \lambda g \in U^X.\, \lambda x \in X.\, f(S(g)(\gamma(x))). \tag{6.33}$$

We show that (Kleisli) identities are preserved, using that comonads on **Sets** are strong, (see Exercise 5.2.14); preservation of composition is proven similarly and is left to the reader.

$$\begin{aligned}\mathcal{L}(X,\gamma)(\mathrm{id}) &= \Lambda(\varepsilon \circ S(\mathrm{ev}) \circ \mathrm{st}' \circ (\mathrm{id} \times \gamma)) \\ &= \Lambda(\mathrm{ev} \circ \varepsilon \circ \mathrm{st}' \circ (\mathrm{id} \times \gamma)) \\ &= \Lambda(\mathrm{ev} \circ (\mathrm{id} \times \varepsilon) \circ (\mathrm{id} \times \gamma)) \\ &= \Lambda(\mathrm{ev}) \\ &= \mathrm{id}.\end{aligned}$$

This $\mathcal{L}$ is also functorial: for a map of coalgebras $(X \xrightarrow{\gamma} S(X)) \xrightarrow{h} (Y \xrightarrow{\beta} S(Y))$ we get a natural transformation $\mathcal{L}(h)\colon \mathcal{L}(Y,\beta) \Rightarrow \mathcal{L}(X,\gamma)$ with components $\mathcal{L}(h)_U = U^h = (-) \circ h\colon U^Y \to U^X$.

Clearly, this yields a faithful functor: if $U^h = U^k\colon U^Y \to U^X$ for each U, then by taking $U = Y$ and precomposing with the identity on Y we get $h = Y^h(\mathrm{id}_Y) = Y^k(\mathrm{id}_Y) = k$. For fullness we have to do more work. Let $\sigma\colon \mathcal{L}(Y,\beta) \Rightarrow \mathcal{L}(X,\gamma)$ be a natural transformation. Applying the component $\sigma_Y\colon Y^Y \to Y^X$ at Y to the identity function yields a map $h = \sigma_Y(\mathrm{id}_Y)\colon X \to Y$. We must show two things: h is a map of coalgebras, and $\sigma_U = U^h\colon U^Y \to U^X$. We start with the latter.

- First we notice that for a function $f\colon U \to V$ we have

$$\mathcal{L}(X,\gamma)(f \circ \varepsilon) = f^X : U^X \longrightarrow V^X.$$

  We need to show $\sigma_U(g) = U^h(g) = g \circ h$ for each $g\colon Y \to U$. But such a function g yields a naturality square:

$$\begin{array}{ccc} \mathcal{L}(Y,\beta)(Y) = Y^Y & \xrightarrow{\quad\sigma_Y\quad} & Y^X = \mathcal{L}(X,\gamma)(Y) \\ {\scriptstyle \mathcal{L}(Y,\beta)(g \circ \varepsilon) = g^Y}\downarrow & & \downarrow{\scriptstyle g^X = \mathcal{L}(X,\gamma)(g \circ \varepsilon)} \\ \mathcal{L}(Y,\beta)(V) = U^Y & \xrightarrow[\quad\sigma_U\quad]{} & U^X = \mathcal{L}(X,\gamma)(U) \end{array}$$

  Now we are done:

$$U^h(g) = g \circ h = g^X(h) = (g^X \circ \sigma_Y)(\mathrm{id}_Y) = (\sigma_U \circ g^Y)(\mathrm{id}_Y) = \sigma_U(g).$$

- In order to show that $h = \sigma_Y(\mathrm{id}_Y)\colon X \to Y$ is a map of coalgebras we use the naturality diagram below; it involves the identity function $S(Y) \to S(Y)$, considered as map $Y \to S(Y)$ in the Kleisli category $\mathcal{K}\ell(S)$:

$$
\begin{array}{ccc}
\mathcal{L}(Y,\beta)(Y) = Y^Y & \xrightarrow{\quad \sigma_Y \quad} & Y^X = \mathcal{L}(X,\gamma)(Y) \\
{\scriptstyle \mathcal{L}(Y,\beta)(\mathrm{id}_{S(Y)})} \big\downarrow & & \big\downarrow {\scriptstyle \mathcal{L}(X,\gamma)(\mathrm{id}_{S(Y)})} \\
\mathcal{L}(Y,\beta)(S(Y)) = S(Y)^Y & \xrightarrow[\quad \sigma_{S(Y)} \quad]{} & U^X = \mathcal{L}(X,\gamma)(S(Y))
\end{array}
$$

The description (6.33) now yields the required equation:

$$
\begin{aligned}
S(h) \circ \gamma &= \mathcal{L}(X,\gamma)(\mathrm{id}_{S(Y)})(h) && \text{by (6.33)} \\
&= (\mathcal{L}(X,\gamma)(\mathrm{id}_{S(Y)}) \circ \sigma_Y)(\mathrm{id}_Y) \\
&= (\sigma_{S(Y)} \circ \mathcal{L}(Y,\beta)(\mathrm{id}_{S(Y)}))(\mathrm{id}_Y) && \text{by naturality} \\
&= \sigma_{S(Y)}(\beta) && \text{by (6.33)} \\
&= S(Y)^h(\beta) && \text{by the previous point} \\
&= \beta \circ h. && \square
\end{aligned}
$$

We turn to assertions for coalgebras. Here we are interested not in the way that assertions are formed syntactically, e.g. via modal operators as in Section 6.5, but in their meaning. There are several significant differences with the approach for algebras.

- In the coalgebraic case we will be using predicates instead of relations (as for algebras).
- We distinguish *axioms* and *axiom systems*. Axioms are simply given by a subset of a final coalgebra. Axiom systems on the other hand involve collections of subsets, given in a functorial manner, as in Definition 6.7.3.

  Axioms are easier to work with in the kind of coalgebraic 'class-style' specifications that we are interested in here – see also the next section – and will thus receive most attention. Following [183] they may be called *behavioural* axioms (or *sinks*, as in [405]).

  In particular, this emphasis on subsets of a final coalgebra as axioms means that arbitrary cofree coalgebras on colours (and 'recolouring'; see Exercise 6.8.8) do not play a big role here – unlike in other work [183, 47, 17].
- We do not describe a deductive calculus for predicates on coalgebras. Such a calculus is described in [17, 422] via a 'child' and a 'recolouring' rule, corresponding to the temporal operators $\square$ and $\boxtimes$ (see Exercise 6.8.8). These rules are defined via closure under *all* operations of some sort (successor and recolouring) and thus involve greatest fixed points. Hence they are not deductive logics in a traditional sense, with finite proof trees.
- As already explained above, before Theorem 6.8.2, we use ordinary (non–finitary) Kleisli categories in the coalgebraic case, since it is natural to have

finitely many variables in (algebraic) terms but not to restrict observations to finite sets.

Definition 6.8.3 1. Let $F\colon \mathbf{Sets} \to \mathbf{Sets}$ be a functor with final coalgebra $Z \xrightarrow{\cong} F(Z)$. **Axioms** for F are given by a (single) subset $\mathcal{A} \subseteq Z$.

A coalgebra $c\colon X \to F(X)$ satisfies $\mathcal{A}$ if the image of the unique coalgebra map $\mathrm{beh}_c\colon X \to Z$ is contained in $\mathcal{A}$; that if there is a map of predicates $\top(X) \to \mathcal{A}$ in

$$\begin{array}{ccc} X & \dashrightarrow & \mathcal{A} \\ {\scriptstyle \top(X)}\Big\| & & \big\downarrow \\ X & \xrightarrow{\ \mathrm{beh}_c\ } & Z \end{array} \qquad \text{or} \qquad \coprod\nolimits_{\mathrm{beh}_c}(\top) \leq \mathcal{A} \qquad \text{or} \qquad \mathrm{beh}_c^{-1}(\mathcal{A}) = \top,$$

where $\top(X) = (X \subseteq X)$ for the truth predicate on X. In that case we write $c \models \mathcal{A}$.

2. Similarly, axioms for a comonad $S : \mathbf{Sets} \to \mathbf{Sets}$ are given by a subset $\mathcal{A} \subseteq S(1)$ of the carrier of the final coalgebra $\delta\colon S(1) \to S^2(1)$, that is, of the cofree coalgebra on a final/singleton set 1.

   An Eilenberg–Moore coalgebra $\gamma\colon X \to S(X)$ satisfies these axioms, written as $\gamma \models \mathcal{A}$, if the image of the unique coalgebra map $S(!_X) \circ \gamma\colon X \to S(1)$ is contained in $\mathcal{A}$.
3. In this way we get full subcategories:

$$\mathbf{CoAlg}(F, \mathcal{A}) \hookrightarrow \mathbf{CoAlg}(F) \qquad \text{and} \qquad \mathcal{EM}(S, \mathcal{A}) \hookrightarrow \mathcal{EM}(S)$$

   of functor and comonad coalgebras satisfying $\mathcal{A}$.

In case the comonad S is a cofree comonad F^∞ on a functor F, then the two ways of describing axioms coincide, because the final F-coalgebra Z is the cofree comonad $F^\infty(1)$ at 1. Moreover, validity for a functor coalgebra $X \to F(X)$ is the same as validity for the associated Eilenberg–Moore coalgebra $X \to F^\infty(X)$; see Exercise 6.8.2.

Axioms as defined above can also be described in a functorial manner. We shall do so for comonads and leave the analogue for functors as an exercise below.

Lemma 6.8.4 *Let $S : \mathbf{Sets} \to \mathbf{Sets}$ be a comonad with axioms $\mathcal{A} \subseteq S(1)$. Then one can define two functors $\mathcal{A}^{\mathcal{K}\ell}$ and $\mathcal{A}^{\mathcal{EM}}$ as below:*

$$\begin{array}{ccc} & & \mathbf{Pred} \\ & {\scriptstyle \mathcal{A}^{\mathcal{K}\ell}}\nearrow & \big\downarrow \\ \mathcal{K}\ell(S) & \longrightarrow & \mathbf{Sets} \end{array} \qquad\qquad \begin{array}{ccc} & & \mathbf{Pred} \\ & {\scriptstyle \mathcal{A}^{\mathcal{EM}}}\nearrow & \big\downarrow \\ \mathcal{EM}(S) & \longrightarrow & \mathbf{Sets} \end{array}$$

Applying these functors to an arbitrary morphism yields a pullback diagram (between predicates).

For an Eilenberg–Moore coalgebra $\gamma\colon X \to S(X)$ *the following statements are then equivalent.*

a. $\gamma \models \mathcal{A}$.
b. $\mathcal{A}^{\mathcal{EM}}(X, \gamma) = \top$.
c. γ *is a map of predicates* $\top(X) \to \mathcal{A}^{\mathcal{K}\ell}(X)$, i.e. $\mathrm{Im}(\gamma) = \coprod_\gamma(\top) \leq \mathcal{A}^{\mathcal{K}\ell}(X)$.
d. *For each* $f\colon X \to U$*, the Kleisli extension* $S(f) \circ \gamma\colon X \to S(U)$ *is a map of predicates* $\top(X) \to \mathcal{A}^{\mathcal{K}\ell}(U)$.

This last formulation is the coalgebraic analogue of validity of axiom systems in algebras from Definition 6.7.3. Hence in the present context we shall call a collection of predicates $\mathcal{B}(U) \subseteq S(U)$ forming a functor $\mathcal{B}\colon \mathcal{K}\ell(S) \to \mathbf{Pred}$ as in the above triangle on the left an **axiom system**. Such a system is automatically closed under substitution – and in particular under recolourings; see Exercise 6.8.8. The system holds in a coalgebra $X \to S(X)$ if the image of the coalgebra is contained in $\mathcal{B}(X) \subseteq S(X)$, as in (c) above. Equivalently, condition (d) may be used.

Proof For an object (set) $U \in \mathcal{K}\ell(S)$ use the unique map $!_U\colon U \to 1$ to define a predicate by pullback:

$$\mathcal{A}^{\mathcal{K}\ell}(U) = S(!_U)^{-1}(\mathcal{A}) \subseteq S(U).$$

We have to show that for each map $g\colon U \to V$ in the Kleisli category $\mathcal{K}\ell(S)$, the resulting Kleisli extension $g^\$ = S(g) \circ \delta\colon S(U) \to S(V)$ is a map of predicates $\mathcal{A}^{\mathcal{K}\ell}(U) \to \mathcal{A}^{\mathcal{K}\ell}(V)$. First we have

$$S(!_V) \circ g^\$ = S(!_V \circ g) \circ \delta = S(!_{S(U)}) \circ \delta = S(!_U \circ \varepsilon) \circ \delta = S(!_U).$$

Next we get the required dashed map in

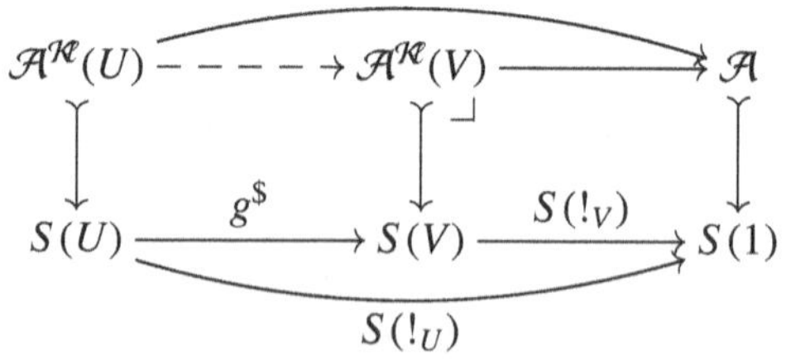

By the pullback lemma (Exercise 4.2.6) the rectangle on the left is a pullback.

Similarly, for an Eilenberg–Moore algebra $\gamma\colon X \to S(X)$ one defines a predicate

$$\mathcal{A}^{\mathcal{EM}}(X, \gamma) = (S(!_X) \circ \gamma)^{-1}(\mathcal{A}) \subseteq X.$$

For a map of coalgebras $(X \xrightarrow{\gamma} S(X)) \xrightarrow{h} (Y \xrightarrow{\beta} S(Y))$ one obtains a pullback on the left below:

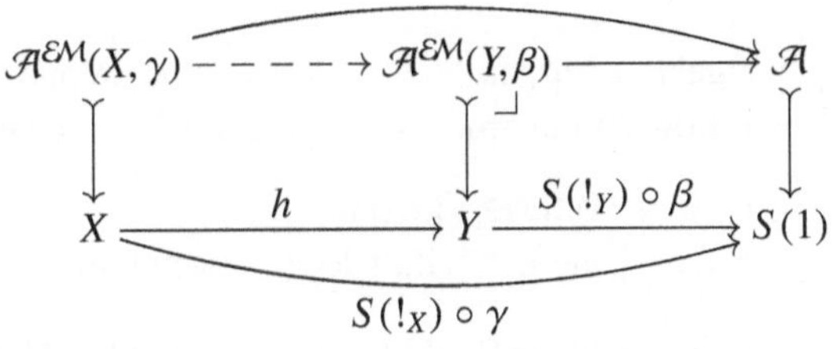

This works since

$$S(!_Y) \circ \beta \circ h = S(!_Y) \circ S(h) \circ \gamma = S(!_X) \circ \gamma.$$

Finally, we prove the equivalence of the statements (a)–(d). Validity $\gamma \models \mathcal{A}$ in (a) means $\coprod_{S(!_X)\circ\gamma}(\top) \leq \mathcal{A}$. Equivalently, $\top = (S(!_X) \circ \gamma)^{-1}(\mathcal{A}) = \mathcal{A}^{\mathcal{EM}}(X, \gamma)$ as in point (b). This is equivalent to the statement in (c), namely $\coprod_\gamma(\top) \subseteq \mathcal{A}^{\mathcal{K}}(X)$, since

$$(S(!_X) \circ \gamma)^{-1}(\mathcal{A}) = \gamma^{-1} S(!_X)^{-1}(\mathcal{A}) = \gamma^{-1}(\mathcal{A}^{\mathcal{K}}(X)).$$

For each function $f : X \to U$ the map $S(f) = (f \circ \varepsilon)^{\$} : S(X) \to S(Y)$ is a map of predicates $\mathcal{A}^{\mathcal{K}}(X) \to \mathcal{A}^{\mathcal{K}}(U)$ by functoriality of $\mathcal{A}^{\mathcal{K}}$. Hence by precomposing this map with the map in (c) we get (d). In order to get from (d) to (a) one instantiates with $f = \mathrm{id}_X$. ❑

Example 6.8.5 We consider a simplified version of the bank account specification from Example 6.5.2, with only a balance $\mathsf{bal} : X \to \mathbb{N}$ and deposit operation $\mathsf{dep} : X \times \mathbb{N} \to X$. The relevant assertion

$$\mathsf{bal}(\mathsf{dep}(x, n)) = \mathsf{bal}(x) + n \tag{6.34}$$

was written in Example 6.5.2 in modal logic style as

$$\mathsf{bal}\downarrow m \vdash [\mathsf{dep}(n)](\mathsf{bal}\downarrow(m+n)).$$

The assertion (6.34) can be interpreted directly in any coalgebra of the form $\langle \mathsf{dep}, \mathsf{bal} \rangle : X \to F(X)$, for $F = (-)^{\mathbb{N}} \times \mathbb{N}$, namely as subset of the state space:

$$\{x \in X \mid \forall n \in \mathbb{N}.\, \mathsf{bal}(\mathsf{dep}(x, n)) = \mathsf{bal}(x) + n\}. \tag{6.35}$$

One expects that assertion (6.34) holds in the coalgebra if this set is the whole of X. We investigate this situation in some detail.

Definition 6.8.3 involves assertions as subsets of final coalgebras. By Proposition 2.3.5 the final F-coalgebra is $\mathbb{N}^{\mathbb{N}^\star}$ with operations

$$\mathsf{bal}(\varphi) = \varphi(\langle\rangle) \quad \text{and} \quad \mathsf{dep}(\varphi, n) = \lambda\sigma \in \mathbb{N}^\star.\, \varphi(n \cdot \sigma).$$

The assertion (6.34) can be described as a subset of this final coalgebra:

$$\mathcal{A} = \{\varphi \in \mathbb{N}^{\mathbb{N}^\star} \mid \forall n \in \mathbb{N}.\, \varphi(\langle n \rangle) = \varphi(\langle\rangle) + n\}.$$

For an arbitrary coalgebra $\langle \mathsf{dep}, \mathsf{bal}\rangle \colon X \to X^{\mathbb{N}} \times \mathbb{N}$ the unique coalgebra map $\mathrm{beh} \colon X \to \mathbb{N}^{\mathbb{N}^\star}$ is, following the proof of Proposition 2.3.5, given by

$$\begin{aligned}\mathrm{beh}(x) &= \lambda\sigma \in \mathbb{N}^\star.\, \mathsf{bal}(\mathsf{dep}^*(x,\sigma)) \\ &= \lambda\langle n_1,\ldots,n_k\rangle \in \mathbb{N}^\star.\, \mathsf{bal}(\mathsf{dep}(\cdots \mathsf{dep}(x,n_1)\cdots, n_k)).\end{aligned}$$

We can now see that the predicate (6.35) is simply $\mathrm{beh}^{-1}(\mathcal{A})$. Hence our earlier statement that validity of the assertion means that the predicate (6.35) is all of X is precisely what is formulated in Definition 6.8.3.

We add two more observations.

1. A simple example of a model of this specification takes the natural numbers $\mathbb{N}$ as state space. The balance operation $\mathsf{bal} \colon \mathbb{N} \to \mathbb{N}$ is the identity function, and the deposit operation $\mathsf{dep} \colon \mathbb{N} \times \mathbb{N} \to \mathbb{N}$ is addition. Clearly, assertion (6.34) holds. It is not hard to see that this model is the final one satisfying this assertion: for an arbitrary coalgebra $c \colon X \to X^{\mathbb{N}} \times \mathbb{N}$ satisfying (6.34), its 'balance' map $\pi_2 \circ c \colon X \to \mathbb{N}$ is a map of coalgebras. Exercise 6.8.4 deals with an alternative 'history' model that keeps track of past transactions.
2. We briefly describe the functorial description $\mathcal{A}^{\mathcal{K}\ell} \colon \mathcal{K}\ell(F^\infty) \to \mathbf{Pred}$ of assertions, from Lemma 6.8.4, for this bank example. An explicit description of the cofree comonad $F^\infty(U) = (U \times \mathbb{N})^{\mathbb{N}^\star}$ is obtained via Propositions 2.5.3 and 2.3.5. Its operations are

$$\mathsf{bal}(\varphi) = \pi_2\varphi(\langle\rangle) \quad \text{and} \quad \mathsf{dep}(\varphi, n) = \lambda\sigma \in \mathbb{N}^\star.\, \varphi(n\cdot\sigma).$$

   The axiom system $\mathcal{A}^{\mathcal{K}\ell} \colon \mathcal{K}\ell(F^\infty) \to \mathbf{Pred}$ from Lemma 6.8.4 determined by (6.34) is given by

$$\begin{aligned}\mathcal{A}^{\mathcal{K}\ell}(U) &= F^\infty(!_U)^{-1}(\mathcal{A}) \\ &= \{\varphi \in F^\infty(U) \mid \forall n \in \mathbb{N}.\, \pi_2\varphi(\langle n\rangle) = \pi_2\varphi(\langle\rangle) + n\}.\end{aligned}$$

This bank account will be investigated further in Example 6.8.10 below.

Recall from Exercise 6.2.5 that for a comonad S on **Sets** the associated predicate lifting $\mathrm{Pred}(S) \colon \mathbf{Pred} \to \mathbf{Pred}$ is also a comonad. Hence we can consider its category of coalgebras. These Eilenberg–Moore coalgebras capture S-invariants.

Lemma 6.8.6 *Let $S \colon$ **Sets** $\to$ **Sets** be a comonad with axiom system $\mathcal{A} \subseteq S(1)$. Recall the induced functor $\mathcal{A}^{\mathcal{K}\ell} \colon \mathcal{K}\ell(S) \to$ **Pred** from Lemma 6.8.4. For* *each set U we define a new predicate $\Box\mathcal{A}^{\mathcal{K}\ell}(U) \subseteq S(U)$ via the Knaster–Tarski fixpoint theorem (see* e.g. *[119, chapter 4]), namely as the greatest fixed point of the following monotone operator:*

$$\mathcal{P}(S(U)) \xrightarrow{\;\mathcal{A}^{\mathcal{K}\ell}(U)\,\wedge\,\delta^{-1}(\mathrm{Pred}(S)(-))\;} \mathcal{P}(S(U)).$$

Hence, $\Box\mathcal{A}^{\mathcal{K}\ell}(U)$ is the greatest invariant for the coalgebra $\delta\colon S(U) \to S^2(U)$, contained in $\mathcal{A}^{\mathcal{K}\ell}(U)$; see Figure 6.4.

This yields a functor (or axiom system) $\Box\mathcal{A}^{\mathcal{K}\ell}$ in a situation

$$\begin{array}{ccc}
 & & \mathcal{EM}(\mathrm{Pred}(S)) \\
 & \overset{\Box\mathcal{A}^{\mathcal{K}\ell}}{\nearrow} & \downarrow \\
 & \overset{\mathcal{A}^{\mathcal{K}\ell}}{\nearrow} & \mathbf{Pred} \circlearrowleft \mathrm{Pred}(S) \\
 & & \downarrow \\
\mathcal{K}\ell(S) & \longrightarrow & \mathbf{Sets} \circlearrowleft S
\end{array}$$

For a coalgebra $\gamma\colon X \to S(X)$ one has

$$\gamma \models \mathcal{A} \iff \gamma \models \Box\mathcal{A}^{\mathcal{K}\ell}.$$

The role of invariants (or subcoalgebras) for validity, as expressed by the last result, is emphasised in [47].

Proof The (greatest) fixed point

$$\Box\mathcal{A}^{\mathcal{K}\ell}(U) = \mathcal{A}^{\mathcal{K}\ell}(U) \wedge \delta^{-1}(\mathrm{Pred}(S)(\Box\mathcal{A}^{\mathcal{K}\ell}(U)))$$

is by construction contained in the subset $\mathcal{A}^{\mathcal{K}\ell}(U) \subseteq S(U)$. The inclusion $\Box\mathcal{A}^{\mathcal{K}\ell}(U) \subseteq \delta^{-1}(\mathrm{Pred}(S)(\Box\mathcal{A}^{\mathcal{K}\ell}(U)))$ gives a factorisation

$$\begin{array}{ccc}
\Box\mathcal{A}^{\mathcal{K}\ell}(U) & \dashrightarrow & \mathrm{Pred}(S)(\Box\mathcal{A}^{\mathcal{K}\ell}(U)) \\
\downarrow & & \downarrow \\
S(U) & \xrightarrow[\delta]{} & S^2(U)
\end{array}$$

By Exercise 6.2.5 this is actually an Eilenberg–Moore coalgebra of the comonad $\mathrm{Pred}(S)$ and thus an S-invariant.

In order to show that $\Box\mathcal{A}^{\mathcal{K}\ell}$ is a functor, assume a map $f\colon S(U) \to V$. We have to show $f^{\$} = S(f) \circ \delta$ is a map of predicates $\Box\mathcal{A}^{\mathcal{K}\ell}(U) \to \Box\mathcal{A}^{\mathcal{K}\ell}(V)$. We have

$$\coprod_{f^{\$}}(\Box\mathcal{A}^{\mathcal{K}\ell}(U)) \subseteq \coprod_{f^{\$}}(\mathcal{A}^{\mathcal{K}\ell}(U)) \subseteq \mathcal{A}^{\mathcal{K}\ell}(V).$$

The second inclusion follows from the fact that $\mathcal{A}^{\mathcal{K}\ell}$ is a functor. Since $\coprod_{f^{\$}}$ preserves invariants, by Exercise 6.2.5, we get $\coprod_{f^{\$}}(\Box\mathcal{A}^{\mathcal{K}\ell}(U)) \subseteq \Box\mathcal{A}^{\mathcal{K}\ell}(V))$, because the latter is the greatest invariant contained in $\mathcal{A}^{\mathcal{K}\ell}(V)$.

Let $\gamma\colon X \to S(X)$ be a coalgebra. If $\gamma \models \Box\mathcal{A}^{\mathcal{K}}$, then $\gamma \models \mathcal{A}$ because of the inclusions $\Box\mathcal{A}^{\mathcal{K}}(1) \subseteq \mathcal{A}^{\mathcal{K}}(1) = \mathcal{A}$, as subsets of $S(1)$. For the other direction, from $\gamma \models \mathcal{A}$ we obtain $\coprod_\gamma(\top) \subseteq \mathcal{A}^{\mathcal{K}}(X)$ as in Lemma 6.8.4(c). But since the image $\coprod_\gamma(\top)$ is an invariant, again by Exercise 6.2.5, we get $\coprod_\gamma(\top) \subseteq \Box\mathcal{A}^{\mathcal{K}}(X)$. ❑

We recall from (6.8) that comprehension $\{-\}$ can be described categorically as a functor **Pred** $\to$ **Sets**, sending a predicate $P \rightarrowtail X$ to its domain P, considered as a set itself. This functor $\{-\}$ is right adjoint to the truth functor $\top$: **Sets** $\to$ **Sets**. Also in the present context comprehension is a highly relevant operation that takes invariants to (sub)coalgebras.

Proposition 6.8.7 *Let S be a comonad on* **Sets***.*

1. *As a functor, S preserves injections.*
2. *Thus* $\{\mathrm{Pred}(S)(P)\} \cong S(\{P\})$*, for each predicate P.*
3. *Comprehension $\{-\}$ lifts to Eilenberg–Moore coalgebras in*

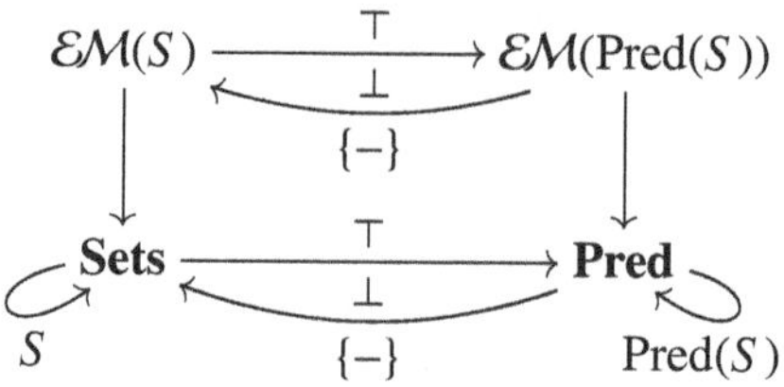

Proof 1. By Lemma 2.1.7 the functor S preserves injections $f\colon X \to Y$ with $X \neq \emptyset$, so we need to consider only the case where $X = \emptyset$. There is a counit map $S(\emptyset) \to \emptyset$, which gives $S(\emptyset) = \emptyset$ since $\emptyset$ is a strict initial object in **Sets**. Hence S preserves all injections.

2. Recall that predicate lifting $\mathrm{Pred}(S)$ applied to a predicate $m\colon P \rightarrowtail X$ is obtained by epi-mono factorisation of $S(m)\colon S(P) \to S(X)$. But since $S(m)$ is injective by (1), we get $\mathrm{Pred}(S)(P) = S(P)$. Or, a bit more formally, $\{\mathrm{Pred}(S)(m)\} = S(\{m\})$.
3. This is proved by Exercise 2.5.13, using the isomorphism from the previous point. ❑

Theorem 6.8.8 *Let S :* **Sets** $\to$ **Sets** *be a comonad with an axiom system $\mathcal{A} \subseteq S(1)$. Then there is a 'subset' comonad $\{S \mid \mathcal{A}\}$:* **Sets** $\to$ **Sets** *with a comonad map $\pi\colon \{S \mid \mathcal{A}\} \Rightarrow S$ and with an isomorphism of categories $\mathcal{EM}(\{S \mid \mathcal{A}\}) \xrightarrow{\cong} \mathcal{EM}(S, \mathcal{A})$ in*

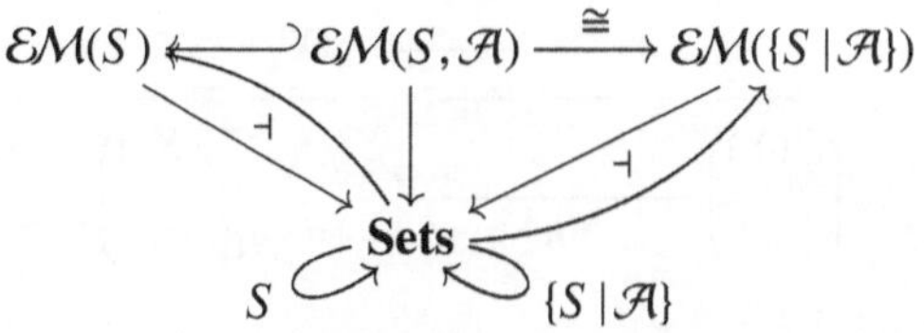

A similar result involving an induced subcomonad occurs in [169], starting from a covariety (a suitably closed class of coalgebras), instead from axioms.

Proof On a set X, define $\{S \mid \mathcal{A}\}(X)$ to be the carrier of the Eilenberg–Moore subcoalgebra obtained by comprehension from the invariant $\Box\mathcal{A}^{\mathcal{K}}(X)$ from Lemma 6.8.6, in

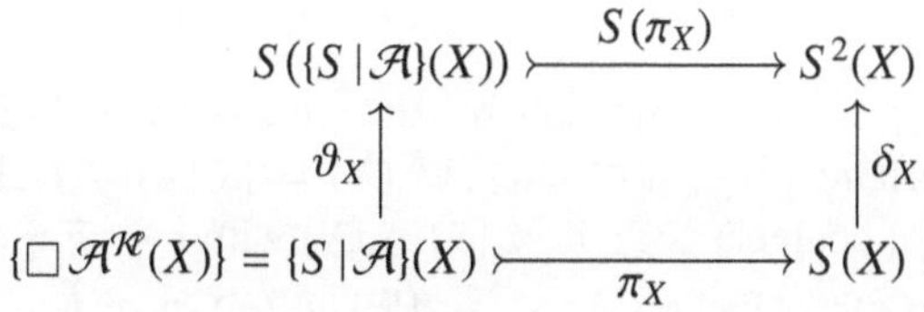

The axioms $\mathcal{A}$ hold by construction in the coalgebra ϑ_X: by functoriality of $\mathcal{A}^{\mathcal{K}}$ we get a map $\{S \mid \mathcal{A}\}(X) \to \mathcal{A}(S(X))$ in

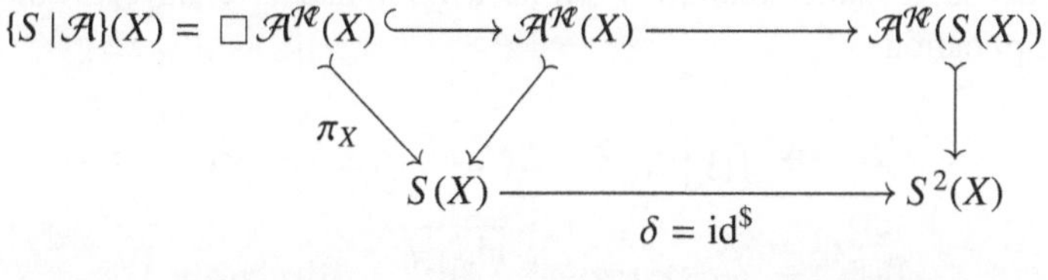

It gives us the required factorisation below, using the pullback property from Lemma 6.8.4:

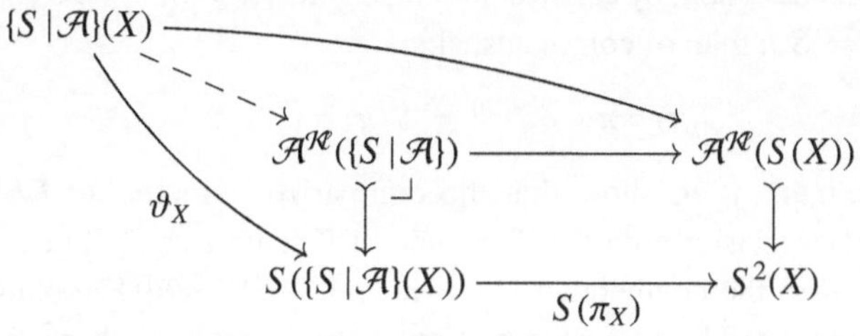

Next we show that the mapping $X \mapsto \vartheta_X$ forms a right adjoint to the forgetful functor $\mathcal{EM}(S,\mathcal{A}) \to$ **Sets**. For a coalgebra $\beta\colon Y \to S(Y)$ satisfying $\mathcal{A}$ there is a bijective correspondence:

$$\frac{Y \xrightarrow{\quad g \quad} X}{\begin{pmatrix} S(Y) \\ \beta\uparrow \\ Y \end{pmatrix} \xrightarrow[\quad h \quad]{} \begin{pmatrix} S(\{S \mid \mathcal{A}\}(X)) \\ \uparrow\vartheta_X \\ \{S \mid \mathcal{A}\}(X) \end{pmatrix}} .$$

This works as follows.

- Given a function $g\colon Y \to X$ we obtain $S(g) \circ \beta\colon Y \to S(X)$, which is a map of coalgebras $\beta \to \delta$. We get $\coprod_{S(g)\circ\beta}(\top) \subseteq \mathcal{A}^{\mathcal{K}}(X)$ in

$$\begin{array}{ccccc} & & \mathcal{A}^{\mathcal{K}}(Y) & \dashrightarrow & \mathcal{A}^{\mathcal{K}}(X) \\ & \nearrow & \downarrow & & \downarrow \\ Y & \xrightarrow[\beta]{} & S(Y) & \xrightarrow[S(g) = (g \circ \varepsilon)^{\$}]{} & S(X) \end{array}$$

  Since $\coprod_{S(g)\circ\beta}(\top)$ is an invariant by Exercise 6.2.5, it is included in the greatest invariant: $\coprod_{S(g)\circ\beta}(\top) \subseteq \Box\mathcal{A}^{\mathcal{K}}(X) = \{S \mid \mathcal{A}\}(X)$. Hence there is a unique map of coalgebras $\overline{g}\colon Y \to \{S \mid \mathcal{A}\}(X)$ with $\pi_X \circ \overline{g} = S(g) \circ \beta$.
- Given a map of coalgebras $h\colon Y \to \{S \mid \mathcal{A}\}(X)$, take $\overline{h} = \varepsilon \circ \pi_X \circ h\colon Y \to X$.

We leave it to the reader to check the remaining details of this correspondence, and proceed to show that $\{S \mid \mathcal{A}\}$ is a comonad. We have a counit and comultiplication:

$$\varepsilon_X^{\{S \mid \mathcal{A}\}} = \Big(\{S \mid \mathcal{A}\}(X) \xrightarrow{\pi_X} S(X) \xrightarrow{\varepsilon_X^S} X\Big)$$

$$\delta_X^{\{S \mid \mathcal{A}\}} = \Big(\{S \mid \mathcal{A}\}(X) \xrightarrow{\quad \overline{\mathrm{id}_{\{S \mid \mathcal{A}\}}} \quad} \{S \mid \mathcal{A}\}^2(X)\Big).$$

The comultiplication is obtained as map of coalgebras via the above adjoint correspondence. Then, by construction it satisfies: $\pi \circ \delta_X^{\{S \mid \mathcal{A}\}} = \vartheta_X$. This makes $\pi\colon \{S \mid \mathcal{A}\} \Rightarrow S$ a map of comonads, since

$$S(\pi) \circ \pi \circ \delta^{\{S \mid \mathcal{A}\}} = S(\pi) \circ \vartheta = \delta^S \circ \pi.$$

What remains is to show that the comparison functor $K\colon \mathcal{EM}(S, \mathcal{A}) \to \mathcal{EM}(\{S \mid \mathcal{A}\})$ is an isomorphism. This functor K sends a coalgebra $\beta\colon Y \to S(Y)$ satisfying $\mathcal{A}$ to the coalgebra $K(\beta) = \overline{\mathrm{id}_Y}\colon Y \to \{S \mid \mathcal{A}\}(Y)$ obtained via the above adjunction. Thus, by construction, $K(\beta)$ is unique with $\pi \circ K(\beta) = \beta$.

In the other direction we have a functor K^{-1} that sends a coalgebra $\gamma\colon X \to \{S \mid \mathcal{A}\}(X)$ to $K^{-1}(\gamma) = \pi \circ \gamma\colon X \to S(X)$. This coalgebra satisfies $\mathcal{A}$ since its image is contained in $\mathcal{A}^{\mathcal{K}}(X)$; see

$$\begin{array}{ccccc} & & & & \mathcal{A}^{\mathcal{K}}(X) \\ & & & \nearrow & \downarrow \\ X & \xrightarrow[\gamma]{} & \{S \mid \mathcal{A}\}(X) & \xrightarrow[\pi]{} & S(X) \end{array}$$

Obviously, $K^{-1}(K(\beta)) = \beta$. And the equation $K(K^{-1}(\beta)) = \beta$ holds since $K(K^{-1}(\beta))$ is the unique map f with $\pi \circ f = K^{-1}(\beta)$. Since $K^{-1}(\beta) = \pi \circ \beta$ this unique map must be β itself. ❑

The next result comes from [236], where it occurs in more limited form, with a direct proof (not via the previous theorem).

Corollary 6.8.9 *Let $F\colon$ **Sets** $\to$ **Sets** be a functor with cofree comonad F^{∞} and axioms $\mathcal{A} \subseteq F^{\infty}(1)$. Then there is a comonad S with natural transformation $S \Rightarrow F$ and an isomorphism of categories*

$$\begin{array}{c} \mathbf{CoAlg}(F, \mathcal{A}) \cong \mathcal{EM}(S) \\ \downarrow \dashv \uparrow \\ \mathbf{Sets} \circlearrowleft S \end{array}$$

Proof Of course one defines S as subset comonad $S = \{F^{\infty} \mid \mathcal{A}\}$ from Theorem 6.8.8, with isomorphism $\mathbf{CoAlg}(F^{\infty}, \mathcal{A}) \cong \mathcal{EM}(S)$ and (composite) natural transformations $S \Rightarrow F^{\infty} \Rightarrow F$. Further, by Exercise 6.8.2 and Proposition 5.4.7 there is an isomorphism $\mathbf{CoAlg}(F, \mathcal{A}) \cong \mathcal{EM}(F^{\infty}, \mathcal{A})$. Hence we are done. ❑

We return to our earlier bank account example, in order to determine the associated comonad that captures the assertion.

Example 6.8.10 In the context of Example 6.8.5 we will explicitly calculate what the comonad is whose Eilenberg–Moore coalgebras are precisely the models of the bank assertion $\mathsf{bal}(\mathsf{dep}(x, n)) = \mathsf{bal}(x) + n$ from (6.34), for coalgebras $\langle \mathsf{dep}, \mathsf{bal} \rangle\colon X \to F(X) = X^{\mathbb{N}} \times \mathbb{N}$. We already saw the cofree comonad comonad F^{∞} is given by $F^{\infty}(X) = (X \times \mathbb{N})^{\mathbb{N}^{\star}}$, with predicates:

$$\mathcal{A}^{\mathcal{K}}(X) = \{\varphi \in F^{\infty}(X) \mid \forall n \in \mathbb{N}.\, \pi_2 \varphi(\langle n \rangle) = \pi_2 \varphi(\langle\rangle) + n\}.$$

It is not hard to see that the greatest invariant contained in it is

$$\Box \mathcal{A}^{\mathcal{K}}(X) = \{\varphi \in F^{\infty}(X) \mid \forall \sigma \in \mathbb{N}^{\star}.\, \forall n \in \mathbb{N}.\, \pi_2 \varphi(n \cdot \sigma) = \pi_2 \varphi(\sigma) + n\}.$$

For elements φ in this subset the function $\pi_2 \circ \varphi\colon \mathbb{N}^{\star} \to \mathbb{N}$ is thus completely determined by the value $\pi_2 \varphi(\langle\rangle) \in \mathbb{N}$ on the empty sequence. Hence this invariant can be identified with

$$S(X) = X^{\mathbb{N}^\star} \times \mathbb{N}.$$

This is the comonad we seek. Its counit and comultiplication are

$$\varepsilon(\varphi, m) = \varphi(\langle\rangle) \qquad \delta(\varphi, m) = \langle\, \lambda\sigma \in \mathbb{N}^\star.\, \langle\, \lambda\tau \in \mathbb{N}^\star.\, \varphi(\sigma \cdot \tau),\ m + \Sigma\sigma \rangle,\ m \,\rangle,$$

where $\Sigma\sigma \in \mathbb{N}$ is the sum of all the numbers in the sequence $\sigma \in \mathbb{N}^\star$.

Now assume we have an Eilenberg–Moore coalgebra $\gamma = \langle \gamma_1, \gamma_2 \rangle \colon X \to S(X) = X^{\mathbb{N}^\star} \times \mathbb{N}$. The coalgebra equations $\varepsilon \circ \gamma = \mathrm{id}$ and $S(\gamma) \circ \gamma = \delta \circ \gamma$ amount to the following two equations:

$$\begin{aligned} \gamma_1(x)(\langle\rangle) &= x \\ \langle\, \lambda\sigma.\, \gamma(\gamma_1(x)(\sigma)),\ \gamma_2(x) \,\rangle &= \langle\, \lambda\sigma.\, \langle\, \lambda\tau.\, \gamma_1(x)(\sigma \cdot \tau),\ \gamma_2(x) + \Sigma\sigma \,\rangle,\ \gamma_2(x) \,\rangle. \end{aligned}$$

The latter equation can be split into two equations: for all $\sigma, \tau \in \mathbb{N}^\star$,

$$\gamma_2(\gamma_1(x)(\sigma)) = \gamma_2(x) + \Sigma\sigma \qquad \gamma_1(\gamma_1(x)(\sigma))(\tau) = \gamma_1(x)(\sigma \cdot \tau).$$

These equations show that γ_2 is the balance operation bal and that γ_1 is the iterated deposit operation dep*, from (2.22), described as monoid action.

We conclude with some additional observations.

Remark 6.8.11 1. In Corollary 6.7.15 we have seen that each finitary monad can be described via operations and assertions. There is no similar result for comonads. The construction for a monad T relies on natural maps of the form $\coprod_{n \in \mathbb{N}} T(n) \times X^n \to T(X)$, from an arity functor to T; they are surjective if and only if T is finitary. Dually, for a comonad S, one may consider maps of the form

$$\begin{array}{ccc} S(X) & \longrightarrow & \prod_{n \in \mathbb{N}} S(n)^{(n^X)} \\ u & \longmapsto & \lambda n.\, \lambda h \in n^X.\, S(h)(u). \end{array}$$

One can require that these maps are injective, and then proceed with the functor $F(X) = \prod_{n \in \mathbb{N}} S(n)^{(n^X)}$ on the right-hand side. However, this functor F does not seem to have cofree coalgebras, so that the subset comonad construction from Theorem 6.8.8 does not work here. Additionally, the functor F is not a standard arity functor of the form that we have worked with in this book. The associated (non-standard) operations have been investigated to some extent in [325], to which we refer for further information. See also [62] for the notion of comodel.

2. Much work has been done towards a 'coalgebraic Birkhoff theorem', characterising a class of coalgebras that is defined by assertions via suitable closure properties; see for instance [183, 411, 168, 405, 321, 47, 29, 17,

169, 422]. Colours and cofree coalgebras play an important role in this work – unlike here.

3. In the end, the categorically inclined reader may wish to try to describe the quotient monad $T/\mathcal{A}$ from Theorem 6.7.11 and the subset comonad $\{S \mid \mathcal{A}\}$ from Theorem 6.8.8 via quotient and comprehension adjunctions between suitable categories of (co)monads.
4. In this book we use quotients of *relations* and comprehension of *predicates*. In a linear (or quantum) setting it is more natural to consider quotients also of predicates. For instance, for a linear subspace $P \subseteq V$ of a vector space V there is the quotient vector space V/P determined by the induced equivalence relation $\sim_P$ on V given by $x \sim_P y \iff y - x \in P$. This setting allows a more uniform logical approach, with predicates only; see [96, 95].

Exercises

6.8.1 Describe the action on morphisms of the functor $O_\#$ from Proposition 6.8.1.

6.8.2 Let $F\colon \mathbf{Sets} \to \mathbf{Sets}$ be a functor with cofree comonad F^∞ and axioms $\mathcal{A} \subseteq F^\infty(1)$, where $F^\infty(1)$ is the final F-coalgebra. Recall from the proof of Proposition 5.4.7 how the Eilenberg–Moore coalgebra $K(c)\colon X \to F^\infty(X)$ associated with the functor coalgebra $c\colon X \to F(X)$ is obtained. Prove $c \models \mathcal{A} \iff K(c) \models \mathcal{A}$.

6.8.3 Let $F\colon \mathbf{Sets} \to \mathbf{Sets}$ be a functor with axioms $\mathcal{A} \subseteq Z$, where $Z \xrightarrow{\cong} F(Z)$ is a final coalgebra. Define, as in Lemma 6.8.4, a functor:

$$\begin{array}{ccc} & & \mathbf{Pred} \\ & \overset{\mathcal{A}}{\nearrow} & \downarrow \\ \mathbf{CoAlg}(F) & \longrightarrow & \mathbf{Sets} \end{array}$$

6.8.4 Consider the bank account specification from Example 6.8.5. Use the set $\mathbb{N}^+$ of non-empty sequences of natural numbers as state space for a 'history' model, with balance operation $\mathsf{bal} = \mathsf{last}\colon \mathbb{N}^+ \to \mathbb{N}$; define a deposit operation $\mathsf{dep}\colon \mathbb{N}^+ \times \mathbb{N} \to \mathbb{N}^+$ such that assertion (6.34) holds.

6.8.5 Prove explicitly that the functor S with ε, δ, as described in Example 6.8.10, is a comonad on **Sets**.

6.8.6 Let $S : \mathbf{Sets} \to \mathbf{Sets}$ be a comonad with axiom system $\mathcal{A}$. Define validity of $\mathcal{A}$ in a functorial model $\mathcal{K}\ell(S) \to \mathbf{Sets}$, in such a way that for a coalgebra $\gamma\colon X \to S(X)$,

$$\gamma \models \mathcal{A} \iff \mathcal{L}(X, \gamma) \models \mathcal{A},$$

where $\mathcal{L}$ is the functor from Theorem 6.8.2.

6.8.7 Consider the more elaborate bank account from Example 6.5.2, with balance, deposit and withdraw operations combined as coalgebra of the functor

$$F(X) = \mathbb{N} \times X^{\mathbb{N}} \times (X + X)^{\mathbb{N}}.$$

1. Prove that this functor can be massaged into the isomorphic form

$$X \longmapsto \mathbb{N} \times \mathcal{P}(\mathbb{N}) \times X^{\mathbb{N}+\mathbb{N}}.$$

2. Use Propositions 2.5.3 and 2.3.5 to determine the cofree coalgebra F^{∞} on F as

$$F^{\infty}(X) = (X \times \mathbb{N} \times \mathcal{P}(\mathbb{N}))^{(\mathbb{N}+\mathbb{N})^{\star}}.$$

Describe the balance, deposit and withdraw operations from Example 6.5.2 explicitly on $F^{\infty}(X)$.

3. Interpret the assertions from Example 6.5.2 as subset $\mathcal{A} \subseteq F^{\infty}(1)$ and as invariant $\Box \mathcal{A}^{\mathcal{K}}(X) \subseteq F^{\infty}(X)$.
4. Prove that the resulting comonad, as in Corollary 6.8.9, is $S(X) = X^{(\mathbb{N}+\mathbb{N})^{\star}} \times \mathbb{N}$. (Thus, the final coalgebra $S(1)$ is $\mathbb{N}$, as for the bank account specification in Example 6.8.5.)

6.8.8 Let S be a comonad on **Sets**. Following [47], we write, for a subset $P \subseteq S(X)$,

$$\boxtimes P = \bigcap \{h^{-1}(P) \mid h \colon S(X) \to S(X) \text{ is a coalgebra map}\}.$$

1. Check that $\boxtimes P \subseteq P$. This $\boxtimes P$ is the greatest subset of P closed under all recolourings h of P.
2. Axiom systems are automatically closed under $\boxtimes$: prove that for an axiom system $\mathcal{A} \colon \mathcal{K}\ell(S) \to$ **Pred**, where $\mathcal{A}(X) \subseteq S(X)$, one has $\boxtimes\mathcal{A}(X) = \mathcal{A}(X)$.

6.9 Coalgebraic Class Specifications

This final section illustrates the use of assertions for coalgebras in the description of state-based systems in computer science, in particular in object–oriented programming. We start with the well-known mathematical structure of Fibonacci numbers, formulated coalgebraically. Figure 6.4 presents a simple illustration of a ‘coalgebraic specification’. It will be explained step by step. A coalgebraic specification is a structured text with a name (here: ‘Fibonacci’) that describes coalgebras with an initial state satisfying assertions.

More formally, a coalgebraic specification consists of three parts or sections, labelled 'operations', 'assertions', 'creation'. Here we give only an informal description; refer to [406, 445] for more details.

The operations section consists of a finite list of coalgebras $c_i\colon X \to F_i(X)$ of simple polynomial functors F_i. Of course, they can be described jointly as a single coalgebra of the product functor $F_1 \times \cdots \times F_n$, but in these specifications it is clearer to describe these operations separately, with their own names. Among the operations one sometimes distinguishes between 'fields' (or 'observers') and 'methods'. Fields are coalgebras of the form $X \longrightarrow A$ whose result type A is a constant that does not contain the state space X. Hence these fields do not change the state, but only give some information about it. In contrast, methods have the state X in their result type and can change the state, i.e. have a side effect. Hence, in Figure 6.4, the operation val is a field and next is a method.

In object-oriented programming languages a class is a basic notion that combines data with associated operations on such data. A coalgebraic specification can be seen as specification of such a class, where the fields capture the data and the methods their operations.

The assertions section contains assertions about the coalgebras in the operations section. They involve a distinguished variable $x\colon X$, so that they can be interpreted as predicates on the state space X, much as in the banking illustration in Example 6.8.5. The assertions are meant to constrain the behaviour of the coalgebras in a suitable manner.

Finally, the creation section of a coalgebraic specification contains assertions about the assumed initial state new. These assertions may involve the coalgebras from the operations section.

A **model** of a coalgebraic specification consists of (1) a coalgebra c of the (combined, product) type described in the operations section of the specification that satisfies the assertions and (2) an initial state that satisfies the creation conditions.

Coalgebraic Specification *Fibonacci*

Operations

  $\mathsf{val}\colon X \longrightarrow \mathbb{N}$

  $\mathsf{next}\colon X \longrightarrow X$

Assertions

  $\mathsf{val}(\mathsf{next}(\mathsf{next}(x))) = \mathsf{val}(\mathsf{next}(x)) + \mathsf{val}(x)$

Creation

  $\mathsf{val}(\mathsf{new}) = 1$

  $\mathsf{val}(\mathsf{next}(\mathsf{new})) = 1$

Figure 6.4 A coalgebraic specification of a Fibonacci system.

Here is a possible model of the *Fibonacci* specification from Figure 6.4. As state space we take

$$X = \{(f, n) \in \mathbb{N}^{\mathbb{N}} \times \mathbb{N} \mid \forall m \geq n.\, f(m+2) = f(m+1) + f(m)\} \qquad (6.36)$$

with operations

$$\mathsf{val}(f, n) = f(n) \quad \text{and} \quad \mathsf{next}(f, n) = (f, n+1).$$

It is clear that the resulting coalgebra $\langle \mathsf{val}, \mathsf{next}\rangle \colon X \to \mathbb{N} \times X$ satisfies the assertion from Figure 6.4. As initial state we can take $\mathsf{new} = (\mathit{fib}, 0) \in X$, where $\mathit{fib} \colon \mathbb{N} \to \mathbb{N}$ is the well-known recursively defined Fibonacci function satisfying $\mathit{fib}(0) = 1$, $\mathit{fib}(1) = 1$, and $\mathit{fib}(m+2) = \mathit{fib}(m+1) + \mathit{fib}(m)$, for all $m \in \mathbb{N}$. Notice that our states $(f, n) \in X$ implicitly keep track of the stage n in the infinite sequence of Fibonacci numbers $\langle \mathit{fib}(0), \mathit{fib}(1), \mathit{fib}(2), \ldots\rangle$. But this stage is not directly visible from the outside. The specification requires only that the current value is available and that a next state can be computed.

Earlier we mentioned that coalgebraic specifications can be understood as specifications of classes in object-oriented programming languages. We shall sketch how this works by describing a class in the object-oriented programming language Java [43] that can be understood as 'implementation' of the *Fibonacci* specification from Figure 6.4. It is presented in Figure 6.5. First we note that this Java implementation uses bounded integers `int`

```
class Fibonacci {

   private int current_value;
   private int previous_value;

   public int val() {
      return current_value;
   }

   public void next() {
      int next_value = current_value + previous_value;
      previous_value = current_value;
      current_value = next_value;
   }

   public Fibonacci() {
      current_value = 1;
      previous_value = 0;
   }
}
```

Figure 6.5 A Java implementation for the Fibonacci specification from Figure 6.4.

where the specification uses (unbounded) natural numbers $\mathbb{N}$, since $\mathbb{N}$ is not available in Java.[2] This already leads to a mismatch. Further, the Java implementation uses an auxiliary field `previous_value` which is not present in the specification. However, since it is `private` and since it has no 'get' method, this `previous_value` is not visible from the outside. Apart from overflow (caused by the bounded nature of `int`), the assertion from Figure 6.4 seems to hold for the implementation. Also, the creation conditions seem to hold for the initial state resulting from the constructor `Fibonacci()` in Figure 6.5.

Continuing the discussion of this Java implementation a bit further, one can ask whether there is a way to make it mathematically precise that the Java implementation from Figure 6.5 yields a model (as defined above) for the coalgebraic specification in Figure 6.4. One way is to give a 'coalgebraic semantics' to Java by interpreting Java programs as suitable coalgebras. This happened for instance in [66, 265, 256]. However, from a programming perspective it makes more sense to incorporate assertions as used in coalgebraic specification into the programming language. This can be done for instance via the specification language JML [91]. It involves assertions, such as class invariants and pre- and post-conditions for methods, that can be checked and verified with the aid of various tools. Figure 6.6 contains an annotated version of the Java Fibonacci class, where logical assertions are preceded by special comment signs `//@` making them recognisable for special JML compilers and tools. The assertions themselves are mostly self-explanatory, except possibly for two keywords: `\old(-)` refers to the value of a field before a method call, and `\result` refers to the outcome of a (non-void) method.

We return to our more mathematically oriented approach to coalgebraic specifications and ask ourselves what the final coalgebra is satisfying the assertion from Figure 6.4 – ignoring the initial state for a moment. The general approach of Theorem 6.8.8, concretely described in Example 6.8.10, tells that we should first look at the final coalgebra of the functor $X \mapsto \mathbb{N} \times X$ – which is $\mathbb{N}^{\mathbb{N}}$ by Proposition 2.3.5 – and consider the greatest invariant $P = \square(\mathit{assertion}) \subseteq \mathbb{N}^{\mathbb{N}}$ as subcoalgebra. It is not hard to see that $P = \{f \in \mathbb{N}^{\mathbb{N}} \mid \forall m.\, f(m+2) = f(m+1) + f(m)\}$. This means that any $f \in P$ is completely determined by its first two values $f(0)$ and $f(1)$. Hence the final coalgebra satisfying the assertions can be identified with $\mathbb{N}^2 = \mathbb{N} \times \mathbb{N}$, with operations $\mathsf{val}\colon \mathbb{N}^2 \to \mathbb{N}$ and $\mathsf{next}\colon \mathbb{N}^2 \to \mathbb{N}^2$ given by

[2] The integral type `int` in Java uses 32-bit 'signed' numbers, which are in the interval $[-2^{31}, 2^{31} - 1] = [-2147483648, 2147483647]$.

```
class Fibonacci {

    //@ invariant previous_value >= 0 &&
    //@           current_value >= previous_value;

    private int current_value;
    private int previous_value;

    //@ ensures \result == current_value;
    public int val() {
       return current_value;
    }

    //@ assignable previous_value, next_value;
    //@ ensures previous_value == \old(current_value) &&
    //@         current_value == \old(current_value) +
    //@                          \old(previous_value);
    public void next() {
       int next_value = current_value + previous_value;
       previous_value = current_value;
       current_value = next_value;
    }

    //@ assignable previous_value, next_value;
    //@ ensures previous_value == 0 && current_value == 1;
    public Fibonacci() {
       current_value = 1;
       previous_value = 0;
    }
}
```

Figure 6.6 The Java *Fibonacci* class from Figure 6.5 with JML annotations.

$$\mathsf{val}(n_1, n_2) = n_2 \quad \text{and} \quad \mathsf{next}(n_1, n_2) = (n_2, n_2 + n_1).$$

It satisfies the assertion from Figure 6.4:

$$\begin{aligned} \mathsf{val}(\mathsf{next}(\mathsf{next}(n_1, n_2))) &= \mathsf{val}(\mathsf{next}(n_2, n_2 + n_1)) \\ &= \mathsf{val}(n_1 + n_2, (n_2 + n_1) + n_2) \\ &= (n_2 + n_1) + n_2 \\ &= \mathsf{val}(\mathsf{next}(n_1, n_2)) + \mathsf{val}(n_1, n_2). \end{aligned}$$

Within this final coalgebra we also find the initial state $\mathsf{new} = (0, 1) \in \mathbb{N}^2$ satisfying the creation condition from Figure 6.4.

Interestingly, this final coalgebra with initial state corresponds closely to the Java implementation from Figure 6.5. It forms the 'minimal realisation' of the required behaviour, which needs to involve only two natural numbers.

Exercises

6.9.1 Describe the unique coalgebra map from the state space (6.36) to the final coalgebra $\mathbb{N}^2$ of the Fibonacci specification in Figure 6.4. Check that it preserves the initial state.

6.9.2 Derive from the Fibonacci specification in Figure 6.4:

$$\Box(\{x \in X \mid \mathsf{val}(\mathsf{next}(x)) \geq \mathsf{val}(x)\})(\mathsf{new}).$$

6.9.3 Consider the functor $F(X) = \mathbb{N} \times X$ for the Fibonacci specification in Figure 6.4.

1. Show that the cofree comonad F^∞ on F is given by $F^\infty(X) = (X \times \mathbb{N})^{\mathbb{N}}$; describe the coalgebra structure $F^\infty(X) \to F(F^\infty(X))$.
2. Interpret the assertion from Figure 6.4 as a subset $\mathcal{A}(X) \subseteq F^\infty(X)$ and determine the greatest invariant $\Box\mathcal{A}(X) \subseteq F^\infty(X)$.
3. Prove that the comonad induced by Corollary 6.8.9 is

$$S(X) = X^{\mathbb{N}} \times \mathbb{N} \times \mathbb{N},$$

with counit and comultiplication

$$\begin{aligned} \varepsilon(\varphi, m, n) &= \varphi(0) \\ \delta(\varphi, m, n) &= \langle \lambda k \in \mathbb{N}. \langle \varphi(k + (-)), \mathit{Fib}(k, m, n) \rangle, \ m, \ n \rangle, \end{aligned}$$

where $\mathit{Fib}(k, -) \colon \mathbb{N}^2 \to \mathbb{N}^2$ is the outcome of the monoid action obtained by doing k Fibonacci steps starting from the input:

$$\mathit{Fib}(0, m, n) = (m, n) \qquad \mathit{Fib}(k + 1, m, n) = \mathit{Fib}(k, n, m + n).$$

Check that the Eilenberg–Moore coalgebras of S correspond to Fibonacci coalgebras (without initial state).

6.9.4 Consider a coalgebra $\langle \mathsf{val}, \mathsf{next} \rangle \colon X \to \mathbb{N} \times X$ satisfying the assertion from the Fibonacci specification in Figure 6.4. Prove that for each $x \in X$ and $\epsilon > 0$,

$$\Diamond\left(\Box\left(\{y \in X \mid \left|\frac{\mathsf{val}(\mathsf{next}(y))}{\mathsf{val}(y)} - \frac{1 + \sqrt{5}}{2}\right| < \epsilon\}\right)\right)(x).$$

(This is coalgebraic/temporal way of saying that the limit of the quotient $\mathsf{val}(\mathsf{next}^{(n+1)}(x))/\mathsf{val}(\mathsf{next}^{(n)}(x))$ is the golden ratio $1 + \sqrt{5}/2$, as n goes to infinity.)

References

[1] A. Abel, B. Pientka, D. Thibodeau, and A. Setzer. Copatterns: Programming infinite structures by observations. In *Principles of Programming Languages*, pages 27–38. ACM Press, New York, 2013.

[2] M. Abott, T. Altenkirch, and N. Ghani. Containers: Constructing strictly positive types. *Theor. Comp. Sci.*, 342:3–27, 2005.

[3] M. Abott, T. Altenkirch, N. Ghani, and C. McBride. Categories of containers. In A.D. Gordon, editor, *Foundations of Software Science and Computation Structures*, number 2620 in Lect. Notes Comp. Sci., pages 23–38. Springer, Berlin, 2003.

[4] M. Abott, T. Altenkirch, N. Ghani, and C. McBride. Derivatives of containers. In M. Hofmann, editor, *Typed Lambda Calculi and Applications*, number 2701 in Lect. Notes Comp. Sci., pages 23–38. Springer, Berlin, 2003.

[5] S. Abramsky. A domain equation for bisimulation. *Inf. & Comp.*, 92:161–218, 1990.

[6] S. Abramsky. Domain theory in logical form. *Ann. Pure & Appl. Logic*, 51(1/2):1–77, 1991.

[7] S. Abramsky. Coalgebras, Chu spaces, and representations of physical systems. *J. Philosophical Logic* 42(3):551–574, 2013.

[8] S. Abramsky and B. Coecke. A categorical semantics of quantum protocols. In K. Engesser, Dov M. Gabbai, and D. Lehmann, editors, *Handbook of Quantum Logic and Quantum Structures: Quantum Logic*, pages 261–323. North Holland, Elsevier, Computer Science Press, 2009.

[9] P. Aczel. *Non-Well-Founded Sets*. Center for the Study of Language and Information (CSLI) Lecture Notes 14, Stanford, CA, 1988.

[10] P. Aczel. Final universes of processes. In S. Brookes, M. Main, A. Melton, M. Mislove, and D. Schmidt, editors, *Mathematical Foundations of Programming Semantics*, number 802 in Lect. Notes Comp. Sci., pages 1–28. Springer, Berlin, 1994.

[11] P. Aczel, J. Adámek, S. Milius, and J. Velebil. Infinite trees and completely iterative theories: A coalgebraic view. *Theor. Comp. Sci.*, 300 (1–3):1–45, 2003.

[12] P. Aczel and N. Mendler. A final coalgebra theorem. In D. Pitt, A. Poigné, and D. Rydeheard, editors, *Category Theory and Computer Science*, number 389 in Lect. Notes Comp. Sci., pages 357–365. Springer, Berlin, 1989.

[13] J. Adámek. Free algebras and automata realizations in the language of categories. *Comment. Math. Univ. Carolinae*, 15:589–609, 1974.

[14] J. Adámek. Observability and Nerode equivalence in concrete categories. In F. Gécseg, editor, *Fundamentals of Computation Theory*, number 117 in Lect. Notes Comp. Sci., pages 1–15. Springer, Berlin, 1981.

[15] J. Adámek. On final coalgebras of continuous functors. *Theor. Comp. Sci.*, 294:3–29, 2003.

[16] J. Adámek. Introduction to coalgebra. *Theor. Appl. Categ.*, 14(8):157–199, 2005.

[17] J. Adámek. A logic of coequations. In L. Ong, editor, *Computer Science Logic*, number 3634 in Lect. Notes Comp. Sci., pages 70–86. Springer, Berlin, 2005.

[18] J. Adámek and V. Koubek. On the greatest fixed point of a set functor. *Theor. Comp. Sci.*, 150:57–75, 1995.

[19] J. Adámek and C. Kupke, editors. *Coalgebraic Methods in Computer Science (CMCS 2008)*, volume 203(5) of Elect. Notes in Theor. Comp. Sci., 2008.

[20] J. Adámek and C. Kupke, editors. *Coalgebraic Methods in Computer Science (CMCS 2008)*, volume 208(12) of *Inf. & Comp.*, 2010.

[21] J. Adámek, D. Lücke, and S. Milius. Recursive coalgebras of finitary functors. *RAIRO-Theor. Inform. and Appl.*, 41:447–462, 2007.

[22] J. Adámek and S. Milius, editors. *Coalgebraic Methods in Computer Science (CMCS'04)*, number 106 in Elect. Notes in Theor. Comp. Sci. Elsevier, Amsterdam, 2004.

[23] J. Adámek and S. Milius, editors. *Coalgebraic Methods in Computer Science (CMCS 2004)*, volume 204(4) of *Inf. & Comp.*, 2006.

[24] J. Adámek, S. Milius, and J. Velebil. A general final coalgebra theorem. *Math. Struct. in Comp. Sci.*, 15(3):409–432, 2005.

[25] J. Adámek, S. Milius, and J. Velebil. Elgot algebras. *Log. Methods Comput. Sci.*, 2(5), 2006.

[26] J. Adámek, S. Milius, and J. Velebil. Algebras with parametrized iterativity. *Theor. Comp. Sci.*, 388:130–151, 2007.

[27] J. Adámek, S. Milius, and J. Velebil. Equational properties of iterative monads. *Inf. & Comp.*, 208(12):1306–1348, 2010.

[28] J. Adámek and H.-E. Porst. From varieties of algebras to varieties of coalgebras. In A. Corradini, M. Lenisa, and U. Montanari, editors, *Coalgebraic Methods in Computer Science*, number 44(1) in Elect. Notes in Theor. Comp. Sci. Elsevier, Amsterdam, 2001.

[29] J. Adámek and H.-E. Porst. On tree coalgebras and coalgebra presentations. *Theor. Comp. Sci.*, 311:257–283, 2004.

[30] J. Adámek and V. Trnková. *Automata and Algebras in Categories*. Kluwer Academic Publishers, Dordrecht, 1990.

[31] J. Adámek and J. Velebil. Analytic functors and weak pullbacks. *Theory and Appl. of Categories*, 21(11):191–209, 2008.

[32] L. Adleman. Computing with DNA. *Scientific American*, 279(2):54–61, 1998.

[33] A. Aho, R. Sethi, and J. Ullman. *Compilers: Principles, Techniques and Tools*. Addison-Wesley, Reading, MA, 1985.

[34] R. Amadio and P.-L. Curien. *Domains and Lambda-Calculi*. Number 46 in Tracts in Theor. Comp. Sci. Cambridge University Press, Cambridge, 1998.

[35] M. Arbib and E. Manes. Foundations of system theory: Decomposable systems. *Automatica*, 10:285–302, 1974.

[36] M. Arbib and E. Manes. Adjoint machines, state-behaviour machines, and duality. *J. Pure & Appl. Algebra*, 6:313–344, 1975.

[37] M. Arbib and E. Manes. *Arrows, Structures and Functors: The Categorical Imperative*. Academic Press, New York, 1975.

[38] M. Arbib and E. Manes. Foundations of system theory: The Hankel matrix. *J. Comp. Syst. Sci*, 20:330–378, 1980.

[39] M. Arbib and E. Manes. Generalized Hankel matrices and system realization. *SIAM J. Math. Analysis*, 11:405–424, 1980.

[40] M. Arbib and E. Manes. Machines in a category. *J. Pure & Appl. Algebra*, 19: 9–20, 1980.

[41] M. Arbib and E. Manes. Parametrized data types do not need highly constrained parameters. *Inf. & Control*, 52:139–158, 1982.

[42] M.A. Arbib. *Theories of Abstract Automata*. Prentice Hall, Englewood Cliffs, NJ, 1969.

[43] K. Arnold and J. Gosling. *The Java Programming Language*. The Java Series. Addison-Wesley, 2nd edition, 1997.

[44] R. Atkey, N. Ghani, B. Jacobs, and P. Johann. Fibrational induction meets effects. In L. Birkedal, editor, *Foundations of Software Science and Computation Structures*, number 7213 in Lect. Notes Comp. Sci., pages 42–57. Springer, Berlin, 2012.

[45] R. Atkey, P. Johann, and N. Ghani. When is a type refinement an inductive type? In M. Hofmann, editor, *Foundations of Software Science and Computation Structures*, number 6604 in Lect. Notes Comp. Sci., pages 72–87. Springer, Berlin, 2011.

[46] S. Awodey. *Category Theory*. Oxford Logic Guides. Oxford University Press, Oxford, 2006.

[47] S. Awodey and J. Hughes. Modal operators and the formal dual of Birkhoff's completeness theorem. *Math. Struct. Comp. Sci.*, 13:233–258, 2003.

[48] E. Bainbridge. A unified minimal realization theory with duality. PhD thesis, University of Michigan, Ann Arbor, 1972. Technical report 140, Department of Computer and Communication Sciences.

[49] E. Bainbridge, P. Freyd, A. Scedrov, and P. Scott. Functorial polymorphism. *Theor. Comp. Sci.*, 70(1):35–64, 1990. Corrigendum in *Theor. Comp. Sci.* 71(3):431, 1990.

[50] J. de Bakker and E. Vink. *Control Flow Semantics*. MIT Press, Cambridge, MA, 1996.

[51] A. Balan and A. Kurz. On coalgebras over algebras. *Theor. Comp. Sci.*, 412(38):4989–5005, 2011.

[52] A-L. Barabási. *Linked: The New Science of Networks*. Perseus Publishing, Cambridge, MA, 2002.

[53] L. Barbosa. Towards a calculus of state-based software components. *J. Universal Comp. Sci.*, 9(8):891–909, 2003.

[54] H. Barendregt. *The Lambda Calculus: Its Syntax and Semantics*. North-Holland, Amsterdam, 2nd rev. edition, 1984.

[55] M. Barr. Terminal coalgebras in well-founded set theory. *Theor. Comp. Sci.*, 114(2):299–315, 1993. Corrigendum in *Theor. Comp. Sci.* 124:189–192, 1994.

[56] M. Barr and Ch. Wells. *Toposes, Triples and Theories*. Springer, Berlin, 1985. Revised and corrected version available at www.cwru.edu/artsci/math/wells/pub/ttt.html.

[57] M. Barr and Ch. Wells. *Category Theory for Computing Science*. Prentice Hall, Englewood Cliffs, NJ, 1990.

[58] F. Bartels. On generalised coinduction and probabilistic specification formats: Distributive laws in coalgebraic modelling. PhD thesis, Free University of Amsterdam, 2004.

[59] F. Bartels, A. Sokolova, and E. de Vink. A hierarchy of probabilistic system types. In H.-P. Gumm, editor, *Coalgebraic Methods in Computer Science*, number 82(1) in Elect. Notes in Theor. Comp. Sci. Elsevier, Amsterdam, 2003.

[60] J. Barwise and L. Moss. *Vicious Circles: On the Mathematics of Non-wellfounded Phenomena*. CSLI Lecture Notes 60, Stanford, CA, 1996.

[61] J. Beck. Distributive laws. In B. Eckman, editor, *Seminar on Triples and Categorical Homology Theory*, number 80 in Lect. Notes Math., pages 119–140. Springer, Berlin, 1969.

[62] M. Behrisch, S. Kerkhoff, and J. Power. Category theoretic understandings of universal algebra and its dual: Monads and Lawvere theories, comonads and what? In U. Berger and M. Mislove, editors, *Mathematical Foundations of Programming Semantics*, number 286 in Elect. Notes in Theor. Comp. Sci., pages 5–16. Elsevier, Amsterdam, 2012.

[63] J. van Benthem. Correspondence theory. In D. Gabbay and F. Guenthner, editors, *Handbook of Philosophical Logic II*, pages 167–247, Reidel, Dordrecht, 1984.

[64] N. Benton, G. Bierman, M. Hyland, and V. de Paiva. Linear lambda calculus and categorical models revisited. In E. Börger, G. Jäger, H. Kleine Büning, S. Martini, and M.M. Richter, editors, *Computer Science Logic*, number 702 in Lect. Notes Comp. Sci., pages 61–84. Springer, Berlin, 1993.

[65] N. Benton, J. Hughes, and E. Moggi. Monads and effects. In G. Barthe, P. Dybjer, L. Pinto, and J. Saraiva, editors, *Applied Semantics*, number 2395 in Lect. Notes Comp. Sci., pages 923–952. Springer, Berlin, 2002.

[66] J. van den Berg and B. Jacobs. The LOOP compiler for Java and JML. In T. Margaria and W. Yi, editors, *Tools and Algorithms for the Construction and Analysis of Systems*, number 2031 in Lect. Notes Comp. Sci., pages 299–312. Springer, Berlin, 2001.

[67] J. Bergstra, A. Ponse, and S.A. Smolka, editors. *Handbook of Process Algebra*. North-Holland, Amsterdam, 2001.

[68] M. Bidoit and R. Hennicker. Proving the correctness of behavioural implementations. In V.S. Alagar and M. Nivat, editors, *Algebraic Methods and Software Technology*, number 936 in Lect. Notes Comp. Sci., pages 152–168. Springer, Berlin, 1995.

[69] M. Bidoit, R. Hennicker, and A. Kurz. On the duality between observability and reachability. In F. Honsell and M. Miculan, editors, *Foundations of Software*

Science and Computation Structures, number 2030 in Lect. Notes Comp. Sci., pages 72–87. Springer, Berlin, 2001.

[70] M. Bílková, A. Kurz, D. Petrişan, and J. Velebil. Relation liftings on preorders and posets. In B. Klin and C. Cîrstea, editors, *Conference on Algebra and Coalgebra in Computer Science (CALCO 2011)*, number 6859 in Lect. Notes Comp. Sci., pages 115–129. Springer, Berlin, 2011.

[71] R. Bird. *Introduction to Functional Programming Using Haskell*. Prentice Hall, Englewood Cliffs, NJ, 2nd edition, 1998.

[72] R. Bird and O. de Moor. *Algebra of Programming*. Prentice Hall Int. Series in Comput. Sci., Englewood Cliffs, NJ, 1996.

[73] P. Blackburn, M. de Rijke, and Y. Venema. *Modal Logic*. Number 53 in Tracts in Theor. Comp. Sci. Cambridge University Press, Cambridge, 2001.

[74] B. Bloom, S. Istrail, and A.R. Meyer. Bisimulation can't be traced. *J. ACM*, 42(1):232–268, 1988.

[75] S.L. Bloom and Z. Ésik. *Iteration Theories: The Equational Logic of Iterative Processes*. EATCS Monographs. Springer, Berlin, 1993.

[76] F. Bonchi, M. Bonsangue, M. Boreale, J. Rutten, and A. Silva. A coalgebraic perspective on linear weighted automata. *Inf. & Comp.*, 211:77–105, 2012.

[77] F. Bonchi, M. Bonsangue, H. Hansen, P. Panangaden, J. Rutten, and A. Silva. Algebra–coalgebra duality in Brzozowski's minimization algorithm. *ACM Trans. Computational Logic*, 15(1), 2014.

[78] F. Bonchi and U. Montanari. Reactive systems, (semi-)saturated semantics and coalgebras on presheaves. *Theor. Comp. Sci.*, 410(41):4044–4066, 2009.

[79] F. Bonchi and D. Pous. Hacking nondeterminism with induction and coinduction. *Communications of the ACM*, 58(2):87–95, 2015.

[80] F. Bonchi and F. Zanasi. Saturated semantics for coalgebraic logic programming. In R. Heckel and S. Milius, editors, *Conference on Algebra and Coalgebra in Computer Science (CALCO 2013)*, number 8089 in Lect. Notes Comp. Sci., pages 80–94. Springer, Berlin, 2013.

[81] F. Bonchi and F. Zanasi. Bialgebraic semantics for logic programming. *Logical Methods in Comp. Sci.*, 11(1):1–47, 2015.

[82] M. Bonsangue, editor. *Coalgebraic Methods in Computer Science (CMCS 2014)*, number 8446 in Lect. Notes Comp. Sci. Springer, Berlin, 2014.

[83] M. Bonsangue, editor. *Coalgebraic Methods in Computer Science 2014*, volume 604 of *Theor. Comp. Sci.*, 2015.

[84] M. Bonsangue, J. Rutten, and A. Silva. Coalgebraic logic and synthesis of Mealy machines. In R. Amadio, editor, *Foundations of Software Science and Computation Structures*, number 4962 in LNCS, pages 231–245. Springer, Berlin, 2008.

[85] F. Borceux. *Handbook of Categorical Algebra*, volumes 50–52 of *Encyclopedia of Mathematics*. Cambridge University Press, Cambridge, 1994.

[86] F. van Breugel and J. Worrell. An algorithm for quantitative verification of probabilistic transition systems in Java for smart cards. In K.G. Larsen and M. Nielsen, editors, *CONCUR 2001 – Concurrency Theory*, number 2154 in Lect. Notes Comp. Sci., pages 336–350. Springer, Berlin, 2001.

[87] R. Brown. *Topology*. John Wiley & Sons, New York, 2nd rev. edition, 1988.

[88] K.B. Bruce, L. Cardelli, G. Castagna, the Hopkins Objects Group (J. Eifrig, S. Smith, V. Trifonov), G. Leavens, and B. Pierce. On binary methods. *Theory & Practice of Object Systems*, 1(3):221–242, 1996.

[89] J.A. Brzozowski. Derivatives of regular expressions. *J. ACM*, 11(4):481–494, 1964.

[90] P. Buchholz. Bisimulation relations for weighted automata. *Theor. Comp. Sci.*, 393(1–3):109–123, 2008.

[91] L. Burdy, Y. Cheon, D. Cok, M. Ernst, J. Kiniry, G. Leavens, K. Leino, and E. Poll. An overview of JML tools and applications. *Int. J. on Software Tools for Technology Transfer*, 7(3):212–232, 2005.

[92] P.J. Cameron. *Sets, Logic and Categories*. Undergraduate Mathematics. Springer, London, 1999.

[93] V. Capretta, T. Uustalu, and V. Vene. Recursive coalgebras from comonads. *Theor. Comp. Sci.*, 204:437–468, 2006.

[94] A. Carboni, M. Kelly, and R. Wood. A 2-categorical approach to change of base and geometric morphisms I. *Cah. de Top. et Géom. Diff.*, 32(1):47–95, 1991.

[95] K. Cho, B. Jacobs, A. Westerbaan, and B. Westerbaan. An introduction to effectus theory. arxiv.org/abs/1512.05813, 2015.

[96] K. Cho, B. Jacobs, A. Westerbaan, and B. Westerbaan. Quotient comprehension chains. In C. Heunen, P. Selinger, and J. Vicary, editors, *Quantum Physics and Logic (QPL) 2015*, number 195 in Elect. Proc. in Theor. Comp. Sci., pages 136–147, 2015.

[97] V. Ciancia. Accessible functors and final coalgebras for named sets. PhD thesis, University of Pisa, 2008.

[98] C. Cîrstea. Integrating observational and computational features in the specification of state-based dynamical systems. *Inf. Théor. et Appl.*, 35(1):1–29, 2001.

[99] C. Cîrstea, A. Kurz, D. Pattinson, L. Schröder, and Y. Venema. Modal logics are coalgebraic. *The Computer Journal*, 54:31–41, 2011.

[100] C. Cîrstea and D. Pattinson. Modular construction of complete coalgebraic logics. *Theor. Comp. Sci.*, 388(1–3):83–108, 2007.

[101] R. Cockett. Introduction to distributive categories. *Math. Struct. in Comp. Sci.*, 3:277–307, 1993.

[102] R. Cockett. Deforestation, program transformation, and cut-elimination. In A. Corradini, M. Lenisa, and U. Montanari, editors, *Coalgebraic Methods in Computer Science*, number 44(1) in Elect. Notes in Theor. Comp. Sci. Elsevier, Amsterdam, 2001.

[103] R. Cockett and T. Fukushima. About Charity. Technical Report 92/480/18, Department of Computer Science University of Calgary, 1992.

[104] R. Cockett and D. Spencer. Strong categorical datatypes I. In R.A.G. Seely, editor, *International Meeting on Category Theory 1991*, Canadian Mathematical Society Proceedings, vol. 13, pp. 141–169, AMS, Montreal, 1992.

[105] R. Cockett and D. Spencer. Strong categorical datatypes II: A term logic for categorical programming. *Theor. Comp. Sci.*, 139:69–113, 1995.

[106] B. Coecke and K. Martin. A partial order on classical and quantum states. In B. Coecke, editor, *New Structures in Physics*, number 813 in Lect. Notes Physics, pages 593–683. Springer, Berlin, 2011.

[107] M. Comini, G. Levi, and M. Meo. A theory of observables for logic programs. *Inf. & Comp.*, 169(1):23–80, 2001.

[108] A. Corradini, B. Klin, and C. Cîrstea, editors. *Coalgebra and Algebra in Computer Science (CALCO'11)*, number 6859 in Lect. Notes Comp. Sci. Springer, Berlin, 2011.

[109] A. Corradini, M. Lenisa, and U. Montanari, editors. *Coalgebraic Methods in Computer Science (CMCS'01)*, number 44(1) in Elect. Notes in Theor. Comp. Sci. Elsevier, Amsterdam, 2001.

[110] A. Corradini, M. Lenisa, and U. Montanari, editors. *Coalgebraic Methods in Computer Science*, volume 13(2) of *Math. Struct. in Comp. Sci.*, 2003. Special issue on CMCS'01.

[111] D. Coumans and B. Jacobs. Scalars, monads and categories. In C. Heunen, M. Sadrzadeh, and E. Grefenstette, editors, *Quantum Physics and Linguistics: A Compositional, Diagrammatic Discourse*, pages 184–216. Oxford University Press, Oxford, 2013.

[112] S. Coupet-Grimal and L. Jakubiec. Hardware verification using co-induction in COQ. In Y. Bertot, G. Dowek, A. Hirschowitz, C. Paulin, and L. Théry, editors, *Theorem Proving in Higher Order Logics*, number 1690 in Lect. Notes Comp. Sci., pages 91–108. Springer, Berlin, 1999.

[113] R. Crole. *Categories for Types*. Cambridge Mathematical Textbooks. Cambridge University Press, 1993.

[114] N.J. Cutland. *Computability*. Cambridge University Press, 1980.

[115] G. D'Agostino and A. Visser. Finality regained: A coalgebraic study of Scott-sets and multisets. *Arch. Math. Log.*, 41:267–298, 2002.

[116] D. van Dalen, C. Doets, and H. de Swart. *Sets: Naive, Axiomatic and Applied.* Number 106 in Pure & Applied Math. Pergamon Press, 1978.

[117] V. Danos, J. Desharnais, F. Laviolette, and P. Panangaden. Bisimulation and cocongruence for probabilistic systems. *Inf. & Comp.*, 204:503–523, 2006.

[118] P. D'Argenio, H. Hermanns, and J.-P. Katoen. On generative parallel composition. In C. Baier, M. Huth, M. Kwiatkowska, and M. Ryan, editors, *Workshop on Probabilistic Methods in Verification (ProbMIV)*, number 22 in Elect. Notes in Theor. Comp. Sci. Elsevier, Amsterdam, 1998.

[119] B. Davey and H. Priestley. *Introduction to Lattices and Order.* Cambridge University Press, 1990.

[120] L. Dennis and A. Bundy. A comparison of two proof critics: Power vs. robustness. In V.A. Carreño, C.A. Muñoz, and S. Tahar, editors, *Theorem Proving in Higher Order Logics*, number 2410 in Lect. Notes Comp. Sci., pages 182–197. Springer, Berlin, 2002.

[121] E. D'Hondt and P. Panangaden. Quantum weakest preconditions. *Math. Struct. in Comp. Sci.*, 16(3):429–451, 2006.

[122] E. Dijkstra and C. Scholten. *Predicate Calculus and Program Semantics.* Springer, Berlin, 1990.

[123] H. Dobbertin. Refinement monoids, Vaught monoids, and Boolean algebras. *Math. Annalen*, 265(4):473–487, 1983.

[124] E.-E. Doberkat. *Stochastic Coalgebraic Logic*. Springer, 2010.

[125] M. Droste and P. Gastin. Weighted automata and weighted logics. In L. Caires, G. Italiano, L. Monteiro, C. Palamidessi, and M. Yung, editors, *International Colloquium on Automata, Languages and Programming*, number 3580 in Lect. Notes Comp. Sci., pages 513–525. Springer, Berlin, 2005.

[126] H. Ehrig and B. Mahr. *Fundamentals of Algebraic Specification I: Equations and Initial Semantics*. Number 6 in EATCS Monographs. Springer, Berlin, 1985.

[127] S. Eilenberg. *Automata, Languages and Machines*. Academic Press, 1974. 2 volumes.

[128] E. Emerson. Temporal and modal logic. In J. van Leeuwen, editor, *Handbook of Theoretical Computer Science*, volume B, pages 995–1072. Elsevier/MIT Press, 1990.

[129] R. Fagin, J.Y. Halpern, Y. Moses, and M.Y. Vardi. *Reasoning about Knowledge*. MIT Press, Cambridge, MA, 1995.

[130] J. Fiadeiro, N. Harman, M. Roggenbach, and J. Rutten, editors. *Coalgebra and Algebra in Computer Science (CALCO'05)*, number 3629 in Lect. Notes Comp. Sci. Springer, Berlin, 2005.

[131] K. Fine. In so many possible worlds. *Notre Dame J. Formal Log.*, 13:516–520, 1972.

[132] M. Fiore. *Axiomatic Domain Theory in Categories of Partial Maps*. Cambridge University Press, Cambridge, 1996.

[133] M. Fiore. A coinduction principle for recursive data types based on bisimulation. *Inf. & Comp.*, 127(2):186–198, 1996.

[134] M. Fiore, N. Gambino, M. Hyland, and G. Winskel. The cartesian closed bicategory of generalised species of structures. *J. London Math. Soc.*, 77(2): 203–220, 2008.

[135] M. Fiore and C.-K. Hur. Equational systems and free constructions (extended abstract). In L. Arge, C. Cachin, T. Jurdzinski, and A. Tarlecki, editors, *International Colloquium on Automata, Languages and Programming*, number 4596 in LNCS, pages 607–618. Springer, Berlin, 2007.

[136] M. Fiore, G. Plotkin, and D. Turi. Abstract syntax and variable binding. In *Logic in Computer Science*, pages 193–202. IEEE, Computer Science Press, Washington, DC, 1999.

[137] M. Fiore and D. Turi. Semantics of name and value passing. In *Logic in Computer Science*, pages 93–104. IEEE, Computer Science Press, Boston, 2001.

[138] M. Fokkinga. Datatype laws without signatures. *Math. Struct. in Comp. Sci.*, 6:1–32, 1996.

[139] W. Fokkink. *Introduction to Process Algebra*. Texts in Theoretical Computer Science. An EATCS Series. Springer, Berlin, 2000.

[140] M. Forti and F. Honsell. Set theory with free construction principles. *Annali Scuola Normale Superiore, Pisa*, X(3):493–522, 1983.

[141] A. Fraenkel, Y. Bar-Hillel, and A. Levy. *Foundations of Set Theory*. North-Holland, Amsterdam, 2nd rev. edition, 1973.

[142] P. Freyd. Aspects of topoi. *Bull. Austr. Math. Soc.*, 7:1–76 and 467–480, 1972.

[143] P. Freyd. Recursive types reduced to inductive types. In *Logic in Computer Science*, pages 498–507. IEEE, Computer Science Press, Philadelphia, 1990.

[144] P. Freyd. Algebraically complete categories. In A. Carboni, M.C. Pedicchio, and G. Rosolini, editors, *Como Conference on Category Theory*, number 1488 in Lect. Notes Math., pages 95–104. Springer, Berlin, 1991.

[145] P. Freyd. Remarks on algebraically compact categories. In M. Fourman, P. Johnstone, and A. Pitts, editors, *Applications of Categories in Computer Science*, number 177 in LMS, pages 95–106. Cambridge University Press, 1992.

[146] P. Freyd and M. Kelly. Categories of continuous functors. *J. Pure & Appl. Algebra*, 2:169–191, 1972.

[147] H. Friedman. Equality between functionals. In *Logic Colloquium. Symposium on Logic Held at Boston 1972–1973*, number 453 in Lect. Notes Math., pages 22–37. Springer, Berlin, 1975.

[148] M. Gabbay and A. Pitts. A new approach to abstract syntax with variable binding. *Formal Aspects Comp.*, 13:341–363, 2002.

[149] N. Ghani, F. Nordvall Forsberg, and A. Simpson. Comprehensive parametric polymorphism: Categorical models and type theory. In B. Jacobs and C. Löding, editors, *Foundations of Software Science and Computation Structures*, number 9634 in Lect. Notes Comp. Sci., pages 3–19. Springer, Berlin, 2016.

[150] N. Ghani, P. Johann, and C. Fumex. Generic fibrational induction. *Logical Methods in Comp. Sci.*, 8(2), 2012.

[151] N. Ghani and J. Power, editors. *Coalgebraic Methods in Computer Science (CMCS 2006)*, volume 164(1) of Elect. Notes in Theor. Comp. Sci., 2006.

[152] V. Giarrantana, F. Gimona, and U. Montanari. Observability concepts in abstract data specifications. In A. Mazurkiewicz, editor, *Mathematical Foundations of Computer Science*, number 45 in Lect. Notes Comp. Sci., pages 576–587. Springer, Berlin, 1976.

[153] J. Gibbons. Origami programming. In J. Gibbons and O. de Moor, editors, *The Fun of Programming*, Cornerstones in Computing, pages 41–60. Palgrave, 2003.

[154] J. Gibbons, G. Hutton, and T. Altenkirch. When is a function a fold or an unfold? In A. Corradini, M. Lenisa, and U. Montanari, editors, *Coalgebraic Methods in Computer Science*, number 44(1) in Elect. Notes in Theor. Comp. Sci. Elsevier, Amsterdam, 2001.

[155] J.-Y. Girard. Linear logic. *Theor. Comp. Sci.*, 50:1–102, 1987.

[156] J.-Y. Girard. Normal functors, power series and λ-calculus. *Ann. Pure & Appl. Logic*, 37:129–177, 1988.

[157] M. Giry. A categorical approach to probability theory. In B. Banaschewski, editor, *Categorical Aspects of Topology and Analysis*, number 915 in Lect. Notes Math., pages 68–85. Springer, Berlin, 1982.

[158] R. van Glabbeek. The linear time-branching time spectrum II. In E. Best, editor, *CONCUR '93. 4th International Conference on Concurrency Theory*, number 715 in Lect. Notes Comp. Sci., pages 66–81. Springer, Berlin, 1993.

[159] R. van Glabbeek, S. Smolka, B. Steffen, and C. Tofts. Reactive, generative, and stratified models of probabilistic processes. In *Logic in Computer Science*, pages 130–141. IEEE, Computer Science Press, Philadelphiaa, 1990.

[160] J. Goguen. Minimal realization of machines in closed categories. *Bull. Amer. Math. Soc.*, 78(5):777–783, 1972.

[161] J. Goguen. Realization is universal. *Math. Syst. Theor.*, 6(4):359–374, 1973.

[162] J. Goguen. Discrete-time machines in closed monoidal categories. I. *J. Comp. Syst. Sci*, 10:1–43, 1975.

[163] J. Goguen, K. Lin, and G. Rosu. Circular coinductive rewriting. In *Automated Software Engineering (ASE'00)*, pages 123–131. IEEE Press, Grenoble, 2000.

[164] J. Goguen and G. Malcolm. A hidden agenda. *Theor. Comp. Sci.*, 245(1):55–101, 2000.

[165] J. Goguen, J. Thatcher, and E. Wagner. An initial algebra approach to the specification, correctness and implementation of abstract data types. In R. Yeh, editor, *Current Trends in Programming Methodology*, pages 80–149. Prentice Hall, Englewood Cliffs, NJ, 1978.

[166] R. Goldblatt. *Topoi: The Categorial Analysis of Logic*. North-Holland, Amsterdam, 2nd rev. edition, 1984.

[167] R. Goldblatt. *Logics of Time and Computation*. CSLI Lecture Notes 7, Stanford, CA, 2nd rev. edition, 1992.

[168] R. Goldblatt. What is the coalgebraic analogue of Birkhoff's variety theorem? *Theor. Comp. Sci.*, 266(1–2):853–886, 2001.

[169] R. Goldblatt. A comonadic account of behavioural covarieties of coalgebras. *Math. Struct. in Comp. Sci.*, 15(2):243–269, 2005.

[170] R. Goldblatt. Final coalgebras and the Hennessy–Milner property. *Ann. Pure & Appl. Logic*, 183:77–93, 2006.

[171] S. Goncharov, S. Milius, and A. Silva. Towards a coalgebraic Chomsky hierarchy. In J. Diaz, I. Lanese, and D. Sangiorgi, editors, *Theoretical Computer Science – 8th IFIP TC 1/WG 2.2 International Conference*, number 8705 in Lect. Notes Comp. Sci., pages 265–280. Springer, Berlin, 2014.

[172] A. Gordon. Bisimilarity as a theory of functional programming. In S. Brookes, M. Main, A. Melton, and M. Mislove, editors, *Mathematical Foundations of Programming Semantics*, number 1 in Elect. Notes in Theor. Comp. Sci. Elsevier, Amsterdam, 1995.

[173] J. Gosling, B. Joy, G. Steele, and G. Bracha. *The Java Language Specification*. The Java Series, 2nd edition. Addison-Wesley, 2000.

[174] S. Gould. What does the dreaded "E" word mean anyway? In *I Have Landed: The End of a Beginning in Natural History*, pages 241–256. Three Rivers Press, New York, 2002.

[175] J.-F. Groote and F. Vaandrager. Structured operational semantics and bisimulation as a congruence. *Inf. & Comp.*, 100(2):202–260, 1992.

[176] H.-P. Gumm. Elements of the general theory of coalgebras. Notes of lectures given at LUATCS'99: Logic, Universal Algebra, Theoretical Computer Science, Johannesburg, 1999.

[177] H.-P. Gumm. Birkhoff's variety theorem for coalgebras. *Contributions to General Algebra*, 13:159–173, 2000.

[178] H.-P. Gumm. Universelle coalgebra, 2001. Appendix in [235].

[179] H.-P. Gumm. Copower functors. *Theor. Comp. Sci.*, 410:1129–1142, 2002.

[180] H.-P. Gumm, editor. *Coalgebraic Methods in Computer Science (CMCS'03)*, number 82(1) in Elect. Notes in Theor. Comp. Sci. Elsevier, Amsterdam, 2003.

[181] H.-P. Gumm, editor. *Coalgebraic Methods in Computer Science*, volume 327 of *Theor. Comp. Sci.*, 2004. Special issue on CMCS'03.

[182] H.-P. Gumm, J. Hughes, and T. Schröder. Distributivity of categories of coalgebras. *Theor. Comp. Sci.*, 308:131–143, 2003.

[183] H.-P. Gumm and T. Schröder. Covarieties and complete covarieties. In B. Jacobs, L. Moss, H. Reichel, and J. Rutten, editors, *Coalgebraic Methods in Computer Science*, number 11 in Elect. Notes in Theor. Comp. Sci. Elsevier, Amsterdam, 1998.

[184] H.-P. Gumm and T. Schröder. Coalgebraic structure from weak limit preserving functors. In H. Reichel, editor, *Coalgebraic Methods in Computer Science*, number 33 in Elect. Notes in Theor. Comp. Sci. Elsevier, Amsterdam, 2000.

[185] H.-P. Gumm and T. Schröder. Monoid-labeled transition systems. In A. Corradini, M. Lenisa, and U. Montanari, editors, *Coalgebraic Methods in Computer Science*, number 44(1) in Elect. Notes in Theor. Comp. Sci., pages 185–204. Elsevier, Amsterdam, 2001.

[186] H.-P. Gumm and T. Schröder. Products of coalgebras. *Algebra Universalis*, 846:163–185, 2001.

[187] H.-P. Gumm and T. Schröder. Coalgebras of bounded type. *Math. Struct. in Comp. Sci.*, 12(5):565–578, 2002.

[188] C. Gunter. *Semantics of Programming Languages: Structures and Techniques*. MIT Press, Cambridge, MA, 1992.

[189] G. Gupta, A. Bansal, R. Min, L. Simon, and A. Mallya. Coinductive logic programming and its applications. In V. Dahl and I. Niemelä, editors, *Logic Programming*, number 4670 in Lect. Notes Comp. Sci., pages 27–44. Springer, Berlin, 2007.

[190] G. Gupta, N. Saeedloei, B. DeVries, R. Min, K. Marple, and F. Kluzniak. Infinite computation, co-induction and computational logic. In A. Corradini, B. Klin, and C. Cîrstea, editors, *Conference on Algebra and Coalgebra in Computer Science (CALCO 2011)*, number 6859 in Lect. Notes Comp. Sci., pages 40–54. Springer, Berlin, 2011.

[191] T. Hagino. A categorical programming language. PhD thesis, University of Edinburgh, 1987. Technical report 87/38.

[192] T. Hagino. A typed lambda calculus with categorical type constructors. In D. Pitt, A Poigné, and D. Rydeheard, editors, *Category and Computer Science*, number 283 in Lect. Notes Comp. Sci., pages 140–157. Springer, Berlin, 1987.

[193] H. H. Hansen, C. Kupke, and R. Leal. Strong completeness for iteration-free coalgebraic dynamic logics. In J. Diaz, I. Lanese, and D. Sangiorgi, editors, *Theoretical Computer Science*, number 8705 in Lect. Notes Comp. Sci., pages 281–295. Springer, Berlin, 2014.

[194] H. H. Hansen, C. Kupke, and E. Pacuit. Neighbourhood structures: Bisimilarity and basic model theory. *Logical Methods in Comp. Sci.*, 5(2), 2009.

[195] H. H. Hansen and J. Rutten. Symbolic synthesis of Mealy machines from arithmetic bitstream functions. *Scientific Annals of Computer Science*, 20: 97–130, 2010.

[196] H.H. Hansen. Coalgebraic modelling: Applications in automata theory and modal logic. PhD thesis, Free University of Amsterdam, 2009.

[197] H. Hansson. *Real-Time Safety Critical Systems: Time and Probability in Formal Design of Distributed Systems*. Elsevier, New York, 1994.

[198] D Harel, D. Kozen, and J. Tiuryn. *Dynamic Logic*. MIT Press, Cambridge, MA, 2000.

[199] R. Hasegawa. Categorical data types in parametric polymorphism. *Math. Struct. in Comp. Sci.*, 4:71–109, 1994.

[200] R. Hasegawa. Two applications of analytic functors. *Theor. Comp. Sci.*, 272(1–2):113–175, 2002.

[201] I. Hasuo. Generic weakest precondition semantics from monads enriched with order. *Theor. Comp. Sci.*, 604:2–29, 2015.

[202] I. Hasuo, C. Heunen, B. Jacobs, and A. Sokolova. Coalgebraic components in a many-sorted microcosm. In A. Kurz and A. Tarlecki, editors, *Conference on Algebra and Coalgebra in Computer Science (CALCO 2009)*, number 5728 in Lect. Notes Comp. Sci., pages 64–80. Springer, Berlin, 2009.

[203] I. Hasuo and B. Jacobs. Context-free languages via coalgebraic trace semantics. In J. Fiadeiro, N. Harman, M. Roggenbach, and J. Rutten, editors, *Conference on Algebra and Coalgebra in Computer Science (CALCO 2005)*, number 3629 in Lect. Notes Comp. Sci., pages 213–231. Springer, Berlin, 2005.

[204] I. Hasuo and B. Jacobs. Traces for coalgebraic components. *Math. Struct. in Comp. Sci.*, 21:267–320, 2011.

[205] I. Hasuo, B. Jacobs, and M. Niqui. Coalgebraic representation theory of fractals. In P. Selinger, editor, *Mathematical Foundations of Programming Semantics*, number 265 in Elect. Notes in Theor. Comp. Sci., pages 351–368. Elsevier, Amsterdam, 2010.

[206] I. Hasuo, B. Jacobs, and A. Sokolova. Generic trace theory via coinduction. *Logical Methods in Comp. Sci.*, 3(4:11), 2007.

[207] I. Hasuo, B. Jacobs, and A. Sokolova. The microcosm principle and concurrency in coalgebra. In R. Amadio, editor, *Foundations of Software Science and Computation Structures*, number 4962 in LNCS, pages 246–260. Springer, Berlin, 2008.

[208] S. Hayashi. Adjunction of semifunctors: Categorical structures in nonextensional lambda calculus. *Theor. Comp. Sci.*, 41:95–104, 1985.

[209] R. Heckel and S. Milius, editors. *Conference on Algebra and Coalgebra in Computer Science (CALCO 2013)*, number 8089 in Lect. Notes Comp. Sci. Springer, Berlin, 2013.

[210] A. Heifetz and P. Mongin. Probability logic for type spaces. *Games and Economic Behavior*, 35(1–2):31–53, 2001.

[211] A. Heifetz and D. Samet. Topology-free typology of beliefs. *J. Economic Theory*, 82(2):324–341, 1998.

[212] M. Hennessy and R. Milner. Algebraic laws for nondeterminism and concurrency. *J. ACM*, 32-1:137–161, 1985.

[213] U. Hensel. Definition and proof principles for data and processes. PhD thesis, Technical University of Dresden, 1999.

[214] U. Hensel and B. Jacobs. Proof principles for datatypes with iterated recursion. In E. Moggi and G. Rosolini, editors, *Category Theory and Computer Science*, number 1290 in Lect. Notes Comp. Sci., pages 220–241. Springer, Berlin, 1997.

[215] U. Hensel and B. Jacobs. Coalgebraic theories of sequences in PVS. *J. Logic and Computation*, 9(4):463–500, 1999.

[216] U. Hensel and D. Spooner. A view on implementing processes: Categories of circuits. In M. Haveraaen, O. Owe, and O.-J. Dahl, editors, *Recent Trends in Data Type Specification*, number 1130 in Lect. Notes Comp. Sci., pages 237–254. Springer, Berlin, 1996.

[217] C. Hermida. Fibrations, logical predicates and indeterminates. PhD thesis, University of Edinburgh, 1993. Technical report LFCS-93-277. Also available as Aarhus Univ. DAIMI technical report PB-462.

[218] C. Hermida and B. Jacobs. Structural induction and coinduction in a fibrational setting. *Inf. & Comp.*, 145:107–152, 1998.

[219] C. Heunen and B. Jacobs. Arrows, like monads, are monoids. In *Mathematical Foundations of Programming Semantics*, number 158 in Elect. Notes in Theor. Comp. Sci. Elsevier, Amsterdam, 2006.

[220] C. Heunen and B. Jacobs. Quantum logic in dagger kernel categories. *Order*, 27(2):177–212, 2010.

[221] W. Hino, H. Kobayashi I. Hasuo, and B. Jacobs. Healthiness from duality. *Logic in Computer Science*. To appear in the 2016 proceedings of http://ieeexplore.ieee.org/xpl/conhome.jsp?reload=true&punumber=1000420.

[222] M. Hirsch and S. Smale. *Differential Equations, Dynamical Systems, and Linear Algebra*. Academic Press, New York, 1974.

[223] C.A.R. Hoare. *Communicating Sequential Processes*. Prentice Hall, 1985. Available at www.usingcsp.com.

[224] D.R. Hofstadter. *Gödel, Escher, Bach: An Eternal Golden Braid*. Basic Books, New York, 1979.

[225] F. Honsell, M. Miculan, and I. Scagnetto. π-calculus in (co)inductive-type theory. *Theor. Comp. Sci.*, 253(2):239–285, 2001.

[226] R. Hoofman and I. Moerdijk. A remark on the theory of semi-functors. *Math. Struct. in Comp. Sci.*, 5(1):1–8, 1995.

[227] R.A. Howard. *Dynamic Probabilistic Systems*. John Wiley & Sons, New York, 1971.

[228] G. Hughes and M. Cresswell. *A New Introduction to Modal Logic*. Routledge, London and New York, 1996.

[229] J. Hughes. Generalising monads to arrows. *Science of Comput. Progr.*, 37:67–111, 2000.

[230] J. Hughes. Modal operators for coequations. In A. Corradini, M. Lenisa, and U. Montanari, editors, *Coalgebraic Methods in Computer Science*, number 44(1) in Elect. Notes in Theor. Comp. Sci. Elsevier, Amsterdam, 2001.

[231] J. Hughes. A study of categories of algebras and coalgebras. PhD thesis, Carnegie Mellon Unive, 2001.

[232] J. Hughes and B. Jacobs. Simulations in coalgebra. *Theor. Comp. Sci.*, 327(1–2):71–108, 2004.

[233] M. Hyland, G. Plotkin, and J. Power. Combining effects: Sum and tensor. *Theor. Comp. Sci.*, 357:70–99, 2006.

[234] M. Hyland and J. Power. The category theoretic understanding of universal algebra: Lawvere theories and monads. In L. Cardelli, M. Fiore, and G. Winskel, editors, *Computation, Meaning, and Logic: Articles Dedicated to Gordon Plotkin*, number 172 in Elect. Notes in Theor. Comp. Sci., pages 437–458. Elsevier, Amsterdam, 2007.

[235] T. Ihringer. *Allgemeine Algebra*, volume 10 of Berliner Studienreihe zur Mathematik. Heldermann Verlag, 2003.

[236] B. Jacobs. Mongruences and cofree coalgebras. In V.S. Alagar and M. Nivat, editors, *Algebraic Methodology and Software Technology*, number 936 in Lect. Notes Comp. Sci., pages 245–260. Springer, Berlin, 1995.

[237] B. Jacobs. Objects and classes, co-algebraically. In B. Freitag, C.B. Jones, C. Lengauer, and H.-J. Schek, editors, *Object-Orientation with Parallelism and Persistence*, pages 83–103. Kluwer Academic, Dordrecht, 1996.

[238] B. Jacobs. Invariants, bisimulations and the correctness of coalgebraic refinements. In M. Johnson, editor, *Algebraic Methodology and Software Technology*, number 1349 in Lect. Notes Comp. Sci., pages 276–291. Springer, Berlin, 1997.

[239] B. Jacobs. *Categorical Logic and Type Theory*. North Holland, Amsterdam, 1999.

[240] B. Jacobs. A formalisation of Java's exception mechanism. In D. Sands, editor, *Programming Languages and Systems (ESOP)*, number 2028 in Lect. Notes Comp. Sci., pages 284–301. Springer, Berlin, 2001.

[241] B. Jacobs. Many-sorted coalgebraic modal logic: A model-theoretic study. *RAIRO-Theor. Inform. and Appl.*, 35(1):31–59, 2001.

[242] B. Jacobs. Comprehension for coalgebras. In L. Moss, editor, *Coalgebraic Methods in Computer Science*, number 65(1) in Elect. Notes in Theor. Comp. Sci. Elsevier, Amsterdam, 2002.

[243] B. Jacobs. The temporal logic of coalgebras via Galois algebras. *Math. Struct. in Comp. Sci.*, 12:875–903, 2002.

[244] B. Jacobs. Trace semantics for coalgebras. In J. Adámek and S. Milius, editors, *Coalgebraic Methods in Computer Science*, number 106 in Elect. Notes in Theor. Comp. Sci. Elsevier, Amsterdam, 2004.

[245] B. Jacobs. Weakest precondition reasoning for Java programs with JML annotations. *J. Logic and Algebraic Programming*, 58:61–88, 2004.

[246] B. Jacobs. A bialgebraic review of deterministic automata, regular expressions and languages. In K. Futatsugi, J.-P. Jouannaud, and J. Meseguer, editors, *Algebra, Meaning and Computation: Essays Dedicated to Joseph A. Goguen on the Occasion of His 65th Birthday*, number 4060 in Lect. Notes Comp. Sci., pages 375–404. Springer, Berlin, 2006.

[247] B. Jacobs. Convexity, duality, and effects. In C. Calude and V. Sassone, editors, *IFIP Theoretical Computer Science 2010*, number 82(1) in IFIP Adv. in Inf. and Comm. Techn., pages 1–19. Springer, Boston, 2010.

[248] B. Jacobs. From coalgebraic to monoidal traces. In B. Jacobs, M. Niqui, J. Rutten, and A. Silva, editors, *Coalgebraic Methods in Computer Science*, volume 264 of Elect. Notes in Theor. Comp. Sci., pages 125–140. Elsevier, Amsterdam, 2010.

[249] B. Jacobs. Coalgebraic walks, in quantum and Turing computation. In M. Hofmann, editor, *Foundations of Software Science and Computation Structures*, number 6604 in Lect. Notes Comp. Sci., pages 12–26. Springer, Berlin, 2011.

[250] B. Jacobs. Probabilities, distribution monads, and convex categories. *Theor. Comp. Sci.*, 412(28):3323–3336, 2011.

[251] B. Jacobs. Bases as coalgebras. *Logical Methods in Comp. Sci.*, 9(3), 2013.

[252] B. Jacobs. Measurable spaces and their effect logic. In *Logic in Computer Science*. IEEE, Computer Science Press, 2013. Available at http://dblp.uni-trier.de/db/conf/lics/lics2013.html#Jacobs13.

[253] B. Jacobs. Dijkstra and Hoare monads in monadic computation. *Theor. Comp. Sci.*, 604:30–45, 2015.

[254] B. Jacobs. A recipe for state and effect triangles. In L. Moss and P. Sobocinski, editors, *Conference on Algebra and Coalgebra in Computer Science (CALCO 2015)*, volume 35 of *LIPIcs*, pages 116–129. Schloss Dagstuhl, 2015.

[255] B. Jacobs, C. Heunen, and I. Hasuo. Categorical semantics for arrows. *J. Funct. Progr.*, 19(3-4):403–438, 2009.

[256] B. Jacobs, J. Kiniry, and M. Warnier. Java program verification challenges. In F. de Boer, M. Bonsangue, S. Graf, and W.-P. de Roever, editors, *Formal Methods for Components and Objects (FMCO 2002)*, number 2852 in Lect. Notes Comp. Sci., pages 202–219. Springer, Berlin, 2003.

[257] B. Jacobs and J. Mandemaker. The expectation monad in quantum foundations. In B. Jacobs, P. Selinger, and B. Spitters, editors, *Quantum Physics and Logic (QPL) 2011*, number 95 in Elect. Proc. in Theor. Comp. Sci., pages 143–182, 2012.

[258] B. Jacobs and J. Mandemaker. Relating operator spaces via adjunctions. In J. Chubb, A. Eskandarian, and V. Harizanov, editors, *Logic and Algebraic Structures in Quantum Computing*, volume 45 of *Lect. Notes in Logic*, pages 123–150. Cambridge University Press, 2016.

[259] B. Jacobs, J. Mandemaker, and R. Furber. The expectation monad in quantum foundations. *Inf. & Comp.*, 2016. Available at http://www.sciencedirect.com/science/article/pii/S0890540116000365.

[260] B. Jacobs, L. Moss, H. Reichel, and J. Rutten, editors. *Coalgebraic Methods in Computer Science (CMCS'98)*, number 11 in Elect. Notes in Theor. Comp. Sci. Elsevier, Amsterdam, 1998.

[261] B. Jacobs, L. Moss, H. Reichel, and J. Rutten, editors. *Coalgebraic Methods in Computer Science*, volume 260(1/2) of *Theor. Comp. Sci.*, 2001. Special issue on CMCS'98.

[262] B. Jacobs, M. Niqui, J. Rutten, and A. Silva, editors. *Coalgebraic Methods in Computer Science*, volume 264(2) of Elect. Notes in Theor. Comp. Sci., 2010. CMCS 2010, Tenth Anniversary Meeting.

[263] B. Jacobs, M. Niqui, J. Rutten, and A. Silva, editors. *Coalgebraic Methods in Computer Science*, volume 412(38) of *Theor. Comp. Sci.*, 2011. CMCS 2010, Tenth Anniversary Meeting.

[264] B. Jacobs and E. Poll. Coalgebras and monads in the semantics of Java. *Theor. Comp. Sci.*, 291(3):329–349, 2003.

[265] B. Jacobs and E. Poll. Java program verification at Nijmegen: Developments and perspective. In K. Futatsugi, F. Mizoguchi, and N. Yonezaki, editors, *Software Security: Theories and Systems*, number 3233 in Lect. Notes Comp. Sci., pages 134–153. Springer, Berlin, 2004.

[266] B. Jacobs and J. Rutten. A tutorial on (co)algebras and (co)induction. *EATCS Bulletin*, 62:222–259, 1997.

[267] B. Jacobs and J. Rutten, editors. *Coalgebraic Methods in Computer Science (CMCS'99)*, number 19 in Elect. Notes in Theor. Comp. Sci. Elsevier, Amsterdam, 1999.

[268] B. Jacobs and J. Rutten, editors. *Coalgebraic Methods in Computer Science*, volume 280(1/2) of *Theor. Comp. Sci.*, 2002. Special issue on CMCS'99.

[269] B. Jacobs and J. Rutten. A tutorial on (co)algebras and (co)induction. In D. Sangiorgi and J. Rutten, editors, *Advanced Topics in Bisimulation and Coinduction*, number 52 in Tracts in Theor. Comp. Sci., pages 38–99. Cambridge University Press, Cambridge, 2011.

[270] B. Jacobs and A. Silva. Initial algebras of terms with binding and algebraic structure. In C. Casadio, B. Coecke, M. Moortgat, and P. Scott, editors, *Categories and Types in Logic, Language, and Physics*, number 8222 in Lect. Notes Comp. Sci., pages 211–234. Springer, Berlin, 2014.

[271] B. Jacobs, A. Sliva, and A. Sokolova. Trace semantics via determinization. In L. Schröder and D. Patinson, editors, *Coalgebraic Methods in Computer Science (CMCS 2012)*, number 7399 in Lect. Notes Comp. Sci., pages 109–129. Springer, Berlin, 2012.

[272] B. Jacobs, A. Sliva, and A. Sokolova. Trace semantics via determinization. *J. Computer and System Sci.*, 81(5):859–879, 2015.

[273] B. Jacobs and A. Sokolova. Exemplaric expressivity of modal logics. *J. Logic and Computation*, 20(5):1041–1068, 2010.

[274] B. Jay. A semantics for shape. *Science of Comput. Progr.*, 25:251–283, 1995.

[275] B. Jay. Data categories. In M. Houle and P. Eades, editors, *Computing: The Australasian Theory Symposium Proceedings*, Melbourne, number 18 in Australian Comp. Sci. Comm., pages 21–28, 1996.

[276] B. Jay and J. Cockett. Shapely types and shape polymorphism. In D. Sannella, editor, *Programming Languages and Systems (ESOP)*, number 788 in Lect. Notes Comp. Sci., pages 302–316. Springer, Berlin, 1994.

[277] J.-B. Jeannin, D. Kozen, and A. Silva. CoCaml: Programming with coinductive types. Technical report http://hdl.handle.net/1813/30798, Cornell University, 2012. Fundamenta Informaticae, to appear.

[278] J.-B. Jeannin, D. Kozen, and A. Silva. Language constructs for non-well-founded computation. In M. Felleisen and P. Gardner, editors, *Programming Languages and Systems (ESOP)*, number 7792 in Lect. Notes Comp. Sci., pages 61–80. Springer, Berlin, 2013.

[279] P. Johnstone. *Topos Theory*. Academic Press, London, 1977.

[280] P. Johnstone. *Stone Spaces*. Number 3 in Cambridge Studies in Advanced Mathematics. Cambridge University Press, Cambridge, 1982.

[281] P. Johnstone. *Sketches of an Elephant: A Topos Theory Compendium*. Number 44 in Oxford Logic Guides. Oxford University Press, Oxford, 2002. 2 volumes.

[282] P. Johnstone, J. Power, T. Tsujishita, H. Watanabe, and J. Worell. An axiomatics for categories of transition systems as coalgebras. In *Logic in Computer Science*. IEEE, Computer Science Press, 1998.

[283] P. Johnstone, J. Power, T. Tsujishita, H. Watanabe, and J. Worrell. On the structure of categories of coalgebras. *Theor. Comp. Sci.*, 260:87–117, 2001.

[284] S. Peyton Jones and P. Wadler. Imperative functional programming. In *Principles of Programming Languages*, pages 71–84. ACM Press, 1993.

[285] A. Joyal. Foncteurs analytiques et espèces de structures. In G. Labelle and P. Leroux, editors, *Combinatoire Enumerative*, number 1234 in Lect. Notes Math., pages 126–159. Springer, Berlin, 1986.

[286] A. Joyal and I. Moerdijk. *Algebraic Set Theory*. Number 220 in London Math. Soc. Lecture Note Series. Cambridge University Press, 1995.

[287] R. Kalman, P. Falb, and M. Arbib. *Topics in Mathematical System Theory*. McGraw-Hill International Series in Pure and Appl. Mathematics, 1969.

[288] B. von Karger. Temporal algebra. *Math. Struct. in Comp. Sci.*, 8:277–320, 1998.

[289] S. Kasangian, M. Kelly, and F. Rossi. Cofibrations and the realization of non-deterministic automata. *Cah. de Top. et Géom. Diff.*, 24:23–46, 1983.

[290] P. Katis, N. Sabadini, and R. Walters. Bicategories of processes. *J. Pure & Appl. Algebra*, 115(2):141–178, 1997.

[291] Y. Kawahara and M. Mori. A small final coalgebra theorem. *Theor. Comp. Sci.*, 233(1–2):129–145, 2000.

[292] K. Keimel, A. Rosenbusch, and T. Streicher. Relating direct and predicate transformer partial correctness semantics for an imperative probabilistic-nondeterministic language. *Theor. Comp. Sci.*, 412:2701–2713, 2011.

[293] J.G. Kemeny and J.L. Snell. *Finite Markov Chains*. Springer-Verlag, New York, 1976.

[294] S.C. Kleene. Representation of events in nerve nets and finite automata. In C. E. Shannon and J. McCarthy, editors, *Automata Studies*, number 34 in Annals of Mathematics Studies, pages 3–41. Princeton University Press, Princeton, NJ, 1956.

[295] A. Klein. Relations in categories. *Illinois J. Math.*, 14:536–550, 1970.

[296] B. Klin. The least fibred lifting and the expressivity of coalgebraic modal logic. In J. Fiadeiro, N. Harman, M. Roggenbach, and J. Rutten, editors, *Conference on Algebra and Coalgebra in Computer Science (CALCO 2005)*, number 3629 in Lect. Notes Comp. Sci., pages 247–262. Springer, Berlin, 2005.

[297] B. Klin. Coalgebraic modal logic beyond sets. In M. Fiore, editor, *Mathematical Foundations of Programming Semantics*, number 173 in Elect. Notes in Theor. Comp. Sci. Elsevier, Amsterdam, 2007.

[298] B. Klin. Bialgebras for structural operational semantics: An introduction. *Theor. Comp. Sci.*, 412(38):5043–5069, 2011.

[299] A. Kock. Monads on symmetric monoidal closed categories. *Arch. Math.*, 21:1–10, 1970.

[300] A. Kock. On double dualization monads. *Math. Scand.*, 27:151–165, 1970.

[301] A. Kock. Bilinearity and cartesian closed monads. *Math. Scand.*, 29:161–174, 1971.

[302] A. Kock. Closed categories generated by commutative monads. *J. Austr. Math. Soc.*, 12:405–424, 1971.

[303] A. Kock. Algebras for the partial map classifier monad. In A. Carboni, M.C. Pedicchio, and G. Rosolini, editors, *Como Conference on Category Theory*, number 1488 in Lect. Notes Math., pages 262–278. Springer, Berlin, 1991.

[304] A. Kock and G.E. Reyes. Doctrines in categorical logic. In J. Barwise, editor, *Handbook of Mathematical Logic*, pages 283–313. North-Holland, Amsterdam, 1977.

[305] E. Komendantskaya and J. Power. Coalgebraic semantics for derivations in logic programming. In B. Klin and C. Cîrstea, editors, *Conference on Algebra and Coalgebra in Computer Science (CALCO 2011)*, number 6859 in Lect. Notes Comp. Sci., pages 268–282. Springer, Berlin, 2011.

[306] E. Komendantskaya, J. Power, and M. Schmidt. Coalgebraic logic programming: from semantics to implementation programming. *J. Logic and Computation*, 26(2):745–783, 2016.

[307] B. König and S. Küpper. Generic partition refinement algorithms for coalgebras and an instantiation to weighted automata. In J. Diaz, I. Lanese, and D. Sangiorgi, editors, *Theoretical Computer Science – 8th IFIP TC 1/WG 2.2 International Conference*, number 8705 in Lect. Notes Comp. Sci., pages 311–325. Springer, Berlin, 2014.

[308] D. Kozen. Semantics of probabilistic programs. *J. Comp. Syst. Sci*, 22(3): 328–350, 1981.

[309] D. Kozen. A completeness theorem for Kleene algebras and the algebra of regular events. *Inf. & Comp.*, 110(2):366–390, 1994.

[310] D. Kozen. Coinductive proof principles for stochastic processes. *Logical Methods in Comp. Sci.*, 3(4):1–14, 2007.

[311] D. Kozen. Optimal coin flipping. Unpublished manuscript, 2009.

[312] D. Kozen and A. Silva. Practical coinduction. *Math. Struct. in Comp. Sci.*, 2016.

[313] M. Kracht. *Tools and Techniques in Modal Logic*. North Holland, Amsterdam, 1999.

[314] S. Krstić, J. Launchbury, and D. Pavlović. Categories of processes enriched in final coalgebras. In F. Honsell and M. Miculan, editors, *Foundations of Software Science and Computation Structures*, number 2030 in Lect. Notes Comp. Sci., pages 303–317. Springer, Berlin, 2001.

[315] C. Kupke, A. Kurz, and Y. Venema. Completeness of the finitary Moss logic. In C. Areces and R. Goldblatt, editors, *Advances in Modal Logic 2008*, volume 7, pages 193–217. King's College Publications, 2003.

[316] C. Kupke, A. Kurz, and Y. Venema. Stone coalgebras. In H.-P. Gumm, editor, *Coalgebraic Methods in Computer Science*, number 82(1) in Elect. Notes in Theor. Comp. Sci. Elsevier, Amsterdam, 2003.

[317] C. Kupke, A. Kurz, and Y. Venema. Stone coalgebras. *Theor. Comp. Sci.*, 327(1–2):109–134, 2004.

[318] C. Kupke and D. Pattinson. Coalgebraic semantics of modal logics: An overview. *Theor. Comp. Sci.*, 412(38):5070–5094, 2011.

[319] C. Kupke and Y. Venema. Coalgebraic automata theory: Basic results. *Logical Methods in Comp. Sci.*, 4:1–43, 2008.

[320] A. Kurz. Coalgebras and modal logic. Notes of lectures given at ESSLLI'01, Helsinki, 1999.

[321] A. Kurz. A covariety theorem for modal logic. In M. Zakharyaschev, K. Segerberg, M. de Rijke, and H. Wansang, editors, *Advances in Modal Logic*, volume 2, pages 367–380. CSLI Publications, Stanford, 2001.

[322] A. Kurz. Specifying coalgebras with modal logic. *Theor. Comp. Sci.*, 260(1–2):119–138, 2001.

[323] A. Kurz and R. Leal. Modalities in the Stone age: A comparison of coalgebraic logics. *Theor. Comp. Sci.*, 430:88–116, 2012.

[324] A. Kurz, D. Petrisan, P. Severi, and F.J. de Vries. Nominal coalgebraic data types with applications to lambda calculus. *Logical Methods in Comp. Sci.*, 9(4):1–52, 2013.

[325] A. Kurz and J. Rosický. Operations and equations for coalgebras. *Math. Struct. in Comp. Sci.*, 15(1):149–166, 2005.

[326] A. Kurz and A. Tarlecki, editors. *Coalgebra and Algebra in Computer Science (CALCO'09)*, number 5728 in Lect. Notes Comp. Sci. Springer, Berlin, 2009.

[327] J. Lambek. A fixed point theorem for complete categories. *Math. Zeitschr.*, 103:151–161, 1968.

[328] K. Larsen and A. Skou. Bisimulation through probabilistic testing. *Inf. & Comp.*, 94:1–28, 1991.

[329] F. Lawvere. Functorial semantics of algebraic theories and some algebraic problems in the context of functorial semantics of algebraic theories. PhD thesis, Columbia University, 1963. Reprinted in *Theory and Applications of Categories*, 5:1–121, 2004.

[330] F. Lawvere and S. Schanuel. *Conceptual Mathematics: A First Introduction to Categories*. Cambridge University Press, Cambridge, 1997.

[331] T. Leinster. A general theory of self-similarity. *Advances in Math.*, 226(4): 2935–3017, 2011.

[332] T. Leinster. *Basic Category Theory*. Cambridge Studies in Advanced Mathematics. Cambridge Univesity Press, Cambridge, 2014.

[333] M. Lenisa, J. Power, and H. Watanabe. Distributivity for endofunctors, pointed and co-pointed endofunctors, monads and comonads. In H. Reichel, editor, *Coalgebraic Methods in Computer Science*, number 33 in Elect. Notes in Theor. Comp. Sci. Elsevier, Amsterdam, 2000.

[334] M. Lenisa, J. Power, and H. Watanabe. Category theory for operational semantics. *Theor. Comp. Sci.*, 327 (1–2):135–154, 2004.

[335] P. Levy. Monads and adjunctions for global exceptions. In *Mathematical Foundations of Programming Semantics*, number 158 in Elect. Notes in Theor. Comp. Sci., pages 261–287. Elsevier, Amsterdam, 2006.

[336] S. Liang, P. Hudak, and M. Jones. Monad transformers and modular interpreters. In *Principles of Programming Languages*, pages 333–343. ACM Press, 1995.

[337] S. Lindley, Ph. Wadler, and J. Yallop. The arrow calculus. *J. Funct. Progr.*, 20(1):51–69, 2010.

[338] D. Lucanu, E.I. Goriac, G. Caltais, and G. Rosu. CIRC: A behavioral verification tool based on circular coinduction. In A. Kurz and A. Tarlecki, editors, *Conference on Algebra and Coalgebra in Computer Science (CALCO 2009)*, number 5728 in Lect. Notes Comp. Sci., pages 433–442. Springer, Berlin, 2009.

[339] G. Malcolm. Behavioural equivalence, bisimulation and minimal realisation. In M. Haveraaen, O. Owe, and O.J. Dahl, editors, *Recent Trends in Data Type Specification*, number 1130 in Lect. Notes Comp. Sci., pages 359–378. Springer, Berlin, 1996.

[340] E. Manes. *Algebraic Theories*. Springer, Berlin, 1974.

[341] E. Manes. *Predicate Transformer Semantics*. Number 33 in Tracts in Theor. Comp. Sci. Cambridge University Press, Cambridge, 1992.

[342] E. Manes and M. Arbib. *Algebraic Appoaches to Program Semantics*. Texts and Monogr. in Comp. Sci.,. Springer, Berlin, 1986.

[343] Z. Manna and A. Pnueli. *The Temporal Logic of Reactive and Concurrent Systems*. Springer-Verlag, Berlin, 1992.

[344] S. Mac Lane. *Categories for the Working Mathematician*. Springer, Berlin, 1971.

[345] S. Mac Lane. *Mathematics: Form and Function*. Springer, Berlin, 1986.

[346] S. Mac Lane and I. Moerdijk. *Sheaves in Geometry and Logic: A First Introduction to Topos Theory*. Springer, New York, 1992.

[347] K. McMillan. *Symbolic Model Checking*. Kluwer Academic, Dordrecht, 1993.

[348] A. Melton, D. Schmidt, and G. Strecker. Galois connections and computer science applications. In D. Pitt, S. Abramsky, A. Poigné, and D. Rydeheard, editors, *Category Theory and Computer Programming*, number 240 in Lect. Notes Comp. Sci., pages 299–312. Springer, Berlin, 1985.

[349] M. Miculan. A categorical model of the fusion calculus. In *Mathematical Foundations of Programming Semantics*, number 218 in Elect. Notes in Theor. Comp. Sci., pages 275–293. Elsevier, Amsterdam, 2008.

[350] T. Miedaner. The soul of the Mark III beast. In D.R. Hofstadter and D.C. Dennet, editors, *The Mind's I*, pages 109–113. Penguin, London, 1981.

[351] S. Milius. A sound and complete calculus for finite stream circuits. In *Logic in Computer Science*, pages 449–458. IEEE, Computer Science Press, 2010.

[352] S. Milius, D. Pattinson, and T. Wißman. A new foundation for finitary corecursion: The locally finite fixpoint and its properties. In B. Jacobs and C. Löding, editors, *Foundations of Software Science and Computation Structures*, number 9634 in Lect. Notes Comp. Sci., pages 107–125. Springer, Berlin, 2016.

[353] R. Milner. An algebraic definition of simulation between programs. In *Second International Joint Conference on Artificial Intelligence*, pages 481–489. British Computer Society Press, London, 1971.

[354] R. Milner. *A Calculus of Communicating Systems*. Lect. Notes Comp. Sci. Springer, Berlin, 1989.

[355] R. Milner. *Communication and Concurrency*. Prentice Hall, Englewood Cliffs, NJ, 1989.

[356] J. Mitchell. *Foundations of Programming Languages*. MIT Press, Cambridge, MA, 1996.

[357] E. Moggi. Notions of computation and monads. *Inf. & Comp.*, 93(1):55–92, 1991.

[358] R. Montague. Universal grammar. *Theoria*, 36:373–398, 1970.

[359] L. Moss. Coalgebraic logic. *Ann. Pure & Appl. Logic*, 96(1–3):277–317, 1999. Erratum in *Ann. Pure & Appl. Logic*, 99(1–3):241–259, 1999.

[360] L. Moss. Parametric corecursion. *Theor. Comp. Sci.*, 260(1–2):139–163, 2001.

[361] L. Moss, editor. *Coalgebraic Methods in Computer Science (CMCS'00)*, number 65(1) in Elect. Notes in Theor. Comp. Sci. Elsevier, Amsterdam, 2002.

[362] L. Moss and I. Viglizzo. Harsanyi type spaces and final coalgebras constructed from satisfied theories. In J. Adámek and S. Milius, editors, *Coalgebraic Methods in Computer Science*, number 106 in Elect. Notes in Theor. Comp. Sci. Elsevier, Amsterdam, 2004.

[363] T. Mossakowski, U. Montanari, and M. Haveraaen, editors. *Coalgebra and Algebra in Computer Science (CALCO'07)*, number 4624 in Lect. Notes Comp. Sci. Springer, Berlin, 2007.

[364] A. Nanevski, G. Morrisett, A. Shinnar, P. Govereau, and L. Birkedal. Ynot: Dependent types for imperative programs. In *International Conference on Functional Programming (ICFP)*, pages 229–240. ACM SIGPLAN Notices, 2008.

[365] M. Nielsen and I. Chuang. *Quantum Computation and Quantum Information*. Cambridge University Press, 2000.

[366] M. Niqui. Formalising exact arithmetic: Representations, algorithms and proofs. PhD thesis, Radboud University Nijmegen, 2004.

[367] P. Odifreddi. *Classical Recursion Theory*. North-Holland, Amsterdam, 1989.

[368] E. Palmgren and I. Moerdijk. Wellfounded trees in categories. *Ann. Pure & Appl. Logic*, 104(1/3):189–218, 2000.

[369] P. Panangaden. *Labelled Markov Processes*. Imperial College Press, London, 2009.

[370] A. Pardo. Combining datatypes and effects. In V. Vene and T. Uustalu, editors, *Advanced Functional Programming*, number 3622 in Lect. Notes Comp. Sci., pages 171–209. Springer, Berlin, 2004.

[371] D. Park. Concurrency and automata on infinite sequences. In P. Deussen, editor, *Proceedings 5th GI Conference on Theoretical Computer Science*, number 104 in Lect. Notes Comp. Sci., pages 15–32. Springer, Berlin, 1981.

[372] R. Paterson. A new notation for arrows. In *International Conference on Functional Programming (ICFP)*, volume 36(10), pages 229–240. ACM, SIGPLAN Notices, 2001.

[373] D. Pattinson. Coalgebraic modal logic: Soundness, completeness and decidability of local consequence. *Theor. Comp. Sci.*, 309(1–3):177–193, 2003.

[374] D. Pattinson. An introduction to the theory of coalgebras. Course notes at the North American Summer School in Logic, Language and Information (NASSLLI), 2003.

[375] D. Pattinson and L. Schröder, editors. *Coalgebraic Methods in Computer Science 2012*, volume 81(5) of *J. Computer and System Sci.*, 2015.

[376] D. Pavlović and M. Escardó. Calculus in coinductive form. In *Logic in Computer Science*, pages 408–417. IEEE, Computer Science Press, 1998.

[377] D. Pavlović, M. Mislove, and J. Worrell. Testing semantics: Connecting processes and process logics. In M. Johnson and V. Vene, editors, *Algebraic Methods and Software Technology*, number 4019 in Lect. Notes Comp. Sci., pages 308–322. Springer, Berlin, 2006.

[378] D. Pavlović and V. Pratt. The continuum as a final coalgebra. *Theor. Comp. Sci.*, 280 (1–2):105–122, 2002.

[379] B. Pierce. *Basic Category Theory for Computer Scientists*. MIT Press, Cambridge, MA, 1991.

[380] A. Pitts. A co-induction principle for recursively defined domains. *Theor. Comp. Sci.*, 124(2):195–219, 1994.

[381] R. Plasmeijer and M. van Eekelen. *Functional Programming and Parallel Graph Rewriting*. Addison-Wesley, Boston, 1993.

[382] G. Plotkin. Lambda definability in the full type hierarchy. In J. Hindley and J. Seldin, editors, *To H.B Curry: Essays on Combinatory Logic, Lambda Calculus and Formalism*, pages 363–373. Academic Press, New York and London, 1980.

[383] G. Plotkin. A structural approach to operational semantics. Report DAIMI FN-19, Aarhus University, reprinted as [384], 1981.

[384] G. Plotkin. A structural approach to operational semantics. *J. Logic and Algebraic Programming*, 60–61:17–139, 2004.

[385] G. Plotkin and M. Abadi. A logic for parametric polymorphism. In M. Bezem and J.-F. Groote, editors, *Typed Lambda Calculi and Applications*, number 664 in Lect. Notes Comp. Sci., pages 361–375. Springer, Berlin, 1993.

[386] G. Plotkin and J. Power. Notions of computation determine monads. In M. Nielsen and U. Engberg, editors, *Foundations of Software Science and Computation Structures*, number 2303 in Lect. Notes Comp. Sci., pages 342–356. Springer, Berlin, 2002.

[387] G. Plotkin and J. Power. Algebraic operations and generic effects. *Appl. Categorical Struct.*, 11(1):69–94, 2003.

[388] G. Plotkin and J. Power. Computational effects and operations: An overview. In *Proceedings of the Workshop on Domains VI*, number 73 in Elect. Notes in Theor. Comp. Sci., pages 149–163. Elsevier, Amsterdam, 2004.

[389] A. Pnueli. The temporal logic of programs. In *Foundations of Computer Science*, pages 46–57. IEEE, 1977.

[390] A. Pnueli. The temporal semantics of concurrent programs. *Theor. Comp. Sci.*, 31:45–60, 1981.

[391] A. Pnueli. Probabilistic verification. *Inf. & Comp.*, 103:1–29, 1993.

[392] E. Poll and J. Zwanenburg. From algebras and coalgebras to dialgebras. In A. Corradini, M. Lenisa, and U. Montanari, editors, *Coalgebraic Methods in Computer Science*, number 44(1) in Elect. Notes in Theor. Comp. Sci. Elsevier, Amsterdam, 2001.

[393] J. Power. Enriched Lawvere theories. *Theory and Appl. of Categories*, 6:83–93, 2000.

[394] J. Power and E. Robinson. Premonoidal categories and notions of computation. *Math. Struct. in Comp. Sci.*, 7(5):453–468, 1997.

[395] J. Power and D. Turi. A coalgebraic foundation for linear time semantics. In M. Hofmann D. Pavlović, and G. Rosolini, editors, *Category Theory and Computer Science 1999*, number 29 in Elect. Notes in Theor. Comp. Sci. Elsevier, Amsterdam, 1999.

[396] S. Pulmannová and S. Gudder. Representation theorem for convex effect algebras. *Commentationes Mathematicae Universitatis Carolinae*, 39(4):645–659, 1998.

[397] H. Reichel. Behavioural equivalence: A unifying concept for initial and final specifications. In *Third Hungarian Computer Science Conference*. Akademiai Kiado, Budapest, 1981.

[398] H. Reichel. *Initial Computability, Algebraic Specifications, and Partial Algebras*. Number 2 in Monographs in Comp. Sci. Oxford University Press, 1987.

[399] H. Reichel. An approach to object semantics based on terminal co-algebras. *Math. Struct. in Comp. Sci.*, 5:129–152, 1995.

[400] H. Reichel, editor. *Coalgebraic Methods in Computer Science (CMCS'00)*, number 33 in Elect. Notes in Theor. Comp. Sci. Elsevier, Amsterdam, 2000.

[401] K. Rosenthal. *Quantales and Their Applications*. Number 234 in Pitman Research Notes in Math. Longman Scientific & Technical, 1990.

[402] M. Rößiger. Languages for coalgebras on datafunctors. In B. Jacobs and J. Rutten, editors, *Coalgebraic Methods in Computer Science*, number 19 in Elect. Notes in Theor. Comp. Sci. Elsevier, Amsterdam, 1999.

[403] M. Rößiger. Coalgebras and modal logic. In H. Reichel, editor, *Coalgebraic Methods in Computer Science*, number 33 in Elect. Notes in Theor. Comp. Sci. Elsevier, Amsterdam, 2000.

[404] M. Rößiger. From modal logic to terminal coalgebras. *Theor. Comp. Sci.*, 260(1–2):209–228, 2001.

[405] G. Roşu. Equational axiomatizability for coalgebra. *Theor. Comp. Sci.*, 260: 229–247, 2001.

[406] J. Rothe, H. Tews, and B. Jacobs. The coalgebraic class specification language CCSL. *J. Universal Comp. Sci.*, 7(2), 2001.

[407] J. Rutten. Processes as terms: Non-well-founded models for bisimulation. *Math. Struct. in Comp. Sci.*, 2(3):257–275, 1992.

[408] J. Rutten. Automata and coinduction (an exercise in coalgebra). In D. Sangiorgi and R. de Simone, editors, *Concur'98: Concurrency Theory*, number 1466 in Lect. Notes Comp. Sci., pages 194–218. Springer, Berlin, 1998.

[409] J. Rutten. Relators and metric bisimulations. In B. Jacobs, L. Moss, H. Reichel, and J. Rutten, editors, *Coalgebraic Methods in Computer Science*, number 11 in Elect. Notes in Theor. Comp. Sci. Elsevier, Amsterdam, 1998.

[410] J. Rutten. Automata, power series, and coinduction: Taking input derivatives seriously (extended abstract). In J. Wiedermann, P. van Emde Boas, and M. Nielsen, editors, *International Colloquium on Automata, Languages and Programming*, number 1644 in Lect. Notes Comp. Sci., pages 645–654. Springer, Berlin, 1999.

[411] J. Rutten. Universal coalgebra: A theory of systems. *Theor. Comp. Sci.*, 249: 3–80, 2000.

[412] J. Rutten. Behavioural differential equations: A coinductive calculus of streams, automata, and power series. *Theor. Comp. Sci.*, 308:1–53, 2003.

[413] J. Rutten. A coinductive calculus of streams. *Math. Struct. in Comp. Sci.*, 15(1):93–147, 2005.

[414] J. Rutten and D. Turi. Initial algebra and final coalgebra semantics for concurrency. In J. de Bakker, W.-P. de Roever, and G. Rozenberg, editors, *A Decade of Concurrency*, number 803 in Lect. Notes Comp. Sci., pages 530–582. Springer, Berlin, 1994.

[415] N. Saeedloei and G. Gupta. Coinductive constraint logic programming. In T. Schrijvers and P. Thiemann, editors, *Functional and Logic Programming*, number 7294 in Lect. Notes Comp. Sci., pages 243–259. Springer, Berlin, 2012.

[416] A. Salomaa. *Computation and Automata*, volume 25 of *Encyclopedia of Mathematics*. Cambridge University Press, 1985.

[417] D. Schamschurko. Modelling process calculi with PVS. In B. Jacobs, L. Moss, H. Reichel, and J. Rutten, editors, *Coalgebraic Methods in Computer Science*, number 11 in Elect. Notes in Theor. Comp. Sci. Elsevier, Amsterdam, 1998.

[418] O. Schoett. Behavioural correctness of data representations. *Science of Comput. Progr.*, 14:43–57, 1990.

[419] L. Schröder. Expressivity of coalgebraic modal logic: The limits and beyond. In V. Sassone, editor, *Foundations of Software Science and Computation Structures*, number 3441 in Lect. Notes Comp. Sci., pages 440–454. Springer, Berlin, 2005.

[420] L. Schröder and D. Patinson, editors. *Coalgebraic Methods in Computer Science (CMCS 2012)*, number 7399 in Lect. Notes Comp. Sci. Springer, Berlin, 2012.

[421] M.P. Schützenberger. On the definition of a family of automata. *Inf. & Control*, 4(2–3):245–270, 1961.

[422] D. Schwencke. Coequational logic for finitary functors. In J. Adámek and C. Kupke, editors, *Coalgebraic Methods in Computer Science*, number 203(5) in Elect. Notes in Theor. Comp. Sci., pages 243–262. Elsevier, Amsterdam, 2008.

[423] D. Scott. Advice on modal logic. In K. Lambert, editor, *Philosophical Problems in Logic: Some Recent Developments*, pages 143–173. Reidel, Dordrecht, 1970.

[424] R. Seely. Linear logic, $*$-autonomous categories and cofree coalgebras. In J. Gray and A. Scedrov, editors, *Categories in Computer Science and Logic*, number 92 in AMS Contemp. Math., pages 371–382, AMS, Providence, RI, 1989.

[425] R. Segala. Modeling and verification of randomized distributed real-time systems. PhD thesis, MIT, 1995.

[426] R. Segala and N. Lynch. Probabilistic simulations for probabilistic processes. In B. Jonsson and J. Parrow, editors, *Concur'94: Concurrency Theory*, number 836 in Lect. Notes Comp. Sci., pages 481–496. Springer, Berlin, 1994.

[427] A. Silva, F. Bonchi, M. Bonsangue, and J. Rutten. Generalizing the powerset construction, coalgebraically. In K. Lodaya and M. Mahajan, editors, *Foundations of Software Technology and Theoretical Computer Science*, volume 8 of *Leibniz International Proceedings in Informatics*, pages 272–283. Schloss Dagstuhl, 2010.

[428] A. Silva, F. Bonchi, M. Bonsangue, and J. Rutten. Quantative Kleene coalgebras. *Inf. & Comp.*, 209(5):822–849, 2011.

[429] A. Silva, F. Bonchi, M. Bonsangue, and J. Rutten. Generalizing determinization from automata to coalgebras. *Logical Methods in Comp. Sci.*, 9(1:09):1–27, 2013.

[430] A. Silva, M. Bonsangue, and J. Rutten. Non-deterministic Kleene coalgebras. *Logical Methods in Comp. Sci.*, 6(3):1–39, 2010.

[431] L. Simon, A. Mallya, A. Bansal, and G. Gupta. Coinductive logic programming. In S. Etalle and M. Truszczynski, editors, *Logic Programming*, number 4079 in Lect. Notes Comp. Sci., pages 330–345. Springer, Berlin, 2006.

[432] M. Smyth. Topology. In S. Abramsky, Dov M. Gabbai, and T. Maibaum, editors, *Handbook of Logic in Computer Science*, volume 1, pages 641–761. Oxford University Press, 1992.

[433] M. Smyth and G. Plotkin. The category theoretic solution of recursive domain equations. *SIAM J. Comput.*, 11:761–783, 1982.

[434] A. Sokolova. Probabilistic systems coalgebraically: A survey. *Theor. Comp. Sci.*, 412(38):5095–5110, 2011.

[435] S. Staton. Relating coalgebraic notions of bisimulation. In A. Kurz and A. Tarlecki, editors, *Conference on Algebra and Coalgebra in Computer Science (CALCO 2009)*, number 5728 in Lect. Notes Comp. Sci., pages 191–205. Springer, Berlin, 2009.

[436] S. Staton. Relating coalgebraic notions of bisimulation. *Logical Methods in Comp. Sci.*, 7(1:13):1–21, 2011.

[437] C. Stirling. *Modal and Temporal Properties of Processes*. Springer, 2001.

[438] M. Stone. Postulates for the barycentric calculus. *Ann. Math.*, 29:25–30, 1949.

[439] N. Swamy, J. Weinberger, C. Schlesinger, J. Chen, and B. Livshits. Verifying higher-order programs with the Dijkstra monad. In *Proceedings of the 34th ACM SIGPLAN Conference on Programming Language Design and Implementation (PLDI)*, pages 387–398. ACM, 2013.

[440] W. Swierstra. A Hoare logic for the state monad. In S. Berghofer, T. Nipkow, C. Urban, and M. Wenzel, editors, *Theorem Proving in Higher Order Logics*, number 5674 in Lect. Notes Comp. Sci., pages 440–451. Springer, Berlin, 2009.

[441] T. Swirszcz. Monadic functors and convexity. *Bull. de l'Acad. Polonaise des Sciences. Sér. des Sciences Math., Astr. et Phys.*, 22:39–42, 1974.

[442] W. Tait. Intensional interpretation of functionals of finite type I. *J. Symb. Logic*, 32:198–212, 1967.

[443] P. Taylor. *Practical Foundations of Mathematics*, number 59 in Cambridge Studies in Advanced Mathematics. Cambridge University Press, 1999.

[444] H. Tews. Coalgebras for binary methods: Properties of bisimulations and invariants. *Inf. Théor. et Appl.*, 35(1):83–111, 2001.

[445] H. Tews. Coalgebraic methods for object-oriented specification. PhD thesis, Technical University Dresden, 2002.

[446] A. Thijs. Simulation and fixpoint semantics. PhD thesis, University of Groningen, 1996.

[447] B. Trancón y Widemann and M. Hauhs. Distributive-law semantics for cellular automata and agent-based models. In A. Corradini, B. Klin, and C. Cîrstea, editors, *Conference on Algebra and Coalgebra in Computer Science (CALCO 2011)*, number 6859 in Lect. Notes Comp. Sci., pages 344–358. Springer, Berlin, 2011.

[448] D. Traytel, A. Popescu, and J. Blanchette. Foundational, compositional (co)datatypes for higher-order logic: Category theory applied to theorem proving. In *Logic in Computer Science*, pages 596–605. IEEE, Computer Science Press, 2012.

[449] V. Trnková. Some properties of set functors. *Comment. Math. Univ. Carolinae*, 10:323–352, 1969.

[450] V. Trnková. Relational automata in a category and their languages. In *Fundamentals of Computation Theory*, number 256 in Lect. Notes Comp. Sci., pages 340–355. Springer, Berlin, 1977.

[451] D. Turi. Functorial operational semantics and its denotational dual. PhD thesis, Free University of Amsterdam, 1996.

[452] D. Turi and G. Plotkin. Towards a mathematical operational semantics. In *Logic in Computer Science*, pages 280–291. IEEE, Computer Science Press, 1997.

[453] D. Turi and J. Rutten. On the foundations of final semantics: Non-standard sets, metric spaces and partial orders. *Math. Struct. in Comp. Sci.*, 8(5):481–540, 1998.

[454] T. Uustalu and V. Vene. Signals and comonads. *J. Universal Comp. Sci.*, 11(7):1310–1326, 2005.

[455] T. Uustalu, V. Vene, and A. Pardo. Recursion schemes from comonads. *Nordic J. Comput.*, 8(3):366–390, 2001.

[456] M. Vardi. Automatic verification of probabilistic concurrent finite state programs. In *Found. Computer Science*, pages 327–338. IEEE, 1985.

[457] Y. Venema. Automata and fixed point logic: A coalgebraic perspective. *Inf. & Comp.*, 204:637–678, 2006.

[458] I. Viglizzo. Final sequences and final coalgebras for measurable spaces. In J. Fiadeiro, N. Harman, M. Roggenbach, and J. Rutten, editors, *Conference on Algebra and Coalgebra in Computer Science (CALCO 2005)*, number 3629 in Lect. Notes Comp. Sci., pages 395–407. Springer, Berlin, 2005.

[459] E. de Vink and J. Rutten. Bisimulation for probabilistic transition systems: A coalgebraic approach. *Theor. Comp. Sci.*, 221:271–293, 1999.

[460] Ph. Wadler. Monads and composable continuations. *Lisp and Symbolic Computation*, 7(1):39–56, 1993.

[461] R. Walters. *Categories and Computer Science*. Carslaw Publications, Sydney, 1991. Also available as: Cambridge Computer Science Text 28, 1992.

[462] M. Wand. Final algebra semantics and data type extension. *J. Comp. Syst. Sci*, 19:27–44, 1979.

[463] W. Wechler. *Universal Algebra for Computer Scientists*, number 25 in EATCS Monographs. Springer, Berlin, 1992.

[464] J. Winter, M. Bonsangue, and J. Rutten. Context-free languages, coalgebraically. In A. Corradini, B. Klin, and C. Cîrstea, editors, *Conference on Algebra and Coalgebra in Computer Science (CALCO 2011)*, number 6859 in Lect. Notes Comp. Sci., pages 359–376. Springer, Berlin, 2011.

[465] M. Wirsing. Algebraic specification. In J. van Leeuwen, editor, *Handbook of Theoretical Computer Science*, volume B, pages 673–788. Elsevier/MIT Press, 1990.

[466] H. Wolff. Monads and monoids on symmetric monoidal closed categories. *Archiv der Mathematik*, 24:113–120, 1973.

[467] U. Wolter. CSP, partial automata, and coalgebras. *Theor. Comp. Sci.*, 280 (1–2):3–34, 2002.

[468] J. Worrell. Toposes of coalgebras and hidden algebras. In B. Jacobs, L. Moss, H. Reichel, and J. Rutten, editors, *Coalgebraic Methods in Computer Science*, number 11 in Elect. Notes in Theor. Comp. Sci. Elsevier, Amsterdam, 1998.

[469] J. Worrell. Terminal sequences for accessible endofunctors. In B. Jacobs and J. Rutten, editors, *Coalgebraic Methods in Computer Science*, number 19 in Elect. Notes in Theor. Comp. Sci. Elsevier, Amsterdam, 1999.

[470] J. Worrell. On the final sequence of a finitary set functor. *Theor. Comp. Sci.*, 338(1–3):184–199, 2005.

[471] G. Wraith. A note on categorical datatypes. In D. Pitt, A. Poigné, and D. Rydeheard, editors, *Category Theory and Computer Science*, number 389 in Lect. Notes Comp. Sci., pages 118–127. Springer, Berlin, 1989.

Definition and Symbol Index

Subject Index